AF669290

*The Most Successful Way of JEE Preparation*

# DPP

*Daily Practice Problems*

## JEE Main & Advanced

# Integral & Its Applications

MATHEMATICS Vol-8

*The Most Successful Way of JEE Preparation*

# DPP

*Daily Practice Problems*

## JEE Main & Advanced

# Integral & Its Applications

MATHEMATICS Vol-8

Amit M Agarwal

arihant
ARIHANT PRAKASHAN (SERIES), MEERUT

DPP
Daily Practice Problems

arihant
ARIHANT PRAKASHAN (SERIES), MEERUT



**Administrative & Production Offices**

**Corporate Office** 'Ramchhaya' 4577/15, Agarwal Road, Darya Ganj, New Delhi -110002
Tele: 011- 47630600, 43518550; Fax: 011- 23280316

**Head Office** Kalindi, TP Nagar, Meerut (UP) - 250002
Tel: 0121-2401479, 2512970, 4004199; Fax: 0121-2401648

**Sales & Support Offices**

Agra, Ahmedabad, Bengaluru, Bareilly, Chennai, Delhi, Guwahati,
Hyderabad, Jaipur, Jhansi, Kolkata, Lucknow, Meerut, Nagpur & Pune

**ISBN :** 978-93-13193-56-2

Published by Arihant Publications (I) Ltd.

**Production Team**

*Publishing Managers* Mahendra Singh Rawat & Keshav Mohan
*Project Head* Sorabh Chaudhary
*Cover Designer* Shanu Mahad
*Inner Designer* Mazher Chaudhary
*Page Layouting* Sunil Sharma, Hemant
*Proof Reading* Himanshu

For further information about the products from Arihant,
log on to www.arihantbooks.com or email to info@arihantbooks.com

/arihantpub /@arihantpub Arihant Publications /arihantpub

# ABOUT DPP

## Daily Practice Problems

Indian Institutes of Technology better known as IITs have always been the dream destination of each science group students who want to pursue their career as Engineers. These are the institutes which have got very high value not even in India but in the whole world. To get the admission in such institutes one need to face entrance Examinations; JEE Main & JEE Advanced, out of which JEE Main is considered as the preliminary exam & JEE Advanced is the final exam to get into the IITs. Approximately per year 12 lakh students enroll themselves for JEE Main and only the top 2.25 lakh in JEE Main merit can attempt for JEE Advanced, the fact clearly express the degree of toughness of JEE Main & JEE Advanced.

To have success in such an exam one needs to be very planned & systematic in its preparation. A very popular wise saying 'Rome was not built in a day' suggests that to achieve a big target like IITs we need to work continuously. For continuous work, one need to have proper planning that how much work he/she has to do on regular basis, i.e., on daily basis. Each student preparing for JEE has many books but unfortunately none of them gives the students a plan for daily studies, which is the most effective success key for JEE.

Keeping the daily practice needs of the students in mind we have come up with DPP Series (Daily Practice Problems) having approximately 26 books. The primary focus of this series is on achieving success through practice & proper planning. For that we have divided the whole syllabus of JEE Main & JEE Advanced; Physics, Chemistry & Mathematics, into different volumes (around 8 volumes of Mathematics & 9 volumes each of Physics & Chemistry), each volume has Daily Practice Problems Sheets having specific questions on various Topics of the individual chapter of the volume ensuring the complete Practice & Assessment of the topic.

In this way we have tried to give you an effective daily planning for JEE keeping in mind the daily hours that a student can devote to his practice, you are just required to practice one sheet of each subject on daily basis. This habit of daily practice will keep you away from exam stress and thoughts like; how much I have prepared/am I preparing properly or not/ do I need to study more on daily basis or what I am doing is ok, etc., which regularly tease your mind and can give you panic at any stage of your preparation.

Some special features that we have tried to incorporate in our Daily Practice Problems (DPP) series are discussed below

1. Micro Level coverage of all the topics of a Chapter/ Unit given in JEE Main & Advanced Syllabus.
2. Each DPP has various types of questions; Subjective Questions, Single/ Multiple Correct Options, Statements type, Comprehension based, Integer Value Answer type, Matching, etc., making you practice & ready for all the question formats of JEE.
3. New & innovative questions in each DPP challenging & improving your problem solving skills.
4. Along with Topical Coverage, Revisal DPPs for JEE Main & JEE Advanced with each Chapter, covering multidisciplinary questions proving comprehensive practice.
5. JEE Main & JEE Advanced Archive (collection of last 13 years' exams questions) with each chapter.
6. Complete Solutions for each DPP.

We hope that DPP series will prove to be very beneficial for the students preparing for JEE, by giving them daily practice in a systematic & planned manner. The suggestions/criticism from the students, teachers & parents about DPP series will be welcomed whole heartedly and we promise to update this series accordingly from time to time.

Daily Practice Problems

# DAILY TASKS...

## 1 INDEFINITE INTEGRAL



## 2 DEFINITE INTEGRAL

DPP
Daily Practice Problems

## 3 DIFFERENTIAL EQUATIONS



## 4 AREA

# 1. Indefinite Integral



DPP

DAILY PRACTICE PROBLEMS

# DPP-1 Standard Forms of Integration

## Subjective Question

Working Space

*Direction* (Q. No. 1) *This question is subjective in nature, need to be solved completely on notebook.*

**1** Integrate the following:

i. $\int (e^{x \log_e a} + e^{a \log_e x} + e^{a \log_e a}) dx$

ii. $\int \frac{1 - \tan^2 x}{1 + \tan^2 x} dx$

iii. $\int \frac{1 + \cos^2 x}{1 + \cos 2x} dx$

iv. $\int 2^x \cdot e^x \, dx$

v. $\int \frac{1 + \tan^2 x}{1 + \cot^2 x} dx$

vi. $\int \frac{e^{5 \log x} - e^{4 \log x}}{e^{3 \log x} - e^{2 \log x}} dx$

vii. $\int \frac{1 + 2x^2}{x^2 (1 + x^2)} dx$

viii. $\int \frac{\cos 2x}{\cos^2 x \sin^2 x} dx$

ix. $\int 4 \cos \frac{x}{2} \cdot \cos x \cdot \cos \frac{21x}{2} dx$

x. $\int \frac{\cos x - \sin x}{\cos x + \sin x} (2 + 2 \sin 2x) dx$

## Only One Option Correct Type

*Direction* (Q. Nos. 2-26) *This section contains 25 multiple choice questions. Each question has four choices (a), (b), (c) and (d), out of which ONLY ONE is correct.*

**2** $\int \frac{\cos^8 x - \sin^8 x}{1 - 2 \sin^2 x \cos^2 x} dx$ is equal to

a. $-\frac{\sin 2x}{2} + C$

b. $\frac{\sin 2x}{2} + C$

c. $\frac{\cos 2x}{2} + C$

d. $-\frac{\cos 2x}{2} + C$

**3** $\int (1 + 2x + 3x^2 + 4x^3 + \cdots) dx$, when $(|x| < 1)$ is equal to

a. $(1 + x)^{-1} + C$

b. $(1 - x)^{-1} + C$

c. $(1 + x)^{-2} + C$

d. None of these

**4** $\int (\sin x \cdot \cos x \cdot \cos 2x \cdot \cos 4x \cdot \cos 8x \cdot \cos 16x) dx$ is equal to

a. $\frac{\sin 16x}{1024} + C$

b. $-\frac{\cos 32x}{1024} + C$

c. $\frac{\cos 32x}{1096} + C$

d. $-\frac{\cos 32x}{1096} + C$

**5** $\int \frac{\sin x + 4 \sin 3x + 6 \sin 5x + 3 \sin 7x}{\sin 2x + 3 \sin 4x + 3 \sin 6x} dx$ is equal to

a. $-2 \sin x + C$

b. $2 \sin x + C$

c. $-2 \cos x + C$

d. $2 \cos x + C$

Working Space

6 $\int (3\sin x\cos^2 x - \sin^3 x)\,dx$ is equal to

a. $\frac{\cos 3x}{3} + C$ b. $-\frac{\cos 3x}{3} + C$

c. $\frac{\sin 3x}{3} + C$ d. $-\frac{\sin 3x}{3} + C$

7 $\int \frac{(1+x)^2}{x(1+x^2)}\,dx$ is equal to

a. $\log x + 2\tan^{-1} x + C$ b. $\log x - 2\tan^{-1} x + C$

c. $-\log x + 2\tan^{-1} x + C$ d. $2\log x + \tan^{-1} x + C$

8 $\int \frac{\cos 2x - \cos 2\alpha}{\cos x - \cos \alpha}\,dx$ is equal to

a. $(\sin x + x\cos\alpha) + C$ b. $(2\sin x + x\cos\alpha) + C$

c. $2(\sin x + x\cos\alpha) + C$ d. None of these

9 $\int \frac{x^4 + x^2 + 1}{2(1+x^2)}\,dx$ is equal to

a. $\left[\frac{x^3}{3} + \tan^{-1} x\right] + C$ b. $\frac{1}{2}[x^3 + \tan^{-1} x] + C$

c. $\frac{1}{2}[x + \tan^{-1} x] + C$ d. $\frac{1}{2}\left[\frac{x^3}{3} + \tan^{-1} x\right] + C$

10 $\int \frac{x^6 - 1}{x^2 + 1}\,dx$ is equal to

a. $\frac{x^5}{5} - \frac{x^3}{3} + x - 2\tan^{-1} x + C$

b. $\frac{x^5}{5} + \frac{x^3}{3} + x - 2\tan^{-1} x + C$

c. $-\frac{x^5}{5} + \frac{x^3}{3} - x + 2\tan^{-1} x + C$

d. $\frac{x^7}{7} + \frac{x^5}{5} - \frac{x^3}{3} + 2\tan^{-1} x + C$

11 $\int \frac{\sec 2x - 1}{\sec 2x + 1}\,dx$ is equal to

a. $(\sec^2 x - x) + C$ b. $(\tan x - x) + C$

c. $(\sec^2 x + x) + C$ d. $(\tan x + x) + C$

12 $\int \left[\sin^2\left(\frac{9\pi}{8} + \frac{x}{4}\right) - \sin^2\left(\frac{7\pi}{8} + \frac{x}{4}\right)\right] dx$ is equal to

a. $2\cos\frac{x}{2} + C$ b. $\sqrt{2}\cos\frac{x}{2} + C$

c. $-\sqrt{2}\cos\frac{x}{2} + C$ d. None of these

13 $\int \frac{\cos 4x + 1}{\cot x - \tan x}\,dx$ is equal to

a. $\frac{\cos 4x}{8} + C$ b. $\frac{-\cos 4x}{4} + C$

c. $\frac{\cos 8x}{8} + C$ d. $\frac{-\cos 4x}{8} + C$

Working Space

**14** $\int \frac{\sin 2x + \sin 5x - \sin 3x}{\cos x + 1 - 2\sin^2 2x} dx$ is equal to

a. $-2\cos x + C$
b. $2\sin x + C$
c. $-2\sin x + C$
d. $2\cos x + C$

**15** $\int \left( \frac{\cot^2 2x - 1}{2\cot 2x} - \cos 8x \cdot \cot 4x \right) dx$ is equal to

a. $\frac{\cos 8x}{8} + C$
b. $\frac{\sin 8x}{8} + C$
c. $-\frac{\cos 8x}{8} + C$
d. $-\frac{\sin 8x}{8} + C$

**16** $\int \frac{\cos^4 x - \sin^4 x}{\sqrt{1 + \cos 4x}} dx$, $(\cos 2x > 0)$ is equal to

a. $-\frac{x}{\sqrt{2}} + C$
b. $2x + C$
c. $\sqrt{2}x + C$
d. $\frac{x}{\sqrt{2}} + C$

**17** $\int \left( \frac{2x^3 + 3x^2 + 4x + 5}{2x + 1} \right) dx$ is equal to

a. $\frac{x^3}{3} + \frac{x^2}{2} + \frac{3x}{2} + \frac{7}{4}\log(2x + 1) + C$
b. $x^3 + \frac{x^2}{2} + \frac{3x}{2} + \frac{7}{4}\log(2x + 1) + C$
c. $-\frac{x^3}{3} + \frac{x^2}{2} + \frac{3x}{2} + \frac{7}{4}\log(2x + 1) + C$
d. $\frac{x^3}{3} + \frac{x^2}{2} + x + \frac{7}{4}\log(2x + 1) + C$

**18** $\int \frac{(x^2 + \sin^2 x)\sec^2 x}{1 + x^2} dx$ is equal to

a. $\tan x - \tan^{-1} x + C$
b. $\tan^{-1} x - \cos x + C$
c. $\tan x + \tan^{-1} x + C$
d. $\tan^{-1} x + \cos x + C$

**19** $\int \frac{dx}{25 + 4x^2}$ is equal to

a. $\frac{1}{5}\tan^{-1}\left(\frac{2x}{5}\right) + C$
b. $\frac{5}{2}\tan^{-1}\left(\frac{2x}{5}\right) + C$
c. $\frac{2}{5}\tan^{-1}\left(\frac{2x}{5}\right) + C$
d. $\frac{1}{10}\tan^{-1}\left(\frac{2x}{5}\right) + C$

**20** $\int \frac{dx}{\sqrt{9 - 16x^2}}$ is equal to

a. $\frac{1}{3}\sin^{-1}\left(\frac{4x}{3}\right) + C$
b. $\frac{1}{4}\sin^{-1}\left(\frac{4x}{3}\right) + C$
c. $\frac{1}{4}\sin^{-1}(4x) + C$
d. $\sin^{-1}\left(\frac{4x}{3}\right) + C$

Working Space

**21** $\int \frac{2x+3}{3x+2}\,dx$ is equal to

a. $\frac{2}{3}x + \frac{5}{9}\log(3x+2) + C$

b. $x + \frac{1}{3}\log(3x+2) + C$

c. $\frac{1}{3}x + \frac{5}{9}\log(3x+2) + C$

d. $2x + \frac{5}{9}\log(3x+2) + C$

**22** $\int \frac{dx}{1+\sin x}$ is equal to

a. $\sec x \tan x + C$

b. $\tan x - \sec x + C$

c. $\sec x - \tan x + C$

d. None of these

**23** $\int \frac{(\sin 2x - \sin 2k)}{\sin x - \sin k + \cos x - \cos k}\,dx$ is equal to

a. $(\sin x - \cos x) + (\sin k + \cos k)x + C$

b. $\sin x \cdot \cos x + (\sin k)x + C$

c. $(\sin x + \cos x) + (\sin k + \cos k)x + C$

d. $\sin x \cdot \cos x + (\cos k)x + C$

**24** $\int \frac{2+3x^2}{x^2(1+x^2)}\,dx$ is equal to

a. $\frac{2}{x} + \tan^{-1}x + C$

b. $-\frac{2}{x} + \tan^{-1}x + C$

c. $\frac{2}{x} + \tan^{-1}x + C$

d. $\frac{1}{x} + \tan^{-1}x + C$

**25** $\int \frac{\cos 8x - \cos 7x}{1+2\cos 5x}\,dx$ is equal to

a. $\frac{\sin 3x}{3} - \frac{\sin 2x}{2} + C$

b. $\cos 3x - \cos 2x + C$

c. $\frac{\cos 3x}{3} + \frac{\cos 2x}{2} + C$

d. $\frac{\sin 3x}{3} + \frac{\sin 2x}{2} + C$

**26** $\int \tan x \cdot \tan 2x \cdot \tan 3x\,dx$ is equal to

a. $\log|\cos x| + \frac{1}{2}\log|\cos 2x| + \frac{1}{3}\log|\cos 3x| + C$

b. $\log|\cos x| - \frac{1}{2}\log|\cos 2x| - \frac{1}{3}\log|\cos 3x| + C$

c. $\log|\cos x| + \frac{1}{2}\log|\cos 2x| - \frac{1}{3}\log|\cos 3x| + C$

d. None of the above

# DPP-2 Problems on Basic Substitution

Working Space

## Subjective Question

*Direction* (Q. No. 1) *This question is subjective in nature, need to be solved completely on notebook.*

**1** Integrate the following:

i. $\int \frac{dx}{(2x-3)^2}$

ii. $\int \frac{x^4}{\sqrt{4+x^5}}dx$

iii. $\int \frac{dx}{(\sin^{-1}x)^3\sqrt{1-x^2}}$

iv. $\int \frac{(\tan^{-1}x)^2}{1+x^2}dx$

v. $\int \frac{dx}{\cos^2 x\sqrt{1+\tan x}}$

vi. $\int \frac{(2x-3)}{x^3-3x+8}dx$

vii. $\int e^{-3x+1}dx$

viii. $\int e^{-x^3}\cdot x^2\,dx$

ix. $\int \frac{dx}{\sqrt{1-25x^2}}$

x. $\int \frac{x}{\sqrt{a^2-x^4}}dx$

xi. $\int \frac{x^2}{x^6+4}dx$

xii. $\int \frac{2x-\sqrt{\sin^{-1}x}}{\sqrt{1-x^2}}dx$

xiii. $\int \frac{\log(x+1)-\log x}{x(x+1)}dx$

xiv. $\int \frac{e^x}{e^x+1}dx$

xv. $\int \frac{dx}{(x+\sqrt{x^2-1})^2}$

xvi. $\int \frac{\cos x}{(\sin^2 x)^{1/3}}dx$

xvii. $\int \frac{(6x-5)}{2\sqrt{3x^2-5x+6}}dx$

xviii. $\int \tan^3 x\cdot d(\tan x)$

xix. $\int \frac{d(1+x^2)}{\sqrt{1+x^2}}$

xx. $\int \frac{d(1+\log_e x)}{\cos^2(1+\log_e x)}$

## Only One Option Correct Type

*Direction* (Q. Nos. 2-15) *This section contains 14 multiple choice questions. Each question has four choices (a), (b), (c) and (d), out of which ONLY ONE is correct.*

**2** Primitive of $\sqrt[3]{\frac{x}{(x^4-1)^4}}$ w.r.t. $x$ is equal to

a. $\frac{3}{4}\left(1+\frac{1}{x^4-1}\right)^{\frac{1}{3}}+C$

b. $-\frac{3}{4}\left(1+\frac{1}{x^4+1}\right)^{\frac{1}{3}}+C$

c. $\frac{4}{3}\left(1+\frac{1}{x^4-1}\right)^{\frac{1}{3}}+C$

d. $-\frac{4}{3}\left(1+\frac{1}{x^4-1}\right)^{\frac{1}{3}}+C$

Working Space

**3** $\int \frac{x}{\sqrt{1+x^2+\sqrt{(1+x^2)^3}}}\,dx$ is equal to

a. $\frac{1}{2}\log(1+\sqrt{1+x^2})+C$

b. $2\sqrt{1+\sqrt{1+x^2}}+C$

c. $2(1+\sqrt{1+x^2})+C$

d. None of these

**4** $\int \frac{\log|x|}{x\sqrt{1+\log|x|}}\,dx$ is equal to

a. $\frac{2}{3}\sqrt{1+\log|x|}\,(\log|x|-5)+C$

b. $\frac{2}{3}\sqrt{1+\log|x|}\,(\log|x|+2)+C$

c. $\frac{1}{3}\sqrt{1+\log|x|}\,(\log|x|-2)+C$

d. $2\sqrt{1+\log|x|}\,(3\log|x|-2)+C$

**5** If $\int \frac{x^4+1}{x(x^2+1)^2}\,dx = A\log|x| + \frac{B}{1+x^2} + C$, then

a. $A=1, B=-1$

b. $A=-1, B=1$

c. $A=1, B=1$

d. $A=-1, B=-1$

**6** $\int \left(\frac{x}{1+x^5}\right)^{3/2} dx$ is equal to

a. $\frac{2}{5}\sqrt{\frac{x^5}{1+x^5}}+C$

b. $\frac{2}{5}\sqrt{\frac{x}{1+x^5}}+C$

c. $\frac{2}{5}\frac{1}{\sqrt{1+x^5}}+C$

d. None of these

**7** $\int x\cdot\frac{\log(x+\sqrt{1+x^2})}{\sqrt{1+x^2}}\,dx$ is equal to

a. $\sqrt{1+x^2}\log(x+\sqrt{1+x^2})+C$

b. $\sqrt{1+x^2}\log(x+\sqrt{1+x^2})-x+C$

c. $\frac{x}{2}\log^2(x+\sqrt{1+x^2})-\frac{x}{\sqrt{1+x^2}}+C$

d. $\frac{x}{2}\log^2(x+\sqrt{1+x^2})+\frac{x}{\sqrt{1+x^2}}+C$

**8** $\int \frac{(x-1)^2}{x^4+2x^2+1}\,dx$ is equal to

a. $\frac{x^3}{3}+x+\frac{x}{x^2+1}+C$

b. $\frac{x^5+x^3+x+3}{3(x^2+1)}+C$

c. $\frac{x^5+4x^3+3x+3}{3(x^2+1)}+C$

d. None of these

**9** $\int \frac{x^2-4}{x^4+24x^2+16}\,dx$ is equal to

a. $\frac{1}{4}\tan^{-1}\left(\frac{x^2+4}{4x}\right)+C$

b. $-\frac{1}{4}\cot^{-1}\left(\frac{x^2+4}{x}\right)+C$

c. $-\frac{1}{4}\cot^{-1}\left[\frac{4(x^2+4)}{x}\right]+C$

d. $\frac{1}{4}\cot^{-1}\left(\frac{x^2+4}{x}\right)+C$

**10** $\int \frac{(x^4-4)}{x^2\sqrt{4+x^2+x^4}}\,dx$ is equal to

a. $\frac{\sqrt{4+x^2+x^4}}{x}+C$

b. $\sqrt{4+x^2+x^4}+C$

c. $\frac{\sqrt{4+x^2+x^4}}{2}+C$

d. $\frac{\sqrt{4+x^2+x^4}}{2x}+C$

Working Space

**11** $\int \frac{x^9}{(x^2+4)^6}\,dx$ is equal to

a. $\frac{1}{5x}\left(4+\frac{1}{x^2}\right)^{-5}+C$
b. $\frac{1}{5}\left(4+\frac{1}{x^2}\right)^{-5}+C$
c. $\frac{1}{10\,x}(1+4x^2)^{-5}+C$
d. $\frac{1}{40}(1+4x^{-2})^{-5}+C$

**12** The value of $\sqrt{2}\int \frac{\sin x}{\sin(x-\pi/4)}\,dx$ is **[AIEEE 2008]**

a. $x+\log\left|\cos\left(x-\frac{\pi}{4}\right)\right|+C$
b. $x-\log\left|\sin\left(x-\frac{\pi}{4}\right)\right|+C$
c. $x+\log\left|\sin\left(x-\frac{\pi}{4}\right)\right|+C$
d. $x-\log\left|\cos\left(x-\frac{\pi}{4}\right)\right|+C$

**13** For any natural number $m$, evaluate $\int (x^{3m}+x^{2m}+x^m)(2x^{2m}+3x^m+6)^{\frac{1}{m}}dx$, where $x>0$. **[IIT JEE 2002]**

a. $\frac{(2x^{3m}+3x^{2m}+6x^m)^{m+1}}{6(m+1)}+C$
b. $\frac{(2x^{3m}+3x^{2m}+6x^m)^{\frac{m+1}{m}}}{6(m+1)}+C$
c. $\frac{(2x^{3m}+3x^{2m}+6x^m)^{\frac{1}{m}}}{6(m+1)}+C$
d. None of these

**14** $\int \frac{(x^2-1)}{x^3\sqrt{2x^4-2x^2+1}}\,dx$ is equal to **[IIT JEE 2006]**

a. $\frac{\sqrt{2x^4-2x^2+1}}{x^2}+C$
b. $\frac{\sqrt{2x^4-2x^2+1}}{x^3}+C$
c. $\frac{\sqrt{2x^4-2x^2+1}}{x}+C$
d. $\frac{\sqrt{2x^4-2x^2+1}}{2x^2}+C$

**15** $\int \frac{\sec^2 x}{(\sec x+\tan x)^{9/2}}\,dx$ is equal to **[IIT JEE 2012]**

a. $-\frac{1}{(\sec x+\tan x)^{11/2}}\left\{\frac{1}{11}-\frac{1}{7}(\sec x+\tan x)^2\right\}+C$
b. $\frac{1}{(\sec x+\tan x)^{11/2}}\left\{\frac{1}{11}-\frac{1}{7}(\sec x+\tan x)^2\right\}+C$
c. $-\frac{1}{(\sec x+\tan x)^{11/2}}\left\{\frac{1}{11}+\frac{1}{7}(\sec x+\tan x)^2\right\}+C$
d. $\frac{1}{(\sec x+\tan x)^{11/2}}\left\{\frac{1}{11}+\frac{1}{7}(\sec x+\tan x)^2\right\}+C$

## One or More than One Options Correct Type

*Direction (Q. Nos. 16-20) This section contains 5 multiple choice questions. Each question has four choices (a), (b), (c) and (d), out of which ONE or MORE THAN ONE are correct.*

**16** Primitive of $\sqrt{1+2\tan x(\sec x+\tan x)}$ w.r.t. $x$ is

a. $\log|\sec x|-\log|\sec x-\tan x|+C$
b. $\log|\sec x+\tan x|+\log|\sec x|+C$
c. $2\log\left|\sec\frac{x}{2}+\tan\frac{x}{2}\right|+C$
d. $\log|1+\tan x(\sec x+\tan x)|+C$

Working Space

17 $\int \frac{dx}{x^3\left(1-\frac{1}{2x^2}\right)}$ is equal to

a. $\log|2x^2-1|+2\log|x|+C$

b. $\log|2x^2-1|-2\log|2x|+C$

c. $\log|2x^2-1|-\log|x^2|-\log 2+C$

d. $\log\left|1-\frac{1}{2x^2}\right|+C$

18 $\int \frac{1}{x^2-1}\log\left|\frac{x-1}{x+1}\right| dx$ is equal to

a. $\frac{1}{2}\log_e^2\left|\frac{x-1}{x+1}\right|+C$

b. $\frac{1}{4}\log^2\left|\frac{x-1}{x+1}\right|+C$

c. $\frac{1}{2}\log^2\left|\frac{x+1}{x-1}\right|+C$

d. $\frac{1}{4}\log^2\left|\frac{x+1}{x-1}\right|+C$

19 $\int \frac{dx}{\sqrt{x-x^2}}$, where $x \in \left(\frac{1}{2}, 1\right)$ is equal to

a. $2\sin^{-1}\sqrt{x}+C$

b. $\sin^{-1}(2x-1)+C$

c. $-\cos^{-1}(2x-1)+C$

d. $\cos^{-1}(2\sqrt{x-x^2})+C$

20 $\int \frac{\sin 2x}{\sin^4 x+\cos^4 x} dx$ is equal to

a. $\cot^{-1}(\cot^2 x)+C$

b. $-\cot^{-1}(\tan^2 x)+C$

c. $\tan^{-1}(\tan^2 x)+C$

d. $-\tan^{-1}(\cos 2x)+C$

## Comprehension Type

**Direction** *(Q. Nos. 21-24) This section contains a paragraph, describing theory, experiments, data, etc. Four questions related to the paragraph have been given. Each question has only one correct answer among the four given options (a), (b), (c) and (d).*

### Passage

Let $\lim_{x\to\infty} \frac{\int (\sin x+\sin ax)^2 dx}{x} = L$

21 For $a=0$, then the value of $L$ is

a. 0

b. $\frac{1}{2}$

c. 1

d. 2

22 For $a=1$, then the value of $L$ is

a. 0

b. $\frac{1}{2}$

c. 1

d. 2

23 For $a=-1$, the value of $L$ is

a. 0

b. $\frac{1}{2}$

c. 1

d. 2

24 For all $a \in R-\{-1, 0, 1\}$, the value of $L$ is

a. 0

b. $\frac{1}{2}$

c. 1

d. 2

# DPP-3 Integration by Parts

## Subjective Question

Working Space

*Direction* (Q. No. 1) *This question is subjective in nature, need to be solved completely on notebook.*

**1** Integrate the following:

i. $\int x \log_e x \, dx$

ii. $\int \frac{x - \sin x}{1 - \cos x} dx$

iii. $\int \tan^{-1}\left(\frac{3x - x^3}{1 - 3x^2}\right) dx$

iv. $\int (\log_e x)^2 \, dx$

v. $\int \sin^{-1} x \, dx$

vi. $\int x^2 \cdot \sin^{-1} x \, dx$

vii. $\int \frac{x \sin^{-1} x^2}{\sqrt{1 - x^2}} dx$

viii. $\int \tan^{-1} \sqrt{x} \, dx$

ix. $\int e^{\sqrt{x}} dx$

x. $\int (\sin^{-1} x)^2 dx$

xi. $\int \sin^{-1}\left(\frac{2x}{1 + x^2}\right) dx$

xii. $\int \tan^{-1}\left(\sqrt{\frac{1 - x}{1 + x}}\right) dx$

xiii. $\int e^{2x} \cdot \sin 3x \, dx$

xiv. $\int \frac{e^{\tan^{-1} x}}{(1 + x^2)^2} dx$

xv. $\int \operatorname{cosec}^3 x \, dx$

xvi. $\int \sin(\log x) dx$

xvii. $\int e^x [\tan x + \log(\sec x)] dx$

xviii. $\int \frac{x e^x}{(x + 1)^2} dx$

xix. $\int e^x \left[\frac{1 + x}{(2 + x)^2}\right] dx$

xx. $\int e^{\tan^{-1} x} \left[\frac{1 + x + x^2}{1 + x^2}\right] dx$

## Only One Option Correct Type

*Direction* (Q. Nos. 2-9) *This section contains 8 multiple choice questions. Each question has four choices (a), (b), (c) and (d), out of which ONLY ONE is correct.*

**2** $\int \sin^2(\log_e x) \, dx$ is equal to

a. $\frac{x}{10}[5 + 2\sin(2\log_e x) + \cos(2\log_e x)] + C$

b. $\frac{x}{10}[5 + 2\sin(2\log x) - \cos(2\log x)] + C$

c. $\frac{x}{10}[5 - 2\sin(2\log_e x) - \cos(2\log_e x)] + C$

d. $\frac{x}{10}[5 - 2\sin(2\log x) + \cos(2\log x)] + C$

Working Space

**3** $\int e^x \left[\dfrac{x+2}{(x+3)^2}\right] dx$ is equal to

a. $e^x \cdot \dfrac{1}{x+3} + C$

b. $e^x \left(2 - \dfrac{6}{x+3}\right) + C$

c. $e^x \left(1 - \dfrac{6}{x+3}\right) + C$

d. $\dfrac{3e^x}{x+3} + C$

**4** $\int e^{\tan^{-1} x} (1 + x + x^2)\, d(\cot^{-1} x)$ is equal to

a. $-e^{\tan^{-1} x} + C$

b. $e^{\tan^{-1} x} + C$

c. $-x \cdot e^{\tan^{-1} x} + C$

d. $x \cdot e^{\tan^{-1} x} + C$

**5** $\int e^x \dfrac{(1 + n \cdot x^{n-1} - x^{2n})}{(1 - x^n)\sqrt{1 - x^{2n}}} dx$ is equal to

a. $e^x \cdot \sqrt{\dfrac{1-x^n}{1+x^n}} + C$

b. $e^x \cdot \sqrt{\dfrac{1+x^n}{1-x^n}} + C$

c. $-e^x \cdot \sqrt{\dfrac{1-x^n}{1+x^n}} + C$

d. None of these

**6** $\int \dfrac{\log x}{x^2} dx$ is equal to **[AIEEE 2002]**

a. $\dfrac{1}{2}(\log x + 1) + C$

b. $-\dfrac{1}{x}(\log x + 1) + C$

c. $\dfrac{1}{x}(\log x + 1) + C$

d. $\dfrac{1}{x}(\log x - 1) + C$

**7** $\int \left[\dfrac{\log x - 1}{1 + (\log x)^2}\right]^2 dx$ is equal to

a. $\dfrac{\log x}{(\log x)^2 + 1} + C$

b. $\dfrac{x}{x^2 + 1} + C$

c. $\dfrac{xe^x}{1 + x^2} + C$

d. $\dfrac{x}{(\log x)^2 + 1} + C$

**8** If $\int f(x)\, dx = \psi(x)$, then $\int x^5 \cdot f(x^3)\, dx$ is equal to

a. $\dfrac{1}{3}\left[x^3 \ \psi(x^3) - \int x^2 \psi(x^3) dx)\right] + C$

b. $\dfrac{1}{3}\left[x^3 \ \psi(x^3) - 3\int x^3 \psi(x^2) dx\right] + C$

c. $\dfrac{1}{3}\left[x^3 \ \psi(x^3) - 3\int x^2 \psi(x^3) dx\right] + C$

d. $\dfrac{1}{3}\left[x^3 \ \psi(x^3) - \int x^3 \psi(x^3) dx\right] + C$

**9** $\int \sin^{-1}\left(\dfrac{2x+2}{\sqrt{4x^2+8x+13}}\right) dx$ is equal to

a. $(x+1)\tan^{-1}\dfrac{2(x+1)}{3} - \dfrac{3}{4}\log(4x^2+8x+13) + C$

b. $(x+1)\tan^{-1}(x+1) - \dfrac{3}{4}\log(4x^2+8x+13) + C$

c. $(x+1)\tan^{-1} 2(x+1) - \dfrac{3}{4}\log(4x^2+8x+13) + C$

d. $(x+1)\tan^{-1}\dfrac{(x+1)}{3} - \dfrac{3}{4}\log(4x^2+8x+13) + C$

## One or More than One Options Correct Type

Working Space

*Direction (Q. Nos. 10-13) This section contains 4 multiple choice questions. Each question has four choices (a), (b), (c) and (d), out of which ONE or MORE THAN ONE are correct.*

**10** $\int e^{3x} \cdot \cos 4x\, dx = e^{3x} (A \sin 4x + B \cos 4x) + C$, then

a. $4A = 3B$ b. $2A = 3B$ c. $3A = 4B$ d. $4A + 3B = 1$

**11** Let $f'(x) = 3x^2 \cdot \sin\left(\frac{1}{x}\right) - x \cos\left(\frac{1}{x}\right)$, $x \neq 0$, $f(0) = 0$, $f\left(\frac{1}{\pi}\right) = 0$, then which of the following is/are not correct?

a. $f(x)$ is continuous at $x = 0$

b. $f(x)$ is not differentiable at $x = 0$

c. $f'(x)$ is discontinuous at $x = 0$

d. $f'(x)$ is differentiable at $x = 0$

**12** $\int x\, e^x \sin x\, dx$ is equal to

a. $\frac{e^x}{2}\{x(\sin x - \cos x) + \cos x\} + C$

b. $\text{Im} \int x\, e^{(1+i)x}\, dx$

c. $\text{Im}\left\{\frac{x\, e^{(1+i)x}}{1+i} - \frac{e^{(1+i)x}}{(1+i)^2}\right\} + C$

d. $\text{Re} \int x\, e^{(1+i)x}\, dx$

**13** If $u = \int e^{ax} \cdot \cos bx\, dx$ and $v = \int e^{ax} \cdot \sin bx\, dx$, then

a. $(a^2 + b^2)(u^2 + v^2) = e^{2ax}$

b. $(a^2 - b^2)(u^2 - v^2) = e^{2ax}$

c. $\tan^{-1}\left(\frac{v}{u}\right) + \tan^{-1}\left(\frac{b}{a}\right) = bx$

d. $\tan^{-1}\left(\frac{v}{u}\right) - \tan^{-1}\left(\frac{b}{a}\right) = ax$

## Comprehension Type

*Direction (Q. Nos. 14 and 15) This section contains a paragraph, describing theory, experiments, data, etc. Two questions related to the paragraph have been given. Each question has only one correct answer among the four given options (a), (b), (c) and (d).*

### Passage

In calculating a number of integrals, we had to use the method of integration by parts several times in sucession. The result could be obtained more rapidly and in a more conside form by using the so called generalised formula for integration by parts.

$$\int u(x)\, v(x)\, dx = u(x)\, v_1(x) - u'(x)\, v_2(x) + u''(x)\, v_3(x) - \cdots + (-1)^{n-1} u^{n-1}(x)\, v_n(x) - (-1)^{n-1} \cdot \int u^n(x) \cdot v_n(x)\, dx$$

where, $v_1(x) = \int v(x)\, dx$, $v_2(x) = \int v_1(x)\, dx$, ..., $v_n = \int v_{n-1}(x)\, dx$ of course, we assume that all derivatives and integrals appearing in this formula exist. The use of the generalised formula for integration by parts is especially useful when calculating $\int P_n(x)\, Q(x)\, dx$, where $P_n(x)$ is polynomial of degree $n$ and the factor of $Q(x)$ is such that it can be integrated successively $(n + 1)$ times.

**14** If $\int (x^3 - 2x^2 + 3x - 1) \cos 2x\, dx = \frac{\sin 2x}{4} u(x) + \frac{\cos 2x}{8} v(x) + C$, then

a. $u(x) = x^3 - 4x^2 + 3x$

b. $u(x) = 2x^3 - 4x^2 + 3x$

c. $v(x) = 3x^2 - 4x + 3$

d. $v(x) = 6x^2 - 8x$

**15** If $\int e^{2x} \cdot x^4 dx = \frac{e^{2x}}{2} f(x) + C$, then $f(x)$ is

a. $\frac{1}{2}\left(x^4 - 2x^3 + 3x^2 - 3x + \frac{3}{2}\right)$

b. $(x^4 - x^3 + 2x^2 - 3x + 2)$

c. $\left(x^4 - 2x^3 + 3x^2 - 3x + \frac{3}{2}\right)$

d. $\left(x^4 - 2x^3 + 2x^2 - 3x + \frac{3}{2}\right)$

# DPP-4 Some Special Substitutions and Integrals

## Subjective Questions

Working Space

*Direction* (Q. Nos. 1 and 2) *These questions are subjective in nature, need to be solved completely on notebook.*

**1** Integrate the following:

i. $\int \frac{dx}{x^2 - x + 1}$ ii. $\int \frac{dx}{2x^2 + x - 1}$ iii. $\int \frac{dx}{\sqrt{x^2 - 2x + 3}}$

iv. $\int \sqrt{2x^2 - 3x + 1}\, dx$ v. $\int \frac{dx}{\sqrt{1 - e^{2x}}}$ vi. $\int \frac{2x}{\sqrt{1 - x^2 - x^4}} dx$

vii. $\int \frac{a^x}{\sqrt{1 - a^{2x}}} dx$ viii. $\int \frac{e^x}{\sqrt{5 - 4e^x - e^{2x}}} dx$ ix. $\int \sqrt{\frac{x}{a^3 - x^3}} dx$

x. $\int \frac{\cos x}{\sqrt{\sin^2 x - 2\sin x - 3}} dx$ xi. $\int \sqrt{\frac{\sin(x - \alpha)}{\sin(x + \alpha)}} dx$ xii. $\int \frac{2\sin 2x - \cos x}{6 - \cos^2 x - 4\sin x} dx$

xiii. $\int \sqrt{\frac{\cos x - \cos^3 x}{1 - \cos^3 x}} dx$ xiv. $\int \sqrt{\frac{a - x}{a + x}} dx$ xv. $\int x \sqrt{\frac{a^2 - x^2}{a^2 + x^2}} dx$

xvi. $\int x \sqrt{1 + x - x^2}\, dx$ xvii. $\int (x + 1)\sqrt{1 - x - x^2}\, dx$ xviii. $\int \frac{2x^2 + 5x + 4}{\sqrt{x^2 + x + 1}} dx$

**2** Integrate the following:

i. $\int \frac{dx}{4\sin^2 x + 9\cos^2 x}$ ii. $\int \frac{\sin x}{\sin 3x} dx$ iii. $\int \frac{dx}{2 + \sin x + \cos x}$

iv. $\int \frac{dx}{\sqrt{3}\sin x + \cos x}$ v. $\int \frac{3\sin x + 2\cos x}{3\cos x + 2\sin x} dx$

## Only One Option Correct Type

*Direction* (Q. Nos. 3-8) *This section contains 6 multiple choice questions. Each question has four choices (a), (b), (c) and (d), out of which ONLY ONE is correct.*

**3** $\int \frac{\sqrt{\cot x} - \sqrt{\tan x}}{\sqrt{2}(\cos x + \sin x)} dx$ is equal to

a. $\sec^{-1}(\sin x + \cos x) + C$

b. $\sec^{-1}(\sin x - \cos x) + C$

c. $\log|(\sin x + \cos x) + \sqrt{\sin 2x}| + C$

d. $\log|(\sin x - \cos x) + \sqrt{\sin 2x}| + C$

Working Space

**4** $\int \frac{\sin x}{\sin 4x} dx$ is equal to

a. $\frac{1}{2\sqrt{2}} \log \left| \frac{1+\sqrt{2}\sin x}{1-\sqrt{2}\sin x} \right| + \frac{1}{8} \log \left| \frac{1+\sin x}{1-\sin x} \right| + C$

b. $\frac{1}{2\sqrt{2}} \log \left| \frac{1+\sqrt{2}\sin x}{1-\sqrt{2}\sin x} \right| - \frac{1}{8} \log \left| \frac{1+\sin x}{1-\sin x} \right| + C$

c. $\frac{1}{4\sqrt{2}} \log \left| \frac{1+\sqrt{2}\sin x}{1-\sqrt{2}\sin x} \right| + \frac{1}{8} \log \left| \frac{1+\sin x}{1-\sin x} \right| + C$

d. $\frac{1}{4\sqrt{2}} \log \left| \frac{1+\sqrt{2}\sin x}{1-\sqrt{2}\sin x} \right| - \frac{1}{8} \log \left| \frac{1+\sin x}{1-\sin x} \right| + C$

**5** $\int \frac{d\theta}{\cos^3\theta \sqrt{\sin 2\theta}}$ is equal to

a. $\frac{\sqrt{2}}{5} (\sqrt{\tan^2\theta + 5}) \tan\theta + C$

b. $\frac{2}{5} (\tan^2\theta + 5)\sqrt{\tan\theta} + C$

c. $\frac{\sqrt{2}}{5} (\tan^2\theta + 5)\sqrt{\tan\theta} + C$

d. $\sqrt{\frac{2}{5}} (\tan^2\theta + 5)\sqrt{\tan\theta} + C$

**6** $\int \frac{dx}{\cos x - \sin x}$ is equal to **[AIEEE 2004]**

a. $\frac{1}{\sqrt{2}} \log \left| \tan\left(\frac{x}{2} - \frac{\pi}{8}\right) \right| + C$

b. $\frac{1}{2} \log \left| \cot\left(\frac{x}{2}\right) \right| + C$

c. $\frac{1}{\sqrt{2}} \log \left| \tan\left(\frac{x}{2} - \frac{3\pi}{8}\right) \right| + C$

d. $\frac{1}{\sqrt{2}} \log \left| \tan\left(\frac{x}{2} + \frac{3\pi}{8}\right) \right| + C$

**7** $\int \frac{dx}{\cos x + \sqrt{3}\sin x}$ is equal to **[AIEEE 2008]**

a. $\frac{1}{2} \log \left| \tan\left(\frac{x}{2} + \frac{\pi}{12}\right) \right| + C$

b. $\frac{1}{2} \log \left| \tan\left(\frac{x}{2} - \frac{\pi}{12}\right) \right| + C$

c. $\log \left| \tan\left(\frac{x}{2} + \frac{\pi}{12}\right) \right| + C$

d. $\log \left| \tan\left(\frac{x}{2} - \frac{\pi}{12}\right) \right| + C$

**8** If $\int \frac{5\tan x}{\tan x - 2} dx = x + a \log|\sin x - 2\sin x| + k$, then $a$ is equal to **[AIEEE 2012]**

a. 2  b. −1  c. −2  d. 1

## Matching List Type

*Direction* (Q. Nos. 9 and 10) *Choices for the correct combination of elements from Column I and Column II are given as options (a), (b), (c) and (d), out of which one is correct.*

**9** Find $\int f(x)dx$ and match the Column I with Column II and mark the correct option from the codes given below.

| | Column I | | Column II |
|---|---|---|---|
| i. | $f(x) = \frac{1}{(a^2+b^2)-(a^2-b^2)\cos x}$ | p. | $\frac{1}{ab}\tan^{-1}\left(\frac{a}{b}\tan\frac{x}{2}\right) + C$ |
| ii. | $f(x) = \frac{1}{a^2\sin^2 x + b^2\cos^2 x}$ | q. | $\frac{1}{a^2\sin\alpha}\tan^{-1}\left(\frac{\tan x}{\sin\alpha}\right) + C, \alpha = \cos^{-1}\left(\frac{b}{a}\right)$ |
| iii. | $f(x) = \frac{1}{a\cos x + b\sin x}$ | r. | $\frac{1}{ab}\tan^{-1}\left(\frac{a}{b}\tan x\right) + C$ |
| iv. | $f(x) = \frac{1}{a^2 - b^2\cos^2 x} (a^2 > b^2)$ | s. | $\frac{1}{\sqrt{a^2+b^2}} \log \left| \tan\frac{1}{2}\left(x + \tan^{-1}\frac{a}{b}\right) \right| + C$ |

**Codes**

| | i | ii | iii | iv |
|---|---|---|---|---|
| a. | p | r | s | q |
| b. | q | r | p | s |
| c. | s | q | p | r |
| d. | q | s | r | p |

10 Find $\int f(x)\,dx$ and then match the following columns and select the correct option from the codes given below.

| | Column I | | Column II |
|---|---|---|---|
| i. | $f(x)=\dfrac{1}{(a^2+x^2)^{3/2}}$ | p. | $C-\dfrac{1}{a}\cdot\sin^{-1}\left(\dfrac{a}{x}\right)$ |
| ii. | $f(x)=\dfrac{x^2}{\sqrt{a^2-x^2}}$ | q. | $\dfrac{a^2}{2}\left\{\sin^{-1}\left(\dfrac{x}{a}\right)-\dfrac{x}{a}\cdot\dfrac{\sqrt{a^2-x^2}}{a}\right\}+C$ |
| iii. | $f(x)=\dfrac{1}{(x^2-a^2)^{3/2}}$ | r. | $C-\dfrac{x}{a^2\sqrt{x^2-a^2}}$ |
| iv. | $f(x)=\dfrac{1}{x\sqrt{x^2-a^2}}$ | s. | $\dfrac{x}{a^2\sqrt{x^2+a^2}}+C$ |

**Codes**

| | i | ii | iii | iv |
|---|---|---|---|---|
| a. | s | q | r | p |
| b. | q | s | p | r |
| c. | r | p | s | r |
| d. | s | r | q | p |

## One Integer Value Correct Type

*Direction (Q. Nos. 11-15) This section contains 5 questions. When worked out will result in an integer from 0 to 9 (both inclusive).*

11 If $\int \dfrac{dx}{5+4\cos x}=a\tan^{-1}[bf(x)]+C$, then $\lim_{x\to 0}\dfrac{2f(x)}{x}$ is equal to

12 If $\int \dfrac{(3x^4-1)}{(x^4+x+1)^2}\,dx=f(x)+C$, then $|\lim_{x\to\infty} x^3\cdot f(x)|$ is equal to

13 If $\int \dfrac{dx}{x^4\,[x\,(x^5-1)]^{1/3}}=a\left(\dfrac{x^\lambda+\mu}{x^\gamma}\right)^{2/3}+C$, then $\left|\dfrac{\lambda}{\gamma}+\mu\right|$ is equal to

14 If $\int(\sqrt{\tan x}+\sqrt{\cot x})\,dx=\sqrt{\lambda}\,\tan^{-1}\left(\dfrac{\sqrt{\tan x}-\sqrt{\cot x}}{\sqrt{\mu}}\right)+C$, then $(\lambda+\mu)$ equals

15 If $\int \dfrac{(ax^2-b)}{x\sqrt{c^2x^2-(ax^2+b)^2}}\,dx=\sin^{-1}[f(x)]+C$, then $\lim_{x\to 0}[2f(x)\cdot x^2]$ is equal to

# DPP-5 $\int \frac{dx}{P\sqrt{Q}}$ Form, Partial Fractions and Reduction Formula

## Subjective Questions

Working Space

*Direction* (Q. Nos. 1-4) *These questions are subjective in nature, need to be solved completely on notebook.*

**1** Integrate the following:

i. $\int \frac{dx}{(x+3)\sqrt{x+1}}$

ii. $\int \frac{dx}{(x+1)\sqrt{x^2+x+1}}$

iii. $\int \sqrt{\frac{2-x}{x-3}}\,dx$

iv. $\int \frac{dx}{(2x^2+3)\sqrt{x^2-4}}$

v. $\int \sqrt{(x-7)(6-x)}\,dx$

**2** Integrate the following:

i. $\int \frac{(2x^3-3x^2-8x-26)}{(2x^2-5x-12)}\,dx$

ii. $\int \frac{(3x+2)}{x^3-6x^2+11x-6}\,dx$

iii. $\int \frac{(3x-2)}{(x-1)^2(x+1)(x+2)}\,dx$

iv. $\int \frac{(2x-1)}{(x+1)(x^2+2)}\,dx$

v. $\int \frac{(x+2)}{(x^2+3x+3)\sqrt{x+1}}\,dx$

vi. $\int \frac{(x^2-4)}{(x^2+1)(x^2+2)(x^2+3)}\,dx$

vii. $\int \frac{x^4-x^2+1}{x^2(x^2+1)^2}\,dx$

viii. $\int \frac{dx}{x(x^5+1)}$

ix. $\int \frac{dx}{x^2(x^7+1)^{6/7}}$

x. $\int \frac{(1-x^2)}{1-x^2+x^4}\,dx$

xi. $\int \frac{dx}{x^4+1}$

xii. $\int x^{-1/2}(2+3x^{1/3})^{-2}dx$

xiii. $\int \frac{(1+x^2)}{(1-x^2)\sqrt{1+x^2+x^4}}\,dx$

xiv. $\int \sqrt{\tan x}\,dx$

xv. $\int \frac{dx}{[(x+2)^8(x-1)^6]^{1/7}}$

**3** If $I_{m,n} = \int (\sin x)^m (\cos x)^n\,dx$, then prove that

$$I_{m,n} = \frac{(\sin x)^{m+1}\cdot(\cos x)^{n-1}}{m+n} + \frac{n-1}{m+n}\cdot I_{m,n-2}.$$

**4** If $I_n = \int \frac{dx}{(x^2+a^2)^n}$, then prove that $I_{n+1} = \frac{1}{2na^2}\cdot\frac{x}{(x^2+a^2)^n} + \frac{(2n-1)}{2n}\cdot\frac{1}{a^2}\cdot I_n$.

Working Space

## Only One Option Correct Type

*Direction* (Q. Nos. 5 and 6) *This section contains 2 multiple choice questions. Each question has four choices (a), (b), (c) and (d), out of which ONLY ONE is correct.*

**5** If $\int \frac{dx}{\sqrt{\sin^3 x\,(\sin x + 2\cos x)}} = f(x) + C$, then $f\left(\frac{\pi}{4}\right)$ is equal to

a. $-1$ b. $-\sqrt{2}$

c. $-\sqrt{3}$ d. $-2$

**6** $\int \frac{(\sqrt{1-x^2} - x)}{\sqrt{1-x^2}\,(1 + x\sqrt{1-x^2})} dx = f(x) + C$, then $f(1)$ is equal to

a. $\pi$ b. $\frac{\pi}{2}$

c. $\frac{\pi}{4}$ d. $\frac{\pi}{8}$

## Comprehension Type

*Direction* (Q. Nos. 7 and 8) *This section contains a paragraph, describing theory, experiments, data, etc. Two questions related to the paragraph have been given. Each question has only one correct answer among the four given options (a), (b), (c) and (d).*

**Passage**

Reduction formula make it possible to reduce an integral dependent on the index, $n > 0$ called the order of the integral, to an integral of the same type with a smaller index. Integration by parts helps us to derive reduction formulas.

**7** If $I_{n,-m} = \int \frac{\sin^n x}{\cos^m x} dx$, then $I_{n,-m} + \frac{n-1}{m-1} I_{n-2,\,2-m}$ is equal to

a. $\frac{\sin^{n-1} x}{\cos^{m-1} x}$ b. $\frac{1}{m-1} \cdot \frac{\sin^{n-1} x}{\cos^{m-1} x}$

c. $\frac{1}{n-1} \cdot \frac{\sin^{n-1} x}{\cos^{m-1} x}$ d. $\frac{n-1}{m-1} \cdot \frac{\sin^{n-1} x}{\cos^{m-1} x}$

**8** If $u_n = \int \frac{x^n}{\sqrt{ax^2 + 2bx + c}} dx$, then $(n+1)\,a\,u_{n+1} + (2n-1)\,b\,u_n + nc \cdot u_{n-1}$ equals

a. $x^{n-1}\sqrt{ax^2 + 2bx + c}$ b. $\frac{x^{n-2}}{\sqrt{ax^2 + 2bx + c}}$

c. $\frac{x^n}{\sqrt{ax^2 + 2bx + c}}$ d. $x^n\sqrt{ax^2 + 2bx + c}$

## One Integer Value Correct Type

*Direction* (Q. Nos. 9 and 10) *This section contains 2 questions. When worked out will result in an integer from 0 to 9 (both inclusive).*

**9** If $\int \frac{dx}{x^{1/2}(1+x^2)^{5/4}} = \frac{A\sqrt{x}}{\sqrt[4]{1+x^2}} + C$, then $A$ equals

**10** If $\int \frac{(2x+3)}{x(x+1)(x+2)(x+3)+1} dx = C - \frac{1}{f(x)}$, where $f(x)$ is of the form of $ax^2 + bx + c$, then $(a+b+c)$ equals

SET - 1

# Revisal Problems for JEE Main

## Only One Option Correct Type

Working Space

**Direction** *(Q. Nos. 1-35) This section contains 35 multiple choice questions. Each question has four choices (a), (b), (c) and (d), out of which ONLY ONE option is correct.*

**1** If $f\left(\dfrac{3x-4}{3x+4}\right)=x+2$, then $\int f(x)\,dx$ is equal to

a. $e^{x+2}\log_e\left|\dfrac{3x-4}{3x+4}\right|+C$

b. $-\dfrac{8}{3}\log_e|1-x|+\dfrac{2}{3}x+C$

c. $\dfrac{8}{3}\log_e|x-1|+\dfrac{x}{3}+C$

d. None of these

**2** $\displaystyle\int\frac{dx}{x^{1/5}(1+x^{4/5})^{1/2}}$ is equal to

a. $\sqrt{1+x^{4/5}}+C$

b. $\dfrac{5}{2}\sqrt{1+x^{4/5}}+C$

c. $x^{4/5}\sqrt{1+x^{4/5}}+C$

d. None of these

**3** If $\int f(x)\sin x\cos x\,dx=\dfrac{1}{2(a^2-b^2)}\log|f(x)|+C$, then $f(x)$ is equal to

a. $\dfrac{1}{a^2\sin^2 x+b^2\cos^2 x}$

b. $\dfrac{1}{a^2\sin^2 x-b^2\cos^2 x}$

c. $\dfrac{1}{a^2\cos^2 x+b^2\sin^2 x}$

d. $\dfrac{1}{a^2\cos^2 x-b^2\sin^2 x}$

**4** If $\displaystyle\int\frac{(\cos x-\sin x)}{\sqrt{8-\sin 2x}}dx=\sin^{-1}\left(\frac{\sin x+\cos x}{a}\right)+C$, then $a$ is equal to

a. 2

b. 3

c. 4

d. None of these

**5** If $\int x\log\left(1+\dfrac{1}{x}\right)dx=f(x)\cdot\log|x+1|+g(x)\log x^2+Lx+C$, then

a. $f(x)=\dfrac{x^2}{2}$

b. $g(x)=\log_e x$

c. $L=1$

d. $L=\dfrac{1}{2}$

**6** $\displaystyle\int\frac{x\sin x^2\cdot e^{\sec x^2}}{\cos^2 x^2}dx$ is equal to

a. $\dfrac{1}{2}e^{\sec x^2}+C$

b. $\dfrac{1}{2}e^{\sin x^2}+C$

c. $\dfrac{1}{2}\sin x^2\, e^{\cos^2 x^2}+C$

d. None of these

**7** $\displaystyle\int x\cdot\sqrt{\frac{2\sin^2(x^2+1)-\sin 2(x^2+1)}{2\sin^2(x^2+1)+\sin 2(x^2+1)}}\,dx$, where $x^2\neq n\pi-1$ and $n\in N$, is equal to

a. $\log\left|\dfrac{1}{2}\sec(x^2+1)\right|+C$

b. $\log\left|\sec\left(\dfrac{x^2+1}{2}\right)\right|+C$

c. $\dfrac{1}{2}\log|\sec(x^2+1)|+C$

d. None of these

*Working Space*

**8** $\int \frac{dx}{\sqrt{\sin^3 x \cdot \sin(x+\alpha)}}$ is equal to

a. $2\,\text{cosec}\,\alpha \cdot \sqrt{\cos\alpha + \sin\alpha \cdot \tan x} + C$

b. $-2\,\text{cosec}\,\alpha \cdot \sqrt{\cos\alpha + \sin\alpha \cot x} + C$

c. $\text{cosec}\,\alpha \cdot \sqrt{\cos\alpha + \sin\alpha \cdot \cot x} + C$

d. None of the above

**9** $\int \frac{f(x)\cdot g'(x) - g(x) f'(x)}{f(x)\cdot g(x)} \cdot \{\log g(x) - \log f(x)\}\, dx$ is equal to

a. $\log_e \left\{\frac{g(x)}{f(x)}\right\} + C$

b. $\frac{1}{2}\left\{\log_e \frac{g(x)}{f(x)}\right\}^2 + C$

c. $\frac{g(x)}{f(x)} \cdot \log_e \left\{\frac{g(x)}{f(x)}\right\} + C$

d. None of these

**10** $\int \frac{dx}{\sin^6 x + \cos^6 x}$ is equal to

a. $\tan^{-1}(\tan x + \cot x) + C$

b. $\tan^{-1}(\cot x - \tan x) + C$

c. $\tan^{-1}(\tan x - \cot x) + C$

d. None of these

**11** $\int e^{\tan^{-1} x} \cdot \left(1 + \frac{x}{1+x^2}\right) dx$ is equal to

a. $\frac{1}{2} x \cdot e^{\tan^{-1} x} + C$

b. $\frac{1}{2} e^{\tan^{-1} x} + C$

c. $x \cdot e^{\tan^{-1} x} + C$

d. $e^{\tan^{-1} x} + C$

**12** Let $f(x)$ be a polynomial satisfying $f(0) = 2$, $f'(0) = 3$ and $f''(x) = f(x)$. Then, $f(4)$ is equal to

a. $\frac{5(e^8+1)}{2e^4}$

b. $\frac{5(e^8-1)}{2e^4}$

c. $\frac{2e^4}{5(e^8-1)}$

d. $\frac{2e^4}{5(e^8+1)}$

**13** $\int \frac{\log_e (x + \sqrt{x^2+1})}{\sqrt{x^2+1}} dx$ is equal to

a. $2\log_e (x + \sqrt{x^2+1}) + C$

b. $\{\log_e (x + \sqrt{x^2+1})\}^2 + C$

c. $\log_e (x + \sqrt{x^2+1}) + C$

d. None of these

**14** $\int \frac{dx}{x^n (1+x^n)^{1/n}}$ $(n \in N)$ is equal to

a. $\frac{1}{1-n} \cdot \left[1 + \frac{1}{x^n}\right]^{1-\frac{1}{n}} + C$

b. $\frac{1}{1+n}\left[1 - \frac{1}{x^n}\right]^{1-\frac{1}{n}} + C$

c. $-\frac{1}{1+n} \cdot \left[1 - \frac{1}{x^n}\right]^{1-\frac{1}{n}} + C$

d. $-\frac{1}{1+n}\left[1 + \frac{1}{x^n}\right]^{1-\frac{1}{n}} + C$

**15** $\int \frac{e^{(x^2 + 4\log x)} - x^3 e^{x^2}}{x-1} dx$ is equal to

a. $\left(\frac{e^{3\log x} - e^{\log x}}{2x}\right) + e^{x^2} + C$

b. $\frac{(x-1)\, x \cdot e^{x^2}}{2} + C$

c. $\frac{(x^2-1)}{2x} e^{x^2} + C$

d. None of these

**16** If $\int \frac{1-x^7}{x(1+x^7)} dx = a \log |x| + b \log |x^7 + 1| + C$, then

a. $a = 1, b = 2/7$

b. $a = -1, b = 2/7$

c. $a = 1, b = -2/7$

d. None of these

Working Space

**17** Given, $f(x)=\begin{vmatrix} 0 & x^2-\sin x & \cos x-2 \\ \sin x - x^2 & 0 & 1-2x \\ 2-\cos x & 2x-1 & 0 \end{vmatrix}$, then $\int f(x)dx$ is equal to

a. $\frac{x^3}{3}-x^2\sin x-\sin 2x+C$
b. $\frac{x^3}{3}-x^2\sin x-\cos 2x+C$
c. $\frac{x^3}{3}-x^2\cos x-\cos 2x+C$
d. None of these

**18** $\int \frac{x^2}{(a+bx^2)^{5/2}}dx$ is equal to

a. $-\frac{1}{3a}\left(\frac{x^2}{a+bx^2}\right)^{3/2}+C$
b. $\frac{1}{3a}\left(\frac{x^2}{a+bx^2}\right)^{3/2}+C$
c. $\frac{1}{2a}\left(\frac{x^2}{a+bx^2}\right)^{2/3}+C$
d. None of these

**19** $\int \frac{\sin^3 x}{(1+\cos^2 x)\cdot\sqrt{1+\cos^2 x+\cos^4 x}}dx$ is equal to

a. $\sec^{-1}(\sec x+\cos x)+C$
b. $\sec^{-1}(\sec x-\cos x)+C$
c. $\sec^{-1}(\sec x-\tan x)+C$
d. None of these

**20** Let $\int e^x\{f(x)-f'(x)\}\,dx=\phi(x)$, then $\int e^x f(x)dx$ is equal to

a. $\phi(x)+e^x f(x)$
b. $\phi(x)-e^x f(x)$
c. $\frac{1}{2}[\phi(x)+e^x f(x)]$
d. $\frac{1}{2}[\phi(x)+e^x f'(x)]$

**21** If $n$ is odd positive integer, then $\int |x^n|\,dx$ is equal to

a. $\left|\frac{x^{n+1}}{n+1}\right|+C$
b. $\frac{x^{n+1}}{n+1}+C$
c. $\frac{|x|^n x}{n+1}+C$
d. None of these

**22** $\int \sin 2x\cdot\log_e(\cos x)dx$ is equal to

a. $\left(\frac{1}{2}+\log_e\cos x\right)\cos^2 x+C$
b. $\cos^2 x\cdot\log_e(\cos x)+C$
c. $\left(\frac{1}{2}-\log_e\cos x\right)\cos^2 x+C$
d. None of these

**23** Let $f(x)=\int \frac{x^2}{(1+x^2)\cdot(1+\sqrt{1+x^2})}dx$ and $f(0)=0$, then $f(1)$ is equal to

a. $\log_e(1+\sqrt{2})$
b. $\log(1+\sqrt{2})-\frac{\pi}{4}$
c. $\log(1+\sqrt{2})+\frac{\pi}{4}$
d. None of these

**24** $\int x\log x\cdot(\log x-1)dx$ is equal to

a. $2(x\log_e x-x)^2+C$
b. $\frac{1}{2}(x\log_e x-x)^2+C$
c. $(x\log_e x)^2+C$
d. $\frac{1}{2}(x\log_e x)^3+C$

**25** $\int \frac{f'(x)}{f(x)\cdot\log\{f(x)\}}dx$ is equal to

a. $\frac{f(x)}{\log\{f(x)\}}+C$
b. $f(x)\cdot\log\{f(x)\}+C$
c. $\log_e|\log f(x)|+C$
d. $\frac{1}{\log|\log f(x)|}+C$

Working Space

**26** $\int x^x (1+\log x)dx$ is equal to

a. $x^x \log_e x + C$ b. $e^{x^x} + C$

c. $x^x + C$ d. None of these

**27** Let $f(x)$ be a polynomical of degree three satisfying $f(0) = -1$ and $f(1) = 0$. Also, 0 is stationary point of $f(x)$. If $f(x)$ does not have an extremum at $x = 0$, then the value of integral $\int \frac{f(x)}{x^3 - 1} dx$ is

a. $\frac{x^2}{2} + C$ b. $x + C$

c. $\frac{x^3}{6} + C$ d. None of these

**28** If $\int \frac{dx}{1-\sin^4 x} = \frac{1}{2} \tan x + A \tan^{-1} \{f(x)\} + C$, then

a. $A = \frac{1}{2\sqrt{2}}$ and $f(x) = \sqrt{2} \tan x$ b. $A = \sqrt{2}$ and $f(x) = \sqrt{2} \tan x$

c. $A = -\sqrt{2}$ and $f(x) = \sqrt{2} \tan x$ d. None of these

**29** Let $\frac{d}{dx} g(x) = g(x)$ and $g(0) = 1$, then $\int g(x) \left( \frac{2 - \sin 2x}{1 - \cos 2x} \right) dx$ is

a. $g(x) \cot x + C$ b. $-g(x) \cot x + C$

c. $\frac{g(x)}{1 - \cos 2x} + C$ d. None of these

**30** $\int \frac{x^2}{(x \sin x + \cos x)^2} dx$ is equal to

a. $\frac{\sin x + x \cos x}{x \sin x + \cos x} + C$ b. $\frac{\sin x - x \cos x}{x \sin x + \cos x} + C$

c. $\frac{\sin x - x \cos x}{x \sin x - \cos x} + C$ d. None of these

**31** If $\int f(x) dx = F(x)$, then $\int x^3 f(x^2) dx$ is equal to

a. $\frac{1}{2}\left[x^2 \{F(x)\}^2 - \int \{F(x)\}^2 dx\right]$ b. $\frac{1}{2}\left[x^2 F(x^2) - \int F(x^2) d(x^2)\right]$

c. $\frac{1}{2}\left[x^2 F(x) - \frac{1}{2} \int \{f(x)\}^2 dx\right]$ d. None of these

**32** $\int \frac{f(x) g'(x) + f'(x) g(x)}{f(x) g(x)} \cdot [\log f(x) + \log g(x)] dx$ is equal to

a. $f(x) g(x) \cdot \log \{f(x) \cdot g(x)\} + C$ b. $\frac{1}{2} [\log \{f(x) \cdot g(x)\}]^2 + C$

c. $[\log \{f(x) \cdot g(x)\}]^2 + C$ d. None of these

**33** $\int \frac{\tan x}{\sqrt{\sin^4 x + \cos^4 x}} dx$ is equal to

a. $\frac{1}{2} \log (\tan^2 x + \sqrt{1 + \tan^4 x}) + C$ b. $\log (\tan^2 x + \sqrt{1 + \tan^4 x}) + C$

c. $\frac{1}{4} \log (\tan^2 x + \sqrt{1 + \tan^4 x}) + C$ d. None of these

**34** If $\int \frac{x^2 + 4}{x^4 + 16} dx = \frac{1}{a} \tan^{-1} \left( \frac{x^2 - 4}{ax} \right) + C$, then $a$ is equal to

a. 4 b. $2\sqrt{2}$

c. 2 d. $\sqrt{2}$

**35** If $I_n = \int (\log_e x)^n dx$, then $I_n + n \cdot I_{n-1}$ is equal to

a. $(x \log_e x)^n$ b. $x (\log_e x)^n$

c. $n (\log_e x)^n$ d. $(\log_e x)^{n-1}$

## Statement Type

Working Space

**Direction** *(Q. Nos. 36-40) This section is based on Statement I and Statement II. Select the correct answer from the codes given below.*

a. Both Statement I and Statement II are correct and Statement II is the correct explanation of Statement I
b. Both Statement I and Statement II are correct but Statement II is not the correct explanation of Statement I
c. Statement I is correct but Statement II is incorrect
d. Statement II is correct but Statement I is incorrect

**36** If $D(x) = \begin{vmatrix} f_1(x) & f_2(x) & f_3(x) \\ a_2 & b_2 & c_2 \\ a_3 & b_3 & c_3 \end{vmatrix}$, where $f_1, f_2, f_3$ are differentiable functions and $a_2, b_2, c_2, a_3, b_3, c_3$ are constants.

**Statement I** $\int D(x)\,dx = \begin{vmatrix} \int f_1(x)\,dx & \int f_2(x)\,dx & \int f_3(x)\,dx \\ a_2 & b_2 & c_2 \\ a_3 & b_3 & c_3 \end{vmatrix} + C$

**Statement II** Integration of sum of the several functions is equal to the sum of integration of individual functions.

**37** **Statement I** If $a > 0$ and $b^2 - 4ac < 0$, then the value of integral $\int \frac{dx}{ax^2 + bx + c}$ will be the type of $\mu \tan^{-1}\left(\frac{x + A}{B}\right) + C$, where $A, B, C$ and $\mu$ are constants.

**Statement II** If $a > 0$, $b^2 - 4ac < 0$, then $ax^2 + bx + c$ can be written as the sum of two squares.

**38** If $y$ is a function of $x$ such that $y(x - y)^2 = x$.

**Statement I** $\int \frac{dx}{x - 3y} = \frac{1}{2}[\log (x - y)^2 - 1]$

**Statement II** $\int \frac{dx}{x - 3y} = \log (x - 3y) + C$

**39** Let $f(x)$ be an indefinite integral of $\sin^2 x$.

**Statement I** The function $f(x)$ satisfies $f(x + \pi) = f(x)$ for all real $x$.
**Statement II** $\sin^2 (x + \pi) = \sin^2 x, \forall\, x$.

**40** **Statement I** If $x > 0, x \neq 1$, then $\int [\log_x e - (\log_x e)^2]\,dx = x \log_x e + C$.

**Statement II** $\int e^x [f(x) + f'(x)]\,dx = e^x f(x) + C$ and $e^t = x$ iff $t = \log x$.

SET - 2

# Revisal Problems for JEE Advanced

## Only One Option Correct Type

Working Space

*Direction* (*Q. Nos. 1-20*) *This section contains 20 multiple choice questions. Each question has four choices (a), (b), (c) and (d), out of which ONLY ONE option is correct.*

**1** Suppose $\begin{vmatrix} f'(x) & f(x) \\ f''(x) & f'(x) \end{vmatrix} = 0$, where $f(x)$ is continuously differentiable function with $f'(x) \neq 0$ and satisfies $f(0) = 1$ and $f'(0) = 2$, then $f(x)$ is equal to

a. $x^2 + x + 1$
b. $2e^x - 1$
c. $e^{2x}$
d. $4e^{x/2} - 3$

**2** If $\int \frac{\cos x - \sin x + 1 - x}{e^x + \sin x + x} dx = \log_e [f(x)] + g(x) + C$, where $C$ is constant of integration and $f(x)$ is positive, then $f(x) + g(x)$ has the value

a. $e^x + \sin x + 2x$
b. $e^x + \sin x$
c. $e^x - \sin x$
d. $e^x + \sin x + x$

**3** $\int x \cdot \frac{\log (x + \sqrt{1 + x^2})}{\sqrt{1 + x^2}} dx$ is equal to

a. $\sqrt{1 + x^2} \cdot \log_e (x + \sqrt{1 + x^2}) + C$
b. $\sqrt{x^2 + 1} \cdot \log (x + \sqrt{1 + x^2}) + x + C$
c. $\sqrt{x^2 + 1} \log_e (x + \sqrt{1 + x^2}) - x + C$
d. $\frac{x}{2} \log^2 (x + \sqrt{1 + x^2}) - \frac{x}{\sqrt{1 + x^2}} + C$

**4** $\int \frac{(2x + 1)}{(x^2 + 4x + 1)^{3/2}} dx$ is equal to

a. $\frac{x}{(x^2 + 4x + 1)^{1/2}} + C$
b. $\frac{x^3}{(x^2 + 4x + 1)^{1/2}} + C$
c. $\frac{x^2}{(x^2 + 4x + 1)^{1/2}} + C$
d. $\frac{1}{(x^2 + 4x + 1)^{1/2}} + C$

**5** $\int \sin (101x) \cdot \sin^{99} x \, dx$ is equal to

a. $\frac{\sin (100x) \cdot (\sin x)^{100}}{100} + C$
b. $\frac{\cos (100x) \cdot (\sin x)^{100}}{100} + C$
c. $\frac{\cos (100x) \cdot (\cos x)^{100}}{100} + C$
d. $\frac{\sin (100x) \cdot (\sin x)^{101}}{101} + C$

**6** Let $f$ be a continuous function satisfying $f'(\log_e x) = \begin{cases} 1, \text{ for } & 0 < x \le 1 \\ x, \text{ for } & x > 1 \end{cases}$ and $f(0) = 0$, then $f(x)$ can be defined as

a. $f(x) = \begin{cases} 1, & \text{if } x \le 0 \\ 1 - e^x, & \text{if } x > 0 \end{cases}$

b. $f(x) = \begin{cases} 1, & \text{if } x \le 0 \\ e^x - 1, & \text{if } x > 0 \end{cases}$

c. $f(x) = \begin{cases} x, & \text{if } x < 0 \\ e^x, & \text{if } x > 0 \end{cases}$

d. $f(x) = \begin{cases} x, & \text{if } x \le 0 \\ e^x - 1, & \text{if } x > 0 \end{cases}$

Working Space

**7** $\int \frac{px^{p+2q-1} - qx^{q-1}}{x^{2p+2q} + 2x^{p+q} + 1} dx$ is equal to

a. $-\frac{x^p}{x^{p+q}+1} + C$ b. $\frac{x^q}{x^{p+q}-1} + C$

c. $-\frac{x^q}{x^{p+q}+1} + C$ d. $\frac{x^p}{x^{p+q}+1} + C$

**8** If $I_n = \int (\sin x)^n \, dx, n \in N$, then $(5I_4 - 6I_6)$ is equal to

a. $\sin x \cdot (\cos x)^5 + C$

b. $\sin 2x \cdot \cos 2x + C$

c. $\frac{\sin 2x}{8} \cdot (\cos^2 2x + 1 - 2\cos 2x) + C$

d. None of the above

**9** $\int \frac{e^{\tan^{-1} x}}{1+x^2} \left[ \left( \sec^{-1} \sqrt{1+x^2} \right)^2 + \cos^{-1} \left( \frac{1-x^2}{1+x^2} \right) \right] dx$, when $x > 0$, is equal to

a. $e^{\tan^{-1} x} \cdot \tan^{-1} x + C$

b. $\frac{e^{\tan^{-1} x} \cdot (\tan^{-1} x)^2}{2} + C$

c. $e^{\tan^{-1} x} \cdot \left[ \sec^{-1} (\sqrt{1+x^2}) \right]^2 + C$

d. $e^{\tan^{-1} x} \cdot \left[ \operatorname{cosec}^{-1} (\sqrt{1+x^2}) \right]^2 + C$

**10** $\int \prod_{r=1}^{n} (x+r) \left( \sum_{k=1}^{n} \frac{1}{x+k} \right) dx$ is equal to

a. $\prod_{r=1}^{n} (x+r)$ b. $\sum_{k=1}^{n} \frac{1}{x+k}$

c. $\sum_{r=1}^{n} \log (x+r)$ d. None of these

**11** If $I_n = \int \cot^n x \, dx$, then $I_0 + I_1 + 2(I_2 + I_3 + \ldots + I_8) + I_9 + I_{10}$ equals

a. $u + \frac{u^2}{2} + \cdots + \frac{u^9}{9}$ b. $-\left( u + \frac{u^2}{2} + \cdots + \frac{u^9}{9} \right)$

c. $-\left( u + \frac{u^2}{2!} + \cdots + \frac{u^9}{9!} \right)$ d. $\frac{u}{2} + \frac{2u^2}{3} + \cdots + \frac{9u^9}{10}$

where, $u = \cot x$

**12** $\int \frac{x^2 (1 - \log_e x) dx}{\log^4 x - x^4}$ is equal to

a. $\frac{1}{2} \log \left( \frac{x}{\log x} \right) - \frac{1}{4} \log (\log^2 x - x^2) + C$

b. $\frac{1}{4} \log \left( \frac{\log x + x}{\log x - x} \right) + \frac{1}{2} \tan^{-1} \left( \frac{\log x}{x} \right) + C$

c. $\frac{1}{4} \log \left( \frac{\log x - x}{\log x + x} \right) - \frac{1}{2} \tan^{-1} \left( \frac{\log x}{x} \right) + C$

d. None of the above

**13** $\int \left( \frac{\sin \theta}{\cos 3\theta} + \frac{\sin 3\theta}{\cos 9\theta} + \frac{\sin 9\theta}{\cos 27\theta} \right) d\theta$ is equal to

a. $\frac{1}{2} \log \left| \frac{\sec 27\theta}{\sec \theta} \right| + C$ b. $\frac{1}{2} \log \left| \frac{\sec \theta}{\sec 27\theta} \right| + C$

c. $\frac{1}{2} \log \left| \frac{\sqrt[27]{\sec 27\theta}}{\sec \theta} \right| + C$ d. None of these

Working Space

**14** If $I = \int \frac{(x^2 + 20)}{(x \sin x + 5\cos x)^2} dx$ is equal to

a. $-\frac{x}{\cos x (x \sin x + 5 \cos x)} + \tan x + C$

b. $\frac{x}{\sin x (x \sin x + 5 \cos x)} + C$

c. $(x \sin x + 5 \cos x)^{-1} \cdot \sin x + 7x + C$

d. None of the above

**15** If $I = \int \frac{x^n}{1 + \frac{x}{1!} + \frac{x^2}{2!} + \cdots + \frac{x^n}{n!}} dx$, then

a. $I = \frac{1}{n!}\left\{x + \log\left(1 + \frac{x}{1!} + \frac{x^2}{2!} + \cdots + \frac{x^n}{n!}\right)\right\} + C$

b. $I = \frac{x^2 - e^x \log x}{n!(e^x - 1)} + C$

c. $I = n! \log\left(\frac{e^x}{1 + \frac{x}{1!} + \frac{x^2}{2!} + \cdots + \frac{x^n}{n!}}\right) + C$

d. None of the above

**16** If $I = \int \frac{dx}{x^r (1 + x^r)^{1/r}}$, $r \in N$ and $I = \alpha (1 + x^{-3})^{\beta} + C$, then the quadratic equation having roots $\alpha$ and $\beta$ is

a. $6x^2 - x - 2 = 0$

b. $6x^2 - x + 2 = 0$

c. $6x^2 + x - 2 = 0$

d. $6x^2 + x + 2 = 0$

**17** $\int \frac{x^6 (x + 1)}{\sqrt{5x^{10} + 6x^9 + x^4}} dx$ is equal to

a. $\frac{\sqrt{5x^6 + 6x^5 + 1}}{15} + C$

b. $\frac{\sqrt{5x^8 + 6x^7 + x^2}}{30} + C$

c. $\frac{\sqrt{5x^8 + 6x^7 + x}}{15} + C$

d. $(5x^{10} + 6x^5 + x^4)^{1/2} + C$

**18** The value of $\int \sqrt{x - 3}\,\{\sin^{-1}(\log_e x) + \cos^{-1}(\log_e x)\}\, dx$ is

a. $\frac{\pi}{3}(x - 3)^{3/2}$

b. $\frac{\pi}{3}(x)^{3/2}$

c. $\frac{\pi}{2}(x - 3)^{3/2}$

d. Does not exist

**19** $\int \frac{\operatorname{cosec}^2 x - 2011}{(\cos x)^{2011}} dx$ is equal to

a. $-\frac{\cot x}{(\cos x)^{2011}} + C$

b. $\frac{\cot x}{(\cos x)^{2010}} + C$

c. $-\frac{\tan x}{(\cos x)^{2011}} + C$

d. $\frac{\cot x}{(\sin x)^{2011}} + C$

**20** Let $P = \int e^{ax} \cdot \cos bx\, dx$ and $Q = \int e^{ax} \cdot \sin bx\, dx$. If $P + iQ = Z + C$, where $Z$ is a complex function and $C$ is constant of integration, then $|Z|$ is equal to

a. $e^{ax}$

b. $\frac{e^{ax}}{a + b}$

c. $\frac{e^{ax}}{\sqrt{a^2 + b^2}}$

d. $\frac{e^{bx}}{\sqrt{a^2 + b^2}}$

## Matching List Type

Working Space

**Direction** (Q. No. 21) *Choices for the correct combination of elements from Column I and Column II are given as options (a), (b), (c) and (d), out of which one is correct.*

**21** Match the Column I and Column II and mark the correct option from the codes given below.

| | Column I | | Column II |
|---|---|---|---|
| i. | Let $f(x) = \int x^{\sin x} \cdot (1 + x\cos x \cdot \log x + \sin x)\,dx$ and $f\left(\frac{\pi}{2}\right) = \frac{\pi^2}{4}$, then the value of $f\left(\frac{\pi}{4}\right)$ is | p. | Rational |
| ii. | Let $g(x) = \int \frac{1 + 2\cos x}{(\cos x + 2)^2}\,dx$ and $g(0) = 0$, then the value of $g\left(\frac{\pi}{2}\right)$ is | q. | Irrational |
| iii. | If real numbers $x$ and $y$ satisfy $(x+5)^2 + (y-12)^2 = (14)^2$. Then, the minimum value of $\sqrt{x^2 + y^2}$ is | r. | Integer |
| iv. | Let $k(x) = \int \frac{(x^2+1)}{\sqrt[3]{x^2 + 3x + 6}}\,dx$ and $k(-1) = \frac{1}{\sqrt[3]{2}}$, then the value of $k(-2)$ is | s. | Prime |

**Codes**

| | i | ii | iii | iv | | i | ii | iii | iv |
|---|---|---|---|---|---|---|---|---|---|
| a. | q | p | p,r | p,r,s | b. | p | q | r | s |
| c. | r,s | p | r | s | d. | r | q | s,p | p |

## Comprehension Type

**Direction** (Q. Nos. 22-26) *This section contains 2 paragraphs, each describing theory, experiments, data, etc. Five questions related to the paragraphs have been given. Each question has only one correct answer among the four given options (a), (b), (c) and (d).*

### Passage I

Let the function $f$ satisfies $f(x)\cdot f'(-x) = f(-x)\cdot f'(x)$ for all $x$ and $f(0) = 3$.

**22** The value of $f(x)\cdot f(-x)$ for all $x$ is

a. 4  
b. 9  
c. 12  
d. 16

**23** The number of roots of $f(x) = 0$ in $[-2, 2]$ is

a. 0  
b. 1  
c. 2  
d. 4

### Passage II

A curve is represented parametrically by the equations $x = e^t \cos t$ and $y = e^t \sin t$, where $t$ is a parameter.

**24** The relation between the parameter $t$ and the angle $\alpha$ between the tangent to the given curve and the $X$-axis is given by

a. $\frac{\pi}{2} - \alpha$  
b. $\frac{\pi}{4} + \alpha$  
c. $\alpha - \frac{\pi}{4}$  
d. $\frac{\pi}{4} - \alpha$

**25** The value of $\frac{d^2y}{dx^2}$ at the point $t = 0$ is

a. 1  
b. 2  
c. $-2$  
d. 3

**26** If $F(t) = \int (x + y)\,dt$, then the value of $F\left(\frac{\pi}{2}\right) - F(0)$ is

a. 1  
b. $-1$  
c. $e^{\pi/2}$  
d. 0

## One Integer Value Correct Type

*Working Space*

**Direction** *(Q. Nos. 27-35) This section contains 9 questions. When worked out will result in an integer from 0 to 9 (both inclusive).*

**27** A curve $g(x) = \int x^{27} \cdot (1 + x + x^2)^6 \cdot (6x^2 + 5x + 4)\,dx$ passing through origin, then $|7g(-1)|$ is equal to

**28** Let $f(xy) = f(x) \cdot f(y), \forall\ x > 0, y > 0$ and $f(x+1) = 1 + x \cdot \{1 + g(x)\}$, where $\lim_{x \to 0} g(x) = 1$. If $\int \frac{f(x)}{f'(x)}\,dx = \frac{x^2}{k} + C$, then $k$ is equal to

**29** If $\int e^{x \sin x + \cos x} \cdot \left( \frac{x^4 \cos^3 x - x \sin x + \cos x}{x^2 \cos^2 x} \right) dx = e^{x \sin x + \cos x} \cdot \frac{1}{x} [f(x)] + C$, then $|f(0)|$ is equal to

**30** Let $A = \int \frac{dx}{x^2 + 6x + 25}$ and $B = \int \frac{dx}{x^2 - 6x - 27}$. If $12(A + B) = \lambda \cdot \tan^{-1}\left(\frac{x+3}{4}\right) + \mu \log\left|\frac{x-9}{x+3}\right| + C$, then the value of $(\lambda + \mu)$ is

**31** Let $\int \frac{dx}{x^{2008} + x} = \frac{1}{p} \log\left(\frac{x^q}{1 + x^r}\right) + C$, where $p, q, r \in N$ and need not to be distinct, then the value of $\left(\frac{p+q}{r}\right)$ is

**32** If $\int \frac{e^x (2 - x^2)}{(1 - x)\sqrt{1 - x^2}}\,dx = \mu e^x \left(\frac{1+x}{1-x}\right)^{\lambda} + C$, then $2\lambda + \mu$ is equal to

**33** Let $f(x)$ be a function such that $f''(x) = f(x)$ and $f'(0) = f(0) = 0$. If the value of $\left|\int \frac{f(x)}{4 - f(x)}\,dx\right| = |\log|k - f(x)|| + C$, then $k$ equals

**34** If $\int \frac{dx}{x^{2010} \cdot (1 + x^{2010})^{1/2010}} = -\frac{1}{a}(1 + x^{-b})^{\frac{c-2}{b}} + k$, then $\left(\frac{a + b - c}{1004}\right)$ is equal to

**35** If $\int \frac{dx}{(1 + \sqrt{x})^{2010}} = 2\left[\frac{1}{\alpha(1 + \sqrt{x})^{\alpha}} - \frac{1}{\beta(1 + \sqrt{x})^{\beta}}\right] + C$, where $\alpha, \beta > 0$, then $|\alpha - \beta|$ is equal to

# JEE Main & AIEEE Archive

(Compilation of Last 13 Years' Questions)

**1** The integral $\int \frac{dx}{x^2(x^4+1)^{\frac{3}{4}}}$ equals **(JEE Main 2015)**

a. $\left(\frac{x^4+1}{x^4}\right)^{\frac{1}{4}}+C$

b. $(x^4+1)^{\frac{1}{4}}+C$

c. $-(x^4+1)^{\frac{1}{4}}+C$

d. $-\left(\frac{x^4+1}{x^4}\right)^{\frac{1}{4}}+C$

**2** The integral $\int\left(1+x-\frac{1}{x}\right)e^{x+\frac{1}{x}}dx$ is equal to **(JEE Main 2014)**

a. $(x-1)e^{x+\frac{1}{x}}+C$ b. $xe^{x+\frac{1}{x}}+C$

c. $(x+1)e^{x+\frac{1}{x}}+C$ d. $-xe^{x+\frac{1}{x}}+C$

**3** If $\int f(x)dx=\psi(x)$, then $\int x^5 f(x^3)dx$ is equal to **(JEE Main 2013)**

a. $\frac{1}{3}[x^3\psi(x^3)-\int x^2\psi(x^3)dx]+C$

b. $\frac{1}{3}x^3\psi(x^3)-3\int x^3\psi(x^3)dx+C$

c. $\frac{1}{3}x^3\psi(x^3)-\int x^2\psi(x^3)dx+C$

d. $\frac{1}{3}[x^3\psi(x^3)-\int x^3\psi(x^3)dx]+C$

**4** If $\int\frac{5\tan x}{\tan x-2}dx=x+a\log|\sin x-2\cos x|+k$, then $a$ is equal to **(AIEEE 2012)**

a. –1 b. –2

c. 1 d. 2

**5** The value of $\sqrt{2}\int\frac{\sin x\,dx}{\sin\left(x-\frac{\pi}{4}\right)}$ is **(AIEEE 2008)**

a. $x-\log\left|\cos\left(x-\frac{\pi}{4}\right)\right|+C$

b. $x+\log\left|\cos\left(x-\frac{\pi}{4}\right)\right|+C$

c. $x-\log\left|\sin\left(x-\frac{\pi}{4}\right)\right|+C$

d. $x+\log\left|\sin\left(x-\frac{\pi}{4}\right)\right|+C$

**6** $\int\frac{dx}{\cos x+\sqrt{3}\sin x}$ is equal to **(AIEEE 2007)**

a. $\frac{1}{2}\log\tan\left(\frac{x}{2}+\frac{\pi}{12}\right)+C$

b. $\frac{1}{2}\log\tan\left(\frac{x}{2}-\frac{\pi}{12}\right)+C$

c. $\log\tan\left(\frac{x}{2}+\frac{\pi}{12}\right)+C$

d. $\log\tan\left(\frac{x}{2}-\frac{\pi}{12}\right)+C$

**7** $\int\left\{\frac{(\log x-1)}{1+(\log x)^2}\right\}^2dx$ is equal to **(AIEEE 2005)**

a. $\frac{x}{(\log x)^2+1}+C$ b. $\frac{xe^x}{1+x^2}+C$

c. $\frac{x}{x^2+1}+C$ d. $\frac{\log x}{(\log x)^2+1}+C$

**8** If $\int\frac{\sin x}{\sin(x-\alpha)}dx=Ax+B\log\sin(x-\alpha)+C$, then the value of $(A, B)$ is **(AIEEE 2004)**

a. $(\sin\alpha, \cos\alpha)$

b. $(\cos\alpha, \sin\alpha)$

c. $(-\sin\alpha, \cos\alpha)$

d. $(-\cos\alpha, \sin\alpha)$

**9** $\int\frac{dx}{\cos x-\sin x}$ is equal to **(AIEEE 2004)**

a. $\frac{1}{\sqrt{2}}\log\left|\tan\left(\frac{x}{2}-\frac{\pi}{8}\right)\right|+C$

b. $\frac{1}{\sqrt{2}}\log\left|\cot\left(\frac{x}{2}\right)\right|+C$

c. $\frac{1}{\sqrt{2}}\log\left|\tan\left(\frac{x}{2}-\frac{3\pi}{8}\right)\right|+C$

d. $\frac{1}{\sqrt{2}}\log\left|\tan\left(\frac{x}{2}+\frac{3\pi}{8}\right)\right|+C$

**10** $\int\frac{dx}{x(x^n+1)}$ is equal to **(AIEEE 2002)**

a. $\frac{1}{n}\log\left(\frac{x^n}{x^n+1}\right)+C$

b. $\frac{1}{n}\log\left(\frac{x^n+1}{x^n}\right)+C$

c. $\log\left(\frac{x^n}{x^n+1}\right)+C$

d. None of the above

# JEE Advanced & IIT JEE Archive

## (Compilation of Last 10 Years' Questions)

**1** The integral $\int \frac{\sec^2 x}{(\sec x + \tan x)^{9/2}} dx$ equals to (for some arbitrary constant $K$)

**(2012, Only One Option Correct Type)**

a. $\frac{-1}{(\sec x + \tan x)^{11/2}} \left\{\frac{1}{11} - \frac{1}{7}(\sec x + \tan x)^2\right\} + K$

b. $\frac{1}{(\sec x + \tan x)^{11/2}} \left\{\frac{1}{11} - \frac{1}{7}(\sec x + \tan x)^2\right\} + K$

c. $\frac{-1}{(\sec x + \tan x)^{11/2}} \left\{\frac{1}{11} + \frac{1}{7}(\sec x + \tan x)^2\right\} + K$

d. $\frac{1}{(\sec x + \tan x)^{11/2}} \left\{\frac{1}{11} + \frac{1}{7}(\sec x + \tan x)^2\right\} + K$

**2** Let $I = \int \frac{e^x}{e^{4x} + e^{2x} + 1} dx$, $J = \int \frac{e^{-x}}{e^{-4x} + e^{-2x} + 1} dx$. Then, for an arbitrary constant $c$, the value of $J - I$ equals

**(2008, Only One Option Correct Type)**

a. $\frac{1}{2}\log\left|\frac{e^{4x} - e^{2x} + 1}{e^{4x} + e^{2x} + 1}\right| + C$

b. $\frac{1}{2}\log\left|\frac{e^{2x} + e^{x} + 1}{e^{2x} - e^{x} + 1}\right| + C$

c. $\frac{1}{2}\log\left|\frac{e^{2x} - e^{x} + 1}{e^{2x} + e^{x} + 1}\right| + C$

d. $\frac{1}{2}\log\left|\frac{e^{4x} + e^{2x} + 1}{e^{4x} - e^{2x} + 1}\right| + C$

**3** Let $f(x) = \frac{x}{(1 + x^n)^{1/n}}$ for $n \geq 2$ and $g(x) = \underbrace{(fofo \ldots of)}_{f \text{ occurs } n \text{ times}}(x)$. Then, $\int x^{n-2} g(x)\, dx$ equals

**(2007, Only One Option Correct Type)**

a. $\frac{1}{n(n-1)}(1 + nx^n)^{1-\frac{1}{n}} + C$

b. $\frac{1}{n-1}(1 + nx^n)^{1-\frac{1}{n}} + C$

c. $\frac{1}{n(n+1)}(1 + nx^n)^{1+\frac{1}{n}} + C$

d. $\frac{1}{n+1}(1 + nx^n)^{1+\frac{1}{n}} + C$

**4** The value of $\int \frac{(x^2 - 1)dx}{x^3\sqrt{2x^4 - 2x^2 + 1}}$ is

**(2006, Only One Option Correct Type)**

a. $2\sqrt{2 - \frac{2}{x^2} + \frac{1}{x^4}} + C$

b. $2\sqrt{2 + \frac{2}{x^2} + \frac{1}{x^4}} + C$

c. $\frac{1}{2}\sqrt{2 - \frac{2}{x^2} + \frac{1}{x^4}} + C$

d. None of these

# 2. Definite Integral

# DPP-1 Basic Definition

Working Space

## Subjective Question

*Direction* (Q. No. 1) *This question is subjective in nature, need to be solved completely on notebook.*

**1** Evaluate the following:

i. $\int_0^{1/2} \frac{x \cdot \sin^{-1} x}{\sqrt{1-x^2}} dx$

ii. $\int_0^{\pi/2} (\sqrt{\tan x} + \sqrt{\cot x}) dx$

iii. $\int_0^{\infty} \frac{dx}{(x + \sqrt{1+x^2})^n}$ for $n > 1$

iv. $\int_0^{\pi/2} \frac{x \sin x \cos x}{(a^2 \cos^2 x + b^2 \sin^2 x)^2} dx$

v. $\int_0^{\pi/2} (\sqrt{\sin x} + \sqrt{\cos x})^{-4} dx$

vi. $\int_0^{\pi/2} \frac{\sin^3 x}{(1+\cos^2 x) \cdot \sqrt{1+\cos^2 x + \cos^4 x}} dx$

## Only One Option Correct Type

*Direction* (Q. Nos. 2-16) *This section contains 15 multiple choice questions. Each question has four choices (a), (b), (c) and (d), out of which ONLY ONE is correct.*

**2** If $\int_0^{\pi/3} \frac{\cos x}{3 + 4\sin x} dx = k \log\left(\frac{3 + 2\sqrt{3}}{3}\right)$, then the value of $k$ is

a. 1/2 b. 1/3
c. 1/4 d. 1/8

**3** $\int_{e^{e}}^{e^{e^{e^{e}}}} \frac{dx}{x \log_e x \log_e(\log_e x) \log[\log_e(\log_e x)]}$ is equal to

a. 1 b. $1/e$
c. $e - 1$ d. $1 + e$

**4** $\int_1^{\infty} (e^{x+1} + e^{3-x})^{-1} dx$ is equal to

a. $\frac{\pi}{4e^2}$ b. $\frac{\pi}{4e}$
c. $\frac{1}{e^2}\left[\frac{\pi}{2} - \tan^{-1}\left(\frac{1}{e}\right)\right]$ d. $\frac{\pi}{2e^2}$

**5** $\int_1^{e} [(x + 1)e^x \cdot \log_e x] dx$ is equal to

a. $e$ b. $e^{e+1}$
c. $e^e(e - 1)$ d. $e^e(e - 1) + e$

Working Space

**6** Let $a, b, c$ be non-zero real numbers such that

$$\int_0^1 (1+\cos^8 x)(ax^2+bx+c)\,dx = \int_0^2 (1+\cos^8 x)\cdot(ax^2+bx+c)dx,$$

then the quadratic equation $ax^2+bx+c=0$ has

a. no root in (0, 2)
b. atleast one root in (0, 2)
c. a double root in (0, 2)
d. None of the above

**7** If $I_n = \int_0^{\pi/4} \tan^n x\,dx$, then $\lim_{n\to\infty} n(I_n + I_{n-2})$ is equal to

a. 1
b. 1/2
c. $\infty$
d. 0

**8** $\int_0^\infty \frac{x\tan^{-1}x}{(1+x^2)^2}dx$ is equal to

a. $\pi/2$
b. $\pi/4$
c. $\pi/6$
d. $\pi/8$

**9** Suppose $f, f', f''$ are continuous on $[0, e]$ and that $f'(e) = f(e) = 1$ and $\int_1^e \frac{f(x)}{x^2}dx = \frac{1}{2}$, then the value of $\int_1^e f''(x)\log x\,dx$ is

a. 0
b. 1
c. 2
d. None of these

**10** $\int_{1/2}^2 \frac{1}{x}\cdot\sin\left(x-\frac{1}{x}\right)dx$ is equal to

a. 0
b. 3/4
c. 5/4
d. 2

**11** $\int_2^4 \left[\log_x 2 - \frac{(\log_x 2)^2}{\log 2}\right]dx$ is equal to

a. 0
b. 1
c. 2
d. 4

**12** Suppose that $F(x)$ is an anti-derivative of $f(x) = \frac{\sin x}{x}$, $x > 0$, then $\int_1^3 \frac{\sin 2x}{x}dx$ can be expressed as

a. $F(6) - F(2)$
b. $\frac{1}{2}[F(6) - F(2)]$
c. $\frac{1}{2}[F(3) - F(1)]$
d. $2[F(6) - F(2)]$

**13** $\int_0^\infty f\left(x+\frac{1}{x}\right)\cdot\frac{\log x}{x}dx$ is equal to

a. 0
b. 1
c. 1/2
d. Cannot be determined

**14** The value of $\int_0^1 \left[\prod_{r=1}^{n}(x+r)\right]\left(\sum_{k=1}^{n}\frac{1}{x+k}\right)dx$ is

a. $n$
b. $n!$
c. $(n+1)!$
d. $n\cdot(n!)$

Working Space

**15** If the value of the integral $\int_1^2 e^{x^2}\,dx$ is $\alpha$, then the value of $\int_e^{e^4} \sqrt{\log_e x}\,dx$ is

a. $e^4 - e - \alpha$
b. $2e^4 - e - \alpha$
c. $2(e^4 - e) - \alpha$
d. $2e^4 - 1 - \alpha$

**16** If for non-zero $x$, $af(x) + bf\left(\frac{1}{x}\right) = \frac{1}{x} - 5$, where $a \neq b$, then $\int_1^2 f(x)\,dx$ is

a. $\frac{1}{a^2+b^2}\left(a\log 2 + 5a + \frac{7b}{2}\right)$
b. $\frac{1}{a^2-b^2}\left(a\log 2 - 5a + \frac{7b}{2}\right)$
c. $\frac{-1}{a^2+b^2}\left(a\log 2 + 5a - \frac{7b}{2}\right)$
d. None of the above

## Comprehension Type

**Direction** (Q. Nos. 17-22) *This section contains 2 paragraphs, each describing theory, experiments, data, etc. Six questions related to the paragraphs have been given. Each question has only one correct answer among the four given options (a), (b), (c) and (d).*

### Passage I

Let $g(x) = \int_0^x f(t)\,dt$, where $f$ is a function whose graph is shown below.

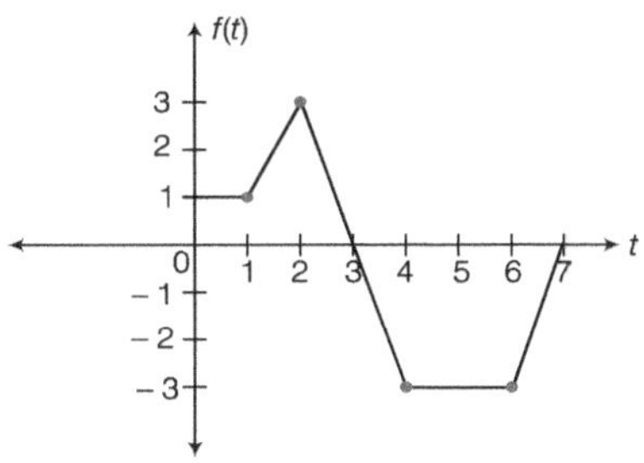


**17** The maximum value of $g(x)$ in $x \in [0, 7]$ is

a. 3
b. 9/2
c. 3/2
d. 6

**18** The value of $x$ at which $g(x)$ becomes zero is

a. 3
b. 4
c. 5
d. 6

**19** The set of values of $x$ in $[0, 7]$ for which $g(x)$ is negative, is

a. (2, 7)
b. (3, 7)
c. (4, 6)
d. (5, 7)

### Passage II

Consider $g(x) = \begin{cases} \left\{\frac{\max [f(t)] + \min [f(t)]}{2}, 0 \le t \le x\right\}, & 0 \le x \le 4 \\ |x-5| + |x-4|, & 4 < x < 5 \\ \tan\left[\sin^{-1}\left(\frac{6-x}{\sqrt{x^2-12x+37}}\right)\right], & x \ge 5 \end{cases}$

where, $f(x) = x^2 - 4x + 3$.

Working Space

20 $\int_2^5 g(x)\,dx$ is equal to

a. 5/3
b. 3
c. 13/3
d. 3/2

21 If $h(x) = \int_0^{x^2} g(t)\,dt$, then complete set of values of $x$ in the interval [0, 7] for which $h(x)$ is decreasing, is

a. (6, 7]
b. (5, 7]
c. $(\sqrt{6}, \sqrt{7}]$
d. $[\sqrt{6}, 7]$

22 $\lim_{x \to 4} \dfrac{g(x) - g(2)}{\log[\cos(4 - x)]}$ is equal to

a. 0
b. 1
c. 2
d. Does not exist

## One Integer Value Correct Type

**Direction** *(Q. Nos. 23-25) This section contains 3 questions. When worked out will result in an integer from 0 to 9 (both inclusive).*

23 The value of $x$ for which $\int_0^x (t^2 - 8t + 13)\,dt = x \cdot \sin\left(\dfrac{a}{x}\right)$ has a solution, $a \neq 0$ is

24 If $\beta + 2\int_0^1 x^2 e^{-x^2}\,dx = \int_0^1 e^{-x^2}\,dx$, then the value of $[\beta]$ (where, $[\cdot]$ denotes the greatest integer function) is

25 For $x > 0$, let $f(x) = \int_1^x \dfrac{\log_e t}{1 + t}\,dt$, then the value of $\dfrac{1}{f(e) + f\left(\dfrac{1}{e}\right)}$ is

# DPP-2 Properties of Definite Integral-I

**Property 1** $\int_a^b f(x)\,dx = \int_a^b f(t)\,dt$ **Property 2** $\int_a^b f(x)\,dx = -\int_b^a f(x)\,dx$

**Property 3** $\int_a^b f(x)\,dx = \int_a^c f(x)\,dx + \int_c^b f(x)\,dx$

## Only One Option Correct Type

Working Space

*Direction* (*Q. Nos.* 1-19) *This section contains* 19 *multiple choice questions. Each question has four choices (a), (b), (c) and (d), out of which ONLY ONE option is correct.*

**1** The value of $\int_0^{\pi} |\cos x|\,dx$ is

a. 2 b. 1
c. 0 d. −1

**2** The value of $\int_0^2 |x^2 + 2x - 3|\,dx$ is

a. 1 b. 2
c. 3 d. 4

**3** The value of $\int_0^3 [x]\,dx$ (where, [·] denotes the greatest integer function) is

a. 1 b. 2
c. 3 d. 4

**4** The value of $\int_0^2 [x^2]\,dx$ (where, [·] denotes the greatest integer function) is

a. $\sqrt{3} - \sqrt{2}$ b. $5 - \sqrt{3} - \sqrt{2}$
c. $\sqrt{5} - \sqrt{3}$ d. None of these

**5** $\int_{-1}^{1} (x - [x])\,dx$ (where, [·] denotes the greatest integer function) is

a. 0 b. 1
c. 2 d. 3

**6** $\int_0^2 \{x\}\,dx$ (where, {} denotes fractional part of $x$) is

a. 0 b. 1
c. 2 d. 3

**7** $\int_0^a \{x^n\}\,dx$ (where, {} denotes fractional part of $x$) is

a. $\sum_{r=1}^{k} (r-1)\{r^{1/n} - (r-1)^{1/n}\} + k(a - k^{1/n})$

b. $\sum_{r=1}^{k} (r-1)\{r^{1/n} - (r-1)^{1/n}\}$

c. $\sum_{r=1}^{k} \{r^{1/n} - (r-1)^{1/n}\} + k(a - k^{1/n})$

d. None of the above

*Working Space*

**8** $\int_0^2 [x^2 - x + 1]\,dx$ (where, [·] denotes the greatest integer function) is equal to

a. $\frac{5-\sqrt{5}}{2}$
b. $\frac{5+\sqrt{5}}{2}$
c. $\frac{1+\sqrt{5}}{2}$
d. $\frac{\sqrt{5}-1}{2}$

**9** $\int_0^{100} [\tan^{-1} x]\,dx$, where [·] denotes the greatest integer function, is equal to

a. 1000
b. $-\tan 1$
c. 0
d. $100 - \tan 1$

**10** $\int_{1/e}^{\tan x} \frac{t}{1+t^2}\,dt + \int_{1/e}^{\cot x} \frac{1}{t(1+t^2)}\,dt$ is equal to

a. $e$
b. $1/e$
c. 1
d. $e - 1$

**11** $\int_0^{2\pi} [2\sin x]\,dx$ (where, [·] denotes the greatest integer function) is equal to

a. $\pi/2$
b. $-\pi/2$
c. $\pi$
d. $-\pi$

**12** $\int_0^{2\pi} [\sqrt{2}\sin x]\,dx$ (where, [·] denotes the greatest integer function) is equal to

a. $\pi/2$
b. $-\pi/2$
c. $\pi$
d. $-\pi$

**13** $\int_0^{\sqrt{2}} [x^2]\,dx$ (where, [·] denotes the greatest integer function) is equal to **[AIEEE 2002]**

a. $\sqrt{2} - 1$
b. $2(\sqrt{2} - 1)$
c. $\sqrt{2}$
d. None of these

**14** $\int_{-2}^{3} |1 - x^2|\,dx$ is equal to **[AIEEE 2004]**

a. 28/3
b. 14/3
c. 7/3
d. 1/3

**15** The value of $\int_1^a [x]\,f'(x)\,dx$, $a > 1$(where, [·] denotes the greatest integer function), is **[AIEEE 2006]**

a. $[a]\,f(a) - \{f(1) + f(2) + \cdots + f([a])\}$
b. $[a]\,f([a]) - \{f(1) + f(2) + \cdots + f(a)\}$
c. $af([a]) - \{f(1) + f(2) + \cdots + f([a])\}$
d. $af(a) - \{f(1) + f(2) + \cdots + f([a])\}$

**16** Let $F(x) = f(x) + f\left(\frac{1}{x}\right)$, where $f(x) = \int_1^x \frac{\log t}{1+t}\,dt$, then $F(e)$ is equal to **[AIEEE 2007]**

a. 1/2
b. 0
c. 1
d. 2

**17** The value of $\int_0^{1.5} x[x^2]\,dx$ (where, [·] denotes the greatest integer function) is **[AIEEE 2011]**

a. 5/4
b. 0
c. 3/2
d. 3/4

**18** $\int_{e^{-1}}^{e^2} \left|\frac{\log_e x}{x}\right| dx$ is equal to

a. 3/2
b. 5/2
c. 3
d. 5

**19** If $f(x) = \begin{cases} e^{\cos x}\cdot \sin x, & \text{for } |x| < 2 \\ 2, & \text{otherwise} \end{cases}$, then $\int_{-2}^{3} f(x)\,dx$ is equal to

a. 0
b. 1
c. 2
d. 3

## Comprehension Type

*Direction (Q. Nos. 20-25) This section contains 2 paragraphs, each describing theory, experiments, data, etc. Six questions related to the paragraphs have been given. Each question has only one correct answer among the four given options (a), (b), (c) and (d).*

### Passage I

If $f(n) = \int_0^{n^2} [\sqrt{x}]\, dx$, where $n \in N$ and [·] denotes the greatest integer function.

**20** The value of $f(n)$ is

a. $\frac{n(n-1)(4n+1)}{6}$ b. $\frac{n(n+1)(4n-1)}{12}$ c. $\frac{n(n^2-1)}{6}$ d. $\frac{n(n+1)(2n+1)}{6}$

**21** The value of $\lim_{n \to \infty} \frac{f(n)}{n^3}$ is

a. $\frac{1}{6}$ b. $\frac{2}{3}$ c. $\frac{1}{3}$ d. $-\frac{1}{3}$

**22** The value of $f(6)$ is

a. 25 b. 125
c. 625 d. None of these

### Passage II

If $F(n) = \int_0^n f(x)dx$, $n \in N$, where $f(x) = \sqrt{x-k+1}$, $k - 1 \le x < k$, $k = 1, 2, 3, \ldots, n$.

**23** The degree of $F(n)$ is

a. 1/2 b. 1 c. 3/2 d. 2

**24** The value of $\lim_{n \to \infty} \frac{F(n)}{n^2}$ is

a. 0 b. $\frac{2}{3}$ c. $\frac{1}{3}$ d. 1

**25** $F(n)$ is

a. continuous for all $n \in N$ b. differentiable for all $n \in N$
c. continuous for all $n \in N - \{1\}$ d. differentiable for all $n \in N - \{1\}$

## One Integer Value Correct Type

*Direction (Q. Nos. 26-30) This section contains 5 questions. When worked out will result in an integer from 0 to 9 (both inclusive).*

**26** If $\int_1^{1.5} [x^2]dx = \sqrt{a} - \sqrt{b}$ (where, [·] denotes the greatest integer function), then find the value of $(a - b)$.

**27** If $\int_0^4 \{\sqrt{x}\}dx = \frac{a}{b}$ (where, {} denotes fractional part of $x$), then the value of $(a - b)$ is

**28** If $\int_0^x \left[\{x\} - \frac{1}{2}\right] dx = a\{x\}^2 - b\{x\}$ (where, {} denotes fractional part of $x$), then $(a + b)$ is equal to

**29** If $\int_0^{2\pi} [\sin x + \cos x]\, dx = |k\pi|$ (where, [·] denotes the greatest integer function), then the value of $k$ is

**30** $\int_{-1}^{1} [x[1 + \sin \pi x]]dx$ (where, [·] denotes the greatest integer function) is equal to

# DPP-3 Properties of Definite Integral-II

**Property 4** $\int_{-a}^{a} f(x)\,dx = \begin{cases} 2\int_0^a f(x)\,dx, & \text{if } f(x) \text{ is even} \\ 0, & \text{if } f(x) \text{ is odd} \end{cases}$

**Property 5** $\int_a^b f(x)\,dx = \int_a^b f(a+b-x)\,dx$

**Property 6** $\int_0^{2a} f(x)\,dx = \begin{cases} 2\int_0^a f(x)\,dx, & \text{if } f(2a-x) = f(x) \\ 0, & \text{if } f(2a-x) = 1-f(x) \end{cases}$

## Subjective Questions

*Working Space*

*Direction* (Q. Nos. 1-3) *These questions are subjective in nature, need to be solved completely on notebook.*

**1** Evaluate the following:

i. $\int_{-a}^{a} x\sqrt{a^2-x^2}\,dx$

ii. $\int_{-1/2}^{1/2} \sec x \cdot \log\left(\frac{1-x}{1+x}\right) dx$

iii. $\int_{-\pi/4}^{\pi/4} \frac{x^7-3x^5+3x^3-x+1}{\cos^2 x}\,dx$

iv. $\int_{-\pi/2}^{\pi/2} \sin^7 x\,dx$

v. $\int_{-\pi/2}^{\pi/2} \sin^2 x\,dx$

vi. $\int_{-\pi/4}^{\pi/4} x^3 \sin^4 x\,dx$

**2** Evaluate the following:

i. $\int_0^{\pi/2} \frac{\sqrt{\sin x}}{\sqrt{\sin x}+\sqrt{\cos x}}\,dx$

ii. $\int_0^{\pi/2} \log(\tan x)\,dx$

iii. $\int_0^{\pi/4} \log(1+\tan x)\,dx$

iv. $\int_1^2 \frac{\sqrt{x}}{\sqrt{3-x}+\sqrt{x}}\,dx$

v. $\int_{\pi/6}^{\pi/3} \frac{1}{1+\sqrt{\cot x}}\,dx$

vi. $\int_a^b \frac{f(x)}{f(x)+f(a+b-x)}\,dx$

vii. $\int_0^{\pi/2} \frac{\sin x-\cos x}{1+\sin x\cos x}\,dx$

viii. $\int_0^{\pi/2} (2\log\sin x - \log\sin 2x)\,dx$

**3** Evaluate the following:

i. $\int_0^{2\pi} \cos^5 x\,dx$

ii. $\int_0^{\pi/2} \log(\sin x)\,dx$

iii. $\int_0^{\pi} \frac{a^2\sin^2 x+b^2\cos x}{a^4\sin^2 x+b^4\cos^2 x}\,dx$

iv. $\int_0^{2\pi} \frac{x\sin^{2n} x}{\sin^{2n} x+\cos^{2n} x}\,dx$

## Only One Option Correct Type

Working Space

**Direction** *(Q. Nos. 4-21) This section contains 18 multiple choice questions. Each question has four choices (a), (b), (c) and (d), out of which ONLY ONE is correct.*

**4** Let $f(x)$ be a function satisfying $f'(x) = f(x)$ with $f(0) = 1$ and $g(x)$ be the function satisfying $f(x) + g(x) = x^2$. Then, the value of $\int_0^1 f(x)\, g(x)\, dx$ is

a. $e - \frac{1}{2}e^2 - \frac{5}{2}$ b. $e - e^2 - 3$

c. $\frac{1}{2}(e - 3)$ d. $e - \frac{e^2}{2} - \frac{3}{2}$

**5** For $f(x) = x^4 + |x|$, let $I_1 = \int_0^{\pi} f(\cos x)\, dx$ and $I_2 = \int_0^{\pi/2} f(\sin x) dx$, then $\frac{I_1}{I_2}$ has the value

a. 1 b. 1/2

c. 2 d. 4

**6** If $f(a + b - x) = f(x)$, then $\int_a^b x f(x)\, dx$ is equal to

a. $\frac{a+b}{2}\int_a^b f(a+b+x) dx$ b. $\frac{a+b}{2}\int_a^b f(b-x) dx$

c. $\frac{a+b}{2}\int_a^b f(x) dx$ d. $\frac{b-a}{2}\int_a^b f(x) dx$

**7** $\int_0^1 x(1-x)^n dx$ is equal to

a. $\frac{1}{n+1} + \frac{1}{n+2}$ b. $\frac{1}{n+1}$

c. $\frac{1}{n+2}$ d. $\frac{1}{n+1} - \frac{1}{n+2}$

**8** If $\int_0^{\pi} x f(\sin x)\, dx = A \int_0^{\pi/2} f(\sin x) dx$, then the value of $A$ is **[AIEEE 2004]**

a. 0 b. $\pi$

c. $\pi/4$ d. $2\pi$

**9** If $f(x) = \frac{e^x}{1+e^x}$, $I_1 = \int_{f(-a)}^{f(a)} x\, g[x(1-x)]\, dx$ and $I_2 = \int_{f(-a)}^{f(a)} g[x(1-x)]\, dx$, then the value of $\frac{I_2}{I_1}$ is **[AIEEE 2004]**

a. 2 b. −3

c. −1 d. 1

**10** The value of $\int_{-\pi}^{\pi} \frac{\cos^2 x}{1+a^x} dx$, $a > 0$, is **[AIEEE 2005, IIT JEE 2001]**

a. $a\pi$ b. $\pi/2$

c. $\pi/a$ d. $2\pi$

**11** $\int_{-3\pi/2}^{-\pi/2} \{(x+\pi)^3 + \cos^2(x+3\pi)\}\, dx$ is equal to **[AIEEE 2006]**

a. $\frac{\pi^4}{32} + \frac{\pi}{2}$ b. $\frac{\pi}{2}$

c. $\frac{\pi}{4} - 1$ d. $\frac{\pi^4}{32}$

Working Space

**12** $\int_0^{\pi} x f(\sin x) dx$ is equal to

a. $\pi \int_0^{\pi} f(\sin x) dx$
b. $\frac{\pi}{2} \int_0^{\pi/2} f(\sin x) dx$
c. $\pi \int_0^{\pi/2} f(\cos x) dx$
d. $\pi \int_0^{\pi} f(\cos x) dx$

**13** The value of $\int_0^{\pi} [\cot x]\, dx$ (where, [·] denotes the greatest integer function) is **[AIEEE 2009]**

a. $-1$
b. $-\pi/2$
c. $\pi/2$
d. $1$

**14** The value of $\int_0^1 \frac{8 \cdot \log(1+x)}{1+x^2} dx$ is **[AIEEE 2011]**

a. $\pi/2 \log 2$
b. $\log 2$
c. $\pi \log 2$
d. $\pi/8 \log 2$

**15** $\int_0^{\pi} e^{\cos^2 x} \cos^3 (2n+1)x \, dx$ is equal to

a. $\pi$
b. $1$
c. $0$
d. None of these

**16** $\int_{-\pi}^{\pi} (\cos ax - \sin bx)^2 dx$, where $a$ and $b$ are integers, is equal to

a. $-\pi$
b. $0$
c. $\pi$
d. $2\pi$

**17** $\int_{-\pi}^{\pi} (1-x^2) \cdot \sin x \cdot \cos^2 x \, dx$ is equal to

a. $0$
b. $\pi - \frac{\pi^3}{3}$
c. $2\pi - \pi^3$
d. $\frac{7}{2} - 2\pi^3$

**18** $\int_{-1/2}^{1/2} \left[ [x] + \log\left(\frac{1+x}{1-x}\right) \right] dx$ is equal to

a. $-1/2$
b. $0$
c. $1$
d. $2\log(1/2)$

**19** $\int_{-2}^{0} [x^3 + 3x^2 + 3x + 3 + (x+1)\cos(x+1)]\, dx$ is **[IIT JEE 2005]**

a. $-4$
b. $0$
c. $4$
d. $6$

**20** $\int_{\sqrt{\log 2}}^{\sqrt{\log 3}} \frac{x \sin x^2}{\sin x^2 + \sin(\log 6 - x^2)} dx$ is **[IIT JEE 2011]**

a. $\frac{1}{4} \log \frac{3}{2}$
b. $\frac{1}{2} \log \frac{3}{2}$
c. $\log \frac{3}{2}$
d. $\frac{1}{6} \log \frac{3}{2}$

**21** $\int_{-\pi/2}^{\pi/2} \left[ x^2 + \log\left(\frac{\pi+x}{\pi-x}\right) \right] \cos x \, dx$ is **[IIT JEE 2012]**

a. $0$
b. $\frac{\pi^2}{2} - 4$
c. $\frac{\pi^2}{2} + 4$
d. $\frac{\pi^2}{2}$

Working Space

## One or More than One Options Correct Type

*Direction* (Q. Nos. 22-26) *This section contains 5 multiple choice questions. Each question has four choices (a), (b), (c) and (d), out of which ONE or MORE THAN ONE are correct.*

**22** $\int_0^{\pi} x f(\sin x)\, dx$ is equal to

a. $\frac{\pi}{2}\int_0^{\pi} f(\sin x)\, dx$
b. $\pi\int_0^{\pi/2} f(\sin x)\, dx$
c. $\pi\int_0^{\pi/2} f(\cos x)\, dx$
d. $\frac{\pi}{2}\int_0^{\pi} f(\cos x)\, dx$

**23** If $I_n = \int_{-\pi}^{\pi} \frac{\sin nx}{(1+\pi^x)\cdot \sin x} dx$, $n = 0, 1, 2, \dots$, then **[IIT JEE 2009]**

a. $I_n = I_{n+2}$
b. $\sum_{m=1}^{10} I_{2m+1} = 10\pi$
c. $\sum_{m=1}^{10} I_{2m} = 0$
d. $I_n = I_{n-1}$

**24** $\int_0^{\pi/2} \frac{\sqrt{\cot x}}{\sqrt{\cot x} + \sqrt{\tan x}} dx$ is equal to

a. $\pi/4$
b. $\pi/2$
c. $\int_{\pi/8}^{3\pi/8} \frac{\sqrt{\cot x}}{\sqrt{\cot x} + \sqrt{\tan x}} dx$
d. $\int_0^{\pi/2} \frac{dx}{1+\tan^3 x}$

**25** $\int_a^b \text{Sgn}\, x\, dx$ (where $a, b \in R$) is equal to

a. $|b| - |a|$
b. $(b-a)\text{Sgn}(b-a)$
c. $b\,\text{Sgn}\,b - a\,\text{Sgn}\,a$
d. $|a| - |b|$

**26** Which of the following is/are true?

a. $\int_a^{\pi-a} x f(\sin x)\, dx = \frac{\pi}{2}\cdot\int_a^{\pi-a} f(\sin x)\, dx$
b. $\int_{-a}^{a} [f(x)]^2\, dx = 2\int_0^a [f(x)]^2 dx$
c. $\int_0^{n\pi} f(\cos^2 x)\, dx = n\cdot\int_0^{\pi} f(\cos^2 x)\, dx$
d. $\int_0^{b-c} f(x+c)\, dx = \int_c^b f(x)\, dx$

## Comprehension Type

*Direction* (Q. Nos. 27-30) *This section contains 2 paragraphs, each describing theory, experiments, data, etc. Four questions related to the paragraphs have been given. Each question has only one correct answer among the four given options (a), (b), (c) and (d).*

### Passage I

Let $f(\alpha) = \int_0^{\pi} \frac{x}{1-\cos\alpha \sin x} dx$

**27** $\lim_{\alpha\to\pi} f(\alpha)$ is equal to

a. $\pi$
b. $2\pi$
c. $-\pi$
d. $-2\pi$

**28** The value of $f'(\pi/2)$ is

a. $\pi$
b. $2\pi$
c. $-\pi$
d. $-2\pi$

Working Space

### Passage II

If $f(a, b) = \int_0^{\pi} \frac{x}{(a^2 \cos^2 x + b^2 \sin^2 x)^2} dx$, $a, b > 0$.

**29** The value of $f(1, 1)$ is

a. $\pi^2$ b. $\pi^3$

c. $\frac{\pi^2}{2}$ d. $\frac{\pi^3}{3}$

**30** The value of $\lim_{a \to \infty} af(a)$ is

a. $\frac{\pi^2}{2b^2}$ b. $\frac{\pi^2}{2b^3}$

c. $\pi^2$ d. None of these

## One Integer Value Correct Type

*Direction* (*Q. Nos.* 31-35) *This section contains* 5 *questions. When worked out will result in an integer from* 0 *to* 9 (*both inclusive*).

**31** The value of $5\int_0^{\pi} \frac{x^3 \cdot \cos^4 x \cdot \sin x}{(\pi^2 - 3\pi x + 3x^2)} dx$ is

**32** If $f$, $g$ and $h$ are continuous on $[0, a]$ such that $f(a - x) = f(x)$, $g(a - x) = -g(x)$ and $3h(x) = 5 + 4h(a - x)$. Then, $\int_0^a f(x)\, g(x)\, h(x)\, dx$ is equal to

**33** $\int_{-\pi}^{\pi} \frac{2x \cdot (1 + \sin x)}{1 + \cos^2 x} dx = k$, then $[k]$ (where, $[\cdot]$ denotes the greatest integer function) is equal to

**34** $\int_0^{\infty} f\left(x + \frac{1}{x}\right) \cdot \frac{\log x}{x} dx$ is equal to

**35** The value of $\frac{1}{1004}\left(\sqrt{\pi \int_0^{2008} x\,|\sin x|\,dx}\right)$ is

# DPP-4 Periodic Functions and Leibnitz Theorem

## Subjective Questions

Working Space

*Direction* (Q. Nos. 1 and 2) *These questions are subjective in nature, need to be solved completely on notebook.*

**1** Evaluate the following:

i. $\int_{-1}^{2} e^{\{x\}} dx$, where {} denotes fractional part.

ii. $\int_{-1}^{2} e^{\{3x\}} dx$

iii. $\int_{0}^{2000\pi} \frac{dx}{1+e^{\sin x}}$

iv. $\int_{-\pi}^{199\pi} \sqrt{\frac{1-\cos 2x}{2}} dx$

**2** Evaluate the following:

i. If $f(x) = \int_{x}^{x^2} \sqrt{\sin t}\, dt$, then find $f'(x)$.

ii. $\lim_{x \to 0} \frac{\int_0^{x^2} \cos t^2 dt}{x \sin x}$

iii. $\lim_{x \to \infty} \frac{\left[\int_0^x e^{t^2} dt\right]^2}{\int_0^x e^{2t^2} dt}$

iv. If $f(x) = e^{g(x)}$ and $g(x) = \int_2^x \frac{t}{1+t^4} dt$, then find $f'(2)$.

## Only One Option Correct Type

*Direction* (Q. Nos. 3-24) *This section contains 22 multiple choice questions. Each question has four choices (a), (b), (c) and (d), out of which ONLY ONE is correct.*

**3** If $\int_0^{f(x)} t^2 dt = x \cos \pi x$, then $f'(a)$ is equal to

a. $-1/9$ b. $-1/3$
c. $1/3$ d. non-existent

**4** Let $f : R \to R$ be a differentiable function and $f(1) = 4$, then the value of $\lim_{x \to 1} \int_4^{f(x)} \frac{2t}{x-1} dt$ is

a. $8f'(1)$ b. $4f'(1)$
c. $2f'(1)$ d. $f'(1)$

**5** For $n \in N$, the value of $\int_0^{n\pi + v} \sqrt{\frac{1+\cos 2x}{2}} dx$, where $\frac{\pi}{2} < v < \pi$ is

a. $2n + 1 - \cos v$ b. $2n - \sin v$
c. $2n + 2 - \sin v$ d. $2n + 1 - \sin v$

Working Space

6 Let $f(x) = \int_2^x \frac{dt}{\sqrt{1+t^4}}$ and $g$ be the inverse of $f$, then the value of $g'(0)$ is

a. 1
b. 17
c. $\sqrt{17}$
d. None of these

7 The value of $\lim_{x \to \infty} \frac{d}{dx} \int_{\sqrt{3}}^{\sqrt{x}} \frac{r^3}{(r+1)(r-1)} dr$ is

a. 0
b. 1
c. 1/2
d. non-existent

8 If $f(x) = \int_0^{\sin^2 x} \sin^{-1} \sqrt{t}\, dt + \int_0^{\cos^2 x} \cos^{-1} \sqrt{t}\, dt$, $x \in [0, \pi/2]$, then $f(x)$ equals

a. $\pi/2$
b. $\pi/4$
c. $\pi$
d. 0

9 The value of $\int_{\pi}^{10\pi} |\sin x| dx$ is

a. 9
b. 10
c. 18
d. 20

10 The value of $\lim_{x \to 0} \frac{\int_0^{x^2} \sec^2 t\, dt}{x \sin x}$ is **[AIEEE 2003]**

a. 0
b. 3
c. 2
d. 1

11 Let $f : R \to R$ be a differentiable function having $f(2) = 6$, $f'(2) = \frac{1}{48}$. Then,

$\lim_{x \to 2} \int_6^{f(x)} \frac{4t^3}{x-2} dt$ equals **[AIEEE 2005]**

a. 24
b. 36
c. 12
d. 18

12 Let $f : (0, \infty) \to R$ and $F(x) = \int_0^x f(t)\, dt$. If $F(x^2) = x^2(1+x)$, then $f(4)$ is equal to

a. 5/4
b. 7
c. 4
d. 2

13 Let $f(x) = \int_1^x \sqrt{2 - t^2}\, dt$, then the real root of $x^2 - f'(x) = 0$ are

a. $\pm 1$
b. $\pm \frac{1}{\sqrt{2}}$
c. $\pm \frac{1}{2}$
d. 0 and 1

14 Let $T > 0$ be a fixed real number. Suppose $f$ is a continuous function such that for all $x \in R$, $f(x+T) = f(x)$. If $I = \int_0^T f(x)\, dx$, then the value of $\int_3^{3+3T} f(2x)\, dx$ is

a. $\frac{3}{2} I$
b. $2I$
c. $3I$
d. $6I$

15 $f(x) = \int_{x^2}^{x^2+1} e^{-t^2} dt$ increases in the interval **[IIT JEE 2003]**

a. no where
b. $x \le 0$
c. $x \in [-2, 2]$
d. $x \ge 0$

Working Space

**16** If $f(x)$ is differentiable and $\int_0^{t^2} x f(x)\, dx = \frac{2}{5}t^5$, then $f\left(\frac{4}{25}\right)$ equals **[IIT JEE 2004]**

a. 2/5
b. −5/2
c. 1
d. 5/2

**17** If $\int_{\sin x}^{1} t^2 f(t)\, dt = (1 - \sin x)$, then the value of $f\left(\frac{1}{\sqrt{3}}\right)$ is **[IIT JEE 2005]**

a. 1/3
b. $1/\sqrt{3}$
c. 3
d. $\sqrt{3}$

**18** $\lim_{x \to \pi/4} \frac{\int_2^{\sec^2 x} f(t)\, dt}{x^2 - \pi^2/16}$ is **[IIT JEE 2007]**

a. $\frac{8}{\pi} f(2)$
b. $\frac{2}{\pi} f(2)$
c. $\frac{2}{\pi} f\left(\frac{1}{2}\right)$
d. $4f(2)$

**19** The value of $\lim_{t \to 0} \log\left[\frac{1}{t}\int_0^t (1 + a\sin bx)^{c/x}\, dx\right]$ is

a. $abc$
b. $\frac{ab}{c}$
c. $\frac{bc}{a}$
d. $\frac{ca}{b}$

**20** Let $y = f(x)$ be a differentiable curve satisfying $\int_2^x f(t)dt = \frac{x^2}{2} + \int_x^2 t^2 f(t)dt$, then

$\int_{-\pi/4}^{\pi/4} \frac{f(x) + x^9 - x^3 + x + 1}{\cos^2 x} dx$ is

a. 0
b. 1
c. 2
d. 4

**21** If $y = f(x)$ is a linear function satisfying. The relation satisfying $f(xy) = f(x) \cdot f(y), \forall x, y \in R$, then the curve $y^2 + \int_0^x (\sin t + a^2 t^3 + bt)\, dt = \alpha, \alpha, b \in R^+$ cuts $y = f^{-1}(x)$ at

a. no point
b. exactly one point
c. atleast two points
d. infinite points

**22** Let $f$ be a non-negative function defined on the interval [0, 1]. If $\int_0^x \sqrt{1 - [f'(t)]^2}\, dt = \int_0^x f(t)\, dt$, $0 \le x \le 1$ and $f(0) = 0$, then **[IIT JEE 2009]**

a. $f(1/2) < 1/2$ and $f(1/3) > 1/3$
b. $f(1/2) > 1/2$ and $f(1/3) > 1/3$
c. $f(1/2) < 1/2$ and $f(1/3) < 1/3$
d. $f(1/2) > 1/2$ and $f(1/3) < 1/3$

**23** The value of $\lim_{x \to 0} \frac{1}{x^3} \int_0^x \frac{t \log(1+t)}{t^4 + 4} dt$ is **[IIT JEE 2010]**

a. 0
b. 1/12
c. 1/24
d. 1/64

**24** Let $f$ be a real-valued function defined on the interval (−1, 1) such that $e^{-x} f(x) = 2 + \int_0^x \sqrt{t^4 + 1}\, dt$, for all $x \in (-1, 1)$ and let $f^{-1}$ be the inverse function of $f$. Then, $(f^{-1})'(2)$ is equal to **[IIT JEE 2010]**

a. 1
b. 1/3
c. 1/2
d. $1/e$

Working Space

## Comprehension Type

*Direction* (Q. Nos. 25-29) *This section contains 2 paragraphs, each describing theory, experiments, data, etc. Five questions related to the paragraphs have been given. Each question has only one correct answer among the four given options (a), (b), (c) and (d).*

### Passage I

Consider the function defined on $[0, 1] \to R$; $f(x) = \dfrac{\sin x - x\cos x}{x^2}$, if $x \neq 0$ and $f(0) = 0$.

**25** $\int_0^1 f(x)dx$ is equal to

a. $1-\sin 1$  b. $\sin 1 - 1$
c. $\sin 1$  d. $-\sin 1$

**26** $\lim_{t \to 0} \dfrac{1}{t^2} \cdot \int_0^t f(x)dx$ is equal to

a. 1/3  b. 1/6
c. 1/12  d. 1/24

### Passage II

Suppose $\lim_{x \to 0} \dfrac{\int_0^x \dfrac{t^2 dt}{(a+t^r)^{1/p}}}{bx - \sin x} = l$, where $p \in N$, $p \geq 2$, $a > 0$, $r > 0$ and $b \neq 0$.

**27** If $l$ exists and is non-zero, then

a. $b > 1$  b. $0 < b < 1$  c. $b < 0$  d. $b = 1$

**28** If $p = 3$ and $l = 1$, then the value of $a$ is

a. 8  b. 3  c. 6  d. 3/2

**29** If $p = 2$, $a = 9$ and $l$ exists, then the value of $l$ is

a. 3/2  b. 2/3  c. 1/3  d. 7/9

## One Integer Value Correct Type

*Direction* (Q. Nos. 30-35) *This section contains 6 questions. When worked out will result in an integer from 0 to 9 (both inclusive).*

**30** If $\int_0^{400\pi} \sqrt{1-\cos 2x}\, dx = k \cdot (200\sqrt{2})$, then the value of $k$ is

**31** If $\int_0^{100} e^{x-[x]} dx = k \cdot [20(e-1)]$, then the value of $k$ is

**32** For any real number $x$, let $[\cdot]$ denotes the greatest integer function. Let $f$ be a real-valued function defined on the interval $[-10, 10]$ by

$$f(x) = \begin{cases} x - [x], & \text{if } [x] \text{ is odd} \\ 1 + [x] - x, & \text{if } [x] \text{ is even} \end{cases},$$

then the value of $\dfrac{\pi^2}{10} \int_{-10}^{10} f(x) \cos \pi x \, dx$ is

**[IIT JEE 2010]**

**33** $\int_{-4}^{-5} e^{(x+5)^2} dx + 3\int_{1/3}^{2/3} e^{9(x-2/3)^2} dx$ is equal to

**34** If $y = \int_0^x f(t) \cdot \sin[2(x-t)]\, dt$ and $\dfrac{d^2y}{dx^2} + 4y = kf(x)$, then the value of $k$ is

**35** The number of points at which $f(x) = \int_0^{x^2} \dfrac{t^2 - 5t + 4}{2 + e^t} dt$ attains local maximum, is

# DPP-5 Estimation, Gamma Function and Limit as the Sum

## Subjective Questions

*Working Space*

**Direction** (Q. Nos. 1-3) *These questions are subjective in nature, need to be solved completely on notebook.*

**1** Prove that

i. $\frac{\pi}{4} < \int_0^1 \frac{1}{\sqrt{1+x^4}}\,dx \le 1$

ii. $1 \le \int_0^1 e^{x^2} dx \le e$

iii. $\log 2 < \int_0^1 \frac{dx}{\sqrt{1+x^6}} < \frac{\pi}{2}$

iv. $1 < \int_0^{\pi/2} \frac{\sin x}{x}\,dx < \pi/2$

**2** Evaluate the following:

i. $\Gamma 2$

ii. $\int_0^{\pi/2} \sin^4 x \cdot \cos^6 x\,dx$

iii. $\int_0^1 x^5 (1-x^2)^4\,dx$

iv. $\int_a^b (x-a)^m \cdot (x-b)^n dx$

**3** Evaluate the following:

i. $\lim_{n\to\infty} \left( \frac{n}{n^2+1^2} + \frac{n}{n^2+2^2} + \frac{n}{n^2+3^2} + \cdots + \frac{n}{n^2+n^2} \right)$

ii. $\lim_{n\to\infty} \left( \frac{1}{n} + \frac{1}{n+1} + \frac{1}{n+2} + \cdots + \frac{1}{3n} \right)$

iii. $\lim_{n\to\infty} \left( \frac{1^2}{1^3+n^3} + \frac{2^2}{2^3+n^3} + \frac{3^2}{3^3+n^3} + \cdots + \frac{n^2}{n^3+n^3} \right)$

iv. $\lim_{n\to\infty} \frac{(n!)^{1/n}}{n}$

## Only One Option Correct Type

**Direction** (Q. Nos. 4-12) *This section contains 9 multiple choice questions. Each question has four choices (a), (b), (c) and (d), out of which ONLY ONE is correct.*

**4** If $I_1 = \int_0^1 2^{x^2} dx$, $I_2 = \int_0^1 2^{x^3} dx$, $I_3 = \int_1^2 2^{x^2} dx$ and $I_4 = \int_1^2 2^{x^3} dx$, then **[AIEEE 2005]**

a. $I_3 > I_4$

b. $I_3 = I_4$

c. $I_1 > I_2$

d. $I_2 > I_1$

Working Space

**5** Let $S_n = \frac{n}{(n+1)(n+2)} + \frac{n}{(n+2)(n+4)} + \frac{n}{(n+3)(n+6)} + \cdots + \frac{1}{(6n)}$,
then $\lim_{n\to\infty} S_n$ is equal to

a. $\log\left(\frac{3}{2}\right)$ b. $\log\left(\frac{9}{2}\right)$

c. 1 d. 0

**6** The value of $\lim_{n\to\infty} \frac{1^p + 2^p + \cdots + n^p}{n^{p+1}}$ is **[AIEEE 2002]**

a. 1 b. $\frac{1}{p+1}$

c. $\frac{1}{p+2}$ d. $p^2$

**7** The value of $\lim_{n\to\infty} \sum_{r=1}^{n} \frac{1}{n} e^{r/n}$ is **[AIEEE 2004]**

a. $e$ b. $e-1$

c. $1-e$ d. $e+1$

**8** The value of $\lim_{n\to\infty} \left(\frac{1}{n^2}\sec^2\frac{1}{n^2} + \frac{2}{n^2}\sec^2\frac{4}{n^2} + \cdots + \frac{1}{n}\sec^2 1\right)$ is **[AIEEE 2005]**

a. $\frac{1}{2}\sec 1$ b. $\frac{1}{2}\operatorname{cosec} 1$

c. $\tan 1$ d. $\frac{1}{2}\tan 1$

**9** Let $I = \int_0^1 \frac{\sin x}{\sqrt{x}}dx$ and $J = \int_0^1 \frac{\cos x}{\sqrt{x}}dx$, then which one of the following is true?

a. $I > 2/3$ and $J > 2$
b. $I < 2/3$ and $J < 2$
c. $I < 2/3$ and $J > 2$
d. $I > 2/3$ and $J < 2$

**10** If $I(m, n) = \int_0^1 t^m (1+t)^n dt$, then $I(m, n)$ in terms of $I(m+1, n-1)$ is

a. $\frac{m}{n+1} I(m+1, n-1)$

b. $\frac{n}{m+1} I(m+1, n-1)$

c. $\frac{2^n}{m+1} - \frac{n}{m+1} I(m+1, n-1)$

d. $\frac{2^n}{m+1} - \frac{n}{n+1} I(m+1, n-1)$

**11** The value of $\lim_{n\to\infty} \sum_{r=1}^{4n} \frac{\sqrt{n}}{\sqrt{r}\,(3\sqrt{r} + 4\sqrt{n})^2}$ is

a. 1/35 b. 1/14

c. 1/10 d. 1/5

**12** The absolute value $\int_{10}^{19} \frac{(\sin x)}{1+x^8}dx$ is less than

a. $10^{-10}$ b. $10^{-11}$

c. $10^{-7}$ d. $10^{-9}$

## One or More than One Options Correct Type

Working Space

**Direction** *(Q. Nos. 13-15) This section contains 3 multiple choice questions. Each question has four choices (a), (b), (c) and (d), out of which ONE or MORE THAN ONE are correct.*

**13** The value(s) of $\int_0^1 \frac{x^4 (1-x)^4}{1+x^2} dx$ is (are) **[IIT JEE 2010]**

a. $\frac{22}{7} - \pi$ b. $\frac{2}{105}$ c. 0 d. $\frac{71}{5} - \frac{3\pi}{2}$

**14** Let $S_n = \sum_{k=0}^{n} \frac{n}{n^2 + kn + k^2}$ and $T_n = \sum_{k=0}^{n-1} \frac{n}{n^2 + kn + k^2}$, for $n = 1, 2, 3, \dots$, then **[IIT JEE 2008]**

a. $S_n < \frac{\pi}{3\sqrt{3}}$
b. $S_n > \frac{\pi}{3\sqrt{3}}$
c. $T_n < \frac{\pi}{3\sqrt{3}}$
d. $T_n > \frac{\pi}{3\sqrt{3}}$

**15** For $a \in R, a \neq 1$, $\lim_{n \to \infty} \frac{(1^a + 2^a + \cdots + n^a)}{(n+1)^{a-1} \cdot [(na+1) + (na+2) + \cdots + (na+n)]} = \frac{1}{60}$, then $a$ is equal to **[JEE Advanced 2013]**

a. 5
b. 7
c. $\frac{-15}{2}$
d. $\frac{-17}{2}$

## Comprehension Type

**Direction** *(Q. Nos. 16-21) This section contains 3 paragraphs, each describing theory, experiments, data, etc. Six questions related to the paragraphs have been given. Each question has only one correct answer among the four given options (a), (b), (c) and (d).*

### Passage I

If $S_n = \int_0^{\pi/2} \frac{\sin(2n-1)x}{\sin x} dx$ and $V_n = \int_0^{\pi/2} \frac{\sin^2 nx}{\sin^2 x} dx$, where $n \in N$.

**16** $S_{n+1} - S_n$ is equal to

a. $\frac{1}{n+1}$
b. $\frac{1}{2n-1}$
c. $\frac{1}{n+1} - \frac{1}{2n-1}$
d. 0

**17** $V_{n+1} - V_n$ is equal to

a. $S_n$
b. $S_{n+1}$
c. $S_{n-1}$
d. $S_{n+1} - S_n$

### Passage II

If $I_n = \int_0^{\pi} \frac{1 - \cos nx}{1 - \cos x} dx$, where $n$ is a non-negative integer.

**18** $I_{n+2} + I_n$ is equal to

a. $I_{n+1}$ b. $2I_{n+1}$ c. $2I_{n-1}$ d. 0

**19** The value of $\int_0^{\pi/2} \frac{\sin^2 nx}{\sin^2 x} dx$ is

a. $n\pi$
b. $\frac{n\pi}{2}$
c. 0
d. None of these

Working Space

## Passage III

Let $u_n = \int_0^{\pi/2} \cos^n x \cdot \cos nx \, dx$, where $n \in N$.

**20** The value of $\frac{u_n}{u_{n-1}}$ is

a. 1
b. 0
c. 1/2
d. 1/4

**21** The value of $u_n$ is

a. $\frac{\pi}{2^n}$
b. $\frac{\pi}{2^{n+1}}$
c. $\frac{\pi}{2^{n-1}}$
d. $\pi$

## One Integer Value Correct Type

*Direction* (*Q. Nos. 22-25*) *This section contains 4 questions. When worked out will result in an integer from 0 to 9 (both inclusive).*

**22** If $\int_0^1 \frac{x^\alpha - 1}{\log x} dx = \frac{1}{\alpha + k}$, then the value of $k$ is

**23** $\lim_{n \to \infty} \frac{\pi}{6n}\left[\sec^2\left(\frac{\pi}{6n}\right) + \sec^2\left(2 \cdot \frac{\pi}{6n}\right) + \cdots + \sec^2 (n-1)\frac{\pi}{6n} + \frac{4}{3}\right]$ has the value $\frac{1}{\sqrt{k}}$, then the value of $k$ is

**24** The interval [0, 4] is divided into $n$ equal subintervals by the points $x_0, x_1, x_2, \ldots, x_{n-1}, x_n$, where $0 = x_0 < x_1 < x_2 < \cdots < x_n = 4$. If $\delta x = x_i - x_{i-1}$ for $i = 1, 2, 3, \ldots, n$. Then, $\lim_{\delta x \to 0} \sum_{i=1}^{n} x_i \cdot \delta x$ is equal to

**25** If $\int_0^\infty x^{2n+1} \cdot e^{-x^2} dx = \frac{n!}{k}$, then the value of $k$ is

SET - 1

# Revisal Problems
## for JEE Main

Working Space

### Only One Option Correct Type

*Direction (Q. Nos. 1-45) This section contains 45 multiple choice questions. Each question has four choices (a), (b), (c) and (d), out of which ONLY ONE option is correct.*

**1** $\int_0^a \frac{g(x)}{f(x)+f(a-x)}dx$ vanishes, if

a. $g(x)$ is odd
b. $g(x)=-g(a-x)$
c. $f(x)=f(a-x)$
d. $f(a-x)=g(x)$

**2** The value of $a$ for which $\int_1^2 \{a^2+(4-4a)x+4x^3\}dx \le 12$, is given by

a. $a=3$
b. $a \le 4$
c. $0 \le a < 3$
d. None of these

**3** If $f(x)$ and $g(x)$ are two continuous functions defined on $[-a, a]$, then the value of

$\int_{-a}^{a} \{f(x)+f(-x)\}\{g(x)+g(-x)\}dx$ is

a. $2a$
b. $f(a)g(a)$
c. $a$
d. 0

**4** If $\int_0^1 \frac{\sin t}{1+t}dt=\alpha$, then the value of $\int_{4\pi-2}^{4\pi} \frac{\sin t/2}{4\pi+2-t}dt$ in terms of $\alpha$, is

a. $2\alpha$
b. $-2\alpha$
c. $\alpha$
d. $-\alpha$

**5** For $x \in R$ and a continuous function $f(x)$, let $I_1=\int_{\sin^2 t}^{1+\cos^2 t} xf[x(2-x)dx]$ and

$I_2=\int_{\sin^2 t}^{1+\cos^2 t} f[x(2-x)]dx$, then $\frac{I_1}{I_2}$ is equal to

a. 0
b. 1
c. 2
d. 3

**6** If $f(a-x)+f(x)=0$ for all $x \in [0, a]$, then the value of $\int_0^a \frac{dx}{1+e^{f(x)}}$ is

a. $a$
b. $a/2$
c. $f(a)$
d. $\frac{1}{2}f(a)$

**7** If $\int_0^a f(x)dx=\lambda$ and $\int_0^a f(2a-x)dx=\mu$, then $\int_0^{2a} f(x)dx$ is equal to

a. $\lambda+\mu$
b. $\lambda-\mu$
c. $2\lambda+\mu$
d. $\lambda+2\mu$

**8** Let $\frac{d}{dx}[F(x)]=\frac{e^{\sin x}}{x}$, $x>0$. If $\int_1^4 \frac{2e^{\sin x^2}}{x}dx=F(k)-F(1)$, then one of the possible values of $k$ is

a. 4
b. 8
c. 16
d. 32

**9** The value of $\int_{-2}^{2} \frac{\sin^2 x}{\left[\frac{x}{\pi}\right]+\frac{1}{2}}dx$ (where, $[\cdot]$ denotes the greatest integer function) is

a. 1
b. 0
c. $4-\sin 4$
d. None of these

Working Space

**10** If [·] denotes greatest integer function, then $\int_{-\pi/2}^{\pi/2}\left(\left[\frac{x}{\pi}\right]+\frac{1}{2}\right)dx$ equals

a. $\pi$ b. $\pi/2$
c. 0 d. $-\pi/2$

**11** $\int_{-\pi/2}^{\pi/2}\frac{1}{e^{\sin x}+1}dx$ is equal to

a. 0 b. 1
c. $-\pi/2$ d. $\pi/2$

**12** If $I_1=\int_0^{\pi/2}\cos\theta\cdot f(\sin\theta+\cos^2\theta)d\theta$ and $I_2=\int_0^{\pi/2}\sin 2\theta\cdot f(\sin\theta+\cos^2\theta)d\theta$, then

a. $I_1=I_2$ b. $I_1=-I_2$
c. $I_1=2I_2$ d. $I_1=-2I_2$

**13** If $I_n=\int_{1/n}^{(an-1)/n}\frac{\sqrt{x}}{\sqrt{a-x}+\sqrt{x}}dx$ for all $n\in N$, then $I_n-nI_{n-1}$ equals

a. $e$ b. $1/e$
c. $-1/e$ d. None of these

**14** $\frac{\int_0^n [x]dx}{\int_0^n \{x\}\,dx}$ (where, [·] and {·} denotes greatest integer function and fractional part of $x$) is equal to

a. $\frac{1}{n-1}$ b. $\frac{1}{n}$
c. $n$ d. $n-1$

**15** Let $f(x)=\max\{x+|x|,\ x-[x]\}$, where [·] denotes greatest integer function, then $\int_{-2}^{2}f(x)dx$ is equal to

a. 3 b. 5
c. 1 d. None of these

**16** If $\int_0^1\frac{e^t}{t+1}dt=a$, then $\int_{b-1}^{b}\frac{e^{-t}}{t-b-1}dt$ is equal to

a. $ae^{-b}$ b. $-ae^{-b}$
c. $be^{-b}$ d. None of these

**17** If $f(x)$ is periodic with period $T$, then

a. $\int_a^b f(x)\,dx=\int_a^{b+T}f(x)\,dx$ b. $\int_a^b f(x)\,dx=\int_{a+T}^{b}f(x)\,dx$
c. $\int_a^b f(x)\,dx=\int_{a+T}^{b+T}f(x)\,dx$ d. $\int_a^b f(x)\,dx=\int_{a+T}^{b+2T}f(x)\,dx$

**18** If $\int_0^x f(t)\,dt=x+\int_x^1 tf(t)\,dt$, then the value of $f(1)$ is

a. 1/2 b. 0
c. 1 d. −1/2

**19** If $I_n=\int_1^e(\log x)^n dx$, then $I_n+nI_{n-1}$ is equal to

a. $1/e$ b. $e$
c. $e-1$ d. None of these

**20** If $\left|\int_a^b f(x)dx\right|=\int_a^b|f(x)|\,dx$, where $a<b$, then $f(x)=0$ has

a. exactly one root in $(a, b)$ b. atleast one root in $(a, b)$
c. no root in $(a, b)$ d. None of these

**21** $\int_0^{100\pi}\sum_{r=1}^{10}\tan rx\,dx$ is equal to

a. $100\pi$ b. $-100\pi$
c. 1 d. None of these

Working Space

**22** If $I_n = \int_0^{\pi/4} \tan^n x\,dx$, then $\frac{1}{I_2 + I_4}, \frac{1}{I_3 + I_5}, \frac{I}{I_4 + I_6}, \dots$ form a/an

a. AP
b. GP
c. HP
d. None of these

**23** $\int_1^{e^6} \left[\frac{\log x}{3}\right] dx$, where [·] denotes greatest integer function, is equal to

a. 0
b. $e^6 - e^3$
c. $e^6 + e^3$
d. $e^3 - e^6$

**24** Let $f(a) > 0$, $f(x)$ be non-decreasing continuous function in $[a, b]$, then $\frac{1}{b-a}\int_a^b f(x)\,dx$ has the

a. maximum value $f(b)$ and minimum value $f(a)$
b. maximum value $f(a)$ and minimum value $f(b)$
c. maximum value $af(a)$ and minimum value $bf(b)$
d. None of the above

**25** Let $f : R \to R$, given by $f(x + y) = f(x) + f(y)$ for all $x, y, \in R$. If $\int_0^2 f(x)\,dx = \alpha$, then $\int_{-2}^2 f(x)\,dx$ is

a. $2\alpha$
b. $\alpha$
c. 0
d. None of these

**26** Let $f : R \to R$ be given by $f(x) = \begin{cases} |x - [x]|, & \text{when } [x] \text{ is odd} \\ |x - [x] - 1|, & \text{when } [x] \text{ is even} \end{cases}$, where [·] denotes greatest integer function, then $\int_{-2}^4 f(x)\,dx$ is equal to

a. 5/2
b. 3/2
c. 5
d. 3

**27** If $\int_{-1}^x f(t)\,dt + f'''(3)\int_x^0 f(t)\,dt = \int_1^x t^3 dt - f'(1)\int_0^x t^2 dt + f''(2)\int_x^3 t\,dt$, then the value of $[1 - f'''(3)] \cdot f'(4)$ is

a. $48 - 8f'(1) + f''(2)$
b. $48 - 8f'(1) - f''(2)$
c. $48 + 8f'(1) + f''(2)$
d. $48 + 8f'(1) - f''(2)$

**28** $\int_{-1}^1 [x^2 + \{x\}]dx$ (where, [·] and {·} denote greatest integer and fractional part of $x$) is equal to

a. $\frac{3-\sqrt{5}}{2}$
b. $\frac{\sqrt{5}}{2}$
c. $\frac{5+\sqrt{5}}{2}$
d. $-\left(\frac{5+\sqrt{5}}{2}\right)$

**29** If $I = \int_0^1 \frac{dx}{1 + x^{\pi/2}}$, then

a. $\log 2 < I < \pi/4$
b. $I < \log 2$
c. $I > \pi/4$
d. None of these

**30** If $f(x) = \int_0^x \{f(t)\}\,dt$ and $\int_0^1 \{f(t)\}^{-1}\,dt = \sqrt{2}$, then

a. $f(x) = \sqrt{2x}$
b. $f(x) = \sqrt{x}$
c. $f(x) = \sqrt{3x - 1}$
d. None of these

**31** If $I_n = \int_0^1 x^n e^{-x}\,dx$, for $n \in N$, then $I_n - nI_{n-1}$ is equal to

a. $e$
b. $1/e$
c. $-1/e$
d. None of these

**32** $\lim_{n \to \infty} \frac{1}{n} \cdot \sum_{r=1}^{2n} \frac{r}{\sqrt{n^2 + r^2}}$ is equal to

a. $1 + \sqrt{5}$
b. $-1 + \sqrt{5}$
c. $-1 + \sqrt{2}$
d. $1 + \sqrt{2}$

**33** The points of extremum of $\phi(x) = \int_1^x e^{-t^2/2} \cdot (1 - t^2)\,dt$ are

a. $x = 1, -1$
b. $x = -1, 2$
c. $x = 2, 1$
d. $x = -2, 1$

Working Space

**34** If $I_n = \int_0^{\pi/2} \frac{\cos^2 nx}{\sin x} dx$, then $I_2 - I_1, I_3 - I_2, I_4 - I_3$ are in

a. GP
b. AP
c. HP
d. None of these

**35** If $\int_n^{n+1} f(x)dx = n^3, n \in Z$, then the value of $\int_{-3}^{3} f(x)dx$ is

a. 9
b. −27
c. −9
d. None of these

**36** If $I = \int_0^1 \frac{x}{8+x^3} dx$, then the smallest interval in which $I$ lies, is

a. (0, 1/8)
b. (0, 1/9)
c. (0, 1/10)
d. (0, 1/7)

**37** If $\int_0^x t\, f(t)dt = \sin x - x\cos x - \frac{x^2}{2}$ for all $x \in R - \{0\}$, then $f\left(\frac{\pi}{6}\right)$ equals

a. 0
b. 1
c. −1/2
d. None of these

**38** If $\int_0^x \frac{bt\cos 4t - a\sin 4t}{t^2} dt = \frac{a\sin 4x}{x}$ for all $x \neq 0$, then

a. $a = 1/4, b = 1$
b. $a = 2, b = 2$
c. $a = -1, b = 4$
d. $a = 2, b = 4$

**39** If $\int_0^{t^2} x f(x)dx = \frac{2}{5}t^5$, then $f\left(\frac{4}{25}\right)$ is equal to

a. 2/5
b. −5/2
c. 1
d. 5/2

**40** $\int_{-2}^{2} |[x]|\, dx$ is equal to

a. 1
b. 2
c. 3
d. 4

**41** The least value of $\phi(x) = \int_{5\pi/4}^{x} (3\sin t + 4\cos t)dt$, when $x \in [5\pi/4, 4\pi/3]$ is

a. $\sqrt{3}$
b. $-2\sqrt{3}$
c. $-2\sqrt{3} + 3/2 + 1/\sqrt{2}$
d. $-3/2 - 1/\sqrt{2}$

**42** If $f\left(\frac{1}{x}\right) + x^2 f(x) = 0$ for all $x > 0$, then $\int_{\sin\theta}^{\operatorname{cosec}\theta} f(x)dx$ is equal to

a. $\sin\theta + \operatorname{cosec}\theta$
b. $\sin^2\theta$
c. $\operatorname{cosec}^2\theta$
d. None of these

**43** If $f\left(\frac{1}{x}\right) + x^2 f(x) = 0$, $x > 0$ and $I = \int_{1/x}^{x} f(t)dt$, $\frac{1}{2} \le x \le 2$, then $I$ is equal to

a. $f(2) - f\left(\frac{1}{2}\right)$
b. $f\left(\frac{1}{2}\right) - f(2)$
c. 0
d. None of these

**44** Let $f(x)$ be a function defined by $f(x) = \int_1^x t(t^2 - 3t + 2)dt$, $1 \le x \le 3$. Then, the range of $f(x)$ is

a. [−1/4, 2]
b. [0, 2]
c. [−1/4, 4]
d. None of these

**45** Let $f(x) = \frac{e^x + 1}{e^x - 1}$ and $\int_0^1 x^3 \cdot \frac{e^x + 1}{e^x - 1} dx = \alpha$, then $\int_{-1}^{1} t^3 f(t)dt$ is equal to

a. 0
b. $\alpha$
c. $2\alpha$
d. None of these

## Statement Type

Working Space

**Direction** **(Q. Nos. 46-50)** ***This section is based on Statement I and Statement II. Select the correct answer from the codes given below.***

a. Both Statement I and Statement II are correct and Statement II is the correct explanation of Statement I
b. Both Statement I and Statement II are correct but Statement II is not the correct explanation of Statement I
c. Statement I is correct but Statement II is incorrect
d. Statement II is correct but Statement I is incorrect

**46** **Statement I** The value of $\int_{\pi/6}^{\pi/3} \frac{dx}{1+\sqrt{\tan x}}$ is $\frac{\pi}{6}$.

**Statement II** $\int_a^b f(x)dx = \int_a^b f(a+b-x)dx$

**47** **Statement I** If $f(x) = \int_0^1 [x\, f(t) + 1]dt$, then $\int_0^3 f(x)dx = 12$.

**Statement II** $f(x) = 3x + 1$

**48** Consider $I = \int_{-\pi/4}^{\pi/4} \frac{dx}{1-\sin x}$

**Statement I** $I = 0$

**Statement II** $\int_{-a}^{a} f(x)dx = 0$, $f(x)$ is odd function.

**49** **Statement I** The function $f(x) = \int_0^x \sqrt{1+t^2}\, dt$ is an odd function and $g(x) = f'(x)$ is an even function.

**Statement II** For a differentiable function $f(x)$, if $f'(x)$ is an even function, then $f(x)$ is odd function.

**50** Given, $f(x) = \sin^3 x$ and $p(x)$ is a quadratic polynomial with leading coefficient unity.

**Statement I** $\int_0^{2\pi} p(x) \cdot f''(x)dx$ vanishes because

**Statement II** $\int_0^{2\pi} f(x)dx$ vanishes.

SET - 2

# Revisal Problems for JEE Advanced

## Only One Option Correct Type

Working Space

*Direction* (Q. Nos. 1-50) *This section contains 50 multiple choice questions. Each question has four choices (a), (b), (c) and (d), out of which ONLY ONE is correct.*

**1** The set of values of $a$ for which the inequality $\int_a^0 (3^{-2x} - 2\cdot 3^{-x})\,dx \ge 0$, is

a. $[0, 1]$
b. $(-\infty, -1]$
c. $[0, \infty)$
d. $(-\infty, -1] \cup [0, \infty)$

**2** The solution set of $\sqrt{5x - 6 - x^2} + \frac{\pi}{2}\left(\int_0^x dz\right) > x\int_0^\pi \sin^2 x\,dx$ is

a. $R$
b. $(1, 6)$
c. $(-6, 1)$
d. $(2, 3)$

**3** $\int_{\pi/4}^{5\pi/4} (|\cos t|\sin t + |\sin t|\cos t)\,dt$ has the value

a. 0
b. 1/2
c. $1/\sqrt{2}$
d. 1

**4** If $\int_{\pi/6}^{\pi/4} \frac{1+\cot x}{e^x \sin x}\,dx = ae^{-\pi/6} + be^{-\pi/6}$, then $(a+b)$ is equal to

a. $2 - \sqrt{2}$
b. $2 + \sqrt{2}$
c. $2\sqrt{2} - 2$
d. $2\sqrt{3} - \sqrt{2}$

**5** For $U_n = \int_0^1 x^n (2-x)^n\,dx$; $V_n = \int_0^1 x^n (1-x)^n\,dx$, $n \in N$, which of the following statement is true?

a. $U_n = 2V_n$
b. $U_n = 2^{-n} V_n$
c. $U_n = 2^{2n} V_n$
d. $U_n = 2^{-2n} V_n$

**6** Let $S(x) = \int_{x^2}^{x^3} \log t\,dt$ $(x > 0)$ and $H(x) = \frac{S'(x)}{x}$, then $H(x)$ is

a. continuous but not differentiable
b. derivable and continuous
c. neither continuous nor differentiable
d. derivable but not continuous

**7** Let $f(x) = \frac{\sin x}{x}$, then $\int_0^{\pi/2} f(x)\cdot f\left(\frac{\pi}{2} - x\right)dx$ is equal to

a. $\frac{2}{\pi}\int_0^\pi f(x)\,dx$
b. $\int_0^\pi f(x)\,dx$
c. $\pi\int_0^\pi f(x)\,dx$
d. $\frac{1}{\pi}\int_0^\pi f(x)dx$

**8** If $\int_0^{f(x)} t^2 dt = x\cos \pi x$, then $f'(9)$ is equal to

a. $-1/9$
b. $-1/3$
c. $1/3$
d. non-existent

**9** The positive value of $a$, so that the definite integral $\int_a^{a^2} \frac{dx}{x + \sqrt{x}}$ achieves smallest value, is

a. $\tan^2 \frac{\pi}{8}$
b. $\tan^2 \frac{3\pi}{8}$
c. $\tan^2 \frac{\pi}{12}$
d. 0

*Working Space*

**10** The value of $f(x) = 1 + x + \int_1^x (\log^2 t + 2\log t)\,dt$, where $f'(x)$ vanishes, is

a. $e^{-1}$
b. 0
c. $2e^{-1}$
d. $1 + 2e^{-1}$

**11** $\lim_{\lambda \to 0}\left[\int_0^1 (1+x)^\lambda dx\right]^{1/\lambda}$ is equal to

a. $2\log 2$
b. $\frac{4}{e}$
c. $\log\frac{4}{e}$
d. 4

**12** The value of $\int_0^1 e^{e^x}(1 + x\cdot e^x)\,dx$ is

a. $e^e$
b. $e^e - e$
c. $e^e - 1$
d. $e$

**13** The value of $\int_{-1}^1 \frac{dx}{(1+e^x)(1+x^2)}$ is

a. $\pi/2$
b. $\pi/4$
c. $\pi/8$
d. $\pi/16$

**14** The set of values of $a$ for which $\int_0^2 (t - \log_2 a)\,dt = \log_2(4/a^2)$ is

a. $a \in R$
b. $a \in R^+$
c. $a < 2$
d. $a > 2$

**15** If $f''(x) = -\frac{1}{x^2} - \pi^2 \sin(\pi x)$; $f'(2) = \pi + \frac{1}{2}$ and $f(1) = 0$, then the value of $f\left(\frac{1}{2}\right)$ is

a. $\log 2$
b. 1
c. $\frac{\pi}{2} - \log 2$
d. $1 - \log 2$

**16** $I = \int_0^1 \frac{\tan^{-1} x}{x}\,dx$ is equal to

a. $\int_0^{\pi/4} \frac{\sin x}{x}\,dx$
b. $\int_0^{\pi/2} \frac{x}{\sin x}\,dx$
c. $\frac{1}{2}\int_0^{\pi/2} \frac{x}{\sin x}\,dx$
d. $\frac{1}{2}\int_0^{\pi/4} \frac{x}{\sin x}\,dx$

**17** Let $I(a) = \int_0^\pi \left(\frac{x}{a} + a\sin x\right)^2 dx$; $a \in R$. Then, the value of $a$ for which $I(a)$ attains its minimum value, is

a. $\sqrt{\pi\sqrt{\frac{2}{3}}}$
b. $\sqrt{\pi\sqrt{\frac{3}{2}}}$
c. $\sqrt{\frac{\pi}{16}}$
d. $\sqrt{\frac{\pi}{13}}$

**18** Suppose $f(x) = ax^2 + bx + c$ is non-negative on $[-1, 1]$. Then, the area under the graph of $f$ over $[-1, 1]$ and the $X$-axis is given by

a. $A = f(-1) + f(1)$
b. $A = f(-1/2) + f(1/2)$
c. $A = \frac{1}{2}[f(-1) + 2f(0) + f(1)]$
d. $A = \frac{1}{3}[f(-1) + 4f(0) + f(1)]$

**19** If $\lim_{a \to \infty} \int_0^\infty \frac{x^2 + ax + 1}{1 + x^4} \cdot \tan^{-1}\left(\frac{1}{x}\right) dx = \frac{\pi^2}{k}$, $k \in N$, then $k$ is equal to

a. 4
b. 8
c. 16
d. 32

**20** $\int_{\pi/2}^{3\pi/2} e^{-x/2} \cdot \frac{\sqrt{1 - \sin x}}{1 + \cos x}\,dx$ is equal to

a. $(2e^{-\pi/3} - \sqrt{2}\,e^{-\pi/4})$
b. $(e^{-\pi/3} - \sqrt{2}\,e^{-\pi/4})$
c. $\left(e^{-\pi/2}\frac{2}{\sqrt{3}} - e^{-\pi/4}\sqrt{2}\right)$
d. None of these

*Working Space*

**21** The value of $x > 1$, satisfying $\int_1^x t \log t \, dt = \frac{1}{4}$ is

a. $\sqrt{e}$
b. $e$
c. $e^2$
d. $e - 1$

**22** The value of $\int_0^{\pi/2} \sin|2x - \alpha| \, dx$, where $\alpha \in [0, \pi]$, is

a. 1
b. $\cos\alpha$
c. $\frac{1+\cos\alpha}{2}$
d. $\frac{1-\cos\alpha}{2}$

**23** $\int_{-1/2}^{1/2} [\sin^{-1}(3x - 4x^3) - \cos^{-1}(4x^3 - 3x)] dx$ is equal to

a. 0
b. $-\pi/2$
c. $7\pi/2$
d. $\pi/2$

**24** Let $f(x) = \int_2^x \frac{dt}{\sqrt{1+t^4}}$ and $g$ be the inverse of $f$, then the value of $g'(0)$ is

a. 1
b. 17
c. $\sqrt{17}$
d. None of these

**25** Let $a_n = \int_0^{\pi/2} (1 - \sin t)^n \sin 2t \, dt$, then $\lim_{n \to \infty} \sum_{n=1}^{n} \frac{a_n}{n}$ is equal to

a. 1/2
b. 1
c. 4/3
d. 3/2

**26** $I = \int_0^{3\pi/4} [(1 + x)\sin x + (1 - x)\cos x] dx$ is equal to

a. $2\tan\frac{3\pi}{8}$
b. $2\tan\frac{\pi}{4}$
c. $2\tan\frac{\pi}{8}$
d. 0

**27** Let $C_n = \int_{1/(n+1)}^{1/n} \frac{\tan^{-1}(nx)}{\sin^{-1}(nx)} dx$, then $\lim_{n \to \infty} n^2 \cdot C_n$ is equal to

a. 1
b. 0
c. $-1$
d. $\frac{1}{2}$

**28** If $f(x) = e^{g(x)}$ and $g(x) = \int_2^x \frac{t}{1+t^4} dt$, then $f'(2)$ is equal to

a. 2/17
b. 0
c. 1
d. Cannot be determined

**29** If $f(x) = \sin x + \int_0^x f'(t)(2\sin t - \sin^2 t) dt$, then $f(x)$ is equal to

a. $\frac{x}{1 - \sin x}$
b. $\frac{\sin x}{1 - \sin x}$
c. $\frac{1 - \cos x}{\cos x}$
d. $\frac{\tan x}{1 - \sin x}$

**30** If $g_n(x) = x^{2n+1} + a_n x + b_n$ $(n \in N)$ satisfies the equation $\int_{-1}^{1} (px + q) g_n(x) dx = 0$ for all linear function $(px + q)$, then

a. $a_n = b_n = 0$
b. $b_n = 0, a_n = -\frac{3}{2n+3}$
c. $a_n = 0, b_n = -\frac{3}{2n+3}$
d. $a_n = b_n = n$

**31** If $F(x) = \int_1^x f(t) dt$, where $f(t) = \int_1^{t^2} \frac{\sqrt{1+u^4}}{u} du$, then the value of $f''(2)$ is

a. $\frac{7}{4\sqrt{17}}$
b. $\frac{15}{\sqrt{17}}$
c. $\sqrt{257}$
d. $\frac{15\sqrt{17}}{68}$

Working Space

**32** Let $f(x)=\int_{-1}^{x} e^{t^2}dt$ and $h(x)=f[1+g(x)]$, where $g(x)$ is defined for all $x$, $g'(x)$ exists for all $x$, and $g(x)<0$ for $x>0$. If $h'(1)=e$ and $g'(1)=1$, then the possible value which $g(1)$ can take, is

a. 0 b. −1
c. −2 d. −4

**33** If $f(2)=3$, $f(5)=7$, given $\int_2^5 f(x)dx=17$, then the value of $\int_3^7 f^{-1}(x)dx$ is

a. 10 b. 11
c. 12 d. 13

**34** Let $f(x)=\int_0^{g(x)} \frac{dt}{\sqrt{1+t^2}}$, where $g(x)=\int_0^{\cos x}(1+\sin t^2)dt$. Also, $h(x)=e^{-|x|}$ and $l(x)=x^2\sin(1/x)$, if $x\neq 0$ and $l(0)=0$, then $f'(\pi/2)$ is equal to

a. $l'(0)$ b. $h'(0^-)$
c. $h'(0^+)$ d. $\lim_{x\to 0}\frac{1-\cos x}{x\sin x}$

**35** $\lim_{t\to 0}\int_0^{2\pi}\frac{|\sin(x+t)-\sin x|}{|t|}dx$ is equal to

a. 0 b. 1
c. 2 d. 4

**36** $\lim_{x\to\infty}\left[x^3\int_{-1/x}^{1/x}\frac{\log(1+t^2)}{1+e^t}dt\right]$ is equal to

a. 1/3 b. 2/3
c. 1 d. 0

**37** If $x=\int_0^y \frac{dt}{\sqrt{1+t^2}}$, then the value of $\frac{d^2y}{dx^2}$ is

a. $\frac{y}{\sqrt{1+y^2}}$ b. $y$
c. $\frac{2y}{\sqrt{1+y^2}}$ d. $4y$

**38** If $\int_1^a x\cdot a^{-[\log_a x]}dx$, where $a>1$ and $[\cdot]$ denotes greatest integer is $\frac{e-1}{2}$, then value of $a$ is

a. $\sqrt{e}$ b. $e$
c. $\sqrt{e+1}$ d. $e-1$

**39** The value of $\lim_{n\to\infty}\sum_{k=1}^{n}\frac{n}{n^2+k^2x^2}$, $x>0$, is

a. $x\tan^{-1}x$ b. $\tan^{-1}x$
c. $\frac{\tan^{-1}x}{x}$ d. $\frac{\tan^{-1}x}{x^2}$

**40** Let $a>0$ and $f(x)$ be monotonically increasing such that $f(0)=0$ and $f(a)=b$, then $\int_0^a f(x)dx+\int_0^b f^{-1}(x)dx$ is equal to

a. $a+b$ b. $ab+b$
c. $ab+a$ d. $ab$

**41** $\int_{19}^{37}[\{x\}^2+3\sin(2\pi x)]dx$, where $\{\cdot\}$ denotes fractional part, is equal to

a. 0 b. 6
c. 9 d. 10

**42** The value of $\lim_{n\to\infty}\int_0^1\frac{n\cdot x^{n-1}}{1+x}dx$ is

a. 0 b. 1/2
c. 1 d. non-existent

Working Space

**43** If $\int_{-1/\sqrt{3}}^{1/\sqrt{3}} \frac{x^4}{1-x^4} \cos^{-1}\left(\frac{2x}{1+x^2}\right) dx = k \int_0^{1/\sqrt{3}} \frac{x^4}{1-x^4} dx$, then the value of $k$ is

a. $\pi$
b. $2\pi$
c. 2
d. 1

**44** If $g(x)$ is the inverse of $f(x)$ and $f(x)$ has domain $x \in [1, 5]$, where $f(1) = 2$ and $f(5) = 10$, then the value of $\int_1^5 f(x)\,dx + \int_2^{10} g(y)\,dy$ is

a. 48
b. 64
c. 71
d. 52

**45** Given a real-valued function $f$ which is monotonic and differentiable, then $\int_a^b [f^2(x) - f^2(a)]\,dx$ is equal to

a. $\int_{f(a)}^{f(b)} x[b - f^{-1}(x)]\,dx$
b. $\int_{f(a)}^{f(b)} x^2 [b - f^{-1}(x)]\,dx$
c. $\int_{f(a)}^{f(b)} 2x[b - f^{-1}(x)]\,dx$
d. None of the above

**46** If $I = \int_0^{1/2} \frac{dx}{\sqrt{1 - x^{2n}}}$, $n \in N$, then

a. $[I] = 1$
b. $0 < I < 1$
c. $\frac{1}{2} \le I \le \frac{\pi}{6}$
d. $0 < \pi \le \frac{1}{2}$

**47** If $f(4 - x) = f(4 + x)$ and $f(8 - x) = f(x + 8)$ and $f(x)$ is function for which $\int_0^8 f(x)\,dx = 5$, then $\int_0^{200} f(x)\,dx$ is equal to

a. 100
b. 25
c. 125
d. 75

**48** If $f[x + f(y)] = f(x) + y, \forall x, y \in R$ and $f(0) = 1$, then $\int_0^{10} f(10 - x)\,dx$ is equal to

a. 1
b. 10
c. $\int_0^1 f(x)\,dx$
d. $10\int_0^1 f(x)\,dx$

**49** The value of $\int_0^\infty \frac{\log_e x}{x^2 + a^2} dx \; (a > 0)$ is

a. $\frac{\log a}{a}$
b. $\frac{\pi \log a}{a}$
c. $\frac{\pi \log a}{4a}$
d. $\frac{\pi \log a}{2a}$

**50** The value of $\int_{-2008}^{2008} \frac{f'(x) + f'(-x)}{(2008)^x + 1} dx$ is

a. 0
b. $f(2008) - f(-2008)$
c. $f(2008) + f(-2008)$
d. None of the above

## One or More than One Options Correct Type

**Direction** *(Q. Nos. 51-60) This section contains 10 multiple choice questions. Each question has four choices (a), (b), (c) and (d), out of which ONE or MORE THAN ONE are correct.*

**51** Let $f : R^+ \to R$ be a differentiable function with $f(1) = 3$ and satisfying $\int_1^{xy} f(t)\,dt = y\int_1^x f(t)\,dt + x\int_1^y f(t)\,dt, \forall x, y \in R^+$, then

a. $f(x) = 3\log_e e(x)$
b. $f(x) = \log_e e(x)$
c. $f'(y) = \frac{3}{y}$
d. $f'(x) = \frac{1}{x}$

Working Space

**52** If $T_n = \sum_{r=2n}^{3n-1} \frac{rn}{r^2+n^2}$ and $S_n = \sum_{r=2n+1}^{3n} \frac{rn}{r^2+n^2}$, then for all $n \in \{1, 2, 3, \ldots\}$

a. $T_n > \frac{1}{2}\log 2$

b. $S_n < \frac{1}{2}\log 2$

c. $T_n < \frac{1}{2}\log 2$

d. $S_n > \frac{1}{2}\log 2$

**53** If $I = \int_0^{\pi/2} \frac{dx}{\sqrt{1+\sin^3 x}}$, then

a. $0 < I < 1$

b. $0 < I < \frac{\pi}{2\sqrt{2}}$

c. $0 < I < \sqrt{2}\pi$

d. $I > 2\pi$

**54** If $\int_0^{\infty} e^{-x^2} dx = \frac{\sqrt{\pi}}{2}$, then

a. $\int_0^{\infty} e^{-2x^2} dx = \frac{\sqrt{\pi}}{2\sqrt{2}}$

b. $\int_0^{\infty} x e^{-x^2} dx = \frac{1}{2}$

c. $\int_0^{\infty} x^2 e^{-x^2} dx = \frac{\sqrt{\pi}}{4}$

d. $\int_0^{\infty} x^2 e^{-x^2} dx = \frac{\pi}{4}$

**55** If $\alpha, \beta\ (\alpha < \beta)$ are two roots of $(6x+1)\,x = 1 + \left[\cos\frac{\pi}{4}\right]$. Then,

$\int_0^{2\alpha} \sin\left(\frac{\pi [x]}{2}\right) dx + \int_0^{3\beta} \cos(\pi [x])\, dx$ (where, $[\cdot]$ denotes greatest integer function), is equal to

a. $\alpha + \beta$

b. 2

c. 0

d. $[2\alpha + 9\beta]$

**56** If $f(x) = \int_1^x \frac{\log t}{1+t} dt$, where $x > 0$, then the value(s) of $x$ satisfying the equation $f(x) + f\left(\frac{1}{x}\right) = 2$ is/are

a. 2

b. $e$

c. $e^{-2}$

d. $e^2$

**57** Let $f(x) = \int_{-1}^{1} (1 - |t|) \cos(xt)\, dt$, then

a. $f(0)$ is not defined

b. $\lim_{x \to 0} f(x) = 2$

c. $\lim_{x \to 0} f(x) = 1$

d. $f(x)$ is continuous at $x = 0$

**58** Let $f(x)$ be real-valued function satisfying, $f(x) = x^2 + x^2 \int_{-1}^{1} t\, f(t)\, dt + x^3 \int_{-1}^{1} f(t)\, dt$, then

a. $\int_{-1}^{1} t \cdot f(t)\, dt = \frac{10}{11}$

b. $f(1) + f(-1) = \frac{30}{11}$

c. $\int_{-1}^{1} t \cdot f(t)\, dt > \int_{-1}^{1} f(t)\, dt$

d. $f(1) - f(-1) = \frac{20}{11}$

**59** If $f[f(x)] = 1 - x$ for all $x \in [0, 1]$ and $J = \int_0^1 f(x)\, dx$, then

a. $f(1/4) + f(3/4) = 1$

b. $J = 1/2$

c. $f(1/3) \cdot f(2/3) = 1$

d. $\int_0^{\pi/2} \frac{\sin x}{(\sin x + \cos x)^3} dx = J$

**60** If $f(x) + \int_0^x g(t)\, dt = \sin x(\cos x - \sin x)$ and $[f'(x)]^2 + [g'(x)]^2 = 1$, then $f(x)$ and $g(x)$ respectively can be

a. $\frac{1}{2}\sin 2x, \sin 2x$

b. $\frac{\cos 2x}{2}, \cos 2x$

c. $\frac{1}{2}\sin 2x, -\sin 2x$

d. $-\sin^2 x, \cos 2x$

Working Space

## Matching List Type

***Direction*** *(Q. Nos. 61 and 62) Choices for the correct combination of elements from Column I and Column II are given as options (a), (b), (c) and (d), out of which one is correct.*

**61** Match the Column I with Column II and mark the correct option from the codes given below.

| | Column I | | Column II |
|---|---|---|---|
| i. | If $f(x)=\int_0^{g(x)}\frac{dt}{\sqrt{1+t^3}}$, where $g(x)=\int_0^{\cos x}(1+\sin t^2)\,dt$, then $f'\left(\frac{\pi}{2}\right)$ is | p. | 3 |
| ii. | If $\int_0^x f(t)\,dt=[f(x)]^2,\ \forall\ x$, then $f(2)$ is | q. | 2 |
| iii. | If $\int_a^b (2+x-x^2)\,dx$ is maximum, then $(a+b)$ is | r. | 1 |
| iv. | If $\lim_{x\to 0}\left(\frac{\sin 2x}{x^3}+a+\frac{b}{x^2}\right)=0$, then $(3a+b)$ is | s. | −1 |

**Codes**

| | i | ii | iii | iv | | i | ii | iii | iv |
|---|---|---|---|---|---|---|---|---|---|
| a. | s | r | r | q | b. | p | s | q | r |
| c. | q | p | r | s | d. | q | p | p | r |

**62** Match the Column I with Column II and mark the correct option from the codes given below.

| | Column I | | Column II |
|---|---|---|---|
| i. | $\int_0^1 [1+(2008)\,x^{2008}]\,e^{x^{2008}}dx$ equals | p. | $e^{-1}$ |
| ii. | $\int_0^1 e^{-x^2}dx+\int_1^{1/e}\sqrt{-\log x}\,dx$ is | q. | $e^{-1/4}$ |
| iii. | $\lim_{n\to\infty}\left[\frac{1^1\cdot 2^3\cdot 3^3\cdots(n-1)^{n-1}\cdot n^n}{n^{1+2+\cdots+n}}\right]^{\frac{1}{n^2}}$ is | r. | $e^{1/2}$ |
| | | s. | $e$ |

**Codes**

| | i | ii | iii | | i | ii | iii |
|---|---|---|---|---|---|---|---|
| a. | s | p | q | b. | p | q | r |
| c. | r | s | q | d. | q | r | p |

## Comprehension Type

***Direction*** *(Q. Nos. 63-68) This section contains 2 paragraphs, each describing theory, experiments, data, etc. Six questions related to the paragraphs have been given. Each question has only one correct answer among the four given options (a), (b), (c) and (d).*

### Passage I

Suppose $f(x)$ and $g(x)$ are two continuous functions defined for $0\le x\le 1$.
Given, $f(x)=\int_0^1 e^{x+t}f(t)\,dt$ and $g(x)=\int_0^1 e^{x+t}g(t)\,dt+x$.

**63** The value of $f(1)$ is

a. 0 b. 1 c. $e^{-1}$ d. $e$

**64** The value of $g(0)-f(0)$ is

a. $\frac{2}{3-e^2}$ b. $\frac{3}{e^2-2}$ c. $\frac{1}{e^2-1}$ d. 0

**65** The value of $\frac{g(0)}{g(2)}$ is

a. 0 b. $\frac{1}{3}$ c. $\frac{1}{e^2}$ d. $\frac{2}{e^2}$

### Passage II

*Working Space*

Suppose $a$ and $b$ are positive real numbers such that $ab = 1$. Let for any real parameter $t$, the distance from the origin to the line $(ae^t)x + (be^{-t})y = 1$ be denoted by $D(t)$.

**66** The value of $I = \int_0^1 \frac{dt}{[D(t)]^2}$ is

a. $\frac{e^2 - 1}{2}\left(b^2 + \frac{a^2}{e^2}\right)$
b. $\frac{e^2 + 1}{2}\left(a^2 + \frac{b^2}{e^2}\right)$
c. $\frac{e^2 - 1}{2}\left(a^2 + \frac{b^2}{e^2}\right)$
d. $\frac{e^2 + 1}{2}\left(b^2 + \frac{a^2}{e^2}\right)$

**67** The value of $b$ at which $I$ is minimum, is

a. $e$
b. $1/e$
c. $1/\sqrt{e}$
d. $\sqrt{e}$

**68** The minimum value of $I$ is

a. $e - 1$
b. $e - 1/e$
c. $e$
d. $e + 1/e$

## One Integer Value Correct Type

***Direction** (Q. Nos. 69-75) This section contains 7 questions. When worked out will result in an integer from 0 to 9 (both inclusive).*

**69** A function $f$ is continuous on the positive real axis, has the property that for all choices of $x > 0$ and $y > 0$. The integral $\int_x^{xy} f(t)dt$ is independent of $x$ (and therefore depends only on $y$). If $f(2) = 2$, then $\int_1^e f(t)dt$ is

**70** Let $I = \int_{k\pi}^{(k+1)\pi} \frac{\sin 2x}{|\sin x| + |\cos x|} dx$, $k \in N$ and $J = \int_0^{\pi/4} \frac{dx}{\sin x + \cos x}$, then $I + 4J$ is equal to

**71** If $f(x)$ is a periodic function with period $T$, so that $f(x + 13) + f(x + 630) = 0$ and $\sum_{p=1}^{777} \int_p^{p+T} f(x)dx = \lambda \int_0^T f(x)dx$, then the value of $\left(\frac{\lambda + T}{2011}\right)$ is

**72** Let $f : R \to R$ and $f(2 - x) = f(2 + x)$ and $f(4 - x) = f(4 + x)$ and given that $\int_0^2 f(x)dx = 5$, then the value of $\frac{1}{25}\int_0^{50} f(x)dx$ is

**73** Let $f : R^+ \to R$ be a differentiable function with $f(1) = 3$ and satisfying $\int_1^{xy} f(t)dt = y\int_1^x f(t)dt + x\int_1^y f(t)dt, \forall x, y \in R^+$, where $f'(y) = \lambda/y$, then $\lambda$ is equal to

**74** If $f(x) = \int_a^x [f(x)]^{-2}dx$, $a \in R$ and $f(10) = 0$, then $[f(x)]^2 \cdot f'(x)$ is equal to

**75** If $\int_0^x \left(x - [x] - \frac{1}{2}\right)dx = \frac{f(x) \cdot g(x)}{2}$ (where, $[\cdot]$ and $\{\cdot\}$ are greatest integer and fractional part of $x$), then the value of $|f(x) - g(x)|$ is

# JEE Main & AIEEE Archive

## (Compilation of Last 13 Years' Questions)

**1** The integral $\int_2^4 \frac{\log x^2}{\log x^2 + \log(36 - 12x + x^2)} dx$ is equal to **(JEE Main 2015)**

a. 2 b. 4 c. 1 d. 6

**2** The integral $\int_0^{\pi} \sqrt{1 + 4\sin^2 \frac{x}{2} - 4\sin\frac{x}{2}}\, dx$ is equal to **(JEE Main 2014)**

a. $\pi - 4$ b. $\frac{2\pi}{3} - 4 - 4\sqrt{3}$

c. $4\sqrt{3} - 4$ d. $4\sqrt{3} - 4 - \frac{\pi}{3}$

**3** **Statement I** The value of the integral $\int_{\pi/6}^{\pi/3} \frac{dx}{1 + \sqrt{\tan x}}$ is equal to $\frac{\pi}{6}$.

**Statement II** $\int_a^b f(x)dx = \int_a^b f(a + b - x)dx$ **(JEE Main 2013)**

a. Both Statement I and Statement II are correct and Statement II is the correct explanation of Statement I

b. Both Statement I and Statement II are correct but Statement II is not the correct explanation of Statement I

c. Statement I is correct but Statement II is incorrect

d. Statement II is correct but Statement I is incorrect

**4** If $g(x) = \int_0^x \cos 4t\, dt$, then $g(x + \pi)$ is equal to **(AIEEE 2012)**

a. $\frac{g(x)}{g(\pi)}$ b. $g(x) + g(\pi)$

c. $g(x) - g(\pi)$ d. $g(x) \cdot g(\pi)$

**5** The value of $\int_0^1 \frac{8\log(1 + x)}{1 + x^2} dx$ is **(AIEEE 2011)**

a. $\frac{\pi}{8}\log 2$ b. $\frac{\pi}{2}\log 2$ c. $\log 2$ d. $\pi \log 2$

**6** Let $[\cdot]$ denotes the greatest integer function, then the value of $\int_0^{1.5} x[x^2]dx$ is **(AIEEE 2011)**

a. $\frac{5}{4}$ b. 0 c. $\frac{3}{2}$ d. $\frac{3}{4}$

**7** Let $p(x)$ be a function defined on $R$ such that $\lim_{x\to\infty} \frac{f(3x)}{f(x)} = 1$, $p'(x) = p'(1 - x)$, for all $x \in [0, 1]$, $p(0) = 1$ and $p(1) = 41$. Then, $\int_0^1 p(x)dx$ is equal to **(AIEEE 2010)**

a. $\sqrt{41}$ b. 21 c. 41 d. 42

**8** $\int_0^{\pi} [\cot x]dx$, where $[\cdot]$ denotes the greatest integer function, is equal to **(AIEEE 2009)**

a. $\frac{\pi}{2}$ b. 1 c. $-1$ d. $-\frac{\pi}{2}$

**9** If $I = \int_0^1 \frac{\sin x}{\sqrt{x}} dx$ and $J = \int_0^1 \frac{\cos x}{\sqrt{x}} dx$. Then, which one of the following is true? **(AIEEE 2008)**

a. $I > \frac{2}{3}$ and $J < 2$ b. $I > \frac{2}{3}$ and $J > 2$

c. $I < \frac{2}{3}$ and $J < 2$ d. $I < \frac{2}{3}$ and $J > 2$

**10** If $F(x) = f(x) + f\left(\frac{1}{x}\right)$, where $f(x) = \int_1^x \frac{\log t}{1 + t} dt$. Then, $F(e)$ is equal to **(AIEEE 2007)**

a. $\frac{1}{2}$ b. 0

c. 1 d. 2

**11** The solution for $x$ of the equation $\int_{\sqrt{2}}^{x} \frac{dt}{t\sqrt{t^2 - 1}} = \frac{\pi}{2}$ is **(AIEEE 2007)**

a. $-\sqrt{2}$ b. $\pi$

c. $\frac{\sqrt{3}}{2}$ d. $2\sqrt{2}$

**12** The value of $\int_3^6 \frac{\sqrt{x}}{\sqrt{9 - x} + \sqrt{x}} dx$ is **(AIEEE 2006)**

a. 3/2 b. 2 c. 1 d. 1/2

**13** $\int_{-3\pi/2}^{-\pi/2} [(x + \pi)^3 + \cos^2(x + 3\pi)]dx$ is equal to **(AIEEE 2006)**

a. $\left(\frac{\pi^4}{32}\right) + \left(\frac{\pi}{2}\right)$ b. $\frac{\pi}{2}$

c. $\left(\frac{\pi}{4}\right) - 1$ d. $\frac{\pi^4}{32}$

**14** $\int_0^{\pi} xf(\sin x)dx$ is equal to **(AIEEE 2006)**

a. $\pi \int_0^{\pi} f(\sin x)dx$ b. $\frac{\pi}{2} \int_0^{\pi/2} f(\sin x)dx$

c. $\pi \int_0^{\pi/2} f(\cos x)dx$ d. $\pi \int_0^{\pi} f(\cos x)dx$

**15** The value of $\int_1^a [x]f'(x)dx$, $a > 1$, where $[x]$ denotes the greatest integer not exceeding $x$, is **(AIEEE 2006)**

a. $[a]f(a) - \{f(1) + f(2) + \ldots + f([a])\}$

b. $[a]f([a]) - \{f(1) + f(2) + \ldots + f(a)\}$

c. $af([a]) - \{f(1) + f(2) + \ldots + f(a)\}$

d. $af(a) - \{f(1) + f(2) + \ldots + f([a])\}$

**16** $\lim_{n\to\infty} \left[\frac{1}{n^2}\sec^2\frac{1}{n^2} + \frac{2}{n^2}\sec^2\frac{4}{n^2} + \ldots + \frac{n}{n^2}\sec^2 1\right]$ is equal to **(AIEEE 2005)**

a. $\frac{1}{2}\tan 1$ b. $\tan 1$

c. $\frac{1}{2}\text{cosec } 1$ d. $\frac{1}{2}\sec 1$

**17** Let $f : R \to R$ be a differentiable function having $f(2) = 6, f'(2) = \left(\frac{1}{48}\right)$. Then, $\lim_{x \to 2} \int_6^{f(x)} \frac{4t^3}{x-2} dt$ is equal to **(AIEEE 2005)**

a. 18 b. 12
c. 36 d. 24

**18** If $I_1 = \int_0^1 2^{x^2} dx, I_2 = \int_0^1 2^{x^3} dx, I_3 = \int_1^2 2^{x^2} dx$ and $I_4 = \int_1^2 2^{x^3} dx$, then **(AIEEE 2005)**

a. $I_3 > I_4$ b. $I_3 = I_4$
c. $I_1 > I_2$ d. $I_2 > I_1$

**19** The value of $\int_{-\pi}^{\pi} \frac{\cos^2 x}{1 + a^x} dx$, where $a > 0$, is **(AIEEE 2005)**

a. $2\pi$ b. $\frac{\pi}{a}$ c. $\frac{\pi}{2}$ d. $a\pi$

**20** The value of $\int_{-2}^{3} |1 - x^2| dx$ is **(AIEEE 2004)**

a. $\frac{28}{3}$ b. $\frac{14}{3}$
c. $\frac{7}{3}$ d. $\frac{1}{3}$

**21** The value of $\int_0^{\pi/2} \frac{(\sin x + \cos x)^2}{\sqrt{1 + \sin 2x}} dx$ is **(AIEEE 2004)**

a. 0 b. 1
c. 2 d. 3

**22** If $\int_0^{\pi} x f(\sin x) dx = A \int_0^{\pi/2} f(\sin x) dx$, then $A$ is equal to **(AIEEE 2004)**

a. 0 b. $\pi$
c. $\frac{\pi}{4}$ d. $2\pi$

**23** If $f(x) = \frac{e^x}{1 + e^x}, I_1 = \int_{f(-a)}^{f(a)} x g[x(1-x)] dx$ and $I_2 = \int_{f(-a)}^{f(a)} g[x(1-x)] dx$, then the value of $\frac{I_2}{I_1}$ is **(AIEEE 2004)**

a. 2 b. $-3$ c. $-1$ d. 1

**24** $\lim_{n \to \infty} \sum_{r=1}^{n} \frac{1}{n} e^{r/n}$ is equal to **(AIEEE 2004)**

a. $e$ b. $e - 1$
c. $1 - e$ d. $e + 1$

**25** If $f(y) = e^y, g(y) = y; y > 0$ and $F(t) = \int_0^t f(t - y) g(y) dy$, then **(AIEEE 2003)**

a. $F(t) = 1 - e^{-t}(1 + t)$ b. $F(t) = e^t - (1 + t)$
c. $F(t) = te^t$ d. $F(t) = te^{-t}$

**26** If $f(a + b - x) = f(x)$, then $\int_a^b x f(x) dx$ is equal to **(AIEEE 2003)**

a. $\frac{a+b}{2} \int_a^b f(b - x) dx$ b. $\frac{a+b}{2} \int_a^b f(x) dx$
c. $\frac{b-a}{2} \int_a^b f(x) dx$ d. $\frac{a+b}{2} \int_a^b f(a + b + x) dx$

**27** The value of $\lim_{x \to 0} \frac{\int_0^{x^2} \sec^2 t \, dt}{x \sin x}$ is **(AIEEE 2003)**

a. 3 b. 2 c. 1 d. $-1$

**28** The value of $I = \int_0^1 x(1 - x)^n dx$ is **(AIEEE 2003)**

a. $\frac{1}{n+1}$ b. $\frac{1}{n+2}$
c. $\frac{1}{n+1} - \frac{1}{n+2}$ d. $\frac{1}{n+1} + \frac{1}{n+2}$

**29** $\lim_{n \to \infty} \frac{\left[\begin{matrix}1 + 2^4 + 3^4 \\ + \ldots + n^4\end{matrix}\right]}{n^5} - \lim_{n \to \infty} \frac{\left[\begin{matrix}1 + 2^3 + 3^3 \\ + \ldots + n^3\end{matrix}\right]}{n^5}$ is equal to **(AIEEE 2003)**

a. 1/30 b. 0
c. 1/4 d. 1/5

**30** Let $\frac{d}{dx} F(x) = \left(\frac{e^{\sin x}}{x}\right), x > 0$.

If $\int_1^4 \frac{3}{x} e^{\sin x^3} dx = F(k) - F(1)$, then one of the possible values of $k$, is **(AIEEE 2003)**

a. 15 b. 16
c. 63 d. 64

**31** If $f(x)$ is a function satisfying $f'(x) = f(x)$ with $f(0) = 1$ and $g(x)$ is a function that satisfies $f(x) + g(x) = x^2$. Then, the value of $\int_0^1 f(x) g(x) dx$ is **(AIEEE 2003)**

a. $e - \frac{e^2}{2} - \frac{5}{2}$ b. $e + \frac{e^2}{2} - \frac{3}{2}$
c. $e - \frac{e^2}{2} - \frac{3}{2}$ d. $e + \frac{e^2}{2} + \frac{5}{2}$

**32** $\int_0^{10\pi} |\sin x| dx$ is equal to **(AIEEE 2002)**

a. 20 b. 8
c. 10 d. 18

**33** If $I_n = \int_0^{\pi/4} \tan^n x \, dx$, then $\lim_{n \to \infty} n[I_n + I_{n+2}]$ is equal to **(AIEEE 2002)**

a. 1/2 b. 1 c. $\infty$ d. 0

**34** $\int_0^2 [x^2] dx$ is equal to **(AIEEE 2002)**

a. $2 - \sqrt{2}$ b. $2 + \sqrt{2}$
c. $\sqrt{2} - 1$ d. $-\sqrt{2} - \sqrt{3} + 5$

**35** $\int_{-\pi}^{\pi} \frac{2x(1 + \sin x)}{1 + \cos^2 x} dx$ is equal to **(AIEEE 2002)**

a. $\frac{\pi^2}{4}$ b. $\pi^2$ c. 0 d. $\frac{\pi}{2}$

**36** $\int_0^{\pi/2} \frac{\sqrt{\sin x}}{\sqrt{\sin x} + \sqrt{\cos x}} dx$ is equal to **(AIEEE 2002)**

a. $\frac{\pi}{4}$ b. $\frac{\pi}{2}$
c. 0 d. 1

# JEE Advanced & IIT JEE Archive

## (Compilation of Last 10 Years' Questions)

**1** Let $f : R \to R$ be a function defined by $f(x) = \begin{cases} [x], & x \le 2 \\ 0, & x > 2 \end{cases}$, where $[x]$ denotes the greatest integer less than or equal to $x$. If $I = \int_{-1}^{2} \frac{xf(x^2)}{2 + f(x+1)} dx$, then the value of $(4I - 1)$ is **(2015 Adv., Integer Type)**

**2** If $\alpha = \int_0^1 (e^{9x + 3\tan^{-1} x}) \left( \frac{12 + 9x^2}{1 + x^2} \right) dx$,

where $\tan^{-1} x$ takes only principal values, then the value of $\left( \log_e |1 + \alpha| - \frac{3\pi}{4} \right)$ is

**(2015 Adv., Integer Type)**

**3** Let $f : R \to R$ be a continuous odd function, which vanishes exactly at one point and $f(1) = \frac{1}{2}$. Suppose that $F(x) = \int_{-1}^{x} f(t)\,dt$ for all $x \in [-1, 2]$ and $G(x) = \int_{-1}^{x} t\,|f\{f(t)\}|\,dt$ for all $x \in [-1, 2]$. If $\lim_{x \to 1} \frac{F(x)}{G(x)} = \frac{1}{14}$, then the value of $f\left(\frac{1}{2}\right)$ is

**(2015 Adv., Integer Type)**

**4** Let $f(x) = 7\tan^8 x + 7\tan^6 x - 3\tan^4 x - 3\tan^2 x$ for all $x \in \left( -\frac{\pi}{2}, \frac{\pi}{2} \right)$. Then, the correct expression(s) is/are

**(2015 Adv., One or More than One Options Correct Type)**

a. $\int_0^{\pi/4} x f(x)\,dx = \frac{1}{12}$

b. $\int_0^{\pi/4} f(x)\,dx = 0$

c. $\int_0^{\pi/4} x f(x)\,dx = \frac{1}{6}$

d. $\int_0^{\pi/4} f(x)\,dx = 1$

**5** Let $f'(x) = \frac{192x^3}{2 + \sin^4 \pi x}$ for all $x \in R$ with $f\left(\frac{1}{2}\right) = 0$. If $m \le \int_{1/2}^{1} f(x)\,dx \le M$, then the possible values of $m$ and $M$ are

**(2015 Adv., One or More than One Options Correct Type)**

a. $m = 13, M = 24$

b. $m = \frac{1}{4}, M = \frac{1}{2}$

c. $m = -11, M = 0$

d. $m = 1, M = 12$

**6** The option(s) with the values of $a$ and $L$ that satisfy the equation

$$\frac{\int_0^{4\pi} e^t (\sin^6 at + \cos^6 at)\,dt}{\int_0^{\pi} e^t (\sin^6 at + \cos^6 at)\,dt} = L,$$ is/are

**(2015 Adv., One or More than One Options Correct Type)**

a. $a = 2, L = \frac{e^{4\pi} - 1}{e^{\pi} - 1}$  b. $a = 2, L = \frac{e^{4\pi} + 1}{e^{\pi} + 1}$

c. $a = 4, L = \frac{e^{4\pi} - 1}{e^{\pi} - 1}$  d. $a = 4, L = \frac{e^{4\pi} + 1}{e^{\pi} + 1}$

**Passage for Q. Nos. (7 and 8)**

Let $F : R \to R$ be a thrice differentiable function. Suppose that $F(1) = 0$, $F(3) = -4$ and $F'(x) < 0$ for all $x \in (1, 3)$. Let $f(x) = xF(x)$ for all $x \in R$.

**(2015 Adv., Comprehension Type)**

**7** The correct statement(s) is/are

a. $f'(1) < 0$

b. $f(2) < 0$

c. $f'(x) \ne 0$ for any $x \in (1, 3)$

d. $f'(x) = 0$ for some $x \in (1, 3)$

**8** If $\int_1^3 x^2 F'(x)\,dx = -12$ and $\int_1^3 x^3 F''(x)\,dx = 40$, then the correct expression(s) is/are

a. $9f'(3) + f'(1) - 32 = 0$  b. $\int_1^3 f(x)\,dx = 12$

c. $9f'(3) - f'(1) + 32 = 0$  d. $\int_1^3 f(x)\,dx = -12$

**9** The following integral $\int_{\pi/4}^{\pi/2} (2\,\text{cosec}\, x)^{17} dx$ is equal to **(2014 Adv., Only One Option Correct Type)**

a. $\int_0^{\log(1+\sqrt{2})} 2(e^u + e^{-u})^{16} du$

b. $\int_0^{\log(1+\sqrt{2})} (e^u + e^{-u})^{17} du$

c. $\int_0^{\log(1+\sqrt{2})} (e^u - e^{-u})^{17} du$

d. $\int_0^{\log(1+\sqrt{2})} 2(e^u - e^{-u})^{16} du$

**10** Let $f : [0, 2] \to R$ be a function which is continuous on $[0, 2]$ and is differentiable on $(0, 2)$ with $f(0) = 1$. Let $F(x) = \int_0^{x^2} f(\sqrt{t})\,dt$, for $x \in [0, 2]$. If $F'(x) = f'(x)$, for all $x \in (0, 2)$, then $F(2)$ equals to

**(2014 Adv., Only One Option Correct Type)**

a. $e^2 - 1$  b. $e^4 - 1$

c. $e - 1$  d. $e^4$

**11** The value of $\int_0^1 4x^3\left\{\frac{d^2}{dx^2}(1-x^2)^5\right\}dx$ is

**(2014 Adv., Integer Type)**

**12** The value of the integral $\int_{-\pi/2}^{\pi/2}\left(x^2+\log\frac{\pi-x}{\pi+x}\right)\cos x\,dx$ is

**(2012, Only One Option Correct Type)**

a. 0 b. $\frac{\pi^2}{2}-4$

c. $\frac{\pi^2}{2}+4$ d. $\frac{\pi^2}{2}$

**13** The value of $\int_{\sqrt{\log 2}}^{\sqrt{\log 3}}\frac{x\sin x^2}{\sin x^2+\sin(\log 6-x^2)}dx$ is

**(2011, Only One Option Correct Type)**

a. $\frac{1}{4}\log\left(\frac{3}{2}\right)$ b. $\frac{1}{2}\log\left(\frac{3}{2}\right)$

c. $\log\left(\frac{3}{2}\right)$ d. $\frac{1}{6}\log\left(\frac{3}{2}\right)$

**14** For $a\in R$ (the set of all real numbers), $a\neq -1$,

$$\lim_{n\to\infty}\frac{(1^a+2^a+\ldots+n^a)}{(n+1)^{a-1}[(na+1)+(na+2)+\ldots+(na+n)]}=\frac{1}{60}.$$

Then, $a$ is equal to

**(2011, One or More than One Options Correct Type)**

a. 5 b. 7

c. $-\frac{15}{2}$ d. $-\frac{17}{2}$

**15** The value(s) of $\int_0^1\frac{x^4(1-x)^4}{1+x^2}dx$ is (are)

**(2010, One or More than One Options Correct Type)**

a. $\frac{22}{7}-\pi$ b. $\frac{2}{105}$

c. 0 d. $\frac{71}{15}-\frac{3\pi}{2}$

**16** For any real number $x$, let $[x]$ denotes the largest integer less than or equal to $x$. Let $f$ be a real-valued function defined on the interval $[-10, 10]$ by $f(x)=\begin{cases}x-[x], & \text{if } [x] \text{ is odd}\\ 1+[x]-x, & \text{if } [x] \text{ is even}\end{cases}$.

Then, the value of $\frac{\pi^2}{10}\int_{-10}^{10}f(x)\cos\pi x\,dx$ is

**(2010, Integer Type)**

**17** If $I_n=\int_{-\pi}^{\pi}\frac{\sin nx}{(1+\pi^x)\sin x}dx,\ n=0,1,2,\ldots,$ then

**(2009, One or More than One Options Correct Type)**

a. $I_n=I_{n+2}$ b. $\sum_{m=1}^{10}I_{2m+1}=10\pi$

c. $\sum_{m=1}^{10}I_{2m}=0$ d. $I_n=I_{n+1}$

**18** Let $f$ be a non-negative function defined on the interval $[0, 1]$. If $\int_0^x\sqrt{1-(f'(t))^2}\,dt=\int_0^x f(t)dt,\ 0\le x\le 1$ and $f(0)=0$, then

**(2009, Only One Option Correct Type)**

a. $f\left(\frac{1}{2}\right)<\frac{1}{2}$ and $f\left(\frac{1}{3}\right)>\frac{1}{3}$

b. $f\left(\frac{1}{2}\right)>\frac{1}{2}$ and $f\left(\frac{1}{3}\right)>\frac{1}{3}$

c. $f\left(\frac{1}{2}\right)<\frac{1}{2}$ and $f\left(\frac{1}{3}\right)<\frac{1}{3}$

d. $f\left(\frac{1}{2}\right)>\frac{1}{2}$ and $f\left(\frac{1}{3}\right)<\frac{1}{3}$

**19** Let $S_n=\sum_{k=0}^{n}\frac{n}{n^2+kn+k^2}$ and $T_n=\sum_{k=0}^{n-1}\frac{n}{n^2+kn+k^2}$, for $n=1,2,3,\ldots,$ then

**(2008, One or More than One Options Correct Type)**

a. $S_n<\frac{\pi}{3\sqrt{3}}$ b. $S_n>\frac{\pi}{3\sqrt{3}}$

c. $T_n<\frac{\pi}{3\sqrt{3}}$ d. $T_n>\frac{\pi}{3\sqrt{3}}$

**20** Match the conditions/expressions in Column I with statement in Column II. **(2007, Matching Type)**

| | Column I | | Column II |
|---|---|---|---|
| i. | $\int_{-1}^{1}\frac{dx}{1+x^2}$ | p. | $\frac{1}{2}\log\left(\frac{2}{3}\right)$ |
| ii. | $\int_0^1\frac{dx}{\sqrt{1-x^2}}$ | q. | $2\log\left(\frac{2}{3}\right)$ |
| iii. | $\int_2^3\frac{dx}{1-x^2}$ | r. | $\frac{\pi}{3}$ |
| iv. | $\int_1^2\frac{dx}{x\sqrt{x^2-1}}$ | s. | $\frac{\pi}{2}$ |

**Codes**

| | i | ii | iii | iv |
|---|---|---|---|---|
| a. | s | s | p | r |
| b. | p | q | r | s |
| c. | q | p | s | r |
| d. | p | r | q | p |

### Passage for Q. Nos. (21-22)

Let $f(x)=(1-x)^2\sin^2 x+x^2$ for all $x\in R$ and let $g(x)=\int_1^x\left[\frac{2(t-1)}{t+1}-\ln t\right]f(t)dt$ for all $x\in(1,\infty)$.

**(2007, Comprehension Type)**

**21** Which of the following is true?

a. $g$ is increasing on $(1,\infty)$

b. $g$ is decreasing on $(1,\infty)$

c. $g$ is increasing on $(1, 2)$ and decreasing on $(2,\infty)$

d. $g$ is decreasing on $(1, 2)$ and increasing on $(2,\infty)$

**22** Consider the statements

P : There exists some $x \in R$ such that

$$f(x) + 2x = 2(1 + x^2)$$

Q : There exists some $x \in R$ such that

$$2f(x) + 1 = 2x(1 + x)$$

Then,

a. both P and Q are correct
b. P is correct and Q is incorrect
c. P is incorrect and Q is correct
d. both P and Q are incorrect

**Passage for Q. Nos. (23-25)**

For every function $f(x)$ which is twice differentiable, these will be good approximation of $\int_a^b f(x)dx = \left(\frac{b-a}{2}\right)\{f(a) + f(b)\}$, for more acurate results for $c \in (a, b)$,

$$F(c) = \frac{c-a}{2}[f(a) - f(c)] + \frac{b-c}{2}[f(b) - f(c)],$$

when $c = \dfrac{a+b}{2}$.

$$\int_a^b f(x)dx = \frac{b-a}{4}\{f(a) + f(b) + 2f(c)\}\,dx$$

**(2006, Comprehension Type)**

**23** Good approximation of $\int_0^{\pi/2} \sin x\,dx$, is

a. $\pi/4$
b. $\pi(\sqrt{2}+1)/4$
c. $\pi(\sqrt{2}+1)/8$
d. $\pi/8$

**24** If $f''(x) < 0, \forall x \in (a, b)$ and $(c, f(c))$ is point of maxima where $c \in (a, b)$, then $f'(c)$ is

a. $\dfrac{f(b) - f(a)}{b-a}$
b. $3\left[\dfrac{f(b) - f(a)}{b-a}\right]$
c. $2\left[\dfrac{f(b) - f(a)}{b-a}\right]$
d. 0

**25** If $\lim_{t \to a} \dfrac{\int_a^t f(x)dx - \frac{(t-a)}{2}\{f(t) + f(a)\}}{(t-a)^3} = 0$, then degree of polynomial function $f(x)$ atmost is

a. 0
b. 1
c. 3
d. 2

**26** The value of

$\int_{-2}^{0} [x^3 + 3x^2 + 3x + 3 + (x+1)\cos(x+1)]\,dx$ is

**(2005, Only One Option Correct Type)**

a. 0
b. 3
c. 4
d. 1

**27** If $\int_{\sin x}^{1} t^2 f(t)\,dt = 1 - \sin x, \forall x \in (0, \pi/2)$, then $f\left(\frac{1}{\sqrt{3}}\right)$ is

**(2005, Only One Option Correct Type)**

a. 3
b. $\sqrt{3}$
c. 1/3
d. None of these

# 3. Differential Equations

# DPP-1 Order, Degree and Formation of Differential Equations

## Subjective Questions

*Direction* (Q. Nos. 1-10) *These questions are subjective in nature, need to be solved completely on notebook.*

Working Space

**1** Find the order and degree of the following differential equations:

i. $\sqrt{\dfrac{d^2y}{dx^2}} = \sqrt[3]{\dfrac{dy}{dx} + 3}$

ii. $\dfrac{d^2y}{dx^2} = \sin\left(\dfrac{dy}{dx}\right)$

iii. $\dfrac{dy}{dx} = \sqrt{3x + 5}$

iv. $\left[1 + \left(\dfrac{dy}{dx}\right)^2\right]^{1/2} = x^2 + y$

v. $\left(1 + \dfrac{dy}{dx}\right)^{1/2} = \dfrac{d^2y}{dx^2}$

vi $\left(\dfrac{dy}{dx}\right) = \sin y$

**2** Find the differential equation of all circles touching the

i. $X$-axis at the origin.

ii. $Y$-axis at the origin.

**3** Find the differential equation of all circles of radius $r$.

**4** Find the differential equation of all circles in the first quadrant, which touch the coordinate axes.

**5** Find the differential equation of all straight lines, which

i. are non-horizontal.

ii. are non-vertical.

**6** Find the differential equation of all conics whose axes coincide with the coordinate axes.

**7** Find the differential equation of all conics whose centre lies at the origin.

**8** Find the differential equation of all straight lines which are at a constant distance $p$ from the origin.

**9** Find the differential equation of all parabola having their axes of symmetry coincident with $X$-axis.

**10** Find the differential equation of all parabolas whose axis of symmetry is parallel to $X$-axis.

## Only One Option Correct Type

Working Space

*Direction* (Q. Nos. 11-20) *This section contains 10 multiple choice questions. Each question has four choices (a), (b), (c) and (d), out of which ONLY ONE is correct.*

**11** The order and degree of the differential equation $\left(1+3\frac{dy}{dx}\right)^{2/3}=4\frac{d^3y}{dx^3}$ are

a. 1 and 2/3
b. 3 and 1
c. 1 and 2
d. 3 and 3

**12** If order and degree of the differential equation $\sqrt[3]{\frac{dy}{dx}}-4\frac{d^2y}{dx^2}-7x=0$ are $a$ and $b$, then the value of $(a+b)$ is

a. 3
b. 4
c. 5
d. 6

**13** The order of differential equation whose general solution is given by $y=(c_1+c_2)\cos(x+c_3)-c_4e^{x+c_5}$ (where, $c_1$, $c_2$, $c_3$, $c_4$ and $c_5$ are arbitrary constants), is

a. 5
b. 4
c. 3
d. 2

**14** The differential equation of the family of curves represented by $y=a+bx+ce^{-x}$ (where, $a$, $b$ and $c$ are arbitrary constants) is

a. $y'''=y'$
b. $y'''+y''=0$
c. $y'''-y''+y'=0$
d. None of these

**15** The differential equation, which represents the family of plane curves $y=e^{cx}$ is

a. $y'=cy$
b. $xy'-\log y=0$
c. $x\log y=yy'$
d. $y\log y=xy'$

**16** The differential equation for the family of curves $x^2+y^2-2ay=0$, where $a$ is an arbitrary constant, is

a. $2(x^2-y^2)y'=xy$
b. $2(x^2+y^2)y'=xy$
c. $(x^2-y^2)y'=2xy$
d. $(x^2+y^2)y'=2xy$

**17** The differential representing the family of the curves $y^2=2c(x+\sqrt{c})$, where $c\ (>0)$, is a parameter, is of order and degree as follows

a. order 1, degree 2
b. order 1, degree 1
c. order 1, degree 3
d. order 2, degree 2

**18** The differential equation whose solution is $Ax^2+By^2=1$, where $A$ and $B$ are arbitrary constants, is of

a. order 1, degree 2
b. order 1, degree 1
c. order 2, degree 1
d. order 2, degree 2

**19** The differential equation of the family of circles with the fixed radius 5 units and centre on the line $y=2$, is

a. $(x-2)(y')^2=25-(y-2)^2$
b. $(y-2)(y')^2=25-(y-2)^2$
c. $(y-2)^2(y')^2=25-(y-2)^2$
d. $(x-2)^2(y')^2=25-(y-2)^2$

**20** The number of values of $m\in N$ for which $y=e^{mx}$ is a solution of the differential equation $D^3y-3D^2y-4Dy+12y=0$, is

a. 0
b. 1
c. 2
d. more than 2

## Comprehension Type

*Direction* (*Q. Nos. 21-25*) *This section contains 2 paragraphs, each describing theory, experiments, data, etc. Five questions related to the paragraphs have been given. Each question has only one correct answer among the four given options (a), (b), (c) and (d).*

Working Space

**Passage I**

If the curve $y = c_1 e^{m_1 x} + c_2 e^{m_2 x} + c_3 e^{m_3 x}$, where $c_1, c_2, c_3$ are arbitrary constants and $m_1, m_2, m_3$ are roots of $m^3 - 7m + 6 = 0$.

**21** The sum of product of two roots at a time is

a. 0
b. 1
c. −7
d. 6

**22** The differential equation corresponding to the given curve, is

a. $y_3 - 7y_1 + 6y = 0$
b. $y_3 - 7y_1 + 6 = 0$
c. $y_3 + 7y_1 + 6y = 0$
d. $y_3 - 7y_1 + 6 = 0$

**Passage II**

All the parabolas each of which has a latusrectum $4a$ and whose axes are parallel to $X$-axis.

**23** The equation is given by

a. $y^2 = 4ax$
b. $(y - k)^2 = 4ax$
c. $(y - k)^2 = 4a(x - h)$
d. $y^2 = 4a(x - h)$

**24** The order and degree are respectively

a. 2, 2
b. 1, 2
c. 1, 1
d. 2, 1

**25** The differential equation is given by

a. $y_1^3 + 2ay_2 = 0$
b. $y_1^3 + a y_2 = 0$
c. $y_1^2 + 2ay_2 = 0$
d. $y_1 + 2a\, y_2^3 = 0$

## One Integer Value Correct Type

*Direction* (*Q. Nos. 26-30*) *This section contains 5 questions. When worked out will result in an integer from 0 to 9 (both inclusive).*

**26** Let the order and degree of $y = Ae^{3x} + Be^{5x}$ be $a$ and $b$, then the value of $a + b$ is

**27** The degree of $\frac{x^2}{a^2} + \frac{y^2}{a^2 + \lambda} = 1$, where $\lambda$ is a parameter, is

**28** The differential equation of the curve $xy = ae^x + be^{-x} + x^2$ is $xy_2 + ay_1 + bxy + x^2 - 2 = 0$, then the value of $a + b$ is

**29** The differential equation of all lines in a plane, which are equally inclined to the coordinate axes is $dy/dx = \lambda$. Then, $|\lambda|$ is equal to

**30** A population grows at the rate of 5% per year. If $x = x(t)$ denotes the number of individuals in the population after $t$ years. The differential equation involving the rate of change of $x$ is given by $\frac{dx}{dt} = (\lambda)\,\%\,(x)$, then the value of $\lambda$ is

DPP-2

# Equation in Variable Separable Form and Reducible to Variable Separation Form

## Only One Option Correct Type

Working Space

*Direction* (Q. Nos. 1-13) *This section contains 13 multiple choice questions. Each question has four choices (a), (b), (c) and (d), out of which ONLY ONE is correct.*

**1** The solution of $xy\frac{dy}{dx}=\frac{1+y^2}{1+x^2}(1+x+x^2)$ is

a. $\sqrt{1+y^2}=Ce^{\tan^{-1}x}$ b. $1+y^2=Cx\,e^{\tan^{-1}x}$

c. $\sqrt{1+y^2}=Cx$ d. $\sqrt{1+y^2}=Cx\,e^{\tan^{-1}x}$

**2** The solution of $(x^3-y^2x^3)\frac{dy}{dx}+y^3+x^2y^3=0$ is

a. $\log\left(\frac{x}{y}\right)=\frac{1}{2}\left(\frac{1}{y^2}+\frac{1}{x^2}\right)+C$ b. $\log\left(\frac{x}{y}\right)=\left(\frac{1}{x^2}+\frac{1}{y^2}\right)+C$

c. $\frac{x}{y}=\log\left(\frac{1}{x^2}+\frac{1}{y^2}\right)+C$ d. $xy=x^2+y^2+C$

**3** The solution of $\sin^{-1}\left(\frac{dy}{dx}\right)=(x+y)$ is

a. $y=\tan(x+y)-\sec(x+y)+C$

b. $x=\tan(x+y)-\sec(x+y)+C$

c. $y=\log[\tan(x+y)]+C$

d. None of the above

**4** The solution of $\frac{dy}{dx}=\cos(x+y)$ is

a. $\frac{x+y}{2}=\tan x+C$ b. $\frac{x-y}{2}=\tan x+C$

c. $\tan\left(\frac{x+y}{2}\right)=x+C$ d. None of these

**5** The solution of differential equation $y\log y+xy'=0$, where $y(1)=e$, is

a. $x(\log y)=1$ b. $xy(\log y)=1$

c. $(\log y)^2=2$ d. None of these

**6** If $\frac{dy}{dx}=y+3>0$ and $y(0)=2$, then $y(\log_e 2)$ is equal to **[AIEEE 2011]**

a. 13 b. −2

c. 7 d. 5

Working Space

**7** If $y = y(x)$ and $\frac{2+\sin x}{y+1} \cdot \frac{dy}{dx} = -\cos x$, $y(0) = 1$, then $y\left(\frac{\pi}{2}\right)$ equals

a. 1/3
b. 2/3
c. −1/3
d. 1

**8** The differential equation $\frac{dy}{dx} = \frac{\sqrt{1-y^2}}{y}$ determines family of circles with **[IIT JEE 2008]**

a. variable radii and fixed centre at (0, 1)
b. variable radii and fixed centre at (0,−1)
c. fixed radius 1 and variable centre along $Y$-axis
d. fixed radius 1 and variable centre along $X$-axis

**9** The general solution of the differential equation $\frac{dy}{dx} = \frac{1-x}{y}$ is a family of curves which looks most likely, as

a. 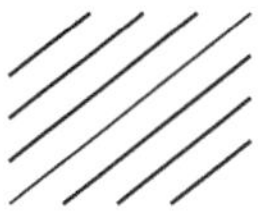

b. 

c. 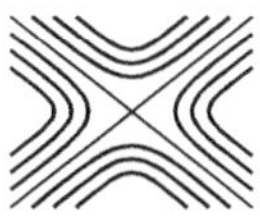

d. 

**10** Which one of the following curves represents the solution of initial value problem $Dy = 100 - y$, where $y(0) = 50$?

a. 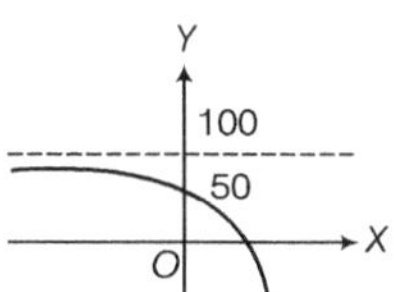


b. 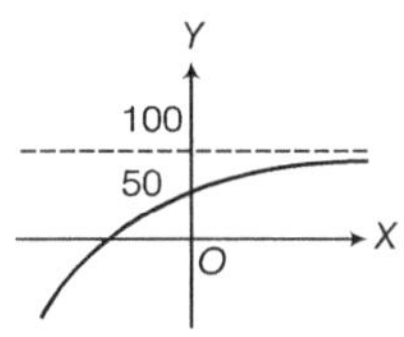


c. 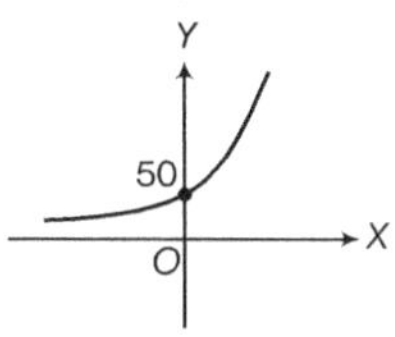


d. 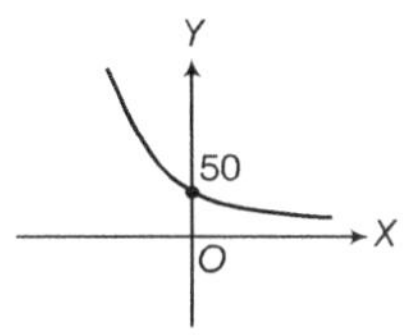


**11** The population $p(t)$ at time $t$ of certain mouse species satisfies the differential equation $\frac{d[p(t)]}{dt} = 0.5p(t) - 450$. If $p(0) = 850$, then the time at which population becomes zero, is **[AIEEE 2012]**

a. $\log 18$
b. $2\log 18$
c. $\log 9$
d. $\frac{1}{2}\log 18$

**12** At present, a firm is manufacturing 2000 items. It is estimated that the rate of change of production $P$ with respect to additional number of workers is given by $\frac{dP}{dx} = 100 - 12\sqrt{x}$. If the firm employs 25 more workers, then new level of production of item is

a. 2500
b. 3000
c. 3500
d. 4500

Working Space

**13** Let $f:\left[\frac{1}{2}, 1\right] \rightarrow R$ (the set of real numbers) be a positive, non-constant and differentiable function such that $f'(x) < 2f(x)$ and $f\left(\frac{1}{2}\right) = 1$. Then, the value of $\int_{1/2}^{1} f(x)dx$ lies in the interval **[JEE Advanced 2013]**

a. $(2e - 1, 2e)$  b. $(e - 1, 2e - 1)$
c. $\left(\frac{e-1}{2}, e - 1\right)$  d. $\left(0, \frac{e-1}{2}\right)$

## One or More than One Options Correct Type

*Direction (Q. Nos. 14-16) This section contains 3 multiple choice questions. Each question has four choices (a), (b), (c) and (d), out of which ONE or MORE THAN ONE are correct.*

**14** The solution of $\frac{dy}{dx} = \frac{ax + h}{by + k}$ represents a parabola, if

a. $a = -2, b = 0$  b. $a = -2, b = 2$
c. $a = 0, b = 2$  d. $a = 0, b = 0$

**15** A normal is drawn at a point $P(x, y)$ of a curve. It meets the $X$-axis and the $Y$-axis in points $A$ and $B$ respectively, such that $\frac{1}{OA} + \frac{1}{OB} = 1$, where $O$ is origin, the equation of such a curve is a circle which passes through (5, 4) and has

a. centre as (1, 1)  b. centre as (2, 1)
c. radius 5  d. radius 4

**16** The graph of function $y = f(x)$ passing through the point (0, 1) and satisfying the differential equation $\frac{dy}{dx} + y\cos x = \cos x$, is such that

a. it is a constant function
b. it is periodic function
c. it is neither an even nor an odd function
d. it is continuous and differentiable function for all $x$

## Comprehension Type

*Direction (Q. Nos. 17-21) This section contains 2 paragraphs, each describing theory, experiments, data, etc. Five questions related to the paragraphs have been given. Each question has only one correct answer among the four given options (a), (b), (c) and (d).*

### Passage I

Differential equation $dy/dx = f(x) \cdot g(y)$ can be solved by seperating variable $\frac{dy}{g(y)} = f(x)\,dx$.

**17** The equation of curve to the point (1, 0) which satisfy the differential equation $(1 + y^2)dx - xy\,dy = 0$ is

a. $x^2 + y^2 = 1$  b. $x^2 - y^2 = 1$
c. $x^2 + y^2 = 2$  d. $x^2 - y^2 = 2$

**18** The solution of the differential equation $\frac{dy}{dx} + \frac{1 + y^2}{\sqrt{1 + x^2}} = 0$ is

a. $\tan^{-1} y + \sin^{-1} x = C$  b. $\tan^{-1} x + \sin^{-1} y = C$
c. $\tan^{-1} y \cdot \sin^{-1} x = C$  d. $\tan^{-1} y - \sin^{-1} x = C$

Working Space

### Passage II

A normal is drawn at a point $P(x, y)$ of a curve. It meets the $X$-axis at $Q$ and $PQ$ is the constant length $k$.

**19** The differential equation described by such curve is

a. $y\dfrac{dy}{dx} = \pm\sqrt{k^2 - y^2}$
b. $\dfrac{dy}{dx} = \pm\sqrt{k^2 - y^2}$
c. $\sqrt{k^2 - y^2}\dfrac{dy}{dx} = \pm y$
d. $(k^2 - y^2)\dfrac{dy}{dx} = \pm y$

**20** The equation of such curve passing through $(0, k)$ is

a. $x^2 - y^2 = k^2$
b. $2x^2 + y^2 = k$
c. $x^2 + y^2 = k^2$
d. None of these

**21** The equation of tangent to the curve passing through (0, 1) at the point (1, 1) is

a. $x - y = 1$
b. $x + y = 1$
c. $2x + y = 1$
d. None of these

## One Integer Value Correct Type

*Direction* (Q. Nos. 22-25) *This section contains 4 questions. When worked out will result in an integer from 0 to 9 (both inclusive).*

**22** If the solution of $\dfrac{dy}{dx} - x\tan(y - x) = 1$ is $\log|\sin(y - x)| = \dfrac{x^2}{k} + C$, then the value of $k$ is

**23** If the solution of $\dfrac{dy}{dx} = (x + y - 1) + \dfrac{x + y}{\log(x + y)}$ is

$1 + \log(x + y) - \log\{k + \log(x + y)\} = x + C$, then $k$ is equal to

**24** If the solution of $\dfrac{dy}{dx} = \cos(x + y) + \sin(x + y)$, is $\log\left|1 + f\left(\dfrac{x + y}{2}\right)\right| = x + C$, then

$\lim\limits_{x \to 0} \dfrac{f(x)}{x}$ is equal to

**25** Let $f$ be a real valued differentiable function on $R$ (the set of all real numbers) such that $f(1) = 1$. If the $y$-intercept of the tangent at any point $P(x, y)$ on the curve $y = f(x)$ is equal to the cube of the abscissa of $P$, then the value of $f(-3)$ is

# DPP-3 Homogeneous Differential Equations and Equation Reducible to Homogeneous Form

## Only One Option Correct Type

*Working Space*

**Direction** (Q. Nos. 1-18) *This section contains 18 multiple choice questions. Each question has four choices (a), (b), (c) and (d), out of which ONLY ONE is correct.*

**1** A curve passes through the point $(1, \pi/4)$ and its slope at any point is given by $\frac{y}{x} - \cos^2\left(\frac{y}{x}\right)$, then the curve has the equation

a. $y = x \tan^{-1}\left(\log \frac{e}{x}\right)$

b. $y = x \tan^{-1}(\log x + 2)$

c. $y = \frac{1}{x} \tan^{-1}\left(\log \frac{e}{x}\right)$

d. None of the above

**2** If $x\frac{dy}{dx} = y(\log y - \log x + 1)$, then the solution of equation is

a. $y \log\left(\frac{x}{y}\right) = Cx$ b. $x \log\left(\frac{y}{x}\right) = Cy$

c. $\log\left(\frac{y}{x}\right) = Cx$ d. $\log\left(\frac{x}{y}\right) = Cy$

**3** The differential equation of all circles passing through the origin and having their centres on the $X$-axis is

a. $x^2 + y^2 = xy\frac{dy}{dx}$ b. $x^2 = y^2 + 3xy\frac{dy}{dx}$

c. $y^2 = x^2 + 2xy\frac{dy}{dx}$ d. $y^2 = x^2 - 2xy\frac{dy}{dx}$

**4** The solution of the differential equation $\frac{dy}{dx} = \frac{x+y}{x}$ satisfying the condition $y(1) = 1$, is **[AIEEE 2008]**

a. $y = \log x + x$ b. $y = x \log x + x^2$

c. $y = xe^{x-1}$ d. $y = x \log x + x$

Working Space

**5** The solution of primitive integral solution $(x^2 + y^2)\,dy = xy\,dx$ is $y = y(x)$. If $y(1) = 1$ and $y(x_0) = e$, then $x_0$ is **[IIT JEE 2005]**

a. $\sqrt{2(e^2 - 1)}$ b. $\sqrt{2(e^2 + 1)}$

c. $\sqrt{3}e$ d. None of these

**6** A curve passes through the point $(1, \pi/6)$. Let the slope of the curve at each point $(x, y)$ be $\frac{y}{x} + \sec\left(\frac{y}{x}\right), x > 0$. Then, the equation of the curve is **[JEE Advanced 2013]**

a. $\sin\left(\frac{y}{x}\right) = \log x + \frac{1}{2}$

b. $\operatorname{cosec}\left(\frac{y}{x}\right) = \log x + 2$

c. $\sec\frac{2y}{x} = \log x + 2$

d. $\cos\frac{2y}{x} = \log x + \frac{1}{2}$

**7** The solution of $(3xy + y^2)dx + (x^2 + xy)\,dy = 0$ is

a. $x^2(y^2 + 2xy) = C$ b. $y^2(y^2 + 2xy) = C$

c. $y^2 + x^3y = C$ d. $y^2(y^2 - 2xy) = C$

**8** The solution of $y \cdot \left\{x\cos\left(\frac{y}{x}\right) + y\sin\left(\frac{y}{x}\right)\right\}dx - x\left\{y \cdot \sin\left(\frac{y}{x}\right) - x\cos\left(\frac{y}{x}\right)\right\}dy = 0$ is

a. $xy^2\cos\left(\frac{y}{x}\right) = C$ b. $x^2y\cos\left(\frac{y}{x}\right) = C$

c. $xy\cos\left(\frac{y}{x}\right) = C$ d. None of these

**9** The solution of $y \cdot \cos(y/x)(x\,dy - y\,dx) + x \cdot \sin(y/x) \cdot (x\,dy + y\,dx) = 0$ is

a. $x^2y \cdot \sin(y/x) = C$ b. $xy^2 \cdot \sin(y/x) = C$

c. $xy\sin(y/x) = C$ d. $xy^2\cos(y/x) = C$

**10** The solution of $\frac{dy}{dx} = \frac{x^2 + y^2 - 2x - 2y + 2}{xy - x - y + 1}$ is

a. $(y - x)^2 = \log|x - 1| + C$

b. $\frac{(y - 1)^2}{(x - 1)^2} = 2\log|x - 1| + C$

c. $(y - x)^2 = \log|x + 1| + C$

d. None of the above

**11** The solution of $xy \cdot \log\left(\frac{x}{y}\right)dx + \left\{y^2 - x^2\log\left(\frac{x}{y}\right)\right\}dy = 0$ is

a. $2x^2\log\left(\frac{x}{y}\right) - x^2 = 4y^2 \cdot \log\left(\frac{C}{x}\right)$

b. $2x^2\log\left(\frac{y}{x}\right) - x^2 = 4y^2 \cdot \log\left(\frac{C}{x}\right)$

c. $x^2\log\left(\frac{y}{x}\right) - x^2 = 4y^2 \cdot \log\left(\frac{C}{x}\right)$

d. None of the above

Working Space

**12** The solution of $(1+e^{x/y})dx + e^{x/y}\cdot\left(1-\frac{x}{y}\right)dy = 0$ is

a. $e^{x/y} - \frac{x}{y} = \frac{C}{y}$

b. $e^{x/y} + \frac{x}{y} = \frac{C}{y}$

c. $\frac{x}{y} - e^{x/y} = Cy^2$

d. None of these

**13** The solution of $\frac{dy}{dx} = \frac{2x - y + 3}{x + 2y + 4}$ is

a. $y^2 - x^2 + xy + 4y - 3x + 1 = C$
b. $y^2 - x^2 - xy - 4y + 3x - 1 = C$
c. $x^2 - y^2 + xy + 4y - 3x - 1 = C$
d. $y^2 - x^2 + xy + 4y - 3x - 1 = C$

**14** The solution of $\frac{dy}{dx} = \frac{x + y - 1}{x - y - 1}$ is

a. $C\sqrt{x^2 + (y-1)^2} = e^{\tan^{-1}\left(\frac{y+1}{x}\right)}$

b. $C\sqrt{x^2 - (y-1)^2} = e^{\tan^{-1}\left(\frac{y+1}{x}\right)}$

c. $C\sqrt{x^2 + (y+1)^2} = e^{\tan^{-1}\left(\frac{y+1}{x}\right)}$

d. None of the above

**15** The solution of $(2x^2 + 3y^2 - 7)\,x \cdot dx - (3x^2 + 2y^2 - 8)\,y \cdot dy = 0$ is
a. $(x^2 + y^2 + 3) = C(x^2 - y^2 - 1)^4$
b. $(x^2 + y^2 - 3)(x^2 + y^2 + 1) = C$
c. $(x^2 + y^2 - 3) = (x^2 - y^2 - 1)^5 \cdot C$
d. None of the above

**16** The solution of $\frac{dy}{dx} + \frac{\cos x\,(3\cos y - 7\sin x - 3)}{\sin y\,(3\sin x - 7\cos y + 7)} = 0$ is

a. $(\cos y - \sin x - 1)^2 (\sin x + \cos y - 1)^5 = C$
b. $(\cos y + \sin x - 1)^2 (\sin x - \cos y + 1)^5 = C$
c. $(\cos y + \sin x + 1)^2 (\sin x + \cos y + 1)^5 = C$
d. None of the above

**17** The solution of $\frac{dy}{dx} = \frac{6x - 4y + 3}{3x - 2y + 1}$ is

a. $(4x - 2y) = 2\log|3x - 2y - 3| + C$
b. $(4x + 2y) = 2\log|3x + 2y + 3| + C$
c. $(4x - 2y) = 2\log|3x - 2y + 3| = C$
d. None of the above

**18** The solution of $(2x + 4y + 3)\cdot\frac{dy}{dx} = 2y + x + 1$ is

a. $\log|4x + 8y + 5| = 4x - 8y + C$
b. $\log|4x - 8y + 5| = 4x + 8y + C$
c. $\log|4x + 8y + 5| = 4x + 8y + C$
d. None of the above

Working Space

## Comprehension Type

*Direction* (Q. Nos. 19 and 20) *This section contains a paragraph, describing theory, experiments, data, etc. Two questions related to the paragraph have been given. Each question has only one correct answer among the four given options (a), (b), (c) and (d).*

**Passage**

The slope of tangent to a curve at any point $(x, y)$ on it is given by $\frac{y}{x} - \cot\left(\frac{y}{x}\right) \cdot \cos\left(\frac{y}{x}\right)$ $(x > 0, y > 0)$ and solution of curve is given by $\sec\left(\frac{y}{x}\right) = f(x) + C$.

**19** The value of $f(x)$ is

a. $\log|x|$

b. $\log|2x|$

c. $\log\left|\frac{1}{x}\right|$

d. None of these

**20** The value of $\lim_{x \to 0} \frac{f(1+x)}{x}$ is

a. $-1$

b. $1$

c. $\frac{1}{2}$

d. $2$

## One Integer Value Correct Type

*Direction* (Q. Nos. 21-25) *This section contains 5 questions. When worked out will result in an integer from 0 to 9 (both inclusive).*

**21** The degree of homogenity of $f(x, y) = \sin\left(\frac{x}{y}\right)$ is

**22** If the solution of $(x + y)dx + (3x + 3y - 4)\,dy = 0$ is $(x + 3y) + 2\log|a + bx + cy| = C_1$, then the value of $(a + b + c)$ is

**23** If the solution of $(2x + 2y + 3)\,dy = (x - 2y + 1)\,dx$ is $y^2 + 2xy + 3y = \frac{x^2}{a} + x + C$, then the value of $a$ is

**24** If the solution of $\frac{dy}{dx} = \frac{\sin y + x}{\sin 2y - x\cos y}$ is $\sin^2 y = x\sin y + \frac{x^2}{a} + C$, then the value of $a$ is

**25** If the solution of $(1 + 2e^{x/y})\,dx + 2e^{x/y} \cdot \left(1 - \frac{x}{y}\right) dy = 0$ is given by $x + 2y\,[f(x)]^{1/y} = C$, then $\lim_{x \to 0} \frac{f(x) - 1}{x}$ is equal to

# DPP-4 Linear Differential Equations and Bernoulli's Equation

## Only One Option Correct Type

Working Space

**Direction** (Q. Nos. 1-10) *This section contains 10 multiple choice questions. Each question has four choices (a), (b), (c) and (d), out of which ONLY ONE is correct.*

**1** If $y(t)$ is a solution of $(1+t)\frac{dy}{dt} - ty = 1$ and $y(0) = -1$, then $y(1)$ is equal to

a. $\frac{1}{2}$ b. $e+\frac{1}{2}$

c. $e-\frac{1}{2}$ d. $-\frac{1}{2}$

**2** Let $p(x)$ be a polynomial such that $p(1) = 0$ and $\frac{d}{dx}[p(x)] > p(x)$ for all $x \geq 1$, then

a. $p(x) > 0, \forall x > 1$ b. $p(x) < 0, \forall x > 1$

c. $p(x) \geq 0, \forall x > 1$ d. $p(x) \leq 0, \forall x > 1$

**3** The solution of differential equation $x^2\frac{dy}{dx}\cdot\cos\left(\frac{1}{x}\right) - y\cdot\sin\left(\frac{1}{x}\right) = -1$, where $y \to -1$, as $x \to \infty$, is

a. $y = \sin\frac{1}{x} - \cos\frac{1}{x}$ b. $y = \frac{x+1}{x\cdot\sin(1/x)}$

c. $y = \cos\frac{1}{x} + \sin\frac{1}{x}$ d. $y = \frac{x+1}{x\cos(1/x)}$

**4** The solution of differential equation $(1+y^2) + (x - e^{\tan^{-1}y})\cdot\frac{dy}{dx} = 0$ is

a. $xe^{2\tan^{-1}y} = e^{\tan^{-1}y} + C$

b. $(x-y) = Ce^{-\tan^{-1}y}$

c. $2xe^{\tan^{-1}y} = e^{2\tan^{-1}y} + C$

d. $xe^{\tan^{-1}y} = \tan^{-1}y + C$

**5** The solution of differential equation $ydx + (x + x^2y)\,dy = 0$ is

a. $-\frac{1}{xy} + C$ b. $-\frac{1}{xy} + \log y = C$

c. $\frac{1}{xy} + \log y = C$ d. $\log y = Cx$

**6** The solution of differential equation $(2x - 10y^3)\frac{dy}{dx} + y = 0$ is

a. $x + y = Ce^{2x}$ b. $y^2 = 2x^3 + C$

c. $xy^2 = 2y^5 + C$ d. $x(y^2 + xy) = 0$

Working Space

**7** The general solution of $y' + y\phi'(x) - \phi(x)\cdot\phi'(x) = 0$, where $\phi(x)$ is known function, is

a. $y = Ce^{-\phi(x)} + \phi(x) - 1$ b. $y = Ce^{\phi(x)} + \phi(x) + k$
c. $y = Ce^{-\phi(x)} - \phi(x) + 1$ d. $y = Ce^{-\phi(x)} + \phi(x) + k$

**8** Consider the differential equation $y^2dx + \left(x - \frac{1}{y}\right)dy = 0$. If $y(1) = 1$, then $x$ is **[AIEEE 2011]**

a. $1 - \frac{1}{y} + \frac{e^{1/y}}{e}$ b. $4 - \frac{2}{y} - \frac{e^{1/y}}{e}$
c. $3 - \frac{1}{y} + \frac{e^{1/y}}{e}$ d. $1 + \frac{1}{y} - \frac{e^{1/y}}{e}$

**9** Let $f(x)$ be differentiable on the interval $(0, \infty)$ such that $f(1) = 1$ and $\lim_{t \to x} \frac{t^2 f(x) - x^2 f(t)}{t - x} = 1$, for each $x > 0$. Then, $f(x)$ is **[IIT JEE 2007]**

a. $\frac{1}{3x} + \frac{2x^2}{3}$ b. $\frac{-1}{3x} + \frac{4x^2}{3}$ c. $-\frac{1}{x} + \frac{1}{x^2}$ d. $\frac{1}{x}$

**10** If $y(x)$ satisfies the differential equation $y' - y\tan x = 2x\sec x$ and $y(0) = 0$, then **[IIT JEE 2012]**

a. $y\left(\frac{\pi}{4}\right) = \frac{\pi^2}{8\sqrt{2}}$ b. $y'\left(\frac{\pi}{4}\right) = \frac{\pi^2}{18}$
c. $y = \left(\frac{\pi}{4}\right) = \frac{\pi^2}{9}$ d. $y'\left(\frac{\pi}{3}\right) = 4\pi + \frac{2\pi^2}{3\sqrt{3}}$

## One or More than One Options Correct Type

*Direction* (Q. Nos. 11-14) *This section contains 4 multiple choice questions. Each question has four choices (a), (b), (c) and (d), out of which ONE or MORE THAN ONE are correct.*

**11** If $y(x)$ satisfies the differential equation $y' + y = \cos x - \sin x$ and $y(\pi/2) = 0$, then

a. $y(0) = 1$ b. $y\left(\frac{\pi}{4}\right) = \frac{1}{\sqrt{2}}$
c. $y\left(\frac{\pi}{3}\right) = 2$ d. $y\left(\frac{\pi}{4}\right) = \sqrt{2}$

**12** If $y(x)$ satisfy the differential equation $y' + y\tan x = 2x + x^2\tan x$ and $y(0) = 0$, then

a. $y\left(\frac{\pi}{4}\right) = \frac{\pi^2}{16}$ b. $y\left(\frac{\pi}{4}\right) = \sqrt{2}\frac{\pi^2}{16}$
c. $y\left(\frac{\pi}{3}\right) = \frac{\pi^2}{9}$ d. $y\left(\frac{\pi}{3}\right) = \sqrt{3}\frac{\pi^2}{9}$

**13** If $y(x)$ satisfy the differential equation $y' + \frac{y}{x} = \cos x + \frac{\sin x}{x}$ and $y(0) = 0$, then

a. $y\left(\frac{\pi}{2}\right) = \frac{1}{2}$ b. $y\left(\frac{\pi}{2}\right) = 1$
c. $y\left(\frac{\pi}{4}\right) = \frac{1}{\sqrt{2}}$ d. $y\left(\frac{\pi}{4}\right) = \frac{1}{2\sqrt{2}}$

**14** If $y(x)$ satisfy the differential equation $y - (x + 2y^2)\cdot y' = 0$ and $y(0) = 1$, then $x$ is

a. $2y(y - 1)$ b. $y(y - 1)$
c. 4, when $y = 2$ d. 1, when $y = 2$

## Comprehension Type

*Direction* (Q. Nos. 15-18) *This section contains 2 paragraphs, each describing theory, experiments, data, etc. Four questions related to the paragraphs have been given. Each question has only one correct answer among the four given options (a), (b), (c) and (d).*

Working Space

**Passage I**

A curve $C$ has the property that the area of the triangle formed by the $X$-axis, tangent to the curve and radius vector of the point of the tangency is constant $k^2$.

**15** The differential equation corresponding to the curve is

a. $x\,dy - y\,dx = -2k^2/y$
b. $x\,dy - y\,dx = -2k^2/y^2$
c. $x\,dy - y\,dx = -2k^2/y^3$
d. $x\,dy + y\,dx = -2k^2/y$

**16** The equation of curve $C$, if it passes through (0, 1) is

a. $y^2x = k^2(1-y^2)$
b. $y^2x = k^2(1+y^2)$
c. $yx = k^2(1-y^2)$
d. $yx = k^2(1+y^2)$

**Passage II**

The length of subnormal at any point $P$ on the curve is directly proportional to $OP^2$, where $O$ is the origin.

**17** The differential equation of the family of curves (where, $\lambda$ is the proportionality constant) is

a. $y\dfrac{dy}{dx} + \lambda y^2 = \lambda x^2$
b. $y\dfrac{dy}{dx} - \lambda y^2 = \lambda x^2$
c. $\dfrac{dy}{dx} + 2\lambda y = 2\lambda x^2$
d. None of these

**18** The family of curves is

a. $y^2 = -x^2 - \dfrac{x}{\lambda} - \dfrac{1}{2\lambda^2} + Ce^{2\lambda x}$
b. $y^2 = x^2 + \dfrac{x}{\lambda} + Ce^{2\lambda x}$
c. $y^2 = x^2 + \dfrac{x}{\lambda} + \dfrac{1}{2\lambda^2} + Ce^{2\lambda x}$
d. None of these

## One Integer Value Correct Type

*Direction* (Q. Nos. 19 and 20) *This section contains 2 questions. When worked out will result in an integer from 0 to 9 (both inclusive).*

**19** Let $f : [1, \infty) \to [2, \infty)$ be a differentiable function such that $f(1) = 2$. If $6\int_1^x f(t)\,dt = 3x\,f(x) - x^3$ for all $x > 1$, then the value of $f(2)$ is

**20** Let $y'(x) + y(x)\,g'(x) = g(x)\cdot g'(x)$, $y(0) = 0$, $x \in R$, where $f'(x)$ denotes $\dfrac{d\,f(x)}{dx}$ and $g(x)$ is a given non-constant differentiable function on $R$ with $g(0) = g(2) = 0$. Then, the value of $y(2)$ is

# DPP-5 Solving by Inspection Method and Orthogonal Trajectory

## Only One Option Correct Type

Working Space

*Direction* (Q. Nos. 1-12) *This section contains 12 multiple choice questions. Each question has four choices (a), (b), (c) and (d), out of which ONLY ONE is correct.*

**1** The solution of $x\,dy + y\,dx + 2x^3dx = 0$ is

a. $xy + x^4 = C$

b. $x^2y + \frac{1}{2}x^4 = e$

c. $xy + \frac{1}{2}x^4 = C$

d. $x^2y + x^4 = C$

**2** The solution of $xdx + y\,dy = a(x^2 + y^2)dy$ is

a. $\log(x^2 + y^2) = 2ay + C$

b. $x^2 + y^2 = 2ay + C$

c. $\log(x^2 - y^2) = 2ay + C$

d. $x^2 - y^2 = 2ay + C$

**3** The solution of $(1 + xy)\,x\,dy + (1 - xy)\,y\,dx = 0$ is

a. $-\frac{1}{xy} + \log y = C$

b. $\frac{1}{xy} + \log y = C$

c. $-\frac{1}{xy} + \log\left(\frac{y}{x}\right) = C$

d. $-\frac{1}{xy} - \log\left(\frac{y}{x}\right) = C$

**4** The solution of $y\,dx - x\,dy + (1 + x^2)dx + x^2 \cdot \sin y\,dy = 0$ is

a. $\frac{y}{x} + \frac{1}{x} - x + \cos y + C = 0$

b. $\frac{y}{x} - \frac{1}{x} - x + \cos y + C = 0$

c. $y + \frac{1}{x} - x + \cos y + C = 0$

d. $y - \frac{1}{x} - x + \cos y + C = 0$

**5** The solution of $(1 + x\sqrt{x^2 + y^2})\,dx + (-1 + \sqrt{x^2 + y^2})\,y\,dy = 0$ is

a. $2x + y^2 + \frac{2}{3}(x^2 + y^2)^{3/2} = C$

b. $2x - y^2 + \frac{2}{3}(x^2 + y^2)^{3/2} = C$

c. $x - y^2 + (x^2 + y^2)^{3/2} = C$

d. $x^2 - y^2 + \frac{2}{3}(x^2 + y^2)^{3/2} = C$

**6** The solution of $\frac{x\,dy}{x^2 + y^2} = \left(\frac{y}{x^2 + y^2} - 1\right)dx$ is

a. $\tan^{-1}(y/x) = -x + C$

b. $\tan^{-1}(y/x) = x + C$

c. $\sin^{-1}(y/x) = -x + C$

d. $\sin^{-1}(y/x) = x + C$

Working Space

**7** The solution of $ye^{x/y}\,dx = (xe^{x/y} + y^2 \sin y)dy$ is

a. $e^{x/y} = -2\cos y + C$
b. $e^{x/y} = \cos y + C$
c. $e^{y/x} = -\cos y + C$
d. $e^{x/y} = -\cos y + C$

**8** The solution of $\{y(1 + x^{-1}) + \sin y\}\,dx + \{x + \log x + x\cos y\}\,dy = 0$ is

a. $xy + x\sin y + y\log y = C$
b. $xy + y\sin y + \log x = C$
c. $xy + x\sin y + y\log x = C$
d. $x + xy\sin y + y\log x = C$

**9** The solution of $x\cos(y/x)\cdot(y\,dx + x\,dy) = y\sin(y/x)(x\,dy - y\,dx)$ is

a. $xy = C\sin(y/x)$
b. $xy = C\cos(y/x)$
c. $xy = C\sec(y/x)$
d. $xy = C\tan(y/x)$

**10** The solution of $x\sin(y/x)\,dy = \{y\sin(y/x) - x\}\,dx$ is

a. $\log x - \cos(y/x) = C$
b. $\log x + \cos(y/x) = C$
c. $\log x - \sin(y/x) = C$
d. $\log x + \sin(y/x) = C$

**11** The orthogonal trajectory of $y = cx^2$, where $c$ is an arbitrary constant, is

a. $x^2 - 2y^2 = C_1$
b. $x^2 + y^2 + x = C_1$
c. $x^2 + 2y^2 = 2C_1$
d. $x^2 - 2y^2 = 2C_1$

**12** The orthogonal trajectory of $x^2 + y^2 - 2cx = 0$ (where, $c$ is a parameter) is

a. $x^2 + y^2 + y = C$
b. $x^2 + y^2 + x = C$
c. $x^2 + 2y^2 + Cy = 0$
d. $x^2 + y^2 + Cy = 0$

## One or More than One Options Correct Type

**Direction** (Q. Nos. 13-15) *This section contains 3 multiple choice questions. Each question has four choices (a), (b), (c) and (d), out of which ONE or MORE THAN ONE are correct.*

**13** If $y(x)$ satisfy the differential equation $\dfrac{x\,dx + y\,dy}{x\,dy - y\,dx} = \sqrt{\dfrac{1 - x^2 - y^2}{x^2 + y^2}}$ and $y(1) = 0$, then

a. $y(0) = 0$
b. $y(0) = -1$
c. $y(0) = 2$
d. $y(0) = -2$

**14** If $y(x)$ satisfy the differential equation $\dfrac{x\,dx - y\,dy}{x\,dy - y\,dx} = \sqrt{\dfrac{1 + x^2 - y^2}{x^2 - y^2}}$, then

a. $\sqrt{x^2 - y^2} + \sqrt{1 + x^2 - y^2} = C\sqrt{\dfrac{x + y}{x - y}}$
b. $\sqrt{x^2 - y^2} - \sqrt{1 + x^2 - y^2} = C\sqrt{\dfrac{x + y}{x - y}}$
c. $\sqrt{x^2 - y^2} + \sqrt{1 + x^2 - y^2} = C\sqrt{\dfrac{x - y}{x + y}}$
d. None of these

**15** A tangent drawn to the curve $y = f(x)$ at $P(x, y)$ cut the $X$-axis and $Y$-axis at $A$ and $B$ respectively such that $BP : AP = 3 : 1$, given that $f(1) = 1$, then

a. equation of the curve is $x\dfrac{dy}{dx} - 3y = 0$
b. normal at (1, 1) is $x + 3y = 4$
c. curve passes through (2, 1/8)
d. equation of the curve is $x\dfrac{dy}{dx} + 3y = 0$

## Comprehension Type

*Direction* (*Q. Nos. 16-23*) *This section contains 3 paragraphs, each describing theory, experiments, data, etc. Eight questions related to the paragraphs have been given. Each question has only one correct answer among the four given options (a), (b), (c) and (d).*

### Passage I

$A$ and $B$ are two separate reservoirs of water, capacity of reservoir $A$ is double the capacity of reservoir $B$. Both the reservoirs are filled completely with water, their inlets are closed and then the water is released simultaneously from both the reservoirs. The rate of flow of water out of each reservoir at any instant of time is proportional to the quantity of water in the reservoir at that time. After one hour the water is released, the quantity of water in reservoir $A$ is 1.5 times the quantity of water in reservoir $B$.

Let $V_A$ and $V_B$ represent volume of reservoir $A$ and $B$ at any time $t$.

**16** If after half an hour $V_A = kV_B$, then $k$ is

a. 3
b. 3/4
c. $\sqrt{3}$
d. None of these

**17** After how many hours, do both the reservoirs have the same quantity of water?

a. $(\log_{4/3} 2)$ h
b. $\log_{3/4} 2$ h
c. 2 h
d. $\dfrac{1}{2-\log_2 3}$ h

**18** If $\dfrac{V_A}{V_B} = f(t)$, where $t$ is time. Then, $f(t)$ is

a. increasing
b. decreasing
c. non-monotonic
d. Data insufficient

### Passage II

If $y = f(x)$ and $y = g(x)$ are the pair of curves such that

i. the tangents at point with equal abscissae intersect on $Y$-axis.
ii. the normals drawn at points with equal abscissae intersect on $X$-axis.
iii. curve $f(x)$ passes through (1, 1) and $g(x)$ passes through (2, 3).

**19** The curve $f(x)$ is given by

a. $\dfrac{2}{x} - x$
b. $2x^2 - \dfrac{1}{x}$
c. $\dfrac{2}{x^2} - x$
d. None of these

**20** The curve $g(x)$ is given by

a. $x - \dfrac{1}{x}$
b. $x + \dfrac{2}{x}$
c. $x^2 - \dfrac{1}{x^2}$
d. None of these

**21** The value of $\int_1^2 [g(x) - f(x)]\,dx$ is

a. 2
b. 3
c. 4
d. $4\log_e 2$

Working Space

### Passage III

Let $f : [0, 1] \to R$ be a function. Suppose the function $f$ is twice differentiable; $f(0) = f(1) = 0$ and satisfies $f''(x) - 2f'(x) + f(x) \geq e^x$, $x \in [0, 1]$.

**22** If the function $e^{-x}f(x)$ assumes its minimum in the interval [0, 1] at $x = 1/4$, then which of the following is true?

a. $f(x) < f(x), \frac{1}{4} < x < \frac{3}{4}$

b. $f'(x) > f(x), 0 < x < \frac{1}{4}$

c. $f'(x) < f(x), 0 < x < \frac{1}{4}$

d. $f'(x) < f(x), \frac{3}{4} < x < 1$

**23** Which of the following is true for $0 < x < 1$? **[JEE Advanced 2013]**

a. $0 < f(x) < \infty$

b. $-\frac{1}{2} < f(x) < \frac{1}{2}$

c. $-\frac{1}{4} < f(x) < 1$

d. $-\infty < f(x) < 0$

## One Integer Value Correct Type

*Direction (Q. Nos. 24 and 25) This section contains 2 questions. When worked out will result in an integer from 0 to 9 (both inclusive).*

**24** If $\int_a^x t \cdot y(t)\, dt = x^2 + y|x|$, then $y$ as a function of $x$ is $y = \lambda - (2 + a^2) e^{\frac{x^2 - a^2}{2}}$, the value of $\lambda$ is

**25** A function $f(x)$ satisfying $\int_0^1 f(tx)\, dt = nf(x)$, where $x > 0$ is $f(x) = ce^{\lambda/\mu}$, then the value of $(\lambda + \mu)$ is

# Revisal Problems for JEE Main

## Only One Option Correct Type

Working Space

*Direction* (Q. Nos. 1-35) *This section contains 35 multiple choice questions. Each question has four choices (a), (b), (c) and (d), out of which ONLY ONE option is correct.*

**1** The differential equation of all parabola whose axis of symmetry is parallel to $X$-axis is of order

a. 2
b. 3
c. 1
d. None of these

**2** The degree of the differential equation $\frac{d^2y}{dx^2}+3\left(\frac{dy}{dx}\right)^2=x^2\log\left(\frac{d^2y}{dx^2}\right)$ is

a. 1
b. 2
c. 3
d. Does not exist

**3** The degree of the differential equation satisfying, $\sqrt{1-x^2}+\sqrt{1-y^2}=a(x-y)$ is

a. 1
b. 2
c. 3
d. 4

**4** The degree of the differential equation $\left(\frac{d^3y}{dx^3}\right)^{2/3}+4-3\frac{d^2y}{dx^2}+5\frac{dy}{dx}=0$ is

a. 1
b. 2
c. 3
d. 4

**5** The degree of the differential equation satisfying the relation

$\sqrt{1+x^2}+\sqrt{1+y^2}=\lambda\left(x\sqrt{1+y^2}-y\sqrt{1+x^2}\right)$ is

a. 1
b. 2
c. 3
d. 4

**6** The degree of the differential equation corresponding to the family of curves $y=a(x+a)^2$, where $a$ is an arbitrary constant, is

a. 2
b. 3
c. 4
d. 5

**7** The order and degree of the differential equation of all tangent lines to the parabola $x^2=4y$, are

a. 1, 2
b. 2, 2
c. 3, 1
d. 4, 1

**8** The degree and order of the differential equation of all parabolas whose axis is $X$-axis, are

a. 2, 1
b. 1, 2
c. 3, 2
d. 1, 1

**9** The solution of $\left(\frac{dy}{dx}\right)^2-x\frac{dy}{dx}+y=0$ is

a. $y=2$
b. $y=2x$
c. $y=2x-4$
d. $y=2x^2-4$

**10** The differential equation representing the family of curves $y^2=2c(x+\sqrt{c})$, where $c$ is a parameter, is of order and degree as

a. 1, 3
b. 2, 3
c. 1, 2
d. 2, 2

Working Space

**11** The degree of the differential equation of all curves having normal of constant length $c$, is

a. 1 b. 3
c. 4 d. None of these

**12** The equation of the curve in which subnormal varies as the square of the ordinate is

a. $y = Ce^{2\lambda x}$ b. $y = Ce^{\lambda x}$
c. $\frac{y^2}{2} + \lambda x = C$ d. $y^2 + \lambda x^2 = C$

**13** The solution of differential equation $xdy - ydx = 0$ represents

a. a parabola whose vertex is at origin
b. a circle whose centre is at origin
c. a rectangular hyperbola
d. straight lines passing through the origin

**14** The degree of the differential equation $x = 1 + \frac{dy}{dx} + \frac{1}{2!}\left(\frac{dy}{dx}\right)^2 + \frac{1}{3!}\left(\frac{dy}{dx}\right)^3 + \cdots \infty$ is

a. 3 b. 1
c. 2 d. not defined

**15** The orthogonal trajectories of the family of curves $a^{n-1} \cdot y = x^n$ (where, $a$ is arbitrary constant) are given by

a. $x^n + n^2y =$ constant b. $ny^2 + x^2 =$ constant
c. $n^2x + y^n =$ constant d. $n^2x - y^n =$ constant

**16** The orthogonal trajectories of the family of circles given by $x^2 + y^2 - 2ay = 0$ (where, $a$ is parameter) is

a. $x^2 + y^2 - kx = 0$ b. $x^2 + y^2 - 2ky = 0$
c. $x^2 + y^2 - 2k_1x - 2k_2y = 0$ d. None of these

**17** The equation of curve in which the portion of the tangent included between the coordinates axes is bisected at the point of contact, is

a. a parabola b. an ellipse
c. a circle d. a hyperbola

**18** The equation of the curve whose subnormal is twice the abscissa, is

a. a circle b. a parabola
c. an ellipse d. a hyperbola

**19** The solution of the differential equation $\frac{x\,dy}{x^2 + y^2} = \left(\frac{y}{x^2 + y^2} - 1\right)dx$ is

a. $y = x\cot(C - x)$ b. $\cos^{-1}(y/x) = (-x + C)$
c. $y = x\tan(C - x)$ d. $y^2/x^2 = x\tan(C - x)$

**20** If $\phi(x)$ is a differentiable function, then the solution of differential equation $dy + \{y\phi'(x) - \phi(x)\,\phi'(x)\}dx = 0$ is

a. $y = \{\phi(x) - 1\} + Ce^{-\phi(x)}$ b. $y\phi(x) = \{\phi(x)\}^2 + C$
c. $ye^{\phi(x)} = \phi(x)e^{\phi(x)} + C$ d. None of these

**21** The solution of the differential equation $\frac{dy}{dx} = \frac{x^2 + y^2 + 1}{2xy}$ satisfying $y(1) = 1$, is

a. a hyperbola b. a circle
c. $y^2 = x(1 + x) - 10$ d. $(x - 2)^2 + (y - 3)^2 = 5$

**22** The solution of the differential equation $x^3\frac{dy}{dx} + 4x^2 \cdot \tan y = e^x \sec y$ satisfying $y(1) = 0$ is

a. $\tan y = (x - 2)e^x \cdot \log 3$
b. $\sin y = e^x(x - 1)x^{-4}$
c. $\tan y = (x - 1)e^x \cdot x^{-3}$
d. $\sin y = e^x(x - 1) \cdot x^{-3}$

Working Space

**23** The solution of differential equation $y(xy + 2x^2y^2)dx + x(xy - x^2y^2)dy = 0$ is given by

a. $2\log|x| - \log|y| - \frac{1}{xy} = C$

b. $2\log|y| - \log|x| - \frac{1}{xy} = C$

c. $2\log|x| + \log|y| + \frac{1}{xy} = C$

d. $2\log|y| + \log|x| + \frac{1}{xy} = C$

**24** The solution of the differential equation $x = 1 + xy\frac{dy}{dx} + \frac{(xy)^2}{2!}\left(\frac{dy}{dx}\right)^2 + \frac{(xy)^3}{3!}\left(\frac{dy}{dx}\right)^3 + \cdots$ is

a. $y = \log_e x + C$

b. $y = (\log_e x)^2 + C$

c. $y = \pm\sqrt{(\log x)^2 + 2C}$

d. $xy = x^y + C$

**25** The solution of the differential equation $\left(\frac{x+y-1}{x+y-2}\right)\frac{dy}{dx} = \left(\frac{x+y+1}{x+y+2}\right)$, given that $y = 1$ when $x = 1$, is

a. $\log\left|\frac{(x-y)^2 - 2}{2}\right| = 2(x + y)$

b. $\log\left|\frac{(x-y)^2 + 2}{2}\right| = 2(x - y)$

c. $\log\left|\frac{(x+y)^2 + 2}{2}\right| = 2(x - y)$

d. None of these

**26** If the solution of differential equation $\frac{dy}{dx} = \frac{ax + 3}{2y + f}$ represents a circle, then the value of $a$ is

a. 2

b. −2

c. 3

d. −4

**27** The curve satisfying $\frac{dy}{dx} = \frac{x^2 + y^2}{x^2 - y^2}$, $y(1) = 2$ has the slope at the point (1, 0) of the curve, is equal to

a. $-\frac{5}{3}$

b. −1

c. 1

d. $\frac{5}{3}$

**28** The equation of the family of curves which intersect the hyperbola $xy = 2$ orthogonally, is

a. $y = \frac{x^3}{6} + C$

b. $y = \frac{x^2}{4} + C$

c. $y = -\frac{x^3}{6} + C$

d. None of these

**29** The general solution of the differential equation $y(x^2y + e^x)dx - e^x dy = 0$ is

a. $x^3y - 3e^x = Cy$

b. $x^3y + 3e^x = 3Cy$

c. $y^3x - 3e^y = Cx$

d. $y^3x + 3e^y = Cx$

**30** If $y(t)$ is a solution of $(1 + t)dy/dt - ty = 1$ and $y(0) = 1$, then $y(1)$ is

a. $-\frac{1}{2}$

b. $e + \frac{1}{2}$

c. $e - \frac{1}{2}$

d. $\frac{1}{2}$

**31** The differential equation of rectangular hyperbolas whose axes are asymptotes of the hyperbola $x^2 - y^2 = a^2$, is

a. $y\frac{dy}{dx} = x$

b. $x\frac{dy}{dx} = -y$

c. $x\frac{dy}{dx} = y$

d. None of these

**32** The differential equation of the family of parabolas having vertex at the origin and the $Y$-axis as the axis of symmetry, is

a. $2y_1 - y = 0$

b. $2y_1 + xy = 0$

c. $xy_1 - 2y = 0$

d. None of these

Working Space

33 The differential equation of the family of ellipses having major and minor axes respectively along the $X$ and $Y$-axes and minor axis is equal to half to the major axis, is

a. $xy' - 4y = 0$
b. $4xy' + y = 0$
c. $4yy' + x = 0$
d. $yy' + 4x = 0$

34 If $\sin x$ is an integrating factor of the differential equation $\frac{dy}{dx} + Py = Q$, then $P$ can be

a. $\log(\sin x)$
b. $\cot x$
c. $\sin x$
d. $\log(\cos x)$

35 A curve having the condition that the slope of tangent at some point is two times the slope of the straight line joining the same point to the origin of coordinates is

a. a circle
b. an ellipse
c. a parabola
d. a hyperbola

## Statement Type

**Direction** (Q. Nos. 36-40) ***This section is based on Statement I and Statement II. Select the correct answer from the codes given below.***

a. Both Statement I and Statement II are correct and Statement II is the correct explanation of Statement I
b. Both Statement I and Statement II are correct but Statement II is not the correct explanation of Statement I
c. Statement I is correct but Statement II is incorrect
d. Statement II is correct but Statement I is incorrect

36 **Statement I** The order of differential equation of all the circles which touches $X$-axis is 2.

**Statement II** The order of differential equation is same as number of independent arbitrary constant in given curve.

37 **Statement I** The order of differential equation whose primitive is $y = A + \log Bx$ is 2.

**Statement II** If there are $n$ arbitrary constants in a family of a curve, then the order of the corresponding differential equation is $n$.

38 **Statement I** The orthogonal trajectory to the curve $(x-a)^2 + (y-b)^2 = r^2$ is $y = mx + b - am$, where $a$ and $b$ are fixed numbers and $r$ and $m$ are parameters.

**Statement II** In a plane, the line that passes through the centre of circle is normal to the circle.

39 **Statement I** $\sin x \cdot \frac{d^2y}{dx^2} + \cos x \frac{dy}{dx} + \tan x = 0$ is not a linear differential equation.

**Statement II** A differential equation is said to be linear, if dependent variable and its differential coefficients occurs in first degree and not multiplied together.

40 Consider the differential equation $(xy - 1)\frac{dy}{dx} + y^2 = 0$.

**Statement I** The solution of the equation is $xy = \log y + C$.

**Statement II** The given differential equation can be expressed as $dx/dy + Px = Q$, where integrating factor is $\log y$.

SET - 2

# Revisal Problems
## for JEE Advanced

### Only One Option Correct Type

Working Space

*Direction* (*Q. Nos.* 1-18) *This section contains* 18 *multiple choice questions. Each question has four choices (a), (b), (c) and (d), out of which ONLY ONE option is correct.*

**1** The area bounded by the curve $y = \int_{1/8}^{\sin^2 x} (\sin^{-1}\sqrt{t})\,dt + \int_{1/8}^{\cos^2 x} (\cos^{-1}\sqrt{t})\,dt$, $0 \le x \le \pi/2$ and the curve satisfying the equation $y(x + y^3)\,dx = x(y^3 - x)\,dy$ by passing through $(4, -2)$ is

a. $\frac{7}{8}\left(\frac{3\pi}{16}\right)^2$　　b. $\frac{1}{8}\left(\frac{3\pi}{16}\right)^3$

c. $\frac{1}{8}\left(\frac{3\pi}{16}\right)^4$　　d. None of these

**2** If $\lim_{x \to a} \frac{\int_a^t f(x)dx - \left(\frac{t-a}{2}\right)\{f(t)+f(a)\}}{(t-a)^3} = 3a$ and $f(1) = -6$, $f'(1) = 42$, then $f(0)$ is

a. $-18$　　b. $-60$

c. $-80$　　d. $20$

**3** If $9 + f''(x) + f'(x) = x^2 + f^2(x)$ is the differential equation of a curve and $P$ is a point of maxima of this curve, then the number of tangents which can be drawn from $P$ to the circle $x^2 + y^2 = 9$ is

a. 2　　b. 1

c. 0　　d. None of these

**4** A function $y = f(x)$ satisfies the differential equation $f(x) \cdot \sin 2x - \cos x + (1 + \sin^2 x) \cdot f'(x) = 0$ with initial condition $y(0) = 0$. The value of $f(\pi/6)$ is equal to

a. 1/5　　b. 3/5

c. 4/5　　d. 2/5

**5** Water is drained from a vertical cylinderical tank by opening a valve at the base of the tank. It is known that the rate at which the water level drops is proportional to the square root of water depth $y$, where the constant of proportionality $k > 0$ depends on the acceleration due to gravity and the geometry of the hole. If $t$ is measured in minutes and $k = 1/15$, then the time to drain the tank, if the water is 4 m deep to start with, is

a. 30 min　　b. 45 min

c. 60 min　　d. 80 min

**6** If the differential equation of the family of curve given by $y = Ax + Be^{2x}$, where $A$ and $B$ are arbitrary constant is of the form $(1-2x) \cdot \frac{d}{dx}\left(\frac{dy}{dx} + ly\right) + k\left(\frac{dy}{dx} + ly\right) = 0$, then the ordered pair $(k, l)$ is

a. $(2, -2)$　　b. $(-2, 2)$

c. $(2, 2)$　　d. $(-2, -2)$

**7** The $x$-intercept of the tangent to a curve is equal to the ordinate of the point of contact. The equation of the curve through the point (1, 1) is

a. $ye^{x/y} = e$　　b. $xe^{x/y} = e$

c. $xe^{y/x} = e$　　d. $ye^{y/x} = e$

**8** A function $y = f(x)$ satisfies the condition $f'(x)\sin x + f(x)\cos x = 1$, $f(x)$ being bounded when $x \to 0$. If $I = \int_0^{\pi/2} f(x)\,dx$, then

a. $\pi/2 < I < \pi^2/4$　　b. $\pi/4 < I < \pi^2/2$

c. $1 < I < \pi/2$　　d. $0 < I < 1$

Working Space

**9** A curve is such that the area of the region bounded by the coordinate axes, the curve and the ordiante of any point on it is equal to the cube of that ordinate. The curve represents

a. a pair of straight lines b. a circle

c. a parabola d. an ellipse

**10** The real value of $m$ for which the substitution, $y = u^m$ will transform the differential equation, $2x^4y\frac{dy}{dx} + y^4 = 4x^6$ into a homogenous equation is

a. $m = 0$ b. $m = 1$

c. $m = 3/2$ d. no value of $m$

**11** Consider the two cases:

**Case I** $y = \sin kt$ satisfies the differential equation $y'' + 9y = 0$.

**Case II** $y = e^{kt}$ satisfy the differential equation $y'' + y' - 6y = 0$.

The value of $k$ for which both the cases are correct, is

a. $-3$ b. 0

c. 2 d. 3

**12** A wet porous substances in the open air loses its moisture at a rate proportional to the moisture content. If a sheet hung in the wind loses half its moisture content during the first hour, then the time when it would have lost 99.9% of its moisture is (weather conditions remains same)

a. more than 100 h b. less than 100 h

c. approximately 10 h d. approximately 9 h

**13** A curve $C$ passes through origin and has the property that at each point $(x, y)$ on it the normal line at that point passes through $(1, 0)$. The equation of a common tangent to the curve $C$ and the parabola $y^2 = 4x$ is

a. $x = 0$ b. $y = 0$

c. $y = x + 1$ d. $x + y + 1 = 0$

**14** A function $y = f(x)$ satisfies $(x + 1)f'(x) - 2(x^2 + x)f(x) = \frac{e^{x^2}}{x+1}$, $\forall\, x > -1$. If $f(0) = 5$, then $f(x)$ is

a. $\left(\frac{3x+5}{x+1}\right)e^{x^2}$ b. $\left(\frac{6x+5}{x+1}\right)e^{x^2}$

c. $\left(\frac{6x+5}{(x+1)^2}\right)e^{x^2}$ d. $\left(\frac{5-6x}{x+1}\right)e^{x^2}$

**15** The curve with the property that the projection of the ordinate on the normal is constant and has a length equal to $a$, is

a. $a \log(\sqrt{y^2 - a^2} + y) = x + C$ b. $x + \sqrt{a^2 - y^2} = C$

c. $(y - a)^2 = Cx$ d. $ay = \tan^{-1}(x + C)$

**16** Consider the differential equation $dy/dx + y \tan x = x \tan x + 1$. Then,

a. the integral curves satisfying the differential equation and given by $y = x + C \sin x$

b. the angle at which the integral curves cut the $Y$-axis is $\pi/2$

c. tangents to all the integral curves at their point of intersections with $Y$-axis are parallel

d. None of the above

**17** The subsitution $y = z^\alpha$ transforms the differential equation $(x^2y^2 - 1)dy + 2xy^3dx = 0$ into a homogenous differential equation for

a. $\alpha = -1$ b. $\alpha = 0$

c. $\alpha = 1$ d. no value of $\alpha$

**18** A curve passing through $(2, 3)$ and satisfying the differential equation $\int_0^x t \cdot y(t)\,dt = x^3 y(x), (x > 0)$ is

a. $xy = 6$ b. $xy = 4$

c. $x^2 + y^2 = 13$ d. $y^2 = \frac{9}{2}x$

Working Space

## One or More than One Options Correct Type

**Direction** (Q. Nos. 19-23) *This section contains 5 multiple choice questions. Each question has four choices (a), (b), (c) and (d), out of which ONE or MORE THAN ONE are correct.*

**19** A curve $y = f(x)$ has the property that the perpendicular distance of the origin from the normal at any point $P$ of the curve is equal to the distance of the point $P$ from the $X$-axis. Then, the differential equation of the curve

a. is homogenous
b. can be converted into linear differential equation with some suitable substitution
c. is the family of circles touching the $X$-axis at origin
d. is the family of circles touching the $Y$-axis at origin

**20** The function $f(x)$ satisfying the equation $f^2(x) + 4f'(x)f(x) + [f'(x)]^2 = 0$ is

a. $f(x) = C \cdot e^{(2-\sqrt{3})x}$
b. $f(x) = C \cdot e^{(2+\sqrt{3})x}$
c. $f(x) = C \cdot e^{(\sqrt{3}-2)x}$
d. $f(x) = C \cdot e^{-(2+\sqrt{3})x}$

**21** A function $y = f(x)$ satisfying the differential equation $\frac{dy}{dx}\sin x - y\cos x + \frac{\sin^2 x}{x^2} = 0$ is such that, $y \to 0$ as $x \to \infty$, then which of the statements is/are correct ?

a. $\lim_{x \to 0} f(x) = 1$
b. $\int_0^{\pi/2} f(x)\,dx < 3/2$
c. $\int_0^{\pi/2} f(x)\,dx > 1$
d. $f(x)$ is odd function

**22** A differential function satisfies $f(x) = \int_0^x [f(t)\cos t - \cos(t - x)]\,dt$, then which of the following hold good?

a. $\min f(x) = 1 - e$
b. $\max f(x) = 1 - e^{-1}$
c. $f''(\pi/2) = e$
d. $f'(0) = 1$

**23** Let $dy/dx + y = f(x)$, where $y$ is a continuous function of $x$ with $y(0) = 1$ and $f(x) = \begin{cases} e^{-x}, & \text{if } 0 \le x < 2 \\ e^{-2}, & \text{if } x > 2 \end{cases}$, then which of the following hold good?

a. $y(1) = 2e^{-1}$
b. $y'(1) = -e^{-1}$
c. $y(3) = -2e^{-3}$
d. $y'(3) = -2e^{-3}$

## Comprehension Type

**Direction** (Q. Nos. 24-29) *This section contains 2 paragraphs, each describing theory, experiments, data, etc. Six questions related to the paragraphs have been given. Each question has only one correct answer among the four given options (a), (b), (c) and (d).*

### Passage I

Let $y = f(x)$ satisfies the equation $f(x) = (e^{-x} + e^{x})\cos x - 2x - \int_0^x (x - t) f'(t)\,dt$.

**24** $y$ satisfies the differential equation

a. $\frac{dy}{dx} = y + e^x \cdot (\cos x - \sin x) - e^{-x}(\cos x + \sin x)$
b. $\frac{dy}{dx} + y = e^x \cdot (\cos x - \sin x) - e^{-x}(\cos x + \sin x)$
c. $\frac{dy}{dx} + y = e^x(\cos x + \sin x) - e^{-x}(\cos x - \sin x)$
d. None of the above

**25** The value of $f'(0) + f''(0)$ equals

a. $-1$
b. 2
c. 1
d. 0

**26** $f(x)$ as a function of $x$ equals

a. $e^{-x}(\cos x - \sin x) + \frac{e^x}{5}(3\cos x - \sin x) + \frac{2}{5}e^{-x}$
b. $e^{-x}(\cos x - \sin x) + \frac{e^x}{5}(3\cos x + \sin x) + \frac{2}{5}e^{-x}$
c. $e^{-x}(\cos x + \sin x) + \frac{e^x}{5}(3\cos x - \sin x) + \frac{2}{5}e^{-x}$
d. None of the above

Working Space

## Passage II

A curve $y = f(x)$ satisfies the differential equation $(1 + x^2) dy/dx + 2xy = 4x^2$ and passes through the origin.

**27** The function $y = f(x)$ is

a. strictly increasing, $\forall\ x \in R$
b. such that it has a minima but no maxima
c. such that it has a maxima but no minima
d. None of the above

**28** The area enclosed by $y = f^{-1}(x)$, the X-axis and the ordinate at $x = 2/3$ is

a. $2 \log 2$
b. $\frac{4}{3} \log 2$
c. $\frac{2}{3} \log 2$
d. $\frac{1}{3} \log 2$

**29** For the function $y = f(x)$, which one of the following does not hold good?

a. $f(x)$ is a rational function
b. $f(x)$ has the same domain and range
c. $f(x)$ is transcidental function
d. $f(x)$ is bijective

## One Integer Value Correct Type

***Direction*** *(Q. Nos. 30-35) This section contains 6 questions. When worked out will result in an integer from 0 to 9 (both inclusive).*

**30** The solution of $y' + \frac{y}{x} = \frac{1}{(1 + \log xy)^2}$ is $xy\,[1 + (\log xy)^2] = \frac{x^\lambda}{\mu} + C$, then the value of $(\lambda + \mu)$ is

**31** The solution of $y'(x^2 - 2x^3y^3) = (3x^2y^4 + 2xy)$ is $x^\lambda y + \frac{x^\mu}{y} = C$, then the value of $(\lambda + \mu)$ is

**32** The solution of $y(2xy + 1)dx + x(1 + 2xy + x^2y^2)dy = 0$ is $-\frac{1}{2(xy)^2} - \frac{\lambda}{xy} + \log y = C$, then the value of $\lambda$ is

**33** The solution of $\cos\left(\frac{x}{y}\right)(y\,dx - x\,dy) = xy^3(x\,dy + y\,dx)$ is $\sin\left(\frac{x}{y}\right) = \frac{(xy)^\lambda}{\mu} + C$, then the value of $(\lambda + \mu)$ is

**34** The solution of $\left\{y\left(1 + \frac{1}{x}\right) + \cos y\right\}dx + \{x + \log x - x \sin y\}dy = 0$ is $xy + y \log x + x^\lambda \cos y = C$, then the value of $\lambda$ is

**35** The solution of $xy' - y = mx^2\sqrt{x^2 - y^2}$ is $\sin^{-1}\left(\frac{y}{x}\right) = m\frac{x^\lambda}{\mu} + C$, then the value of $(\lambda - \mu)$ is

# JEE Main & AIEEE Archive

(Compilation of Last 13 Years' Questions)

**1** Let $y(x)$ be the solution of the differential equation $(x\log x)\frac{dy}{dx} + y = 2x\log x, (x \geq 1)$. Then, $y(e)$ is equal to **(JEE Main 2015)**

a. $e$ b. 0

c. 2 d. $2e$

**2** Let the population of rabbits surviving at a time $t$ be governed by the differential equation $\frac{dp(t)}{dt} = \frac{1}{2}p(t) - 200$. If $p(0) = 100$, then $p(t)$ is equal to **(JEE Main 2014)**

a. $400 - 300e^{t/2}$ b. $300 - 200e^{-t/2}$

c. $600 - 500e^{t/2}$ d. $400 - 300e^{-t/2}$

**3** At present, a firm is manufacturing 2000 items. It is estimated that the rate of change of production $P$ with respect to additional number of workers $x$ is given by $\frac{dP}{dx} = 100 - 12\sqrt{x}$. If the firm employees 25 more workers, then the new level of production of items is **(JEE Main 2013)**

a. 2500 b. 3000

c. 3500 d. 4500

**4** The population $p(t)$ at time $t$ of a certain mouse species satisfies the differential equation $\frac{dp(t)}{dt} = 0.5(t) - 450$. If $p(0) = 850$, then the time at which the population becomes zero is **(AIEEE 2012)**

a. 2 log 18 b. log 9

c. $\frac{1}{2}\log 18$ d. log 18

**5** If $\frac{dy}{dx} = y + 3 > 0$ and $y(0) = 2$, then $y(\log 2)$ is equal to **(AIEEE 2011)**

a. 5 b. 13

c. $-2$ d. 7

**6** Let $I$ be the purchase value of an equipment and $V(t)$ be the value after it has been used for $t$ years. The value $V(t)$ depreciates at a rate given by differential equation $\frac{dV(t)}{dt} = -k(T-t)$, where $k > 0$ is a constant and $T$ is the total life in years of the equipment. Then, the scrap value $V(T)$ of the equipment is **(AIEEE 2011)**

a. $I - \frac{k}{2}T^2$ b. $I - \frac{k(T-t)^2}{2}$

c. $e^{-kT}$ d. $T^2 - \frac{1}{k}$

**7** Consider the differential equation $y^2dx + \left(x - \frac{1}{y}\right)dy = 0$. If $y(1) = 1$, then $x$ is given by **(AIEEE 2011)**

a. $1 - \frac{1}{y} + \frac{e^{1/y}}{e}$ b. $4 - \frac{2}{y} - \frac{e^{1/y}}{e}$

c. $3 - \frac{1}{y} + \frac{e^{1/y}}{e}$ d. $1 + \frac{1}{y} - \frac{e^{1/y}}{e}$

**8** Solution of the differential equation $\cos x\,dy = y(\sin x - y)dx$, $0 < x < \frac{\pi}{2}$, is **(AIEEE 2010)**

a. $\sec x = (\tan x + C)y$

b. $y\sec x = \tan x + C$

c. $y\tan x = \sec x + C$

d. $\tan x = (\sec x + C)y$

**9** The differential equation which represents the family of curves $y = c_1e^{c_2x}$, where $c_1$ and $c_2$ are arbitrary constants, is **(AIEEE 2009)**

a. $y' = y^2$ b. $y'' = y'y$

c. $yy'' = y'$ d. $yy'' = (y')^2$

**10** The differential equation of the family of circles with fixed radius 5 units and centre on the line $y = 2$ is **(AIEEE 2008)**

a. $(x-2)^2y'^2 = 25 - (y-2)^2$

b. $(x-2)y'^2 = 25 - (y-2)^2$

c. $(y-2)y'^2 = 25 - (y-2)^2$

d. $(y-2)^2y'^2 = 25 - (y-2)^2$

**11** The solution of the differential equation $\frac{dy}{dx} = \frac{x+y}{x}$ satisfying the condition $y(1) = 1$ is

a. $y = x\log x + x$ b. $y = \log x + x$

c. $y = x\log x + x^2$ d. $y = xe^{(x-1)}$

**12** The differential equation of all circles passing through the origin and having their centres on the $X$-axis is **(AIEEE 2007)**

a. $x^2 = y^2 + xy\frac{dy}{dx}$ b. $x^2 = y^2 + 3xy\frac{dy}{dx}$

c. $y^2 = x^2 + 2xy\frac{dy}{dx}$ d. $y^2 = x^2 - 2xy\frac{dy}{dx}$

**13** The differential equation whose solution is $Ax^2 + By^2 = 1$, where $A$ and $B$ are arbitrary constants, is of **(AIEEE 2006)**

a. first order and second degree

b. first order and first degree

c. second order and first degree

d. second order and second degree

**14** The differential equation representing the family of curves $y^2 = 2c\,(x + \sqrt{c})$, where $c > 0$, is a parameter, is of order and degree as follows **(AIEEE 2005)**

a. order 2, degree 2 b. order 1, degree 3
c. order 1, degree 1 d. order 1, degree 2

**15** If $x\dfrac{dy}{dx} = y(\log y - \log x + 1)$, then the solution of the equation is **(AIEEE 2005)**

a. $\log\left(\dfrac{x}{y}\right) = Cy$ b. $\log\left(\dfrac{y}{x}\right) = Cx$
c. $x\log\left(\dfrac{y}{x}\right) = Cy$ d. $y\log\left(\dfrac{x}{y}\right) = Cx$

**16** The differential equation for the family of curves $x^2 + y^2 - 2ay = 0$, where $a$ is an arbitrary constant, is **(AIEEE 2004)**

a. $2(x^2 - y^2)y' = xy$ b. $2(x^2 + y^2)y' = xy$
c. $(x^2 - y^2)y' = 2xy$ d. $(x^2 + y^2)y' = 2xy$

**17** The solution of the differential equation $y\,dx + (x + x^2y)\,dy = 0$ is **(AIEEE 2004)**

a. $-\dfrac{1}{xy} = C$ b. $-\dfrac{1}{xy} + \log y = C$
c. $\dfrac{1}{xy} + \log y = C$ d. $\log y = Cx$

**18** The degree and order of the differential equation of the family of all parabolas whose axis is $X$-axis, are respectively **(AIEEE 2003)**

a. 2, 1 b. 1, 2 c. 3, 2 d. 2, 3

**19** The solution of the differential equation

$$(1 + y^2) + (x - e^{\tan^{-1} y})\frac{dy}{dx} = 0, \text{ is}$$ **(AIEEE 2003)**

a. $(x - 2) = Ce^{-2\tan^{-1} y}$
b. $2xe^{\tan^{-1} y} = e^{2\tan^{-1} y} + C$
c. $xe^{\tan^{-1} y} = \tan^{-1} y + C$
d. $xe^{2\tan^{-1} y} = e^{\tan^{-1} y} + C$

**20** The order and degree of the differential equation $\left(1 + 3\dfrac{dy}{dx}\right)^{\frac{2}{3}} = 4\dfrac{d^3y}{dx^3}$ are **(AIEEE 2002)**

a. $\left(1, \dfrac{2}{3}\right)$ b. (3, 1)
c. (3, 3) d. (1, 2)

**21** The solution of the equation $\dfrac{d^2y}{dx^2} = e^{-2x}$ is **(AIEEE 2002)**

a. $\dfrac{e^{-2x}}{4}$ b. $\dfrac{e^{-2x}}{4} + Cx + D$
c. $\dfrac{1}{4}e^{-2x} + Cx^2 + D$ d. $\dfrac{1}{4}e^{-2x} + C + D$

**22** The differential equation of all non-vertical lines in a plane is **(AIEEE 2002)**

a. $\dfrac{d^2y}{dx^2} = 0$ b. $\dfrac{d^2x}{dy^2} = 0$
c. $\dfrac{dy}{dx} = 0$ d. $\dfrac{dx}{dy} = 0$

# JEE Advanced & IIT JEE Archive

(Compilation of Last 10 Years' Questions)

**1** Let $F(x)=\int_x^{x^2+\frac{\pi}{6}} 2\cos^2 t\, dt$ for all $x \in R$ and $f:\left[0,\frac{1}{2}\right]\to[0,\infty)$ be a continuous function. For $a \in\left[0,\frac{1}{2}\right]$, if $F'(a)+2$ is the area of the region bounded by $x=0$, $y=0$, $y=f(x)$ and $x=a$, then $f(0)$ is **(2015 Adv., Integer Type)**

**2** Let $y(x)$ be a solution of the differential equation $(1+e^x)y'+ye^x=1$. If $y(0)=2$, then which of the following statement(s) is/are true?
**(2015 Adv., One or More than One Options Correct Type)**

a. $y(-4)=0$
b. $y(-2)=0$
c. $y(x)$ has a critical point in the interval $(-1, 0)$
d. $y(x)$ has no critical point in the interval $(-1, 0)$

**3** Consider the family of all circles whose centres lie on the straight line $y=x$. If this family of circles is represented by the differential equation $Py''+Qy'+1=0$, where $P, Q$ are the functions of $x, y$ and $y'$ (here, $y'=\frac{dy}{dx}$, $y''=\frac{d^2y}{dx^2}$), then which of the following statement(s) is/are true?
**(2015 Adv., One or More than One Options Correct Type)**

a. $P=y+x$
b. $P=y-x$
c. $P+Q=1-x+y+y'+(y')^2$
d. $P-Q=x+y-y'-(y')^2$

**4** The function $y=f(x)$ is the solution of the differential equation $\frac{dy}{dx}+\frac{xy}{x^2-1}=\frac{x^4+2x}{\sqrt{1-x^2}}$ in $(-1,1)$ satisfying $f(0)=0$. Then, $\int_{-\sqrt{3}/2}^{\sqrt{3}/2} f(x)dx$ is
**(2014 Adv., Only One Option Correct Type)**

a. $\frac{\pi}{3}-\frac{\sqrt{3}}{2}$ b. $\frac{\pi}{3}-\frac{\sqrt{3}}{4}$
c. $\frac{\pi}{6}-\frac{\sqrt{3}}{4}$ d. $\frac{\pi}{6}-\frac{\sqrt{3}}{2}$

**5** Let $f:[1/2, 1]\to R$ (the set of all real numbers) be a positive, non-constant and differentiable function such that $f'(x)<2f(x)$ and $f(1/2)=1$. Then, the value of $\int_{1/2}^{1} f(x)dx$ lies in the interval
**(2013 Adv., Only One Option Correct Type)**

a. $(2e-1, 2e)$
b. $(e-1, 2e-1)$
c. $\left(\frac{e-1}{2}, e-1\right)$
d. $\left(0, \frac{e-1}{2}\right)$

**6** A curve passes through the point $\left(1,\frac{\pi}{6}\right)$. Let the slope of the curve at each point $(x, y)$ be $\frac{y}{x}+\sec\left(\frac{y}{x}\right)$, $x>0$.
Then, the equation of the curve is
**(2013 Adv., Only One Option Correct Type)**

a. $\sin\left(\frac{y}{x}\right)=\log x+\frac{1}{2}$ b. $\operatorname{cosec}\left(\frac{y}{x}\right)=\log x+2$
c. $\sec\left(\frac{2y}{x}\right)=\log x+2$ d. $\cos\left(\frac{2y}{x}\right)=\log x+\frac{1}{2}$

**Passage for Q. Nos. (7 and 8)**

Let $f:[0, 1]\to R$ (the set of all real numbers) be a function. Suppose the function $f$ is twice differentiable, $f(0)=f(1)=0$ and satisfies $f''(x)-2f'(x)+f(x)\ge e^x$, $x\in[0, 1]$.
**(2013 Adv., Comprehension Type)**

**7** Which of the following is true for $0<x<1$?

a. $0<f(x)<\infty$ b. $-\frac{1}{2}<f(x)<\frac{1}{2}$
c. $-\frac{1}{4}<f(x)<1$ d. $-\infty<f(x)<0$

**8** If the function $e^{-x}f(x)$ assumes its minimum in the interval $[0, 1]$ a $x=1/4$, then which of the following is true?

a. $f'(x)<f(x)$, $\frac{1}{4}<x<\frac{3}{4}$
b. $f'(x)>f(x)$, $0<x<\frac{1}{4}$
c. $f'(x)<f(x)$, $0<x<\frac{1}{4}$
d. $f'(x)<f(x)$, $\frac{3}{4}<x<1$

**9** If $y(x)$ satisfies the differential equation $y'-y\tan x=2x\sec x$ and $y(0)$, then
**(2012, One or More than One Options Correct Type)**

a. $y\left(\frac{\pi}{4}\right)=\frac{\pi^2}{8\sqrt{2}}$ b. $y'\left(\frac{\pi}{4}\right)=\frac{\pi^2}{18}$
c. $y\left(\frac{\pi}{3}\right)=\frac{\pi^2}{9}$ d. $y'\left(\frac{\pi}{3}\right)=\frac{4\pi}{3}+\frac{2\pi^2}{3\sqrt{3}}$

**10** Let $y'(x)+y(x)g'(x)=g(x)g'(x)$, $y(0)=0$, $x\in R$, where $f'(x)$ denotes $\frac{d\,f(x)}{dx}$ and $g(x)$ is a given non-constant differentiable function on $R$ with $g(0)=g(2)=0$. Then, the value of $y(2)$ is
**(2011, Integer Type)**

**11** If the function $f(x)=x^3+e^{x/2}$ and $g(x)=f^{-1}(x)$, then the value of $g'(1)$ is **(2009, Integer Type)**

**12** Let $f : R \rightarrow R$ be a continuous function, which satisfies $f(x) = \int_0^x f(t)\,dt$. Then, the value of $f(\ln 5)$ is **(2009, Integer Type)**

**13** Let a solution $y = y(x)$ of the differential equation $x\sqrt{x^2-1}\,dy - y\sqrt{y^2-1}\,dx = 0$ satisfy $y(2) = \frac{2}{\sqrt{3}}$

**Statement I** $y(x) = \sec\left(\sec^{-1} x - \frac{\pi}{6}\right)$ and

**Statement II** $y(x)$ is given by

$$\frac{1}{y} = \frac{2\sqrt{3}}{x} - \sqrt{1 - \frac{1}{x^2}}.$$

**(2008, Statement Type)**

a. Both Statement I and Statement II are correct and Statement II is the correct explanation of Statement I
b. Both Statement I and Statement II are correct but Statement II is not the correct explanation of Statement I
c. Statement I is correct but Statement II is incorrect
d. Statement II is correct but Statement I is incorrect

**14** Let $f(x)$ be differentiable on the interval $(0, \infty)$ such that $f(1) = 1$, and $\lim_{t \to x} \frac{t^2 f(x) - x^2 f(t)}{t - x} = 1$ for each $x > 0$. Then, $f(x)$ is **(2007, Only One Option Correct Type)**

a. $\frac{1}{3x} + \frac{2x^2}{3}$  b. $-\frac{1}{3x} + \frac{4x^2}{3}$
c. $-\frac{1}{x} + \frac{2}{x^2}$  d. $\frac{1}{x}$

**15** The differential equation $\frac{dy}{dx} = \frac{\sqrt{1-y^2}}{y}$ determines a family of circles with **(2007, Only One Option Correct Type)**

a. variable radii and a fixed centre at (0, 1)
b. variable radii and a fixed centre at (0, – 1)
c. fixed radius 1 and variable centres along the X-axis
d. fixed radius 1 and variable centres along the Y-axis

**16** Tangent is drawn at any point $P$ of a curve which passes through (1, 1) cutting X-axis and Y-axis at $A$ and $B$ respectively. If $BP : AP = 3 : 1$, then **(2006 Adv., One or More than One Options Correct Type)**

a. differential equation of the curve is $3x\frac{dy}{dx} + y = 0$
b. differential equation of the curve is $3x\frac{dy}{dx} - y = 0$
c. curve is passing through $\left(\frac{1}{8}, 2\right)$
d. normal at (1, 1) is $x + 3y = 4$

**17** If length of tangent at any point on the curve $y = f(x)$ intercepted between the point and the X-axis is of length 1. Find the equation of the curve. **(2005, Subjective Type)**

**18** If $x\,dy = y\,(dx + y\,dy)$, $y(1) = 1$ and $y(x) > 0$. Then, $y(-3)$ is equal to **(2005, Only One Option Correct Type)**

a. 3  b. 2
c. 1  d. 0

# 4. Area

# DPP-1 Area Under the Curves

## Subjective Questions

Working Space

*Direction* (Q. Nos. 1 and 2) *These questions are subjective in nature, need to be solved completely on notebook.*

**1** Find the area bounded by

i. $y = \sec^2 x$, $x = \pi/6$, $x = \pi/3$ and $X$-axis.

ii. $y = \sin 2x$, $X$-axis and lines $x = \pi/4$ and $x = 3\pi/4$.

iii. $y = 4x^2$, $x = 0$, $y = 1$ and $y = 4$ in the first quadrant.

iv. $y = x^2 + 2$ above $X$-axis between $x = 2$ and $x = 3$.

v. $y = \sqrt{1 - x^2}$ with coordinate axes bounded in the first quadrant.

vi. $y = x|x|$, $X$-axis and the abscissae $x = -1/2$ and $x = 1$.

**2** Find the area bounded by the curves.

i. $x = y^2$ and $x = 3 - 2y^2$.

ii. $y = \sqrt{x}$ and $y = x$.

iii. $x = 1/2$, $x = 2$, $y = \log_e x$ and $y = 2^x$.

iv. $y = 2^x$, $y = 2x - x^2$ bounded by the straight lines $x = 0$ and $x = 2$.

v. $y = (x-1)(x-2)(x-3)$ lying between abscissae $x = 0$, $x = 3$ and $X$-axis.

vi. $x = -2y^2$ and $x = 1 - 3y^2$.

## Only One Option Correct Type

*Direction* (Q. Nos. 3-22) *This section contains 20 multiple choice questions. Each question has four choices (a), (b), (c) and (d), out of which ONLY ONE is correct.*

**3** If the area bounded by the $X$-axis, curve $y = f(x)$ and the lines $x = 1$, $x = b$ is equal to $\sqrt{b^2 + 1} - \sqrt{2}$ for all $b > 1$, then $f(x)$ is equal to

a. $\sqrt{x+1}$ b. $\sqrt{x-1}$

c. $\sqrt{x^2+1}$ d. $x/\sqrt{1+x^2}$

**4** The area of greatest rectangle that can be inscribed in the ellipse $\dfrac{x^2}{a^2} + \dfrac{y^2}{b^2} = 1$, is

a. $2ab$ b. $ab$

c. $\sqrt{ab}$ d. $a/b$

**5** The area enclosed by the curve $y = \log_e(x + e)$ and the coordinate axes is

a. 1 sq unit b. 2 sq units

c. 3 sq units d. 4 sq units

**6** The area of the region bounded by the curves $y = |x - 1|$ and $y = 3 - |x|$, is

a. 6 sq units b. 2 sq units c. 3 sq units d. 4 sq units

Working Space

**7** The area of region bounded by the curves $y = |x - 2|$, $x = 1$, $x = 3$ and the $X$-axis is

a. 1 sq unit
b. 2 sq units
c. 3 sq units
d. 4 sq units

**8** The parabolas $y^2 = 4x$ and $x^2 = 4y$ divide the square region bounded by $x = 4$, $y = 4$ and the coordinate axes. If $S_1$, $S_2$ and $S_3$ are respectively the area of these parts numbered from top to bottom, then $S_1 : S_2 : S_3$ is

a. 1 : 2 : 1
b. 1 : 2 : 3
c. 2 : 1 : 2
d. 1 : 1 : 1

**9** The area of region enclosed between the curves $y^2 = x$ and $y = |x|$ is **[AIEEE 2007]**

a. 2/3 sq unit
b. 1 sq unit
c. 1/6 sq unit
d. 1/3 sq unit

**10** The area of region bounded by the parabola $(y - 2)^2 = (x - 1)$, the tangent to the parabola at the point (2, 3) and the $X$-axis, is **[AIEEE 2009]**

a. 9 sq units
b. 12 sq units
c. 3 sq units
d. 6 sq units

**11** Let $f(x)$ be a non-negative continuous function such that the area bounded by the curve $y = f(x)$, $X$-axis and the ordinates $x = \pi/4$ and $x = \beta > \pi/4$, is $(\beta \sin \beta + \pi/4 \cos \beta + \sqrt{2}\beta)$. Then, $f(\pi/2)$ is equal to

a. $\left(\frac{\pi}{4} + \sqrt{2} - 1\right)$
b. $\left(\frac{\pi}{4} - \sqrt{2} + 1\right)$
c. $\left(1 - \frac{\pi}{4} - \sqrt{2}\right)$
d. $\left(1 - \frac{\pi}{4} + \sqrt{2}\right)$

**12** The area of the plane region bounded by the curves $x + 2y^2 = 0$ and $x + 3y^2 = 1$, is

a. $\frac{4}{3}$ sq units
b. $\frac{5}{3}$ sq units
c. $\frac{2}{3}$ sq units
d. $\frac{1}{3}$ sq units

**13** The area (in sq units) bounded by the curves $y = \cos x$, $y = \sin x$ between the ordinates $x = 0$ and $x = 3\pi/2$, is **[AIEEE 2010]**

a. $4\sqrt{2} - 2$
b. $4\sqrt{2} + 2$
c. $4\sqrt{2} - 1$
d. $4\sqrt{2} + 1$

**14** The area (in sq units) bounded by the curves $y = \sqrt{x}$, $2y - x + 3 = 0$, $X$-axis and lying in the first quadrant, is **[JEE Main 2013]**

a. 9
b. 36
c. 18
d. 27/4

**15** The triangle formed by the tangent to the curve $f(x) = x^2 + bx - b$ at the point (1, 1) and the coordinate axes lie in the first quadrant. If the area is 2 sq units, then the value of $b$ is

a. $-1$
b. $-2$
c. $-3$
d. 1

**16** The area (in sq units) of region bounded by the curves $y = x^2$, $y = |2 - x^2|$ and $y = 2$, which lies to the right of line $x = 1$, is

a. $\frac{20}{3} - 4\sqrt{2}$
b. $\frac{20}{3} + 4\sqrt{2}$
c. $\frac{10}{3} - 4\sqrt{2}$
d. $\frac{10}{3} - 2\sqrt{2}$

Working Space

**17** The area (in sq units) enclosed by the curves $y = \sin x + \cos x$ and $y = |\cos x - \sin x|$ over the interval $[0, \pi/2]$, is **[JEE Advanced 2013]**

a. $4\sqrt{2} - 1$
b. $2\sqrt{2}(\sqrt{2} - 1)$
c. $2(\sqrt{2} + 1)$
d. $2\sqrt{2}(\sqrt{2} + 1)$

**18** The area enclosed between the curves $y = ax^2$ and $x = ay^2$ $(a > 0)$ is 1 sq unit, then the value of $a$ is

a. $1/\sqrt{3}$
b. $1/2$
c. 1
d. $1/3$

**19** The area (in sq units) of the region bounded by the curves $y = \log_e x$ and $y = (\log_e x)^2$, is

a. $e + 1$
b. $e - 1$
c. $3 - e$
d. 1

**20** The area bounded by the curves $y = -\sqrt{-x}$ and $x = -\sqrt{-y}$, where $x, y \le 0$ is

a. $\frac{4}{3}$ sq units
b. $\frac{1}{3}$ sq unit
c. $\frac{2}{3}$ sq unit
d. Cannot be determined

**21** The area of the region(s) enclosed by $y = x^2$ and $y = \sqrt{|x|}$ is

a. 1/3 sq unit
b. 2/3 sq unit
c. 1/6 sq unit
d. 1 sq unit

**22** The area (in sq units) bounded by the curves $y = x(1 - \log_e x)$ and positive $X$-axis between $x = 1/e$ and $x = e$, is

a. $\frac{e^2 - 4e^{-2}}{5}$
b. $\frac{e^2 - 5e^{-2}}{4}$
c. $\frac{4e^2 - e^{-2}}{5}$
d. None of these

## One or More than One Options Correct Type

*Direction* (Q. Nos. 23-26) *This section contains 4 multiple choice questions. Each question has four choices (a), (b), (c) and (d), out of which ONE or MORE THAN ONE are correct.*

**23** The area enclosed by the curve $y = \sin x$ between $x = 2n\pi$ and $x = 2(n + 1)\pi$, is

a. $\int_0^{2\pi} \sin x \, dx$
b. $2\int_0^{\pi} \sin x \, dx$
c. $4\int_0^{\pi/2} \sin x \, dx$
d. 4

**24** The area of region bounded by the curve $y = x - x^2$ and line $y = mx$ equals 9/2 sq units, then the value(s) of $m$ is/are

a. $-4$
b. $-2$
c. 2
d. 4

**25** If $C_1 : y = \frac{1}{1 + x^2}$ and $C_2 : y = \frac{x^2}{2}$ are two curves lying in the $XY$-plane, then the area (in sq units) bounded by

a. the curve $C_1$ and $y = 0$ is $\pi$
b. the $C_1$ and $C_2$ is $\left(\frac{\pi}{2} - \frac{1}{3}\right)$
c. $C_1$ and $C_2$ is $\left(1 - \frac{\pi}{2}\right)$
d. $C_1$ and $X$-axis is $\pi/2$

**26** The area (in sq units) of the region of plane bounded by $(|x|, |y|) \le 1$ and $xy \le 1/2$ is

a. less than $4 \log 3$
b. $15/4$
c. $2 + 2\log 2$
d. $3 + \log 2$

*Working Space*

## Matching List Type

*Direction (Q. No. 27) Choices for the correct combination of elements from Column I and Column II are given as options (a), (b), (c) and (d), out of which one is correct.*

**27** Let $f(x) = |x|$, $g(x) = |x-1|$ and $h(x) = |x+1|$. Match the Column I with Column II and mark the correct option from the codes given below.

| | Column I | | Column II |
|---|---|---|---|
| i. | The area bounded by min $\{f(x), g(x)\}$ and X-axis is | p. | 1/8 sq unit |
| ii. | The area bounded by min $\{f(x), h(x)\}$ and X-axis is | q. | 1/4 sq unit |
| iii. | The area bounded by min $\{f(x), g(x), h(x)\}$ and X-axis is | r. | 1/2 sq unit |
| iv. | The area bounded by min $\{f(x), g(x), h(x)\}$ and $y = 1/2$ is | s. | 3/4 sq unit |

**Codes**

| | i | ii | iii | iv |
|---|---|---|---|---|
| a. | p | r | q | s |
| b. | r | q | p | s |
| c. | q | q | r | s |
| d. | s | p | q | r |

## Comprehension Type

*Direction (Q. Nos. 28-30) This section contains a paragraph, describing theory, experiments, data, etc. Three questions related to the paragraph have been given. Each question has only one correct answer among the four given options (a), (b), (c) and (d).*

### Passage

Five curves are defined as follows:

$C_1 : |x+y| \le 1,$ $C_2 : |x-y| \le 1$

$C_3 : |x| \le 1/2,$ $C_4 : |y| \le 1/2$

$C_5 : 3x^2 + 3y^2 = 1,$

**28** The area (in sq units) bounded by $C_1$ and $C_2$ which does not contain the area bounded by $C_5$, is

a. $2 - \frac{\pi}{4}$
b. $2 - \frac{\pi}{6}$
c. $2 - \frac{\pi}{3}$
d. 2

**29** The part of area of curve $C_5$, which does not contain the area of $C_3$ and $C_4$, is

a. $\frac{\pi}{3} - \frac{1}{2}$
b. $\frac{\pi}{3} - 1$
c. $\frac{\pi}{3} - \frac{1}{6}$
d. $\frac{2\pi}{9} - \frac{1}{\sqrt{3}}$

**30** The part of area which is bounded by $C_1$ and $C_2$ but not bounded by $C_3$ and $C_4$, is

a. 1
b. $\frac{1}{2}$
c. $\frac{1}{3}$
d. None of these

# DPP-2 Curve Sketching and Its Applications in Finding the Area

## Only One Option Correct Type

Working Space

**Direction** *(Q. Nos. 1-10) This section contains 10 multiple choice questions. Each question has four choices (a), (b), (c) and (d), out of which ONLY ONE is correct.*

**1** The area contained between the two arms of curve $(y - x)^2 = x^3$ between $x = 0$ and $x = 1$, is

a. $\frac{3}{5}$ sq unit b. $\frac{4}{5}$ sq unit

c. $\frac{2}{5}$ sq unit d. $\frac{1}{5}$ sq unit

**2** The area enclosed by $2x^2 + 6xy + 5y^2 = 1$ is

a. $\frac{\pi}{4}$ sq unit b. $\frac{\pi}{2}$ sq units

c. $\pi$ sq units d. $\frac{3\pi}{2}$ sq units

**3** Let $A(m)$ be area bounded by parabola $y = x^2 + 2x - 3$ and the line $y = mx + 1$, then the least area of $A(m)$ is

a. $\frac{8}{3}$ sq units b. $\frac{16}{3}$ sq units

c. $\frac{32}{3}$ sq units d. None of these

**4** The area enclosed by the curves $y = 2x^2$ and $y = x^4 - 2x^2$ is

a. $\frac{128}{11}$ sq units b. $\frac{128}{15}$ sq units

c. 128 sq units d. None of these

**5** The area (in sq units) of region bounded by the curves $y = x^2$ and $y = \sec^{-1}[-\sin^2 x]$, where $[\cdot]$ denotes the greatest integer function, is

a. $4\pi\frac{\sqrt{\pi}}{3}$ b. $\frac{4\sqrt{\pi}}{3}$

c. $\frac{4\pi^2}{3}$ d. None of these

**6** Suppose $y = f(x)$ and $y = g(x)$ are two functions whose graphs intersect at three points (0, 4), (2, 2) and (4, 0) with $f(x) > g(x)$ for $0 < x < 2$ and $f(x) < g(x)$ for $2 < x < 4$. If $\int_0^4 [f(x) - g(x)]\, dx = 10$ and $\int_2^4 [g(x) - f(x)]\, dx = 5$. The area between two curves for $0 < x < 2$ is

a. 5 sq units b. 10 sq units

c. 15 sq units d. 20 sq units

Working Space

**7** The area (in sq units) bounded by the curves $y = xe^{-x}$, $xy = 0$ and $x = c$, where $c$ is the $x$-coordinate of the curve's inflection point, is

a. $3e^{-2}$ b. $1-2e^{-2}$

c. $1-e^{-2}$ d. $1-3e^{-2}$

**8** Let $a$ be a positive constant number. Consider two curves $C_1 : y = e^x$, $C_2 : y = e^{a-x}$. Let $S$ be the area of the part surrounding by $C_1$, $C_2$ and $Y$-axis, then $\lim_{a \to 0} \frac{S}{a^2}$ equals

a. 4 b. $\frac{1}{2}$

c. 0 d. $\frac{1}{4}$

**9** The area (in sq units) enclosed by graph of function $y = \log^2 x - 1$ lying in the 4th quadrant is

a. $\left(4e - \frac{1}{e}\right)$ b. $2\left(e + \frac{1}{e}\right)$

c. $\frac{2}{e}$ d. $\frac{4}{e}$

**10** The area of region enclosed between the curves $x = y^2 - 1$ and $x = |y|\sqrt{1-y^2}$ is

a. 1 sq unit b. $\frac{4}{3}$ sq units

c. $\frac{2}{3}$ sq unit d. 2 sq units

## Statement Type

**Direction** *(Q. Nos. 11 and 12) This section is based on Statement I and Statement II. Select the correct answer from the codes given below.*

a. Both Statement I and Statement II are correct and Statement II is the correct explanation of Statement I
b. Both Statement I and Statement II are correct but Statement II is not the correct explanation of Statement I
c. Statement I is correct but Statement II is incorrect
d. Statement II is correct but Statement I is incorrect

**11** **Statement I** The area bounded by the curve $|x| + |y| = a\,(a > 0)$ is $2a^2$ and area bounded by $|px + qy| + |qx - py| = a$, where $p^2 + q^2 = 1$, is also $2a^2$.

**Statement II** Since, $\alpha x + \beta y = 0$ is perpendicular to $\beta x - \alpha y = 0$. We can take one as $X$-axis and another as $Y$-axis and therefore, the area of bounded by $|\alpha x + \beta y| + |\beta x - \alpha y| = a$ is $2a^2$, $\forall\, \alpha, \beta \in R$, $a \neq 0$, $b \neq 0$.

**12** **Statement I** The area bounded by parabola $y = x^2 - 4x + 3$ and $y = 0$ is 4/3 sq units.

**Statement II** The area bounded by curves $y = f(x) \geq 0$ and $y = 0$ between the ordinates $x = a$ and $x = b\,(b > a)$ is $\int_a^b f(x)\,dx$.

## Comprehension Type

**Direction** (Q. Nos. 13-21) *This section contains 3 paragraphs, each describing theory, experiments, data, etc. Nine questions related to the paragraphs have been given. Each question has only one correct answer among the four given options (a), (b), (c) and (d).*

### Passage I

Consider two curves $y = \frac{1}{x^2}$ and $y = \frac{1}{4(x-1)}$. $a$ is a number such that $a > 2$ and reciprocal of the area of figure bounded by the curves, the lines $x = 2$ and $x = a$, is $a$ itself. $b$ is the number such that $1 < b < 2$ and the area bounded by the curves and the lines $x = b$, $x = 2$ is equal to $(1 - 1/b)$.

**13** The value of $(\log_e a - \log_e b)$ is a/an

a. positive integer
b. negative integer
c. rational number
d. irrational number

**14** If $A = \begin{bmatrix} \log(a-1) & 0 \\ 0 & \log(b-1) \end{bmatrix}$, then $A^{-1}$ is equal to

a. $-A/4$
b. $A$
c. $4A$
d. $A/4$

**15** If $z$ is a complex number such that $z = \log(a-1) + i\log(b-1)$, then $\arg(z)$ is

a. $\pi/4$
b. $3\pi/4$
c. $-\pi/4$
d. $-3\pi/4$

### Passage II

Consider the function defined implicity by the equation $y^3 - 3y + x = 0$ on various intervals in the real line. If $x \in (-\infty, -2) \cup (2, \infty)$. The equation implicity defines a unique real valued differentiable function $y = f(x)$. If $x \in (-2, 2)$, the equation implicity defines a unique real valued differentiable function $y = g(x)$ satisfying $g(0) = 0$. **[IIT JEE 2008]**

**16** If $f(-10\sqrt{2}) = 2\sqrt{2}$, then $f''(-10\sqrt{2})$ is equal to

a. $\frac{4\sqrt{2}}{7^3 \cdot 3^2}$
b. $-\frac{4\sqrt{2}}{7^3 \cdot 3^2}$
c. $\frac{4\sqrt{2}}{7^3 \cdot 3}$
d. $\frac{-4\sqrt{2}}{7^3 \cdot 3}$

**17** The area of the region bounded by the curve $y = f(x)$, the $X$-axis and the lines $x = a$, $x = b$, where $-\infty < a < b < -2$, is

a. $\int_a^b \frac{x}{3[\{f(x)\}^2 - 1]} dx + bf(b) - af(a)$

b. $-\int_a^b \frac{x}{3[\{f(x)\}^2 - 1]} dx + bf(b) - af(a)$

c. $\int_a^b \frac{x}{3[\{f(x)\}^2 - 1]} dx - bf(b) + af(a)$

d. $-\int_a^b \frac{x}{3[\{f(x)\}^2 - 1]} dx - bf(b) + af(a)$

**18** $\int_{-1}^{1} g'(x) dx$ is equal to

a. $2g(-1)$
b. $0$
c. $-2g(1)$
d. $2g(1)$

Working Space

## Passage III

Consider the polynomial $f(x) = 1 + 2x + 3x^2 + 4x^3$. Let $S$ be the sum of all distinct real roots of $f(x)$ and let $t = |S|$.

**19** The real number $S$ lies in the interval

a. $\left(-\frac{1}{4}, 0\right)$
b. $\left(-11, -\frac{3}{4}\right)$
c. $\left(-\frac{3}{4}, \frac{-1}{2}\right)$
d. $\left(0, \frac{1}{4}\right)$

**20** The area bounded by the curve $y = f(x)$ and the lines $x = 0$, $y = 0$ and $x = t$, lies in the interval

a. $\left(\frac{15}{16}, 3\right)$
b. $\left(\frac{21}{64}, \frac{11}{16}\right)$
c. $(9, 10)$
d. $\left(0, \frac{21}{64}\right)$

**21** The function $f'(x)$ is

a. increasing in $\left(-t, -\frac{1}{4}\right)$ and decreasing in $\left(-\frac{1}{4}, t\right)$
b. decreasing in $\left(-t, -\frac{1}{4}\right)$ and increasing in $\left(-\frac{1}{4}, t\right)$
c. increasing in $(-t, t)$
d. decreasing in $(-t, t)$

## One Integer Value Correct Type

*Direction* (Q. Nos. 22-25) *This section contains 4 questions. When worked out will result in an integer from 0 to 9 (both inclusive).*

**22** Let $f(x) = \text{maximum } \{x^2, (1-x)^2, 2x(1-x)\}$, where $0 \le x \le 1$. If the area of the region bounded by the curves $y = f(x)$, $X$-axis, $x = 0$ and $x = 1$ is $\left(\frac{m}{n}\right)$, then $(2m - n)$ is equal to

**23** The area enclosed by curves $y = \cos x - \cos^2 x$ and $y = x^2\left(x^2 - \frac{\pi^2}{4}\right)$ is $\left(a - \frac{\pi}{b} + \frac{\pi^5}{120}\right)$ sq units. Then, the value of $(a + b)$ is

**24** If $f(x) = [x] + \{x\}^2$ (where, $[\cdot]$ and $\{\cdot\}$ denote greatest integer function and fractional part) and the area bounded by $y = f(x)$, $y = f^{-1}(x)$, $x = 0$ and $x = \frac{3}{2}$ is $\left(\frac{a + 7\sqrt{2}}{b\sqrt{c}}\right)$. Then, the value of $(b - 2a - c)$ is

**25** Let $f(x)$ be a function which satisfy the equation $f(xy) = f(x) + f(y)$, $\forall\ x > 0, y > 0$, such that $f'(1) = 2$ and the area of the region bounded by the curves $y = f(x)$, $y = |x^3 - 6x^2 + 11x - 6|$ and $x = 0$ is $\frac{2}{\lambda}$. Then, the value of $\lambda$ is

SET - 1

# Revisal Problems
## for JEE Main

### Only One Option Correct Type

Working Space

*Direction* (*Q. Nos. 1-17*) *This section contains 17 multiple choice questions. Each question has four choices (a), (b), (c) and (d), out of which ONLY ONE option is correct.*

**1** The area bounded by the curve $y=\left[\dfrac{x^2}{64}+2\right]$, (where $[\cdot]$ denotes greatest integer function), and lines $y = x - 1$, $x = 0$ above the $X$-axis, is

a. 2 sq units b. 3 sq units c. 4 sq units d. None of these

**2** The area bounded by the curve $y = xe^{|x|}$ and lines $|x| = 1$, $y = 0$, is

a. 4 sq units b. 6 sq units c. 1 sq unit d. 2 sq units

**3** The area bounded by the curves $y = \log x$, $y = \log|x|$, $y = |\log x|$ and $y = |\log|x||$ is

a. 5 sq units b. 2 sq units c. 4 sq units d. None of these

**4** Let $f(x)$ be a continuous function such that the area bounded by the curve $y = f(x)$. If area bounded by $X$-axis and the lines $x = 0$, $x = a$ is $\left(1+\dfrac{a^2}{2}\sin a\right)$. Then,

a. $f\left(\dfrac{\pi}{2}\right)=1+\dfrac{\pi^2}{8}$ b. $f(a)=1+\dfrac{a^2}{2}\sin a$

c. $f(a)=a\sin a+\dfrac{1}{2}a^2\cos a$ d. None of these

**5** The area in sq units bounded by $|x-1|\le 2$ and $x^2-y^2=1$ is

a. $6\sqrt{2}+\dfrac{1}{2}\log|3+2\sqrt{2}|$ b. $6\sqrt{2}+\dfrac{1}{2}\log|3-2\sqrt{2}|$

c. $6\sqrt{2}-\log|3+2\sqrt{2}|$ d. None of these

**6** The area bounded by the curve $f(x) = x + \sin x$ and its inverse function between the ordinates $x = 0$ to $x = 2\pi$, is

a. $4\pi$ sq units b. $8\pi$ sq units

c. 4 sq units d. 8 sq units

**7** The area bounded by $f(x)=\dfrac{(x-1)(x+1)}{(x-2)}$, $X$-axis and ordinates $x = 0$ and $x=\dfrac{3}{2}$, is

a. $\dfrac{4}{5}$ sq unit b. $\dfrac{7}{8}$ sq unit c. 1 sq unit d. None of these

**8** If $f(x) = \max\{\sin x, \cos x, 1/2\}$, then area of the region (in sq units) bounded by the curve $y = f(x)$, $X$ and $Y$-axes and $x = 5\pi/3$, is

a. $\sqrt{2}-\sqrt{3}+\dfrac{5\pi}{12}$ b. $\sqrt{2}+\dfrac{\sqrt{3}}{2}+\dfrac{5\pi}{12}$ c. $\sqrt{2}+\dfrac{\sqrt{3}}{2}-\dfrac{5\pi}{12}$ d. None of these

**9** If $y = f(x)$ makes positive intercepts of 2 units and 1 unit on $X$ and $Y$-axes and enclose an area of $\dfrac{3}{4}$ sq unit with the axes, then $\int_0^2 x f'(x)\,dx$ is equal to

a. $\dfrac{3}{2}$ b. 1 c. $\dfrac{5}{4}$ d. $\dfrac{-3}{4}$

**10** The area (in sq unit) of the region $\{(x, y): |x| \le |y| \text{ and } x^2 + y^2 \le 1\}$ is

a. $\dfrac{3\pi}{8}$ b. $\dfrac{5\pi}{8}$

c. $\dfrac{\pi}{2}$ d. $\dfrac{\pi}{8}$

Working Space

**11** If $\begin{bmatrix} 4a^2 & 4a & 1 \\ 4b^2 & 4b & 1 \\ 4c^2 & 4c & 1 \end{bmatrix} \begin{bmatrix} f(-1) \\ f(1) \\ f(2) \end{bmatrix} = \begin{bmatrix} 3a^2+3a \\ 3b^2+3b \\ 3c^2+3c \end{bmatrix}$, $f(x)$ is a quadratic function and its maximum value occurs at a point $V$. A point of intersection of $y = f(x)$ with $X$-axis and point $B$ is such that chord $AB$ subtends right angle at $V$, then the area enclosed by $f(x)$ and chord $AB$ is

a. $\frac{125}{3}$ sq units b. $\frac{1}{3}$ sq unit c. $\frac{25}{3}$ sq units d. $\frac{5}{3}$ sq units

**12** Let the straight line $x = b$ divides the area enclosed by $y = (1-x)^2$, $y = 0$ and $x = 0$ into two parts $R_1$ $(0 \le x \le b)$ and $R_2$ $(b \le x \le 1)$ such that $R_1 - R_2 = 1/4$, then $b$ equals

a. $\frac{3}{4}$ b. $\frac{1}{2}$ c. $\frac{1}{3}$ d. $\frac{1}{4}$

**13** Let $f : [-1, 2] \to [0, \infty)$ be a continuous function such that $f(x) = f(1-x)$ for all $x \in [1, -2]$. Let $R_1 = \int_{-1}^{2} x f(x) dx$ and $R_2$ be the area of the region bounded by $y = f(x)$, $x = -1$, $x = 2$ and the $X$-axis. Then,

a. $R_1 = 2R_2$ b. $R_1 = 3R_2$ c. $2R_1 = R_2$ d. $3R_1 = R_2$

**14** The curve $f(x) = Ax^2 + Bx + C$ passes through the point (1, 3) and the line $4x + y = 8$ is tangent to it at the point (2, 0). The area enclosed by $y = f(x)$, the tangent line and the $Y$-axis, is

a. $\frac{4}{3}$ sq units b. $\frac{8}{3}$ sq units c. $\frac{16}{3}$ sq units d. $\frac{32}{3}$ sq units

**15** The area of the region(s) enclosed by the curves $y = x^2$ and $y = \sqrt{|x|}$ is

a. $\frac{1}{3}$ sq unit b. $\frac{2}{3}$ sq unit c. $\frac{9}{4}$ sq units d. $\frac{13}{4}$ sq units

**16** The area enclosed by $y = x^3$, its normal at (1, 1) and $X$-axis, is

a. $\frac{7}{4}$ sq units b. $\frac{9}{4}$ sq units
c. $\frac{5}{4}$ sq units d. 2 sq units

**17** If $f(x) = \sin x$, $\forall x \in (0, \pi/2]$, $f(x) + f(\pi - x) = 2$ for all $x \in (\pi/2, \pi]$ and $f(x) = f(2\pi - x)$ for all $x \in (\pi, 2\pi]$, then the area enclosed by $y = f(x)$ and $X$-axis is

a. $\pi$ sq units b. $2\pi$ sq units
c. 2 sq units d. 4 sq units

## Statement Type

**Direction** *(Q. Nos. 18-20) This section is based on Statement I and Statement II. Select the correct answer from the codes given below.*

a. Both Statement I and Statement II are correct and Statement II is the correct explanation of Statement I
b. Both Statement I and Statement II are correct but Statement II is not the correct explanation of Statement I
c. Statement I is correct but Statement II is incorrect
d. Statement II is correct but Statement I is incorrect

**18** **Statement I** The area of the curve $y = \sin^2 x$ from 0 to $\pi$ will be more than that of curve $y = \sin x$ from 0 to $\pi$.

**Statement II** $t^2 > t$, if $t \in R - [0, 1]$.

**19** **Statement I** The area bounded by the curves $y = x^2 - 3$ and $y = kx + 2$ is the least, if $k = 0$.

**Statement II** The area bounded by $y = x^2 - 3$ and $y = kx + 2$ is $\sqrt{k^2 + 20}$.

**20** Consider the curves $y = x - \lambda x^2$ and $y = x^2/\lambda$, $(\lambda > 0)$.

**Statement I** The area bounded between the curves is maximum, when $\lambda = 1$.

**Statement II** The area bounded between the curves is $\frac{\lambda^2}{(1+\lambda^2)^2}$ sq units.

SET - 2

# Revisal Problems
## for JEE Advanced

## Only One Option Correct Type

Working Space

*Direction* (*Q. Nos.* 1-13) *This section contains 13 multiple choice questions. Each question has four choices (a), (b), (c) and (d), out of which ONLY ONE option is correct.*

**1** The point $P(x, y)$ moves in $XY$-plane in such a way that $[|x|]+[|y|]=1$, where $[\cdot]$ denotes the greatest integer function. Area of the region representing all possible positions of the point $P$, is

a. 4 sq units  b. 16 sq units
c. $2\sqrt{2}$ sq units  d. 8 sq units

**2** Let $A(x)=a\sin x\int_0^x \cos t\,dt+2\int_0^x t\,dt+a\cos^2 x-x^2$. If $x^2-2x+3>A(x), \forall x\in R$. Then, the greatest area bounded by $x\,A(x)$, $y=0$ and $x=5$, is

a. 16 sq units  b. 25 sq units
c. 8 sq units  d. $\frac{35}{2}$ sq units

**3** The area (in sq units) bounded by the curves $y=\sin^{-1}|\sin x|$ and $y=(\sin^{-1}|\sin x|)^2$, $0\le x\le 2\pi$, is

a. $\frac{1}{3}+\frac{\pi^2}{4}$  b. $\frac{1}{6}+\frac{\pi^3}{8}$
c. 2  d. None of these

**4** The area (in sq units) of the region $\{(x, y): x^2+y^2\le 5, \{||x|-|y||\ge 1\}$, is

a. $4\left[\pi-\tan^{-1}\left(\frac{24}{7}\right)\right]-4$  b. $5\left[\pi-\tan^{-1}\left(\frac{24}{7}\right)\right]-4$
c. $3\left[\pi-\tan^{-1}\left(\frac{24}{7}\right)\right]-4$  d. $2\left[\pi-\tan^{-1}\left(\frac{24}{7}\right)\right]-1$

**5** The area bounded by the minimum $\{|x|, |y|\}=1$ and the maximum of $\{|x|, |y|\}=2$ is

a. 4 sq units  b. 8 sq units
c. 16 sq units  d. 9 sq units

**6** A square $ABCD$ is inscribed in a circle of radius 4 and a point $P$ moves inside the circle such that $d(P, AB)\le\min\{d(P, BC), d(P, DA)\}$, where $d(P, AB)$ is the distance of a point $P$ from the line $AB$. The area of region (in sq units) covered by the moving point $P$ is

a. $4\pi$  b. $8\pi$
c. $8\pi-16$  d. $4\pi-4$

**7** The area of the loop of the curve $ay^2=x^2(a-x)$ is

a. $4a^2$ sq units  b. $\frac{8a^2}{15}$ sq units
c. $\frac{16a^2}{9}$ sq units  d. None of these

**8** If $f(x)=\begin{cases}\sqrt{\{x\}}, & \text{for } x\notin Z\\ 1, & \text{for } x\in Z\end{cases}$ and $g(x)=\{x\}^2$, where $\{\cdot\}$ denotes fractional part of $x$, then the area bounded by $f(x)$ and $g(x)$ for $x\in[0, 6]$, is

a. $\frac{2}{3}$ sq unit  b. 2 sq units
c. $\frac{10}{3}$ sq units  d. 6 sq units

Working Space

9 Let $ABC$ be a triangle with vertices $A(b, 2\sqrt{3}+2)$, $B(4, 2)$ and $C(8, 2)$. Let $R$ be the region consisting of all points $P$ inside $\Delta ABC$, which satisfy $d(P, BC) \geq \max\{d(P, AB), d(P, AC)\}$, where $d(P, C, L)$ denotes distance of the point $P$ from the line $L$. Then, the area (in sq units) of the region $R$ is

a. $\frac{2\sqrt{3}}{3}$
b. $\frac{4\sqrt{3}}{3}$
c. $2\sqrt{3}$
d. $4\sqrt{3}$

10 Three points $O(0, 0)$, $P(a, a^2)$ and $Q(-b, b^2)$ $(a > 0, b > 0)$ are on the parabola $y = x^2$. Let $S_1$ be the area bounded by the line $PQ$ and the parabola and let $S_2$ be the area of the $\Delta OPQ$, then the minimum value of $S_1/S_2$, is

a. $\frac{4}{3}$
b. $\frac{5}{3}$
c. 2
d. $\frac{7}{3}$

11 $y = f(x)$ is a function which satisfy

(i) $f(0) = 0$
(ii) $f''(x) = f'(x)$
(iii) $f'(0) = 1$

Then, the area (in sq units) bounded by the graph of $y = f(x)$, the lines $x = 0$, $x - 1 = 0$ and $y + 1 = 0$, is

a. $e$
b. $e - 2$
c. $e - 1$
d. $e + 1$

12 The curve $y = ax^2 + bx + c$ passes through the point (1, 2) and its tangent at origin is the line $y = x$. The area bounded by the curve, the ordinate of the curve at minima and the tangent line, is

a. $\frac{1}{24}$ sq unit
b. $\frac{1}{12}$ sq unit
c. $\frac{1}{8}$ sq unit
d. $\frac{1}{6}$ sq unit

13 A function $y = f(x)$ satisfies the differential equation $dy/dx - y = (\cos x - \sin x)$ with initial condition that $y$ is bounded, when $x \to \infty$. The area enclosed by $y = f(x)$, $y = \cos x$ and $Y$-axis in the first quadrant is

a. $\sqrt{2} - 1$
b. $\sqrt{2}$
c. 1
d. $\frac{1}{\sqrt{2}}$

## One or More than One Options Correct Type

**Direction** (Q. Nos. 14-20) *This section contains 7 multiple choice questions. Each question has four choices (a), (b), (c) and (d), out of which ONE or MORE THAN ONE are correct.*

14 Let $T$ be the triangle with vertices $(0, 0)$, $(0, c^2)$, $(c, c^2)$ and $R$ be the region between $y = cx$ and $y = x^2$, where $c > 0$, then

a. area $(R) = \frac{c^3}{6}$
b. area $(R) = \frac{c^3}{3}$
c. $\lim_{c \to 0^+} \frac{\text{area }(T)}{\text{area }(R)} = 3$
d. $\lim_{c \to 0^+} \frac{\text{area }(T)}{\text{area }(R)} = \frac{3}{2}$

15 Suppose $f$ is defined from $R \to [-1, 1]$ as $f(x) = \frac{x^2 - 1}{x^2 + 1}$, where $R$ is the set of real numbers.

Then, which statements does not hold?

a. $f$ is many-one and onto
b. $f$ increases for $x > 0$ and decreases for $x < 0$
c. Minimum value is not attained even though $f$ is bounded
d. The area included by the curve $y = f(x)$ and the line $y = 1$ is $\pi$ sq units

Working Space

**16** Consider $f(x)=\begin{cases}\cos x, & 0\le x<\dfrac{\pi}{2}\\ \left(\dfrac{\pi}{2}-x\right)^2, & \dfrac{\pi}{2}\le x<\pi\end{cases}$, such that $f$ is periodic with period $\pi$, then

a. the range of $f$ is $[0, \pi^2/4)$

b. $f$ is continuous for all real $x$, but not differentiable for some real $x$

c. $f$ is continuous for all real $x$

d. the area bounded by $y = f(x)$ and the $X$-axis from $x = -n\pi$ to $x = n\pi$ is $2n\left(1+\dfrac{\pi^3}{24}\right)$ for a given $n \in N$

**17** Which of the following statement(s) is/are true for the function $f(x) = (x-1)^2(x-2)+1$ defined on $[0, 2]$?

a. Range of $f$ is $\left[\dfrac{23}{27}, 1\right]$

b. The coordinates of the turning point of the graph of $y = f(x)$ occur at (1, 1) and $\left(\dfrac{5}{3}, \dfrac{23}{27}\right)$

c. The value of $p$ for which the equation $f(x) = p$ has 3 distinct solutions lie in interval $\left(\dfrac{23}{27}, 1\right)$

d. The area enclosed by $y = f(x)$, the line $x = 0$ and $y = 1$ as $x$ varies from 0 to 1 is 7/12 sq unit

**18** Consider the functions $f(x)$ and $g(x)$, both defined from $R \to R$ and are defined as $f(x) = 2x - x^2$ and $g(x) = x^n$, where $n \in N$. If the area between $f(x)$ and $g(x)$ is 1/2, then $n$ is a divisor of

a. 12
b. 15
c. 20
d. 30

**19** If area of the region bounded by the curve $y = x^2 - x$ and the lines $y = mx$ and $y = nx$ is $\dfrac{37}{6}$ sq units

$(m > n, m, n, \in N)$, then

a. $m = 4$
b. $n = 2$
c. $m = 3$
d. $n = 3$

**20** If $A_i$ is the area bounded by $|x - a_i| + |y| = b_i$, $i \in N$, where $a_{i+1} = a_i + \dfrac{3}{2}b_i$ and $b_{i+1} = \dfrac{b_i}{2}$, $a_1 = 0$, $b_1 = 32$, then

a. $A_3 = 128$
b. $A_3 = 256$
c. $\lim\limits_{n\to\infty}\sum\limits_{i=1}^{n} A_i = \dfrac{8}{3}(32)^2$
d. $\lim\limits_{n\to\infty}\sum\limits_{i=1}^{n} A_i = \dfrac{4}{3}(16)^2$

## Comprehension Type

**Direction** (Q. Nos. 21-32) *This section contains 4 paragraphs, each describing theory, experiments, data, etc. Twelve questions related to the paragraphs have been given. Each question has only one correct answer among the four given options (a), (b), (c) and (d).*

### Passage I

Consider the function $f(x) = x^3 - 8x^2 + 20x - 13$.

**21** The number of positive integers $x$ for which $f(x)$ is a prime number, is

a. 1
b. 2
c. 3
d. 4

**22** The function $f(x)$ defined for $R \to R$

a. is one-one and onto

b. is many-one and onto

c. has 3 real roots

d. is such that $f(x_1)\cdot f(x_2) < 0$, where $x_1$ and $x_2$ are the roots of $f'(x) = 0$

**23** The area enclosed by $y = f(x)$ and the coordinates axes is

a. $\dfrac{65}{12}$ sq units
b. $\dfrac{13}{12}$ sq units
c. $\dfrac{71}{12}$ sq units
d. None of these

Working Space

### Passage II

Let $f(x) = 3 + \int_0^x \frac{1 + \sin t}{t^2 + 2} dt$, $\forall\, x \in R$ and $p(x)$ is a second degree polynomial function of $x$ such that $p(0) = f(0)$; $p'(0) = f'(0)$ and $p''(0) = f''(0)$.

**24** The least value of $P(x)$ is

a. 7/2 b. 9/5 c. 12/5 d. 11/4

**25** Slope of the tangent line at the point $\{-2, p(-2)\}$ on the curve $y = p(x)$ is

a. −1/2 b. −2/3 c. −3/2 d. −5/4

**26** The area enclosed by $y = p(x)$ and the $Y$-axis and the tangent at the point $\{-2, p(-2)\}$ is

a. 3/4 b. 8/3 c. 12/7 d. 2/3

### Passage III

Let $f : A \to A$, where $A \subseteq (-1, \infty)$ and $\{f(x + f(y) + x\, f(y)\} = y + f(x) + y\, f(x)$, $\forall x, y \in A$ and $\frac{f(x)}{x}$ be strictly increasing, $\forall\, x \in (-1, -0) \cup (0, \infty)$.

**27** $f(x)$ is equal to (where, $f(x) = x$ has 1 solution)

a. $xe^{1+x}$ b. $x \log(1 + x)$
c. $\frac{x}{1+x}$ d. None of these

**28** The area bounded by $f(x)$ and its horizontal asymptote, $Y$-axis and $x = 2$, is

a. log2 b. 1
c. log3 d. infinite

**29** For $y = f(x)$, $x = 0$ is the point of

a. maxima b. minima
c. inflection d. None of these

### Passage IV

Let $f(x) = x^2 - 3x + 2$ be a function, $\forall x \in R$.

**30** The area bounded by the curve $f(|x|)$ and $X$-axis, is

a. 2 sq units b. 5/2 sq units
c. 3/5 sq unit d. 5/3 sq units

**31** The area bounded by the curve $|f(|x|)|$ between $1 \le |x| \le 2$ and the $X$-axis is

a. 1/5 sq unit b. 1/4 sq unit
c. 1/3 sq unit d. 1/2 sq unit

**32** If $|f(|x|)| = a$ has 6 real roots, then

a. $a \in (0, 1/4)$ b. $a \in (1/4, 2)$
c. $a \in (-1/4, 0)$ d. None of these

## One Integer Value Correct Type

***Direction** (Q. Nos. 33-35) This section contains 3 questions. When worked out will result in an integer from 0 to 9 (both inclusive).*

**33** Let $f$ be differentiable function such that $f(x) = x^2 + \int_0^x e^{-t} f(x - t)\, dt$, where $f(1) + f(2) + f(3) + \cdots + f(9) = \lambda \cdot (480)$, then $\lambda$ is equal to

**34** If $A$ is the area between the curve $y = x^6 (1 - x)^7$ and $X$-axis, then the number of prime factors of $A^{-1}$ is

**35** Area of the figure contained between the parabola $x^2 = 4y$ and the curve $y = \frac{8}{x^2 + 4}$ is $\left(4\pi - \frac{k}{3}\right)$, then the value of $k$ is

# JEE Main & AIEEE Archive

## (Compilation of Last 13 Years' Questions)

**1** The area (in sq units) of the region described by $\{(x, y): y^2 \le 2x \text{ and } y \ge 4x - 1\}$ is **(JEE Main 2015)**

a. $\frac{7}{32}$ b. $\frac{5}{64}$
c. $\frac{15}{64}$ d. $\frac{9}{32}$

**2** The area of the region described by $A = \{(x, y): x^2 + y^2 \le 1 \text{ and } y^2 \le 1 - x\}$ is **(JEE Main 2014)**

a. $\frac{\pi}{2} + \frac{4}{3}$ b. $\frac{\pi}{2} - \frac{4}{3}$
c. $\frac{\pi}{2} - \frac{2}{3}$ d. $\frac{\pi}{2} + \frac{2}{3}$

**3** The area (in sq units) bounded by the curves $y = \sqrt{x}$, $2y - x + 3 = 0$, $X$-axis and lying in the first quadrant is **(JEE Main 2013)**

a. 9 b. 36 c. 18 d. $\frac{27}{4}$

**4** The area (in sq units) bounded between the parabolas $x^2 = \frac{y}{4}$ and $x^2 = 9y$ and the straight line $y = 2$, is **(AIEEE 2012)**

a. $20\sqrt{2}$ b. $\frac{10\sqrt{2}}{3}$
c. $\frac{20\sqrt{2}}{3}$ d. $10\sqrt{2}$

**5** The area of the region enclosed by the curves $y = x$, $x = e$, $y = \frac{1}{x}$ and the positive $X$-axis is **(AIEEE 2011)**

a. 1 sq unit b. $\frac{3}{2}$ sq units
c. $\frac{5}{2}$ sq units d. $\frac{1}{2}$ sq unit

**6** The area bounded by the curves $y^2 = 4x$ and $x^2 = 4y$ is **(AIEEE 2011)**

a. 0 b. $\frac{32}{3}$
c. $\frac{16}{3}$ d. $\frac{8}{3}$

**7** The area bounded by the curves $y = \cos x$ and $y = \sin x$ between the abscissae $x = 0$ and $x = \frac{3\pi}{2}$, is **(AIEEE 2009)**

a. $(4\sqrt{2} - 2)$ sq units b. $(4\sqrt{2} + 2)$ sq units
c. $(4\sqrt{2} - 1)$ sq units d. $(4\sqrt{2} + 1)$ sq units

**8** The area of the region bounded by the parabola $(y - 2)^2 = x - 1$, the tangent to the parabola at the point $(2, 3)$ and the $X$-axis is **(AIEEE 2009)**

a. 6 sq units b. 9 sq units
c. 12 sq units d. 3 sq units

**9** The area of the plane region bounded by the curves $x + 2y^2 = 0$ and $x + 3y^2 = 1$ is equal to **(AIEEE 2008)**

a. $\frac{4}{3}$ sq units b. $\frac{5}{3}$ sq units
c. $\frac{1}{3}$ sq unit d. $\frac{2}{3}$ sq unit

**10** The area enclosed between the curves $y^2 = x$ and $y = |x|$ is **(AIEEE 2007)**

a. $\frac{2}{3}$ sq unit b. 1 sq unit
c. $\frac{1}{6}$ sq unit d. $\frac{1}{3}$ sq unit

**11** If $f(x)$ is a non-negative continuous function such that the area bounded by the curve $y = f(x)$, $X$-axis and the ordinates $x = \frac{\pi}{4}$ and $x = \beta > \frac{\pi}{4}$ is $\left(\beta \sin \beta + \frac{\pi}{4} \cos \beta + \sqrt{2}\beta\right)$. Then, $f\left(\frac{\pi}{2}\right)$ is equal to **(AIEEE 2005)**

a. $\left(1 - \frac{\pi}{4} + \sqrt{2}\right)$ b. $\left(1 - \frac{\pi}{4} - \sqrt{2}\right)$
c. $\left(\frac{\pi}{4} - \sqrt{2} + 1\right)$ d. $\left(\frac{\pi}{4} + \sqrt{2} - 1\right)$

**12** The area enclosed between the curve $y = \log_e (x + e)$ and the coordinate axes is **(AIEEE 2005)**

a. 4 sq units b. 3 sq units
c. 2 sq units d. 1 sq unit

**13** The parabolas $y^2 = 4x$ and $x^2 = 4y$ divide the square region bounded by the lines $x = 4$, $y = 4$ and the coordinate axes. If $S_1, S_2$ and $S_3$ are respectively the areas of these parts numbered from top to bottom, then $S_1 : S_2 : S_3$ is equal to **(AIEEE 2005)**

a. 1 : 1 : 1 b. 2 : 1 : 2
c. 1 : 2 : 3 d. 1 : 2 : 1

**14** The area of the region bounded by the curves $y = |x - 2|$, $x = 1$, $x = 3$ and the $X$-axis is **(AIEEE 2004)**

a. 1 sq unit b. 2 sq units
c. 3 sq units d. 4 sq units

**15** The area of the region bounded by the curves $y = |x - 1|$ and $y = 3 - |x|$ is **(AIEEE 2003)**

a. 2 sq units b. 3 sq units
c. 4 sq units d. 6 sq units

**16** The area bounded by the curve $y = 2x - x^2$ and the straight line $y = -x$ is given by **(AIEEE 2002)**

a. $\frac{9}{2}$ sq units b. $\frac{43}{6}$ sq units
c. $\frac{35}{6}$ sq units d. None of these

# JEE Advanced & IIT JEE Archive

## (Compilation of Last 10 Years' Questions)

**1** The area enclosed by the curves $y = \sin x + \cos x$ and $y = |\cos x - \sin x|$ over the interval $\left[0, \frac{\pi}{2}\right]$ is

**(2014 Adv., Only One Option Correct Type)**

a. $4(\sqrt{2} - 1)$ b. $2\sqrt{2}(\sqrt{2} - 1)$
c. $2(\sqrt{2} + 1)$ d. $2\sqrt{2}(\sqrt{2} + 1)$

**2** The common tangents to the circle $x^2 + y^2 = 2$ and the parabola $y^2 = 8x$ touch the circle at the points $P, Q$ and the parabola at the points $R, S$. Then, the area of the quadrilateral $PQRS$ is

**(2014 Adv., Only One Option Correct Type)**

a. 3 b. 6
c. 9 d. 15

**3** If $S$ is the area of the region enclosed by $y = e^{-x^2}$, $y = 0$, $x = 0$ and $x = 1$. Then,

**(2012, One or More than One Options Correct Type)**

a. $S \geq \frac{1}{e}$

b. $S \geq 1 - \frac{1}{e}$

c. $S \leq \frac{1}{4}\left(1 + \frac{1}{\sqrt{e}}\right)$

d. $S \leq \frac{1}{\sqrt{2}} + \frac{1}{\sqrt{e}}\left(1 - \frac{1}{\sqrt{2}}\right)$

**4** Let the straight line $x = b$ divides the area enclosed by $y = (1 - x)^2$, $y = 0$ and $x = 0$ into two parts $R_1 (0 \leq x \leq b)$ and $R_2 (b \leq x \leq 1)$ such that $R_1 - R_2 = \frac{1}{4}$. Then, $b$ equals

**(2011, Only One Option Correct Type)**

a. $\frac{3}{4}$ b. $\frac{1}{2}$
c. $\frac{1}{3}$ d. $\frac{1}{4}$

**5** Let $f : [-1, 2] \to [0, \infty)$ be a continuous function such that $f(x) = f(1 - x)$ for all $x \in [-1, 2]$. Let $R_1 = \int_{-1}^{2} x f(x) dx$ and $R_2$ be the area of the region bounded by $y = f(x)$, $x = -1$, $x = 2$ and the $X$-axis. Then,

**(2011, Only One Option Correct Type)**

a. $R_1 = 2R_2$ b. $R_1 = 3R_2$
c. $2R_1 = R_2$ d. $3R_1 = R_2$

**6** Area of the region bounded by the curve $y = e^x$ and lines $x = 0$ and $y = e$ is

**(2009, One or More than One Options Correct Type)**

a. $e - 1$ b. $\int_1^e \ln(e + 1 - y) dy$
c. $e - \int_0^1 e^x dx$ d. $\int_1^e \ln y \, dy$

**7** The area of the region between the curves $y = \sqrt{\frac{1 + \sin x}{\cos x}}$ and $y = \sqrt{\frac{1 - \sin x}{\cos x}}$ and bounded by the lines $x = 0$ and $x = \frac{\pi}{4}$ is

**(2008, Only One Option Correct Type)**

a. $\int_0^{\sqrt{2}-1} \frac{t}{(1 + t^2)\sqrt{1 - t^2}} dt$

b. $\int_0^{\sqrt{2}-1} \frac{4t}{(1 + t^2)\sqrt{1 - t^2}} dt$

c. $\int_0^{\sqrt{2}+1} \frac{4t}{(1 + t^2)\sqrt{1 - t^2}} dt$

d. $\int_0^{\sqrt{2}+1} \frac{t}{(1 + t^2)\sqrt{1 - t^2}} dt$

**8** The area of the equilateral triangle, in which three coins of radius 1 cm are placed, as shown in the figure, is

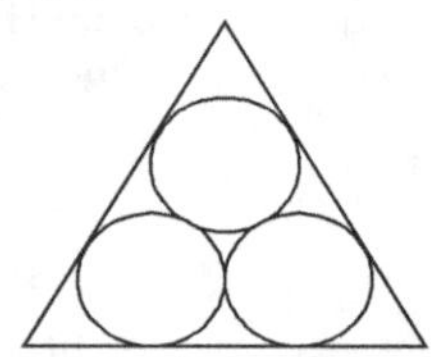

**(2005, Only One Option Correct Type)**

a. $(6 + 4\sqrt{3})$ sq cm b. $(4\sqrt{3} - 6)$ sq cm
c. $(7 + 4\sqrt{3})$ sq cm d. $4\sqrt{3}$ sq cm

**9** The area bounded by the curves $y = (x - 1)^2$, $y = (x + 1)^2$ and $y = \frac{1}{4}$ is

**(2005, Only One Option Correct Type)**

a. $\frac{1}{3}$ sq unit b. $\frac{2}{3}$ sq unit
c. $\frac{1}{4}$ sq unit d. $\frac{1}{5}$ sq unit

**10** Find the area bounded by the curves

$$x^2 = y, \; x^2 = -y \text{ and } y^2 = 4x - 3.$$

**(2005, Subjective Type)**

**11** If $\begin{bmatrix} 4a^2 & 4a & 1 \\ 4b^2 & 4b & 1 \\ 4c^2 & 4c & 1 \end{bmatrix} \begin{bmatrix} f(-1) \\ f(1) \\ f(2) \end{bmatrix} = \begin{bmatrix} 3a^2 + 3a \\ 3b^2 + 3b \\ 3c^2 + 3c \end{bmatrix}$,

$f(x)$ is a quadratic function and its maximum value occurs at a point $V$. $A$ is a point of intersection of $y = f(x)$ with $X$-axis and point $B$ is such that chord $AB$ subtends a right angled at $V$. Find the area enclosed by $f(x)$ and chord $AB$.

**(2005, Subjective Type)**

# Answers *with* Explanations

# Answers

## 1. Indefinite Integral

### DPP-1 Standard Forms of Integration

**1.** (i) $\frac{a^x}{\log_e a}+\frac{x^{a+1}}{a+1}+a^a x+C$ (ii) $\frac{\sin 2x}{2}+C$

(iii) $\frac{1}{2}(\tan x+x)+C$ (iv) $\frac{2^x e^x}{1+\log 2}+C$

(v) $\tan x-x+C$ (vi) $\frac{x^3}{3}+C$

(vii) $\frac{-1}{x}+\tan^{-1}x+C$ (viii) $-\cot x-\tan x+C$

(ix) $\frac{\sin 12x}{12}+\frac{\sin 10x}{10}+\frac{\sin 11x}{11}+\frac{\sin 9x}{9}+C$ (x) $\sin 2x+C$

**2.** *b* **3.** *b* **4.** *b* **5.** *b* **6.** *b* **7.** *a* **8.** *c* **9.** *d* **10.** *a* **11.** *b*
**12.** *c* **13.** *d* **14.** *a* **15.** *c* **16.** *d* **17.** *a* **18.** *a* **19.** *d* **20.** *b* **21.** *a*
**22.** *b* **23.** *a* **24.** *b* **25.** *a* **26.** *c*

### DPP-2 Problems on Basic Substitution

**1.** (i) $-\frac{1}{2(2x-3)}+C$ (ii) $\frac{2}{5}\sqrt{4+x^5}+C$

(iii) $-\frac{1}{2}\frac{1}{(\sin^{-1}x)^2}+C$ (iv) $\frac{(\tan^{-1}x)^3}{3}+C$

(v) $2\sqrt{1+\tan x}+C$ (vi) $\log|x^2-3x+8|+C$

(vii) $-\frac{1}{3}e^{-3x+1}+C$ (viii) $-\frac{1}{3}e^{-x^3}+C$

(ix) $\frac{1}{5}\sin^{-1}(5x)+C$ (x) $\frac{1}{2}\sin^{-1}\left(\frac{x^2}{a}\right)+C$

(xi) $\frac{1}{6}\tan^{-1}\left(\frac{x^2}{2}\right)+C$ (xii) $-2\sqrt{1-x^2}-\frac{2}{3}(\sin^{-1}x)^{3/2}+C$

(xiii) $-\frac{1}{2}\left[\log\left(1+\frac{1}{x}\right)\right]^2+C$ (xiv) $\log|e^x+1|+C$

(xv) $\frac{2}{3}[x^3-\sqrt{(x^3-1)^3}]-x+C$ (xvi) $3(\sin x)^{1/3}+C$

(xvii) $\sqrt{3x^2-5x+6}+C$ (xviii) $\frac{\tan^4 x}{4}+C$

(xix) $2\sqrt{1+x^2}+C$ (xx) $\tan(1+\log x)+C$

**2.** *a* **3.** *b* **4.** *a* **5.** *c* **6.** *a* **7.** *b* **8.** *d* **9.** *a* **10.** *a* **11.** *d*
**12.** *c* **13.** *b* **14.** *d* **15.** *c* **16.** *a,b,d* **17.** *b,c,d* **18.** *b,d* **19.** *a,b,c,d* **20.** *a,b,c,d*
**21.** *b* **22.** *d* **23.** *a* **24.** *c*

### DPP-3 Integration by Parts

**1.** (i) $\log x\cdot\frac{x^2}{2}-\frac{x^2}{4}+C$ (ii) $-x\cot\frac{x}{2}+C$

(iii) $3x\tan^{-1}x-\frac{3}{2}\log|1+x^2|+C$ (iv) $x(\log^2 x+2\log x+2)+C$

(v) $\sin^{-1}x\cdot x+\sqrt{1-x^2}+C$ (vi) $\frac{x^3}{3}\sin^{-1}x+\frac{\sqrt{1-x^2}}{3}-\frac{1}{9}(1-x^2)^3+C$

(vii) $-\sqrt{1-x^2}\sin^{-1}x^2+2\sqrt{1+x^2}+C$ (viii) $(x+1)\tan^{-1}\sqrt{x}-\sqrt{x}+C$

(ix) $2e^{\sqrt{x}}(\sqrt{x}-1)+C$ (x) $[(\sin^{-1}x)^2-2]\sin(\sin^{-1}x)+2\sin^{-1}x\cos(\sin^{-1}x)+C$

(xi) $2[x\tan^{-1}x-\log|\sec(\tan^{-1}x)|]+C$ (xii) $\frac{1}{2}\left[x\cos^{-1}x-\sqrt{1-x^2}\right]+C$

(xiii) $\frac{e^{2x}}{13}[2\sin 3x - 3\cos 3x] + C$ (xiv) $\frac{e^t}{10}[(2\sin 2t + \cos 2t) + 5] + C$, where $t = \tan^{-1} x$

(xv) $-\frac{1}{2}\text{cosec}\, x \cdot \cot x + \frac{1}{2}\log(\text{cosec}\, x - \cot x) + C$

(xvi) $\frac{x}{2}[\sin(\log x) - \cos(\log x)] + C$ (xvii) $\log(\sec x) \cdot e^x + C$

(xviii) $\frac{e^x}{x+1} + C$ (xix) $\frac{e}{2+x} + C$ (xx) $x \cdot e^{\tan^{-1} x} + C$

**2.** *c* **3.** *a* **4.** *c* **5.** *b* **6.** *b* **7.** *d* **8.** *c* **9.** *a* **10.** *c,d*
**11.** *b,c,d* **12.** *a,b,c,d* **13.** *a,c* **14.** *b* **15.** *c*

## DPP-4 Some Special Substitutions and Integrals

**1.** (i) $\frac{2}{\sqrt{3}}\tan^{-1}\left(\frac{2x-1}{\sqrt{3}}\right) + C$ (ii) $\frac{1}{3}\log\left|\frac{2x-1}{2(x+1)}\right| + C$

(iii) $\log(x-1) + \sqrt{x^2 - 2x + 3} + C$

(iv) $\sqrt{2}\left[\frac{1}{8}(4x-3)\sqrt{x^2 - \frac{3x}{2} + \frac{1}{2}} - \frac{1}{32}\log\left|\left(x - \frac{3}{4}\right) + \sqrt{x^2 - \frac{3x}{2} + \frac{1}{2}}\right|\right] + C$

(v) $\frac{1}{2}\log\left|\frac{\sqrt{1-e^{2x}} - 1}{\sqrt{1-e^{2x}} + 1}\right| + C$ (vi) $\sin^{-1}\left(\frac{2x^2+1}{\sqrt{5}}\right) + C$

(vii) $\frac{1}{\log a}\sin^{-1}(a^x) + C$ (viii) $\sin^{-1}\left(\frac{e^x+2}{3}\right) + C$

(ix) $\frac{2}{3}\sin^{-1}\left(\frac{x^{3/2}}{a^{3/2}}\right) + C$ (x) $\log|(\sin x - 1) + \sqrt{\sin^2 x - 2\sin x - 3}| + C$

(xi) $-\cos\alpha \cdot \sin^{-1}\left(\frac{\cos x}{\cos\alpha}\right) - \sin\alpha \cdot \log\left|\sin x - \sqrt{\sin^2 x - \sin^2\alpha}\right| + C$

(xii) $2\log|\sin^2 x - 4\sin x + 5| + 7\tan^{-1}(\sin x - 2) + C$

(xiii) $\frac{-2}{3}\sin^{-1}(\cos^{3/2} x) + C$ (xiv) $a\sin^{-1}\frac{x}{a} + \sqrt{a^2 - x^2} + C$

(xv) $\frac{1}{2}a^2\sin^{-1}\left(\frac{x^2}{a^2}\right) + \frac{1}{4}\sqrt{a^4 - x^4} + C$

(xvi) $-\frac{1}{3}(1 + x - x^2)^{3/2} + \frac{1}{2}\left[\frac{x - \frac{1}{2}}{2}\sqrt{1 + x - x^2} + \frac{5}{8}\sin^{-1}\frac{2x-1}{\sqrt{5}}\right] + C$

(xvii) $-\frac{1}{3}(1 - x - x^2)^{3/2} + \frac{1}{8}(2x+1)\sqrt{1 - x - x^2} + \frac{5}{16}\sin^{-1}\frac{2x+1}{\sqrt{5}} + C$

(xviii) $\left(x + \frac{2}{7}\right)\sqrt{x^2 + x - 1} + \frac{5}{4}\log\left\{\left(x + \frac{1}{2}\right) + \sqrt{x^2 + x + 1}\right\} + C$

**2.** (i) $\frac{1}{6}\tan^{-1}\left(\frac{2\tan x}{3}\right) + C$ (ii) $\frac{1}{2\sqrt{3}}\log\left|\frac{\sqrt{3} + \tan x}{\sqrt{3} - \tan x}\right| + C$ (iii) $\sqrt{2}\tan^{-1}\left[\frac{\tan\frac{x}{2} + 1}{\sqrt{2}}\right] + C$

(iv) $\frac{1}{2}\log\left|\frac{2 - \sqrt{3} + \tan\frac{x}{2}}{2 + \sqrt{3} - \tan\frac{x}{2}}\right| + C$ (v) $\frac{12}{13}x - \frac{5}{13}\log|3\cos x + 2\sin x| + C$

**3.** *a* **4.** *d* **5.** *c* **6.** *d* **7.** *a* **8.** *a* **9.** *a* **10.** *a* **11.** 1 **12.** 1
**13.** 0 **14.** 4 **15.** 0

## DPP-5 $\int \frac{dx}{P\sqrt{Q}}$ Form, Partial Fractions and Reduction Formula

**1.** (i) $\sqrt{2}\tan^{-1}\sqrt{\frac{x+1}{2}} + C$ (ii) $-\frac{1}{\sqrt{3}}\log\left|\frac{(\sqrt{3}+1)x + 3 + \sqrt{3}}{(1-\sqrt{3})x + 3 - \sqrt{3}}\right| + C$ (iii) $-\sin^{-1}(2x-5) + \sqrt{(2-x)(x-3)}$

(iv) $\frac{1}{2\sqrt{33}}\log\left|\frac{\sqrt{11}x + \sqrt{3x^2 - 12}}{\sqrt{11}x - \sqrt{3x^2 - 12}}\right| + C$ (v) $\frac{2x-13}{4}\sqrt{(x-7)(6-x)} + \frac{1}{8}\sin^{-1}(2x-13) + C$

**2.** (i) $\frac{x^2}{2}+x+\frac{9}{4}\log(2x^2-5x-12)+\frac{29}{56}\log\left|\frac{2x-12}{2x+2}\right|+C$ (ii) $\frac{5}{2}\log|(x-1)|-8\log|x-2|+\frac{11}{2}\log|x-3|+C$

(iii) $\frac{13}{36}\log(x-1)-\frac{1}{6(x-1)}-\frac{5}{4}\log(x+1)+\frac{8}{9}\log(x+2)+C$

(iv) $-\log|x+1|-\frac{1}{2}\log|x^2+2|+\frac{1}{\sqrt{2}}\tan^{-1}\frac{x}{\sqrt{2}}+C$ (v) $\frac{2}{\sqrt{3}}\tan^{-1}\left[\frac{x}{\sqrt{3}(x+1)}\right]+C$

(vi) $-\frac{5}{2}\tan^{-1}x+\frac{6}{\sqrt{2}}\tan^{-1}\frac{x}{\sqrt{2}}-\frac{7}{2\sqrt{3}}\tan^{-1}\frac{x}{\sqrt{3}}+C$ (vii) $-\frac{1}{x}-\frac{3}{2}\cdot\frac{x}{(x^2+1)^2}-\frac{3}{2}\tan^{-1}x+C$

(viii) $-\frac{1}{5}\log|1+x^{-5}|+C$ (ix) $-(1+x^{-7})^{1/7}+C$ (x) $-\frac{1}{2\sqrt{3}}\log\left|\frac{x+\frac{1}{x}-\sqrt{3}}{x+\frac{1}{x}+\sqrt{3}}\right|+C$

(xi) $\frac{1}{2\sqrt{2}}\tan\left(\frac{x^2-1}{x\sqrt{2}}\right)-\frac{1}{4\sqrt{2}}\log\left|\frac{x^2+1-\sqrt{2}x}{x^2+1+\sqrt{2}x}\right|+C$ (xii) $\frac{1}{\sqrt{6}}\tan^{-1}\left(\sqrt{\frac{3}{2}}x^{1/6}\right)-\frac{x^{1/6}}{(2+3x^{1/3})}+C$

(xiii) $-\frac{1}{2\sqrt{3}}\log\left|\frac{\sqrt{x^2+\frac{1}{x^2}+1}-\sqrt{3}}{\sqrt{x^2+\frac{1}{x^2}+1}+\sqrt{3}}\right|+C$ (xiv) $\frac{1}{\sqrt{2}}\left[\tan^{-1}\left(\frac{t-\frac{1}{t}}{\sqrt{2}}\right)+\frac{1}{2}\log\left(\frac{t+\frac{1}{t}-\sqrt{2}}{t+\frac{1}{t}+\sqrt{2}}\right)\right]+C$, where $t=\sqrt{\tan x}$

(xv) $\frac{7}{3}\left(\frac{x-1}{x+2}\right)^{1/7}+C$

**5.** *c* **6.** *b* **7.** *b* **8.** *d* **9.** 2 **10.** 5

### Revisal Problems (JEE Main)

**1.** *b* **2.** *b* **3.** *a* **4.** *b* **5.** *d* **6.** *a* **7.** *b* **8.** *b* **9.** *b* **10.** *c*
**11.** *c* **12.** *b* **13.** *d* **14.** *a* **15.** *d* **16.** *c* **17.** *d* **18.** *b* **19.** *a* **20.** *c*
**21.** *c* **22.** *c* **23.** *b* **24.** *b* **25.** *c* **26.** *c* **27.** *b* **28.** *a* **29.** *b* **30.** *d*
**31.** *b* **32.** *b* **33.** *a* **34.** *b* **35.** *b* **36.** *a* **37.** *a* **38.** *c* **39.** *d* **40.** *a*

### Revisal Problems (JEE Advanced)

**1.** *c* **2.** *b* **3.** *c* **4.** *a* **5.** *a* **6.** *d* **7.** *c* **8.** *c* **9.** *c* **10.** *a*
**11.** *b* **12.** *c* **13.** *c* **14.** *a* **15.** *c* **16.** *a* **17.** *a* **18.** *d* **19.** *a* **20.** *c*
**21.** *a* **22.** *b* **23.** *a* **24.** *c* **25.** *b* **26.** *c* **27.** 1 **28.** 4 **29.** 1 **30.** 4
**31.** 2 **32.** 2 **33.** 4 **34.** 2 **35.** 1

### JEE Main & AIEEE Archive

**1.** *d* **2.** *b* **3.** *c* **4.** *d* **5.** *d* **6.** *a* **7.** *a* **8.** *b* **9.** *d* **10.** *a*

### JEE Advanced & IIT JEE Archive

**1.** *c* **2.** *c* **3.** *a* **4.** *c*

## 2. Definite Integral

### DPP-1 Basic Definition

**1.** (i) $\frac{1}{2}-\frac{\sqrt{3}\pi}{12}$ (ii) $\sqrt{2}\pi$ (iii) $\frac{n}{n^2-1}$ (iv) $\frac{\pi}{4ab^2(a+b)}$ (v) $\frac{1}{3}$ (vi) $\frac{\pi}{6}$

**2.** *c* **3.** *a* **4.** *a* **5.** *d* **6.** *b* **7.** *a* **8.** *d* **9.** *d* **10.** *a* **11.** *a*
**12.** *a* **13.** *a* **14.** *d* **15.** *b* **16.** *b* **17.** *b* **18.** *c* **19.** *d* **20.** *b* **21.** *c*
**22.** *a* **23.** 6 **24.** 0 **25.** 2

### DPP-2 Properties of Definite Integral-I

**1.** *a* **2.** *d* **3.** *c* **4.** *b* **5.** *b* **6.** *b* **7.** *a* **8.** *a* **9.** *d* **10.** *c*
**11.** *d* **12.** *d* **13.** *a* **14.** *a* **15.** *a* **16.** *a* **17.** *d* **18.** *b* **19.** *c* **20.** *a*
**21.** *b* **22.** *b* **23.** *b* **24.** *a* **25.** *a,b* **26.** 2 **27.** 4 **28.** 1 **29.** 1 **30.** 0

## DPP-3 Properties of Definite Integral-II

1. (i) 0 (ii) 0 (iii) 2 (iv) 0 (v) $\frac{\pi}{2}$ (vi) 0

2. (i) $\frac{\pi}{4}$ (ii)0 (iii) $\frac{\pi}{8}\log 2$ (iv) $\frac{1}{2}$ (v) $\frac{\pi}{12}$ (vi) $\frac{b-a}{2}$ (vii) 0 (viii) $-\frac{\pi}{2}\log 2$

3. (i) 0 (ii) $-\frac{\pi}{2}\log 2$ (iii) $\frac{\pi}{a^2+b^2}$ (iv) $\pi^2$

4. *d* 5. *c* 6. *c* 7. *d* 8. *b* 9. *a* 10. *b* 11. *b* 12. *c* 13. *b*
14. *c* 15. *c* 16. *d* 17. *a* 18. *a* 19. *c* 20. *a* 21. *b* 22. *a,b,c*
23. *a,b,c* 24. *a,d* 25. *a,c* 26. *a,b,c,d* 27. *a* 28. *c* 29. *c* 30. *d* 31. 1
32. 0 33. 9 34. 0 35. 2

## DPP-4 Periodic Functions and Leibnitz Theorem

1. (i) $3(e-1)$ (ii) $3(e-1)$ (iii) $1000\pi$ (iv) 400

2. (i) $2x\sqrt{\sin x^2}-\sqrt{\sin x}$ (ii) 1 (iii) 0 (iv) 2 /17

3. *a* 4. *a* 5. *b* 6. *c* 7. *c* 8. *b* 9. *c* 10. *d* 11. *d* 12. *c*
13. *a* 14. *c* 15. *b* 16. *a* 17. *c* 18. *a* 19. *a* 20. *c* 21. *c* 22. *c*
23. *b* 24. *b* 25. *a* 26. *b* 27. *d* 28. *a* 29. *b* 30. 4 31. 5 32. 4
33. 0 34. 2 35. 2

## DPP-5 Estimation, Gamma Function and Limit as the Sum

2. (i) 1 (ii) $\frac{3\pi}{512}$ (iii) $\frac{1}{210}$ (iv) $\frac{(b-a)^{m+n+1}\cdot m!\cdot n!}{(m+n+1)!}$ 3. (i) $\frac{\pi}{4}$ (ii) $\log 3$ (iii) $\frac{1}{3}\log 2$ (iv) $\frac{1}{e}$

4. *c* 5. *a* 6. *b* 7. *b* 8. *d* 9. *b* 10. *c* 11. *c* 12. *c* 13. *a*
14. *a,d* 15. *b,d* 16. *d* 17. *b* 18. *b* 19. *b* 20. *c* 21. *b* 22. 1 23. 3
24. 8 25. 2

## Revisal Problems (JEE Main)

1. *b* 2. *a* 3. *d* 4. *d* 5. *b* 6. *b* 7. *a* 8. *c* 9. *b* 10. *c*
11. *d* 12. *a* 13. *d* 14. *d* 15. *b* 16. *b* 17. *c* 18. *a* 19. *b* 20. *c*
21. *d* 22. *a* 23. *b* 24. *a* 25. *c* 26. *d* 27. *b* 28. *a* 29. *a* 30. *a*
31. *c* 32. *b* 33. *a* 34. *c* 35. *b* 36. *b* 37. *c* 38. *a* 39. *a* 40. *d*
41. *c* 42. *d* 43. *c* 44. *a* 45. *c* 46. *d* 47. *c* 48. *d* 49. *c* 50. *a*

## Revisal Problems (JEE Advanced)

1. *d* 2. *d* 3. *a* 4. *a* 5. *c* 6. *b* 7. *a* 8. *a* 9. *a* 10. *d*
11. *b* 12. *a* 13. *b* 14. *b* 15. *d* 16. *c* 17. *a* 18. *d* 19. *c* 20. *d*
21. *a* 22. *c* 23. *b* 24. *c* 25. *a* 26. *a* 27. *d* 28. *a* 29. *b* 30. *b*
31. *c* 32. *c* 33. *c* 34. *c* 35. *d* 36. *a* 37. *b* 38. *a* 39. *c* 40. *d*
41. *b* 42. *b* 43. *a* 44. *a* 45. *c* 46. *c* 47. *c* 48. *d* 49. *d* 50. *b*
51. *a,c* 52. *a,b* 53. *b,c* 54. *a,b,c* 55. *b,d* 56. *c,d* 57. *c,d* 58. *b,d* 59. *a,b,d* 60. *c,d*
61. *a* 62. *a* 63. *a* 64. *a* 65. *b* 66. *c* 67. *d* 68. *b* 69. 4 70. 4
71. 1 72. 5 73. 3 74. 1 75. 1

## JEE Main & AIEEE Archive

1. *c* 2. *d* 3. *d* 4. *b,c* 5. *d* 6. *d* 7. *b* 8. *d* 9. *c* 10. *a*
11. *a* 12. *a* 13. *b* 14. *c* 15. *a* 16. *a* 17. *a* 18. *c* 19. *c* 20. *a*
21. *c* 22. *b* 23. *a* 24. *b* 25. *b* 26. *b* 27. *c* 28. *c* 29. *d* 30. *d*
31. *c* 32. *a* 33. *b* 34. *d* 35. *b* 36. *a*

## JEE Advanced & IIT JEE Archive

1. 0 2. 9 3. 7 4. *a,b* 5. *d* 6. *a,c* 7. *a,b,c* 8. *c,d* 9. *a* 10. *b*
11. 2 12. *b* 13. *a* 14. *b,d* 15. *a* 16. 4 17. *a,b,c* 18. *c* 19. *a,d* 20. *a*
21. *b* 22. *c* 23. *c* 24. *a* 25. *b* 26. *c* 27. *a*

# 3. Differential Equations

## DPP-1 Order, Degree and Formation of Differential Equations

**1.** (i) Order = 2, Degree = 3 (ii) Order = 2, Degree = Not defined (iii) Order = 1, Degree = 2
(iv) Order = 1, Degree = 2 (v) Order = 2, Degree = 2 (vi) Order = 1, Degree = 1

**2.** (i) $(x^2 - y^2)\frac{dy}{dx} = 2xy$ (ii) $(y^2 - x^2) = 2xy\frac{dy}{dx}$ **3.** $\left[1+\left(\frac{dy}{dx}\right)^2\right]^3 = r^2\left[\frac{d^2y}{dx^2}\right]^2$

**4.** $(x-y)^2\left[1+\left(\frac{dy}{dx}\right)^2\right] = \left[x + y\frac{dy}{dx}\right]^2$

**5.** (i) $\frac{dx}{dy} = m$ and $\frac{d^2x}{dy^2} = 0$ (ii) $\frac{dy}{dx} = m$ and $\frac{d^2y}{dx^2} = 0$ **6.** $y\frac{dy}{dx} = x\left[\left(\frac{dy}{dx}\right)^2 + y\frac{d^2y}{dx^2}\right]$

**7.** $\left(3x\frac{d^2y}{dx^2} + x^2\frac{d^3y}{dx^3}\right)\left(y - x\frac{dy}{dx}\right) = 3x\frac{d^2y}{dx^2}\left(y - x\frac{dy}{dx} - x^2\frac{d^2y}{dx^2}\right)$

**8.** $\left(y - x\frac{dy}{dx}\right)^2 = p^2\left[1+\left(\frac{dy}{dx}\right)^2\right]$ **9.** $y\left(\frac{d^2y}{dx^2}\right) + \left(\frac{dy}{dx}\right)^2 = 0$ **10.** $\frac{d^3x}{dy^3} = 0$

**11.** *d* **12.** *c* **13.** *c* **14.** *b* **15.** *d* **16.** *c* **17.** *c* **18.** *c* **19.** *c* **20.** *c*
**21.** *c* **22.** *a* **23.** *c* **24.** *d* **25.** *a* **26.** 3 **27.** 1 **28.** 1 **29.** 1 **30.** 5

## DPP-2 Equation in Variable Separable Form and Reducible to Variable Separation Form

**1.** *d* **2.** *a* **3.** *b* **4.** *c* **5.** *a* **6.** *c* **7.** *a* **8.** *d* **9.** *b* **10.** *b*
**11.** *b* **12.** *c* **13.** *d* **14.** *a,c* **15.** *a,c* **16.** *a,b,d* **17.** *b* **18.** *a* **19.** *a* **20.** *c*
**21.** *b* **22.** 2 **23.** 1 **24.** 1 **25.** 9

## DPP-3 Homogeneous Differential Equations and Equation Reducible to Homogeneous Form

**1.** *a* **2.** *c* **3.** *c* **4.** *d* **5.** *c* **6.** *a* **7.** *a* **8.** *c* **9.** *c* **10.** *b*
**11.** *a* **12.** *b* **13.** *d* **14.** *c* **15.** *c* **16.** *a* **17.** *c* **18.** *a* **19.** *c* **20.** *a*
**21.** 0 **22.** 0 **23.** 2 **24.** 2 **25.** 1

## DPP-4 Linear Differential Equations and Bernoulli's Equation

**1.** *d* **2.** *a* **3.** *a* **4.** *c* **5.** *b* **6.** *c* **7.** *a* **8.** *d* **9.** *a* **10.** *a*
**11.** *a,b* **12.** *a,c* **13.** *b,c* **14.** *a,c* **15.** *a* **16.** *c* **17.** *b* **18.** *a* **19.** 6 **20.** 0

## DPP-5 Solving by Inspection Method and Orthogonal Trajectory

**1.** *c* **2.** *a* **3.** *c* **4.** *a* **5.** *b* **6.** *a* **7.** *d* **8.** *c* **9.** *c* **10.** *a*
**11.** *c* **12.** *d* **13.** *a* **14.** *a* **15.** *c,d* **16.** *c* **17.** *a,d* **18.** *b* **19.** *a* **20.** *b*
**21.** *b* **22.** *c* **23.** *d* **24.** 2 **25.** 1

## Revisal Problems (JEE Main)

**1.** *b* **2.** *d* **3.** *a* **4.** *b* **5.** *a* **6.** *b* **7.** *a* **8.** *b* **9.** *c* **10.** *a*
**11.** *d* **12.** *b* **13.** *d* **14.** *b* **15.** *b* **16.** *a* **17.** *d* **18.** *d* **19.** *c* **20.** *a*
**21.** *a* **22.** *b* **23.** *a* **24.** *c* **25.** *d* **26.** *b* **27.** *c* **28.** *a* **29.** *b* **30.** *a*
**31.** *b* **32.** *c* **33.** *c* **34.** *b* **35.** *c* **36.** *a* **37.** *d* **38.** *a* **39.** *d* **40.** *c*

## Revisal Problems (JEE Advanced)

**1.** *c* **2.** *b* **3.** *c* **4.** *d* **5.** *c* **6.** *a* **7.** *a* **8.** *a* **9.** *c* **10.** *c*
**11.** *a* **12.** *c* **13.** *a* **14.** *b* **15.** *a* **16.** *c* **17.** *a* **18.** *a* **19.** *a,b,d* **20.** *c,d*
**21.** *a,b,c* **22.** *a,b,c* **23.** *a,b,d* **24.** *b* **25.** *d* **26.** *a* **27.** *a* **28.** *c* **29.** *c* **30.** 4
**31.** 5 **32.** 2 **33.** 4 **34.** 1 **35.** 0

### JEE Main & AIEEE Archive

| | | | | | | | | | |
|---|---|---|---|---|---|---|---|---|---|
| 1. c | 2. a | 3. c | 4. a | 5. d | 6. a | 7. d | 8. a | 9. d | 10. d |
| 11. a | 12. c | 13. c | 14. b | 15. b | 16. c | 17. b | 18. b | 19. b | 20. c |
| 21. b | 22. a | | | | | | | | |

### JEE Advanced & IIT JEE Archive

| | | | | | | | | | |
|---|---|---|---|---|---|---|---|---|---|
| 1. 3 | 2. a,c | 3. b,c | 4. b | 5. d | 6. a | 7. b | 8. c | 9. a,d | 10. 0 |
| 11. 2 | 12. 0 | 13. c | 14. a | 15. c | 16. a,c | | | | |

17. $\sqrt{1-y^2}-\log\left|\dfrac{1+\sqrt{1-y^2}}{1-\sqrt{1-y^2}}\right|=\pm x+C$ 18. a

# 4. Area

## DPP-1 Area Under the Curves

1. (i) $\dfrac{2}{\sqrt{3}}$ sq unit (ii) 1 sq unit (iii) $\dfrac{7}{3}$ sq units (iv) $\dfrac{25}{3}$ sq units (v) $\dfrac{\pi}{4}$ sq unit (vi) $\dfrac{7}{24}$ sq unit

2. (i) 4 sq units (ii) $\dfrac{1}{6}$ sq unit (iii) $\left(\dfrac{4-\sqrt{2}}{\log 2}-\dfrac{5}{2}\log 2+\dfrac{3}{2}\right)$ sq units (iv) $\left(\dfrac{3}{\log 2}-\dfrac{4}{3}\right)$ sq unit (v) $\dfrac{11}{4}$ sq units (vi) $\dfrac{4}{3}$ sq units

| | | | | | | | | | |
|---|---|---|---|---|---|---|---|---|---|
| 3. d | 4. a | 5. a | 6. d | 7. a | 8. d | 9. c | 10. a | 11. d | 12. a |
| 13. a | 14. a | 15. c | 16. a | 17. b | 18. a | 19. c | 20. b | 21. b | 22. b |
| 23. a,b,c,d | | 24. b,d | 25. a,b | 26. a,d | 27. c | 28. c | 29. d | 30. a | |

## DPP-2 Curve Sketching and Its Applications in Finding the Area

| | | | | | | | | | |
|---|---|---|---|---|---|---|---|---|---|
| 1. b | 2. c | 3. c | 4. b | 5. a | 6. c | 7. d | 8. d | 9. d | 10. d |
| 11. c | 12. b | 13. a | 14. d | 15. c | 16. b | 17. a | 18. d | 19. c | 20. a |
| 21. b | 22. 7 | 23. 4 | 24. 6 | 25. 8 | | | | | |

### Revisal Problems (JEE Main)

| | | | | | | | | | |
|---|---|---|---|---|---|---|---|---|---|
| 1. c | 2. d | 3. b | 4. c | 5. c | 6. d | 7. b | 8. b | 9. d | 10. c |
| 11. a | 12. b | 13. c | 14. b | 15. b | 16. a | 17. b | 18. d | 19. c | 20. c |

### Revisal Problems (JEE Advanced)

| | | | | | | | | | |
|---|---|---|---|---|---|---|---|---|---|
| 1. d | 2. b | 3. d | 4. b | 5. a | 6. a | 7. b | 8. b | 9. b | 10. a |
| 11. c | 12. a | 13. a | 14. a,c | 15. a,c,d | 16. a,d | 17. b,c,d | 18. b,c,d | 19. b,c | 20. a,c |
| 21. c | 22. b | 23. a | 24. d | 25. a | 26. d | 27. d | 28. a | 29. d | 30. a |
| 31. c | 32. d | 33. 2 | 34. 5 | 35. 4 | | | | | |

### JEE Main & AIEEE Archive

| | | | | | | | | | |
|---|---|---|---|---|---|---|---|---|---|
| 1. d | 2. a | 3. a | 4. c | 5. b | 6. c | 7. a | 8. b | 9. a | 10. c |
| 11. a | 12. d | 13. a | 14. a | 15. c | 16. a | | | | |

### JEE Advanced & IIT JEE Archive

| | | | | | | | | |
|---|---|---|---|---|---|---|---|---|
| 1. b | 2. d | 3. a,b,d | 4. b | 5. c | 6. b,c,d | 7. b | 8. a | 9. a |

10. $\dfrac{1}{3}$ sq unit 11. $\dfrac{125}{3}$ sq units

# 1. Indefinite Integral

## DPP-1 Standard Forms of Integration

**1.** (i) Let $I = \int (e^{x\log_e a} + e^{a\log_e x} + e^{a\log_e a})dx$

$= \int e^{\log_e a^x}dx + \int e^{\log_e x^a}dx + \int e^{\log_e a^a}dx$

$= \int a^x\, dx + \int x^a dx + \int a^a dx$

$= \dfrac{a^x}{\log_e a} + \dfrac{x^{a+1}}{a+1} + a^a x + C$

$\left[\because \int a^x dx = \dfrac{a^x}{\log a} + C, \int x^n dx = \dfrac{x^{n+1}}{n+1} + C \text{ and } \int k dx = kx + C\right]$

(ii) Let $I = \int \dfrac{1-\tan^2 x}{1+\tan^2 x}dx$

$= \int \dfrac{1-\tan^2 x}{\sec^2 x}dx$ $\quad [\because \sec^2 x = 1 + \tan^2 x]$

$= \int (\cos^2 x - \sin^2 x)\, dx$

$= \int \cos 2x\, dx$ $\quad [\because \cos 2x = \cos^2 x - \sin^2 x]$

$= \dfrac{\sin 2x}{2} + C$

(iii) Let $I = \int \dfrac{1+\cos^2 x}{1+\cos 2x}dx$

$= \int \dfrac{1+\cos^2 x}{2\cos^2 x}dx$ $\quad [\because 1 + \cos 2x = 2\cos^2 x]$

$= \dfrac{1}{2}\int (\sec^2 x + 1)dx$

$= \dfrac{1}{2}[\int \sec^2 x\, dx + \int dx]$ $\quad [\because \int \sec^2 x\, dx = \tan x + C]$

$= \dfrac{1}{2}(\tan x + x) + C$

(iv) Let $I = \int 2^x \cdot e^x dx = \int (2e)^x dx$

$= \dfrac{(2e)^x}{\log(2e)} + C$ $\quad \left[\because \int a^x\, dx = \dfrac{a^x}{\log_e a}\right]$

$= \dfrac{2^x \cdot e^x}{\log 2 + \log e} + C = \dfrac{2^x e^x}{1 + \log 2} + C$

(v) Let $I = \int \dfrac{1+\tan^2 x}{1+\cot^2 x}dx$

$= \int \dfrac{\sec^2 x}{\operatorname{cosec}^2 x}dx$ $\quad \left[\because 1+\tan^2 x = \sec^2 x, \; 1+\cot^2 x = \operatorname{cosec}^2 x\right]$

$= \int \tan^2 x\, dx = \int (\sec^2 x - 1)\, dx$

$= (\tan x - x) + C$

(vi) Let $I = \int \dfrac{e^{5\log x} - e^{4\log x}}{e^{3\log x} - e^{2\log x}}dx$

$= \int \dfrac{x^5 - x^4}{x^3 - x^2}dx$ $\quad [\because e^{\log x} = x]$

$= \int \dfrac{x^4(x-1)}{x^2(x-1)}dx = \int x^2 dx = \dfrac{x^3}{3} + C$

(vii) Let $I = \int \dfrac{1+2x^2}{x^2(1+x^2)}dx = \int \dfrac{(1+x^2)+x^2}{x^2(1+x^2)}dx$

$= \int \left[\dfrac{1}{x^2} + \dfrac{1}{1+x^2}\right]dx = \int \dfrac{1}{x^2}dx + \int \dfrac{1}{1+x^2}dx$

$= -\dfrac{1}{x} + \tan^{-1} x + C$ $\quad \left[\because \int \dfrac{1}{x^2+a^2}dx = \dfrac{1}{a}\tan^{-1}\dfrac{x}{a} + C\right]$

(viii) Let $I = \int \dfrac{\cos 2x}{\cos^2 x \sin^2 x}dx$

$= \int \dfrac{\cos^2 x - \sin^2 x}{\cos^2 x \sin^2 x}$ $\quad [\because \cos 2x = \cos^2 x - \sin^2 x]$

$= \int \left(\dfrac{1}{\sin^2 x} - \dfrac{1}{\cos^2 x}\right)dx$

$= \int (\operatorname{cosec}^2 x - \sec^2 x)dx$ $\quad \left[\because \int \operatorname{cosec}^2 x\, dx = -\cot x, \; \int \sec^2 x dx = \tan x\right]$

$= -\cot x - \tan x + C$

**Alternate Method**

$I = \int \dfrac{\cos 2x}{\cos^2 x \sin^2 x}dx = 4\int \dfrac{\cos 2x}{(2\cos x \sin x)^2}dx$

$= 4\int \dfrac{\cos 2x}{(\sin 2x)^2}dx = 4\int \cot 2x \operatorname{cosec} 2x dx$

$= -2\operatorname{cosec} 2x - C$

(ix) Let $I = \int 4\cos\dfrac{x}{2}\cdot\cos x\cdot\cos\dfrac{21x}{2}dx$

$= \int 2\left(2\cos\dfrac{x}{2}\cos\dfrac{21x}{2}\right)\cos x\, dx$

$= 2\int (\cos 11x + \cos 10x)\cos x\, dx$

$= \int (2\cos 11x \cos x + 2\cos 10x \cos x)\, dx$

$= \int (\cos 12x + \cos 10x + \cos 11x + \cos 9x)dx$

$= \dfrac{1}{12}\sin 12x + \dfrac{1}{10}\sin 10x + \dfrac{1}{11}\sin 11x + \dfrac{1}{9}\sin 9x + C$

(x) Let $I = \int \dfrac{\cos x - \sin x}{\cos x + \sin x}(2 + 2\sin 2x)dx$

$= \int \dfrac{(\cos x - \sin x)\, 2\, (1 + \sin 2x)}{\cos x + \sin x}dx$

$= 2\int \dfrac{(\cos x - \sin x)(\cos x + \sin x)^2}{(\cos x + \sin x)}dx$

$[\because 1 + \sin 2x = (\cos x + \sin x)^2]$

$= 2\int (\cos x - \sin x)(\cos x + \sin x)dx$

$= 2\int (\cos^2 x - \sin^2 x)dx = 2\int \cos 2x\, dx$

$= \dfrac{2}{2}\sin 2x + C = \sin 2x + C$

**2.** (b) Let $I = \int \dfrac{\cos^8 x - \sin^8 x}{1 - 2\sin^2 x \cos^2 x}dx$

$= \int \dfrac{(\cos^4 x)^2 - (\sin^4 x)^2}{(\cos^2 x + \sin^2 x)^2 - 2\sin^2 x \cos^2 x}dx$

$= \int \dfrac{(\cos^4 x + \sin^4 x)(\cos^4 x - \sin^4 x)}{(\cos^4 x + \sin^4 x)}dx$

$= \int (\cos^4 x - \sin^4 x)\, dx$

$= \int (\cos^2 x + \sin^2 x)(\cos^2 x - \sin^2 x)dx$

$= \int (\cos^2 x - \sin^2 x)dx = \int \cos 2x\, dx$

$= \dfrac{1}{2}\sin 2x + C$

**3.** (b) Since, $(1-x)^{-1} = 1 + x + x^2 + x^3 + x^4 + \cdots$

$\Rightarrow \dfrac{d}{dx}(1-x)^{-1} = 1 + 2x + 3x^2 + 4x^3 + \cdots$

Since, derivative of $(1-x)^{-1} = 1 + 2x + 3x^2 + 4x^3 + \cdots$

$\therefore$ Anti-derivative of $1 + 2x + 3x^2 + 4x^3 + \cdots$ is $(1-x)^{-1}$.

i.e. $\int (1 + 2x + 3x^2 + 4x^3 + \cdots \infty)\, dx = (1-x)^{-1} + C$

**4.** (b) Let $I = \int (\sin x \cdot \cos x \cdot \cos 2x \cdot \cos 4x \cdot \cos 8x \cdot \cos 16x)\, dx$

$$= \int \sin x \cdot \left(\frac{\sin 2^5 x}{2^5 \sin x}\right) dx$$

$$\left[\because \cos\theta \cdot \cos 2\theta \cdot \cos 2^2\theta \cdot \cos 2^4\theta \cdots \cos 2^{n-1}\theta = \frac{\sin 2^n\theta}{2^n \sin\theta}\right]$$

$$= \frac{1}{32}\int \sin 32x\, dx = \frac{-\cos 32x}{32 \times 32} + C$$

$$= -\frac{\cos 32x}{1024} + C$$

**5.** (b) Let $I = \int \frac{\sin x + 4\sin 3x + 6\sin 5x + 3\sin 7x}{\sin 2x + 3\sin 4x + 3\sin 6x} dx$

$$= \int \frac{(\sin x + \sin 3x) + 3(\sin 3x + \sin 5x) + 3(\sin 5x + \sin 7x)}{\sin 2x + 3\sin 4x + 3\sin 6x} dx$$

$$= \int \frac{2\sin 2x \cos x + 3 \times 2\sin 4x \cos x + 3 \times 2\sin 6x \cos x}{\sin 2x + 3\sin 4x + 3\sin 6x} dx$$

$$\left[\because \sin C + \sin D = 2\sin\frac{C+D}{2}\cos\frac{C-D}{2}\right]$$

$$= \int \frac{2\cos x[\sin 2x + 3\sin 4x + 3\sin 6x]}{\sin 2x + 3\sin 4x + 3\sin 6x} dx$$

$$= 2\int \cos x\, dx = 2\sin x + C$$

**6.** (b) Let $I = \int (3\sin x \cos^2 x - \sin^3 x) dx$

$$= \int [3\sin x(1 - \sin^2 x) - \sin^3 x]\, dx$$

$$= \int (3\sin x - 3\sin^3 x - \sin^3 x) dx$$

$$= \int (3\sin x - 4\sin^3 x)\, dx = \int \sin 3x\, dx$$

$$= \frac{-\cos 3x}{3} + C$$

**7.** (a) Let $I = \int \frac{(1+x)^2}{x(1+x^2)} dx = \int \frac{1 + x^2 + 2x}{x(1+x^2)} dx$

$$= \int \left[\frac{1+x^2}{x(1+x^2)} + \frac{2x}{(1+x^2)x}\right] dx = \int \left[\frac{1}{x} + \frac{2}{1+x^2}\right] dx$$

$$= \log x + 2\tan^{-1} x + C$$

**8.** (c) Let $I = \int \frac{\cos 2x - \cos 2\alpha}{\cos x - \cos\alpha} dx$

$$= \int \frac{2\cos^2 x - 1 - 2\cos^2\alpha + 1}{\cos x - \cos\alpha} dx$$

$$= \int \frac{2(\cos^2 x - \cos^2\alpha)}{\cos x - \cos\alpha} dx$$

$$= 2\int (\cos x + \cos\alpha) dx$$

$$= 2(\sin x + x\cos\alpha) + C$$

**9.** (d) Let $I = \int \frac{x^4 + x^2 + 1}{2(1+x^2)} dx$

$$= \int \frac{x^2(1+x^2) + 1}{2(1+x^2)} dx = \int \left[\frac{x^2}{2} + \frac{1}{2(1+x^2)}\right] dx$$

$$= \frac{1}{2}\int \left[x^2 + \frac{1}{1+x^2}\right] dx = \frac{1}{2}\left[\frac{x^3}{3} + \tan^{-1} x\right] + C$$

**10.** (a) Let $I = \int \frac{x^6 - 1}{x^2 + 1} dx$

$$= \int \left[(x^4 - x^2 + 1) - \frac{2}{x^2 + 1}\right] dx \quad \text{[by long division]}$$

$$= \frac{x^5}{5} - \frac{x^3}{3} + x - 2\tan^{-1} x + C$$

**11.** (b) Let $I = \int \frac{\sec 2x - 1}{\sec 2x + 1} dx$

$$= \int \frac{1 - \cos 2x}{1 + \cos 2x} dx = \int \frac{1 - 1 + 2\sin^2 x}{1 + 2\cos^2 x - 1} dx$$

$$= \int \frac{\sin^2 x}{\cos^2 x} dx = \int \tan^2 x\, dx$$

$$= \int (\sec^2 x - 1)\, dx = \tan x - x + C$$

**12.** (c) Let $I = \int \left[\sin^2\left(\frac{9\pi}{8} + \frac{x}{4}\right) - \sin^2\left(\frac{7\pi}{8} + \frac{x}{4}\right)\right] dx$

$$= \int \sin\left(\frac{9\pi}{8} + \frac{x}{4} + \frac{7\pi}{8} + \frac{x}{4}\right) \cdot \sin\left(\frac{8\pi}{8} + \frac{x}{4} - \frac{7\pi}{8} - \frac{x}{4}\right) dx$$

$$[\because (\sin^2 A - \sin^2 B) = \sin(A+B)\sin(A-B)]$$

$$= \int \sin\left(2\pi + \frac{x}{2}\right) \cdot \sin\frac{\pi}{4} dx$$

$$= \frac{1}{\sqrt{2}}\int \sin\frac{x}{2} dx \quad [\because \sin(2\pi + \theta) = \sin\theta]$$

$$= \frac{1}{\sqrt{2}} \times 2\left(-\cos\frac{x}{2}\right) + C$$

$$= -\sqrt{2}\cos\frac{x}{2} + C$$

**13.** (d) Let $I = \int \frac{\cos 4x + 1}{\cot x - \tan x} dx$

$$= \int \frac{2\cos^2 2x - 1 + 1}{\frac{\cos x}{\sin x} - \frac{\sin x}{\cos x}} dx \quad [\because \cos 2A = 2\cos^2 A - 1]$$

$$= \int \frac{2\cos^2 2x(\sin x \cdot \cos x)}{\cos^2 x - \sin^2 x} dx = \int \frac{\cos^2 2x \cdot \sin 2x}{\cos 2x} dx$$

$$= \int \cos 2x \sin 2x\, dx = \frac{1}{2}\int \sin 4x\, dx$$

$$= \frac{1}{2}\left(-\frac{\cos 4x}{4}\right) = -\frac{1}{8}\cos 4x + C$$

**14.** (a) Let $I = \int \frac{\sin 2x + \sin 5x - \sin 3x}{\cos x + 1 - 2\sin^2 2x} dx$

$$= \int \frac{\sin 2x + 2\cos 4x \cdot \sin x}{\cos x + \cos 4x} dx$$

$$\left[\begin{array}{l}\because (\sin C - \sin D) = 2\cos\frac{C+D}{2}\sin\frac{C-D}{2} \\ \text{and } 1 - 2\sin^2 A = \cos 2A\end{array}\right]$$

$$= \int \frac{2\sin x \cos x + 2\cos 4x \cdot \sin x}{\cos x + \cos 4x}$$

$$= \int \frac{2\sin x[\cos x + \cos 4x]}{(\cos x + \cos 4x)} dx$$

$$= 2\int \sin x\, dx = -2\cos x + C$$

**15.** (c) Let $I = \int \left(\frac{\cot^2 2x - 1}{2\cot 2x} - \cos 8x \cdot \cot 4x\right) dx$

$$= \int (\cot 4x - \cos 8x \cdot \cot 4x)\, dx \quad \left[\because \cot 2A = \frac{2\cot^2 A - 1}{2\cot A}\right]$$

$$= \int \cot 4x\,(1 - \cos 8x) dx$$

$$= \int \cot 4x\, 2\sin^2 4x\, dx \quad [\because (1 - \cos 2A) = 2\sin^2 A]$$

$$= \int 2\sin 4x \cos 4x\, dx$$

$$= \int \sin 8x\, dx \quad [\because 2\sin A\cos A = \sin 2A]$$

$$= -\frac{1}{8}\cos 8x + C$$

**16.** (d) Let $I = \int \frac{\cos^4 x - \sin^4 x}{\sqrt{1+\cos 4x}} dx$ [$\cos 2x > 0$]

$$= \int \frac{(\cos^2 x)^2 - (\sin^2 x)^2}{\sqrt{1+\cos 4x}} dx$$

$$= \int \frac{(\cos^2 x + \sin^2 x)(\cos^2 x - \sin^2 x)}{\sqrt{2\cos^2 2x}} dx$$

[$\because 1 + \cos 2A = 2\cos^2 A$]

$$= \int \frac{\cos^2 x - \sin^2 x}{\sqrt{2}\cos 2x} dx = \frac{1}{\sqrt{2}} \int \frac{\cos 2x}{\cos 2x} dx$$

$$= \frac{1}{\sqrt{2}} \int 1 dx = \frac{1}{\sqrt{2}} x + C$$

**17.** (a) Let $I = \int \left( \frac{2x^3 + 3x^2 + 4x + 5}{2x+1} \right) dx$

$$= \int \left[ x^2 + x + \frac{3}{2} + \frac{7}{2(2x+1)} \right] dx$$

$$= \frac{x^3}{3} + \frac{x^2}{2} + \frac{3x}{2} + \frac{7}{2} \times \frac{1}{2} \log(2x+1) + C$$

$$= \frac{x^3}{3} + \frac{x^2}{2} + \frac{3x}{2} + \frac{7}{4} \log(2x+1) + C$$

**18.** (a) Let $I = \int \frac{(x^2 + \sin^2 x)\sec^2 x}{1+x^2} dx$

$$= \int \frac{x^2 \sec^2 x + \sin^2 x \sec^2 x}{(1+x^2)} dx$$

$$= \int \frac{x^2 \sec^2 x + \tan^2 x}{(1+x^2)} dx = \int \frac{x^2(1+\tan^2 x) + \tan^2 x}{(1+x^2)} dx$$

$$= \int \frac{x^2 + (x^2+1)\tan^2 x}{(1+x^2)} dx$$

$$= \left( \frac{x^2}{1+x^2} + \tan^2 x \right) dx$$

$$= \int \left[ 1 - \frac{1}{1+x^2} + \sec^2 x - 1 \right] dx = \int \left( \sec^2 x - \frac{1}{1+x^2} \right) dx$$

$$= \tan x - \tan^{-1} x + C$$

**19.** (d) Let $I = \int \frac{dx}{25 + 4x^2} = \frac{1}{4} \int \frac{dx}{\left(\frac{5}{2}\right)^2 + x^2}$

$$= \frac{1}{4} \times \frac{1}{\frac{5}{2}} \tan^{-1} \left( \frac{x}{\frac{5}{2}} \right) + C \quad \left[ \because \int \frac{dx}{a^2 + x^2} = \frac{1}{a} \tan^{-1} \frac{x}{a} + C \right]$$

$$= \frac{1}{10} \tan^{-1} \left( \frac{2x}{5} \right) + C$$

**20.** (b) Let $I = \int \frac{dx}{\sqrt{9 - 16x^2}}$

$$= \frac{1}{4} \int \frac{dx}{\sqrt{\frac{9}{16} - x^2}} = \frac{1}{4} \int \frac{dx}{\sqrt{\left(\frac{3}{4}\right)^2 - x^2}}$$

$$= \frac{1}{4} \sin^{-1} \left( \frac{x}{\frac{3}{4}} \right) + C \quad \left[ \because \int \frac{dx}{\sqrt{a^2 - x^2}} = \sin^{-1} \frac{x}{a} + C \right]$$

$$= \frac{1}{4} \sin^{-1} \left( \frac{4x}{3} \right) + C$$

**21.** (a) Let $I = \int \frac{2x+3}{3x+2} dx$

$$= \int \left[ \frac{2}{3} + \frac{5}{3(3x+2)} \right] dx \quad \text{[by long division]}$$

$$= \frac{1}{3} \left[ 2x + \frac{5}{3} \log(3x+2) \right] + C$$

$$\left[ \because \int \frac{1}{a+b} dx = \frac{-1}{a} \log(ax+b) + C \right]$$

$$= \frac{2}{3} x + \frac{5}{9} \log(3x+2) + C$$

**22.** (b) Let $I = \int \frac{dx}{1 + \sin x}$

$$= \int \frac{1 - \sin x}{1 - \sin^2 x} dx = \int \frac{1 - \sin x}{\cos^2 x} dx$$

$$= \int (\sec^2 x - \tan x \cdot \sec x) dx$$

$$= \tan x - \sec x + C$$

**23.** (a) Let $I = \int \frac{\sin 2x - \sin 2k}{\sin x - \sin k + \cos x - \cos k} dx$

$$= \int \frac{\sin 2x - \sin 2k}{(\sin x + \cos x) - (\sin k + \cos k)} dx$$

$$= \int \frac{(\sin 2x - \sin 2k)[(\sin x + \cos x) + (\sin k + \cos k)]}{(\sin x + \cos)^2 - (\sin k + \cos k)^2} dx$$

$$= \int \frac{(\sin 2x - \sin 2k)[(\sin x + \cos x) + (\sin k + \cos k)]}{1 + \sin 2x - 1 - \sin 2k} dx$$

$$= \int \frac{(\sin 2x - \sin 2k)[(\sin x + \cos x) + (\sin k + \cos k)]}{(\sin 2x - \sin 2k)} dx$$

$$= \int [\sin x + \cos x + (\sin k + \cos k)] dx$$

$$= -\cos x + \sin x + (\sin k + \cos k)x + C$$

**24.** (b) Let $I = \int \frac{2 + 3x^2}{x^2(1+x^2)} dx$

$$= \int \frac{(1+2x^2) + (1+x^2)}{x^2(1+x^2)} dx = \int \left[ \frac{1+2x^2}{x^2(1+x^2)} + \frac{1}{x^2} \right] dx$$

$$= \int \left[ \frac{(1+x^2) + x^2}{x^2(1+x^2)} + \frac{1}{x^2} \right] dx = \int \left[ \frac{1}{x^2} + \frac{1}{1+x^2} + \frac{1}{x^2} \right] dx$$

$$= -\frac{1}{x} + \tan^{-1} x - \frac{1}{x} + C = -\frac{2}{x} + \tan^{-1} x + C$$

**25.** (a) Let $I = \int \frac{\cos 8x - \cos 7x}{1 + 2\cos 5x} dx = \int \frac{-2\sin\frac{15x}{2}\sin\frac{x}{2}}{1 + 2\cos 5x} dx$

On multiplying and dividing by $\sin\frac{5x}{2}$, we have

$$= -\int \frac{2\sin\frac{15x}{2}\sin\frac{x}{2}\sin\frac{5x}{2}}{\sin\frac{5x}{2} + 2\sin\frac{5x}{2}\cos 5x}$$

$$= -\int \frac{2\sin\frac{15x}{2}\sin\frac{x}{2}\sin\frac{5x}{2}}{\sin\frac{5x}{2} + \sin\frac{15}{2} - \sin\frac{5x}{2}}$$

$$= -\int 2\sin\frac{x}{2} \cdot \sin\frac{5x}{2} dx = -\int (\cos 2x - \cos 3x) dx$$

$$= \int (\cos 3x - \cos 2x) dx = \frac{1}{3}\sin 3x - \frac{1}{2}\sin 2x + C$$

**26.** (c) We have, $\tan 3x = \tan(2x + x)$

$\tan 3x = \tan(2x + x)$

$$\Rightarrow \tan 3x = \frac{\tan 2x + \tan x}{1 - \tan 2x \tan x} \quad \left[ \because \tan(A+B) = \frac{\tan A + \tan B}{1 - \tan A \tan B} \right]$$

$\Rightarrow \quad \tan 3x - \tan 3x \tan 2x \tan x = \tan 2x + \tan x$

$\Rightarrow \quad \tan x \tan 2x \tan 3x = \tan 3x - \tan 2x - \tan x$

$\therefore \quad \int \tan x \tan 2x \tan 3x dx = \int (\tan 3x - \tan 2x - \tan x) dx$

$$= -\frac{1}{3}\log|\cos 3x| + \frac{1}{2}\log|\cos 2x| + \log|\cos x| + C$$

$$= \log|\cos x| + \frac{1}{2}\log|\cos 2x| - \frac{1}{3}\log|\cos 3x| + C$$

# DPP-2 Problems on Basic Substitution

**1.** (i) Let $I = \int \frac{dx}{(2x-3)^2}$

Put $2x - 3 = t$

$\Rightarrow \quad 2dx = dt \quad \Rightarrow \quad dx = \frac{dt}{2}$

$\therefore \quad I = \frac{1}{2}\int \frac{dt}{t^2} = -\frac{1}{2t} + C = -\frac{1}{2(2x-3)} + C$

(ii) Let $I = \int \frac{x^4}{\sqrt{4+x^5}}dx$

Put $(4 + x^5) = t$

$\Rightarrow \quad 5x^4 dx = dt$

$\Rightarrow \quad x^4 dx = \frac{dt}{5}$

$\therefore \quad I = \frac{1}{5}\int \frac{dt}{\sqrt{t}} = \frac{1}{5} \times 2t^{1/2} + C$

$= \frac{2}{5}\sqrt{4+x^5} + C$

(iii) Let $I = \int \frac{dx}{(\sin^{-1} x)^3 \sqrt{1-x^2}}$

Put $\quad \sin^{-1} x = t$

$\Rightarrow \quad \frac{1}{\sqrt{1-x^2}}dx = dt$

$\therefore \quad I = \int \frac{dt}{t^3} = -\frac{1}{2t^2} + C = -\frac{1}{2}\frac{1}{(\sin^{-1} x)^2} + C$

(iv) Let $I = \int \frac{(\tan^{-1} x)^2}{1+x^2}dx$

Put $\tan^{-1} x = t$

$\Rightarrow \frac{1}{1+x^2}dx = dt$

$\therefore \quad I = \int t^2 dt = \frac{t^3}{3} + C = \frac{(\tan^{-1} x)^3}{3} + C$

(v) Let $I = \int \frac{dx}{\cos^2 x\sqrt{1+\tan x}}$

$I = \int \frac{\sec^2 x}{\sqrt{1+\tan x}}dx$

Put $1 + \tan x = t \Rightarrow \sec^2 x\, dx = dt$

$\therefore \quad I = \frac{dt}{\sqrt{t}} = 2\sqrt{t} + C = 2\sqrt{1+\tan x} + C$

(vi) Let $I = \int \frac{(2x-3)}{x^2-3x+8}dx$

Put $x^2 - 3x + 8 + t$

$\Rightarrow \quad (2x-3)\, dx = dt$

$\therefore \quad I = \int \frac{dt}{t} = \log|t| + C$

$= \log|x^2 - 3x + 8| + C$

(vii) Let $I = \int e^{-3x+1}dx$

Put $-3x + 1 = t$

$\Rightarrow \quad -3dx = dt$

$\Rightarrow \quad dx = \frac{dt}{-3}$

$\therefore \quad I = \int \frac{e^t}{-3}dt$

$= -\frac{1}{3}e^t + C$

$= -\frac{1}{3}e^{-3x+1} + C$

(viii) Let $I = \int e^{-x^3} \cdot x^2 dx$

Put $x^3 = t \Rightarrow \quad 3x^2 dx = dt$

$\Rightarrow \quad x^2\, dx = \frac{dt}{3}$

$\therefore \quad I = \frac{1}{3}\int e^{-t}dt = \frac{-1}{3}e^{-t} + C$

$= -\frac{1}{3}e^{-x^3} + C$

(ix) Let $I = \int \frac{dx}{\sqrt{1-25x^2}} = \frac{1}{5}\int \frac{dx}{\sqrt{\frac{1}{25} - x^2}}$

$= \frac{1}{5}\int \frac{dx}{\sqrt{\left(\frac{1}{5}\right)^2 - x^2}}$

$= \frac{1}{5}\sin^{-1}\left(\frac{x}{1/5}\right) + C$

$= \frac{1}{5}\sin^{-1}(5x) + C$

(x) Let $I = \int \frac{x}{\sqrt{a^2-x^4}}dx = \int \frac{x}{\sqrt{a^2-(x^2)^2}}dx$

Put $x^2 = t \quad \Rightarrow \quad 2x\, dx = dt$

$\Rightarrow \quad x\, dx = \frac{dt}{2}$

$\therefore \quad I = \frac{1}{2}\int \frac{dt}{\sqrt{a^2-t^2}} = \frac{1}{2}\sin^{-1}\left(\frac{t}{a}\right) + C$

$= \frac{1}{2}\sin^{-1}\left(\frac{x^2}{a}\right) + C$

(xi) Let $I = \int \frac{x^2}{x^6+4}dx \quad \Rightarrow \quad I = \int \frac{x^2}{(x^3)^2 + (2)^2}dx$

Put $x^3 = t$, then $3x^2\, dx = dt$

$\Rightarrow \quad x^2\, dx = \frac{dt}{3}$

$\therefore \quad I = \frac{1}{3}\int \frac{dt}{t^2 + (2)^2}$

$= \frac{1}{3} \times \frac{1}{2}\tan^{-1}\left(\frac{t}{2}\right) + C = \frac{1}{6}\tan^{-1}\left(\frac{x^2}{2}\right) + C$

(xii) Let $I = \int \frac{2x - \sqrt{\sin^{-1} x}}{\sqrt{1-x^2}}dx$

$= 2\int \frac{x}{\sqrt{1-x^2}}dx - \int \frac{\sqrt{\sin^{-1} x}}{\sqrt{1-x^2}}dx$

Put $1 - x^2 = t$ in the first integral

$\Rightarrow \quad -2x dx = dt$

$\Rightarrow \quad x dx = -\frac{1}{2}dt$

and $\quad \sin^{-1} x = u$ in the second integral

$\Rightarrow \quad \frac{1}{\sqrt{1-x^2}}dx = du$

$\therefore \quad I = 2\int \frac{dt}{-2\sqrt{t}} - \int \sqrt{u}\, du$

$= -\int \frac{dt}{\sqrt{t}} - \int \sqrt{u}\, du$

$= -2\sqrt{t} - \frac{2u^{3/2}}{3} + C$

$= -2\sqrt{1-x^2} - \frac{2}{3}(\sin^{-1} x)^{3/2} + C$

(xiii) Let $I = \int \frac{\log(x+1) - \log x}{x(x+1)} dx$

$$= \int \frac{\log\left(\frac{x+1}{x}\right)}{x(x+1)} dx = \int \frac{\log\left(1+\frac{1}{x}\right)}{x(x+1)} dx$$

Put $\log\left(1+\frac{1}{x}\right) = t$

$\Rightarrow \quad \frac{1}{1+\frac{1}{x}} \cdot \left(-\frac{1}{x^2}\right) dx = dt$

$\Rightarrow \quad -\frac{dx}{x(x+1)} = dt$

$\therefore \quad I = \int -t\, dt$

$$= \frac{-\left[\log\left(1+\frac{1}{x}\right)\right]^2}{2} + C$$

(xiv) Let $I = \int \frac{e^x}{e^x + 1} dx$

Put $e^x + 1 = t$

$\Rightarrow \quad e^x dx = dt$

$\therefore \quad I = \int \frac{dt}{t}$

$= \log|t| + C$

$= \log|e^x + 1| + C$

(xv) Let $I = \int \frac{dx}{(x + \sqrt{x^2 - 1})^2}$

Put $x = \sec\theta \Rightarrow dx = \sec\theta \tan\theta d\theta$

$\therefore \quad I = \int \frac{\sec\theta \tan\theta}{(\sec\theta + \tan\theta)^2} d\theta$

$$= \int \frac{\sec\theta\tan\theta \cdot (\sec\theta - \tan\theta)^2}{(\sec^2\theta - \tan^2\theta)^2} d\theta$$

$$= \int \sec\theta\tan\theta\, (\sec^2\theta + \tan^2\theta - 2\sec\theta\tan\theta)\, d\theta$$

$$= \int \sec\theta\tan\theta\, (2\sec^2\theta - 1 - 2\sec\theta\tan\theta)\, d\theta$$

$$= \int (2\sec^3\theta\tan\theta - \sec\theta\tan\theta - 2\sec^2\theta\tan^2\theta) d\theta$$

$$= \frac{2\sec^3\theta}{3} - \sec\theta - \frac{2\tan^3\theta}{3} + C$$

$$= \frac{2}{3}(\sec^3\theta - \tan^3\theta) - \sec\theta + C$$

$$= \frac{2}{3}[x^3 - \sqrt{(x^3-1)^3}] - x + C$$

(xvi) Let $I = \int \frac{\cos x}{(\sin^2 x)^{1/3}} dx$

Put $\sin x = t$

$\Rightarrow \quad \cos x dx = dt$

$\therefore \quad I = \int \frac{dt}{t^{2/3}} = 3t^{1/3} + C$

$= 3(\sin x)^{1/3} + C$

(xvii) Let $I = \int \frac{(6x-5)dx}{2\sqrt{3x^2 - 5x + 6}}$

Put $\sqrt{3x^2 - 5x + 6} = t$

$\frac{(6x-5)}{2\sqrt{3x^2-5x+6}} dx = dt$

$\therefore \quad I = \int dt$

$= t + C$

$= \sqrt{3x^2 - 5x + 6} + C$

(xviii) Let $I = \int \tan^3 x\, d(\tan x) = \frac{\tan^4 x}{4} + C$

**Alternate Method**

Let $I = \tan^3 x\, d\, (\tan x)$

$= \int \tan^3 x \cdot \sec^2 x\, dx \quad [\because d(\tan x) = \sec^2 x]$

Put $\tan x = t$

$\Rightarrow \quad \sec^2 x dx = dt$

$\therefore \quad I = \int t^3 dt = \frac{t^4}{4} + C = \frac{\tan^4 x}{4} + C$

(xix) Let $I = \int \frac{d(1+x^2)}{\sqrt{1+x^2}} = 2\sqrt{1+x^2} + C$

**Alternate Method**

Let $\quad I = \int \frac{d(1+x^2)}{\sqrt{1+x^2}}$

$= \int \frac{2x}{\sqrt{1+x^2}} dx$

Put $1 + x^2 = t$

$\Rightarrow \quad 2x dx = dt$

$\therefore \quad I = \int \frac{dt}{\sqrt{t}} = 2\sqrt{t} + C$

$= 2\sqrt{1+x^2} + C$

(xx) Let $I = \int \frac{d(1+\log_e x)}{\cos^2(1+\log_e x)}$

$= \int \sec^2(1+\log x) d(1+\log x)$

$= \tan(1+\log x) + C$

**Alternate Method**

Let $\; I = \int \frac{d(1+\log x)}{\cos^2(1+\log x)} = \int \frac{\frac{1}{x}}{\cos^2(1+\log x)} dx$

$= \int \frac{\sec^2(1+\log x)}{x} dx$

Put $1 + \log x = t$

$\Rightarrow \quad \frac{1}{x} dx = dt$

$\therefore \quad I = \int \sec^2 + dt$

$= \tan t + C = \tan(1+\log x) + C$

**2.** (a) Let $I = \int \left[\frac{x}{(x^4-1)^4}\right]^{1/3} dx$

$$= \int \left[\frac{x}{x^{16}\left(1-\frac{1}{x^4}\right)^4}\right]^{1/3} dx = \int \left[\frac{1}{x^{15}\left(1-\frac{1}{x^4}\right)^4}\right]^{1/3} dx$$

$$= \int \frac{dx}{x^5\left(1-\frac{1}{x^4}\right)^{4/3}}$$

Put $1 - \frac{1}{x^4} = t$

$\Rightarrow \quad \frac{4}{x^5} dx = dt$

$\therefore \quad I = \frac{1}{4}\int \frac{dt}{t^{4/3}} = \frac{3}{4t^{1/3}} + C$

$$= \frac{3}{4\left(1-\frac{1}{x^4}\right)^{1/3}} + C = \frac{3}{4}\left(1 + \frac{1}{x^4-1}\right)^{\frac{1}{3}} + C$$

**3.** (b) Let $I = \int \frac{x}{\sqrt{1+x^2+\sqrt{(1+x^2)^3}}}dx$

$$= \int \frac{x}{\sqrt{(1+x^2)+(1+x^2)^{3/2}}}dx$$

$$= \int \frac{x}{\sqrt{(1+x^2)[1+(1+x^2)^{1/2}]}}dx$$

Put $1+x^2 = t \Rightarrow 2x\,dx = dt$

$\Rightarrow \quad x\,dx = \frac{dt}{2}$

$\therefore \quad I = \frac{1}{2}\int \frac{dt}{\sqrt{t(1+t^{1/2})}}$

Again, put $1+t^{1/2} = u$

$$= \frac{1}{2\sqrt{t}}dt = du$$

Now, $\quad I = \int \frac{du}{\sqrt{u}} = 2\sqrt{u} + C = 2\sqrt{1+(1+x^2)^{1/2}} + C$

**4.** (a) Let $I = \int \frac{\log|x|}{x\sqrt{1+\log|x|}}dx$

Put $1+\log|x| = t$

$\Rightarrow \quad \frac{1}{x}dx = dt$

$\therefore \quad I = \frac{1}{2}+1\int \frac{t-1}{\sqrt{t}}dt = \int\left(\sqrt{t}-\frac{1}{\sqrt{t}}\right)dt$

$$= \frac{t^{\frac{1}{2}+1}}{3/2} - 2\sqrt{t} + C = \frac{2}{3}t^{3/2} - 2\sqrt{t} + C$$

$$= 2\sqrt{t}\left[\frac{1}{3}t - 2\right] + C$$

$$= 2\sqrt{1+\log|x|}\left[\frac{1}{3}(1+\log|x|) - 2\right] + C$$

$$= 2\sqrt{1+\log|x|}\left[\frac{1}{3}\log|x| - \frac{5}{3}\right] + C$$

**5.** (c) Let $I = \int \frac{(x^4+1)}{x(x^2+1)^2}dx$

$$= \int\left\{\frac{x^4+2x^2+1}{x(x^2+1)^2} - \frac{2x^2}{x(x^2+1)^2}\right\}dx$$

$$= \int\left\{\frac{1}{x} - \frac{2x}{(x^2+1)^2}\right\}dx$$

$$= \log|x| - \int \frac{2x}{(x^2+1)^2}dx$$

Put $x^2+1 = t \Rightarrow 2dx = dt$

$\therefore \quad I = \log|x| - \int \frac{dt}{t^2}$

$$= \log|x| + \frac{1}{t} + C$$

$$= \log|x| + \frac{1}{x^2+1} + C$$

Hence, $\quad A = 1$ and $B = 1$

**6.** (a) We have, $\quad I = \int\left(\frac{x}{1+x^5}\right)^{3/2}dx$

$$= \int\left[\frac{x}{x^5\left(1+\frac{1}{x^5}\right)}\right]^{3/2}dx = \int\left[\frac{1}{x^4\left(1+\frac{1}{x^5}\right)}\right]^{3/2}dx$$

$$= \int \frac{dx}{x^6\left(1+\frac{1}{x^5}\right)^{3/2}}$$

Put $1+\frac{1}{x^5} = t$

$\Rightarrow \quad -\frac{5}{x^6}dx = dt$

$\therefore \quad I = -\frac{1}{5}\int \frac{dt}{t^{3/2}}$

$$= -\frac{1}{5}\left(\frac{-2}{\sqrt{t}}\right) + C = \frac{2}{5}\frac{1}{\sqrt{1+\frac{1}{x^5}}} + C$$

$$= \frac{2}{5}\sqrt{\frac{x^5}{x^5+1}} + C$$

**7.** (b) Let $I = \int x\cdot\frac{\log(x+\sqrt{1+x^2})}{\sqrt{1+x^2}}dx$

Put $\sqrt{1+x^2} = t \Rightarrow x = \sqrt{t^2-1}$

$\Rightarrow \quad \frac{2x}{2\sqrt{1+x^2}}dx = dt$

$\Rightarrow \quad \frac{x}{\sqrt{1+x^2}}dx = dt$

$\therefore \quad I = \int \log(\sqrt{t^2-1}+t)dt$

$$= \log(t+\sqrt{t^2-1})t - \int \frac{\left(1+\frac{t}{\sqrt{t^2-1}}\right)t}{t+\sqrt{t^2-1}}dt$$

$$= t\log(t+\sqrt{t^2-1}) - \frac{1}{2}\int \frac{2t}{\sqrt{t^2-1}}dt$$

$$= t\log\left(t+\sqrt{t^2-1}\right) - \sqrt{t^2-1} + C$$

$$= \sqrt{1+x^2}\log(x+\sqrt{1+x^2}) - x + C$$

**8.** (d) Let $I = \int \frac{(x-1)^2}{x^4+2x^2+1}dx$

$$= \int\left\{\frac{x^2+1}{(x^2+1)^2} - \frac{2x}{(x^2+1)^2}\right\}dx$$

$$= \int \frac{1}{x^2+1}dx - \int \frac{2x}{(x^2+1)^2}dx$$

$$= \tan^{-1}x - \int \frac{dt}{t^2}$$

Put $t = x^2+1 \Rightarrow dt = 2x\,dx$

$\therefore \quad I = \tan^{-1}x + \frac{1}{t} + C$

$$= \tan^{-1}x + \frac{1}{x^2+1} + C$$

**9.** (a) Let $I = \int \frac{x^2-4}{x^4+24x^2+16}dx$

$$= \int \frac{\left(1-\frac{4}{x^2}\right)}{x^2+24+\frac{16}{x^2}}dx = \int \frac{\left(1-\frac{4}{x^2}\right)}{x^2+\frac{16}{x^2}+8+16}dx$$

$$= \int \frac{\left(1-\frac{4}{x^2}\right)}{(x+4/x)^2+4^2}dx$$

Put $x+\frac{4}{x} = t \Rightarrow \left(1-\frac{4}{x^2}\right)dx = dt$

$\therefore \quad I = \int \frac{dt}{t^2+4^2} = \frac{1}{4}\tan^{-1}\left(\frac{t}{4}\right) + C$

$$= \frac{1}{4}\tan^{-1}\left(\frac{x^2+4}{4x}\right) + C$$

**10.** (a) Let $I = \int \frac{(x^4 - 4)}{x^2\sqrt{4 + x^2 + x^4}}dx$

$= \int \frac{(x^4 - 4)}{x^3\sqrt{\frac{4}{x^2} + 1 + x^2}}dx$

Put $\frac{4}{x^2} + 1 + x^2 = t$

$\Rightarrow \left(2x - \frac{8}{x^3}\right)dx = dt$

$\therefore \quad I = \int \frac{\left(x - \frac{4}{x^3}\right)dx}{\sqrt{x^2 + \frac{4}{x^2} + 1}}$

$= \frac{1}{2}\int \frac{dt}{\sqrt{t}} = \frac{1}{2}\cdot\frac{t^{1/2}}{1/2} + C$

$= \sqrt{\frac{4}{x^2} + 1 + x^2} + C = \frac{\sqrt{4 + x^2 + x^4}}{x} + C$

**11.** (d) Let $I = \int \frac{x^9}{(x^2 + 4)^6}dx = \int \frac{x^9}{x^{12}\left(1 + \frac{4}{x^2}\right)^6}dx$

Put $1 + \frac{4}{x^2} = t$

$\Rightarrow \quad \frac{-8}{x^3}dx = dt$

$\therefore \quad I = -\frac{1}{8}\int \frac{dt}{t^6} = -\frac{1}{8}\cdot\frac{1}{(-5)t^5} + C$

$= \frac{1}{40t^5} + C$

$= \frac{1}{40}(1 + 4x^{-2})^{-5} + C$

**12.** (c) Let $I = \sqrt{2}\int \frac{\sin x}{\sin\left(x - \frac{\pi}{4}\right)}dx$

Put $x - \frac{\pi}{4} = t$

$\Rightarrow \quad dx = dt$

$\therefore \quad I = \sqrt{2}\int \frac{\sin\left(\frac{\pi}{4} + t\right)dt}{\sin t}$

$= \sqrt{2}\int\left(\frac{1}{2}\cot t + \frac{1}{\sqrt{2}}\right)dt$

$= \log|\sin t| + t + C$

$= x + \log\left|\sin\left(x - \frac{\pi}{4}\right)\right| + C$

**13.** (b) Let $I = \int (x^{3m} + x^{2m} + x^m)(2x^{2m} + 3x^m + 6)^{1/m}dx$

$= \int\left(\frac{x^{3m} + x^{2m} + x^m}{x}\right)(x)(2x^{2m} + 3x^m + 6)^{1/m}dx$

$= \int (x^{3m-1} + x^{2m-1} + x^{m-1})(2x^{3m} + 3x^{2m} + 6x^m)^{1/m}dx$

Let $2x^{3m} + 3x^{2m} + 6x^m = t$

$\Rightarrow \quad dt = 6m(x^{3m-1} + x^{2m-1} + x^{m-1})dx$

$\therefore \quad I = \int t^{\frac{1}{m}}\frac{dt}{6m} = \frac{1}{6m}\frac{t^{1+\frac{1}{m}}}{\frac{1}{m} + 1} + C$

$= \frac{1}{6(m+1)}(2x^{3m} + 3x^{2m} + 6x^m)^{\frac{m+1}{m}} + C$

**14.** (d) Let $I = \int \frac{(x^2 - 1)}{x^3\sqrt{2x^4 - 2x^2 + 1}}dx = \frac{1}{4}\int \frac{\frac{4}{x^3} - \frac{4}{x^5}}{\sqrt{2 - \frac{2}{x^2} + \frac{1}{x^4}}}dx$

Put $2 - \frac{2}{x^2} + \frac{1}{x^4} = t$

$\Rightarrow \quad \left(\frac{4}{x^3} - \frac{4}{x^5}\right)dx = dt$

$\therefore \quad I = \frac{1}{4}\int \frac{dt}{\sqrt{t}} = \frac{2\sqrt{t}}{4} + C$

$= \frac{\sqrt{2 - \frac{2}{x^2} + \frac{1}{x^4}}}{2} + C$

$= \frac{\sqrt{2x^4 - 2x^2 + 1}}{2x^2} + C$

**15.** (c) Let $I = \frac{\sec^2 x}{(\sec x + \tan x)^{9/2}}dx$

Put $\quad \sec x + \tan x = t$

$\Rightarrow \quad (\sec x\tan x + \sec^2 x)dx = dt$

$\Rightarrow \quad \sec x\, dx = \frac{dt}{t}$

$\therefore \quad \sec x - \tan x = \frac{1}{t} \Rightarrow \sec x = \frac{1}{2}\left(t + \frac{1}{t}\right)$

$\therefore \quad I = \int \frac{\sec x\cdot\sec x}{(\sec x + \tan x)^{9/2}}dx$

$= \int \frac{\frac{1}{2}\left(t + \frac{1}{t}\right)\frac{dt}{t}}{t^{9/2}}$

$= \frac{1}{2}\int\left(\frac{1}{t^{9/2}} + \frac{1}{t^{13/2}}\right)dt$

$= -\frac{1}{2}\left[\frac{2}{7t^{7/2}} + \frac{2}{11t^{11/2}}\right] + C$

$= -\left[\frac{1}{7(\sec x + \tan x)^{7/2}} + \frac{1}{11(\sec x + \tan x)^{11/2}}\right] + C$

$= \frac{-1}{(\sec x + \tan x)^{11/2}}\left[\frac{1}{11} + \frac{1}{7}(\sec x + \tan x)^2\right] + C$

**16.** (a,b,d) Let $I = \int \sqrt{1 + 2\tan x(\sec x + \tan x)}\,dx$

$= \int \sqrt{1 + 2\tan x\sec x + 2\tan^2 x}\,dx$

$= \int \sqrt{[(1 + \tan^2 x) + \tan^2 x + 2\tan x\sec x]}\,dx$

$= \int \sqrt{[\sec^2 x + \tan^2 x + 2\tan x\sec x]}\,dx$

$= \int \sqrt{(\sec x + \tan x)^2}\,dx = \int (\sec x + \tan x)\,dx$

$= \log|\sec x + \tan x| + \log|\sec x| + C$

So, option (b) is correct.

$= \log\frac{1}{\sec x - \tan x} + \log|\sec x| + C$

$= -\log|\sec x - \tan x| + \log|\sec x| + C$

So, option (a) is correct.

Again, $\log|\sec x + \tan x| + \log|\sec x| + C$

$= \log|\sec x(\sec x + \tan x)| + C$

$= \log|\sec^2 x + \sec x\tan x| + C$

$= \log|1 + \tan^2 x + \sec x\tan x| + C$

$= \log|1 + \tan x(\tan x + \sec x)| + C$

So, option (d) is correct.

**17.** (b,c,d) Let $I = \int \frac{dx}{x^3\left(1-\frac{1}{2x^2}\right)}$

Put $1 - \frac{1}{2x^2} = t$

$\Rightarrow \quad \frac{1}{x^3}dx = dt$

$\therefore \quad I = \int \frac{dt}{t} = \log|t| + C$

$= \log\left|1 - \frac{1}{2x^2}\right| + C$

So, option (d) is correct.

$= \log\left|\frac{2x^2-1}{2x^2}\right| + C$

$= \log|2x^2 - 1| - \log|2x^2|$

So, option (b) is correct.

Again, $\log|2x^2 - 1| - \log|2x^2|$

$= \log|2x^2 - 1| - \log 2 - \log|x^2| + C$

So, option (c) is correct.

**18.** (b,d) Let $I = \int \frac{1}{x^2-1}\log\left|\frac{x-1}{x+1}\right| dx$

Put $\quad \log\left|\frac{x-1}{x+1}\right| = t$

$\Rightarrow \quad \frac{x+1}{x-1}\cdot\frac{2}{(x+1)^2}dx = dt$

$\Rightarrow \quad \left(\frac{2}{x^2-1}\right)dx = dt$

$\therefore \quad I = \frac{1}{2}\int t\cdot dt = \frac{1}{2}\frac{t^2}{2} + C$

$= \frac{1}{4}\log^2\left|\frac{x-1}{x+1}\right| + C$

So, option (b) is correct.

$= \frac{1}{4}\left\{-\log\left|\frac{x+1}{x-1}\right|\right\}^2 + C$

$= \frac{1}{4}\log^2\left|\frac{x+1}{x-1}\right| + C$

So, option (d) is correct.

**19.** (a,b,c,d) Let $I = \int \frac{dx}{\sqrt{x-x^2}} = \int \frac{dx}{\sqrt{-\left(x^2 - x + \frac{1}{4} - \frac{1}{4}\right)}}$

$= \int \frac{dx}{\sqrt{-\left[\left(x-\frac{1}{2}\right)^2 - \left(\frac{1}{2}\right)^2\right]}}$

$= \int \frac{dx}{\sqrt{\left(\frac{1}{2}\right)^2 - \left(x-\frac{1}{2}\right)^2}}$

$= \sin^{-1}\frac{\left(x-\frac{1}{2}\right)}{1/2} + C$

$= \sin^{-1}(2x-1) + C$ [option b]

$= \cos^{-1}(2\sqrt{x-x^2}) + C$ [option d]

Again, $\quad I = \int \frac{dx}{\sqrt{x-x^2}} = \int \frac{dx}{\sqrt{x}\sqrt{1-x}}$

$\Rightarrow \quad I = \int \frac{dx}{\sqrt{x}\sqrt{(1)^2 - (\sqrt{x})^2}}$

Put $\sqrt{x} = 1 \Rightarrow \frac{1}{2\sqrt{x}}dx = dt$

$\therefore \quad I = 2\int \frac{dt}{\sqrt{1^2 - t^2}}$

$= 2\sin^{-1}t + C$

$= 2\cdot\sin^{-1}\sqrt{x} + C$ [option a]

$= \cos^{-1}(1-2x) + C$

$= -\cos^{-1}(2x-1) + C$ [option c]

**20.** (a,b,c,d) Let $I = \int \frac{\sin 2x}{(\sin^4 x + \cos^4 x)}dx$

$= \int \frac{\frac{2\sin x\cos x}{\cos^4 x}}{\tan^4 x + 1}dx$

$= \int \frac{2\tan x\cdot\sec^2 x}{(\tan^2 x)^2 + 1}dx$

Put $\tan^2 x = t$

$\Rightarrow \quad 2\tan x\cdot\sec^2 x\,dx = dt$

$\therefore \quad I = \int \frac{dt}{t^2+1} = \tan^{-1}t + C$

$= \tan^{-1}(\tan^2 x) + C$ [option c]

$= \frac{\pi}{2} - \cot^{-1}(\tan^2 x) + C$

[using $\tan^{-1}x + \cot^{-1}x = \pi/2$]

$= -\cot^{-1}(\tan^2 x) + C$ [option b]

Also, $\quad \tan^{-1}(x) = \cot^{-1}(1/x)$

$\Rightarrow \quad \tan^{-1}(\tan^2 x) = \cot^{-1}(\cot^2 x) + C$ [option a]

and $\quad -\tan^{-1}(\cos 2x) + C = -\tan^{-1}\left(\frac{1-\tan^2 x}{1+\tan^2 x}\right) + C$

$= -\tan^{-1}1 + \tan^{-1}(\tan^2 x) + C$

$= \tan^{-1}(\tan^2 x) + C$ [option d]

**21.** (b) For $a = 0$, $L = \lim_{x\to\infty}\frac{\int \sin^2 x\,dx}{x}$

$= \lim_{x\to\infty}\frac{\int(1-\cos 2x)dx}{2x} = \lim_{x\to\infty}\left[\frac{x}{2x} - \frac{\sin 2x}{4x}\right]$

$= \lim_{x\to\infty}\left[\frac{1}{2} - \frac{\sin 2x}{4x}\right] = \frac{1}{2}$

**22.** (d) For $a = 1$, $L = \lim_{x\to\infty}\frac{\int 4\sin^2 x dx}{x}$

$= \lim_{x\to\infty}\frac{2\int(1-\cos 2x)\,dx}{x} = \frac{2\left(x - \frac{\sin 2x}{2}\right)}{x}$

$= \lim_{x\to\infty} 2 - \frac{\sin 2x}{x} = 2$

**23.** (a) For $a = -1$, $L = \lim_{x\to\infty}\frac{0}{x} = 0$

**24.** (c) For $a \in R$, $L = \lim_{x\to\infty}\frac{\int(\sin^2 x + \sin^2 ax + 2\sin x\cdot\sin a x)\,dx}{x}$

$= \lim_{x\to\infty}\frac{\int\left\{\frac{1-\cos 2x}{2} + \frac{1-\cos 2ax}{2} + \cos(a-1)x - \cos(a+1)x\right\}dx}{x}$

$= \lim_{x\to\infty}\left[1 - \frac{1}{x}\cdot\left\{\frac{\sin 2x}{4} - \frac{\sin 2ax}{4a} + \frac{\sin(a-1)x}{a-1} - \frac{\sin(a+1)x}{a+1}\right\}\right]$

$= 1$

# DPP-3 **Integration by Parts**

**1.** (i) Let $I = \int \underset{II}{x} \cdot \underset{I}{\log_e x}\, dx$

Using integration by parts, we have

$$I = \log x \cdot \frac{x^2}{2} - \int \left(\frac{1}{x} \cdot \frac{x^2}{2}\right) dx$$

$$= \log x \cdot \frac{x^2}{2} - \frac{1}{2}\frac{x^2}{2} + C$$

$$= \log x \cdot \frac{x^2}{2} - \frac{1}{4}x^2 + C$$

(ii) Let $I = \int \frac{x - \sin x}{1 - \cos x} dx$

$$= -\int \frac{x}{1 - \cos x} dx - \int \frac{\sin x}{1 - \cos x} dx$$

$$= \int \frac{x}{2\sin^2 x/2} dx - \int \frac{2\sin x/2 \cos x/2}{2\sin^2 x/2} dx$$

$$= \int \underset{I}{\frac{x}{2}} \underset{II}{\operatorname{cosec}^2 \frac{x}{2}}\, dx - \int \cot\frac{x}{2}\, dx$$

$$= \frac{1}{2}\int \underset{I}{x} \underset{II}{\operatorname{cosec}^2 \frac{x}{2}}\, dx - \int \cot\frac{x}{2}\, dx$$

$$= \frac{1}{2}\left[x\left(-2\cot\frac{x}{2}\right) - \int \left(1 \cdot -2\cot\frac{x}{2}\right)dx\right] - \int \cot\frac{x}{2}\, dx + C$$

$$= -\,x\cot\frac{x}{2} + \int \cot\frac{x}{2}\, dx - \int \cot\frac{x}{2} + C$$

$$= -\,x\cot\frac{x}{2} + C$$

(iii) Let $I = \int \tan^{-1}\left(\frac{3x - x^3}{1 - 3x^2}\right) dx$

But $x = \tan\theta$

$\Rightarrow \quad \theta = \tan^{-1} x$

$\therefore \quad I = \int \tan^{-1}(\tan 3\theta)\, d\theta$

$$= \int 3\theta\, d\theta$$

$$= 3\int \tan^{-1} x\, dx$$

$$= 3\int \underset{I}{\tan^{-1} x} \cdot \underset{II}{1}\, dx$$

$$= 3\int \tan^{-1} x \cdot x - \int \frac{1}{1 + x^2} \cdot x\, dx$$

$$= 3\left\{x\tan^{-1} x - \int \frac{x}{1 + x^2} dx\right\}$$

Put $1 + x^2 = t$

$\Rightarrow \quad 2x\, dx = dt$

$\Rightarrow \quad x\, dx = \frac{dt}{2}$

$\therefore \quad I = 3\left[x\tan^{-1} x - \frac{1}{2}\int \frac{dt}{t}\right]$

$$= 3\left[x\tan^{-1} x - \frac{1}{2}\log|t|\right] + C$$

$$= 3x\tan^{-1} x - \frac{3}{2}\log|1 + x^2| + C$$

(iv) Let $I = \int (\log_e x)^2\, dx$

Put $\log x = t \quad \Rightarrow \quad x = e^t$

$\Rightarrow \quad \frac{1}{x} dx = dt$

$\Rightarrow \quad dx = e^t dt$

$\therefore \quad I = -\int \underset{I}{t^2} \cdot \underset{II}{e^t}\, dt$

$$= t^2 e^t - \int \underset{I}{2t} \cdot \underset{II}{e^t} + C$$

$$= t^2 e^t - 2\left[t e^t - \int 1 \cdot e^t dt\right] + C$$

$$= t^2 \cdot e^t - 2t e^t + 2e^t + C$$

$$= e^t (t^2 - 2t + 2) + C$$

$$= e^{\log x}[(\log x)^2 + 2\log x + 2] + C$$

$$= x(\log^2 x + 2\log x + 2) + C$$

(v) Let $I = \int \sin^{-1} x\, dx$

$$= \int \underset{I}{(\sin^{-1} x)} \cdot \underset{II}{1}\, dx$$

$$= \sin^{-1} x \cdot x - \int \frac{1}{\sqrt{1 - x^2}} \cdot x\, dx$$

$$= \sin^{-1} x \cdot x - \int \frac{x}{\sqrt{1 - x^2}} dx$$

$$= \sin^{-1} x \cdot x + \frac{1}{2}\int \frac{dt}{\sqrt{t}}$$

$\left[\text{put } 1 - x^2 = t \Rightarrow -2x\,dx = dt \Rightarrow x\,dx = \frac{-1}{2} dt\right]$

$$= \sin^{-1} x \cdot x + \frac{1}{2} \times 2\sqrt{t} + C$$

$$= \sin^{-1} x \cdot x + \sqrt{1 - x^2} + C$$

(vi) Let $I = \int \underset{II}{x^2} \cdot \underset{I}{\sin^{-1} x}\, dx$

$$= \sin^{-1} x\, \frac{x^3}{3} - \int \frac{1}{\sqrt{1 - x^2}} \cdot \frac{x^3}{3} dx$$

$$= \sin^{-1} x \cdot \frac{x^3}{3} - \frac{1}{3}\int \frac{x^3}{\sqrt{1 - x^2}} dx + C$$

Put $1 - x^2 = t^2 \quad \Rightarrow \quad -2x\, dx = 2t\, dt$

$\Rightarrow \quad x\, dx = -t\, dt$

$\therefore \quad I = \sin^{-1} x \cdot \frac{x^3}{3} + \frac{1}{3}\left[\int \frac{(1 - t^2) \cdot t dt}{t}\right] + C$

$$= \sin^{-1} x \cdot \frac{x^3}{3} + \frac{1}{3}\left[\int (1 - t^2)\, dt\right] + C$$

$$= \sin^{-1} x \cdot \frac{x^3}{3} + \frac{1}{3}\left[t - \frac{t^3}{3}\right] + C$$

$$= \sin^{-1} x\, \frac{x^3}{3} + \frac{1}{3}\left[\sqrt{1 - x^2} - \frac{(1 - x^2)^3}{3}\right] + C$$

$$= \frac{x^3}{3} \cdot \sin^{-1} x + \frac{\sqrt{1 - x^2}}{3} - \frac{1}{9}(1 - x^2)^3 + C$$

(vii) Let $I = \int \frac{x\sin^{-1} x^2}{\sqrt{1-x^2}} dx$

Put $x^2 = t$

$\Rightarrow \quad 2x dx = dt$

$\Rightarrow \quad x dx = \frac{dt}{2}$

$\therefore \quad I = \frac{1}{2}\int \frac{\sin^{-1} t}{\sqrt{1-t}} dt = \frac{1}{2}\int \frac{\sin^{-1} t}{\sqrt{1-t}} dt$

$= \frac{1}{2}\int \left[\underset{\text{I}}{\sin^{-1} t} \times \underset{\text{II}}{\frac{1}{\sqrt{1-t}}}\right] dt$

$= \frac{1}{2}\left[\sin^{-1} t \cdot \int \frac{dt}{\sqrt{1-t}} - \int \frac{1}{\sqrt{1-t}}\left(\int \frac{dt}{\sqrt{1-t}}\right) dt\right]$

$= \frac{1}{2}\left[\sin^{-1} t(-2\sqrt{1-t}) - \int \frac{1}{\sqrt{1-t^2}} \times (-2\sqrt{1-t}) dt\right]$

$= \frac{1}{2}\left[\sin^{-1} t(-2\sqrt{1-t}) + 2\int \frac{dt}{\sqrt{1+t}}\right]$

$= \frac{1}{2}[-2\sin^{-1} t\sqrt{1-t} + 2(2\sqrt{1+t})] + C$

$= -\sqrt{1-t}\sin^{-1} t + 2\sqrt{1+t} + C$

$= -\sqrt{1-x^2}\sin^{-1} x^2 + 2\sqrt{1+x^2} + C$

(viii) Let $I = \int \tan^{-1}\sqrt{x} dx$

Put $\sqrt{x} = t$

$\Rightarrow \quad \frac{1}{2\sqrt{x}} dx = dt$

$\Rightarrow \quad \frac{1}{2t} dx = dt$

$\Rightarrow \quad dx = 2t dt$

$\therefore \quad I = 2\int \underset{\text{I}}{(\tan^{-1} t)}\underset{\text{II}}{t}\, dt$

$= 2\left[\tan^{-1} t\frac{t^2}{2} - \int \frac{1}{1+t^2}\cdot\frac{t^2}{2} dt\right] + C$

$= \left[t^2\tan^{-1} t - \int \frac{t^2}{1+t^2} dt\right] + C$

$= \left[t^2\tan^{-1} t - \int\left(1 - \frac{1}{1+t^2}\right) dt\right] + C$

$= t^2\tan^{-1} t - \int dt + \int \frac{dt}{1+t^2} + C$

$= t^2\tan^{-1} t - t + \tan^{-1} t + C$

$= (t^2+1)\tan^{-1} t - t + C$

$= (x+1)\tan^{-1}\sqrt{x} - \sqrt{x} + C$

(ix) Let $I = \int e^{\sqrt{x}} dx$

Put $\sqrt{x} = t$

$\Rightarrow \quad \frac{1}{2\sqrt{x}} dx = dt$

$\Rightarrow \quad dx = 2t\, dt$

$\therefore \quad I = 2\int \underset{\text{I}}{t}\cdot \underset{\text{II}}{e^t} dt$

$= 2[t\cdot e^t - \int 1\cdot e^t dt] + C$

$= 2[t\cdot e^t - e^t] + C$

$= 2e^t(t-1) + C$

$= 2e^{\sqrt{x}}(\sqrt{x}-1) + C$

(x) Let $I = \int (\sin^{-1} x)^2 dx$

Put $\sin^{-1} x = t \quad \Rightarrow \quad x = \sin t$

$\frac{1}{\sqrt{1-x^2}} dx = dt$

$\Rightarrow \quad dx = \sqrt{1-\sin^2 t}\, dt$

$\Rightarrow \quad dx = \cos t\, dt$

$\therefore \quad I = \int \underset{\text{I}}{t^2}\cdot \underset{\text{II}}{\cos t}\, dt$

$= t^2\cdot\sin t - \int \underset{\text{I}}{2t}\cdot \underset{\text{II}}{\sin t}\, dt$

$= t^2\sin t - 2[-t\cos t + \int 1\cdot\cos t\, dt] + C$

$= t^2\sin t + 2t\cos t - 2\sin t + C$

$= t^2\sin t + 2t\cos t - 2\sin t + C$

$= (t^2 - 2)\sin t + 2t\cos t + C$

$= [(\sin^{-1} x)^2 - 2]\sin(\sin^{-1} x) + 2\sin^{-1} x\cos(\sin^{-1} x) + C$

(xi) Let $I = \int \sin^{-1}\left(\frac{2x}{1+x^2}\right) dx$

Put $x = \tan\theta \quad \Rightarrow \quad \theta = \tan^{-1} x$

$\Rightarrow \quad x = \sec^2\theta\, d\theta$

$\therefore \quad I = \int \sin^{-1}\left(\frac{2\tan\theta}{1+\tan^2\theta}\right)\sec^2\theta\, d\theta$

$= \int \sin^{-1}(\sin 2\theta)\sec^2\theta\, d\theta$

$= \int 2\theta\sec^2\theta\, d\theta$

$= 2\int \underset{\text{I}}{\theta}\,\underset{\text{II}}{\sec^2\theta}\, d\theta$

$= 2[\theta\tan\theta - \int 1\cdot\tan\theta\, d\theta] + C$

$= 2[\theta\tan\theta - \log|\sec\theta| + C$

$= 2[\tan^{-1} x\tan(\tan^{-1} x) - \log|\sec(\tan^{-1} x)|] + C$

$= 2[x\tan^{-1} x - \log|\sec(\tan^{-1} x)|] + C$

(xii) Let $I = \int \tan^{-1}\left(\sqrt{\frac{1-x}{1+x}}\right) dx$

We have, $\quad 2\tan^{-1} x = \cos^{-1}\left(\frac{1-x^2}{1+x^2}\right)$

$\therefore \quad \tan^{-1}\sqrt{\frac{1-x}{1+x}} = \frac{1}{2}\cos^{-1}\left(\frac{1-\frac{1-x}{1+x}}{1+\frac{1-x}{1+x}}\right)$

$= \frac{1}{2}\cos^{-1} x$

Thus, we have

$I = \int \tan^{-1}\sqrt{\frac{1-x}{1+x}}\, dx$

$= \frac{1}{2}\int \cos^{-1} x\, dx$

$= \frac{1}{2}\left[x\cos^{-1} x - \int \frac{-1}{\sqrt{1-x^2}}\cdot x dx\right]$

$= \frac{1}{2}\left[x\cos^{-1} x - \frac{1}{2}\int \frac{-2x}{\sqrt{1-x^2}} dx\right]$

$= \frac{1}{2}[x\cos^{-1} x - \sqrt{1-x^2}] + C$

(xiii) Let $= \int e^{2x} \cdot \sin 3x\, dx$
II I

$$= \sin 3x \left(\frac{e^{2x}}{2}\right) - 3\cos 3x \cdot \left(\frac{e^{2x}}{2}\right) dx$$

$$= \frac{1}{2}\sin 3x\, e^{2x} - \frac{3}{2}\int e^{2x} \cos 3x dx$$
II I

$$= \frac{1}{2}\sin 3x e^{2x} - \frac{3}{2}\left[\cos 3x \cdot \frac{e^{2x}}{2} + \int 3\sin 3x \left(\frac{e^{2x}}{2}\right) dx\right]$$

$$= \frac{1}{2}\sin 3x \cdot e^{2x} - \frac{3}{2}\left[\frac{1}{2} e^{2x} \cos 3x + \frac{3}{2}\int e^{2x} \sin 3x\, dx\right]$$

$$= \frac{1}{2}\sin 3x\, e^{2x} - \frac{3}{4} e^{2x} \cos 3x - \frac{9}{4} I$$

$$\Rightarrow (1 + 9/4)\, I = \frac{1}{2} e^{2x} \sin 3x - \frac{3}{4} e^{2x} \cos 3x + C$$

$$\therefore \quad I = \frac{4}{13} e^{2x} \left[\frac{1}{2}\sin 3x - \frac{3}{4}\cos 3x\right] + C$$

$$= \frac{2e^{2x}}{13}\left[\sin 3x - \frac{3}{2}\cos 3x\right] + C$$

$$= \frac{e^{2x}}{13}[2\sin 3x - 3\cos 3x] + C$$

**Alternate Method**

We know that

$$\int e^{ax} \sin bx dx = \frac{e^{ax}}{a^2 + b^2}[a \sin bx - b\cos bx] + C$$

$$\Rightarrow \int e^{2x} \sin 3x dx = \frac{e^{2x}}{(2)^2 + (3)^2}[2\sin 3x - 3\cos 3x] + C$$

$$= \frac{e^{2x}}{13}[2\sin 3x - 3\cos 3x] + C$$

(xiv) Let $I = \int \frac{e^{\tan^{-1} x}}{(1 + x^2)^2} dx$

Put $\tan^{-1} x = t$

$$\Rightarrow \quad x = \tan t$$

$$\Rightarrow \quad \frac{1}{1 + x^2} dx = dt$$

$$\therefore \quad I = \int \frac{e^t}{1 + \tan^2 t} dt$$

$$= \int \frac{e^t}{\sec^2 t} dt = \int e^t \cos^2 t\, dt$$

$$= \frac{1}{2}\int e^t (\cos 2t + 1)\, dt = \frac{1}{2}\left[\int e^t \cos 2t\, dt + \int e^t\, dt\right]$$

$$= \frac{1}{2}\left[\frac{e^t}{5}(2\sin 2t + \cos 2t) + e^t\right] + C$$

$$= \frac{1}{2}\left[\frac{e^t}{5}\{(2\sin 2t + \cos 2t) + 5\}\right] + C$$

$$= \frac{e^t}{10}[(2\sin 2t + \cos 2t) + 5] + C, \text{ where } t = \tan^{-1} x$$

(xv) Let $I = \int \operatorname{cosec}^3 x dx$

$$= \int \operatorname{cosec}^2 x \cdot \operatorname{cosec} x\, dx$$
II I

$$= \operatorname{cosec} x\,(-\cot x) - \int (-\operatorname{cosec} x \cot x) \cdot (-\cot x) dx$$

$$= -\operatorname{cosec} x \cdot \cot x - \int \operatorname{cosec} x \cdot \cot^2 x dx$$

$$= -\operatorname{cosec} x \cdot \cot x - \int \operatorname{cosec} x\,(\operatorname{cosec}^2 x - 1)\, dx$$

$$= -\operatorname{cosec} x \cdot \cot x - \int (\operatorname{cosec}^3 x - \operatorname{cosec} x) dx$$

$$= -\operatorname{cosec} x \cdot \cot x - \int \operatorname{cosec}^3 x\, dx + \int \operatorname{cosec} x\, dx$$

$$I = -\operatorname{cosec} x \cdot \cot x - I + \int \operatorname{cosec} x\, dx$$

$$2I = -\operatorname{cosec} x \cdot \cot x + \log(\operatorname{cosec} x - \cot x) + C$$

$$I = -\frac{1}{2}\operatorname{cosec} x \cdot \cot x + \frac{1}{2}\log(\operatorname{cosec} x - \cot x) + C$$

(xvi) Let $I = \int \sin(\log x)\, dx$

Put $\log x = t$

$$\Rightarrow \quad x = e^t$$

$$\Rightarrow \quad \frac{1}{x} dx = dt$$

$$\Rightarrow \quad dx = e^t dt$$

$$\therefore \quad I = \int e^t \sin t\, dt$$

$$= \frac{e^t}{1^2 + 1^2}[\sin t - \cos t] + C$$

$$= \frac{e^t}{2}[\sin t - \cos t] + C$$

$$= \frac{e^{\log x}}{2}[\sin(\log x) - \cos(\log x)] + C$$

$$= \frac{x}{2}[\sin(\log x) - \cos(\log x)] + C$$

(xvii) Let $I = \int e^x [\tan x + \log(\sec x)] dx$

$$= \int e^x \tan x dx + \int e^x \log(\sec x)\, dx$$
II I

$$= \int e^x \tan x dx + \log(\sec x)\, e^x - \int \tan x \cdot e^x dx + C$$

$$= \log(\sec x) \cdot e^x + C$$

**Alternate Method**

Let $I = \int e^x [\tan x + \log(\sec x)] dx$

Again, $f(x) = \log(\sec x)$

$$\Rightarrow \quad f'(x) = \frac{1}{\sec x} \cdot \sec x \cdot \tan x$$

$$\Rightarrow \quad f'(x) = \tan x$$

$$\therefore \quad I = \int e^x [f'(x) + f(x)] dx$$

$$= e^x f(x) + C$$

$$= e^x \log(\sec x) + C \qquad \left[\because \int e^x [f(x) + f'(x)] dx = e^x f(x) + C\right]$$

(xviii) Let $I = \int \frac{x e^x}{(x + 1)^2} dx$

$$= \int \frac{\{(x + 1) - 1\} e^x}{(x + 1)^2} dx$$

$$= \int e^x \left[\frac{1}{x + 1} - \frac{1}{(x + 1)^2}\right] dx$$

Let $f(x) = \frac{1}{x + 1}$

$$\therefore \quad f'(x) = -\frac{1}{(x + 1)^2} = \int e^x [f(x) + f'(x)]\, dx$$

$$= e^x f(x) + C \left[\because \int e^x [f(x) + f'(x)]\, dx = e^x f(x) + C\right]$$

$$= e^x \left(\frac{1}{x + 1}\right) + C$$

(xix) Let $I = \int e^x \left[\frac{1 + x}{(2 + x)^2}\right] dx$

$$= \int e^x \left[\frac{(2 + x)}{(2 + x)^2} - \frac{1}{(2 + x)^2}\right] dx$$

$$= \int e^x \left[\frac{1}{(2 + x)} - \frac{1}{(2 + x)^2}\right] dx$$

Again, let $f(x) = \frac{1}{2+x}$

$$f'(x) = -\frac{1}{(2+x)^2} = e^x \times \frac{1}{2+x} + C = \frac{e^x}{2+x} + C$$

$$[\because \int e^x[(f(x) + f'(x)]\,dx = e^x f(x) + C]$$

(xx) Let $I = \int e^{\tan^{-1}x}\left[\frac{1+x+x^2}{1+x^2}\right]dx$

Put $\tan^{-1}x = t \Rightarrow x = \tan t$

$\Rightarrow \quad \frac{1}{1+x^2}dx = dt$

$$\therefore \quad I = \int e^t(1+\tan t+\tan^2 t)\,dt$$

$$= \int e^t(\sec^2 t + \tan t)\,dt$$

Again, let $f(x) = \tan t$

$$f'(x) = \sec^2 t = e^t \tan t + C$$

$$[\because e^x\int [f(x) + f'(x)]\,dx = e^x f(x) + C]$$

$$= e^{\tan^{-1}x}\tan(\tan^{-1}x) + C$$

$$= x \cdot e^{\tan^{-1}x} + C$$

**2.** (c) Let $I = \int \sin^2(\log x)\,dx$

Put $\log x = t \Rightarrow x = e^t$

$\Rightarrow \quad \frac{1}{x}dx = dt$

$\Rightarrow \quad dx = e^t dt$

$$\therefore \quad I = \int \sin^2 t \cdot e^t dt$$

$$= \frac{1}{2}\int e^t(1-\cos 2t)\,dt$$

$$= \frac{1}{2}\left[\int e^t\,dt - \int e^t \cos 2t\,dt\right]$$

$$= \frac{1}{2}\left[e^t - \frac{e^t}{5}(2\sin 2t + \cos 2t)\right] + C$$

$$= \frac{x}{10}[5 - 2\sin(2\log x) - \cos(2\log x)] + C$$

**3.** (a) Let $I = \int e^x\left(\frac{x+2}{(x+3)^2}\right)dx$

$$= \int e^x\left[\frac{x+3}{(x+3)^2} - \frac{1}{(x+3)^2}\right]dx$$

$$= \int e^x \cdot \left[\frac{1}{x+3} - \frac{1}{(x+3)^2}\right]dx$$

$$= e^x \cdot \frac{1}{x+3} + C$$

**4.** (c) Let $I = \int e^{\tan^{-1}x}(1+x+x^2)\,d(\cot^{-1}x)$

$$= -\int \frac{e^{\tan^{-1}x}(1+x+x^2)}{1+x^2}dx$$

Put $\tan^{-1}x = t \Rightarrow x = e^t$

$\Rightarrow \quad \frac{1}{1+x^2}dx = dt$

$$\therefore \quad I = -\int e^t(1+\tan t+\tan^2 t)\,dt$$

$$= -\int e^t(\sec^2 + \tan t)\,dt$$

$$= -e^t \tan t + C$$

$$= -e^{\tan^{-1}x}\tan(\tan^{-1}x) + C$$

$$= -xe^{\tan^{-1}x} + C$$

**5.** (b) Let $I = \int e^x \frac{(1+nx^{n-1}-x^{2n})}{(1-x^n)\sqrt{1-x^{2n}}}dx = \int e^x \frac{(1-2^{2n})+nx^{n-1}}{(1-x^n)\sqrt{1-x^{2n}}}dx$

$$= \int e^x\left[\frac{1-x^{2n}}{(1-x^n)\sqrt{1-x^{2n}}} - \frac{nx^{n-1}}{(1-x^n)\sqrt{1-x^{2n}}}\right]dx$$

$$= \int e^x\left[\frac{1+x^n}{\sqrt{1-x^{2n}}} - \frac{nx^{n-1}}{(1-x^n)\sqrt{(1-x^{2n})}}\right]dx$$

$$= \int e^x\left[\sqrt{\frac{1+x^n}{1-x^n}} - \frac{nx^{n-1}}{(1-x^n)\sqrt{1-x^{2n}}}\right]dx$$

Let $f(x) = \sqrt{\frac{1+x^n}{1-x^n}}$, then

$$f'(x) = \frac{-nx^{n-1}}{(1-x^n)\sqrt{1-x^{2n}}}$$

$$\therefore \quad I = e^x\sqrt{\frac{1+x^n}{1-x^n}} + C \qquad \left[\because \int e^x[f(x)+f'(x)] = e^x f(x) + C\right]$$

**6.** (b) Let $I = \int \frac{\log x}{x^2}dx$

Put $\log x = t \Rightarrow \frac{1}{x}dx = dt$

$$\therefore \quad I = \int \frac{t}{e^t}dt = \int e^{-t} \cdot t\,dt$$

$$\qquad\qquad\quad \text{II} \quad \text{I}$$

$$= -t \cdot e^{-t} - \int 1 \cdot (-e^{-t})\,dt$$

$$= -t \cdot e^{-t} + \int e^{-t}\,dt$$

$$= -t \cdot e^{-t} - e^{-t} + C$$

$$= -\log x \cdot e^{-\log x} - e^{-\log x} + C$$

$$= -e^{-\log x}(\log x + 1) + C$$

$$= -\frac{1}{x}(\log x + 1) + C$$

**7.** (d) Let $I = \int\left[\frac{\log x - 1}{1+(\log x)^2}\right]^2 dx$

Put $\log x = t \Rightarrow x = e^t$

$\Rightarrow \quad dx = e^t dt$

$$\therefore \quad I = \int\left[\frac{(t-1)}{1+t^2}\right]^2 e^t dt = \int e^t\left[\frac{1}{1+t^2} - \frac{2t}{(1+t^2)^2}\right]dt$$

$$= e^t\frac{1}{1+t^2} + C = \frac{x}{1+(\log x)^2} + C$$

**8.** (c) Given, $\int f(x)\,dx = \psi(x)$

Now, $\quad I = \int x^5 f(x^3)\,dx$

Put $x^3 = t \Rightarrow 3x^2 dx = dt$

$\Rightarrow \quad x^2 dx = \frac{dt}{3}$

$$\therefore \quad I = \frac{1}{3}\int t\ f(t)\,dt$$

$$\qquad\qquad\quad \text{I} \quad \text{II}$$

$$= \frac{1}{3}[t \cdot \int f(t)dt - \int 1 \cdot \{\int f(t)dt\}\,dt] + C$$

$$= \frac{1}{3}[t \cdot \psi(t) - \int \psi(t)\,dt] + C$$

$$= \frac{1}{3}[x^3\psi(x^3) - \int \psi(x^3)\,d(x^3)] + C$$

$$= \frac{1}{3}[x^3\psi(x^3) - 3\int \psi(x^3)\,x^2 dx] + C$$

$$= \frac{1}{3}[x^3\psi(x^3) - 3\int x^2\psi(x^3)\,dx] + C$$

**9.** (a) Let $I = \int \sin^{-1}\left(\frac{2x+2}{\sqrt{4x^2+8x+13}}\right)dx$

$$= \int \sin^{-1}\left(\frac{2x+2}{\sqrt{(2x+2)^2+3^2}}\right)dx$$

Put $2x + 2 = 3\tan\theta$

$\Rightarrow \quad 2x\,dx = 3\sec^2\theta\,d\theta$

$\therefore \quad I = \int \sin^{-1}\left(\frac{3\tan\theta}{3\sec\theta}\right)\frac{3}{2}\sec^2\theta d\theta$

$$= \frac{3}{2}\int \theta\sec^2\theta\,d\theta$$

$$= \frac{3}{2}(\theta\tan\theta - \int \tan\theta\,d\theta)$$

$$= \frac{3}{2}(\theta\tan\theta - \log|\sec\theta|) + C$$

$$= \frac{3}{2}\left[\frac{2x+2}{3}\tan^{-1}\left(\frac{2x+2}{3}\right) - \log\sqrt{1+\left(\frac{2x+2}{3}\right)^2}\right] + C$$

$$= \frac{3}{2}\left[\frac{2}{3}(x+1)\tan^{-1}\left\{\frac{2}{3}(x+1) - \log\sqrt{4x^2+8x+13}\right\}\right] + C$$

$$= (x+1)\tan^{-1}\frac{2(x+1)}{3} - \frac{3}{4}\log(4x^2+8x+13) + C$$

**10.** (c,d) Given, $\int e^{3x}\cos 4x dx = e^{3x}(A\sin 4x + B\cos 4x) + C$

Now, $\int e^{3x}\cos 4x = \frac{e^{3x}}{(3)^2+(4)^2}[4\sin 4x + 3\cos 4x] + C$

$$= \frac{e^{3x}}{25}[4\sin 4x + 3\cos 4x] + C$$

$$= e^{3x}\left[\frac{4}{25}\sin 4x + \frac{3}{25}\cos 4x\right] + C$$

$\Rightarrow \quad A = \frac{4}{25}, B = \frac{3}{25}$

$\therefore \quad 3A = 3\times\frac{4}{25} = \frac{3}{25}\times 4 = 4B$

and $\quad 4A + 3B = 4\times\frac{4}{25} + 3\times\frac{3}{25}$

$$= \frac{16}{25} + \frac{9}{25} = \frac{25}{25} = 1$$

**11.** (b,c,d) $f'(x) = 3x^2\cdot\sin(1/x) - x\cos(1/x),\ x \neq 0$

On integrating, $f(x) = \begin{cases} x^3\cdot\sin(1/x), & x \neq 0 \\ 0, & x = 0 \end{cases}$

$\therefore$ $f(x)$ is continuous at $x = 0$.

$\Rightarrow f(x)$ is differentiable at $x = 0$.

Again, $f'(x) = 3x^2\cdot\sin(1/x) - x\cos(1/x)$

does not exists at $x = 0$.

$\Rightarrow$ Continuous at $x = 0$.

$\therefore$ $f'(x)$ is not differentiable at $x = 0$.

**12.** (a,b,c,d) We have,

$\int xe^x\cdot\sin x dx$ = Imaginary part of $\int xe^{(1+i)x}dx$

$= \text{Im}\int xe^{(1+i)x}dx$, using integration by parts

$$= \text{Im}\left\{\frac{xe^{(1+i)x}}{(1+i)} - \frac{e^{(1+i)x}}{(1+i)^2}\right\}$$

$$= \frac{xe^x}{2}\cdot\text{Im}(\cos x + i\sin x)(1-i) - \frac{e^x}{4}\text{Im}[\cos x + i\sin x\cdot x(-2i)]$$

$$= \frac{e^x}{2}[x\cdot(\sin x - \cos x) + \cos x] + C$$

**13.** (a,c) Given, $u = \int e^{ax}\cos bx\,dx = \frac{e^{ax}}{a^2+b^2}[b\sin bx + a\cos bx]$

and $\quad v = \int e^{ax}\sin bx\,dx = \frac{e^{ax}}{a^2+b^2}[a\sin bx - b\cos bx]$

Now, $\quad u^2 = \frac{e^{2ax}}{(a^2+b^2)^2}(b\sin bx + a\cos bx)^2$

and $\quad v^2 = \frac{e^{2ax}}{(a^2+b^2)^2}(a\sin bx - b\cos bx)^2$

Then, $u^2 + v^2 = \frac{e^{2ax}}{(a^2+b^2)^2}(a^2+b^2)$

$\Rightarrow \quad u^2 + v^2 = \frac{e^{2ax}}{a^2+b^2}$

$\Rightarrow \quad (a^2+b^2)(u^2+v^2) = e^{2ax}$

Again, $\quad u^2 - v^2$

$$= \frac{e^{2ax}\left[\begin{matrix}(b\sin bx + a\cos bx + a\sin bx - b\cos bx)\\(b\sin bx + a\cos bx - a\sin bx + b\cos bx)\end{matrix}\right]}{(a^2+b^2)^2}$$

$$= \frac{e^{2ax}}{(a^2+b^2)^2}[\{(a+b)\sin bx + (a-b)\cos bx\}\ (b-a)\sin bx + (a+b)\cos bx]$$

$$= \frac{e^{2ax}\left[\begin{matrix}-(a^2-b^2)\sin^2 bx + (a+b)^2\sin bx\cos bx\\-(a-b)^2\sin bx\cos bx + (a^2-b^2)\cos^2 bx\end{matrix}\right]}{(a^2+b^2)^2}$$

$\Rightarrow \quad u^2 - v^2 \neq e^{2ax}$

Now, $\quad \frac{v}{u} = \frac{a\sin bx - b\cos bx}{a\cos bx + b\sin bx}$

$\Rightarrow \quad \frac{v}{u} = \frac{\tan bx - b/a}{1 + (b/a)\tan bx}$

$\Rightarrow \quad \frac{v}{u} = \tan\left[bx - \tan^{-1}\left(\frac{b}{a}\right)\right]$

$\therefore \quad \tan^{-1}\left(\frac{v}{u}\right) = bx - \tan^{-1}\left(\frac{b}{a}\right)$

i.e. $\quad \tan^{-1}\left(\frac{v}{u}\right) + \tan^{-1}\left(\frac{b}{a}\right) = bx$

**14.** (b) We have,

$\int (x^3 - 2x^2 + 3x - 1)\cos 2x dx$ (I, II)

$$= (x^3 - 2x^2 + 3x - 1)\cdot\left(\frac{\sin 2x}{2}\right) - (3x^2 - 4x + 3)\cdot\left(-\frac{\cos 2x}{4}\right) + (6x-4)\cdot\left(\frac{-\sin 2x}{8}\right) - (6)\cdot\left(\frac{\cos 2x}{16}\right) + C$$

$$= \frac{\sin 2x}{4}(2x^3 - 4x^2 + 6x - 2 - 3x + 2) + \frac{\cos 2x}{8}(6x^2 - 8x + 6 - 3) + C$$

$$= \frac{\sin 2x}{4}(2x^3 - 4x^2 + 3x) + \frac{\cos 2x}{8}(6x^2 - 8x + 3) + C$$

$\therefore \quad u(x) = 2x^3 - 4x^2 + 3x$ and $v(x) = 6x^2 - 8x + 3$

**15.** (c) We have,

$$\int x^4\cdot e^{2x}dx = (x^4)\cdot\left(\frac{e^{2x}}{2}\right) - (4x^3)\left(\frac{e^{2x}}{4}\right) + (12x^2)\cdot\left(\frac{e^{2x}}{8}\right) - (24x)\left(\frac{e^{2x}}{16}\right) + (24)\left(\frac{e^{2x}}{24}\right) + C$$

$$= \frac{e^{2x}}{2}\left[x^4 - 2x^3 + 3x^2 - 3x + \frac{3}{2}\right] + C$$

# DPP-4 Some Special Substitutions and Integrals

**1.** (i) Let $I = \int \frac{dx}{x^2 - x + 1}$

$$= \int \frac{dx}{x^2 - x + \frac{1}{4} - \frac{1}{4} + 1} = \int \frac{dx}{\left(x - \frac{1}{2}\right)^2 + \frac{3}{4}}$$

$$= \int \frac{dx}{\left(x - \frac{1}{2}\right)^2 + \left(\frac{\sqrt{3}}{2}\right)^2}$$

Put $x - \frac{1}{2} = t \Rightarrow dx = dt$

$$\therefore I = \int \frac{dt}{t^2 + \left(\frac{\sqrt{3}}{2}\right)^2}$$

$$= \frac{2}{\sqrt{3}} \tan^{-1}\left(\frac{2t}{\sqrt{3}}\right) + C \left[\because \int \frac{1}{x^2 + a^2} dx = \frac{1}{a} \tan^{-1}\left(\frac{x}{a}\right) + C\right]$$

$$= \frac{2}{\sqrt{3}} \tan^{-1}\left(\frac{2x - 1}{\sqrt{3}}\right) + C$$

(ii) Let $I = \frac{dx}{2x^2 + x - 1}$

$$= \frac{1}{2} \int \frac{dx}{x^2 + \frac{x}{2} - \frac{1}{2}}$$

$$= \frac{1}{2} \int \frac{dx}{x^2 + \frac{x}{2} + \frac{1}{16} - \frac{1}{16} - \frac{1}{2}}$$

$$= \frac{1}{2} \int \frac{dx}{\left(x + \frac{1}{4}\right)^2 - \frac{9}{16}}$$

$$= \frac{1}{2} \int \frac{dx}{\left(x + \frac{1}{4}\right)^2 - \left(\frac{3}{4}\right)^2}$$

Put $x + \frac{1}{4} = t \Rightarrow dx = dt$

$$\therefore \quad I = \frac{1}{2} \int \frac{dt}{t^2 - \left(\frac{3}{4}\right)^2}$$

$$= \frac{1}{2} \cdot \frac{1}{2 \cdot (3/4)} \cdot \log \left|\frac{t - 3/4}{t + 3/4}\right| + C$$

$$\left[\because \int \frac{dx}{x^2 - a^2} = \frac{1}{2a} \log \left|\frac{x - a}{x + a}\right| + C\right]$$

$$= \frac{1}{2} \cdot \frac{3}{2} \cdot \log \left|\frac{x + \frac{1}{4} - \frac{3}{4}}{x + \frac{1}{4} + \frac{3}{4}}\right| + C$$

$$= \frac{1}{3} \log \left|\frac{x - \frac{1}{2}}{x + 1}\right| + C = \frac{1}{3} \log \left|\frac{2x - 1}{2(x + 1)}\right| + C$$

(iii) Let $I = \int \frac{dx}{\sqrt{x^2 - 2x + 3}}$

$$= \int \frac{dx}{\sqrt{x^2 - 2x + 1 - 1 + 3}} = \int \frac{dx}{\sqrt{(x - 1)^2 + (\sqrt{2})^2}}$$

$$= \log |(x - 1) + \sqrt{(x - 1)^2 + (\sqrt{2})^2}$$

$$\left[\because \int \frac{dx}{\sqrt{x^2 + a^2}} = \log |x + \sqrt{x^2 + a^2}| + C\right]$$

$$= \log |(x - 1) + \sqrt{x^2 - 2x + 3}| + C$$

(iv) Let $I = \int \sqrt{2x^2 - 3x + 1}\, dx$

$$= \sqrt{2} \int \left(\sqrt{x^2 - \frac{3x}{2} + \frac{1}{2}}\right) dx$$

$$= \sqrt{2} \int \sqrt{\left(x^2 - \frac{3x}{2} + \frac{9}{16} - \frac{9}{16} + \frac{1}{2}\right)} dx$$

$$= \sqrt{2} \int \sqrt{\left(x - \frac{3}{4}\right)^2 - \frac{1}{16}}\, dx$$

$$= \sqrt{2} \int \sqrt{\left(x - \frac{3}{4}\right)^2 - \left(\frac{1}{4}\right)^2}\, dx$$

$$= \left[\sqrt{2}\, \frac{1}{2}\left(x - \frac{3}{4}\right)\sqrt{\left(x - \frac{3}{4}\right)^2 - \frac{1}{16}}\right.$$

$$\left. - \frac{1}{2 \times 16} \log \left|\left(x - \frac{3}{4}\right) + \sqrt{\left(x - \frac{3}{4}\right)^2 - \frac{1}{16}}\right|\right] + C$$

$$\left[\because \int \sqrt{x^2 - a^2}\, dx = \frac{x}{2}\sqrt{x^2 - a^2} - \frac{1}{2a^2} \log \left| x + \sqrt{x^2 - a^2}\right| + C\right]$$

$$= \sqrt{2} \left[\frac{1}{8}(4x - 3)\sqrt{x^2 - \frac{3x}{2} + \frac{1}{2}}\right.$$

$$\left. - \frac{1}{32} \log \left|\left(x - \frac{3}{4}\right) + \sqrt{x^2 - \frac{3x}{2} + \frac{1}{2}}\right|\right. + C$$

(v) Let $I = \int \frac{1}{\sqrt{1 - e^{2x}}} dx$

Put $1 - e^{2x} = t^2 \Rightarrow -2e^{2x} dx = 2t\, dt \Rightarrow dx = \frac{t}{t^2 - 1} dt$

$$\therefore \quad I = \int \frac{1}{\sqrt{t^2}} \cdot \frac{t}{t^2 - 1} dt$$

$$= \int \frac{dt}{t^2 - 1} = \frac{1}{2} \log \left|\frac{t - 1}{t + 1}\right| + C$$

$$= \frac{1}{2} \log \left|\frac{\sqrt{1 - e^{2x}} - 1}{\sqrt{1 - e^{2x}} + 1}\right| + C$$

(vi) Let $I = \int \frac{2x}{\sqrt{1 - x^2 - x^4}} dx$

Put $x^2 = t \Rightarrow 2x\, dx = dt$

$$\therefore \quad I = \int \frac{dt}{\sqrt{1 - t - t^2}} = \int \frac{dt}{\sqrt{1 + \frac{1}{4} - \frac{1}{4} - t - t^2}}$$

$$= \int \frac{dt}{\sqrt{\frac{5}{4} - \left(t + \frac{1}{2}\right)^2}} = \int \frac{dt}{\left(\frac{\sqrt{5}}{2}\right)^2 - \left(t + \frac{1}{2}\right)^2}$$

$$= \sin^{-1}\left(\frac{t + \frac{1}{2}}{\sqrt{5/2}}\right) + C = \sin^{-1}\left(\frac{2x^2 + 1}{\sqrt{5}}\right) + C$$

(vii) Let $I = \int \frac{a^x}{\sqrt{1 - a^{2x}}} dx$

Put $a^x = t \Rightarrow a^x \log a\, dx = dt$

$$\Rightarrow \quad a^x dx = \frac{dt}{\log a}$$

$$\therefore \quad I = \int \frac{1}{\sqrt{1 - t^2}} \cdot \frac{dt}{\log a} = \frac{1}{\log a} \sin^{-1}(t) + C$$

$$= \frac{1}{\log a} \sin^{-1}(a^x) + C$$

(viii) Let $I = \int \frac{e^x}{\sqrt{5 - 4e^x - e^{2x}}} dx$

Put $e^x = t \Rightarrow e^x\, dx = dt$

$$\therefore \quad I = \int \frac{e^x}{\sqrt{5 - 4t - t^2}} dx$$

$$= \int \frac{dt}{\sqrt{5 + 4 - 4 - 4t - t^2}}$$

$$= \int \frac{dt}{\sqrt{(3)^2 - (t + 2)^2}}$$

$$= \sin^{-1}\left(\frac{t + 2}{3}\right) + C$$

$$= \sin^{-1}\left(\frac{e^x + 2}{3}\right) + C$$

(ix) Let $I = \int \sqrt{\frac{x}{a^3 - x^3}}\, dx = \int \frac{x^{1/2}}{\sqrt{(a^{3/2})^2 - (x^{3/2})^2}} dx$

Put $x^{3/2} = t \Rightarrow \frac{3}{2} x^{1/2}\, dx = dt \Rightarrow x^{1/2}\, dx = \frac{2}{3} dt$

$$\therefore \quad I = \frac{2}{3} \int \frac{dt}{\sqrt{(a^{3/2})^2 - (t)^2}}$$

$$= \frac{2}{3} \sin^{-1}\left(\frac{t}{a^{3/2}}\right) + C$$

$$\left[\because \int \frac{dx}{\sqrt{a^2 - x^2}} = \sin^{-1}\left(\frac{x}{a}\right) + C\right]$$

$$= \frac{2}{3} \sin^{-1}\left(\frac{x^{3/2}}{a^{3/2}}\right) + C$$

(x) Let $I = \int \frac{\cos x}{\sqrt{\sin^2 x - 2\sin x - 3}} dx$

Put $\sin x = t \Rightarrow \cos x\, dx = dt$

$$\therefore \quad I = \int \frac{dt}{\sqrt{t^2 - 2t - 3}} = \int \frac{dt}{\sqrt{t^2 - 2t + 1 - 1 - 3}}$$

$$= \int \frac{dt}{\sqrt{(t - 1)^2 - (2)^2}} = \log|(t - 1) + \sqrt{(t - 1)^2 - (2)^2}| + C$$

$$= \log|(\sin x - 1) + \sqrt{\sin^2 x - 2\sin x - 3}| + C$$

(xi) Let $I = \int \sqrt{\frac{\sin(x - \alpha)}{\sin(x + \alpha)}}\, dx$

$$= \int \sqrt{\frac{\sin(x - \alpha)}{\sin(x + \alpha)} \times \frac{\sin(x - \alpha)}{\sin(x - \alpha)}}\, dx$$

$$= \int \frac{\sin(x - \alpha)}{\sqrt{\sin^2 x - \sin^2 \alpha}} dx = \int \frac{\sin x \cos \alpha - \cos x \sin \alpha}{\sqrt{\sin^2 x - \sin^2 \alpha}} dx$$

$$= \cos \alpha \int \frac{\sin x}{\sqrt{\sin^2 x - \sin^2 \alpha}} dx - \sin \alpha \int \frac{\cos x}{\sqrt{\sin^2 x - \sin^2 \alpha}} dx$$

$$= \cos \alpha \int \frac{\sin x}{\sqrt{1 - \cos^2 x - \sin^2 \alpha}} dx - \sin \alpha \int \frac{\cos x}{\sqrt{\sin^2 x - \sin^2 \alpha}} dx$$

$$= \cos \alpha \int \frac{\sin x}{\sqrt{\cos^2 \alpha - \cos^2 x}} dx - \sin \alpha \int \frac{\cos x}{\sqrt{\sin^2 x - \sin^2 \alpha}} dx$$

In the first part, put $\cos x = t$, so that $-\sin x\, dx = dt$
and in second part put $\sin x = u$, so that $\cos x\, dx = dt$

$$\therefore I = -\cos \alpha \int \frac{dt}{\sqrt{\cos^2 \alpha - t^2}} - \sin \alpha \int \frac{du}{\sqrt{u^2 - \sin^2 \alpha}}$$

$$= -\cos \alpha \cdot \sin^{-1}\left(\frac{t}{\cos \alpha}\right) - \sin \alpha \cdot \log\left|u - \sqrt{u^2 - \sin^2 \alpha}\right| + C$$

$$= -\cos \alpha \cdot \sin^{-1}\left(\frac{\cos x}{\cos \alpha}\right)$$

$$- \sin \alpha \cdot \log|\sin x - \sqrt{\sin^2 x - \sin^2 \alpha}| + C$$

(xii) Let $I = \int \frac{2\sin 2x - \cos x}{6 - \cos^2 x - 4\sin x} dx$

$$= \int \frac{(4\sin x - 1)\cos x}{6 - (1 - \sin^2 x) - 4\sin x} dx$$

$$= \int \frac{(4\sin x - 1)\cos x}{\sin^2 x - 4\sin x + 5} dx$$

Put $\sin x = t \Rightarrow \cos x\, dx = dt$

$$\therefore \quad I = \int \frac{(4t - 1)}{(t^2 - 4t + 5)} dt \quad \ldots(i)$$

Now, let $(4t - 1) = \lambda(2t - 4) + \mu$

On comparing the coefficients of like powers of $t$, we get

$$2\lambda = 4, -4\lambda + \mu = -1$$

$$\Rightarrow \quad \lambda = 2, \mu = 7 \quad \ldots(ii)$$

$$\therefore \quad I = \int \frac{2(2t - 4) + 7}{(t^2 - 4t + 5)} dt \quad \text{[using Eqs. (i) and (ii)]}$$

$$= 2\int \frac{2t - 4}{t^2 - 4t + 5} dt + 7\int \frac{dt}{t^2 - 4t + 5}$$

$$= 2\log|t^2 - 4t + 5| + 7\int \frac{dt}{t^2 - 4t + 4 - 4 + 5}$$

$$= 2\log|t^2 - 4t + 5| + 7\int \frac{dt}{(t - 2)^2 + 1^2}$$

$$= 2\log|t^2 - 4t + 5| + 7\tan^{-1}(t - 2) + C$$

$$= 2\log|\sin^2 x - 4\sin x + 5| + 7\tan^{-1}(\sin x - 2) + C$$

(xiii) Let $I = \int \sqrt{\frac{\cos x - \cos^3 x}{(1 - \cos^3 x)}}\, dx$

$$= \int \frac{\sqrt{\cos x}\,(\sqrt{1 - \cos^2 x})}{\sqrt{1 - (\cos^{3/2} x)^2}} dx$$

$$= \int \frac{\sqrt{\cos x} \cdot \sin x}{\sqrt{1 - (\cos^{3/2} x)^2}} dx$$

Put $\cos^{3/2} x = t \Rightarrow \frac{3}{2}\cos^{1/2} x\,(-\sin x)\, dx = dt$

$$\therefore \quad I = -\frac{2}{3}\int \frac{dt}{\sqrt{1 - t^2}}$$

$$= -\frac{2}{3}\sin^{-1}(t) + C$$

$$= -\frac{2}{3}\sin^{-1}(\cos^{3/2} x) + C$$

(xiv) Let $I = \int \sqrt{\frac{a - x}{a + x}}\, dx$

$$= \int \sqrt{\frac{a - x}{a + x} \times \frac{a - x}{a - x}}\, dx$$

$$= \int \frac{a - x}{\sqrt{a^2 - x^2}} dx$$

$$= \int \frac{a}{\sqrt{a^2 - x^2}} dx - \int \frac{x}{\sqrt{a^2 - x^2}} dx$$

$$= 2\sin^{-1}\left(\frac{x}{a}\right) - \int \frac{x}{\sqrt{a^2 - x^2}} dx$$

Put $a^2 - x^2 = t^2$

$$\Rightarrow \quad -2x\,dx = 2t\,dt$$

$$\Rightarrow \quad x\,dx = -t\,dt$$

$$\therefore \quad I = a\sin^{-1}\left(\frac{x}{a}\right) + t + C$$

$$= a\sin^{-1}\frac{x}{a} + \sqrt{a^2 - x^2} + C$$

(xv) Let $I = \int x \cdot \sqrt{\frac{a^2 - x^2}{a^2 + x^2}}\, dx$

Put $x^2 = t \Rightarrow 2x\, dx = dt$

$$\therefore \quad I = \int \sqrt{\frac{a^2 - t}{a^2 + t}} \cdot \frac{dt}{2}$$

$$= \frac{1}{2}\int \sqrt{\frac{a^2 - t}{a^2 + t} \cdot \frac{a^2 - t}{a^2 - t}}\, dt$$

$$= \frac{1}{2}\int \frac{a^2 - t}{\sqrt{a^4 - t^2}}\, dt$$

$$= \frac{1}{2} a^2 \int \frac{dt}{\sqrt{(a^2)^2 - t^2}} - \frac{1}{2}\int \frac{t}{\sqrt{a^4 - t^2}}\, dt$$

$$= \frac{1}{2} a^2 \sin^{-1}\left(\frac{t}{a^2}\right) + \frac{1}{4}\int \frac{du}{\sqrt{u}}$$

where, $a^4 - t^2 = u \Rightarrow 2t\, dt = du$

$$= \frac{1}{2} a^2 \sin^{-1}\left(\frac{t}{a^2}\right) + \frac{1}{4}\left(\frac{u^{1/2}}{1/2}\right) + C$$

$$= \frac{1}{2} a^2 \sin^{-1}\left(\frac{x^2}{a^2}\right) + \frac{1}{4}\sqrt{a^4 - x^4} + C$$

(xvi) Let $I = \int x\sqrt{1 + x - x^2}\, dx$

Put $x = \lambda\left[\frac{d}{dx}(1 + x - x^2)\right] + \mu$

$\Rightarrow \quad x = \lambda(1 - 2x) + \mu$

On comparing the coefficients of like powers of $x$, we get

$$1 = -2\lambda$$

and $\quad \lambda + \mu = 0$

$$\Rightarrow \quad \lambda = -\frac{1}{2}, \mu = \frac{1}{2}$$

$$\therefore \quad I = \int\left[-\frac{1}{2}(1 - 2x) + \frac{1}{2}\right]\sqrt{1 + x - x^2}\, dx$$

$$= -\frac{1}{2}\int (1 - 2x)\sqrt{1 + x - x^2}\, dx + \frac{1}{2}\int \sqrt{1 + x - x^2}\, dx$$

$$= -\frac{1}{2}\int \sqrt{t}\, dt + \frac{1}{2}\int \sqrt{-\left(x^2 - x + \frac{1}{4} - \frac{1}{4} - 1\right)}\, dx$$

[where, $t = 1 + x - x^2$]

$$= -\frac{1}{2}\left(\frac{t^{3/2}}{3/2}\right) + \frac{1}{2}\int \sqrt{\left(\frac{\sqrt{5}}{2}\right)^2 - \left(x - \frac{1}{2}\right)^2}\, dx$$

$$= -\frac{1}{3}t^{3/2} + \frac{1}{2}\left[\left(x - \frac{1}{2}\right)\sqrt{\left(\frac{\sqrt{5}}{2}\right)^2 - \left(x - \frac{1}{2}\right)^2} + \frac{1}{2}\left(\frac{\sqrt{5}}{2}\right)^2 \sin^{-1}\left(\frac{x - \frac{1}{2}}{\frac{\sqrt{5}}{2}}\right)\right] + C$$

$$= -\frac{1}{3}(1 + x - x^2)^{3/2} + \frac{1}{2}\left[\frac{x - \frac{1}{2}}{2}\sqrt{1 + x - x^2} + \frac{5}{8}\sin^{-1}\frac{2x - 1}{\sqrt{5}}\right] + C$$

(xvii) Let $I = \int (x + 1)\sqrt{1 - x - x^2}\, dx$

Put $x + 1 = \lambda \cdot \left[\frac{d}{dx}(1 - x - x^2)\right] + \mu$

$\Rightarrow \quad x + 1 = \lambda(-1 - 2x) + \mu$

On comparing the coefficients of like powers of $x$, we get

$$-2\lambda = 1 \text{ and } \mu - \lambda = 1$$

$$\Rightarrow \quad \lambda = -\frac{1}{2} \text{ and } \mu = \frac{1}{2}$$

$$\therefore \quad (x + 1) = -\frac{1}{2}(-1 - 2x) + \frac{1}{2}$$

$$\therefore \quad I = \int\left[-\frac{1}{2}(-1 - 2x) + \frac{1}{2}\right]\sqrt{1 - x - x^2}\, dx$$

$$= -\frac{1}{2}\int (-1 - 2x)\sqrt{1 - x - x^2}\, dx + \frac{1}{2}\int \sqrt{1 - x - x^2}\, dx$$

$$= -\frac{1}{2}\int (-1 - 2x)\sqrt{1 - x - x^2}\, dx + \frac{1}{2}\int \sqrt{1 - \left(x^2 + x + \frac{1}{4} - \frac{1}{4}\right)}\, dx$$

Put $t = 1 - x - x^2$ in first integral, then

$dt = (-1 - 2x)\, dx$

$$\therefore \quad I = -\frac{1}{2}\int \sqrt{t}\, dt + \frac{1}{2}\int \sqrt{\left(\frac{\sqrt{5}}{2}\right)^2 - \left(x + \frac{1}{2}\right)^2}\, dx$$

$$= -\frac{1}{2}\left(\frac{t^{3/2}}{3/2}\right) + \left[\frac{1}{2}\left(x + \frac{1}{2}\right)\sqrt{1 - x - x^2} + \frac{1}{2} \times \frac{5}{4}\sin^{-1}\frac{x + \frac{1}{2}}{\frac{\sqrt{5}}{2}}\right] + C$$

$$= -\frac{1}{3}(1 - x - x^2)^{3/2} + \frac{1}{8}(2x + 1)\sqrt{1 - x - x^2} + \frac{5}{16}\sin^{-1}\frac{2x + 1}{\sqrt{5}} + C$$

(xviii) Let $I = \int \frac{2x^2 + 5x + 4}{\sqrt{x^2 + x + 1}}\, dx$

Put $2x^2 + 5x + 4 = \lambda(x^2 + x + 1) + \mu\left\{\frac{d}{dx}(x^2 + x + 1)\right\} + \gamma$

$\Rightarrow 2x^2 + 5x + 4 = \lambda(x^2 + x + 1) + \mu(2x + 1) + \gamma$

On comparing the coefficients of like terms, we get

$$\lambda = 2, \lambda + 2\mu = 5, \lambda + \mu + \gamma = 4$$

$$\Rightarrow \quad \lambda = 2, \mu = \frac{3}{2}, \gamma = \frac{1}{2}$$

$$\therefore I = \int\left[2 \cdot \frac{(x^2 + x + 1)}{\sqrt{x^2 + x + 1}} + \frac{3}{2} \cdot \frac{(2x + 1)}{\sqrt{x^2 + x + 1}} + \frac{1}{2} \cdot \frac{1}{\sqrt{x^2 + x + 1}}\right] dx$$

$$= 2\int \frac{x^2 + x + 1}{\sqrt{x^2 + x + 1}}\, dx + \frac{3}{2}\int \frac{(2x + 1)}{\sqrt{x^2 + x + 1}}\, dx + \frac{1}{2}\int \frac{dx}{\sqrt{x^2 + x + 1}}$$

Put $x^2 + x + 1 = t$ in second integral, then

$(2x + 1)\, dx = dt$

$$\therefore I = 2\int \sqrt{x^2 + x + 1}\, dx + \frac{3}{2}\int \frac{dt}{\sqrt{t}} + \frac{1}{2}\int \frac{dx}{\sqrt{x^2 + x + 1}}$$

$$= 2\int \sqrt{\left(x + \frac{1}{2}\right)^2 + \left(\frac{\sqrt{3}}{2}\right)^2}\, dx + \frac{3}{2} \cdot \frac{t^{1/2}}{1/2} + \frac{1}{2}\int \frac{dx}{\sqrt{\left(x + \frac{1}{2}\right)^2 + \left(\frac{\sqrt{3}}{2}\right)^2}}$$

$$= 2\left[\frac{1}{2}\left(x+\frac{1}{2}\right)\sqrt{x^2+x+1} + \frac{1}{2}\cdot\frac{3}{4}\log\left(x+\frac{1}{2}\right)\sqrt{x^2+x+1}\right] + 3\sqrt{x^2+x+1} + \frac{1}{2}\log\left|\left(x+\frac{1}{2}\right)+\sqrt{x^2+x+1}\right| + C$$

$$= \left(x+\frac{1}{2}\right)\sqrt{x^2+x+1} + \frac{3}{4}\log\left\{\left(x+\frac{1}{2}\right)+\sqrt{x^2+x+1}\right\} + 3\sqrt{x^2+x+1} + \frac{1}{2}\log\left\{\left(x+\frac{1}{2}\right)\right\} + \sqrt{x^2+x+1} + C$$

$$= \left(x+\frac{7}{2}\right)\sqrt{x^2+x-1} + \frac{5}{4}\log\left\{\left(x+\frac{1}{2}\right)+\sqrt{x^2+x+1}\right\} + C$$

**2.** (i) Let $I = \int \frac{dx}{4\sin^2 x + 9\cos^2 x}$

On dividing numerator and denominator by $\cos^2 x$, we get

$$I = \int \frac{\sec^2 x}{4\tan^2 x + 9}dx$$

Put $\tan x = t \Rightarrow \sec^2 x\, dx = dt$

$$\therefore \quad I = \int \frac{dt}{4t^2+9} = \frac{1}{4}\int \frac{dt}{t^2 + \left(\frac{3}{2}\right)^2}$$

$$= \frac{1}{4}\cdot\frac{1}{3/2}\tan^{-1}\left(\frac{t}{3/2}\right) + C$$

$$= \frac{1}{6}\tan^{-1}\left(\frac{2\tan x}{3}\right) + C$$

(ii) Let $I = \int \frac{\sin x}{\sin 3x}dx$

$$= \int \frac{\sin x}{3\sin x - 4\sin^3 x}dx = \int \frac{dx}{3 - 4\sin^2 x}$$

On dividing numerator and denominator by $\cos^2 x$, we get

$$I = \int \frac{\sec^2 x}{3\sec^2 x - 4\tan^2 x}dx$$

$$= \int \frac{\sec^2 x}{3(1+\tan^2 x) - 4\tan^2 x}dx$$

$$= \int \frac{\sec^2 x}{3 - \tan^2 x}dx$$

Put $\tan x = t \Rightarrow \sec^2 x\, dx = dt$

$$\therefore \quad I = \int \frac{dt}{(\sqrt{3})^2 - t^2} = \frac{1}{2\sqrt{3}}\log\left|\frac{\sqrt{3}+t}{\sqrt{3}-t}\right| + C$$

$$= \frac{1}{2\sqrt{3}}\log\left|\frac{\sqrt{3}+\tan x}{\sqrt{3}-\tan x}\right| + C$$

(iii) Let $I = \int \frac{dx}{2+\sin x + \cos x}$

$$= \int \frac{dx}{2 + \frac{2\tan\frac{x}{2}}{1+\tan^2\frac{x}{2}} + \frac{1-\tan^2\frac{x}{2}}{1+\tan^2\frac{x}{2}}}$$

$$\left[\because \sin x = \frac{2\tan\frac{x}{2}}{1+\tan^2\frac{x}{2}}, \cos x = \frac{1-\tan^2\frac{x}{2}}{1+\tan^2\frac{x}{2}}\right]$$

$$= \int \frac{\sec^2\frac{x}{2}}{2 + 2\tan^2\frac{x}{2} + 2\tan\frac{x}{2} + 1 - \tan^2\frac{x}{2}}dx$$

$$= \int \frac{\sec^2\frac{x}{2}}{\tan^2\frac{x}{2} + 2\tan\frac{x}{2} + 3}dx$$

Put $\tan\frac{x}{2} = t \Rightarrow \frac{1}{2}\sec^2\frac{x}{2}dx = dt$

$$\therefore \quad I = \int \frac{2}{t^2+2t+3}dt$$

$$= 2\int \frac{dt}{t^2+2t+1+2}$$

$$= 2\int \frac{dt}{(t+1)^2 + (\sqrt{2})^2}$$

$$= 2\cdot\frac{1}{\sqrt{2}}\tan^{-1}\left(\frac{t+1}{\sqrt{2}}\right) + C$$

$$= \sqrt{2}\tan^{-1}\left[\frac{\tan\frac{x}{2}+1}{\sqrt{2}}\right] + C$$

(iv) Let $I = \int \frac{dx}{\sqrt{3}\sin x + \cos x}$

$$= \int \frac{dx}{\sqrt{3}\cdot\frac{2\tan\frac{x}{2}}{1+\tan^2\frac{x}{2}} + \frac{1-\tan^2\frac{x}{2}}{1+\tan^2\frac{x}{2}}}$$

$$= \int \frac{\sec^2\frac{x}{2}}{2\sqrt{3}\tan\frac{x}{2} + 1 - \tan^2\frac{x}{2}}dx$$

Put $\tan\frac{x}{2} = t \Rightarrow \frac{1}{2}\sec^2\frac{x}{2}dx = dt$

$$\therefore \quad I = \int \frac{2}{-t^2 + 2\sqrt{3}t + 1}dt$$

$$= 2\int \frac{dt}{-t^2 + 2\sqrt{3}t - 3 + 3 + 1}$$

$$= 2\int \frac{dt}{4 - (t-\sqrt{3})^2}$$

$$= 2\int \frac{dt}{(2)^2 - (t-\sqrt{3})^2}$$

$$= 2\cdot\frac{1}{2\times 2}\log\left|\frac{2+t-\sqrt{3}}{2-t+\sqrt{3}}\right| + C$$

$$= \frac{1}{2}\log\left|\frac{2-\sqrt{3}+\tan\frac{x}{2}}{2+\sqrt{3}-\tan\frac{x}{2}}\right| + C$$

(v) Let $I = \int \frac{3\sin x + 2\cos x}{3\cos x + 2\sin x}dx$

Put $3\sin x + 2\cos x = \lambda(3\cos x + 2\sin x) + \mu\left\{\frac{d}{dx}\{3\cos x + 2\sin x\}\right\}$

$\Rightarrow \quad 3\sin x + 2\cos x = \lambda(3\cos x + 2\sin x) + \mu(-3\sin x + 2\cos x)$

On comparing the coefficients of $\sin x$ and $\cos x$ on both sides, we get

$$3 = 2\lambda - 3\mu \text{ and } 2 = 3\lambda + 2\mu$$

$\Rightarrow \quad \lambda = \frac{12}{13}$ and $\mu = -\frac{5}{13}$

$\therefore \quad I = \lambda \int dx + \mu \int \frac{-3\sin x + 2\cos x}{3\cos x + 2\sin x} dx$

$= \frac{12}{13} x - \frac{5}{13} \log|3\cos x + 2\sin x| + C$

**3.** (a) Let $I = \int \frac{(\cos x - \sin x)}{\sqrt{2\cos x \sin x} \cdot (\sin x + \cos x)} dx$

$= \int \frac{(\cos x - \sin x)}{(\sin x + \cos x)\sqrt{(\sin x + \cos x)^2 - 1}} dx$

Put $(\cos x + \sin x) = t \Rightarrow (\cos x - \sin x)\, dx = dt$

$\therefore \quad I = \int \frac{dt}{t\sqrt{t^2 - 1}} = \sec^{-1}(t) + C$

$= \sec^{-1}(\sin x + \cos x) + C$

**4.** (d) Let $I = \int \frac{\sin x}{\sin 4x} dx = \int \frac{\sin x}{2\sin 2x \cos 2x} dx$

$= \int \frac{\sin x}{4\sin x \cdot \cos x \cdot \cos 2x} dx$

$= \frac{1}{4}\int \frac{\cos x}{\cos^2 x \cdot (\cos 2x)} dx$

$= \frac{1}{4}\int \frac{\cos x}{(1 - \sin^2 x)\cdot(1 - 2\sin^2 x)} dx$

Put $\sin x = t \quad \Rightarrow \quad \cos x\, dx = dt$

$\therefore \quad I = \frac{1}{4}\int \frac{dt}{(1 - t^2)(1 - 2t^2)}$

$= \frac{1}{4}\int \left(\frac{-1}{1 - t^2} + \frac{2}{1 - 2t^2}\right) dt$ [using partial fractions]

$= -\frac{1}{4}\int \frac{dt}{1 - t^2} + \frac{1}{4}\cdot\int \frac{dt}{\left(\frac{1}{\sqrt{2}}\right)^2 - t^2}$

$= \frac{-1}{4} \times \frac{1}{2}\cdot\log\left|\frac{1 + t}{1 - t}\right| + \frac{1}{4} \times \frac{1}{2\cdot\frac{1}{\sqrt{2}}}\log\left|\frac{\frac{1}{\sqrt{2}} + t}{\frac{1}{\sqrt{2}} - t}\right| + C$

$= \frac{1}{4\sqrt{2}}\log\left|\frac{1 + \sqrt{2}t}{1 - \sqrt{2}t}\right| - \frac{1}{8}\log\left|\frac{1 + t}{1 - t}\right| + C$

$= \frac{1}{4\sqrt{2}}\log\left|\frac{1 + \sqrt{2}\sin x}{1 - \sqrt{2}\sin x}\right| - \frac{1}{8}\log\left|\frac{1 + \sin x}{1 - \sin x}\right| + C$

**5.** (c) Let $I = \int \frac{d\theta}{\cos^3\theta \sqrt{\sin 2\theta}}$

$= \frac{1}{\sqrt{2}}\int \frac{d\theta}{\sin^{1/2}\theta \cos^{7/2}\theta}$

$= \frac{1}{\sqrt{2}}\int \sin^{-1/2}\theta \cos^{-7/2}\theta\, d\theta$

$= \frac{1}{\sqrt{2}}\int \tan^{-1/2}\theta\, (\sec^2\theta)^2 d\theta$

Put $\tan\theta = t \Rightarrow \sec^2\theta\, d\theta = dt$

$= \frac{1}{\sqrt{2}}\int t^{-1/2}(1 + t^2)\, dt$

$= \frac{1}{\sqrt{2}}\int (t^{-1/2} + t^{3/2})\, dt$

$= \frac{1}{\sqrt{2}}\left[\frac{t^{1/2}}{1/2} + \frac{2}{5}t^{5/2}\right] + C$

$= \frac{1}{\sqrt{2}}\left[2\tan^{1/2}\theta + \frac{2}{5}\tan^{5/2}\theta\right] + C$

$= \frac{\sqrt{2}}{5}(\tan^2\theta + 5)\sqrt{\tan\theta} + C$

**6.** (d) Let $I = \int \frac{dx}{\cos x - \sin x}$

$= \frac{1}{\sqrt{2}}\int \frac{dx}{\cos x \cos\frac{\pi}{4} - \sin x \sin\frac{\pi}{4}}$

$= \frac{1}{\sqrt{2}}\int \frac{dx}{\cos\left(x + \frac{\pi}{4}\right)}$

$= \frac{1}{\sqrt{2}}\int \sec\left(x + \frac{\pi}{4}\right) dx$

$= \frac{1}{\sqrt{2}}\log\left|\tan\left(\frac{\pi}{4} + \frac{x}{2}\right) + \frac{\pi}{8}\right| + C$

$= \frac{1}{\sqrt{2}}\log\left|\tan\left(\frac{x}{2} + \frac{3\pi}{8}\right)\right| + C$

**7.** (a) Let $I = \int \frac{\cos x}{\cos x + \sqrt{3}\sin x} dx$

$= \frac{1}{2}\int \frac{dx}{\frac{1}{2}\cos x + \frac{\sqrt{3}}{2}\sin x}$

$= \frac{1}{2}\int \frac{dx}{\cos x \cos\frac{\pi}{3} + \sin x \sin\frac{\pi}{3}}$

$= \frac{1}{2}\int \frac{dx}{\cos\left(x - \frac{\pi}{3}\right)} = \frac{1}{2}\int \sec\left(x - \frac{\pi}{3}\right) dx$

$= \frac{1}{2}\log\left|\tan\left(\frac{\pi}{4} + \frac{x}{2} - \frac{\pi}{6}\right)\right| + C$

$= \frac{1}{2}\log\left|\tan\left(\frac{x}{2} + \frac{\pi}{12}\right)\right| + C$

**8.** (a) Let $I = \int \frac{5\tan x}{\tan x - 2} dx$

On multiplying by $\cos x$ in numerator and denominator, we get

$I = \int \frac{5\sin x}{\sin x - 2\cos x} dx$

This special integration requires special substitutation of type

$N^r = A(D^r) + B\frac{d}{dx}(D^r)$

$\Rightarrow \quad 5\sin x = A(\sin x - 2\cos x) + B(\cos x + 2\sin x)$

Comparing coefficients of $\sin x$ and $\cos x$ on both sides, we get

$A + 2B = 5$

$B - 2A = 0$

$\Rightarrow \quad A = 1, B = 2$

$\Rightarrow \quad 5\sin x = (\sin x - 2\cos x) + 2(\cos x + 2\sin x)$

$\therefore \quad I = \int \frac{(\sin x - 2\cos x) + 2(\cos x + 2\sin x)}{(\sin x - 2\cos x)} dx$

$= \int \frac{\sin x - 2\cos x}{\sin x - 2\cos x} dx + 2\int \frac{\cos x + 2\sin x}{\sin x - 2\cos x} dx$

$= \int dx + 2\int \frac{\cos x + 2\sin x}{\sin x - 2\cos x} dx$

Put $\sin x - 2\cos x = t$

$\Rightarrow (\cos x + 2\sin x)\, dx = dt$

$\therefore \quad I = x + 2\int \frac{dt}{t}$

$= x + 2\log|t| + k$

$= x + 2\log|\sin x - 2\cos x| + k$

But $\int \frac{5\tan x}{\tan x - 2} dx = x + a\log|\sin x - 2\sin x| + k$

$\Rightarrow \quad a = 2$

**9.** (a)

(i) We have,

$$f(x) = \frac{1}{(a^2+b^2)-(a^2-b^2)\cos x}$$

$$= \frac{1}{(a^2+b^2)-(a^2-b^2)\dfrac{\left(1-\tan^2\frac{x}{2}\right)}{1+\tan^2\frac{x}{2}}}$$

$$= \frac{\sec^2\frac{x}{2}}{(a^2+b^2)\left(1+\tan^2\frac{x}{2}\right)-(a^2-b^2)\left(1-\tan^2\frac{x}{2}\right)}$$

$$= \frac{\sec^2\frac{x}{2}}{a^2+b^2+(a^2+b^2)\tan^2\frac{x}{2}-a^2+b^2+(a^2-b^2)\tan^2\frac{x}{2}}$$

$$= \frac{\sec^2\frac{x}{2}}{2b^2+2a^2\tan^2\frac{x}{2}}$$

$$= \frac{\sec^2\frac{x}{2}}{2a^2\left(\frac{b^2}{a^2}+\tan^2\frac{x}{2}\right)}$$

Now, $\int f(x)dx = \frac{1}{2a^2}\int \frac{\sec^2\frac{x}{2}}{\frac{b^2}{a^2}+\tan^2\frac{x}{2}}dx$

Put $\tan\frac{x}{2} = t \Rightarrow \sec^2\frac{x}{2}dx = 2dt$

$$= \frac{1}{a^2}\int \frac{dt}{\frac{b^2}{a^2}+t^2}$$

$$= \frac{1}{a^2}\cdot\frac{a}{b}\tan^{-1}\left(\frac{t}{b/a}\right)+C$$

$$= \frac{1}{ab}\tan^{-1}\left(\frac{\tan\frac{x}{2}}{\frac{b}{a}}\right)+C$$

$$= \frac{1}{ab}\tan^{-1}\left(\frac{a}{b}\tan\frac{x}{2}\right)+C$$

(ii) We have,

$$f(x) = \frac{1}{a^2\sin^2 x+b^2\cos^2 x}$$

$$= \frac{\sec^2 x}{a^2\tan^2 x+b^2}$$

Now, $\int f(x)dx = \int \frac{\sec^2 x}{a^2\tan^2 x+b^2}dx$

Put $\tan x = t \Rightarrow \sec^2 x\,dx = dt$

$$= \int \frac{dt}{a^2t^2+b^2}$$

$$= \frac{1}{a^2}\int \frac{dt}{t^2+\left(\frac{b}{a}\right)^2}$$

$$= \frac{1}{a^2}\times\frac{a}{b}\tan^{-1}\left(\frac{t}{\frac{b}{a}}\right)+C$$

$$= \frac{1}{ab}\tan^{-1}\left(\frac{a\tan x}{b}\right)+C$$

(iii) We have, $f(x) = \frac{1}{a\cos x+b\sin x}$

On substituting $a = r\sin\theta$ and $b = r\cos\theta$, we get

$$r = \sqrt{a^2+b^2} \text{ and } \theta = \tan^{-1} a/b$$

$$\therefore \int f(x)\,dx = \int \frac{dx}{r\sin\theta + r\cos\theta}$$

$$= \frac{1}{r}\int \frac{dx}{\sin(x+\theta)}$$

$$= \frac{1}{r}\int \text{cosec}\,(x+\theta)\,dx$$

$$= \frac{1}{r}\log\left|\tan\left(\frac{x}{2}+\frac{\theta}{2}\right)\right|+C$$

$$= \frac{1}{\sqrt{a^2+b^2}}\log\left|\tan\left(\frac{x}{2}+\frac{1}{2}\tan^{-1}\frac{a}{b}\right)\right|+C$$

(iv) We have,

$$f(x) = \frac{1}{a^2-b^2\cos^2 x} = \frac{\sec^2 x}{a^2\sec^2 x-b^2}$$

Now, $\int f(x)\,dx = \int \frac{\sec^2 x}{a^2\sec^2 x-b^2}dx$

$$= \int \frac{\sec^2 x}{a^2\tan^2 x+(a^2-b^2)}dx$$

$$= \int \frac{dt}{a^2t^2+(a^2-b^2)}$$

$$= \frac{1}{a^2}\cdot\int \frac{dt}{t^2+\left[\frac{\sqrt{a^2-b^2}}{a}\right]^2}$$

$$= \frac{1}{a^2}\cdot\frac{1}{\frac{\sqrt{a^2-b^2}}{a}}\cdot\tan^{-1}\left[\frac{at}{\sqrt{a^2-b^2}}\right]+C$$

$$= \frac{1}{a^2\cdot\sqrt{1-\frac{b^2}{a^2}}}\cdot\tan^{-1}\left(\frac{t}{\sqrt{1-\frac{b^2}{a^2}}}\right)+C$$

$$= \frac{1}{a^2\cdot\sin\alpha}\cdot\tan^{-1}\left[\frac{\tan x}{\sin\alpha}\right]+C, \text{ where } \alpha = \cos^{-1}\left(\frac{b}{a}\right)$$

Hence, (i) → (p), (ii) → (r), (iii) → (s), (iv) → (q)

**10.** (a)

(i) Let $I = \int \frac{dx}{(a^2+x^2)^{3/2}} = \int \frac{dx}{x^3\cdot\left(\frac{a^2}{x^2}+1\right)^{3/2}}$

Put $a^2x^{-2}+1 = t \Rightarrow -\frac{2a^2}{x^3}dx = dt$

$$\therefore \quad I = \frac{-1}{2a^2}\int\frac{dt}{t^{3/2}} = \frac{-1}{2a^2}\cdot\frac{1}{t^{1/2}\cdot(-1/2)}+C$$

$$= \frac{1}{a^2}\cdot\frac{x}{\sqrt{a^2+x^2}}+C = \frac{x}{a^2\sqrt{x^2+a^2}}+C$$

(ii) Let $I = \int \frac{x^2}{\sqrt{a^2-x^2}}dx$

Put $x = a\sin\theta \Rightarrow dx = a\cos\theta\,d\theta$

$$\therefore I = \int \frac{a^2\sin^2\theta\cdot a\cos\theta}{a\cdot\cos\theta}d\theta = \frac{1}{2}a^2\int(1-\cos 2\theta)d\theta$$

$$= \frac{a^2}{2}\cdot\left\{\theta-\frac{\sin 2\theta}{2}\right\}+C$$

$$= \frac{a^2}{2}\left\{\sin^{-1}\left(\frac{x}{a}\right)-\frac{x}{a}\cdot\frac{\sqrt{a^2-x^2}}{a}\right\}+C$$

(iii) Let $I = \int \frac{dx}{(x^2 - a^2)^{3/2}} = \int \frac{dx}{x^3\left(1 - \frac{a^2}{x^2}\right)^{3/2}}$

Put $1 - a^2x^{-2} = t \Rightarrow 2a^2x^{-3}dx = dt \Rightarrow \frac{dx}{x^3} = \frac{dt}{2a^2}$

$\therefore \quad I = \int \frac{dt}{2a^2 \cdot t^{3/2}} = \frac{1}{2a^2} \cdot \frac{1}{t^{1/2} \cdot (-1/2)} + C$

$= -\frac{1}{a^2 \cdot \sqrt{1 - \frac{a^2}{x^2}}} + Cx = C - \frac{x}{a^2 \cdot \sqrt{x^2 - a^2}}$

(iv) Let $I = \int \frac{dx}{x\sqrt{x^2 - a^2}}$

Put $x = a\operatorname{cosec}\theta \Rightarrow dx = -a\operatorname{cosec}\theta \cdot \cot\theta d\theta$

$\therefore \quad I = \frac{-a\operatorname{cosec}\theta\cot\theta}{a\operatorname{cosec}\theta \cdot a\cot\theta} d\theta = \frac{-1}{a} \cdot \theta + C$

$= \frac{-1}{a} \cdot \sin^{-1}\left(\frac{a}{x}\right) + C$

Hence, (i) → (s), (ii) → (q), (iii) → (r), (iv) → (p)

**11.** (1) Given, $\int \frac{dx}{5 + 4\cos x} = a\tan^{-1}[bf(x)] + C$

Let $\quad I = \int \frac{dx}{5 + 4\cos x} = \int \frac{dx}{5 + 4\left(\frac{1 - \tan^2\frac{x}{2}}{1 + \tan^2\frac{x}{2}}\right)}$

$= \int \frac{\sec^2\frac{x}{2}}{9 + \tan^2\frac{x}{2}} dx$

Put $\tan\frac{x}{2} = t \Rightarrow \frac{1}{2}\sec^2\frac{x}{2} dx = dt$

$\therefore \quad I = 2\int \frac{dt}{9 + t^2}$

$= 2\int \frac{dt}{(3)^2 + t^2} = \frac{2}{3}\tan^{-1}\left(\frac{t}{3}\right) + C$

$= \frac{2}{3}\tan^{-1}\left(\frac{1}{3}\tan\frac{x}{2}\right) + C$

$\Rightarrow \quad f(x) = \tan\frac{x}{2}$

Now, $\lim_{x \to 0} \frac{2f(x)}{x} = 2\lim_{x \to 0} \frac{\tan\frac{x}{2}}{x} = \lim_{x \to 0} \frac{\tan\frac{x}{2}}{\frac{x}{2}} = 1$

**12.** (1) Let $I = \int \frac{(3x^4 - 1)}{(x^4 + x + 1)^2} dx = \int \frac{(3x^4 - 1)}{x^2(x^3 + 1 + 1/x)^2} dx$

$= \int \frac{(3x^2 - x^{-2})}{(x^3 + 1 + 1/x)^2} dx$

Put $\left(x^3 + 1 + \frac{1}{x}\right) = t \Rightarrow \left(3x^2 - \frac{1}{x^2}\right)dx = dt$

$\therefore \quad I = \int \frac{dt}{t^2} = -\frac{1}{t} + C$

$= -\frac{1}{x^3 + 1 + \frac{1}{x}} + C = -\frac{x}{x^4 + x + 1} + C$

$\Rightarrow \quad f(x) = -\frac{x}{x^4 + x + 1}$

$\therefore \quad x^3 \cdot f(x) = -\frac{x^4}{x^4 + x + 1}$

Now, $\left|\lim_{x \to \infty} x^3 \cdot f(x)\right| = \left|\lim_{x \to \infty} \frac{-x^4}{x^4 + x + 1}\right| = 1$

**13.** (0) Let $I = \int \frac{dx}{x^6(1 - x^{-5})^{1/3}}$

Put $1 - x^{-5} = t \Rightarrow \frac{5}{x^6}dx = dt$

$\therefore \quad I = \frac{1}{5}\int \frac{dt}{t^{1/3}} = \frac{1}{5} \cdot \frac{t^{2/3}}{2/3} + C$

$= \frac{3}{10}\left(\frac{x^5 - 1}{x^5}\right)^{2/3} + C$

On comparing with $a\left(\frac{x^\lambda + \mu}{x^\gamma}\right)^{2/3} + C$, we get

$\lambda = 5, \mu = -1, \gamma = 5$

$\therefore \quad \left|\frac{\lambda}{\gamma} + \mu\right| = 0$

**14.** (4) $\int (\sqrt{\tan x} + \sqrt{\cot x})dx$

$= \sqrt{\lambda}\tan^{-1}\left(\frac{\sqrt{\tan x} - \sqrt{\cot x}}{\sqrt{\mu}}\right) + C$

Now, $\int (\sqrt{\tan x} + \sqrt{\cot x})dx = \int \frac{\sin x + \cos x}{\sqrt{\sin x \cos x}} dx$

$= \sqrt{2}\int \frac{\sin x + \cos x}{\sqrt{\sin 2x}} dx = \sqrt{2}\int \frac{\sin x + \cos x}{\sqrt{1 - (\sin x - \cos x)^2}} dx$

$= \sqrt{2}\int \frac{dt}{\sqrt{1 - t^2}}$, where $t = \sin x - \cos x$

$= \sqrt{2}\sin^{-1}(t) + C$

$= \sqrt{2}\tan^{-1}\frac{t}{\sqrt{1 - t^2}} + C$

$= \sqrt{2}\tan^{-1}\left(\frac{\sin x - \cos x}{\sqrt{\sin 2x}}\right) + C$

$= \sqrt{2}\tan^{-1}\left(\frac{\sqrt{\tan x} - \sqrt{\cot x}}{\sqrt{2}}\right) + C$

$\Rightarrow \quad \lambda = 2, \mu = 2$

$\therefore \quad \lambda + \mu = 2 + 2 = 4$

**15.** (0) Given, $\int \frac{(ax^2 - b)}{x\sqrt{c^2x^2 - (ax^2 + b)^2}} dx = \sin^{-1}[f(x)] + C$

Now, $\quad I = \int \frac{(ax^2 - b)}{x\sqrt{c^2x^2 - (ax^2 + b)^2}} dx$

$= \int \frac{\left(a - \frac{b}{x^2}\right)}{\sqrt{c^2 - \left(ax + \frac{b}{x}\right)^2}} dx$

Put $ax + \frac{b}{x} = t \Rightarrow \left(a - \frac{b}{x^2}\right)dx = dt$

$\therefore \quad I = \int \frac{dt}{\sqrt{c^2 - t^2}} = \sin^{-1}\left(\frac{t}{c}\right) + C$

$= \sin^{-1}\left(\frac{ax + \frac{b}{x}}{c}\right) + C$

$\Rightarrow \quad f(x) = \frac{ax^2 + b}{cx}$

Now, $\lim_{x \to 0} [2f(x) \cdot x^2] = \lim_{x \to 0} 2 \cdot \frac{ax^2 + b}{cx} \cdot x^2$

$= 2\lim_{x \to 0} \frac{(ax^3 + bx)}{c} = 0$

## DPP-5 $\int \frac{dx}{P\sqrt{Q}}$ Form, Partial Fractions and Reduction Formula

**1.** (i) Let $I = \int \frac{dx}{(x+3)\sqrt{x+1}}$

Put $x + 1 = t^2 \Rightarrow dx = 2t\,dt$

$\therefore \quad I = 2\int \frac{dt}{t^2+2} = 2\int \frac{dt}{t^2+(\sqrt{2})^2}$

$= 2\cdot\frac{1}{\sqrt{2}}\tan^{-1}\frac{t}{\sqrt{2}} + C = \sqrt{2}\tan^{-1}\sqrt{\frac{x+1}{2}} + C$

(ii) Let $I = \int \frac{dx}{(x+1)\sqrt{x^2+x+1}}$

Put $x + 1 = \frac{1}{t} \Rightarrow dx = -\frac{1}{t^2}dt$

$\therefore \quad I = \int \frac{-1/t^2}{\frac{1}{t}\sqrt{\frac{1}{t^2}+\frac{1}{t}+1}}dt = \int \frac{-1/t^2}{\frac{1}{t^2}\sqrt{1+t+t^2}}dt$

$= -\int \frac{dt}{\sqrt{t^2+t+1}} = -\int \frac{dt}{\sqrt{\left(t+\frac{1}{2}\right)^2+\frac{3}{4}}}$

$= -\int \frac{dt}{\sqrt{\left(t+\frac{1}{2}\right)^2+\left(\frac{\sqrt{3}}{2}\right)^2}}$

$= -\frac{1}{2\times\frac{\sqrt{3}}{2}}\log\left|\frac{t+\frac{1}{2}+\frac{\sqrt{3}}{2}}{t+\frac{1}{2}-\frac{\sqrt{3}}{2}}\right| + C$

$= -\frac{1}{\sqrt{3}}\log\left|\frac{\frac{1}{x+1}+\frac{1+\sqrt{3}}{2}}{\frac{1}{x+1}+\frac{1-\sqrt{3}}{2}}\right| + C$

$= -\frac{1}{\sqrt{3}}\log\left|\frac{2+(1+\sqrt{3})(x+1)}{2+(1-\sqrt{3})(x+1)}\right| + C$

$= -\frac{1}{\sqrt{3}}\log\left|\frac{2+x+1+\sqrt{3}x+\sqrt{3}}{2+x+1-\sqrt{3}x-\sqrt{3}}\right| + C$

$= -\frac{1}{\sqrt{3}}\log\left|\frac{(\sqrt{3}+1)x+3+\sqrt{3}}{(1-\sqrt{3})x+3-\sqrt{3}}\right| + C$

(iii) Let $I = \int \sqrt{\frac{2-x}{x-3}}\,dx$

$= \int \frac{2-x}{\sqrt{(2-x)(x-3)}}dx$

$= \int \frac{2}{\sqrt{(2-x)(x-3)}}dx - \int \frac{x}{\sqrt{(2-x)(x-3)}}dx$

$= \int \frac{2}{\sqrt{5x-x-6}}dx - \int \frac{x}{\sqrt{5x-x^2-6}}dx$

$= 2\int \frac{dx}{\sqrt{x^2-5x+6-\frac{25}{4}+\frac{25}{4}}} + \frac{1}{2}\int \frac{-2x+5}{\sqrt{-\left(x-\frac{5}{2}\right)^2+\left(\frac{1}{2}\right)^2}}$

$- \int \frac{5\,dx}{\sqrt{\left(x-\frac{5}{2}\right)^2\left(\frac{1}{2}\right)^2}}$

$= -3\int \frac{dx}{\sqrt{\left(x-\frac{5}{2}\right)^2-\left(\frac{1}{2}\right)^2}} + \frac{1}{2}\int \frac{(-2x+5)\,dx}{\sqrt{\left(\frac{1}{2}\right)^2\left(x-\frac{5}{2}\right)^2}}$

$= -\sin^{-1}(2x-5) + \sqrt{(2-x)(x-3)}$

(iv) Let $I = \int \frac{dx}{(2x^2+3)\sqrt{x^2-4}}$

Put $x = \frac{1}{t} \Rightarrow dx = -\frac{1}{t^2}dt$

$\therefore \quad I = \int \frac{-1/t^2}{\left(\frac{2}{t^2}+3\right)\sqrt{\frac{1}{t^2}-4}}dt$

$= \int \frac{-1/t^2}{\left(\frac{2}{t^2}+3\right)\sqrt{\frac{1}{t^2}-4}}dt$

$= \int \frac{-1/t^2}{\frac{(2+3t^2)\sqrt{1-4t^2}}{t^3}}dt$

$= \int \frac{-t}{(2+3t^2)\sqrt{1-4t^2}}dt$

Again, put $t^2 = u \Rightarrow 2t\,dt = du$

Now, $\quad I = -\frac{1}{2}\int \frac{du}{(2+3u)\sqrt{1-4u}}$

Put $1-4u = z^2 \Rightarrow -4\,du = 2z\,dz$

$\therefore \quad I = \frac{1}{4}\int \frac{z}{2+3\left(\frac{1-z^2}{4}\right)\times z}dz$

$= \frac{1}{4}\int \frac{dz}{\frac{8+3-3z^2}{4}} = \int \frac{dz}{11-3z^2}$

$= \frac{1}{3}\int \frac{dz}{\frac{11}{3}-z^2} = \frac{1}{3}\int \frac{dz}{\left(\sqrt{\frac{11}{3}}\right)^2-z^2}$

$= \frac{1}{3}\times\frac{1}{2\sqrt{\frac{11}{3}}}\log\left|\frac{\sqrt{\frac{11}{3}}+z}{\sqrt{\frac{11}{3}}-z}\right| + C$

$= \frac{1}{2\sqrt{33}}\log\left|\frac{\sqrt{\frac{11}{3}}+\sqrt{\frac{x^2-4}{x^2}}}{\sqrt{\frac{11}{3}}-\sqrt{\frac{x^2-4}{x^2}}}\right| + C$

$= \frac{1}{2\sqrt{33}}\log\left|\frac{\sqrt{11}x+\sqrt{3x^2-12}}{\sqrt{11}x-\sqrt{3x^2-12}}\right| + C$

(v) Let $I = \int \sqrt{(x-7)(6-x)}\,dx$

$= \int \sqrt{6x-x^2-42+7x}\,dx = \int \sqrt{13x-x^2-42}\,dx$

$= \int \sqrt{-(x^2-13x+42)}\,dx$

$= \int \sqrt{-\left[x^2-13x+\left(\frac{13}{2}\right)^2-\left(\frac{13}{2}\right)^2+42\right]}\,dx$

$= \int \sqrt{-\left[\left(x-\frac{13}{2}\right)^2-\frac{1}{4}\right]}\,dx = \int \sqrt{\left(\frac{1}{2}\right)^2-\left(x-\frac{13}{2}\right)^2}\,dx$

$= \frac{\left(x-\frac{13}{2}\right)}{2}\sqrt{(x-7)(6-x)} + \frac{1}{8}\sin^{-1}\left(\frac{x-\frac{13}{2}}{\frac{1}{2}}\right) + C$

$= \frac{2x-13}{4}\sqrt{(x-7)(6-x)} + \frac{1}{8}\sin^{-1}(2x-13) + C$

**2.** (i) Let $I = \int \frac{2x^3 - 3x^2 - 8x - 26}{(2x^2 - 5x - 12)} dx$

$$= \int \left[ (x+1) + \frac{9x - 14}{2x^2 - 5x - 12} \right] dx$$

$$= \int (x+1)\, dx + \int \frac{9x - 14}{2x^2 - 5x - 12} dx$$

$$= \frac{x^2}{2} + x + \int \frac{9x - 14}{2x^2 - 5x - 12} dx$$

Put $9x - 14 = A\frac{d}{dx}(2x^2 - 5x - 12) + B$

$\Rightarrow 9x - 14 = A(4x - 5) + B$

$\Rightarrow 9x - 14 = 4Ax - 5A + B$

On comparing the coefficients of like terms, we get

$$4A = 9, \ -5A + B = -14$$

$$\Rightarrow \quad A = \frac{9}{4}, B = -4 + \frac{45}{4} = \frac{29}{4}$$

$$\therefore I = \frac{x^2}{2} + x + \int \frac{\left[ \frac{9}{4}(4x - 5) + \frac{29}{4} \right]}{2x^2 - 5x - 12} dx$$

$$= \frac{x^2}{2} + x + \frac{9}{4} \int \frac{4x - 5}{2x^2 - 5x - 12} dx + \frac{29}{4} \int \frac{dx}{2x^2 - 5x - 12}$$

$$= \frac{x^2}{2} + x + \frac{9}{4} \log (2x^2 - 5x - 12) + \frac{29}{8} \int \frac{dx}{\left(x - \frac{5}{2}\right)^2 - \frac{49}{4}} + C$$

$$= \frac{x^2}{2} + x + \frac{9}{4} \log (2x^2 - 5x - 12)$$

$$+ \frac{29}{8} \times \frac{1}{2 \times \frac{7}{2}} \log \left| \frac{x - \frac{5}{2} - \frac{7}{2}}{x - \frac{5}{2} + \frac{7}{2}} \right| + C$$

$$= \frac{x^2}{2} + x + \frac{9}{4} \log (2x^2 - 5x - 12) + \frac{29}{56} \log \left| \frac{2x - 12}{2x + 2} \right| + C$$

(ii) Let $I = \int \frac{(3x + 2)}{x^3 - 6x^2 + 11x - 6} dx$

We have, $\frac{3x + 2}{x^3 - 6x^2 + 11x - 6} = \frac{3x + 2}{(x-1)(x-2)(x-3)}$

Again, let $\frac{3x + 2}{(x-1)(x-2)(x-3)} = \frac{A}{x-1} + \frac{B}{x-2} + \frac{C}{x-3}$

$$\Rightarrow \frac{3x + 2}{(x-1)(x-2)(x-3)} = \frac{A(x-2)(x-3) + B(x-1)(x-3) + C(x-1)(x-2)}{(x-1)(x-2)(x-3)}$$

$\Rightarrow \quad 3x + 2 = A(x-2)(x-3) + B(x-1)(x-3) + C(x-1)(x-2)$ ...(i)

On putting $x = 1$ in Eq. (i), we get

$$5 = A(1-2)(1-3) \Rightarrow A = 5/2$$

On putting $x = 2$ in Eq. (i), we get

$$8 = B(2-1)(2-3) \Rightarrow B = -8$$

On putting $x = 3$ in Eq. (i), we get

$$11 = C(3-1)(3-2) \Rightarrow C = \frac{11}{2}$$

$$\therefore \quad \frac{3x + 2}{x^3 - 6x^2 + 11x - 6} = \frac{5}{2(x-1)} - \frac{8}{x-2} + \frac{11}{2(x-3)}$$

Now, $I = \int \frac{3x + 2}{x^3 - 6x^2 + 11x - 6}$

$$= \frac{5}{2} \int \frac{dx}{x-1} - 8 \int \frac{dx}{x-2} + \frac{11}{2} \int \frac{dx}{x-3}$$

$$= \frac{5}{2} \log |(x-1)| - 8 \log |x-2| + \frac{11}{2} \log |x-3| + C$$

(iii) Let $I = \int \frac{3x - 2}{(x-1)^2 (x+1)(x+2)} dx$

We have,

$$\frac{3x - 2}{(x-1)^2(x+1)(x+2)} = \frac{A}{x-1} + \frac{B}{(x-1)^2} + \frac{C}{x+1} + \frac{D}{x+2}$$

$\Rightarrow \quad (3x - 2) = A(x-1)(x+1)(x+2) + B(x+1)(x+2) + C(x-1)^2(x+2) + D(x-1)^2(x+1)$ ...(i)

On putting $x = 1$ in Eq. (i), we get

$$1 = B(1+1)(1+2) \Rightarrow B = \frac{1}{6}$$

On putting $x = -1$ in Eq. (i), we get

$$-5 = C(-2)^2(-1+2) \Rightarrow C = -\frac{5}{4}$$

On putting $x = -2$ in Eq. (i), we get

$$-8 = D(-3)^2(-1) \Rightarrow \quad D = \frac{8}{9}$$

On equating the coefficients of $x^3$ both sides, we get

$$0 = A + C + D$$

$$\Rightarrow \quad A = -C - D = \frac{5}{4} - \frac{8}{9} = \frac{13}{36}$$

$$\therefore I = \int \frac{3x - 2}{(x-1)^2(x+1)(x+2)}$$

$$= \int \frac{13}{36(x-1)} dx + \frac{1}{6} \int \frac{dx}{(x-1)^2} - \frac{5}{4} \int \frac{dx}{x+1} + \frac{8}{9} \int \frac{dx}{x+2}$$

$$= \frac{13}{36} \log (x-1) - \frac{1}{6(x-1)} - \frac{5}{4} \log (x+1) + \frac{8}{9} \log (x+2) + C$$

(iv) Let $I = \int \frac{(2x - 1)}{(x+1)(x^2 + 2)} dx$

We have, $\frac{2x - 1}{(x+1)(x^2 + 2)} = \frac{A}{x+1} + \frac{Bx + C}{x^2 + 2}$

$$\Rightarrow \quad \frac{2x - 1}{(x+1)(x^2 + 2)} = \frac{A(x^2 + 2) + (Bx + C)(x+1)}{(x+1)(x^2 + 2)}$$

$\Rightarrow \quad 2x - 1 = A(x^2 + 2) + (Bx + C)(x+1)$ ...(i)

On putting $x = -1$ in Eq. (i), we get

$$-3 = A(3) \Rightarrow A = -1$$

Now, comparing the coefficient of like powers of $x$ on both sides of Eq. (i), we get

$$A + B = 0, C + 2A = -1 \text{ and } C + B = 2$$

$$-1 + B = 0, C - 2 = -1 \quad \Rightarrow B = 1, C = 1$$

$$\therefore \quad \frac{2x - 1}{(x+1)(x^2 + 2)} = -\frac{1}{(x+1)} + \frac{x + 1}{(x^2 + 2)}$$

Now, $I = \int \frac{2x - 1}{(x+1)(x^2 + 2)} = -\int \frac{dx}{x+1} + \int \frac{x + 1}{x^2 + 2} dx$

$$= -\log |x + 1| - \int \frac{x}{x^2 + 2} dx + \int \frac{dx}{x^2 + 2}$$

$$= -\log |x + 1| - \frac{1}{2} \log |x^2 + 2| + \frac{1}{\sqrt{2}} \tan^{-1} \frac{x}{\sqrt{2}} + C$$

(v) Let $I = \int \frac{(x + 2)}{(x^2 + 3x + 3)\sqrt{x + 1}} dx$

Put $x + 1 = t^2 \quad \Rightarrow \quad dx = 2t\, dt$

$$\therefore I = \int \frac{(t^2 + 1)\, 2t}{\{(t^2 - 1)^2 + 3(t^2 - 1) + 3\} t} dt = \int \frac{t^2 + 1}{t^4 + t^2 + 1} dt$$

$$= 2 \int \frac{1 + 1/t^2}{t^2 + \frac{1}{t^2} + 1} dt = 2 \int \frac{1 + \frac{1}{t^2}}{\left(t - \frac{1}{t}\right)^2 + (\sqrt{3})^2} dt$$

$$= 2 \int \frac{du}{u^2 + (\sqrt{3})^2}, \text{ where } t - \frac{1}{t} = u$$

$$= \frac{2}{\sqrt{3}} \tan^{-1} \frac{u}{\sqrt{3}} + C = \frac{2}{\sqrt{3}} \tan^{-1} \left( \frac{t - (1/t)}{\sqrt{3}} \right) + C$$

$$= \frac{2}{\sqrt{3}} \tan^{-1} \left( \frac{t^2 - 1}{t\sqrt{3}} \right) + C = \frac{2}{\sqrt{3}} \tan^{-1} \left[ \frac{x}{\sqrt{3(x + 1)}} \right] + C$$

(vi) Let $I = \int \frac{x^2 - 4}{(x^2 + 1)(x^2 + 2)(x^2 + 3)} dx$

We have, $\frac{x^2 - 4}{(x^2 + 1)(x^2 + 2)(x^2 + 3)}$

Put $x^2 = t$, then

$$\frac{t - 4}{(t + 1)(t + 2)(t + 3)} = \frac{A}{t + 1} + \frac{B}{t + 2} + \frac{C}{t + 3}$$

$$\Rightarrow \frac{t - 4}{(t + 1)(t + 2)(t + 3)} = \frac{A(t + 2)(t + 3) + B(t + 1)(t + 3) + C(t + 1)(t + 2)}{(t + 1)(t + 2)(t + 3)}$$

$$\Rightarrow (t - 4) = A(t + 2)(t + 3) + B(t + 1)(t + 3) + C(t + 1)(t + 2) \quad \dots(i)$$

On putting $t = -1$ in Eq. (i), we get

$$-5 = A(-1 + 2)(-1 + 3) \Rightarrow A = -\frac{5}{2}$$

On putting $t = -2$ in Eq. (i), we get

$$-6 = B(-2 + 1)(-2 + 3) \Rightarrow B = 6$$

On putting $t = -3$ in Eq. (i), we get

$$-7 = C(-3 + 1)(-3 + 2) \Rightarrow C = -7/2$$

$$\therefore \frac{t - 4}{(t + 1)(t + 2)(t + 3)} = -\frac{5}{2(t - 1)} + \frac{6}{(t - 2)} - \frac{7}{2(t - 3)}$$

$$\Rightarrow \frac{x^2 - 4}{(x^2 + 1)(x^2 + 2)(x^2 + 3)} = -\frac{5}{2(x^2 + 1)} + \frac{6}{x^2 + 2} - \frac{7}{2(x^2 + 3)}$$

$$\therefore \quad I = \int \frac{(x^2 - 4)}{(x^2 - 1)(x^2 - 2)(x^2 - 3)} dx$$

$$= -\frac{5}{2}\int \frac{dx}{x^2 + 1} + 6\int \frac{dx}{x^2 + 2} - \frac{7}{2}\int \frac{dx}{x^2 + 3}$$

$$= -\frac{5}{2}\tan^{-1} x + \frac{6}{\sqrt{2}}\tan^{-1}\frac{x}{\sqrt{2}} - \frac{7}{2\sqrt{3}}\tan^{-1}\frac{x}{\sqrt{3}} + C$$

(vii) Let $I = \int \frac{x^4 - x^2 + 1}{x^2(x^2 + 1)^2} dx$

We have $\frac{x^4 - x^2 + 1}{x^2(x^2 + 1)^2}$

Put $x^2 = t$, then

$$\frac{t^2 - t + 1}{t(t + 1)^2} = \frac{A}{t} + \frac{B}{t + 1} + \frac{C}{(t + 1)^2}$$

$$\Rightarrow \quad t^2 - t + 1 = A(t + 1)^2 + Bt(t + 1) + C(t) \quad \dots(i)$$

On putting $t = 0$ in Eq. (i), we get $1 = A$

On putting $t = -1$ in Eq. (i), we get $3 = -C \Rightarrow C = -3$

On putting $t = 2$ in Eq. (i), we get $3 = 9A + 6B + 2C$

$$\Rightarrow \quad 6B + 3 = 3$$

$$B = 0$$

$$\therefore I = \int \frac{x^4 - x^2 + 1}{x^2(x^2 + 1)^2} = \int \frac{dx}{x^2} - \int \frac{3}{(x^2 + 1)^2} dx = -\frac{1}{x} - I_1 + C \quad \dots(ii)$$

Now, $\quad I_1 = 3\int \frac{dx}{(x^2 + 1)^2}$

We know that, $\int \frac{dx}{x^2 + a^2} = \frac{1}{a}\tan^{-1}\frac{x}{a}$

$$\Rightarrow \quad \int \frac{dx}{x^2 + 1} = \tan^{-1} x$$

Also, $\int \frac{1}{x^2 + 1} \cdot 1\, dx = \frac{1}{x^2 + 1} \cdot x - \int \frac{-2x^2}{(x^2 + 1)^2} dx$

$$= \frac{x}{x^2 + 1} + 2\int \frac{x^2}{(x^2 + 1)^2} dx$$

$$= \frac{x}{x^2 + 1} + 2\int \frac{x^2 + 1 - 1}{(x^2 + 1)^2} dx$$

$$\Rightarrow \quad \int \frac{dx}{x^2 + 1} = \frac{x}{x^2 + 1} + 2\int \frac{dx}{x^2 + 1} - 2\int \frac{dx}{(x^2 + 1)^2}$$

$$\Rightarrow \quad \tan^{-1} x = \frac{x}{x^2 + 1} + 2\tan^{-1} x - 2\int \frac{dx}{(x^2 + 1)^2}$$

$$\Rightarrow \quad \int \frac{dx}{(x^2 + 1)^2} = \frac{x}{2(x^2 + 1)} + \frac{3}{2}\tan^{-1} x$$

$$\therefore \quad I_1 = 3\int \frac{dx}{(x^2 + 1)^2} = \frac{3}{2} \cdot \frac{x}{(x^2 + 1)^2} + \frac{3}{2}\tan^{-1} x$$

On substituting the value of $I_1$ in Eq. (ii), we get

$$\int \frac{x^4 - x^2 + 1}{x^2(x^2 + 1)^2} = -\frac{1}{x} - \frac{3}{2} \cdot \frac{x}{(x^2 + 1)^2} - \frac{3}{2}\tan^{-1} x + C$$

(viii) Let $I = \int \frac{dx}{x(x^5 + 1)} = \int \frac{dx}{x^6\left(1 + \frac{1}{x^5}\right)}$

Put $1 + \frac{1}{x^5} = t \Rightarrow -\frac{5}{x^6}dx = dt$

$$\therefore \quad I = -\frac{1}{5}\int \frac{dt}{t} = -\frac{1}{5}\log|t| + C$$

$$= -\frac{1}{5}\log\left|1 + \frac{1}{x^5}\right| + C = -\frac{1}{5}\log|1 + x^{-5}| + C$$

(ix) Let $I = \int \frac{dx}{x^2(x^7 + 1)^{6/7}} = \int \frac{dx}{x^8\left(1 + \frac{1}{x^7}\right)^{6/7}}$

Put $1 + \frac{1}{x^7} = t \quad \Rightarrow \quad -\frac{7}{x^8}dx = dt$

$$\therefore \quad I = -\frac{1}{7}\int \frac{dt}{t^{6/7}} = -\frac{1}{7} \cdot \frac{t^{-6/7 + 1}}{-\frac{6}{7} + 1} + C$$

$$= -t^{1/7} + C = -(1 + x^{-7})^{1/7} + C$$

(x) Let $I = \int \frac{1 - x^2}{1 - x^2 + x^4} dx = \int \frac{x^2\left(\frac{1}{x^2} - 1\right)dx}{x^2\left(\frac{1}{x^2} - 1 + x^2\right)}$

$$= -\int \frac{\left(1 - \frac{1}{x^2}\right)dx}{x^2 + \frac{1}{x^2} - 1} = -\int \frac{\left(1 - \frac{1}{x^2}\right)dx}{\left(x + \frac{1}{x}\right)^2 - 3}$$

Put $x + \frac{1}{x} = t \Rightarrow \left(1 - \frac{1}{x^2}\right)dx = dt$

$$\therefore \; I = -\int \frac{dt}{t^2 - 3} = -\int \frac{dt}{t^2 - (\sqrt{3})^2} = -\frac{1}{2\sqrt{3}}\log\left|\frac{t - \sqrt{3}}{t + \sqrt{3}}\right| + C$$

$$= -\frac{1}{2\sqrt{3}}\log\left|\frac{x + \frac{1}{x} - \sqrt{3}}{x + \frac{1}{x} + \sqrt{3}}\right| + C$$

(xi) Let $I = \int \frac{dx}{x^4 + 1} = \frac{1}{2}\int \frac{2}{x^4 + 1} dx$

$$= \frac{1}{2}\int \frac{1 + x^2 + 1 - x^2}{x^4 + 1} dx = \frac{1}{2}\left[\int \frac{1 + x^2}{1 + x^4}dx + \int \frac{1 - x^2}{1 + x^4}dx\right]$$

$$= \frac{1}{2}\left[\int \frac{x^2 + 1}{x^4 + 1}dx - \int \frac{x^2 - 1}{x^4 + 1}dx\right]$$

$$= \frac{1}{2}\left[\int \frac{1 + \frac{1}{x^2}}{x^2 + \frac{1}{x^2}}dx - \int \frac{1 - \frac{1}{x^2}}{x^2 + \frac{1}{x^2}}dx\right]$$

$$= \frac{1}{2}\left[\int \frac{\left(1 + \frac{1}{x^2}\right)}{\left(x - \frac{1}{x}\right)^2 + 2}dx - \int \frac{\left(1 - \frac{1}{x^2}\right)}{\left(x + \frac{1}{x}\right)^2 - 2}dx\right]$$

$$= \frac{1}{2}\left[\int \frac{dt}{t^2 + (\sqrt{2})^2} - \int \frac{du}{u^2 - (\sqrt{2})^2}\right]$$

where, $t = x - \frac{1}{x}, u = x + \frac{1}{x}$

$$\therefore \quad I = \frac{1}{2}\left[\frac{1}{\sqrt{2}}\tan^{-1}\left(\frac{t}{\sqrt{2}}\right) - \frac{1}{2\sqrt{2}}\log\left|\frac{u - \sqrt{2}}{u + \sqrt{2}}\right|\right] + C$$

$$= \frac{1}{2\sqrt{2}}\tan^{-1}\left(\frac{x - \frac{1}{x}}{\sqrt{2}}\right) - \frac{1}{4\sqrt{2}}\log\left|\frac{x + \frac{1}{x} - \sqrt{2}}{x + \frac{1}{x} + \sqrt{2}}\right| + C$$

$$= \frac{1}{2\sqrt{2}}\tan\left(\frac{x^2 - 1}{x\sqrt{2}}\right) - \frac{1}{4\sqrt{2}}\log\left|\frac{x^2 + 1 - \sqrt{2}x}{x^2 + 1 + \sqrt{2}x}\right| + C$$

(xii) Let $I = \int \frac{dx}{x^{1/2}\cdot(2 + 3x^{1/3})^2}$

Put $x = t^6 \Rightarrow dx = 6t^5 dt$

$$\therefore \quad I = \int \frac{6t^5}{t^3\cdot(2 + 3t^2)^2}dt = 6\int \frac{t^2}{(2 + 3t^2)^2}dt$$

$$= 2\int \left[\frac{1}{(2 + 3t^2)} - \frac{2}{(2 + 3t^2)^2}\right]dt$$

$$= 2\int \frac{dt}{3t^2 + 2} - 4\int \frac{dt}{9t^4 + 12t^2 + 4}$$

$$= 2\cdot\int \frac{dt}{3t^2 + 2} - \int \frac{\{(2 + 3t^2) + (2 - 3t^2)\}dt}{(3t^2 + 2)^2}$$

$$= 2\int \frac{dt}{3t^2 + 2} - \int \frac{dt}{3t^2 + 2} - \int \frac{2 - 3t^2}{(3t^2 + 2)^2}dt$$

$$= \int \frac{dt}{3t^2 + 2} + \int \frac{(3 - 2/t^2)}{(3t + 2/t)^2}dt$$

Put $3t + \frac{2}{t} = z \Rightarrow \left(3 - \frac{1}{t^2}\right)dt = dz$

$$\therefore \quad I = \frac{1}{3}\int \frac{dt}{t^2 + \left(\sqrt{\frac{2}{3}}\right)^2} + \int \frac{dz}{z^2}$$

$$= \frac{1}{3}\cdot\frac{1}{\sqrt{\frac{2}{3}}}\cdot\tan^{-1}\left(\frac{t}{\sqrt{\frac{2}{3}}}\right) - \frac{1}{z} + C$$

$$= \frac{1}{\sqrt{6}}\tan^{-1}\left(\sqrt{\frac{3}{2}}t\right) - \frac{t}{(2 + 3t^2)} + C$$

$$= \frac{1}{\sqrt{6}}\tan^{-1}\left(\sqrt{\frac{3}{2}}x^{1/6}\right) - \frac{x^{1/6}}{(2 + 3x^{1/3})} + C$$

(xiii) Let $I = \int \frac{(1 + x^2)}{(1 - x^2)\sqrt{1 + x^2 + x^4}}dx$

$$= \int \frac{x^2(1/x^2 + 1)\,dx}{x^2\left(\frac{1}{x} - x\right)\sqrt{\frac{1}{x^2} + 1 + x^2}} = \int \frac{\left(\frac{1}{x^2} + 1\right)dx}{\left(\frac{1}{x} - x\right)\sqrt{\frac{1}{x^2} + x^2 + 1}}$$

$$= -\int \frac{\left(1 + \frac{1}{x^2}\right)dx}{\left(x - \frac{1}{x}\right)\sqrt{\left(x - \frac{1}{x}\right)^2 + 3}}$$

Put $x - \frac{1}{x} = t$

$\Rightarrow \left(1 + \frac{1}{x^2}\right)dx = dt$

Now, $I = -\int \frac{dt}{t\sqrt{t^2 + 3}}$

Again, put $t^2 + 3 = s^2 \Rightarrow 2t\ dt = 2s ds$

$$\therefore \quad I = -\int \frac{s}{s(s^2 - 3)}ds = -\int \frac{ds}{s^2 - (\sqrt{3})^2}$$

$$= -\frac{1}{2\sqrt{3}}\log\left|\frac{s - \sqrt{3}}{s + \sqrt{3}}\right| + C$$

$$= -\frac{1}{2\sqrt{3}}\log\left|\frac{\sqrt{t^2 + 3} - \sqrt{3}}{\sqrt{t^2 + 3} + \sqrt{3}}\right| + C$$

$$= -\frac{1}{2\sqrt{3}}\log\left|\frac{\sqrt{\left(x - \frac{1}{x}\right)^2 + 3} - \sqrt{3}}{\sqrt{\left(x - \frac{1}{x}\right)^2 + 3} + \sqrt{3}}\right| + C$$

$$= -\frac{1}{2\sqrt{3}}\log\left|\frac{\sqrt{x^2 + \frac{1}{x^2} + 1} - \sqrt{3}}{\sqrt{x^2 + \frac{1}{x^2} + 1} + \sqrt{3}}\right| + C$$

(xiv) Let $I = \int \sqrt{\tan x}\,dx$

Put $\tan x = t^2 \Rightarrow \sec^2 x dx = 2t\ dt$

$$\therefore I = \int t\cdot\frac{2t}{1 + t^4}dt = \int \frac{2t^2}{t^4 + 1}dt$$

$$= \int \frac{t^2 + 1}{t^4 + 1}dt + \int \frac{t^2 - 1}{t^4 + 1}dt = \int \frac{1 + \frac{1}{t^2}}{t^2 + \frac{1}{t^2}}dt + \int \frac{1 - \frac{1}{t^2}}{t^2 + \frac{1}{t^2}}dt$$

$$= \int \frac{1 + \frac{1}{t^2}}{\left(t^2 + \frac{1}{t}\right)^2 + 2}dt + \int \frac{1 + \frac{1}{t^2}}{\left(t^2 + \frac{1}{t}\right)^2 - 2}dt$$

$\therefore = \int \frac{ds}{s^2 + (\sqrt{2})^2} + \int \frac{dr}{r^2 - (\sqrt{2})^2}$, where $s = t - \frac{1}{t}$ and $r = t + \frac{1}{t}$

$$= \frac{1}{\sqrt{2}}\tan^{-1}\left(\frac{s}{\sqrt{2}}\right) + \frac{1}{2\sqrt{2}}\log\left|\frac{r - \sqrt{2}}{r + \sqrt{2}}\right| + C$$

$$= \frac{1}{\sqrt{2}}\left[\tan^{-1}\left(\frac{t - \frac{1}{t}}{\sqrt{2}}\right) + \frac{1}{2}\log\left(\frac{t + \frac{1}{t} - \sqrt{2}}{t + \frac{1}{t} + \sqrt{2}}\right)\right] + C,$$

where $t = \sqrt{\tan x}$

(xv) Let $I = \int \frac{dx}{(x + 2)^{8/7}(x - 1)^{6/7}} = \int \frac{dx}{(x + 2)^2\cdot\left(\frac{x - 1}{x + 2}\right)^{6/7}}$

Put $\frac{x - 1}{x + 2} = t \Rightarrow \frac{3}{(x + 2)^2}dx = dt$

$$\therefore \quad I = \frac{1}{3}\int \frac{dt}{t^{6/7}} = \frac{1}{3}\cdot\frac{t^{1/7}}{1/7} + C = \frac{7}{3}\left(\frac{x - 1}{x + 2}\right)^{1/7} + C$$

**3.** Let $A = \sin^{m+1}x\cos^{n-1}x$

$$\therefore \quad \frac{dA}{dx} = (m + 1)\sin^m x\cos x\cdot\cos^{n-1}x$$
$$+ \sin^{m+1}x(n - 1)\cos^{n-2}x(-\sin x)$$
$$= (m + 1)\sin^m x\cos^n x - \sin^{m+2}x(n - 1)\cos^{n-2}x$$
$$= (m + 1)\sin^m x\cos^n x - \sin^m x(1 - \cos^2 x)(n - 1)\cos^{n-2}x$$
$$= (m + 1)\sin^m x\cos^n x - \sin^m x\cos^{n-2}x(n - 1)$$
$$+ (n - 1)\sin^m x\cos^n x$$
$$= (m + n)\sin^m x\cos^n x - (n - 1)\sin^m x\cos^{n-2}x$$

On integrating both sides w.r.t. $x$, we get

$$A = \int (m + n)\sin^m x\cos^n x\,dx - (n - 1)\int \sin^m x\cos^{n-2}x\,dx$$

$$\Rightarrow \quad A = (m + n)\int \sin^m x\cos^n x\,dx - (n - 1)\int \sin^m x\cos^{n-2}x\,dx$$

$$\Rightarrow (m + n)\int \sin^m x\cos^n x\,dx = (n - 1)\int \sin^m x\cos^{n-2}x\,dx + A$$

$\Rightarrow (m+n)\int \sin^m x \cos^n x dx$

$= (n-1)\int \sin^m x \cos^{n-2} x\, dx + \sin^{m+1} x \cos^{n-1} x$

$\Rightarrow \int \sin^m x \cos^n x\, dx = \frac{n-1}{m+n}\int \sin^m x \cos^{n-2} x\, dx + \frac{\sin^{m+1} x \cos^{n-1} x}{m+n}$

$\therefore \quad I_{m,n} = \frac{\sin^{m+1} x \cos^{n-1} x}{m+n} + \frac{n-1}{m+n} I_{m,n-2}$

**Hence proved.**

**4.** Given, $I_n = \int \frac{dx}{(x^2+a^2)^n} = \int \underset{\text{I}}{\frac{1}{(x^2+a^2)^n}} \cdot \underset{\text{II}}{1}\, dx$

On applying integration by parts, we get

$I_n = \frac{1}{(x^2+a^2)^n}\cdot x - \int \frac{2x}{(x^2+a^2)^{n+1}}\cdot(-n)\cdot(x)\, dx$

$= \frac{x}{(x^2+a^2)^n} + 2n\int \frac{x^2}{(x^2+a^2)^{n+1}} dx$

$= \frac{x}{(x^2+a^2)^n} + 2n\int \frac{x^2+a^2-a^2}{(x^2+a^2)^{n+1}} dx$

$\therefore \quad I_n = \frac{x}{(x^2+a^2)^n} + 2n\int \frac{1}{(x^2+a^2)^n} dx - 2a^2 n\int \frac{dx}{(x^2+a^2)^{n+1}}$

$I_n = \frac{x}{(x^2+a^2)^n} + 2n\, I_n - 2na^2\, I_{n+1}$

$\Rightarrow \quad 2na^2\, I_{n+1} = \frac{x}{(x^2+a^2)^n} + (2n-1)\, I_n$

$\Rightarrow \quad I_{n+1} = \frac{1}{2na^2}\cdot\frac{x}{(x^2+a^2)^n} + \frac{(2n-1)}{2n}\cdot\frac{1}{a^2}\cdot I_n$

**Hence proved.**

**5.** (c) Let $I = \int \frac{dx}{\sqrt{\sin^3 x\cdot(\sin x + 2\cos x)}}$

$= \int \frac{\text{cosec}^2 x}{\sqrt{1+2\cot x}} dx$

Put $1 + 2\cot x = t^2 \Rightarrow -2\,\text{cosec}^2 x\, dx = 2t\, dt$

$\therefore \quad I = -\int \frac{t}{t} dt = -t + C$

$= -\sqrt{1+2\cot x} + C$

Now, $\quad f(x) = -\sqrt{1+2\cot x}$

$\therefore \quad f\left(\frac{\pi}{4}\right) = -\sqrt{3}$

**6.** (b) Let $I = \int \frac{(\sqrt{1-x^2}-x)}{\sqrt{1-x^2}(1+x\sqrt{1-x^2})} dx$

Put $\quad x = \sin\theta \Rightarrow dx = \cos\theta\, d\theta$

$\therefore \quad I = \int \frac{(\cos\theta - \sin\theta)\cdot\cos\theta}{\cos\theta(1+\sin\theta\cos\theta)} d\theta = \int \frac{2(\cos\theta-\sin\theta)}{2+\sin 2\theta} d\theta$

$= 2\int \frac{(\cos\theta - \sin\theta)}{1+(\sin\theta+\cos\theta)^2} d\theta$

Again, put $\sin\theta + \cos\theta = t$

$\Rightarrow \quad (\cos\theta - \sin\theta)d\theta = dt$

$\therefore \quad I = 2\int \frac{dt}{1+t^2} = 2\tan^{-1}(t) + C$

$= 2\tan^{-1}(\sin\theta + \cos\theta) + C$

$= 2\tan^{-1}(x + \sqrt{1-x^2}) + C$

$\therefore \quad f(x) = 2\tan^{-1}(x+\sqrt{1-x^2})$

$\Rightarrow \quad f(1) = 2\tan^{-1}(1) = \pi/2$

**7.** (b) $I_{n,-m} = \int \frac{\sin^n x}{\cos^m x} dx$

$= \int \underset{\text{I}}{\sin^{n-1} x}\cdot \underset{\text{II}}{\frac{\sin x}{\cos^m x}} dx$

$= \sin^{n-1} x\cdot\frac{(\cos x)^{-m+1}}{m-1}$

$- \int (n-1)\sin^{n-2} x\cos x\frac{(\cos x)^{-m+1}}{m-1} dx$

$= \frac{1}{m-1}\cdot\frac{\sin^{n-1} x}{\cos^{m-1} x} - \frac{n-1}{m-1}\int \frac{\sin^{n-2} x}{\cos^{m-2} x} dx$

$I_{n,-m} = \frac{1}{m-1}\cdot\frac{\sin^{n-1} x}{\cos^{m-1} x} - \frac{n-1}{m-1} I_{n-2,\,2-m}$

$\Rightarrow \quad I_{n,-m} + \frac{n-1}{m-1} I_{n-2,\,2-m} = \frac{1}{m-1}\cdot\frac{\sin^{n-1} x}{\cos^{m-1} x}$

**8.** (d) Let $u_n = \int \frac{x^n}{\sqrt{ax^2+2bx+c}}$

$\therefore \quad u_{n+1} = \int \frac{x^{n+1}}{\sqrt{ax^2+2bx+c}} = \int \frac{x^n\cdot x}{\sqrt{ax^2+2bx+c}}$

$\Rightarrow \quad u_{n+1} = \frac{1}{2a}\int \frac{x^n(2ax+2b-2b)}{ax^2+bx+c} dx$

$\Rightarrow 2a\, u_{n+1} = \int \frac{x^n(2ax+2b)}{\sqrt{ax^2+2bx+c}} dx - 2b\int \frac{x^n}{\sqrt{ax^2+2bx+c}} dx$

$= x^n\cdot 2\sqrt{ax^2+2bx+c} - 2n\int x^{n-1}\sqrt{ax^2+2bx+c} - 2bu_n$

$= 2x^n\sqrt{ax^2+2bx+c} - 2n\int \frac{x^{n-1}(ax^2+2bx+c)}{\sqrt{ax^2+2bx+c}} - 2b\, u_n$

$= 2x^n\sqrt{ax^2+2bx+c} - 2n(au_{n-1} - 2b\, u_n + cu_{n-1}) - 2bu_n$

$\Rightarrow \quad 2a(n+1)u_{n+1} = 2x^n\sqrt{ax^2+2bx+c}$

$- 4n\, bu_n - 2n\, Cu_n - 1 - 2bu_n$

$\Rightarrow a(n+1)u_{n+1} + b(2n-1)u_n + nc\, u_{n-1}$

$= x^n\sqrt{ax^2+2bx+c}$

**9.** (2) Let $I = \int \frac{dx}{x^{1/2}\cdot(1+x^2)^{5/4}}$

$= \frac{dx}{x^{1/2}\cdot x^{5/2}\cdot(x^{-2}+1)^{5/4}} = \int \frac{dx}{x^3\cdot(1+x^{-2})^{5/4}}$

Put $1 + x^{-2} = t \Rightarrow \frac{-2}{x^3} dx = dt$

$\therefore \quad I = \frac{-1}{2}\int \frac{dt}{t^{5/4}}$

$= -\frac{1}{2}\cdot\frac{1}{t^{1/4}\cdot(-1/4)} = \frac{2}{(1+x^{-2})^{1/4}}$

$= \frac{2\sqrt{x}}{(1+x^2)^{1/4}} + C$

$\therefore \quad A = 2$

**10.** (5) Let $I = \int \frac{2x+3}{(x^2+3x)(x^2+3x+2)+1} dx$

Put $x^2 + 3x = t \Rightarrow (2x+3)dx = dt$

$\therefore \quad I = \int \frac{dt}{t(t+2)+1} = \int \frac{dt}{(t+1)^2}$

$= C - \frac{1}{t+1} = C - \frac{1}{x^2+3x+1}$

$\Rightarrow \quad a = 1, b = 3, c = 1$

$\therefore \quad a + b + c = 5$

## Revisal Problems (JEE Main)

**1.** (b) Given, $f\left(\frac{3x-4}{3x+4}\right)=x+2$

Let $\frac{3x-4}{3x+4}=\alpha$

$\Rightarrow \frac{(3x-4)+(3x+4)}{(3x-4)-(3x+4)}=\frac{\alpha+1}{\alpha-1}$

[by componendo and dividendo]

$\Rightarrow \frac{6x}{-8}=\frac{\alpha+1}{\alpha-1}$

$\Rightarrow x=-\frac{4}{3}\left(\frac{\alpha+1}{\alpha-1}\right)$

$\Rightarrow x+2=-\frac{4\alpha+4}{3\alpha-3}+2$

$=\frac{-4\alpha-4+6\alpha-6}{3\alpha-3}$

$=\frac{2\alpha-10}{3\alpha-3}$

$\therefore f\left(\frac{3x-4}{3x+4}\right)=x+2$

$\Rightarrow f(\alpha)=\frac{2\alpha-10}{3\alpha-3}$

$=\frac{2}{3}\left(\frac{\alpha-5}{\alpha-1}\right)$

$=\frac{2}{3}\left(\frac{\alpha-1-4}{\alpha-1}\right)$

$=\frac{2}{3}\left(1-\frac{4}{\alpha-1}\right)$

$=\frac{2}{3}-\frac{8}{3(\alpha-1)}$

$\Rightarrow f(x)=\frac{2}{3}-\frac{8}{3(x-1)}$

$\therefore \int f(x)\,dx=\int\left[\frac{2}{3}-\frac{8}{3(x-1)}\right]dx$

$\Rightarrow \int f(x)\,dx=\frac{2}{3}x-\frac{8}{3}\log|x-1|+C$

**2.** (b) Let $I=\int\frac{dx}{x^{1/5}(1+x^{4/5})^{1/2}}=\int x^{-1/5}(1+x^{4/5})^{-1/2}dx$

Put $1+x^{4/5}=t^2$

$\Rightarrow \frac{4}{5}x^{-1/5}dx=2t\,dt$

$\therefore I=\frac{5}{2}\int\frac{t}{t}dt$

$=\frac{5}{2}\int dt=\frac{5}{2}t+C$

$=\frac{5}{2}\sqrt{(1+x^{4/5})}+C$

**3.** (a) Let $I=\int\frac{\sin x\cos x}{a^2\sin^2x+b^2\cos^2x}dx$

$=\frac{1}{2(a^2-b^2)}\int\frac{(a^2-b^2)\sin 2x}{a^2\sin^2x+b^2\cos^2x}dx$

$=\frac{1}{2(a^2-b^2)}\int\frac{1}{a^2\sin^2x+b^2\cos^2x}d(a^2\sin^2x+b^2\cos^2x)$

$=\frac{1}{2(a^2-b^2)}\log|a^2\sin^2x+b^2\cos^2x|+C$

Hence, $f(x)=\frac{1}{a^2\sin^2x+b^2\cos^2x}$

**4.** (b) Let $I=\int\frac{\cos x-\sin x}{\sqrt{8-\sin 2x}}dx$

$=\int\frac{1}{\sqrt{(3)^2-(\sin x+\cos x)^2}}d(\sin x+\cos x)$

$=\sin^{-1}\left(\frac{\sin x+\cos x}{3}\right)+C$

Hence, $a=3$

**5.** (d) Let $I=\int x\log\left(1+\frac{1}{x}\right)dx$

$=\int \underset{\text{II}}{x}\,\underset{\text{I}}{\log(1+x)}\,dx-\int \underset{\text{II}}{x}\,\underset{\text{I}}{\log x}\,dx$

$=\frac{x^2}{2}\log(1+x)-\frac{1}{2}\int\frac{x^2}{x+1}dx-\left\{\frac{x^2}{2}\log x-\int\frac{1}{x}\times\frac{x^2}{2}dx\right\}$

$\Rightarrow I=\frac{x^2}{2}\log(1+x)-\frac{1}{2}\int x-1+\frac{1}{x+1}dx-\frac{x^2}{2}\log x+\frac{x^2}{4}+C$

$\Rightarrow I=\frac{x^2}{2}\log(1+x)-\frac{1}{2}\left(\frac{x^2}{2}-x\right)-\frac{1}{2}\log(x+1)$

$-\frac{x^2}{2}\log x+\frac{x^2}{4}+C$

$\Rightarrow I=\left(\frac{x^2-1}{2}\right)\log(1+x)-\frac{x^2}{2}\log x+\frac{x}{2}+C$

Hence, $L=\frac{1}{2}$

**6.** (a) Let $I=\int\frac{x\sin x^2e^{\sec x^2}}{\cos^2x^2}dx$

$=\frac{1}{2}\int e^{\sec x^2}\cdot 2x\tan x^2\sec x^2\,dx$

$=\frac{1}{2}\int e^{\sec x^2}d(\sec x^2)=\frac{1}{2}e^{\sec x^2}+C$

**7.** (b) Let $I=\int x\sqrt{\frac{2\sin^2(x^2+1)-\sin 2(x^2+1)}{2\sin^2(x^2+1)+\sin 2(x^2+1)}}dx$

$=\int x\sqrt{\frac{2\sin(x^2+1)-2\sin(x^2+1)\cos(x^2+1)}{2\sin(x^2+1)+2\sin(x^2+1)\cos(x^2+1)}}dx$

$=\int x\sqrt{\frac{1-\cos(x^2+1)}{1+\cos(x^2+1)}}dx=\int x\cdot\tan\left(\frac{x^2+1}{2}\right)dx$

$=\int\tan\left(\frac{x^2+1}{2}\right)d\left(\frac{x^2+1}{2}\right)$

$=\log\left|\sec\left(\frac{x^2+1}{2}\right)\right|+C$

**8.** (b) Let $I=\int\frac{dx}{\sqrt{\sin^3x\sin(x+\alpha)}}$

$=\int\frac{dx}{\sin^3x\{\sin x\cos\alpha+\cos x\sin\alpha\}}$

$=\int\frac{dx}{\sin^4x\{\cos\alpha+\cot x\sin\alpha\}}$

$=\int\frac{\operatorname{cosec}^2x}{\sqrt{\cos\alpha+\sin\alpha\cot x}}dx$

Put $\cos\alpha+\sin\alpha\cot x=t^2$

$\Rightarrow -\sin\alpha\operatorname{cosec}^2x\,dx=2t\,dt$

$\therefore I=-\frac{2}{\sin\alpha}\int\frac{t}{t}dt=\frac{2}{\sin\alpha}\cdot t+C$

$=-2\operatorname{cosec}\alpha\sqrt{\cos\alpha+\sin\alpha\cot x}+C$

9. (b) Let $I = \int \frac{f(x)g'(x) - g(x)f'(x)}{f(x)\cdot g(x)}[\log g(x) - \log f(x)]\,dx$

$$= \int \log\left\{\frac{g(x)}{f(x)}\right\}\cdot d\left\{\log\frac{g(x)}{f(x)}\right\}$$

$$= \frac{1}{2}\left\{\log\frac{g(x)}{f(x)}\right\}^2 + C$$

10. (c) Let $I = \int \frac{dx}{\sin^6 x + \cos^6 x}$

On dividing numerator and denominator by $\cos^6 x$, we get

$$I = \int \frac{\sec^6 x}{1+\tan^6 x}dx$$

$$= \int \frac{(1+\tan^2 x)^2 \sec^2 x}{1+\tan^6 x}dx$$

Put $\tan x = t$

$\Rightarrow$ $\sec^2 x\,dx = dt$

$\therefore$ $$I = \int \frac{(1+t^2)^2}{1+t^6}dt$$

$$= \int \frac{(1+t^2)^2}{(1+t^2)(1-t^2+t^4)}dt$$

$$= \int \frac{1+t^2}{(1-t^2+t^4)}dt = \int \frac{(1+1/t^2)\,dt}{\left(\frac{1}{t^2}-1+t^2\right)}$$

$$= \int \frac{\left(1+\frac{1}{t^2}\right)dt}{\left(t-\frac{1}{t}\right)^2+1}$$

Put $t - \frac{1}{t} = z$

$\Rightarrow$ $1 + \frac{1}{t^2}dt = dz$

$$= \int \frac{dz}{z^2+1} = \tan^{-1}(z) + C$$

$$= \tan^{-1}\left(\frac{t^2-1}{t}\right) + C$$

$$= \tan^{-1}\left(\frac{\tan^2 x - 1}{\tan x}\right) + C$$

$$= \tan^{-1}(\tan x - \cot x) + C$$

11. (c) Let $I = \int e^{\tan^{-1}x}\left(1 + \frac{x}{1+x^2}\right)dx$

$$= \int e^{\tan^{-1}x}\left(\frac{1+x+x^2}{1+x^2}\right)dx$$

Put $x = \tan\theta \Rightarrow dx = \sec^2\theta\,d\theta$

$\therefore$ $I = \int e^{\theta}(1+\tan\theta+\tan^2\theta)\,d\theta$

$$= \int e^{\theta}(\sec^2\theta + \tan\theta)\,d\theta$$

$= e^{\theta}\tan\theta + C$ $\quad [\because \int e^x[f(x)+f'(x)]\,dx = e^x \tan x + C]$

$$= x\cdot e^{\tan^{-1}x} + C$$

12. (b) We have,

$$f''(x) = f(x)$$

$\Rightarrow$ $2f'(x)f''(x) = 2f(x)f'(x)$

$\Rightarrow$ $\frac{d}{dx}[f'(x)]^2 = \frac{d}{dx}[f(x)]^2$

$\Rightarrow$ $[f'(x)]^2 \;\; + [f(x)]^2 + C$ $\quad \ldots$(i)

Now, $f(0) = 2$ and $f'(0) = 3$

From Eq. (i), we get

$$[f'(0)]^2 = [f(0)]^2 + C$$

$\Rightarrow$ $9 = 4 + C \Rightarrow C = 5$

$\therefore$ $[f'(x)]^2 = [f(x)]^2 + 5$

$\Rightarrow$ $f'(x) = \sqrt{5+[f(x)]^2}$

$\Rightarrow$ $\int \frac{1}{(\sqrt{5})^2+[f(x)]^2}d[f(x)] = \int dx$

$\Rightarrow$ $\log\left[f(x) + \sqrt{5+[f(x)^2]}\right] = x + C_1$

$\therefore$ $f(0) = 2 \Rightarrow \log(2+3) = C_1 \Rightarrow C_1 = \log 5$

$\therefore$ $\log\{f(x) + \sqrt{5+[f(x)]^2}\} = x + \log 5$

$\Rightarrow$ $\log\left\{\frac{f(x)+\sqrt{5+[f(x)]^2}}{5}\right\} = x$

$\Rightarrow$ $f(x) + \sqrt{5+[f(x)]^2} = 5e^x$

$\Rightarrow$ $\sqrt{5+[f(x)]^2} + f(x) = 5e^x$

and $\sqrt{5+[f(x)]^2} - f(x) = 5e^{-x}$

$\Rightarrow$ $2f(x) = 5(e^x - e^{-x})$

$\Rightarrow$ $f(x) = \frac{5}{2}(e^x - e^{-x})$

$\therefore$ $f(4) = \frac{5}{2}(e^4 - e^{-4}) = \frac{5(e^8-1)}{2e^4}$

13. (d) Let $I = \int \frac{\log(x+\sqrt{x^2+1})}{\sqrt{x^2+1}}dx$

$$= \int \log(x+\sqrt{x^2+1})\,d\left[\log(x+\sqrt{x^2+1})\right]$$

$$= \frac{\left[\log(x+\sqrt{x^2+1})\right]^2}{2} + C$$

14. (a) Let $I = \int \frac{dx}{x^n(1+x^n)^{1/n}}$

$$= \int \frac{dx}{x^{n+1}\left(1+\frac{1}{x^n}\right)^{1/n}}$$

$$= \int (1+x^{-n})^{-1/n}x^{-n-1}dx$$

$$= -\frac{1}{n}\int (1+x^{-n})^{-1/n}\,d(1+x^n)$$

$$= -\frac{1}{n}\frac{(1+x^{-n})^{-1/n+1}}{\left(-\frac{1}{n}+1\right)} + C$$

$$= \frac{1}{1-n}\left[1+\frac{1}{x^n}\right]^{1-\frac{1}{n}} + C$$

15. (d) Let $I = \int \frac{e^{x^2+4\log x} - x^3e^{x^2}}{x-1}dx$

$$= \int \frac{e^{x^2}\cdot x^4 - x^3e^{x^2}}{x-1}dx$$

$$= \int x^3\cdot e^{x^2}dx$$

$$= \frac{1}{2}\int t\,e^t dt, \text{ where } t = x^2$$

$\therefore$ $I = \frac{1}{2}(t-1)e^t + C$

$$= \frac{1}{2}(x^2-1)e^{x^2} + C$$

**16.** (c) Given,

$$\int \frac{1-x^7}{x(1+x^7)}dx = a\log|x| + b\log|x^7+1| + C$$

On differentiating both sides, we get

$$\frac{1-x^7}{x(1+x^7)} = \frac{a}{x} + b\frac{7x^6}{1+x^7}$$

$$\Rightarrow \quad 1-x^7 = a(1+x^7) + 7bx^7$$

On comparing, we have

$$a = 1 \text{ and } a + 7b = -1$$

$$\Rightarrow \quad b = -2/7$$

**17.** (d) We have,

$$f(x) = \begin{vmatrix} 0 & x^2-\sin x & \cos x - 2 \\ \sin x - x^2 & 0 & 1-2x \\ 2-\cos x & 2x-1 & 0 \end{vmatrix}$$

$$= \begin{vmatrix} 0 & \sin x - x^2 & 2-\cos x \\ x^2 - \sin x & 0 & 2x-1 \\ \cos x - 2 & 1-2x & 0 \end{vmatrix}$$

[interchanging row and column]

$$= -\begin{vmatrix} 0 & x^2-\sin x & \cos x - 2 \\ \sin x - x^2 & 0 & 1-2x \\ 2-\cos x & 2x-1 & 0 \end{vmatrix}$$

[taking (– 1) common from each column]

$$\Rightarrow \quad f(x) = -f(x) \Rightarrow f(x) = 0$$

$$\therefore \quad \int f(x)\,dx = 0$$

**18.** (b) Let $I = \int \frac{x^2}{(a+bx^2)^{5/2}}dx$

$$= \int \frac{x^2}{x^5\left(\frac{a}{x^2}+b\right)^{5/2}}dx$$

$$= \int \frac{dx}{x^3\left(\frac{a}{x^2}+b\right)^{5/2}}$$

Put $\frac{a}{x^2} + b = t^2$

$$\Rightarrow \quad -\frac{2a}{x^3}dx = 2t\,dt$$

$$\therefore \quad I = -\frac{1}{a}\int \frac{t}{t^5}dt$$

$$= -\frac{1}{a}\int t^{-4}dt$$

$$= -\frac{1}{a}\frac{t^{-3}}{-3}$$

$$= \frac{1}{3a} \times \frac{1}{\left(\frac{a}{x^2}+b\right)^{3/2}} + C$$

$$= \frac{1}{3a} \cdot \frac{x^3}{(a+bx^2)^{3/2}} + C$$

**19.** (a) We have,

$$I = \int \frac{\sin^3 x}{(1+\cos^2 x)\sqrt{1+\cos^2 x + \cos^4 x}}dx$$

$$= -\int \frac{(1-\cos^2 x)\,d(\cos x)}{(1+\cos^2 x)\sqrt{1+\cos^2 x+\cos^4 x}}$$

$$= -\int \frac{1-t^2}{(1+t^2)\sqrt{1+t^2+t^4}}dt, \text{ where } t = \cos x$$

$$= \int \frac{1-\frac{1}{t^2}}{\left(t+\frac{1}{t}\right)\sqrt{\left(t+\frac{1}{t}\right)^2-1}}dt$$

$$= \int \frac{1\,d\left(t+\frac{1}{t}\right)}{\left(1+\frac{1}{t}\right)\sqrt{\left(t+\frac{1}{t}\right)^2-1^2}}$$

$$= \sec^{-1}\left(t+\frac{1}{t}\right) + C$$

$$= \sec^{-1}(\cos x + \sec x) + C$$

**20.** (c) We know that

$$\int e^x[f(x) + f'(x)]dx = e^x f(x) \quad \dots\text{(i)}$$

It is given that

$$\int e^x[f(x) - f'(x)]\,dx = \phi(x) \quad \dots\text{(ii)}$$

On adding Eqs. (i) and (ii), we get

$$2\int e^x f(x)\,dx = \phi(x) + e^x f(x)$$

$$\Rightarrow \quad \int e^x f(x)\,dx = \frac{1}{2}[\phi(x) + e^x f(x)]$$

**21.** (c) Let $I = \int |x^n|dx$

Since, integrand in modulus function, therefore following cases arises.

***Case* I** When $x \geq 0$

In this case, we have

$$I = \int |x^n|\,dx = \int |x|^n\,dx$$

$$\Rightarrow \quad I = \int x^n dx \quad [\because |x| = n]$$

$$= \frac{x^n+1}{n+1} + C$$

$$= \frac{|x|^n x}{n+1} + C \quad [\because x \geq 0, |x| = x]$$

***Case* II** When $x < 0$

In this case, we have

$$|x| = -x$$

$$\therefore \quad I = \int |x^n|\,dx = \int |x|^n dx = \int (-x)^n dx$$

$$\Rightarrow \quad \int |x^n|dx = -\int x^n dx \quad [\because n \text{ is odd}]$$

$$\Rightarrow \quad I = -\frac{x^{n+1}}{n+1} + C$$

$$= \frac{(-x)^n x}{n+1} + C \quad [\because n \text{ is odd}, (-x)^n = -x^n]$$

$$= \frac{|x|^n x}{n+1} + C$$

Hence, $\int |x^n|dx = \frac{|x|^n x}{n+1} + C$

**22.** (c) Let $I = \int \sin 2x \log(\cos x)\,dx$

$$= 2\int \cos x \log \cos x \sin x\,dx$$

$$= -2\int t \log t\,dt, \text{ where } t = \cos x$$

$$= -2\left[\frac{t^2}{2}\log t - \frac{t^2}{4}\right] + C$$

$$= -t^2 \log t + \frac{t^2}{2} + C$$

$$= \left[\frac{1}{2} - \log \cos x\right]\cos^2 x + C$$

**23.** (b) Let $f(x) = \int \frac{x^2}{(1+x^2)(1+\sqrt{1+x^2})} dx$

$$= \int \frac{\sqrt{1+x^2}-1}{x^2+1} dx$$

$$= \int \frac{dx}{\sqrt{x^2+1}} - \int \frac{dx}{x^2+1}$$

$\Rightarrow \quad f(x) = \log(x + \sqrt{x^2+1}) - \tan^{-1} x + C$

Now, $f(0) = 0$

$0 = \log 1 - \tan^{-1}(0) + C$

$\Rightarrow \quad C = 0$

$\therefore \quad f(x) = \log(x + \sqrt{x^2+1}) - \tan^{-1} x$

$\Rightarrow \quad f(1) = \log(\sqrt{2}+1) - \frac{\pi}{4}$

**24.** (b) Let $I = \int x \log x (\log x - 1) dx = \int \log x (x \log x - x) dx$

$$= \int (x \log x - x) \, d(x \log x - x)$$

$$= \frac{(x \log x - x)^2}{2} + C$$

**25.** (c) Let $I = \int \frac{f'(x)}{f(x) \log [f(x)]} dx$

Put $\log f(x) = t$

$\Rightarrow \quad \frac{1}{f(x)} f'(x)\, dx = dt$

$\therefore \quad I = \int \frac{dt}{t} = \log |t| + C$

$= \log |\log f(x)| + C$

**26.** (c) Let $I = \int x^x (1 + \log x)\, dx$

Now, let $A = x^x$

$\Rightarrow \quad \log A = x \log x$

$\Rightarrow \quad \frac{1}{A} \frac{dA}{dx} = x \times \frac{1}{x} + \log x$

$\Rightarrow \quad \frac{1}{A} \frac{dA}{dx} = 1 + \log x$

$\Rightarrow \quad \frac{dA}{dx} = A(1 + \log x)$

$\Rightarrow \quad \frac{dA}{dx} = x^x (1 + \log x)$

On integrating both sides, we get

$$\int x^x (1 + \log x)\, dx = A + C = x^x + C$$

**27.** (b) Let $f(x) = ax^3 + bx^2 + cx + d$, then

$f(0) = -1$ and $f(1) = 0$

$\Rightarrow \quad d = -1$ and $a + b + c + d = 0$

$\Rightarrow \quad d = -1$ and $a + b + c = 1$ ...(i)

It is given that $x = 0$ is a stationary point of $f(x)$ but it is not a point of extremum.

Therefore, $f'(0) = 0, f''(0) = 0$ and $f'''(0) \neq 0$

Now, $f(x) = ax^3 + bx^2 + cx + d$

$\Rightarrow \quad f'(x) = 3ax^2 + 2bx + c$

$f''(x) = 6ax + 2b$ and $f'''(x) = 6a$

$\therefore \quad f'(0) = 0, f''(0) = 0$ and $f'''(0) \neq 0$

$\Rightarrow \quad c = 0, b = 0$ and $a \neq 0$ ...(ii)

From Eqs. (i) and (ii), we get

$a = 1, b = c = 0$ and $d = -1$

$\therefore \quad f(x) = x^3 - 1$

Hence, $\int \frac{f(x)}{x^3-1} = \int 1 \cdot dx = x + C$

**28.** (a) Let $I = \int \frac{1}{1-\sin^4 x} dx$

$$= \int \frac{1}{(1-\sin^2 x)(1+\sin^2 x)} dx$$

$$= \int \frac{\sec^2 x}{1+\sin^2 x} dx = \int \frac{1+\tan^2 x}{1+2\tan^2 x} d(\tan x)$$

$\Rightarrow \quad I = \frac{1}{2} \int \frac{\tan^2 x + \frac{1}{2} + \frac{1}{2}}{\tan^2 x + \frac{1}{2}} d(\tan x)$

$\Rightarrow \quad I = \frac{1}{2} \int \left[ 1 + \frac{1}{2} \frac{1}{\left(\tan^2 x + \frac{1}{2}\right)} \right] d(\tan x)$

$$= \frac{1}{2} \left[ \tan x + \frac{\sqrt{2}}{2} \tan^{-1}(\sqrt{2} \tan x) \right] + C$$

$$I = \frac{1}{2} \tan x + \frac{\sqrt{2}}{4} \tan^{-1}(\sqrt{2} \tan x) + C$$

Hence, $A = \frac{\sqrt{2}}{4}$ and $f(x) = \sqrt{2} \tan x$

$\Rightarrow \quad A = \frac{1}{2\sqrt{2}}$

**29.** (b) We have, $\frac{d}{dx}\{g(x)\} = g(x)$

$\Rightarrow \quad g'(x) = g(x) \Rightarrow \frac{g'(x)}{g(x)} = 1$

$\Rightarrow \quad \int \frac{g'(x)}{g(x)} dx = \int 1 \cdot dx$

$\Rightarrow \quad \log [g(x)] = x$

$\Rightarrow \quad g(x) = C \cdot e^x$

Now, $g(0) = 1 \Rightarrow C = 1$

$\therefore \quad g(x) = e^x$

$\therefore \quad \int g(x) \frac{2 - \sin 2x}{1 - \cos 2x} dx$

$$= \int e^x (\operatorname{cosec}^2 x - \cot x)\, dx$$

$$= -e^x \cot x + C$$

$$= -g(x) \cot x + C$$

**30.** (d) Let $I = \int \frac{x^2}{(x \sin x + \cos x)^2} dx$

We know, $\frac{d}{dx}\left[\frac{1}{x \sin x + \cos x}\right]$

$$= -\frac{(\sin x + x \cos x - \sin x)}{(x \sin x + \cos x)^2}$$

$$= \frac{-x \cos x}{(x \sin x + \cos x)^2}$$

$\therefore \int \frac{-x \cos x}{(x \sin x + \cos x)^2} dx = \frac{1}{x \sin x + \cos x}$

$\Rightarrow I = \int \underset{\text{II}}{\frac{-x \cos x}{(x \sin x + \cos x)^2}} \cdot \underset{\text{I}}{\frac{-x}{\cos x}} dx$

$$= \left(\frac{-x}{\cos x}\right) \frac{1}{x \sin x + \cos x} - \int \left\{ \frac{-\cos x - x \sin x}{\cos^2 x} \times \frac{1}{x \sin x + \cos x} \right\} dx$$

$$= \frac{-x}{\cos x (x \sin x + \cos x)} + \int \sec^2 x \, dx$$

$$= \frac{-x}{\cos x (x \sin x + \cos x)} + \tan x + C$$

**31.** (b) We have,

$$\int f(x)\,dx = F(x)$$

$$\therefore \quad \int x^3 f(x^2)\,dx = \frac{1}{2}\int \underset{\text{I}}{x^2}\, \underset{\text{II}}{f(x^2)}\,d(x^2)$$

$$= \frac{1}{2}[x^2 F(x^2) - \int F(x^2)\,d(x^2)]$$

**32.** (b) Let $I = \int \frac{f(x)g'(x) + f'(x)g(x)}{f(x)g(x)}[\log f(x) + \log g(x)]\,dx$

$$= \int \log[f(x)\cdot g(x)] \times \frac{1}{f(x)\,g(x)}\,d[f(x)\,g(x)]$$

$$= \int \log[f(x)\,g(x)]\,d[\log f(x)\,g(x)]$$

$$= \frac{1}{2}[\log\{f(x)g(x)\}]^2 + C$$

**33.** (a) Let $I = \int \frac{\tan x}{\sqrt{\sin^4 x + \cos^4 x}}\,dx$

$$= \int \frac{\tan x \sec^2 x}{\sqrt{\tan^4 x + 1}}\,dx$$

$$= \frac{1}{2}\int \frac{1}{\sqrt{(\tan^2 x)^2 + 1}}\,d(\tan^2 x)$$

$$= \frac{1}{2}\log(\tan^2 x + \sqrt{1 + \tan^4 x}) + C$$

**34.** (b) Let $I = \int \frac{x^2 + 4}{x^4 + 16}\,dx$

$$= \int \frac{\left(1 + \frac{4}{x^2}\right)dx}{x^2 + \frac{16}{x^2} - 8 + 8}$$

$$= \int \frac{\left(1 + \frac{4}{x^2}\right)}{\left(x - \frac{4}{x}\right)^2 + 8}\,dx$$

$$= \int \frac{dt}{t^2 + 8}, \text{ where } t = x - \frac{4}{x}$$

$$= \frac{1}{2\sqrt{2}}\tan^{-1}\left(\frac{t}{2\sqrt{2}}\right) + C$$

$$= \frac{1}{2\sqrt{2}}\tan^{-1}\left(\frac{x - \frac{4}{x}}{2\sqrt{2}}\right) + C$$

$$= \frac{1}{2\sqrt{2}}\tan^{-1}\left(\frac{x^2 - 4}{2\sqrt{2}x}\right) + C$$

**35.** (b) Let $I_n = \int (\log x)^n \cdot 1\,dx$

$$\Rightarrow \quad I_n = x(\log x)^n - \int x \cdot n(\log x)^{n-1}\frac{1}{x}\,dx$$

$$\Rightarrow \quad I_n = x(\log x)^n - \int x \cdot n(\log x)^{n-1}\frac{1}{x}\,dx$$

$$\Rightarrow \quad I_n = x(\log x)^n - n\,I_{n-1}$$

$$\therefore \quad I_n + nI_{n-1} = x(\log x)^n$$

**36.** (a) Statement II is correct by definition.

$\Rightarrow$ Statement I is correct.

i.e.
$$\int D(x)\,dx = \begin{vmatrix} \int f_1(x)dx & \int f_2(x)dx & \int f_3(x)dx \\ a_2 & b_2 & c_2 \\ a_3 & b_3 & c_3 \end{vmatrix} + C$$

**37.** (a) If $a > 0$ and $b^2 - 4ac < 0$, then

$$ax^2 + bx + c = a\left(x + \frac{b}{2a}\right)^2 + \frac{4ac - b^2}{4a}$$

$$\Rightarrow \quad \int \frac{dx}{ax^2 + bx + c} = \int \frac{dx}{a\left(x + \frac{b}{2a}\right)^2 + k^2}$$

where, $k^2 = \frac{4ac - b^2}{4a} > 0$

which will have the answer of the type

$$\frac{1}{a}\cdot\frac{1}{k/\sqrt{a}}\tan^{-1}\frac{x + b/2a}{k/\sqrt{a}} + C \text{ or } u \tan^{-1}\frac{x + A}{B} + C$$

Hence, Statement I and Statement II are correct and Statement II is the correct explanation of Statement I.

**38.** (c) The Statement II is incorrect, since while writing

$$\int \frac{dx}{x - 3y} = \log(x - 3y) + C$$

We are assuming that $y$ is a constant.

We will know prove the Statement I.

From the given relation, $(x - y)^2 = \frac{x}{y}$

$$\Rightarrow \quad 2\log(x - y) = \log x - \log y$$

Also, $\frac{dy}{dx} = \left(-\frac{y}{x}\right)\frac{x + y}{x - 3y}$

To prove that integral relation it is sufficient to show that $\frac{d}{dx}(\text{RHS}) = \frac{1}{x - 3y}$

Now, $\text{RHS} = \frac{1}{2}\log\left[\frac{x}{y} - 1\right] \quad \left[\because (x - y)^2 = \frac{x}{y}\right]$

$$= \frac{1}{2}[\log(x - y) - \log y]$$

$$= \frac{1}{2}\left[\frac{\log x - \log y}{2} - \log y\right]$$

$$= \frac{1}{4}[\log x - 3\log y]$$

$$\Rightarrow \quad \frac{d}{dx}(\text{RHS}) = \frac{1}{4}\left[\frac{1}{x} - \frac{3}{y}\frac{dy}{dx}\right]$$

$$= \frac{1}{4}\left[\frac{1}{x} - \frac{3}{y}\left(-\frac{y}{x}\right)\frac{x + y}{x - 3y}\right] = \frac{1}{x - 3y}$$

Thus, Statement I is correct.

**39.** (d) $f(x) = \int \sin^2 x\,dx = \int \frac{1 - \cos 2x}{2}\,dx$

$$\Rightarrow \quad f(x) = \frac{1}{4}(2x - \sin 2x) + C$$

Since, $f(x + \pi) \neq f(x)$

$\therefore$ Statement I is incorrect.

But Statement II is correct as $\sin^2 x$ is periodic with period $\pi$.

**40.** (a) We have, $\int [\log_x e - (\log_x e)^2]\,dx$

$$= \int \left[\frac{1}{\log x} - \frac{1}{(\log x)^2}\right]dx$$

$$= \int \left(\frac{1}{t} - \frac{1}{t^2}\right)e^t\,dt, \text{ where } t = \log x$$

$$= \frac{e^t}{t} + C = \frac{x}{\log x} + C = x\log_x e + C$$

Hence, Statement I and Statement II are correct and Statement II is the correct explanation of Statement I.

## Revisal Problems (JEE Advanced)

**1.** (c) Given, $\begin{vmatrix} f'(x) & f(x) \\ f''(x) & f'(x) \end{vmatrix} = 0$

$\Rightarrow \quad f'(x)\cdot f'(x) - f(x) f''(x) = 0$

$\Rightarrow \quad \dfrac{[f'(x)]^2 - f(x)f''(x)}{[f'(x)]^2} = 0 \quad \Rightarrow \quad \dfrac{d}{dx}\left[\dfrac{f(x)}{f'(x)}\right] = 0$

On integrating $\quad \dfrac{f(x)}{f'(x)} = C \quad$ ...(i)

On putting $x = 0$, $\dfrac{f(0)}{f'(0)} = C \Rightarrow C = \dfrac{1}{2}$

Hence, $\quad \dfrac{f(x)}{f'(x)} = \dfrac{1}{2} \quad$ ...(ii)

From Eq. (i), $2f(x) = f'(x)$

$\dfrac{f'(x)}{f(x)} = 2$

Again integrating, we have

$\log [f(x)] = 2x + K$

On putting $x = 0$, $K = 0$

$\therefore \quad f(x) = e^{2x}$

**2.** (b) Given,

$\displaystyle\int \frac{\cos x - \sin x + 1 - x}{e^x + \sin x + x} dx = \log [f(x)] + g(x) + C$

Now, let $\quad I = \displaystyle\int \frac{\cos x - \sin x + 1 - x}{e^x + \sin x + x} dx$

$= \displaystyle\int \frac{(e^x + \cos x + 1) - (e^x + \sin x + x)}{e^x + \sin x + x} dx$

$= \log (e^x + \sin x + x) - x + C$

$\therefore \quad f(x) = e^x + \sin x + x$ and $g(x) = -x$

Now, $f(x) + g(x) = e^x + \sin x$

**3.** (c) Let $I = \displaystyle\int x \cdot \frac{\log (x + \sqrt{x^2 + 1})}{\sqrt{x^2 + 1}} dx$

Put $\quad t = \sqrt{x^2 + 1}$

$\Rightarrow \quad dt = \dfrac{x}{\sqrt{x^2 + 1}} dx$

$\therefore \quad I = \displaystyle\int \log (t + \sqrt{t^2 - 1})\, dx$

$= \log (t + \sqrt{t^2 - 1})\cdot t - \displaystyle\int \frac{1 + \frac{t}{\sqrt{t^2 - 1}}}{t + \sqrt{t^2 - 1}} \cdot t\, dt$

$= t \cdot \log(t + \sqrt{t^2 - 1}) - \dfrac{1}{2} \displaystyle\int \frac{2t}{\sqrt{t^2 - 1}} dt$

$= t \cdot \log (t + \sqrt{t^2 - 1}) - \sqrt{t^2 - 1} + C$

$= \sqrt{1 + x^2} \log \left( x + \sqrt{1 + x^2} \right) - x + C$

**4.** (a) Let $I = \displaystyle\int \frac{(2x + 1)}{(x^2 + 4x + 1)^{3/2}} dx$

$= \displaystyle\int \frac{(2x + 1)}{x^3\left(1 + \frac{4}{x} + \frac{1}{x^2}\right)^{3/2}} dx = \int \frac{2x^{-2} + x^{-3}}{\left(1 + \frac{4}{x} + \frac{1}{x^2}\right)^{3/2}} dx$

Put $\dfrac{1}{x^2} + \dfrac{4}{x} + 1 = t^2$

$\Rightarrow \left(\dfrac{-2}{x^3} - \dfrac{4}{x^2}\right) dx = 2t\, dt \Rightarrow \left(\dfrac{2}{x^2} + \dfrac{1}{x^3}\right) dx = -t\, dt$

$\therefore \quad I = \displaystyle\int \frac{-t}{t^3} dt = \frac{1}{t} + C = \frac{x}{\sqrt{x^2 + 4x + 1}} + C$

**5.** (a) Let $I = \displaystyle\int \sin (101x) \sin^{99} x dx$

$= \displaystyle\int \sin (100x + x) (\sin x)^{99} dx$

$= \displaystyle\int \{\sin (100x) \cos x + \cos 100\, x \sin x\} (\sin x)^{99} dx$

$= \displaystyle\int \underset{\text{I}}{\sin (100 x)} \underset{\text{II}}{\cos x \cdot (\sin x)^{99}} dx + \int \cos (100x) (\sin x)^{100} dx$

$= \dfrac{\sin (100x) (\sin x)^{100}}{100} - \dfrac{100}{100} \displaystyle\int \cos (100x) (\sin x)^{100} dx$

$+ \displaystyle\int \cos (100x) (\sin x)^{100} dx$

$= \dfrac{\sin (100x)(\sin x)^{100}}{100} + C$

**6.** (d) Given, $\quad f'(\log x) = \begin{cases} 1, & 0 < x \le 1 \\ x, & x > 1 \end{cases}$

Put $\log x = t \quad \Rightarrow \quad x = e^t$

$\therefore \quad f'(t) = \begin{cases} 1, & 0 < e^t \le 1 \\ e^t, & e^t > 1 \end{cases}$

On integrating, we get

$f(t) = \begin{cases} t + C, & 0 < e^t \le 1 \\ e^t + C, & e^t > 1 \end{cases}$

$f(t) = t + C, 0 < e^t \le 1$

$f(0) = 0 + C \Rightarrow C = 0 \quad [\because f(0) = 0]$

$\therefore \quad f(t) = t, t \le 0$

and $\quad f(t) = e^t + C, e^t > 1$

$f(t) = e^t = C, t > 0$

$f(0) = e^0 + C$

$0 = 1 + C \quad [\because f(0) = 0]$

$\Rightarrow \quad C = -1$

$\therefore \quad f(t) = e^t - 1$

$\therefore \quad f(t) = \begin{cases} t, & t \le 0 \\ e^t - 1, & t > 0 \end{cases}$

$\Rightarrow \quad f(x) = \begin{cases} x, & x \le 0 \\ e^x - 1, & x > 0 \end{cases}$

**7.** (c) Let $I = \displaystyle\int \frac{px^{p + 2q - 1} - qx^{q - 1}}{(x^{p+q} + 1)^2} dx$

$= \displaystyle\int \frac{px^{p-1} - q\, x^{-q-1}}{(x^p + x^{-q})^2} dx$

On taking $x^q$ as $x^{2q}$ common from denominator and take it in numerator.

Put $x^p + x^{-q} = t$

$\Rightarrow \quad (px^{p-1} - qx^{-q-1})\, dx = dt$

$\therefore \quad I = \displaystyle\int \frac{dt}{t^2} = -\frac{1}{t} + C = -\left(\frac{x^q}{x^{p+q} + 1}\right) + C$

**8.** (c) Let $I_n = \displaystyle\int \sin^n x dx$

$= \displaystyle\int \underset{\text{I}}{\sin^{n-1} x} \underset{\text{II}}{\sin x}\, dx$

$= -\sin^{n-1} x \cos x + \displaystyle\int (n - 1) \sin^{n-2} x \cos^2 x\, dx$

$= -\sin^{n-1} x \cos x + (n - 1) \displaystyle\int (\sin^{n-2} x - \sin^n x)\, dx$

$= -\sin^{n-1} x \cos x + (n - 1)I_{n-2} - (n - 1)I_n$

$\therefore \quad nI_n = -\sin^{n-1} x \cos x + (n - 1)I_{n-2} + C \quad$ ...(i)

Put $n = 6$ in Eq. (i), we get

$6I_6 = -\sin^5 x \cos x + 5I_4 + C$

$\Rightarrow \quad 5I_4 - 6I_6 = \sin^5 x \cos x + C \quad$ ...(ii)

Now, $\sin^5 x \cos x = \frac{1}{2}\sin^4 x\,(2\sin x\cos x)$

$$= \frac{1}{2}\sin^4 x \sin 2x = \frac{1}{2}\left[\frac{1-\cos 2x}{2}\right]^2 \sin 2x$$

$$\Rightarrow \quad \sin^5 x\cdot\cos x = \frac{\sin 2x}{8}[1+\cos^2 2x - 2\cos 2x] \quad \ldots\text{(iii)}$$

From Eqs. (ii) and (iii), we get

$$5I_4 - 6I_6 = \frac{\sin 2x}{8}[1+\cos^2 2x - 2\cos 2x] + C$$

**9.** (c) We know that

$$\sec^{-1}\sqrt{1+x^2} = \tan^{-1} x$$

$$\cos^{-1}\left(\frac{1-x^2}{1+x^2}\right) = 2\tan^{-1} x, \text{ for } x>0$$

$$I = \int \frac{e^{\tan^{-1} x}}{1+x^2}\{(\tan^{-1} x)^2 + 2\tan^{-1} x\}dx$$

Put $\tan^{-1} x = t \Rightarrow \frac{1}{1+x^2}dx = dt$

$$\therefore \quad I = \int e^t (t^2 + 2t)\,dt = e^t\cdot t^2 + C$$

$$= e^{\tan^{-1} x}(\tan^{-1} x)^2 + C$$

$$= e^{\tan^{-1} x}\cdot\left[(\sec^{-1}\sqrt{1+x^2})\right]^2 + C$$

**10.** (a) Let $I = \int \prod_{r=1}^{n}(x+r)\left(\sum_{k=1}^{n}\frac{1}{x+k}\right)dx$

Now, consider $A = (x+1)(x+2)(x+3)\cdots(x+n)$

$$A = \prod_{r=1}^{n}(x+r) \quad \ldots\text{(i)}$$

On taking log both sides, we have

$$\log A = \log(x+1) + \log(x+2) + \cdots + \log(x+n)$$

On differentiating w.r.t. $x$, we get

$$\frac{1}{A}\frac{dA}{dx} = \frac{1}{x+1} + \frac{1}{x+2} + \frac{1}{x+3} + \cdots + \frac{1}{x+n}$$

$$\Rightarrow \quad \frac{dA}{dx} = A\left[\frac{1}{x+1}+\frac{1}{x+2}+\frac{1}{x+3}+\cdots+\frac{1}{x+n}\right] = A\left[\sum_{k=1}^{n}\frac{1}{x+k}\right]$$

$$\Rightarrow \quad \frac{dA}{dx} = \prod_{r=1}^{n}(x+r)\left[\sum_{k=1}^{n}\frac{1}{x+k}\right] \quad \text{[using Eq. (i)]}$$

On integrating both sides, we have,

$$A = \int \prod_{r=1}^{n}(x+r)\left(\sum_{k=1}^{n}\frac{1}{x+k}\right)dx$$

$$\Rightarrow \quad \int \prod_{r=1}^{n}(x+r)\left(\sum_{k=1}^{n}\frac{1}{x+k}\right)dx = \prod_{r=1}^{n}(x+r)$$

**11.** (b) Let $I_n = \int \cot^n x\,dx$

$$= \int \cot^{n-2} x(\operatorname{cosec}^2 x - 1)\,dx = -\frac{u^{n-1}}{n-1} - I_{n-2}$$

$$\Rightarrow \quad I_n + I_{n-2} = -\frac{u^{n-1}}{n-1} \quad \text{[put } n = 2, 3, 4, \ldots, 10]$$

$$I_2 + I_0 = -\frac{u}{1}$$

$$I_3 + I_1 = -\frac{u^2}{2}$$

$$I_4 + I_2 = -\frac{u^3}{3}$$

$$\vdots \qquad \vdots$$

$$I_{10} + I_8 = -\frac{u^9}{9}$$

On adding all the above, we get

$$I_0 + I_1 + 2(I_2 + I_3 + \cdots I_8) + I_5 + I_{10} = -\left(u + \frac{u^2}{2} + \cdots + \frac{u^9}{9}\right)$$

**12.** (c) Let $I = \int \frac{x^2(1-\log x)}{\log^4 x - x^4}dx$

$$= \int \frac{x^2(1-\log x)}{x^4\left[\left(\frac{\log x}{x}\right)^4 - 1\right]}dx = \int \frac{1-\log x}{x^2\left[\left(\frac{\log x}{x}\right)^4 - 1\right]}dx$$

Put $\frac{\log x}{x} = t$

$$\Rightarrow \left(\frac{1-\log x}{x^2}\right)dx = dt$$

$$\therefore \quad I = \int \frac{dt}{(t^4-1)} = \int \frac{dt}{(t^2+1)(t^2-1)}$$

$$= \frac{1}{2}\int \frac{(t^2+1)-(t^2-1)}{(t^2+1)(t^2-1)}dt$$

$$= \frac{1}{2}\left[\int\frac{dt}{t^2-1} - \int\frac{dt}{t^2+1}\right]$$

$$= \frac{1}{2}\left(\frac{1}{2}\log\frac{t-1}{t+1} - \tan^{-1} t\right) + C$$

$$= \frac{1}{4}\log\left(\frac{\log x - x}{\log x + x}\right) - \frac{1}{2}\tan^{-1}\left(\frac{\log x}{x}\right) + C$$

**13.** (c) On solving

$$\frac{\sin\theta}{\cos 3\theta} = \frac{1}{2}[\tan 3\theta - \tan\theta]$$

$$\frac{\sin 3\theta}{\cos 9\theta} = \frac{1}{2}[\tan 9\theta - \tan 3\theta]$$

$$\frac{\sin 9\theta}{\cos 27\theta} = \frac{1}{2}[\tan 27\theta - \tan 9\theta]$$

$$\therefore \int\left(\frac{\sin\theta}{\cos 3\theta} + \frac{\sin 3\theta}{\cos 9\theta} + \frac{\sin 9\theta}{\cos 27\theta}\right)d\theta$$

$$= \frac{1}{2}\int(\tan 27\theta - \tan\theta)\,d\theta$$

$$= \frac{1}{2}\left[\frac{1}{27}\log(\sec 27\theta - \log\sec\theta\right] + C$$

$$= \frac{1}{2}\log\left|\frac{\sqrt[27]{\sec 27\theta}}{\sec\theta}\right| + C$$

**14.** (a) Let $I = \int \frac{x^2+20}{(x\sin x + 5\cos x)^2}dx$

$$= \int \frac{x^8(x^2+20)}{x^8(x\sin x + 5\cos x)^2}dx$$

$$= \int \frac{x^8(x^2+20)}{(x^5\sin x + 5x^4\cos x)^2}dx$$

$$= \int \frac{(x^5+20x^3)\cos x}{(x^5\sin x + 5x^4\cos x)^2}\cdot\frac{x^5}{\cos x}dx$$

Note that $\frac{d}{dx}[x^5\sin x + 5x^4\cos x] = (x^5+20x^3)\cos x$

Integrating by parts, we have

$$I = -\frac{x^5}{\cos x(x^5\sin x + 5x^4\cos x)} + \int \frac{5x^4\cos x + x^5\sin x}{x^5\sin x + 5x^4\cos x}\sec^2 x\,dx$$

$$= -\frac{x}{\cos x(x\sin x + 5\cos x)} + \tan x + C$$

**15.** (c) Let $I = n!\int dx - n!\int \dfrac{\left(1+\frac{x}{1!}+\frac{x^2}{2!}+\cdots+\frac{x^{n-1}}{(n-1)!}\right)}{1+\frac{x}{1!}+\frac{x^2}{2!}+\cdots+\frac{x^n}{n!}}dx$

$$= n!(x) - n!\log\left(1+\frac{x}{1!}+\frac{x^2}{2!}+\cdots+\frac{x^n}{n!}\right)+C$$

$$= n!\left[\log_e e^x - \log_e\left(1+\frac{x}{1!}+\frac{x^2}{2!}+\cdots+\frac{x^n}{n!}\right)\right]+C$$

$$= n!\log_e\left(\frac{e^x}{1+\frac{x}{1!}+\frac{x^2}{2!}+\cdots+\frac{x^n}{n!}}\right)+C$$

**16.** (a) Let $I = \int \dfrac{dx}{x^r(1+x^r)^{1/r}} = \int \dfrac{dx}{x^{r+1}\left(1+\frac{1}{x^r}\right)^{1/r}}$

$$= \int (1+x^{-r})^{-1/r}\cdot x^{-r-1}\,dx = -\frac{1}{r}\int (1+x^{-r})^{\frac{-1}{r}} d(1+x^{-r})$$

$$= -\frac{1}{r}\frac{(1+x^{-r})^{-\frac{1}{r}+1}}{\left(-\frac{1}{r}+1\right)}+C = \frac{1}{1-r}\left(1+\frac{1}{x^r}\right)^{\frac{r-1}{r}}+C \quad \ldots\text{(i)}$$

Now, put $r = 3$ in Eq. (i), we get

$$I = \frac{1}{-2}(1+x^{-3})^{2/3}+C$$

But $\quad I = \alpha(1+x^{-3})^{\beta}+C \Rightarrow \alpha = -\frac{1}{2}, \beta = \frac{2}{3}$

$\therefore$ Required equation is $x^2 - \left(-\frac{1}{2}+\frac{2}{3}\right)x + \left(-\frac{1}{2}\right)\left(\frac{2}{3}\right) = 0$

$$\Rightarrow \quad x^2 - \frac{1}{6}x - \frac{1}{3} = 0$$

$$\Rightarrow \quad 6x^2 - x - 2 = 0$$

**17.** (a) Let $I = \int \dfrac{(x^5+x^4)}{\sqrt{5x^6+6x^5+1}}dx$

Put $\quad 5x^6+6x^5+1 = t$

$\Rightarrow \quad (30x^5+30x^4)\,dx = dt \Rightarrow (x^5+x^4)dx = \dfrac{dt}{30}$

$$\therefore \quad I = \frac{1}{30}\int \frac{dt}{t^{1/2}} = \frac{1}{30}\cdot\frac{t^{1/2}}{1/2}+C$$

$$= \frac{1}{15}\sqrt{5x^6+6x^5+1}+C$$

**18.** (d) We know that

$$\sin^{-1}x + \cos^{-1}x = \frac{\pi}{2}, \text{ if } -1 \le x \le 1 \text{ and } x \ge 3$$

$\therefore \sin^{-1}(\log x) + \cos^{-1}(\log x)$ does not exist.

Hence, the given integral does not exist.

**19.** (a) Let $I = \int \dfrac{\operatorname{cosec}^2 x - 2011}{(\cos x)^{2011}}dx$

$$= \int \underset{\text{I}}{(\cos x)^{-2011}}\,\underset{\text{II}}{\operatorname{cosec}^2 x}\,dx - \int \frac{2011}{(\cos x)^{2011}}dx$$

$$= (\cos x)^{-2011}(-\cot x) - \int (-2011)(\cos x)^{-2012}(-\sin x)$$

$$(-\cot x)dx - 2011\int \frac{dx}{(\cos x)^{2011}}$$

$$= (\cos x)^{-2011}(-\cot x) + 2011\int \frac{dx}{(\cos x)^{2011}}$$

$$-2011\int \frac{dx}{(\cos x)^{2011}}+C$$

$$= \frac{-\cot x}{(\cos x)^{2011}}+C$$

**20.** (c) Given,

$$P = \int e^{ax}\cos bx\,dx = \frac{e^{ax}}{a^2+b^2}(b\sin bx + a\cos bx)+C_1$$

$$Q = \int e^{ax}\sin bx\,dx = \frac{e^{ax}}{a^2+b^2}(a\sin bx - b\cos bx)+C_2$$

Now, $P + iQ$

$$= \frac{e^{ax}}{a^2+b^2}$$

$$[(b\sin bx + a\cos bx) + i\,(a\sin bx - b\cos bx)] + C_1 + C_2$$

$$= \frac{e^{ax}}{a^2+b^2}[(b\sin bx + a\cos bx) + i(a\sin bx - b\cos bx)]+C$$

[where, $C = C_1 + C_2$]

Given that, $P + iQ = Z + C$

$$\Rightarrow \quad Z = \frac{e^{ax}}{a^2+b^2}[(b\sin bx + a\cos bx) + i(a\sin bx - b\cos bx)]$$

$$\therefore \quad |Z| = \frac{e^{ax}}{a^2+b^2}\sqrt{(b\sin bx + a\cos bx)^2 + (a\sin bx - b\cos bx)^2}$$

$$= \frac{e^{ax}}{a^2+b^2}\sqrt{\begin{bmatrix} b^2\sin^2 bx + a^2\cos^2 bx \\ +2ab\sin bx\cos bx + a^2\sin^2 bx \\ +b^2\cos^2 bx - 2ab\sin bx\cos bx\end{bmatrix}}$$

$$= \frac{e^{ax}}{a^2+b^2}\sqrt{a^2+b^2} = \frac{e^{ax}}{\sqrt{a^2+b^2}}$$

**21.** (a)

(i) $f(x) = \int x^{\sin x}(1 + x\cos x\cdot\log x + \sin x)\,dx$

If $F(x) = x^{\sin x} = e^{\sin x\log x}$

$$\therefore \quad xF'(x) = x^{\sin x}(x\cos x\log x + \sin x)$$

$$\therefore \quad f(x) = \int [F(x) + xF'(x)] = xF(x) + C$$

$$f(x) = x\cdot x^{\sin x} + C$$

$$f\left(\frac{\pi}{2}\right) = \frac{\pi}{2} - \frac{\pi}{2} + C \Rightarrow C = 0$$

$$\therefore \quad f(x) = x\,x^{\sin x}$$

$$\Rightarrow \quad f\left(\frac{\pi}{4}\right) = \frac{\pi}{4}\left(\frac{\pi}{4}\right)^{1/\sqrt{2}} = \left(\frac{\pi}{4}\right)^{1+\frac{1}{\sqrt{2}}} \quad \text{[irrational]}$$

(ii) $g(x) = \int \dfrac{\cos x(\cos x + 2) + \sin^2 x}{(\cos x + 2)^2}dx$

$$= \int \underset{\text{II}}{\cos x}\cdot\underset{\text{I}}{\frac{1}{\cos x + 2}}dx + \int \frac{\sin^2 x}{(\cos x + 2)^2}dx$$

$$= \frac{1}{\cos x + 2}\sin x - \int \frac{\sin^2 x}{(\cos x + 2)^2}dx + \int \frac{\sin^2 x}{(\cos x + 2)^2}dx + C$$

$$= \frac{\sin x}{\cos x + 2} + C$$

**Alternate Method**

$$\frac{d}{dx}\left(\frac{\sin x}{\cos x + 2}\right) = \frac{(\cos x + 2)\cos x + \sin^2 x}{(\cos x + 2)^2} = \frac{2\cos x + 1}{(\cos x + 2)^2}$$

$$\therefore \int \frac{2\cos x + 1}{(\cos x + 2)^2}dx = \frac{\sin x}{\cos x + 2} + C$$

$$g(x) = \frac{\sin x}{\cos x + 2} + C$$

$$g(0) = 0 \Rightarrow C = 0$$

$$\therefore \quad g(x) = \frac{\sin x}{\cos x + 2} \Rightarrow g\left(\frac{\pi}{2}\right) = \frac{1}{2} \quad \text{[rational]}$$

(iii) Let $\quad x + 5 = 14\cos\theta$

$$y - 12 = 14\sin\theta$$

$$\therefore \quad x^2 + y^2 = (14\cos\theta - 5)^2 + (14\sin\theta + 12)^2$$

$$= 196 + 25 + 144 + 28(12\sin\theta - 5\cos\theta)$$

$$= 365 + 28(12\sin\theta - 5\cos\theta)$$

$$\therefore \quad \left|\sqrt{x^2 - y^2}\right|_{\min} = \sqrt{365 - 28\times 13} = \sqrt{365 - 364} = 1$$

(iv) $k(x) = \int \frac{(x^2+1)}{(x^3+3x+6)^{1/3}} dx$

Put $x^3 + 3x + 6 = t^3$

$3(x^2+1)dx = 3t^2\,dt$

$\Rightarrow \quad K(x) = \int \frac{t^2}{t} dt = \frac{t^2}{2} + C$

$\Rightarrow \quad K(x) = \frac{1}{2}(x^3 + 3x + 6)^{2/3} + C$

$\Rightarrow \quad K(-1) = \frac{1}{2}(2)^{2/3} + C \Rightarrow C = 0$

$\Rightarrow \quad K(-1) = \frac{1}{2}(x^3 + 3x + 6)^{2/3}$

$\therefore \quad K(-2) = \frac{1}{2}(-8)^{2/3} = 2$

Hence, (i) $\rightarrow$ (q), (ii) $\rightarrow$ (p), (iii) $\rightarrow$ (p,r), (iv) $\rightarrow$ (p,r,s)

**22.** (b) Given, $f(x) f'(-x) = f(-x) f'(x)$

$$\frac{f'(x)}{f(x)} = \frac{f'(-x)}{f(-x)}$$

On integrating, $\log f(x) = \log f(-x) + \log C$

$\log f(x) + \log f(-x) = C$

$f(x) \cdot f(-x) = C$

But $f(0) = 3$

$\therefore \quad f^2(0) = C \Rightarrow C = 9$

$\therefore \quad f(x) f(-x) = 9$

**23.** (a) Let $x = \alpha$ be the root of $f(x) = 0$.

$f(\alpha) = 0$

$f(x)\,f(-x) = 9$

Put $x = \alpha \quad \Rightarrow \quad 0 = 9$ [impossible]

Therefore, $f(x)$ has no root but $f(0) = 3$.

Therefore, $f(x) > 0, \forall\ x \in R$ as $f$ is continuous possible function $f(x) = 3e^{-x}$.

**24.** (c) Given, $\quad x = e^t \cos t$ and $y = e^t \sin t$

$\Rightarrow \quad \frac{dx}{dt} = e^t \cos t + (-\sin t) e^t$

$= e^t (\cos t - \sin t)$

and $\quad \frac{dy}{dt} = e^t \sin t + \cos t\, e^t$

$= e^t (\cos t + \sin t)$

Now, $\quad \frac{dy}{dx} = \frac{dy/dt}{dx/dt} = \frac{\cos t + \sin t}{\cos t - \sin t}$

$\Rightarrow \quad \frac{dy}{dx} = \frac{1 + \tan t}{1 - \tan t} = \tan\left(\frac{\pi}{4} + t\right)$

Since, $\alpha$ is the angle between the tangent to the curve and $X$-axis, then

$\frac{dy}{dx} = \tan \alpha$

$\Rightarrow \quad \tan \alpha = \tan\left(\frac{\pi}{4} + t\right) \Rightarrow \alpha = \frac{\pi}{4} + t$

$\therefore \quad t = \alpha - \frac{\pi}{4}$

**25.** (b) $\frac{d^2y}{dx^2} = \frac{d}{dx}\left(\frac{dy}{dx}\right) = \frac{d}{dt}\left(\frac{\cos t + \sin t}{\cos t - \sin t}\right) \cdot \frac{dt}{dx}$

$$= \frac{\left[\begin{array}{l}(\cos t - \sin t)(-\sin t + \cos t) \\ -(\cos t + \sin t)(-\sin t - \cos t)\end{array}\right]}{(\cos t - \sin t)^2} \times \frac{1}{e^t(\cos t - \sin t)}$$

$$\Rightarrow \frac{d^2y}{dx^2} = \frac{(\cos t - \sin t)^2 + (\cos t + \sin t)^2}{(\cos t - \sin t)^2} \times \frac{1}{e^t(\cos t - \sin t)}$$

$$= \frac{2}{(\cos t - \sin t)^2} \times \frac{1}{e^t(\cos t - \sin t)}$$

$$\Rightarrow \quad \frac{d^2y}{dx^2} = \frac{2}{e^t} \times \frac{1}{(\cos t - \sin t)^3}$$

$$\therefore \quad \left[\frac{d^2y}{dx^2}\right]_{t=0} = 2$$

**26.** (c) $F(t) = \int (e^t \cos t + e^t \sin t)\,dt$

$\Rightarrow F(t) = \int e^t (\cos t + \sin t)\,dt$

$F(t) = e^t \sin t + C$

$\left[\text{as } \frac{d}{dx} \sin x = \cos x, \int e^x [f(x) + f'(x)]\,dx = e^x f(x) + C\right]$

Now, $F\left(\frac{\pi}{2}\right) - F(0) = e^{\pi/2} \sin \frac{\pi}{2} + C - e^0 \sin 0 - C = e^{\pi/2}$

**27.** (1) Given, $g(x) = \int x^{27}(1 + x + x^2)^6 (6x^2 + 5x + 4)\,dx$

$= \int (x^4 + x^5 + x^6)^6 (6x^5 + 5x^4 + 4x^3)\,dx$

Let $\quad x^6 + x^5 + x^4 = t$

$\Rightarrow \quad (6x^5 + 5x^4 + 4x^3)\,dx = dt$

$\therefore \quad g(x) = \int t^6\,dt = \frac{t^7}{7} + C$

$= \frac{1}{7}(x^4 + x^5 + x^6)^7 + C$

$g(0) = 0$ [since, $g(x)$ is pass through the origin]

$\Rightarrow \quad 0 = 0 + C \Rightarrow C = 0$

Now, $\quad g(x) = \frac{1}{7}(x^4 + x^5 + x^6)^7$

$\therefore \quad g(-1) = \frac{1}{7}(1 - 1 + 1)^7 = \frac{1}{7}$

Hence, $\quad |7g(-1)| = 7 \times \frac{1}{7} = 1$

**28.** (4) Given, $f(xy) = f(x) \cdot f(y)$ ...(i)

On putting, $x = y = 1$, $f(1) = 1$

On differentiating w.r.t. $x$ and $y$ as constant

$f'(xy) \cdot y = f'(x) f(y)$

When put $x = 1$, then $\quad f'(y) y = f'(1) \cdot f(y)$

$\Rightarrow \quad \frac{f'(y)}{f(y)} = \frac{f'(1)}{y}$

$\therefore \quad \frac{f(x)}{f'(x)} = \frac{x}{f'(1)}$ ...(ii)

Here, $f'(1) = \lim_{h \to 0} \frac{f(1+h) - f(1)}{h} = \lim_{h \to 0} \frac{1 + h\{1 + g(h)\} - 1}{h}$

$= \lim_{h \to 0} \{1 + g(h)\} = 2$

From Eq. (ii), we get

$\frac{f(x)}{f'(x)} = \frac{x}{2}$

$\therefore \quad \int \frac{f(x)}{f'(x)} dx = \frac{x^2}{4} + C$

On comparing, $k = 4$

**29.** (1) Given,

$$\int e^{x \sin x + \cos x} \left(\frac{x^4 \cos^3 x - x \sin x + \cos x}{x^2 \cos^2 x}\right) dx$$

$$= e^{x \sin x + \cos x} \frac{f(x)}{x} + C$$

Now, $\quad \int e^{x \sin x + \cos x} \left(\frac{x^4 \cos^3 x - x \sin x + \cos x}{x^2 \cos^2 x}\right) dx$

$$= \int x \cdot e^{(x \sin x + \cos x)} \cdot x \cos x\, dx$$

$$- \int \underset{\text{I}}{e^{(x \sin x + \cos x)}} \cdot \underset{\text{II}}{\frac{x \sin x - \cos x}{(x \cos x)^2}}\, dx$$

On applying integration by parts, we get

$$= \{x \cdot e^{x \sin x - \cos x} - \int e^{(x \sin x + \cos x)} dx$$

$$- \left\{e^{(x \sin x + \cos x)} \frac{1}{x \cos x} - \int e^{(x \sin x + \cos x)} dx\right\}$$

$$= e^{x \sin + \cos x}\left(x - \frac{1}{x \cos x}\right) + C$$

$$\Rightarrow \quad f(x) = \frac{x^2 \cos x - 1}{\cos x}$$

$$\therefore \quad f(0) = 1 \Rightarrow |f(0)| = 1$$

**30.** (4) Given, $A = \int \frac{dx}{x^2 + 6x + 25}$ and $B = \int \frac{dx}{x^2 - 6x - 27}$

$$\Rightarrow \quad A = \int \frac{dx}{x^2 + 6x + 9 - 9 + 25}$$

and $$B = \int \frac{dx}{x^2 - 6x + 9 - 9 - 27}$$

$$\Rightarrow \quad A = \int \frac{dx}{(x+3)^2 + 16} \text{ and } B = \int \frac{dx}{(x-3)^2 - 36}$$

$$\Rightarrow \quad A = \int \frac{dx}{(x+3)^2 + (4)^2} \text{ and } B = \int \frac{dx}{(x-3)^2 - (6)^2}$$

$$\Rightarrow \quad A = \frac{1}{4} \tan^{-1}\left(\frac{x+3}{4}\right) \text{ and } B = \frac{1}{2 \times 6} \log \left|\frac{x-3-6}{x-3+6}\right|$$

$$\Rightarrow \quad A = \frac{1}{4} \tan^{-1}\left(\frac{x+3}{4}\right) \text{ and } B = \frac{1}{12} \log \left|\frac{x-9}{x+3}\right|$$

Now, $12(A + B) = 12\left[\frac{1}{4} \tan^{-1}\left(\frac{x+3}{4}\right) + \frac{1}{12} \log \frac{x-9}{x+3}\right]$

$$= 3 \tan^{-1}\left(\frac{x+3}{4}\right) + \log \left|\frac{x-9}{x+3}\right|$$

$$\therefore \quad \lambda = 3, \mu = 1 \Rightarrow \lambda + \mu = 4$$

**31.** (2) Given, $\int \frac{dx}{x^{2008} + x} = \frac{1}{p} \log\left(\frac{x^q}{1 + x^r}\right) + C$

Let $I = \int \frac{dx}{x^{2008} + x} = \int \frac{dx}{x^{2008}\left(1 + \frac{1}{x^{2007}}\right)}$

Put $1 + \frac{1}{x^{2007}} = t \Rightarrow -\frac{2007}{x^{2008}} dx = dt$

$$\therefore \quad I = -\frac{1}{2007} \int \frac{dt}{t}$$

$$= -\frac{1}{2007} \log |t| + C$$

$$= -\frac{1}{2007} \log \left|1 + \frac{1}{x^{2007}}\right| + C$$

$$= -\frac{1}{2007} \log \left|\frac{x^{2007} + 1}{x^{2007}}\right| + C$$

$$= \frac{1}{2007} \log \left|\frac{x^{2007}}{1 + x^{2007}}\right| + C$$

$\Rightarrow p = 2007, q = 2007$ and $r = 2007$

Hence, $\frac{p+q}{r} = \frac{2007 + 2007}{2007} = \frac{2 \times 2007}{2007} = 2$

**32.** (2) Given, $\int \frac{e^x (2 - x^2)}{(1-x)\sqrt{1-x^2}} dx = \mu e^x \left(\frac{1+x}{1-x}\right)^{\lambda} + C$

Let $$I = \int \frac{e^x (2 - x^2)}{(1-x)\sqrt{1-x^2}} dx$$

$$\Rightarrow \quad I = \int e^x \left[\frac{1}{(1-x)\sqrt{1-x^2}} + \sqrt{\frac{1+x}{1-x}}\right] dx$$

As, $\frac{d}{dx}\left[\sqrt{\frac{1+x}{1-x}}\right] = \frac{1}{(1-x)\sqrt{1-x^2}}$

$$\therefore \quad I = e^x \sqrt{\frac{1+x}{1-x}} + C$$

$$\Rightarrow \quad I = e^x \left(\frac{1+x}{1-x}\right)^{1/2} + C$$

$$\Rightarrow \quad \mu = 1, \lambda = \frac{1}{2}$$

$$\therefore \quad 2\lambda + \mu = \left(2 \times \frac{1}{2} + 1\right) = 2$$

**33.** (4) $\because f''(x) = f(x)$

$$\Rightarrow \quad \int f''(x) \cdot f'(x) dx = \int f(x) \cdot f'(x) dx$$

$$\Rightarrow \quad [f'(x)]^2 = [f(x)]^2 + C$$

At $x = 0$, $C = 0$

$$\Rightarrow \quad f'(x) = \pm f(x)$$

$$\therefore \quad \left|\int \frac{f(x)}{4 - f(x)} dx\right| = \left|\int \frac{\pm f'(x)}{4 - f(x)} dx\right|$$

$$= |\pm \log[4 - f(x)] + C|$$

On comparing, $k = 4$

**34.** (2) Consider the integral

$$I = \int \frac{dx}{x^n (1 + x^n)^{1/n}}, n \in N = \int \frac{dx}{x^{n+1}\left(1 + \frac{1}{x^n}\right)^{1/n}}$$

$$= \int (1 + x^{-n})^{-\frac{1}{n}} x^{-n-1} dx = \int -\frac{1}{n} \int (1 + x^{-n})^{-1/n} d(1 + x^{-n})$$

$$= -\frac{1}{n} \frac{(1 + x^{-n})^{-\frac{1}{n}+1}}{\left(-\frac{1}{n} + 1\right)} + K = \frac{1}{1-n}\left(1 - \frac{1}{x^n}\right)^{1 - \frac{1}{n}} + K$$

$$= \frac{1}{1-n} (1 + x^{-n})^{\frac{n-1}{n}} + K$$

Now, $\int \frac{dx}{x^n (1 + x^n)^{1/n}} = \frac{1}{1-n} (1 + x^{-n})^{\frac{n-1}{n}} + K$

Put $n = 2010$, then

$$\int \frac{dx}{x^{2010} (1 + x^{2010})^{\frac{1}{2010}}} = \frac{1}{1 - 2010} (1 + x^{-2010})^{\frac{2010-1}{2010}} + K$$

$$= \frac{1}{-2009} (1 + x^{-2010})^{\frac{2009}{2010}} + K$$

$\Rightarrow a = 2009, b = 2010, c = 2011$

Hence, $\frac{a + b - c}{1004} = \frac{2009 + 2010 - 2011}{1004} = \frac{2008}{1004} = 2$

**35.** (1) Let $I = \int \frac{dx}{(1 + \sqrt{x})^{2010}} = \int \frac{\sqrt{x}}{\sqrt{x}(1 + \sqrt{x})^{2010}} dx$

Put $1 + \sqrt{x} = t \Rightarrow \frac{1}{2\sqrt{x}} dx = dt$

$$\Rightarrow \quad dx = 2(t-1) dt$$

$$\therefore \quad I = \int \frac{2(t-1)}{t^{2010}} dt$$

$$= 2 \int \left[\frac{1}{t^{2009}} - \frac{1}{t^{2010}}\right] dt$$

$$= 2\left[\frac{1}{2009 \cdot t^{2009}} - \frac{1}{2008 \cdot t^{2008}}\right] + C$$

$$\therefore \quad \alpha = 2009, \beta = 2008$$

Then, $|\alpha - \beta| = 1$

## JEE Main & AIEEE Archive

**1.** (d) Let $I = \int \frac{dx}{x^2(x^4+1)^{\frac{3}{4}}} = \int \frac{dx}{x^5\left(1+\frac{1}{x^4}\right)^{\frac{3}{4}}}$

Put $1 + \frac{1}{x^4} = t^4$

$\Rightarrow \quad -\frac{4}{x^5}dx = 4t^3dt$

$\Rightarrow \quad \frac{dx}{x^5} = -t^3dt$

$\therefore \quad I = \int \frac{-t^3}{t^3}dt = -\int dt = -t + C$

$= -\left(1+\frac{1}{x^4}\right)^{\frac{1}{4}} + C$

**2.** (b) $\int \left(1 + x - \frac{1}{x}\right)e^{x+\frac{1}{x}}dx$

$= \int e^{x+\frac{1}{x}}dx + \int x\left(1-\frac{1}{x^2}\right)e^{x+\frac{1}{x}}dx$

$= \int e^{x+\frac{1}{x}}dx + xe^{x+\frac{1}{x}} - \int \frac{d}{dx}(x)e^{x+\frac{1}{x}}dx$

$= \int e^{x+\frac{1}{x}}dx + xe^{x+\frac{1}{x}} - \int e^{x+\frac{1}{x}}dx$

$\left[\because \int\left(x-\frac{1}{x^2}\right)e^{x+\frac{1}{x}}dx = e^{x+\frac{1}{x}}\right]$

$= xe^{x+\frac{1}{x}} + C$

**3.** (c) Given, $\int f(x)\,dx = \psi(x)$

Let $\quad I = \int x^5 f(x^3)\,dx$

Put $\quad x^3 = t \Rightarrow x^2dx = \frac{dt}{3} \quad$ ...(i)

$\therefore I = \frac{1}{3}\int t\,f(t)\,dt$

$= \frac{1}{3}[t\,\psi(t) - \int \psi(t)\,dt]$ [using integration by parts]

$= \frac{1}{3}[x^3\psi(x^3) - 3\int x^2\psi(x^3)\,dx] + C$ [from Eq. (i)]

$= \frac{1}{3}x^3\psi(x^3) - \int x^2\psi(x^3)\,dx + C$

**4.** (d) **Given Integral** $\int \frac{5\tan x}{\tan x - 2}dx$

**To find** The value of '$a$',

if $\quad \int \frac{5\tan x}{\tan x - 2}dx = x + a\log|\sin x - 2\cos x| + k \quad$ ...(i)

Now, let us assume that

$$I = \int \frac{5\tan x}{\tan x - 2}dx$$

On multiplying by $\cos x$ in numerator and denominator, we get

$$I = \int \frac{5\sin x}{\sin x - 2\cos x}dx$$

This special integration requires special substitution of type

$N^r = A(D^r) + B\left(\frac{dD^r}{dx}\right)$, $A$ and $B$ are constants.

Let $\quad 5\sin x = A(\sin x - 2\cos x) + B(\cos x + 2\sin x)$

$\Rightarrow 0\cos x + 5\sin x = (A+2B)\sin x + (B-2A)\cos x$

On comparing the coefficients of $\sin x$ and $\cos x$, we get

$A + 2B = 5$

and $\quad B - 2A = 0$

On solving the above two equations in $A$ and $B$, we get

$A = 1$

and $\quad B = 2$

$\Rightarrow 5\sin x = (\sin x - 2\cos x) + 2(\cos x + 2\sin x)$

$\Rightarrow \quad I = \int \frac{5\sin x}{\sin x - 2\cos x}dx$

$= \int \frac{(\sin x - 2\cos x) + 2(\cos x + 2\sin x)}{(\sin x - 2\cos x)}dx$

$= \int \frac{\sin x - 2\cos x}{\sin x - 2\cos x}dx + 2\int \frac{(\cos x + 2\sin x)}{(\sin x - 2\cos x)}dx$

$= \int 1\,dx + 2\int \frac{d(\sin x - 2\cos x)}{(\sin x - 2\cos x)}$

$= x + 2\log|(\sin x - 2\cos x)| + k \quad$ ...(ii)

where, $k$ is the constant of integration.

Now, by comparing the value of $I$ in Eqs. (i) and (ii), we get

$a = 2$

**5.** (d) Let $\quad I = \sqrt{2}\int \frac{\sin x}{\sin\left(x - \frac{\pi}{4}\right)}dx$

Put $x - \frac{\pi}{4} = t \Rightarrow dx = dt$

$\therefore \quad I = \sqrt{2}\int \frac{\sin\left(\frac{\pi}{4}+t\right)dt}{\sin t}$

$= \sqrt{2}\int\left[\frac{1}{\sqrt{2}}\cot t + \frac{1}{\sqrt{2}}\right]dt$

$= \log|\sin t| + t + C_1$

$= x + \log\left|\sin\left(x - \frac{\pi}{4}\right)\right| + C \quad \left[\because C = C_1 - \frac{\pi}{4}\right]$

**6.** (a) $\int \frac{dx}{\cos x + \sqrt{3}\sin x}$

$= \int \frac{dx}{2\left(\frac{1}{2}\cos x + \frac{\sqrt{3}}{2}\sin x\right)}$

$= \frac{1}{2}\int \sec\left(x - \frac{\pi}{3}\right)dx$

$= \frac{1}{2}\log\tan\left(\frac{x}{2} - \frac{\pi}{6} + \frac{\pi}{4}\right) + C$

$= \frac{1}{2}\log\tan\left(\frac{x}{2} + \frac{\pi}{12}\right) + C$

**7.** (a) $\int\left\{\frac{(\log x - 1)}{1 + (\log x)^2}\right\}^2dx$

$= \int \frac{(\log x)^2 + 1 - 2\log x}{[(\log x)^2 + 1]^2}dx$

$= \int \frac{(\log x)^2 + 1 - 2x\left(\log x \cdot \frac{1}{x}\right)}{[(\log x)^2 + 1]^2}dx$

$= \int \frac{d}{dx}\left[\frac{x}{(\log x)^2 + 1}\right]dx$

$= \frac{x}{(\log x)^2 + 1} + C$

**Alternate Method**

Let $I = \int \left\{\frac{(\log x - 1)}{1 + (\log x)^2}\right\}^2 dx$

Put $t = \log x$

$\Rightarrow \quad x = e^t$

$\Rightarrow \quad dx = e^t dt$

$\therefore \quad I = \int \left\{\frac{t-1}{1+t^2}\right\}^2 e^t \, dt$

$= \int \frac{(1+t^2) - 2t}{(1+t^2)^2} \cdot e^t dt$

$= \int \frac{e^t}{1+t^2} \cdot dt - \int \frac{2te^t}{(1+t^2)^2} \cdot dt$

Using by parts,

$I = \frac{1}{1+t^2} \cdot e^t - \int \frac{-1}{(1+t^2)^2} \cdot 2t \cdot e^t \, dt - \int \frac{2t\, e^t}{(1+t^2)^2} dt$

$= \frac{e^t}{1+t^2} + \int \frac{2te^t}{(1+t^2)^2} dt - \int \frac{2te^t}{(1+t^2)^2} dt$

$\therefore \quad I = \frac{e^t}{1+t^2} + C$

$= \frac{x}{(\log x)^2 + 1} + C$

**8.** (b) Let $I = \int \frac{\sin x}{\sin (x - \alpha)} dx$

Put $x - \alpha = t$

$\Rightarrow \quad dx = dt$

$\therefore \quad I = \int \frac{\sin (t + \alpha)}{\sin t} dt$

$\Rightarrow \quad I = \int \frac{\sin t \cos \alpha + \cos t \sin \alpha}{\sin t} dt$

$\Rightarrow \quad I = \cos \alpha \int 1\, dt + \sin \alpha \int \frac{\cos t}{\sin t} dt$

$\Rightarrow \quad I = \cos \alpha (t) + \sin \alpha \log \sin t + C_1$

$\Rightarrow \quad I = \cos \alpha (x - \alpha) + \sin \alpha \log \sin (x - \alpha) + C_1$

$\Rightarrow \quad I = x \cos \alpha + \sin \alpha \log \sin (x - \alpha) + C$

[let $C = -\alpha \cos \alpha + C_1$]

But $\quad I = Ax + B \log \sin (x - \alpha) + C$

$\therefore \quad x \cos \alpha + \sin \alpha \log \sin (x - \alpha) + C = Ax + B \log \sin (x - \alpha) + C$

On comparing both sides, we get

$A = \cos \alpha$

and $\quad B = \sin \alpha$

**Alternate Method**

Given that,

$$\int \frac{\sin x}{\sin (x - \alpha)} dx = Ax + B \log \sin (x - \alpha) + C$$

On differentiating both sides w.r.t. $x$, we get

$$\frac{\sin x}{\sin (x - \alpha)} = A + B \frac{\cos (x - \alpha)}{\sin (x - \alpha)}$$

$\Rightarrow \quad \sin x = A \sin (x - \alpha) + B \cos (x - \alpha)$

$\Rightarrow \quad \sin x = A(\sin x \cos \alpha - \cos x \sin \alpha) + B(\cos x \cos \alpha + \sin x \sin \alpha)$

$\Rightarrow \quad \sin x = \sin x (A \cos \alpha + B \sin \alpha) + \cos x (B \cos \alpha - A \sin \alpha)$

On comparing the coefficients of $\sin x$ and $\cos x$ both sides, we get

$A \cos \alpha + B \sin \alpha = 1 \quad \ldots \text{(i)}$

and $\quad B \cos \alpha - A \sin \alpha = 0 \quad \ldots \text{(ii)}$

On solving Eqs. (i) and (ii), we get

$A = \cos \alpha$ and $B = \sin \alpha$

**9.** (d) Let $I = \int \frac{dx}{\cos x - \sin x}$

$= \frac{1}{\sqrt{2}} \int \frac{dx}{\left(\frac{1}{\sqrt{2}} \cos x - \frac{1}{\sqrt{2}} \sin x\right)}$

$= \frac{1}{\sqrt{2}} \int \frac{dx}{\cos \left(x + \frac{\pi}{4}\right)}$

$= \frac{1}{\sqrt{2}} \int \sec \left(x + \frac{\pi}{4}\right) dx$

$= \frac{1}{\sqrt{2}} \log \left| \tan \left(\frac{\pi}{4} + \frac{x}{2} + \frac{\pi}{8}\right) \right| + C$

$= \frac{1}{\sqrt{2}} \log \left| \tan \left(\frac{x}{2} + \frac{3\pi}{8}\right) \right| + C$

**10.** (a) Let $I = \int \frac{dx}{x(x^n + 1)} = \int \frac{x^{n-1}}{x^n(x^n + 1)} dx$

Put $\quad x^n + 1 = t$

$\Rightarrow \quad nx^{n-1} dx = dt$

$\therefore \quad I = \frac{1}{n} \int \frac{dt}{t(t-1)}$

$= \frac{1}{n} \int \left(\frac{1}{t-1} - \frac{1}{t}\right) dt$

$= \frac{1}{n} \log \left(\frac{t-1}{t}\right) + C$

$= \frac{1}{n} \log \left(\frac{x^n}{x^n + 1}\right) + C$

## JEE Advanced & IIT JEE Archive

**1.** (c) **Concept Involved**

Integration by substitution

i.e. $$I=\int f(g(x))\cdot g'(x)\,dx$$

Put $g(x)=t \Rightarrow g'(x)\,dx=dt$

$\therefore$ $$I=\int f(t)\,dt$$

**Description of Situation** Generally, students gets confused after substitution, i.e.

$$\sec x+\tan x=t$$

Now, for $\sec x$ we should use

$$\sec^2 x-\tan^2 x=1$$

$\Rightarrow$ $$(\sec x-\tan x)(\sec x+\tan x)=1$$

$\Rightarrow$ $$\sec x-\tan x=\frac{1}{t}$$

Here, $$I=\int\frac{\sec^2 x}{(\sec x+\tan x)^{9/2}}dx$$

Put $\sec x+\tan x=t$

$\Rightarrow$ $$(\sec x\tan x+\sec^2 x)\,dx=dt$$

$\Rightarrow$ $$\sec x\cdot t\,dx=dt$$

$\Rightarrow$ $$\sec x\,dx=\frac{dt}{t}$$

$\therefore$ $$\sec x-\tan x=\frac{1}{t}$$

$\Rightarrow$ $$\sec x=\frac{1}{2}\left(t+\frac{1}{t}\right)$$

$\therefore$ $$I=\int\frac{\sec x\cdot\sec x}{(\sec x+\tan x)^{9/2}}dx$$

$\Rightarrow$ $$I=\int\frac{\frac{1}{2}\left(t+\frac{1}{t}\right)\cdot\frac{dt}{t}}{t^{9/2}}=\frac{1}{2}\int\left(\frac{1}{t^{9/2}}+\frac{1}{t^{13/2}}\right)dt$$

$$=-\frac{1}{2}\left\{\frac{2}{7t^{7/2}}+\frac{2}{11t^{11/2}}\right\}+K$$

$$=-\left[\frac{1}{7(\sec x+\tan x)^{7/2}}+\frac{1}{11(\sec x+\tan x)^{11/2}}\right]+K$$

$$=\frac{-1}{(\sec x+\tan x)^{11/2}}\left\{\frac{1}{11}+\frac{1}{7}(\sec x+\tan x)^2\right\}+K$$

**2.** (c) Since, $$I=\int\frac{e^x}{e^{4x}+e^{2x}+1}dx$$

$$J=\int\frac{e^{3x}}{1+e^{2x}+e^{4x}}dx$$

$\therefore$ $$J-I=\int\frac{(e^{3x}-e^x)}{1+e^{2x}+c^{4x}}dx$$

Put $e^x=u \Rightarrow e^x dx=du$

$\therefore$ $$J-I=\int\frac{(u^2-1)}{1+u^2+u^4}du$$

$$=\int\frac{\left(1-\frac{1}{u^2}\right)}{1+\frac{1}{u^2}+u^2}du=\int\frac{\left(1-\frac{1}{u^2}\right)}{\left(u+\frac{1}{u}\right)^2-1}du$$

Put $u+\frac{1}{u}=t$

$\Rightarrow$ $$\left(1-\frac{1}{u^2}\right)du=dt$$

$\therefore$ $$J-I=\int\frac{dt}{t^2-1}$$

$$=\frac{1}{2}\log\left|\frac{t-1}{t+1}\right|+C$$

$$=\frac{1}{2}\log\left|\frac{u^2-u+1}{u^2+u+1}\right|+C$$

$$=\frac{1}{2}\log\left|\frac{e^{2x}-e^x+1}{e^{2x}+e^x+1}\right|+C$$

**3.** (a) Given, $$f(x)=\frac{x}{(1+x^n)^{1/n}} \text{ for } n\ge 2$$

$\therefore$ $$ff(x)=\frac{f(x)}{[1+f(x)^n]^{1/n}}=\frac{x}{(1+2x^n)^{1/n}}$$

and $$fff(x)=\frac{x}{(1+3x^n)^{1/n}}$$

$\therefore$ $$g(x)=\underbrace{(fofo\ldots of)}_{n\text{ times}}(x)=\frac{x}{(1+nx^n)^{1/n}}$$

Let $I=\int x^{n-2}g(x)\,dx$

$$=\int\frac{x^{n-1}}{(1+nx^n)^{1/n}}dx$$

$$=\frac{1}{n^2}\int\frac{n^2x^{n-1}}{(1+nx^n)^{1/n}}dx$$

$$=\frac{1}{n^2}\int\frac{\frac{d}{dx}(1+nx^n)}{(1+nx^n)^{1/n}}dx$$

$\therefore$ $$I=\frac{1}{n(n-1)}(1+nx^n)^{1-\frac{1}{n}}+C$$

**4.** (c) Let $$I=\int\frac{(x^2-1)\,dx}{x^3\sqrt{2x^4-2x^2+1}}$$

On dividing numerator and denominator by $x^5$,

$$=\int\frac{\left(\frac{1}{x^3}-\frac{1}{x^5}\right)dx}{\sqrt{2-\frac{2}{x^2}+\frac{1}{x^4}}}$$

Put $2-\frac{2}{x^2}+\frac{1}{x^4}=t$

$\Rightarrow$ $$\left(\frac{4}{x^3}-\frac{4}{x^5}\right)dx=dt$$

$\therefore$ $$I=\frac{1}{4}\int\frac{dt}{\sqrt{t}}$$

$$=\frac{1}{4}\cdot\frac{t^{1/2}}{1/2}+C$$

$$=\frac{1}{2}\sqrt{2-\frac{2}{x^2}+\frac{1}{x^4}}+C$$

# 2. Definite Integral

## DPP-1 Basic Definition

**1.** (i) Let $I = \int_0^{1/2} \frac{x \cdot \sin^{-1} x}{\sqrt{1-x^2}} dx$

Put $\sin^{-1} x = t \Rightarrow x = \sin t$

$$\Rightarrow \quad \frac{1}{\sqrt{1-x^2}} dx = dt$$

When $x = 0$, then $t = 0$

and when $x = \frac{1}{2}$, then $t = \frac{\pi}{6}$

$$\therefore \quad I = \int_0^{\pi/6} \underset{\text{I}}{t} \cdot \underset{\text{II}}{\sin t}\, dt$$

$$= [t\,(-\cos t)]_0^{\pi/6} - \int_0^{\pi/6} 1 \cdot (-\cos t)\, dt$$

$$= [-t \cos t]_0^{\pi/6} + \int_0^{\pi/6} \cos t\, dt$$

$$= [-t \cos t]_0^{\pi/6} + [\sin t]_0^{\pi/6}$$

$$= -\frac{\sqrt{3}}{2} \times \frac{\pi}{6} + \frac{1}{2} = \frac{1}{2} - \frac{\sqrt{3}\pi}{12}$$

(ii) Let $I = \int_0^{\pi/2} (\sqrt{\tan x} + \sqrt{\cot x})\, dx$

$$= \int_0^{\pi/2} \frac{\sin x + \cos x}{\sqrt{\sin x \cos x}} dx = \sqrt{2} \int_0^{\pi/2} \frac{\sin x + \cos x}{\sqrt{\sin 2x}} dx$$

$$= \sqrt{2} \int_0^{\pi/2} \frac{d(-\cos x + \sin x)}{\sqrt{1 - (\sin x - \cos x)^2}}$$

$$= \sqrt{2}\, [\sin^{-1}(\sin x - \cos x)]_0^{\pi/2}$$

$$= \sqrt{2}[\sin^{-1}(1) - \sin^{-1}(-1)] = \sqrt{2}\, [\sin^{-1}(1) + \sin^{-1}(1)]$$

$$= \sqrt{2}\, [2 \sin^{-1}(1)] = \sqrt{2} \times 2 \times \frac{\pi}{2} = \sqrt{2}\pi$$

(iii) Let $I = \int_0^{\infty} \frac{dx}{(x + \sqrt{1+x^2})^n}$ for $n > 1$

Put $x + \sqrt{1+x^2} = t$

$$\Rightarrow \quad \left[1 + \frac{1}{2\sqrt{1+x^2}}(2x)\right] dx = dt$$

$$\Rightarrow \quad \left[1 + \frac{x}{\sqrt{1+x^2}}\right] dx = dt$$

$$\Rightarrow \quad \left[\frac{\sqrt{1+x^2} + x}{\sqrt{1+x^2}}\right] dx = dt \quad \Rightarrow \quad \frac{t}{\sqrt{1+x^2}} dx = dt$$

$$\Rightarrow \quad dx = \frac{\sqrt{1+x^2}}{t} dt$$

Now, $x + \sqrt{1+x^2} = t$ and $\sqrt{1+x^2} - x = \frac{1}{t}$

$$\therefore \quad t + \frac{1}{t} = 2\sqrt{1+x^2}$$

$$\Rightarrow \quad \sqrt{1+x^2} = \frac{t^2+1}{2t} \quad \Rightarrow \quad dx = \left(\frac{t^2+1}{2t^2}\right) dt$$

When $x = 0$, then $t = 1$ and when $x = \infty$, then $t = \infty$

$$\therefore \quad I = \int_1^{\infty} \frac{t^2+1}{2t^2 \cdot t^n} dt = \int_1^{\infty} \frac{t^2+1}{2t^{n+2}} dt = \frac{1}{2} \int_1^{\infty} \left(\frac{1}{t^n} + \frac{1}{t^{n+2}}\right) dt$$

$$= \frac{1}{2} \lim_{a \to +\infty} \left[\frac{-1}{(n-1)t^{n-1}} - \frac{1}{(n+1)t^{n+1}}\right]_1^a$$

$$= \frac{1}{2}\left[-0 - 0 - \left(-\frac{1}{n-1} - \frac{1}{n+1}\right)\right] = \frac{n}{n^2-1}$$

(iv) Let $I = \int_0^{\pi/2} \frac{x \sin x \cos x}{(a^2 \cos^2 x + b^2 \sin^2 x)^2} dx$

$$= \int_0^{\pi/2} \frac{x}{2} \cdot \frac{2 \sin x \cos x}{(a^2 \cos^2 x + b^2 \sin^2 x)^2} dx$$

$$= \frac{1}{2(b^2 - a^2)} \cdot \int_0^{\pi/2} \underset{\text{I}}{x} \cdot \underset{\text{II}}{\left\{\frac{d(a^2 \cos^2 x + b^2 \sin^2 x)}{(a^2 \cos^2 x + b^2 \sin^2 x)^2}\right\}} dx$$

Using integration by parts

$$= \frac{1}{2(b^2 - a^2)}\left[\left(x \cdot \frac{-1}{(a^2 \cos^2 x + b^2 \sin^2 x)}\right)_0^{\pi/2} - \int_0^{\pi/2} \frac{-1}{(a^2 \cos^2 x + b^2 \sin^2 x)} dx\right]$$

$$= \frac{1}{2(b^2 - a^2)}\left[\left(-\frac{\pi}{2b^2} - 0\right) + \frac{1}{b} \int_0^{\pi/2} \frac{b \sec^2 x}{a^2 + b^2 \tan^2 x} dx\right]$$

$$= \frac{1}{2(b^2 - a^2)}\left[-\frac{\pi}{2b^2} + \frac{1}{b} \int_0^{\infty} \frac{dt}{a^2 + t^2}\right]$$

$$= \frac{1}{2(b^2 - a^2)}\left[-\frac{\pi}{2b^2} + \frac{1}{b} \cdot \frac{1}{a} \cdot \left(\tan^{-1} \frac{t}{a}\right)_0^{\infty}\right]$$

$$= \frac{-\pi}{4b^2(b^2 - a^2)} + \frac{1}{2ab(b^2 - a^2)} \cdot \left(\frac{\pi}{2}\right)$$

$$= \frac{\pi}{4ab^2(b^2 - a^2)}[b - a] = \frac{\pi}{4ab^2(b + a)}$$

(v) Let $I = \int_0^{\pi/2} (\sqrt{\sin x} + \sqrt{\cos x})^{-4} dx$

$$\therefore \quad I = \int_0^{\pi/2} \frac{dx}{(\sqrt{\sin x} + \sqrt{\cos x})^4} = \int_0^{\pi/2} \frac{dx}{\cos^2 x \cdot (1 + \sqrt{\tan x})^4}$$

$$= \int_0^{\pi/2} \frac{\sec^2 x}{(1 + \sqrt{\tan x})^4} dx$$

Put $\tan x = t \Rightarrow \sec^2 x\, dx = dt$

$$\Rightarrow \quad I = \int_0^{\infty} \frac{dt}{(1 + \sqrt{t})^4}$$

Let $t = u^2 \Rightarrow dt = 2u\, du$

$$= \int_0^{\infty} \frac{2u}{(1+u)^4} du = 2 \int_0^{\infty} \frac{u + 1 - 1}{(u+1)^4} du$$

$$= 2 \int_0^{\infty} \left[\frac{1}{(u+1)^3} - \frac{1}{(u+1)^4}\right] du$$

$$= 2\left[\frac{(u+1)^{-2}}{-2} - \frac{(u+1)^{-3}}{-3}\right]_0^{\infty}$$

$$= 2\left[0 - \left(-\frac{1}{2} + \frac{1}{3}\right)\right] = \frac{1}{3}$$

(vi) Let $I = \int_0^{\pi/2} \frac{\sin^3 x}{(1 + \cos^2 x)\sqrt{1 + \cos^2 x + \cos^4 x}} dx$

$$= \int_0^{\pi/2} \frac{(1 - \cos^2 x) \sin x}{(1 + \cos^2 x)\sqrt{1 + \cos^2 x + \cos^4 x}} dx$$

$$= -\int_0^{\pi/2} \frac{(1 - \cos^2 x)\, d(\cos x)}{(1 + \cos^2 x)\sqrt{1 + \cos^2 x + \cos^4 x}}$$

$$= \int_1^0 \frac{(1 - t^2)}{(1 + t^2)\sqrt{1 + t^2 + t^4}} dt, \text{ where } t = \cos x$$

$$= -\int_0^1 \frac{\left(1-\frac{1}{t^2}\right)dt}{\left(t+\frac{1}{t}\right)\sqrt{\left(t+\frac{1}{t}\right)^2-1}}$$

$$= -\int_0^1 \frac{1}{\left(t+\frac{1}{t}\right)\sqrt{\left(t-\frac{1}{t}\right)^2-1}}\, d\left(t+\frac{1}{t}\right)$$

$$= -\left[\sec^{-1}\left(t+\frac{1}{t}\right)\right]_0^1 = -\left(\frac{\pi}{3}-\frac{\pi}{2}\right) = \frac{\pi}{6}$$

**2.** (c) Let $I = \int_0^{\pi/3} \frac{\cos x}{3+4\sin x}dx$

Put $3 + 4\sin x = t \Rightarrow 4\cos x\, dx = dt$

When, $x = \frac{\pi}{3}$, then $t = 3 + 2\sqrt{3}$

and when $x = 0$, then $t = 3$

$$\therefore \quad I = \frac{1}{4}\int_3^{3+2\sqrt{3}} \frac{dt}{t} = \frac{1}{4}[\log(t)]_3^{3+2\sqrt{3}}$$

$$= \frac{1}{4}[\log(3+2\sqrt{3}) - \log 3]$$

$$= \frac{1}{4}\log\left(\frac{3+2\sqrt{3}}{3}\right) \quad \ldots(i)$$

But it is given that,

$$\int_0^{\pi/3} \frac{\cos x}{3+4\sin x}dx = k\log\left(\frac{3+2\sqrt{3}}{3}\right) \quad \ldots(ii)$$

From Eqs. (i) and (ii), we get

$$k = \frac{1}{4}$$

**3.** (a) Let $I = \int_{e^{e^e}}^{e^{e^{e^e}}} \frac{dx}{x\log x \log(\log x)\cdot\log[\log(\log x)]}$

Put $\log\{\log(\log x)\} = t$

$$\Rightarrow \frac{1}{\log(\log x)} \times \frac{1}{\log x} \times \frac{1}{x}\, dx = dt$$

$$\therefore \quad I = \int_1^e \frac{dt}{t} = [\log|t|]_1^e = [1-0] = 1$$

**4.** (a) Let $I = \int_1^\infty (e^{x+1} + e^{3-x})^{-1}dx$

$$= \int_1^\infty \frac{dx}{(e^{x+1}+e^{3-x})} = \int_1^\infty \frac{dx}{(ee^x + e^3 e^{-x})}$$

$$= \int_1^\infty \frac{e^x}{e(e^{2x}+e^2)}dx$$

[multiplying numerator and denominator by $e^x$]

Put $e^x = t \Rightarrow e^x dx = dt$

When $x = 1$, then $t = e$

and when $x = \infty$, then $t = \infty$

$$\therefore \quad I = \frac{1}{e}\int_e^\infty \frac{dt}{t^2+e^2} = \left[\frac{1}{e^2}\tan^{-1}\left(\frac{t}{e}\right)\right]_e^\infty$$

$$= \frac{1}{e^2}\left[\frac{\pi}{2}-\frac{\pi}{4}\right] = \frac{\pi}{4e^2}$$

**5.** (d) Let $I = \int_1^e [(x+1)e^x \log x]dx$

$$= \int_1^e (xe^x \log x + e^x \log x)dx$$

$$= \int_1^e e^x(x\log x + \log x)dx$$

$$= \int_1^e e^x(x\log x + \log x + 1 - 1)dx$$

$$= \int_1^e e^x(x\log x + \log x + 1)dx - \int_1^e e^x dx$$

$$= [e^x x\log x - e^x]_1^e = e^{e+1} - e^e + e$$

$$= e^e(e-1) + e$$

**6.** (b) Given, $\int_0^1 (1+\cos^8 x)(ax^2+bx+c)dx$

$$= \int_0^2 (1+\cos^8 x)(ax^2+bx+c)dx$$

$$\Rightarrow \int_0^1 (1+\cos^8 x)(ax^2+bx+c)dx$$

$$= \int_0^1 (1+\cos^8 x)(ax^2+bx+c)dx + \int_1^2 (1+\cos^8 x)(ax^2+bx+c)dx$$

$$\Rightarrow \quad \int_1^2 (1+\cos^8 x)(ax^2+bx+c)dx = 0$$

$\because$ $(1+\cos^8 x)(ax^2+bx+c)$ must have atleast one root in (1, 2).

$\Rightarrow ax^2+bx+c = 0$ must have atleast one root $\in (1, 2)$.

$\therefore$ Atleast one root $\in (0, 2)$.

**7.** (a) Given, $I_n = \int_0^{\pi/4} \tan^n x\, dx$

$$= \int_0^{\pi/4} \tan^{n-2}x \tan^2 x\, dx$$

$$= \int_0^{\pi/4} \sec^2 x\cdot\tan^{n-2}x\, dx - \int_0^{\pi/4}\tan^{n-2}x\, dx$$

$$\Rightarrow \quad I_n = \int_0^1 t^{n-2}dt - I_{n-2}, \text{ where } t = \tan x$$

$$\therefore \quad I_n + I_{n-2} = \left[\frac{t^{n-1}}{n-1}\right]_0^1 \Rightarrow I_n + I_{n-2} = \frac{1}{n-1}$$

Now, $\lim_{n\to\infty} n(I_n + I_{n-2}) = \lim_{n\to\infty} n\left(\frac{1}{n-1}\right)$

$$= \lim_{n\to\infty}\frac{n}{n\left(1-\frac{1}{n}\right)} = \lim_{n\to\infty}\frac{1}{1-\frac{1}{n}} = 1$$

**8.** (d) Let $I = \int_0^\infty \frac{x\tan^{-1}x}{(1+x^2)^2}dx$

Put $\tan^{-1}x = t \Rightarrow x = \tan t$

$$\Rightarrow \quad \frac{1}{1+x^2}dx = dt$$

When $x = 0$, then $t = 0$

and when $x = \infty$, then $t = \frac{\pi}{2}$

$$\therefore \quad I = \int_0^{\pi/2} \frac{t\tan t}{1+\tan^2 t}dt$$

$$= \int_0^{\pi/2} t\sin t\cos t\, dt = \frac{1}{2}\int_0^{\pi/2} \underset{\text{I}}{t}\, \underset{\text{II}}{\sin 2t}\, dt$$

$$= \frac{1}{2}\left[t\cdot\frac{(-\cos 2t)}{2} - \int_0^{\pi/2}\frac{-\cos 2t}{2}dt\right]$$

$$= \frac{1}{2}\left[-\frac{t}{2}\cos 2t + \frac{1}{2}\int_0^{\pi/2}\cos 2t\, dt\right]$$

$$= \frac{1}{4}\left[-t\cos 2t + \frac{\sin 2t}{2}\right]_0^{\pi/2}$$

$$= \frac{1}{4}\left[\left(\frac{\pi}{2}+0\right) - (-0+0)\right] = \frac{\pi}{8}$$

**9.** (d) Given, $f'(e) = f(e) = 1$

$$\int_1^e \frac{f(x)}{x^2}dx = \frac{1}{2}$$

Now, $\int_1^e \underset{\text{II}}{f''(x)}\, \underset{\text{I}}{\log x}\, dx = [\log x\cdot(f'(x))]_1^e - \int_1^e \underset{\text{I}}{\frac{1}{x}}\underset{\text{II}}{f'(x)}dx$

$$= (1-0) - \left[\left(\frac{1}{x}f(x)\right)_1^e - \int_1^e\left(-\frac{1}{x^2}\right)f(x)dx\right]$$

$$= 1 - \left[\left(\frac{1}{e}-f(1)\right)+\frac{1}{2}\right] \quad \left[\because \int_1^e \frac{f(x)}{x^2} = \frac{1}{2}\right]$$

$$= \frac{1}{2} - \frac{1}{e} + f(1)$$

**10.** (a) Let $I = \int_{1/2}^{2} \frac{1}{x} \sin\left(x - \frac{1}{x}\right) dx$

Put $x = \frac{1}{t} \Rightarrow dx = -\frac{1}{t^2} dt$

$$\therefore \quad I = \int_{2}^{1/2} t \sin\left(\frac{1}{t} - t\right)\left(-\frac{1}{t^2}\right) dt = \int_{2}^{1/2} \frac{1}{t} \sin\left(t - \frac{1}{t}\right) dt$$

$$= -\int_{1/2}^{2} \frac{1}{t} \sin\left(t - \frac{1}{t}\right) dt = -\int_{1/2}^{2} \frac{1}{x} \sin\left(x - \frac{1}{x}\right) dx$$

$$\Rightarrow \quad I = -I \Rightarrow 2I = 0$$

$$\Rightarrow \quad I = 0$$

**11.** (a) Let $I = \int_{2}^{4} \left[\frac{1}{\log_2 x} - \frac{1}{\log 2} \cdot \frac{1}{(\log_2 x)^2}\right] dx$

$$= \int_{2}^{4} 1 \cdot (\log_2 x)^{-1} dx - \frac{1}{\log 2} \int_{2}^{4} \frac{1}{(\log_2 x)^2} dx$$

$$= [(\log_2 x)^{-1} \cdot x]_2^4 - \int_{2}^{4} \left[-\frac{1}{(\log_2 x)^2} \cdot \frac{1}{x} \cdot \frac{1}{\log 2} \cdot x\right] dx$$

$$- \frac{1}{\log 2} \int_{2}^{4} \frac{dx}{(\log_2 x)^2}$$

$$= \left(\frac{4}{2} - \frac{2}{1}\right) + \int_{2}^{4} \frac{dx}{\log 2 \cdot (\log_2 x)^2} - \int_{2}^{4} \frac{dx}{\log 2 \cdot (\log_2 x)^2}$$

$$= 0$$

**12.** (a) Given, $f(x) = \frac{\sin x}{x}, x > 0$

$$\int f(x) dx = \int \frac{\sin x}{x} dx = F(x) \qquad \text{...(i)}$$

Let $\quad I = \int_{1}^{3} \frac{\sin 2x}{x} dx$

Put $2x = t \Rightarrow dx = \frac{dt}{2}$

$$\therefore \quad I = \int_{2}^{6} \frac{\sin t}{t} dt = \int_{2}^{6} f(t) dt \qquad \text{[by Eq. (i)]}$$

$$= F(6) - F(2)$$

**13.** (a) Let $I = \int_{0}^{\infty} f\left(x + \frac{1}{x}\right) \frac{\log x}{x} dx$

Put $\quad \log x = t \Rightarrow x = e^t \Rightarrow dx = e^t dt$

$$\therefore \quad I = \int_{-\infty}^{\infty} f(e^t + e^{-t}) \frac{t}{e^t} e^t dt$$

$$= \int_{-\infty}^{\infty} f(e^t + e^{-t}) \cdot t \, dt \qquad [\because \text{function is odd}]$$

$$= 0$$

**Alternate Method**

Put $x = \frac{1}{t} \Rightarrow dx = -\frac{1}{t^2} dt$

$$\therefore \quad I = \int_{\infty}^{0} f\left(\frac{1}{t} + t\right) \frac{\log(1/t)}{1/t} \left(-\frac{1}{t^2}\right) dt$$

$$= -\int_{\infty}^{0} f\left(t + \frac{1}{t}\right)\left(\frac{-\log t}{t}\right) dt$$

$$= -\int_{0}^{\infty} f\left(t + \frac{1}{t}\right) \frac{\log t}{t} dt = -\int_{0}^{\infty} f\left(t + \frac{1}{t}\right) \frac{\log t}{t} dt$$

$$= -\int_{0}^{\infty} f\left(x + \frac{1}{x}\right) \frac{\log x}{x} dx$$

$$\Rightarrow \quad I = -I \Rightarrow 2I = 0$$

$$\Rightarrow \quad I = 0$$

**14.** (d) Let $I = \int_{0}^{1} \prod_{r=1}^{n} (x + r) \left(\sum_{k=1}^{n} \frac{1}{x + k}\right) dx$

The given integrand is differential coefficient of $\prod_{r=1}^{n} (x + r)$.

$$\Rightarrow \quad I = \left[\prod_{r=1}^{n} (x + r)\right]_0^1 = (n + 1)! - n! = n \cdot n!$$

**Alternate Method**

Consider $\quad (x + 1)(x + 2)(x + 3) \cdots (x + n) = e^t$

So that when $x = 0$, then $t = \log(n!)$

and when $x = 1$, then $t = \log(n + 1)!$

$$\log(x + 1) + \log(x + 2) + \log(x + 3) + \cdots + \log(x + n) = t$$

$$\therefore \quad \left[\frac{1}{x + 1} + \frac{1}{x + 2} + \frac{1}{x + 3} + \cdots + \frac{1}{x + n}\right] dx = dt$$

$$\therefore I = \int_{\ln n!}^{\ln (n + 1)!} e^t dt = [e^t]_{\ln n!}^{\ln (n + 1)!}$$

$$= e^{\ln (n + 1)!} - e^{\ln n!} = (n + 1)! - n! = n \cdot n!$$

**15.** (b) Given, $\int_{1}^{2} e^{x^2} dx = \alpha \qquad \text{...(i)}$

Let $\quad I_1 = \int_{e}^{e^4} \sqrt{\log_e x} \, dx$

Put $\log_e x = t^2 \Rightarrow x = e^{t^2}$

$$\Rightarrow \quad dx = e^{t^2} \cdot 2t \, dt$$

$$\therefore \quad I = \int_{1}^{2} \underset{\text{I}}{t} \cdot \underset{\text{II}}{(e^{t^2} \cdot 2t)} dt$$

$$= [t \cdot e^{t^2}]_1^2 - \int_{1}^{2} 1 \cdot e^{t^2} dt$$

$$= 2e^4 - e - \alpha \qquad \text{[by Eq. (i)]}$$

**16.** (b) Let us first find the function satisfying

$$af(x) + bf\left(\frac{1}{x}\right) = \frac{1}{x} - 5 \qquad \text{...(i)}$$

On replacing $x$ by $\frac{1}{x}$, we have

$$af\left(\frac{1}{x}\right) + bf(x) = x - 5 \qquad \text{...(ii)}$$

Now, eliminating $f\left(\frac{1}{x}\right)$ from Eqs. (i) and (ii), we get

$$f(x) = \frac{\frac{a}{x} - 5a - bx + 5b}{a^2 - b^2}$$

$$\Rightarrow \int_{1}^{2} f(x) dx = \int_{1}^{2} \frac{\frac{a}{x} - 5a - bx + 5b}{a^2 - b^2} dx$$

$$= \frac{1}{a^2 - b^2}\left[a \log x - b\frac{x^2}{2} + 5(b - a)x\right]_1^2$$

$$= \frac{1}{a^2 - b^2}\left[a \log 2 - 2b + 10(b - a) + \frac{b}{2} - 5(b - a)\right]$$

$$= \frac{1}{a^2 - b^2}\left[a \log 2 - 5a + \frac{7b}{2}\right]$$

**Explanations for Q. Nos. (17-19)**

$g(x)$ is maximum, when $x = 3$

(as below $X$-axis, area would be subtracted)

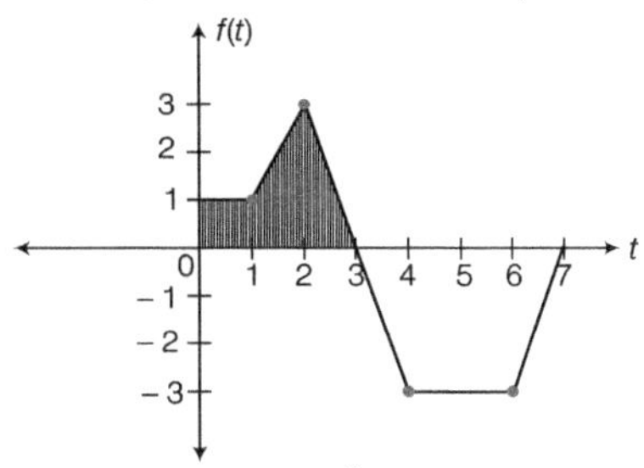


$$\therefore \quad g(x) = \int_{0}^{x} f(t) dt$$

Maximum, when $x = 3$

$\Rightarrow \quad [g(x)]_{x=3} = \int_0^3 f(t)\,dt$

$$= (1 \times 1) + \frac{1}{2}(1+3) \times 1 + \frac{1}{2} \times 1 \times 3$$

$$= 1 + \frac{4}{2} + \frac{3}{2}$$

$$= 3 + \frac{3}{2} = \frac{9}{2}$$

The value of $x$ at which $g(x)$ becomes zero is $x = 5$.

$\Rightarrow \quad \int_0^3 f(t)\,dt = -\int_3^5 f(t)\,dt$

$\Rightarrow \quad \int_0^5 f(t)\,dt = 0$

$\therefore \quad x = 5$

As $g(x)$ is zero, when $x = 5$

$\therefore$ $g(x)$ is negative, when $5 < x < 7$

**17.** (b) **18.** (c) **19.** (d)

**Explanations for Q. Nos. (20-22)**

$$g(x) = \begin{cases} \dfrac{\max[f(t)] + \min[f(t)]}{2}, & 0 \le t \le x,\ 0 \le x \le 4 \\ |x-5| + |x-4|, & 4 < x < 5 \\ \tan\left[\sin^{-1}\left(\dfrac{6-x}{\sqrt{x^2-12x+37}}\right)\right], & x \ge 5 \end{cases}$$

where, $f(x) = x^2 - 4x + 3$

Now, $f(x) = x^2 - 4x + 3 \Rightarrow f'(x) = 2x - 4$

$\Rightarrow \quad f'(x) = 0 \Rightarrow x = 2$

$\Rightarrow \quad f(0) = 3$

$f(2) = -1,\ f(4) = 3$

$\therefore$ Max of $f(x) = 3$ and min $f(x) = -1$

$$\therefore \quad g(x) = \begin{cases} \dfrac{1+1}{2} = 1, & 0 \le x \le 4 \\ x - 4 - x + 5 = 1, & 4 < x < 5 \\ \tan\left[\sin^{-1}\left(\dfrac{6-x}{\sqrt{x^2-12x+37}}\right)\right], & x \ge 5 \end{cases}$$

Now, $\int_2^5 g(x)\,dx = \int_2^4 g(x)\,dx + \int_4^5 g(x)\,dx$

$$= \int_2^4 1\,dx + \int_4^5 1\,dx = [x]_2^4 + [x]_4^5$$

$$= (4-2) + (5-4) = 2 + 1 = 3$$

and $\quad h(x) = \int_0^{x^2} g(t)\,dt \Rightarrow h'(x) = g(x^2)2x$

If $h(x)$ is decreasing, then $h'(x) \le 0$

$$2xg(x^2) \le 0 \Rightarrow x < 0 \text{ or } g(x^2) < 0$$

But $\quad x \nless 0 \Rightarrow g(x^2) < 0$

$\Rightarrow \quad \tan\left[\sin^{-1}\left(\dfrac{6-x^2}{\sqrt{x^4-12x^2+37}}\right)\right] < 0$

$\Rightarrow \quad \sin^{-1}\left(\dfrac{6-x^2}{\sqrt{x^4-12x^2+37}}\right) < 0$

$\Rightarrow \quad \dfrac{6-x^2}{\sqrt{x^4-12x^2+37}} < 0$

$\Rightarrow \quad 6 - x^2 < 0 \quad [\because x^4 - 12x^2 + 37 > 0;\ a > 0,\ D < 0,\ f(x) > 0]$

$\Rightarrow \quad x^2 > 6 \Rightarrow x > \sqrt{6}$

$\therefore \quad x \in (\sqrt{6}, 7]$

$\Rightarrow \quad \lim_{x \to 4} \dfrac{g(x) - g(2)}{\log[\cos(4-x)]} \qquad \left[\dfrac{0}{0} \text{ form}\right]$

$$= \lim_{x \to 4} \frac{g'(x)}{\dfrac{-\sin(4-x)(-1)}{\cos(4-x)}} \qquad \text{[by L' Hospital's rule]}$$

$$= \lim_{x \to 4} \frac{g'(x)}{\tan(4-x)} \qquad \left[\frac{0}{0} \text{ form}\right]$$

$$= \lim_{x \to 4} \frac{g''(x)}{-\sec^2(4-x)} \qquad \text{[by L'Hospital's rule]}$$

$$= \lim_{x \to 4} \frac{g''(x)}{-1} \qquad \left[\begin{array}{l}\because g(x) = 1 \text{ for } 0 \le x \le 4,\ 4 < x \le 5 \\ \therefore g'(x) = 0 \text{ and } g''(x) = 0\end{array}\right]$$

$$= 0$$

**20.** (b) **21.** (c) **22.** (a)

**23.** (6) Given, $\int_0^x (t^2 - 8t + 13)\,dt = x\sin\left(\dfrac{a}{x}\right)$

$\Rightarrow \quad \dfrac{x^3}{3} - 4x^2 + 13x = x\sin\left(\dfrac{a}{x}\right)$

Now, $\quad x^2 - 12x + 39 = 3 \cdot \sin\left(\dfrac{a}{x}\right)$

$\Rightarrow \quad (x-6)^2 + 3 = 3\sin\left(\dfrac{a}{x}\right)$

Since, $(x-6)^2 + 3 \ge 3$ for all $x \in R$ and $3\sin\left(\dfrac{a}{x}\right) \le 3$ for all $x \in R - \{0\}$.

$\therefore$ The given equation is meaningful, if each side is equal to 3.

$\therefore \quad (x-6)^2 + 3 = 3 \Rightarrow x = 6$

**24.** (0) Consider the integral $\int_0^1 x^2 e^{-x^2}\,dx$

$$= -\frac{1}{2}\int_0^1 x(-2x)e^{-x^2}\,dx$$

$$= -\frac{1}{2}\left\{\left[xe^{-x^2}\right]_0^1 - \int_0^1 e^{-x^2}\,dx\right\}$$

[integration by parts]

Now, $\quad \int_0^1 x^2 e^{-x^2}\,dx = -\dfrac{1}{2e} + \dfrac{1}{2}\int_0^1 e^{-x^2}\,dx$

$\Rightarrow \quad \dfrac{1}{2e} + \int_0^1 x^2 e^{-x^2}\,dx = \dfrac{1}{2}\int_0^1 e^{-x^2}\,dx$

$\Rightarrow \quad \dfrac{1}{e} + 2\int_0^1 x^2 e^{-x^2}\,dx = \int_0^1 e^{-x^2}\,dx \Rightarrow \beta = \dfrac{1}{e}$

$\therefore \quad [\beta] = \left[\dfrac{1}{e}\right] = 0$

**25.** (2) Let $F(x) = f(x) + f\left(\dfrac{1}{x}\right) = \int_1^x \dfrac{\log t}{1+t}\,dt + \int_1^x \dfrac{\log t}{1+t}\,dt$

In second intergral, let $t = \dfrac{1}{y} \Rightarrow dt = -\dfrac{1}{y^2}dy$

Now, $F(x) = \int_1^x \dfrac{\log t}{1+t}dt + \int_1^x \dfrac{-\log y}{y(1+y)}dy$

$$= \int_1^x \frac{\log t}{1+t}dt + \int_1^x \frac{\log y}{y(1+y)}dy$$

$$= \int_1^x \frac{\log t}{1+t}dt + \int_1^x \frac{\log t}{t(1+t)}dt = \int_1^x \frac{\log t}{t}dt$$

$$= \frac{1}{2}(\log x)^2 = \frac{1}{2}(\log x)^2$$

$\Rightarrow \quad F(e) = \dfrac{1}{2}$

$\therefore \quad \dfrac{1}{f(e) + f\left(\dfrac{1}{e}\right)} = \dfrac{1}{F(e)} = \dfrac{1}{1/2} = 2$

# DPP-2 **Properties of Definite Integral**

**1.** (a) Let $I = \int_0^{\pi} |\cos x|\, dx$

$$= \int_0^{\pi/2} |\cos x|\, dx + \int_{\pi/2}^{\pi} |\cos x|\, dx$$
$$= \int_0^{\pi/2} (\cos x)\, dx + \int_{\pi/2}^{\pi} |\cos x|\, dx$$
$$= \int_0^{\pi/2} \cos x dx - \int_{\pi/2}^{\pi} (\cos x)\, dx$$
$$= [\sin x]_0^{\pi/2} - [\sin x]_{\pi/2}^{\pi} = (1 - 0) - (0 - 1) = 2$$

**2.** (d) Let $I = \int_0^2 | x^2 + 2x - 3|\, dx$

$$= \int_0^1 (x^2 + 2x - 3)\, dx + \int_1^2 | x^2 + 2x - 3|\, dx \qquad \ldots (i)$$

We have, $x^2 + 2x - 3 = (x + 3)(x - 1)$

$x^2 + 2x - 3 > 0$ for $x < -3$ or $x > 1$

$x^2 + 2x - 3 < 0$ for $-3 < x < 1$

$$\Rightarrow \; |x^2 + 2x - 3| = \begin{cases} (x^2 + 2x - 3), & x < -3 \text{ or } x > 1 \\ -(x^2 + 2x - 3), & -3 < x < 1 \end{cases}$$

Eq. (i) becomes,

$$I = \int_0^1 -(x^2 + 2x - 3)\, dx + \int_1^2 (x^2 + 2x - 3)\, dx$$
$$= -\left[\frac{x^3}{3} + x^2 - 3x\right]_0^1 + \left[\frac{x^3}{3} + x^2 - 3x\right]_1^2 = 4$$

**3.** (c) Let $I = \int_0^3 [x]\, dx$

$$= \int_0^1 [x]\, dx + \int_1^2 [x]\, dx + \int_2^3 [x]\, dx$$
$$\Rightarrow \int_0^3 [x]\, dx = \int_0^1 0\, dx + \int_1^2 1\, dx + \int_2^3 2\, dx$$
$$= (2 - 1) + 2(3 - 2) = 3$$

**4.** (b) Let $I = \int_0^2 [x^2]\, dx$

$$= \int_0^1 [x^2]\, dx + \int_1^{\sqrt{2}} [x^2]\, dx + \int_{\sqrt{2}}^{\sqrt{3}} [x^2]\, dx + \int_{\sqrt{3}}^2 [x^2]\, dx$$
$$= \int_0^1 0 dx + \int_1^{\sqrt{2}} 1\, dx + \int_{\sqrt{2}}^{\sqrt{3}} 2\, dx + \int_{\sqrt{3}}^2 3\, dx$$
$$= 0 + 1(\sqrt{2} - 1) + 2(\sqrt{3} - \sqrt{2}) + 3(2 - \sqrt{3})$$
$$= 5 - \sqrt{3} - \sqrt{2}$$

**5.** (b) Let $I = \int_{-1}^1 (x - [x])\, dx = \int_{-1}^1 x\, dx - \int_{-1}^1 [x]\, dx$

$$= \left[\frac{x^2}{2}\right]_{-1}^1 - \left(\int_{-1}^0 [x]\, dx + \int_0^1 [x]\, dx\right)$$
$$= \frac{1}{2}(1 - 1) - \left(\int_{-1}^0 -1\, dx + \int_0^1 0\, dx\right)$$
$$= 0 + [x]_{-1}^0 = 1$$

**Alternate Method**

The graph of $x - [x]$ is as follows

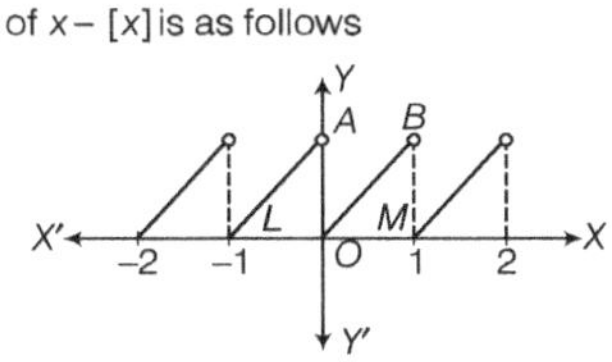


It is evident from the graph of $y = f(x)$ that

$$\int_{-1}^1 f(x)\, dx = \text{Area of } \Delta OLA + \text{Area of } \Delta OMB$$
$$\Rightarrow \int_{-1}^1 f(x)\, dx = 2 \text{ (Area of } \Delta OMB)$$
$$\Rightarrow \int_{-1}^1 f(x)\, dx = 2 \times \frac{1}{2}(OM \times MB) = 1 \times 1 = 1$$

**6.** (b) Let $I = \int_0^2 \{x\}\, dx$

$$= \int_0^2 (x - [x])dx = \int_0^2 x\, dx - \int_0^2 [x]\, dx$$
$$= \left[\frac{x^2}{2}\right]_0^2 - \left[\int_0^1 [x]\, dx + \int_1^2 [x]\, dx\right]$$
$$= \frac{1}{2}(4 - 0) - \left(\int_0^1 0\, dx + \int_1^2 1\, dx\right)$$
$$= 2 - [x]_1^2 = 2 - 1 = 1$$

**7.** (a) Let $I = \int_0^a [x^n]dx$

Let $k$ be a natural number such that $k \le a^n < k + 1$,

then
$$\int_0^a [x^n]dx = \int_0^1 [x^n]\, dx + \int_1^{2^{1/n}} [x^n]\, dx + \int_{2^{1/n}}^{3^{1/n}} [x^n]\, dx + \cdots + \int_{(k-1)^{1/n}}^{k^{1/n}} [x^n]\, dx + \int_{k^{1/n}}^{(a^n)^{1/n}} x^n dx$$
$$= \int_0^1 0\, dx + \int_1^{2^{1/n}} 1\, dx + \int_{2^{1/n}}^{3^{1/n}} 2\, dx + \cdots + \int_{(k-1)^{1/n}}^{k^{1/n}} (k - 1)\, dx + \int_{k^{1/n}}^{(a^n)^{1/n}} k dx$$
$$= 0 + 1\cdot(2^{1/n} - 1) + 2\cdot(3^{1/n} - 2^{1/n}) + \cdots + (k - 1)[k^{1/n} - (k - 1)^{1/n}] + k(a - k^{1/n})$$
$$\Rightarrow \sum_{r=1}^{k} (r - 1)[r^{1/n} - (r - 1)^{1/n}] + k(a - k^{1/n})$$

**8.** (a) Here, $f(x) = x^2 - x + 1$ is a non-monotonic function, such problems should be solved by graphical method. Now, $g(x) = [x^2 - x + 1]$, $\forall x \in [0,2]$ could be plotted as

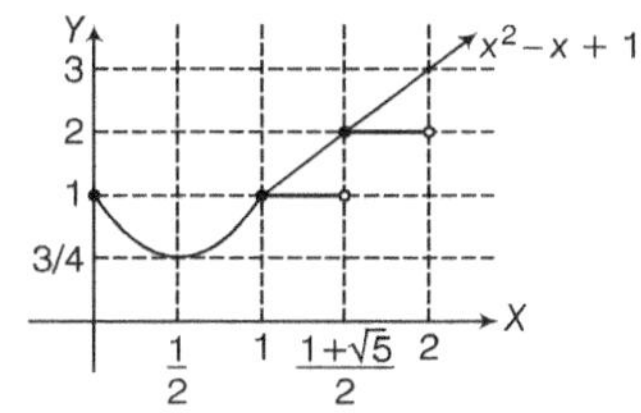


$$\therefore \quad I = \int_0^2 [x^2 - x + 1]\, dx$$
$$= \int_0^1 0\, dx + \int_1^{\frac{1+\sqrt{5}}{2}} 1\, dx + \int_{\frac{1+\sqrt{5}}{2}}^2 2\, dx$$
$$= \left(\frac{1 + \sqrt{5}}{2} - 1\right) + 2\left(2 - \frac{1 + \sqrt{5}}{2}\right) = \frac{5 - \sqrt{5}}{2}$$

**9.** (d) Let $I = \int_0^{100} [\tan^{-1} x]dx$

The graph of $\tan^{-1} x$ is shown as

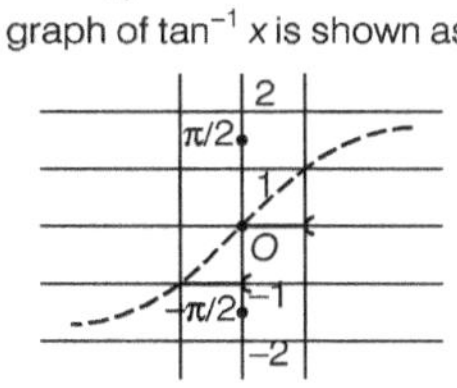


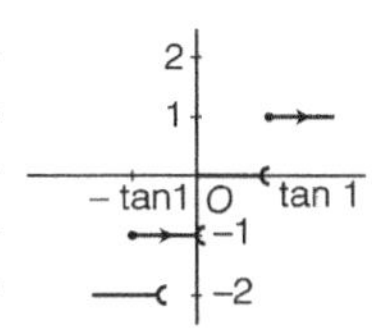


$\therefore \int_0^{100} [\tan^{-1} x]\, dx$ is shown as

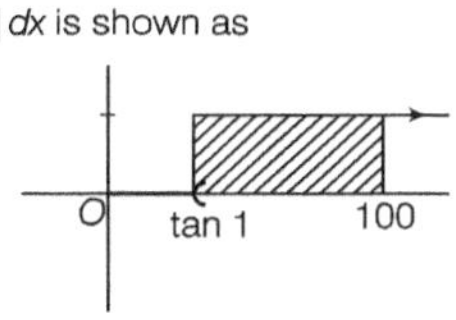


$$= 1 \times (100 - \tan 1)$$

**10.** (c) Consider the integral, $\int_{1/e}^{\cot x} \frac{dt}{t(1+t^2)}$

On putting $t = \frac{1}{u}$, we get

$$\int_{1/e}^{\cot x}\frac{dt}{t(1+t^2)} = \int_e^{\tan x} \frac{-1/u^2}{\frac{1}{u}\left(1+\frac{1}{u^2}\right)}du = -\int_e^{\tan x}\frac{u}{1+u^2}du$$

$$\Rightarrow \int_{1/e}^{\cot x}\frac{dt}{t(1+t^2)} = -\int_e^{\tan x}\frac{t}{1+t^2}dt$$

$$= -\left[\int_e^{1/e}\frac{t}{1+t^2}dt + \int_{1/e}^{\tan x}\frac{t}{1+t^2}dt\right]$$

$$\Rightarrow \int_{1/e}^{\cot x}\frac{dt}{t(1+t^2)} = -\int_e^{1/e}\frac{t}{1+t^2}dt - \int_{1/e}^{\tan x}\frac{t}{1+t^2}dt$$

$$\therefore \int_{1/e}^{\tan x}\frac{t}{1+t^2}dt + \int_{1/e}^{\cot x} t(1+t^2)\,dt$$

$$= -\int_e^{1/e}\frac{t}{1+t^2}dt = -\frac{1}{2}[\log(1+t^2)]_e^{1/e}$$

$$= -\frac{1}{2}\left[\log\left(1+\frac{1}{e^2}\right) - \log(1+e^2)\right]$$

$$= -\frac{1}{2}[\log(e^2+1) - 2\log e - \log(1+e^2)]$$

$$= \log e = 1$$

**11.** (d) Let $I = \int_0^{2\pi}[2\sin x]dx$

The graph of $2\sin x$ is shown below

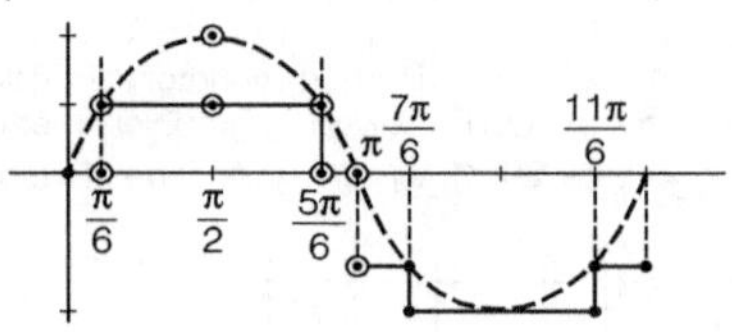


From the graph,

$$\therefore I = \int_{\pi/6}^{5\pi/6} 1\,dx + \int_{\pi}^{7\pi/6} -1\,dx + \int_{7\pi/6}^{11\pi/6} -2\,dx + \int_{11\pi/6}^{2\pi} -1\,dx$$

$$= \left(\frac{5\pi}{6}-\frac{\pi}{6}\right) + \left(\frac{-7\pi}{6}+\pi\right) + 2\left(\frac{-11\pi}{6}+\frac{7\pi}{6}\right) + \left(-2\pi + \frac{11\pi}{6}\right)$$

$$= \frac{2\pi}{3} - \frac{\pi}{6} - \frac{8\pi}{6} - \frac{\pi}{6} = -\pi$$

**12.** (d) Let $I = \int_0^{2\pi}[\sqrt{2}\sin x]\,dx$

Since, $0 \le x \le 2\pi \Rightarrow -\sqrt{2} \le \sqrt{2}\sin x \le \sqrt{2}$

$\therefore \sqrt{2}\sin x = -\sqrt{2}, \sqrt{2}\sin x = -1, \sqrt{2}\sin x = 0, \sqrt{2}\sin x = 1,$
$\sqrt{2}\sin x = \sqrt{2}$

$\Rightarrow \sin x = -1, \sin x = -\frac{1}{\sqrt{2}}, \sin x = 0, \sin x = \frac{1}{\sqrt{2}}, \sin x = 1$

Now,
$\sin x = -1 \Rightarrow x = \frac{3\pi}{2}$

$\sin x = -\frac{1}{\sqrt{2}} \Rightarrow x = \frac{5\pi}{4}, \frac{7\pi}{4}$

$\sin x = 0 \Rightarrow x = 0, \pi, 2\pi$

$\sin x = \frac{1}{\sqrt{2}} \Rightarrow x = \frac{\pi}{4}, \frac{3\pi}{4}$

$\sin x = 1 \Rightarrow x = \frac{\pi}{2}$

$$\therefore I = \int_0^{2\pi}[\sqrt{2}\sin x]\,dx$$

$$= \int_0^{\pi/4} 0\,dx + \int_{\pi/4}^{\pi/2} 1\,dx + \int_{\pi/2}^{3\pi/4} 1\,dx + \int_{3\pi/4}^{\pi} 0\,dx + \int_{\pi}^{5\pi/4}(-1)\,dx + \int_{5\pi/4}^{3\pi/2}(-2)\,dx + \int_{3\pi/2}^{7\pi/4}(-2)\,dx + \int_{7\pi/4}^{2\pi}(-1)\,dx$$

$$= 0 + \left(\frac{\pi}{2}-\frac{\pi}{4}\right) + \left(\frac{3\pi}{4}-\frac{\pi}{2}\right) + 0 - \left(\frac{5\pi}{4}-\pi\right) - 2\left(\frac{3\pi}{2}-\frac{5\pi}{4}\right) - 2\left(\frac{7\pi}{4}-\frac{3\pi}{2}\right) - \left(2\pi - \frac{7\pi}{4}\right)$$

$$\frac{\pi}{4}+\frac{\pi}{4}-\frac{\pi}{4}-2\left(\frac{\pi}{4}\right)-2\left(\frac{\pi}{4}\right)-\left(\frac{\pi}{4}\right) = -\pi$$

**13.** (a) Let $I = \int_0^{\sqrt{2}}[x^2]\,dx = \int_0^1[x^2]\,dx + \int_1^{\sqrt{2}}[x^2]\,dx$

$$= \int_0^1 0\,dx + \int_1^{\sqrt{2}} 1\,dx = [x]_1^{\sqrt{2}} = \sqrt{2} - 1$$

**14.** (a) Let $I = \int_{-2}^{3}|1 - x^2|\,dx$

$$= \int_{-2}^{-1}(x^2 - 1)\,dx + \int_{-1}^{1}(1 - x^2)\,dx + \int_1^3 (x^2 - 1)\,dx$$

$$= \left[\frac{x^3}{3} - x\right]_{-2}^{-1} + \left[x - \frac{x^3}{3}\right]_{-1}^{1} + \left[\frac{x^3}{3} - x\right]_1^3$$

$$= \left[-\frac{1}{3} + 1 + \frac{8}{3} - 2\right] + \left[1 - \frac{1}{3} + 1 - \frac{1}{3}\right] + \left[9 - 3 - \frac{1}{3} + 1\right]$$

$$= \frac{4}{3} + \frac{4}{3} + \frac{20}{3} = \frac{28}{3}$$

**15.** (a) Since, $\int_1^a [x]\,f'(x)\,dx$

$$= \int_1^2 f'(x)\,dx + \int_2^3 2f'(x)\,dx + \cdots + \int_{[a]}^a [a]\,f'(x)\,dx$$

$$= [f(x)]_1^2 + 2[f(x)]_2^3 + \cdots + [a]\,[f(x)]_{[a]}^a$$

$$= f(2) - f(1) + 2f(3) - 2f(2) + \cdots + [a]\,f(a) - [a]\,f([a])$$

$$= [a]\,f(a) - \{f(1) + f(2) + \cdots + f([a])\}$$

**16.** (a) Since, $f(x) = \int_1^x \frac{\log t}{1+t}dt$ and $F(e) = f(e) + \frac{1}{f(e)}$

$$\Rightarrow F(e) = \int_1^e \frac{\log t}{1+t}dt + \int_1^{1/e}\frac{\log t}{1+t}dt$$

On replacing $t$ by $\frac{1}{t}$ in second integration

$$\therefore F(e) = \int_1^e \frac{\log t}{1+t}dt + \int_1^e \frac{\log t}{t(1+t)}dt = \int_1^e \frac{\log t}{t}dt = \left[\frac{(\log t)^2}{2}\right]_1^e$$

$$= \frac{1}{2}[(\log e)^2 - (\log 1)^2] = \frac{1}{2}$$

**17.** (d) Let $I = \int_0^{1.5} x[x^2]\,dx$

$$= \int_0^1 x\cdot 0\,dx + \int_1^{\sqrt{2}} x\cdot 1\,dx + \int_{\sqrt{2}}^{1.5} x\cdot 2\,dx$$

$$= 0 + \left[\frac{x^2}{2}\right]_1^{\sqrt{2}} + [x^2]_{\sqrt{2}}^{1.5}$$

$$= \frac{1}{2}(2 - 1) + [(1.5)^2 - 2]$$

$$= \frac{1}{2} + 2.25 - 2 = \frac{1}{2} + \frac{1}{4} = \frac{3}{4}$$

**18.** (b) We have,

$$I = \int_{1/e}^{e^2}\left|\frac{\log x}{x}\right|dx = \int_{1/e}^{e^2}\frac{|\log x|}{x}dx$$

$$= \int_{1/e}^{1}\frac{|\log x|}{x}dx + \int_1^{e^2}\frac{\log x}{x}dx \quad \left[\because \log x < 0 \text{ for } \frac{1}{e} < x < 1;\ \log x > 0 \text{ for } x > 1\right]$$

$$= -\int_{1/e}^{1}\log x\, d(\log x) + \int_1^{e^2}\log x\, d(\log x)$$

$$= -\left[\frac{1}{2}(\log x)^2\right]_{1/e}^{1} + \left[\frac{1}{2}(\log x)^2\right]_1^{e^2}$$

$$= -\left[\frac{1}{2}\times 0 - \frac{1}{2}\right] + \left[\frac{1}{2}\times 4 - 0\right]$$

$$= \frac{1}{2} + 2 = \frac{5}{2}$$

**19.** (c) We have, $I = \int_{-2}^{3} f(x)\,dx$

$$= \int_{-2}^{2} e^{\cos x}\sin x\,dx + \int_{2}^{3} 2\,dx$$

$$= 0 + 2(3-2) \quad \left[\begin{array}{l}\because g(x) = e^{\cos x}\sin x \text{ is odd function}\\ \therefore \int_{-2}^{3} g(x)\,dx = 0\end{array}\right]$$

$$= 2$$

**20.** (a) We have, $f(n) = \int_0^{n^2}[\sqrt{x}]\,dx = \sum_{r=1}^{n}\int_{(r-1)^2}^{r^2}[\sqrt{x}]\,dx$

$$= \sum_{r=1}^{n}\int_{(r-1)^2}^{r^2}(r-1)\,dx = \sum_{r=1}^{n}(r-1)(2r-1)$$

$$= \sum_{r=1}^{n}(2r^2 - 3r + 1)$$

$$= 2\times\frac{n(n+1)(2n+1)}{6} - 3\times\frac{n(n+1)}{2} + n$$

$$= \frac{n}{6}[2(2n^2+3n+1) - 9n - 9 + 6]$$

$$= \frac{n}{6}(4n^2 - 3n - 1) = \frac{n}{6}(n-1)(4n+1)$$

**21.** (b) We have, $\lim_{n\to\infty}\frac{f(n)}{n^3}$

$$= \lim_{n\to\infty}\frac{\frac{n}{6}(n-1)(4n+1)}{n^3}$$

$$= \frac{1}{6}\lim_{n\to\infty}\frac{n^3\left(1-\frac{1}{n}\right)\left(4+\frac{1}{n}\right)}{n^3}$$

$$= \frac{1}{6}\lim_{n\to\infty}\left(1-\frac{1}{n}\right)\left(4+\frac{1}{n}\right) = \frac{4}{6} = \frac{2}{3}$$

**22.** (b) Since, $f(n) = \frac{n}{6}(n-1)(4n+1)$

$$\Rightarrow \quad f(6) = \frac{6}{6}(6-1)(4\times 6+1) = 5\times 25 = 125$$

**Explanations for Q. Nos. (23-25)**

We have, $\int_0^n f(x)\,dx = \sum_{r=1}^{n}\int_{r-1}^{r} f(x)\,dx$

$$= \sum_{r=1}^{n}\int_{r-1}^{r}\sqrt{x-r+1}\,dx$$

$$= \sum_{r=1}^{n}\int_0^1 \sqrt{t}\,dt, \text{ where } t = x - r + 1$$

$$= n\int_0^1 \sqrt{t}\,dt = n\cdot\frac{[t^{3/2}]_0^1}{3/2}$$

$$\therefore \quad F(n) = n\cdot\frac{2}{3}(1-0) = \frac{2n}{3}$$

$\therefore$ Degree of $F(n)$ is 1.

Now, value of $\lim_{n\to\infty}\frac{F(n)}{n^2} = \lim_{n\to\infty}\frac{2n}{3n^2} = 0$

Also, $F(n)$ is continuous and differentiable for all $n \in N$.

**23.** (b) **24.** (a) **25.** (a,b)

**26.** (2) Let $\int_1^{1.5}[x^2]\,dx = \int_1^{\sqrt{2}} 1\,dx + \int_{\sqrt{2}}^{1.5} 2\,dx$

$$= [x]_1^{\sqrt{2}} + [2x]_{\sqrt{2}}^{1.5}$$

$$= [\sqrt{2}-1] + 2(1.5-\sqrt{2})$$

$$= (\sqrt{2}-1) + (3-2\sqrt{2})$$

$$= -\sqrt{2} + 3 - 1 = 2 - \sqrt{2}$$

$$= \sqrt{4} - \sqrt{2}$$

But $\int_1^{1.5}[x^2]\,dx = \sqrt{a} - \sqrt{b} \Rightarrow a = 4, b = 2$

$$\therefore \quad a - b = 4 - 2 = 2$$

**27.** (4) $\int_0^4 \{\sqrt{x}\}\,dx = \int_0^1 \sqrt{x}\,dx + \int_1^4 (\sqrt{x}-1)\,dx$

$$= \left[\frac{2}{3}x^{3/2}\right]_0^1 + \left[\frac{2}{3}x^{3/2} - x\right]_1^4$$

$$= \frac{2}{3} + \frac{2}{3}(8-1) - 3 = \frac{7}{3}$$

But $\int_0^4 \{x\}\,dx = \frac{a}{b} \Rightarrow a = 7, b = 3$

$$\therefore \quad a - b = 7 - 3 = 4$$

**28.** (1) We have, $\{x\} = x - [x]$

Let $k \le x \le k+1$, where $k \in N$.

Then, $I = \int_0^x\left(\{x\} - \frac{1}{2}\right)dx = \int_0^x\left(x - [x] - \frac{1}{2}\right)dx$

$$= \int_0^x (x - [x])\,dx - \int_0^x \frac{1}{2}\,dx$$

$$= \int_0^k (x-[x])\,dx + \int_k^x (x-[x])\,dx - \frac{x}{2}$$

$$= \frac{k}{2} + \int_k^x (x-k)\,dx - \frac{x}{2}$$

$$= \frac{k}{2} + \left(\frac{x^2}{2} - kx\right) - \left(\frac{k^2}{2} - k^2\right) - \frac{x}{2}$$

$$= \frac{x^2}{2} - kx - \frac{k(k+1)}{2} - \frac{x}{2}$$

$$= \frac{1}{2}(x^2 - 2kx + k^2) + \frac{1}{2}(k - x)$$

$$= \frac{1}{2}(x-k)^2 - \frac{1}{2}(x-k) = \frac{1}{2}(x-[x])^2 - \frac{1}{2}(x-[x])$$

$$I = \frac{1}{2}\{x\}^2 - \frac{1}{2}\{x\}$$

But $\int_0^x\left[\{x\} - \frac{1}{2}\right]dx = a\{x\}^2 - b\{x\} \Rightarrow a = \frac{1}{2}, b = \frac{1}{2}$

$$\therefore \quad a + b = \frac{1}{2} + \frac{1}{2} = 1$$

**29.** (1) Let $I = \int_0^{2\pi}[\sin x + \cos x]\,dx$

We know $\sin x + \cos x = \sqrt{2}\sin\left(x + \frac{\pi}{4}\right)$ ...(i)

$$\therefore \quad \sin x + \cos x = \begin{cases} 1, & 0 \le x \le \frac{\pi}{2} \\ 0, & \frac{\pi}{2} < x < \frac{3\pi}{4} \\ -1, & \frac{3\pi}{4} < x \le \pi \\ -2, & \pi < x < \frac{3\pi}{2} \\ -1, & \frac{3\pi}{2} \le x \le \frac{7\pi}{4} \\ 0, & \frac{7\pi}{4} \le x \le 2\pi \end{cases}$$

$$\therefore \int_0^{2\pi}[\sin x + \cos x]\,dx = \int_0^{\pi/2} 1\,dx + \int_{\pi/2}^{3\pi/4} 0\,dx + \int_{3\pi/4}^{\pi}(-1)\,dx + \int_{\pi}^{3\pi/2}(-2)\,dx + \int_{3\pi/2}^{7\pi/4}(-1)\,dx + \int_{7\pi/4}^{2\pi} 0\,dx$$

$$= \frac{\pi}{2} + 0 - \left(\pi - \frac{3\pi}{4}\right) - 2\left(\frac{3\pi}{2} - \pi\right) - \left(\frac{7\pi}{4} - \frac{3\pi}{2}\right) + 0$$

$$= -\pi$$

$$\therefore \quad |k\pi| = \pi$$

$$\Rightarrow \quad k = 1$$

**30.** (0) Let $I = \int_{-1}^{1}[x[1 + \sin \pi x]]\,dx$

$$= \int_{-1}^{0}[x[1+\sin \pi x]]\,dx + \int_0^1 [x[1+\sin x]]\,dx$$

$$= \int_{-1}^{0}[x\times 0]\,dx + \int_0^1 [x\times 1]\,dx$$

$$= 0$$

# DPP-3 Properties of Definite Integral

**1.** (i) Let $I = \int_{-a}^{a} x\sqrt{a^2 - x^2}\,dx$

Again, let $f(x) = x\sqrt{a^2 - x^2}$

$f(-x) = -x\sqrt{a^2 - x^2}$

$\Rightarrow \quad f(-x) = -f(x)$

$\therefore$ $f(x)$ is odd function.

Hence, $I = \int_{-a}^{a} x\sqrt{a^2 - x^2}\,dx = 0$

$$\left[\because \int_{-a}^{a} f(x)\,dx = 0, \text{ if } f(x) \text{ is odd.}\right]$$

(ii) Let $I = \int_{-1/2}^{1/2} \sec x \cdot \log\left(\frac{1-x}{1+x}\right)dx$

Again, let $f(x) = \sec x \log\left(\frac{1-x}{1+x}\right)$

$\Rightarrow f(-x) = \sec(-x)\log\left(\frac{1+x}{1-x}\right)$

$= \sec x \cdot \log\left(\frac{1+x}{1-x}\right)$

Now, $f(x) + f(-x) = \sec x\left[\log\left(\frac{1-x}{1+x}\right) + \log\left(\frac{1+x}{1-x}\right)\right]$

$= \sec x\,[\log(1-x) - \log(1+x) + \log(1+x) - \log(1-x)]$

$= 0$

$\Rightarrow f(x)$ is odd function.

$\therefore \quad -\int_{-1/2}^{1/2} \sec x \log\left(\frac{1-x}{1+x}\right)dx = 0$

$$\left[\because \int_{-a}^{a} f(x)\,dx = 0, \text{ if } f(x) \text{ is odd}\right]$$

(iii) Let $I = \int_{-\pi/4}^{\pi/4} \left(\frac{x^7 - 3x^5 + 3x^3 - x + 1}{\cos^2 x}\right)dx$

$= \int_{-\pi/4}^{\pi/4} \{\sec^2 x\,(x^7 - 3x^5 + 3x^3 - x) + \sec^2 x\}\,dx$

$= \int_{-\pi/4}^{\pi/4} \sec^2 x(x^7 - 3x^5 + 3x^3 - x)\,dx + \int_{-\pi/4}^{\pi/4} \sec^2 x\,dx$

$= 0 + \int_{-\pi/4}^{\pi/4} \sec^2 x\,dx$

[$\because \sec^2 x\,(x^7 - 3x^5 + 3x^3 - x)$ is odd function.]

$= [\tan x]_{-\pi/4}^{\pi/4}$

$= \left[\tan\frac{\pi}{4} - \tan\left(-\frac{\pi}{4}\right)\right]$

$= \left[\tan\frac{\pi}{4} + \tan\frac{\pi}{4}\right]$

$= 1 + 1 = 2$

(iv) Let $I = \int_{-\pi/2}^{\pi/2} \sin^7 x\,dx$

Again, let $f(x) = \sin^7 x$

$f(-x) = \sin^7(-x)$

$= -\sin^7 x$

$= -f(x)$

$\Rightarrow f(x)$ is odd function.

$\therefore \quad I = \int_{-\pi/2}^{\pi/2} \sin^7 x\,dx = 0 \quad \left[\because \int_{-a}^{a} f(x)\,dx = 0, \text{ if } f(x) \text{ is odd.}\right]$

(v) Let $I = \int_{-\pi/2}^{\pi/2} \sin^2 x\,dx$

$= \frac{1}{2}\int_{-\pi/2}^{\pi/2} (1 - \cos 2x)\,dx$

$= \frac{1}{2}\left[\int_{-\pi/2}^{\pi/2} 1\,dx - \int_{-\pi/2}^{\pi/2} \cos 2x\,dx\right]$

$= \frac{1}{2}\left[[x]_{-\pi/2}^{\pi/2} - \int_{0}^{\pi/2} \cos 2x\,dx\right]$ [$\because \cos 2x$ is even function]

$= \frac{1}{2}\left[\left(\frac{\pi}{2} + \frac{\pi}{2}\right) - \left[\frac{\sin 2x}{2}\right]_0^{\pi/2}\right]$

$= \frac{1}{2}[\pi - 0] = \frac{\pi}{2}$

(vi) Let $I = \int_{-\pi/4}^{\pi/4} x^3 \sin^4 x\,dx$

Again, let $f(x) = x^3 \sin^4 x$

$\Rightarrow \quad f(-x) = (-x)^3 \sin^4(-x) = -x^3 \sin^4 x$

$\Rightarrow \quad f(-x) = -f(x)$

$\therefore$ $f(x)$ is odd function.

Now, $\quad I = \int_{-\pi/4}^{\pi/4} x^3 \sin^4 x\,dx = 0$

$$\left[\because \int_{-a}^{a} f(x)\,dx = 0, \text{ if } f(x) \text{ is odd.}\right]$$

**2.** (i) Let $I = \int_0^{\pi/2} \frac{\sqrt{\sin x}}{\sqrt{\sin x} + \sqrt{\cos x}}\,dx \quad \ldots(i)$

$= \int_0^{\pi/2} \frac{\sqrt{\sin\left(\frac{\pi}{2} - x\right)}}{\sqrt{\sin\left(\frac{\pi}{2} - x\right)} + \sqrt{\cos\left(\frac{\pi}{2} - x\right)}}\,dx$

$$\left[\because \int_0^a f(x)\,dx = \int_0^a f(a-x)\,dx\right]$$

$= \int_0^{\pi/2} \frac{\sqrt{\cos x}}{\sqrt{\cos x} + \sqrt{\sin x}}\,dx \quad \ldots(ii)$

On adding Eqs. (i) and (ii), we get

$2I = \int_0^{\pi/2} \frac{\sqrt{\sin x} + \sqrt{\cos x}}{\sqrt{\sin x} + \sqrt{\cos x}}\,dx$

$\Rightarrow \quad 2I = \int_0^{\pi/2} dx = [x]_0^{\pi/2} = \frac{\pi}{2}$

$\Rightarrow \quad 2I = \frac{\pi}{2}$

$\Rightarrow \quad I = \frac{\pi}{4}$

(ii) Let $I = \int_0^{\pi/2} \log(\tan x)\,dx \quad \ldots(i)$

$= \int_0^{\pi/2} \log\left[\tan\left(\frac{\pi}{2} - x\right)\right]dx \quad \left[\because \int_0^a f(x)dx = \int_0^a f(a-x)dx\right]$

$= \int_0^{\pi/2} \log(\cot x)\,dx \quad \ldots(ii)$

On adding Eqs. (i) and (ii), we get

$2I = \int_0^{\pi/2} \log(\tan x \times \cot x)\,dx$

$\Rightarrow \quad 2I = \int_0^{\pi/2} \log(1)\,dx$

$\Rightarrow \quad 2I = 0$

$\therefore \quad I = 0$

(iii) Let $I = \int_0^{\pi/4} \log(1 + \tan x)\,dx$

$$= \int_0^{\pi/4} \log\left[1 + \tan\left(\frac{\pi}{4} - x\right)\right]dx$$

$$\left[\because \int_0^a f(x)\,dx = \int_0^a f(a-x)\,dx\right]$$

$$= \int_0^{\pi/4} \log\left(1 + \frac{\tan\frac{\pi}{4} - \tan x}{1 + \tan\frac{\pi}{4}\tan x}\right)dx$$

$$= \int_0^{\pi/4} \log\left(1 - \frac{1-\tan x}{1+\tan x}\right)dx = \int_0^{\pi/4} \log\left(\frac{2}{1+\tan x}\right)dx$$

$$= \int_0^{\pi/4} [\log 2 - \log(1 + \tan x)]\,dx$$

$$= \int_0^{\pi/4} \log 2\,dx - \int_0^{\pi/4} \log(1 + \tan x)\,dx$$

$\Rightarrow \quad I = \log 2[x]_0^{\pi/4} - I$

$\Rightarrow \quad 2I = \log 2\left(\frac{\pi}{4}\right)$

$\therefore \quad I = \frac{\pi}{8}\log 2$

(iv) Let $I = \int_1^2 \frac{\sqrt{x}}{\sqrt{3-x} + \sqrt{x}}dx$ ...(i)

$$= \int_1^2 \frac{\sqrt{1+2-x}}{\sqrt{3-3+x} + \sqrt{3-x}}dx$$

$$\left[\because \int_a^b f(x)\,dx = \int_a^b f(a+b-x)\,dx\right]$$

$$I = \int_1^2 \frac{\sqrt{3-x}}{\sqrt{x} + \sqrt{3-x}}dx \qquad \text{...(ii)}$$

On adding Eqs. (i) and (ii), we get

$$2I = \int_1^2 \frac{\sqrt{x} + \sqrt{3-x}}{\sqrt{3-x} + \sqrt{x}}dx = \int_1^2 dx$$

$\Rightarrow \quad 2I = [x]_1^2 = (2-1) = 1$

$\therefore \quad I = \frac{1}{2}$

(v) Let $I = \int_{\pi/6}^{\pi/3} \frac{dx}{1 + \sqrt{\cot x}}$

$$= \int_{\pi/6}^{\pi/3} \frac{\sqrt{\sin x}}{\sqrt{\sin x} + \sqrt{\cos x}}dx \qquad \text{...(i)}$$

$$= \int_{\pi/6}^{\pi/3} \frac{\sqrt{\sin\left(\frac{\pi}{6} + \frac{\pi}{3} - x\right)}}{\sqrt{\sin\left(\frac{\pi}{6} + \frac{\pi}{3} - x\right)} + \sqrt{\cos\left(\frac{\pi}{6} + \frac{\pi}{3} - x\right)}}dx$$

$$\left[\because \int_a^b f(x)\,dx = \int_a^b f(a+b-x)\,dx\right]$$

$$= \int_{\pi/6}^{\pi/3} \frac{\sqrt{\sin\left(\frac{\pi}{2} - x\right)}}{\sqrt{\sin\left(\frac{\pi}{2} - x\right)} + \sqrt{\cos\left(\frac{\pi}{2} - x\right)}}dx$$

$$= \int_{\pi/6}^{\pi/3} \frac{\sqrt{\cos x}}{\sqrt{\cos x} + \sqrt{\sin x}}dx \qquad \text{...(ii)}$$

On adding Eqs. (i) and (ii), we get

$$2I = \int_{\pi/6}^{\pi/3} \frac{\sqrt{\sin x} + \sqrt{\cos x}}{\sqrt{\sin x} + \sqrt{\cos x}}dx$$

$\Rightarrow \quad 2I = \int_{\pi/6}^{\pi/3} dx \quad \Rightarrow \quad 2I = [x]_{\pi/6}^{\pi/3}$

$\Rightarrow \quad 2I = \frac{\pi}{3} - \frac{\pi}{6} \quad \Rightarrow \quad 2I = \frac{\pi}{6}$

$\therefore \quad I = \frac{\pi}{12}$

(vi) Let $I = \int_a^b \frac{f(x)}{f(x) + f(a+b-x)}dx$ ...(i)

$$= \int_a^b \frac{f(a+b-x)}{f(a+b-x) + f(x)}dx \qquad \text{...(ii)}$$

$$\left[\because \int_a^b f(x)\,dx = \int_a^b f(a+b-x)\,dx\right]$$

On adding Eqs. (i) and (ii), we get

$$2I = \int_a^b \frac{f(x) + f(a+b-x)}{f(x) + f(a+b-x)}dx = \int_a^b 1\,dx$$

$\Rightarrow \quad 2I = [x]_a^b = b - a$

$\Rightarrow \quad 2I = b - a$

$\therefore \quad I = \left(\frac{b-a}{2}\right)$

(vii) Let $I = \int_0^{\pi/2} \frac{\sin x - \cos x}{1 + \sin x \cos x}dx$ ...(i)

$$= \int_0^{\pi/2} \frac{\sin\left(\frac{\pi}{2} - x\right) - \cos\left(\frac{\pi}{2} - x\right)}{1 + \sin\left(\frac{\pi}{2} - x\right)\cos\left(\frac{\pi}{2} - x\right)}dx$$

$$= \int_0^{\pi/2} \frac{\cos x - \sin x}{1 + \sin x \cos x}dx \qquad \text{...(ii)}$$

On adding Eqs. (i) and (ii), we get

$$2I = \int_0^{\pi/2} \frac{\sin x - \cos x + \cos x - \sin x}{1 + \sin x \cos x}dx$$

$\Rightarrow \quad 2I = 0$

$\therefore \quad I = 0$

(viii) Let $I = \int_0^{\pi/2} (2\log \sin x - \log \sin 2x)\,dx$

$$= \int_0^{\pi/2} \log\left(\frac{\sin^2 x}{\sin 2x}\right)dx$$

$$= \int_0^{\pi/2} \log\left(\frac{\sin x}{2\cos x}\right)dx$$

$$= \int_0^{\pi/2} (\log \sin x - \log 2 - \log \cos x)\,dx$$

$$= \int_0^{\pi/2} \log \sin x\,dx - \int_0^{\pi/2} \log \cos x\,dx - \int_0^{\pi/2} \log 2\,dx$$

$$= \int_0^{\pi/2} \log\left[\sin\left(\frac{\pi}{2} - x\right)\right]dx - \int_0^{\pi/2} \log \cos x\,dx - \int_0^{\pi/2} \log 2\,dx$$

$$\left[\because \int_0^a f(x)\,dx = \int_0^a f(a-x)\,dx\right]$$

$$= \int_0^{\pi/2} \log \cos x\,dx - \int_0^{\pi/2} \log \cos x\,dx - \int_0^{\pi/2} \log 2\,dx$$

$$= -\int_0^{\pi/2} \log 2\,dx$$

$$= -\log 2[x]_0^{\pi/2}$$

$$= -\log 2\left(\frac{\pi}{2} - 0\right)$$

$$= -\frac{\pi}{2}\log 2$$

**3.** (i) Let $I = \int_0^{2\pi} \cos^5 x\,dx$

Let $f(x) = \cos^5 x$

Then, $f(2\pi - x) = \cos^5(2\pi - x)$

$= \cos^5 x = f(x)$

$\therefore I = 2\int_0^{\pi} \cos^5 x\,dx \quad \left[\because \int_0^{2a} f(x)dx = 2\int_0^a f(x)dx, \text{ if } f(2a-x) = f(x)\right]$

$= 2\int_0^{\pi} \cos^5(\pi - x)\,dx$

$= 2\int_0^{\pi} -\cos^5 x\,dx \quad \left[\because \int_0^{2a} f(x)dx = 0, \text{ if } f(2a-x) = -f(x)\right]$

$= 0$

(ii) Let $I = \int_0^{\pi/2} \log(\sin x)dx$ ...(i)

$= \int_0^{\pi/2} \log \sin\left(\frac{\pi}{2} - x\right)dx$ $\left[\because \int_0^a f(x)\,dx = \int_0^a f(a-x)dx\right]$

$I = \int_a^{\pi/2} \log\cos x dx$ ...(ii)

On adding Eqs. (i) and (ii), we get

$$2I = \int_0^{\pi/2} (\log \sin x + \log \cos x)dx$$

$$= \int_0^{\pi/2} \log(\sin x \cos x)dx$$

$$= \int_0^{\pi/2} \log\left(\frac{2\sin x\cos x}{2}\right)dx$$

$$= \int_0^{\pi/2} \log\left(\frac{\sin 2x}{2}\right)dx$$

$$= \int_0^{\pi/2} \log \sin 2x\,dx - \int_0^{\pi/2} \log 2\,dx$$

$$= I_1 - \int_0^{\pi/2} \log 2 dx$$

$$= I_1 - \log 2\,[x]_0^{\pi/2}$$

$$= I_1 - \frac{\pi}{2}\log 2 \quad \text{...(iii)}$$

Now, $I_1 = \int_0^{\pi/2} \log \sin 2x\,dx$

Put $2x = t$

$\Rightarrow \quad dx = \frac{dt}{2}$

Then, $I_1 = \frac{1}{2}\int_0^{\pi} \log \sin t\,dt$

$$= \frac{1}{2} \times 2 \int_0^{\pi/2} \log \sin t\,dt$$

$\left[\because \int_0^{2a} f(x)\,dx = \int_0^a f(x)\,dx, \text{ if } f(2a-x) = f(x)\right]$

$$= \int_0^{\pi/2} \log \sin t dt$$

$$= \int_0^{\pi/2} \log \sin x dx \quad \text{[by property 1]}$$

$\Rightarrow \quad = I$ ...(iv)

Using Eq. (iv) in Eq. (iii), we get

$$2I = I - \frac{\pi}{2}\log 2$$

$$I = -\frac{\pi}{2}\log 2$$

(iii) Let $I = \int_0^{\pi} \frac{a^2 \sin^2 x + b^2 \cos x}{a^4 \sin^2 x + b^4 \cos^2 x}dx$

$$= 2a^2 \int_0^{\pi/2} \frac{\sin^2 x}{a^4 \sin^2 x + b^4 \cos^2 x}dx + b^2 \int_0^{\pi} \frac{\cos x}{a^4 \sin^2 x + b^4 \sin^2 x}dx$$

$$= 2a^2 \int_0^{\pi/2} \frac{\tan^2 x}{a^4 \tan^2 x + b^4}dx + 0$$

$$= 2a^2 \int_0^{\pi/2} \frac{\tan^2 x \cdot \sec^2 x}{(a^4 \tan^2 x + b^4)(1 + \tan^2 x)}dx$$

Put $\tan x = t \Rightarrow \sec^2 x\,dx = dt$

$\therefore \quad I = 2a^2 \int_0^{\infty} \frac{t^2}{(a^4t^2 + b^4)(1+t^2)}dt$

$$= \frac{2a^2}{b^4 - a^4}\int_0^{\infty} \frac{t^2(b^4 - a^4)}{(a^4t^2 + b^4)(1+t^2)}dt$$

$$= \frac{2a^2}{b^4 - a^4}\int_0^{\infty} \frac{b^4t^2 + b^4 - a^4t^2 - b^4}{(a^4t^2 + b^4)(1+t^2)}dt$$

$$= \frac{2a^2}{b^4 - a^4}\int_0^{\infty}\left(\frac{b^4}{a^4t^2 + b^4} - \frac{1}{t^2+1}\right)dt$$

$$= \frac{2a^2}{b^4 - a^4}\left[\frac{b^4}{a^2b^2}\tan^{-1}\left(\frac{a^2t}{b^2}\right) - \tan^{-1}t\right]_0^{\infty}$$

$$= \frac{2a^2}{b^4 - a^4}\left[\frac{\pi}{2}\cdot\frac{b^2}{a^2} - \frac{\pi}{2}\right]$$

$$= \frac{2a^2}{b^4 - a^4}\cdot\frac{\pi}{2}\cdot\frac{(b^2 - a^2)}{a^2}$$

$$= \frac{\pi}{b^2 + a^2}$$

(iv) Let $I = \int_0^{2\pi} \frac{x \sin^{2n} x}{\sin^{2n} x + \cos^{2n} x}dx$ ...(i)

$$= \int_0^{2\pi} \frac{(2\pi - x)\sin^{2n} x}{\sin^{2n} x + \cos^{2n} x}dx \quad \text{...(ii)}$$

$\left[\because \int_0^a f(x)\,dx = \int_0^a f(a-x)\,dx\right]$

On adding Eqs. (i) and (ii), we get

$$2I = \int_0^{2\pi} \frac{2\pi \sin^{2n} x}{\sin^{2n} x + \cos^{2n} x}dx$$

$$I = \pi\int_0^{2\pi} \frac{\sin^{2n} x}{\cos^{2n} x + \sin^{2n} x}dx$$

$$= 2\pi\int_0^{\pi} \frac{\sin^{2n} x}{\sin^{2n} x + \cos^{2n} x}dx$$

$\left[\because \int_a^{2a} f(x)\,dx = 2\int_0^a f(x)\,dx, \text{ if } f(2a - x) = f(x)\right]$

$$I = 4\pi\int_0^{\pi/2} \frac{\sin^{2n} x}{\sin^{2n} x + \cos^{2n} x}dx \quad \text{...(iii)}$$

$$I = 4\pi\int_0^{\pi/2} \frac{\sin^{2n}\left(\frac{\pi}{2} - x\right)}{\sin^{2n}\left(\frac{\pi}{2} - x\right) + \cos^{2n}\left(\frac{\pi}{2} - x\right)}dx$$

$$I = 4\pi\int_0^{\pi/2} \frac{\cos^{2n} x}{\cos^{2n} x + \sin^{2n} x}dx \quad \text{...(iv)}$$

On adding Eqs. (iii) and (iv), we get

$$2I = 4\pi\int_0^{\pi/2} \frac{\sin^{2n} x + \cos^{2n} x}{\sin^{2n} x + \cos^{2n} x}dx$$

$$= 4\pi\int_0^{\pi/2} dx = 4\pi[x]_0^{\pi/2}$$

$\Rightarrow \quad 2I = 4\pi \times \frac{\pi}{2}$

$\therefore \quad I = \pi^2$

**4.** (d) Given, $f'(x) = f(x)$

$\Rightarrow \quad f(x) = ce^x$ and since $f(0) = 1$

$\therefore \quad 1 = f(0) = c$

$\therefore \quad f(x) = e^x$ and hence $g(x) = x^2 - e^x$

Thus, $\int_0^1 f(x)g(x)\,dx = \int_0^1 (x^2e^x - e^{2x})dx$

$$= [x^2e^x]_0^1 - 2\int_0^1 xe^x dx - \left[\frac{e^{2x}}{2}\right]_0^1$$

$$= (e - 0) - 2\{[xe^x]_0^1 - [e^x]_0^1\} - \frac{1}{2}(e^2 - 1)$$

$$= (e - 0) - 2[(e - 0) - (e - 1)] - \frac{1}{2}(e^2 - 1)$$

$$= e - \frac{1}{2}e^2 - \frac{3}{2}$$

**5.** (c) Given, $f(x) = x^4 + |x|$

$$I_1 = \int_0^{\pi} f(\cos x)dx$$

$$I_2 = \int_0^{\pi/2} f(\sin x)dx$$

Clearly, $f(x)$ is an even function.

Hence, $I_1 = \int_0^{\pi} f[\cos(\pi - x)]dx$

$$= \int_0^{\pi} f(-\cos x)dx$$

$$= \int_0^{\pi} f(\cos x)dx$$

$\Rightarrow$ $I_1 = 2\int_0^{\pi/2} f(\cos x)\,dx$

$$= 2\int_0^{\pi/2} f(\sin x)dx = 2I_2$$

$\therefore$ $\dfrac{I_1}{I_2} = 2$

**Alternate Method**

Let $u = \cos x \Rightarrow du = -\sin x\,dx$

$\therefore$ $I_1 = \int_{-1}^{1} \dfrac{f(u)}{\sqrt{1-u^2}}du$

$\Rightarrow$ $I_1 = 2\int_0^{1} \dfrac{f(u)}{\sqrt{1-u^2}}du$ ...(i)

Similarly, with $\sin x = t$ in $I_2$, we have

$\cos x\,dx = dt$

$\Rightarrow$ $\sqrt{1-t^2}dx = dt$

$\Rightarrow$ $dx = \dfrac{dt}{\sqrt{1-t^2}}$

$\therefore$ $I_2 = \int \dfrac{f(t)}{\sqrt{1-t^2}}dt$ ...(ii)

From Eqs. (i) and (ii), we get

$$\frac{I_1}{I_2} = 2$$

**6.** (c) Let $I = \int_a^b xf(x)dx$ ...(i)

$$= \int_a^b (a+b-x)f(a+b-x)dx$$

$\left[\because \int_a^b f(x)\,dx = \int_a^b f(a+b-x)dx\right]$

$= \int_a^b (a+b-x)\,f(x)dx$ ...(ii)

$[\because f(a+b-x) = f(x)]$

On adding Eqs. (i) and (ii), we get

$$2I = \int_a^b (a+b)f(x)dx$$

$\Rightarrow$ $2I = (a+b)\int_a^b f(x)dx$

$\therefore$ $I = \dfrac{a+b}{2}\int_a^b f(x)\,dx$

**7.** (d) Let $I = \int_0^1 x(1-x)^n dx$

On replacing $x$ by $(1-x)$, we get

$$I = \int_0^1 (1-x)x^n\,dx$$

$$= \int_0^1 (x^n - x^{n+1})dx$$

$$= \left[\frac{x^{n+1}}{n+1} - \frac{x^{n+2}}{n+2}\right]_0^1$$

$$= \left[\frac{1}{n+1} - \frac{1}{n+2}\right]$$

**8.** (b) Let $I = \int_0^{\pi} xf(\sin x)dx$ ...(i)

$I = \int_0^{\pi} (\pi - x)f(\sin x)dx$ ...(ii)

On adding Eqs. (i) and (ii), we get

$$2I = \int_0^{\pi} \pi f(\sin x)dx$$

$$I = \frac{\pi}{2}\int_0^{\pi} f(\sin x)dx$$

$$I = \pi\int_0^{\pi/2} f(\sin x)dx$$

$\Rightarrow$ $\int_0^{\pi} x f(\sin x)dx = \pi\int_0^{\pi/2} f(\sin x)dx$

But it is $A\int_0^{\pi/2} f(\sin x)\,dx$

$\therefore$ $A = \pi$

**9.** (a) Given, $f(x) = \dfrac{e^x}{1+e^x}$

$$I_1 = \int_{f(-a)}^{f(a)} x\,g[x(1-x)]dx$$

$$I_2 = \int_{f(-a)}^{f(a)} g\{x(1-x)\}dx$$

$\therefore$ $f(a) = \dfrac{e^a}{1+e^a}$ ...(i)

and $f(-a) = \dfrac{e^{-a}}{1+e^{-a}} = \dfrac{1}{1+e^a}$ ...(ii)

On adding Eqs. (i) and (ii), we get

$f(a) + f(-a) = 1$

$\Rightarrow$ $f(a) = 1 - f(-a)$

Let $f(-a) = t$

$\Rightarrow$ $f(a) = 1 - t$

Now, $I_1 = \int_t^{1-t} x\,g[x(1-x)]dx$ ...(iii)

$\Rightarrow$ $I_1 = \int_t^{1-t} (1-x)g[x(1-x)]dx$ ...(iv)

$\left[\because \int_a^b f(x)dx = \int_a^b f(a+b-x)dx\right]$

On adding Eqs. (iii) and (iv), we get

$$2I_1 = \int_t^{1-t} g\{x(1-x)\}dx = I_2$$

$\therefore$ $\dfrac{I_2}{I_1} = 2$

**10.** (b) Let $I = \int_{-\pi}^{\pi} \dfrac{\cos^2 x}{1+a^x}dx,\ a > 0$ ...(i)

On replacing $x$ by $-x$, we get

$I = \int_{-\pi}^{\pi} \dfrac{a^x\cos^2 x}{1+a^x}dx$ ...(ii)

On adding Eqs. (i) and (ii), we get

$$2I = \int_{-\pi}^{\pi} \frac{(1+a^x)\cos^2 x}{(1+a^x)}dx$$

$$= \int_{-\pi}^{\pi} \cos^2 x\,dx = \int_{-\pi}^{\pi}\left(\frac{\cos 2x + 1}{2}\right)dx$$

$$= 2\int_0^{\pi}\left(\frac{\cos 2x + 1}{2}\right)dx$$

$$= \int_0^{\pi}(\cos 2x + 1)\,dx$$

$$= \left[\frac{\sin 2x}{2} + x\right]_0^{\pi}$$

$\Rightarrow$ $2I = \pi$

$\therefore$ $I = \pi/2$

**11.** (b) Let $I = \int_{-3\pi/2}^{-\pi/2} \{(x+\pi)^3 + \cos^2(x+3\pi)\}dx$

$$= \int_{-3\pi/2}^{-\pi/2} [(x+\pi)^3 + \cos^2 x]dx \quad \ldots(i)$$

$$= \int_{-3\pi/2}^{-\pi/2} \left[\left(-\frac{\pi}{2} - \frac{3\pi}{2} - x + \pi\right)^3 + \cos^2\left(-\frac{3\pi}{2} - \frac{\pi}{2} - x\right)\right]dx$$

$$= \int_{-3\pi/2}^{-\pi/2} [(-\pi - x)^3 + \cos^2(-2\pi - x)]dx$$

$$= \int_{-3\pi/2}^{-\pi/2} [-(x+\pi)^3 + \cos^2 x]dx \quad \ldots(ii)$$

On adding Eqs. (i) and (ii), we get

$$2I = \int_{-3\pi/2}^{-\pi/2} 2\cos^2 x\, dx$$

$$= 2\int_{-3\pi/2}^{-\pi/2} \left(\frac{\cos 2x + 1}{2}\right)dx$$

$$= \int_{-3\pi/2}^{-\pi/2} (\cos 2x + 1)dx$$

$$= \left(\frac{\sin 2x}{2} + x\right)_{-3\pi/2}^{-\pi/2}$$

$$= \left(0 - \frac{\pi}{2}\right) - \left(0 - \frac{3\pi}{2}\right)$$

$$= -\frac{\pi}{2} + \frac{3\pi}{2}$$

$\Rightarrow \quad 2I = \pi$

$\therefore \quad I = \frac{\pi}{2}$

**12.** (c) Let $I = \int_0^\pi x\, f(\sin x)dx \quad \ldots(i)$

$$I = \int_0^\pi (\pi - x) f[\sin(\pi - x)]dx \quad \left[\because \int_0^a f(x)\,dx = \int_0^a f(a-x)dx\right]$$

$$I = \int_0^\pi (\pi - x)\, f(\sin x)\, dx \quad \ldots(ii)$$

On adding Eqs. (i) and (ii), we get

$$2I = \int_0^\pi \pi f(\sin x)dx$$

$\Rightarrow \quad 2I = 2\int_0^{\pi/2} \pi f(\sin x)dx$

$$\left[\because \int_0^{2a} f(x)dx = 2\int_0^a f(x)\,dx, \text{ if } f(2a - x) = f(x)\right]$$

$\Rightarrow \quad 2I = 2\pi \int_0^{\pi/2} f(\sin x)dx$

$\Rightarrow \quad I = \pi \int_0^{\pi/2} f\left[\sin\left(\frac{\pi}{2} - x\right)\right]dx$

$\Rightarrow \quad I = \pi \int_0^{\pi/2} f(\cos x)dx \quad \left[\because \int_0^a f(x)dx = \int_0^a f(a-x)\,dx\right]$

**13.** (b) Let $I = \int_0^\pi [\cot x]dx \quad \ldots(i)$

$$= \int_0^\pi [\cot(\pi - x)]dx \quad \left[\because \int_0^a f(x)dx = \int_0^a f(a-x)dx\right]$$

$$I = \int_0^\pi [-\cot x]\,dx \quad \ldots(ii)$$

On adding Eqs. (i) and (ii), we get

$$2I = \int_0^\pi \{[\cot x] + [-\cot x]\}dx$$

$$= \int_0^\pi -1\,dx \quad \left[\because [x] + [-x] = \begin{cases} -1, & x \notin Z \\ 0, & x \in Z \end{cases}\right]$$

$$= -[x]_0^\pi$$

$\Rightarrow \quad 2I = -\pi$

$\therefore \quad I = -\frac{\pi}{2}$

**14.** (c) Let $I = \int_0^1 \frac{8 \cdot \log(1+x)}{1+x^2}dx$

Put $x = \tan\theta \Rightarrow dx = \sec^2\theta\, d\theta$

when, $x = 0$, then $\theta = 0$

and when $x = 1$, then $\theta = \frac{\pi}{4}$

$\therefore \quad I = \int_0^{\pi/4} \frac{8\log(1+\tan\theta)}{1+\tan^2\theta}\sec^2\theta\, d\theta$

$$= 8\int_0^{\pi/4} \log(1+\tan\theta)d\theta$$

$$= 8 \times \frac{\pi}{8}\log 2$$

$$\left[\because \int_0^{\pi/4} \log(1+\tan x)dx = \frac{\pi}{8}\log 2, \text{see Q. 2 part (iii)}\right]$$

$$= \pi \log 2$$

**15.** (c) Let $I = \int_0^\pi e^{\cos^2 x}\cos^3(2n+1)x\,dx$

$$= \int_0^\pi e^{\cos^2(\pi - x)}\cos^3(2n+1)(\pi - x)dx$$

$$= \int_0^\pi e^{\cos^2 x}\cos^3\{(2n+1)\pi - (2n+1)x\}dx$$

$$= \int_0^\pi -e^{\cos^2 x}\cos^3(2n+1)x\,dx$$

$$= -\int_0^\pi e^{\cos^2 x}\cos^3(2n+1)x\,dx$$

$\Rightarrow \quad I = -I \Rightarrow 2I = 0$

$\therefore \quad I = 0$

**16.** (d) Let $I = \int_{-\pi}^{\pi} (\cos ax - \sin bx)^2\,dx$

$$= \int_{-\pi}^{\pi} (\cos^2 ax + \sin^2 bx - 2\cos ax \cdot \sin bx)\,dx$$

$$= 2\int_0^\pi (\cos^2 ax + \sin^2 bx)\,dx - 0$$

$$= \int_0^\pi (1 + \cos 2ax + 1 - \cos 2bx)\,dx$$

$$= \left[2x + \frac{\sin 2ax}{2a} - \frac{\sin 2bx}{2b}\right]_0^\pi = 2\pi$$

**17.** (a) Let $I = \int_{-\pi}^{\pi} (1 - x^2)\sin x \cdot \cos^2 x\,dx$

Let $f(x) = (1 - x^2)\sin x\cos^2 x$

$f(-x) = (1 - x^2)(-\sin x)\cos^2 x$

$= -(1 - x^2)\sin x\cos^2 x = -f(x)$

$\Rightarrow f(x)$ is an odd function.

Hence, $I = \int_{-\pi}^{\pi} (1 - x^2)\sin x \cdot \cos^2 x\,dx$

$$\left[\because \int_{-a}^{a} f(x)\,dx = 0, \text{ if } f(x) \text{ is odd.}\right]$$

$= 0$

**18.** (a) Let $I = \int_{-1/2}^{1/2} \left[[x] + \log\left(\frac{1+x}{1-x}\right)\right]dx$

$$= \int_{-1/2}^{1/2} [x]dx + \int_{-1/2}^{1/2} \log\left(\frac{1+x}{1-x}\right)dx$$

$$= \int_{-1/2}^{0} [x]dx + \int_0^{1/2} [x]\,dx + \int_{-1/2}^{1/2} \log\left(\frac{1+x}{1-x}\right)dx$$

$$= \int_{-1/2}^{0} -1\,dx + \int_0^{1/2} 0\,dx + 0$$

$$\left[\because \log\left(\frac{1+x}{1-x}\right) \text{ is an odd function.}\right]$$

$$= [-x]_{-1/2}^{0}$$

$$= -\left[0 + \frac{1}{2}\right] = -\frac{1}{2}$$

**19.** (c) Let $I = \int_{-2}^{0}[x^3 + 3x^2 + 3x + 3 + (x+1)\cos(x+1)]dx$

$$= \int_{-2}^{0}\{x^3 + 1 + 3x^2 + 3x + 2 + (x+1)\cos(x+1)\}dx$$

$$= \int_{-2}^{0}\{(x+1)^3 + 2 + (x+1)\cos(x+1)\}dx$$

Put $x + 1 = t \Rightarrow dx = dt$

when $x = -2 \Rightarrow t = -1$ and when $x = 0 \Rightarrow t = 1$

$\therefore \quad I = \int_{-1}^{1}(t^3 + 2 + t\cos t)dt$

$$= \int_{-1}^{1} t^3\, dt + 2\int_{-1}^{1} dt + \int_{-1}^{1} t\cos t\, dt$$

$= 0 + [2t]_{-1}^{1} + 0 \quad$ [$\because t^3$ and $t\cos t$ are odd functions.]

$= [2 - (-2)]$

$= 4$

**20.** (a) Let $I = \int_{\sqrt{\log 2}}^{\sqrt{\log 3}} \frac{x\sin x^2}{\sin x^2 + \sin(\log 6 - x^2)}dx$

Put $x^2 = t \Rightarrow 2x\,dx = dt$

$$= \frac{1}{2}\int_{\log 2}^{\log 3}\frac{\sin t}{\sin t + \sin(\log 6 - t)}dt \quad \ldots(i)$$

$$= \frac{1}{2}\int_{\log 2}^{\log 3}\frac{\sin(\log 2 + \log 3 - t)}{\sin(\log 2 + \log 3 - t) + \sin t}dt$$

$$\left[\because \int_a^b f(x)\,dx = \int_a^b f(a + b - x)\,dx\right]$$

$$= \frac{1}{2}\int_{\log 2}^{\log 3}\frac{\sin(\log 6 - t)}{\sin(\log 6 - t) + \sin t}dt \quad \ldots(ii)$$

On adding Eqs. (i) and (ii), we get

$$2I = \frac{1}{2}\int_{\log 2}^{\log 3} 1\, dt$$

$$2I = \frac{1}{2}[t]_{\log 2}^{\log 3} = \frac{1}{2}[\log 3 - \log 2]$$

$$2I = \frac{1}{2}\log\frac{3}{2}$$

$\therefore \quad I = \frac{1}{4}\log\frac{3}{2}$

**21.** (b) Let $I = \int_{-\pi/2}^{\pi/2}\left[x^2 + \log\left(\frac{\pi + x}{\pi - x}\right)\right]\cos x\,dx$

$$= \int_{-\pi/2}^{\pi/2} x^2\cos x\,dx + \int_{-\pi/2}^{\pi/2}\log\left(\frac{\pi + x}{\pi - x}\right)\cos x\,dx$$

$$= \int_{-\pi/2}^{\pi/2} x^2\cos x\,dx + 0$$

$$\left[\because \log\left(\frac{\pi + x}{\pi - x}\right)\cos x \text{ is an odd function.}\right]$$

$$= \int_{-\pi/2}^{\pi/2} \underset{\text{I}}{x^2}\, \underset{\text{II}}{\cos x}\,dx$$

$$= [x^2\sin x]_{-\pi/2}^{\pi/2} - \int_{-\pi/2}^{\pi/2} \underset{\text{I}}{2x}\cdot \underset{\text{II}}{\sin x}\,dx$$

$$= [x^2\sin x]_{-\pi/2}^{\pi/2} - 2\left[-x\cos x - \int -\cos x\,dx\right]_{-\pi/2}^{\pi/2}$$

$$= [x^2\sin x]_{-\pi/2}^{\pi/2} - 2[-x\cos x + \sin x]_{-\pi/2}^{\pi/2}$$

$$= [x^2\sin x + 2x\cos x - 2\sin x]_{-\pi/2}^{\pi/2}$$

$$= \left(\frac{\pi^2}{4} + 0 - 2\right) - \left(-\frac{\pi^2}{4} + 0 + 2\right)$$

$$= \frac{\pi^2}{4} - 2 + \frac{\pi^2}{4} - 2$$

$$= \frac{2\pi^2}{4} - 4$$

$$= \frac{\pi^2}{2} - 4$$

**22.** (a,b,c) Let $I = \int_0^{\pi} x\, f(\sin x)dx \quad \ldots(i)$

$$I = \int_0^{\pi}(\pi - x)\, f(\sin x)dx \quad \ldots(ii)$$

On adding Eqs. (i) and (ii), we get

$$2I = \int_0^{\pi}\pi\, f(\sin x)dx$$

$\Rightarrow \quad I = \frac{\pi}{2}\int_0^{\pi} f(\sin x)dx$

Again, $\quad 2I = \int_0^{\pi}\pi\, f(\sin x)dx$

$$2I = 2\pi\int_0^{\pi/2} f(\sin x)dx$$

$\Rightarrow \quad I = \pi\int_0^{\pi/2} f(\sin x)dx = \pi\int_0^{\pi/2} f\left(\sin\frac{\pi}{2} x\right)dx$

$\therefore \quad I = \pi\int_0^{\pi/2} f(\cos x)\,dx$

**23.** (a,b,c) Given, $I_n = \int_{-\pi}^{\pi}\frac{\sin nx}{(1 + \pi^x)\sin x}dx$

Using $\int_{-a}^{a} f(x)dx = \int_0^a [f(x) + f(-x)]\,dx$, we have

$$I_n = \int_0^{\pi}\left\{\frac{\sin nx}{(1 + \pi^x)\sin x} + \frac{\sin nx}{(1 + \pi^{-x})\sin x}\right\}dx$$

$$= \int_0^{\pi}\frac{\sin nx}{\sin x}\left\{\frac{1}{1 + \pi^x} + \frac{\pi^x}{1 + \pi^x}\right\}dx$$

$$= \int_0^{\pi}\frac{\sin nx}{\sin x}\left\{\frac{1 + \pi^x}{1 + \pi^x}\right\}dx = \int_0^{\pi}\frac{\sin nx}{\sin x}dx$$

Now, $I_{n+2} - I_n = \int_0^{\pi}\left[\frac{\sin(n+2)x}{\sin x} - \frac{\sin nx}{\sin x}\right]dx$

$$= \int_0^{\pi}\left[\frac{\sin(n+2)x - \sin nx}{\sin x}\right]dx$$

$\Rightarrow \quad I_{n+2} - I_n = 2\int_0^{\pi}\cos(n+1)x\,dx$

$\Rightarrow \quad I_{n+2} - I_n = 2\left[\frac{\sin(n+1)x}{n+1}\right]_0^{\pi}$

$\Rightarrow \quad I_{n+2} - I_n = 0$

$\Rightarrow \quad I_{n+2} = I_n$

for $n = 1, 2, 3, \ldots$

$\Rightarrow \quad I_n = I_{n-2} = I_{n-4} = \cdots = I_2 = I_0$, if $n$ is even

and $\quad I_n = I_{n-2} = I_{n-4} = \cdots = I_3 = I_1$, if $n$ is odd

$\Rightarrow \quad I_n = \begin{cases} 0, \text{ if } n \text{ is even} \\ \pi, \text{ if } n \text{ is odd}\end{cases} \quad \left[\because I_0 = 0,\ I_1 = \int_0^{\pi}\frac{\sin x}{\sin x}dx = \pi\right]$

$\Rightarrow \quad \sum_{m=1}^{10} I_{2m} = 0$ and $\sum_{m=1}^{10} I_{2m+1} = 10\pi$

**24.** (a,d) Let $I = \int_0^{\pi/2}\frac{\sqrt{\cot x}}{\sqrt{\cot x} + \sqrt{\tan x}}dx \quad \ldots(i)$

$$I = \int_0^{\pi/2}\frac{\sqrt{\cot\left(\frac{\pi}{2} - x\right)}}{\sqrt{\cot\left(\frac{\pi}{2} - x\right)} + \sqrt{\tan\left(\frac{\pi}{2} - x\right)}}dx$$

$$I = \int_0^{\pi/2}\frac{\sqrt{\tan x}}{\sqrt{\tan x} + \sqrt{\cot x}}dx \quad \ldots(ii)$$

On adding Eqs. (i) and (ii), we get

$$2I = \int_0^{\pi/2}\frac{\sqrt{\cot x} + \sqrt{\tan x}}{\sqrt{\tan x} + \sqrt{\cot x}}dx = \int_0^{\pi/2} dx$$

$\Rightarrow \quad 2I = [x]_0^{\pi/2} \Rightarrow 2I = \frac{\pi}{2}$

$\Rightarrow \quad I = \frac{\pi}{4} \quad$ [option (a)]

Again, $I = \int_0^{\pi/2} \frac{dx}{1 + \tan^3 x}$

$$I = \int_0^{\pi/2} \frac{\cos^3 x}{\cos^3 x + \sin^3 x} dx \quad \text{...(i)}$$

$$\Rightarrow \quad I = \int_0^{\pi/2} \frac{\cos^3\left(\frac{\pi}{2} - x\right)}{\cos^3\left(\frac{\pi}{2} - x\right) + \sin^3\left(\frac{\pi}{2} - x\right)} dx$$

$$\Rightarrow \quad I = \int_0^{\pi/2} \frac{\sin^3 x}{\sin^3 x + \cos^3 x} dx \quad \text{...(ii)}$$

On adding Eqs. (i) and (ii), we get

$$2I = \int_0^{\pi/2} dx$$

$$\Rightarrow \quad 2I = [x]_0^{\pi/2} = \frac{\pi}{2}$$

$$\therefore \quad I = \frac{\pi}{4}$$

**25.** (a,c) Let $I = \int_a^b \text{Sgn } x \, dx$

$$\Rightarrow \quad \text{Sgn } x = \begin{cases} 1, & x > 0 \\ -1, & x < 0 \\ 0, & x = 0 \end{cases}$$

$$\Rightarrow \quad I = \int_a^b dx = [|x|]_a^b = |b| - |a|$$

Again, $I = \int_a^b \text{Sgn } x \, dx$

$= \text{Sgn } x \int_a^b dx$

$= [\text{Sgn } x \cdot x]_a^b$ [$\because$ Sgn $x$ = constant value for all $x$]

$= b \text{ Sgn}(b) - a\text{Sgn}(a)$

**26.** (a,b,c,d) (a) Let $I = \int_a^{\pi - a} x f(\sin x) dx$ ...(i)

$$= \int_a^{\pi - a} (a + \pi - a - x) f(\sin x) dx$$

$$I = \int_a^{\pi - a} (\pi - x) f(\sin x) dx \quad \text{...(ii)}$$

On adding Eqs. (i) and (ii), we get

$$2I = \int_a^{\pi - a} \pi f(\sin x) dx$$

$$I = \frac{\pi}{2} \int_a^{\pi - a} f(\sin x) dx$$

(b) $\int_{-a}^{a} [f(x)]^2 dx = 2\int_0^a [f(x)]^2 dx$ [$\because [f(x)]^2$ is an even function.]

(c) $\int_0^{n\pi} f(\cos^2 x) dx$

Since, $\cos^2 x$ is periodic function wtih period $\pi$.

$$\therefore \quad \int_0^{n\pi} f(\cos^2 x) dx = n\int_0^{\pi} f(\cos^2 x) dx$$

(d) Let $I = \int_0^{b-c} f(x + c) dx$ ...(i)

Put $x + c = t \Rightarrow dx = dt$

$$\therefore \quad I = \int_c^b f(t) dt$$

$$I = \int_c^b f(x) dx \quad \text{...(ii)}$$

From Eqs. (i) and (ii), we get

$$\int_0^{b-c} f(x + c) dx = \int_c^b f(x) dx$$

**27.** (a) Given, $f(\alpha) = \int_0^{\pi} \frac{x}{1 - \cos\alpha \sin x} dx$ ...(i)

$$= \int_0^{\pi} \frac{(\pi - x)}{1 - \cos\alpha \sin(\pi - x)} dx$$

$$f(\alpha) = \int_0^{\pi} \frac{(\pi - x)}{1 - \cos\alpha \sin x} dx \quad \text{...(ii)}$$

On adding Eqs. (i) and (ii), we get

$$2f(\alpha) = \int_0^{\pi} \frac{\pi}{1 - \cos\alpha \sin x} dx$$

$$= \pi \int_0^{\pi} \frac{dx}{1 - \cos\alpha \sin x}$$

On dividing numerator and denominator by $\cos^2 \frac{x}{2}$, we get

$$2f(\alpha) = \pi \int_0^{\pi} \frac{\sec^2 \frac{x}{2}}{1 + \tan^2 \frac{x}{2} - 2\cos\alpha \tan\frac{x}{2}} dx$$

Put $\tan\frac{x}{2} = t$

$$\Rightarrow \quad \frac{1}{2}\sec^2\frac{x}{2} dx = dt$$

$$\therefore \quad 2f(\alpha) = 2\pi \int_0^{\infty} \frac{dt}{1 + t^2 - 2\cos\alpha t}$$

$$\Rightarrow \quad 2f(\alpha) = 2\pi \int_0^{\infty} \frac{dt}{\cos^2\alpha + \sin^2\alpha + t^2 - 2t\cos\alpha}$$

$$\Rightarrow \quad 2f(\alpha) = 2\pi \int_0^{\infty} \frac{dt}{(t - \cos\alpha)^2 + \sin^2\alpha}$$

$$\Rightarrow \quad f(\alpha) = \left[\frac{\pi}{\sin\alpha} \tan^{-1}\left(\frac{t - \cos\alpha}{\sin\alpha}\right)\right]_0^{\infty}$$

$$f(\alpha) = \frac{\pi}{\sin\alpha} [\tan^{-1}(\infty) - \tan^{-1}(-\cot\alpha)]$$

$$= \frac{\pi}{\sin\alpha}\left[\frac{\pi}{2} + \tan^{-1}(\cot\alpha)\right]$$

$$= \frac{\pi}{\sin\alpha}\left[\frac{\pi}{2} + \left(\frac{\pi}{2} - \alpha\right)\right]$$

Now, $f(\alpha) = \frac{\pi(\pi - \alpha)}{\sin\alpha}$

$$\therefore \quad \lim_{\alpha \to \pi} f(\alpha) = \lim_{\alpha \to \pi} \frac{\pi(\pi - \alpha)}{\sin\alpha} \quad \left[\frac{0}{0} \text{ form}\right]$$

$$= \lim_{\alpha \to \pi} \frac{-\pi}{\cos\alpha}$$

$$= \frac{-\pi}{-1} = \pi$$

**28.** (c) $f(\alpha) = \frac{\pi(\pi - \alpha)}{\sin\alpha}$

$$f'\left(\frac{\pi}{2}\right) = \frac{\sin\alpha(-\pi) - \pi(\pi - \alpha)\cos\alpha}{\sin^2\alpha}$$

$$f'(\alpha) = \frac{-\pi\sin\alpha - \pi\cos\alpha(\pi - \alpha)}{\sin^2\alpha}$$

$$f'\left(\frac{\pi}{2}\right) = \frac{-\pi\sin\frac{\pi}{2} - \pi\cos\frac{\pi}{2}\left(\pi - \frac{\pi}{2}\right)}{\sin^2\frac{\pi}{2}} = \frac{-\pi - 0}{(1)^2}$$

$$\therefore \quad f'\left(\frac{\pi}{2}\right) = -\pi$$

**29.** (c) $f(a, b) = \int_0^{\pi} \frac{x}{(a^2\cos^2 x + b^2\sin^2 x)^2} dx$ ...(i)

$$f(a, b) = \int_0^{\pi} \frac{(\pi - x)}{(a^2\cos^2 x + b^2\sin^2 x)^2} dx \quad \text{...(ii)}$$

$$\left[\because \int_0^a f(x) dx = \int_0^a f(a - x) dx\right]$$

On adding Eqs. (i) and (ii), we get

$$2f(a, b) = \pi \int_0^{\pi} \frac{dx}{(a^2\cos^2 x + b^2\sin^2 x)^2}$$

$$2f(a, b) = 2\pi \int_0^{\pi/2} \frac{dx}{(a^2\cos^2 x + b^2\sin^2 x)^2} \quad \text{...(iii)}$$

Consider the integral

$$I=\int_0^{\pi/2}\frac{dx}{a^2\cos^2 x+b^2\sin^2 x}$$

On dividing numerator and denominator by $\cos^2 x$, we get

$$I=\int_0^{\pi/2}\frac{\sec^2 x}{a^2+b^2\tan^2 x}dx$$

Put $\tan x=t \Rightarrow \sec^2 x\,dx=dt$

when $x=0$, then $t=0$

and when $x=\frac{\pi}{2}$, then $t=\infty$

$$\therefore \quad I=\int_0^{\infty}\frac{dt}{a^2+b^2t^2}=\frac{1}{b^2}\int_0^{\infty}\frac{dt}{\left(\frac{a}{b}\right)^2+t^2}$$

$$=\frac{1}{b^2}\cdot\frac{1}{a/b}\left[\tan^{-1}\left(\frac{bt}{a}\right)\right]_0^{\infty}$$

$$=\frac{1}{ab}\left[\frac{\pi}{2}-0\right]$$

$$=\frac{\pi}{2ab}$$

$$\Rightarrow \quad \int_0^{\pi/2}\frac{dx}{a^2\cos^2 x+b^2\sin^2 x}=\frac{\pi}{2ab}$$

On differentiating w.r.t. $a$ both sides, we get

$$\Rightarrow \quad -\int_0^{\pi/2}\frac{2a\cos^2 x}{(a^2\cos^2 x+b^2\sin^2 x)}dx=-\frac{\pi}{2b}\left(\frac{1}{a^2}\right)$$

$$\Rightarrow \quad \int_0^{\pi/2}\frac{\cos^2 x}{(a^2\cos^2 x+b^2\sin^2 x)^2}dx=\frac{\pi}{4a^3b} \quad \ldots\text{(iv)}$$

Similarly, on differentiating Eq. (i) w.r.t. $b$, we get

$$\int_0^{\pi/2}\frac{\sin^2 x}{(a^2\cos^2 x+b^2\sin^2 x)^2}dx=\frac{\pi}{4ab^3} \quad \ldots\text{(v)}$$

On adding Eqs. (iv) and (v), we get

$$\int_0^{\pi/2}\frac{dx}{(a^2\cos^2 x+b^2\sin^2 x)^2}=\frac{\pi}{4ab}\left(\frac{1}{a^2}+\frac{1}{b^2}\right) \quad \ldots\text{(vi)}$$

Using Eq. (vi) in Eq. (iii), we get

$$f(a,b)=\pi\times\frac{\pi}{4ab}\left(\frac{1}{a^2}+\frac{1}{b^2}\right)$$

$$f(a,b)=\frac{\pi^2}{4ab}\left(\frac{1}{a^2}+\frac{1}{b^2}\right)$$

$$\therefore \quad f(1,1)=\frac{\pi^2}{2}$$

**30.** (d) We have, $\lim_{a\to\infty} a f(a)$

$$=\lim_{a\to\infty}\frac{a\pi^2(a^2+b^2)}{4a^3b^3}$$

$$=\frac{\pi^2}{4b^3}$$

**31.** (1) Let $I=5\int_0^{\pi}\frac{x^3\cdot\cos^4 x\cdot\sin x}{(\pi^2-3\pi x+3x^2)}dx \quad \ldots\text{(i)}$

$$\Rightarrow \quad I=5\int_0^{\pi}\frac{(\pi-x)^3\cdot\cos^4(\pi-x)\cdot\sin(\pi-x)}{\pi^2-3\pi(\pi-x)+3(\pi-x)^2}dx \quad \ldots\text{(ii)}$$

On adding Eqs. (i) and (ii), we get

$$2I=5\int_0^{\pi}\frac{(\pi^2-3\pi x+3x^2)\cdot\cos^4 x\cdot\sin x}{(\pi^2-3\pi x+3x^2)}dx$$

$$\Rightarrow \quad 2I=5\int_0^{\pi}\cos^4 x\cdot\sin x\,dx=5(-1)\int_1^{-1}t^4 dt$$

$$\therefore \quad 2I=2\Rightarrow I=1$$

**32.** (0) Let $I=\int_0^a f(x)g(x)h(x)\,dx$

$$=\int_0^a f(a-x)g(a-x)h(a-x)\,dx$$

$$=-\int_0^a f(x)g(x)h(a-x)\,dx$$

$\because \quad 7I=3I+4I$

$$=\int_0^a\{f(x)g(x)\}\{3h(x)-4h(a-x)\}dx$$

$$=5\int_0^a f(x)g(x)\,dx=0$$

Since, $f(a-x)\cdot g(a-x)=-f(x)g(x)$

$$\Rightarrow \quad I=0$$

**33.** (9) Let $I=\int_{-\pi}^{\pi}\frac{2x(1+\sin x)}{(1+\cos^2 x)}dx$

$$=\int_{-\pi}^{\pi}\frac{2x}{(1+\cos^2 x)}dx+\int_{-\pi}^{\pi}\frac{2x\sin x}{(1+\cos^2 x)}dx$$

$$I=0+4\int_0^{\pi}\frac{x\sin x}{(1+\cos^2 x)}dx \quad \ldots\text{(i)}$$

$$I=4\int_0^{\pi}\frac{(\pi-x)\sin x}{(1+\cos^2 x)}dx \quad \ldots\text{(ii)}$$

On adding Eqs. (i) and (ii), we get

$$2I=4\pi\int_0^{\pi}\frac{\sin x}{1+\cos^2 x}dx$$

Put $\cos x=t \Rightarrow -\sin x\cdot dx=dt$

$$\therefore \quad 2I=-4\pi\int_1^{-1}\frac{dt}{1+t^2}=-4\pi\,[\tan^{-1}t]_1^{-1}$$

$$=-4\pi\left(-\frac{\pi}{4}-\frac{\pi}{4}\right)$$

$$\Rightarrow \quad 2I=2\pi^2$$

$$\therefore \quad I=\pi^2=k\Rightarrow[k]=9$$

**34.** (0) Let $I=\int_0^{\infty}f\left(x+\frac{1}{x}\right)\frac{\log x}{x}dx$

Put $\log x=t \Rightarrow x=e^t$

$$\Rightarrow \quad dx=e^t dt$$

$$\therefore \quad I=\int_{-\infty}^{\infty}f(e^t+e^{-t})\frac{t}{e^t}e^t dt$$

$$=\int_{-\infty}^{\infty}f(e^t+e^{-t})t\,dt$$

$$=0 \quad [\because \text{function is odd}]$$

**35.** (2) Let $I=\pi\int_0^{2008}x|\sin x|\,dx$

Put $\pi x=t \Rightarrow dx=\frac{dt}{\pi}$

$$\therefore \quad I=\frac{1}{\pi}\cdot\frac{\pi}{\pi}\int_0^{2008\pi}t\,|\sin t|\,dt$$

$$=\frac{1}{\pi}\int_0^{2008\pi}t\,|\sin t|\,dt \quad \ldots\text{(i)}$$

$$I=\frac{1}{\pi}\int_0^{2008\pi}(2008\pi-t)|\sin t|\,dt \quad \ldots\text{(ii)}$$

On adding Eqs. (i) and (ii), we get

$$2I=\frac{2008\pi}{\pi}\int_0^{2008\pi}|\sin t|\,dt=(2008)^2\int_0^{\pi}|\sin t|\,dt$$

$$\Rightarrow \quad I=(2008)^2$$

Now, $\frac{1}{1004}\sqrt{\pi\int_0^{2008}x|\sin x|dx}$

$$=\frac{1}{1004}\sqrt{(2008)^2}=\frac{1}{1004}\times 2008=2$$

# DPP-4 Periodic Functions and Leibnitz Theorem

**1.** (i) Let $I = \int_{-1}^{2} e^{\{x\}} dx$

$= \int_{-1}^{2} e^{x-[x]} dx$

We know that $x - [x]$ is periodic with period 1.

$= \int_{-1\times 1}^{2\times 1} e^{x-[x]} dx$

$= [2-(-1)] \int_0^1 e^{x-[x]} dx$

$= 3\int_0^1 e^{x-0} dx$

$= 3\int_0^1 e^x dx$

$= 3[e^x]_0^1 = 3(e-1)$

(ii) Let $I = \int_{-1}^{2} e^{\{3x\}} dx$

Put $3x = t$

$\Rightarrow \quad 3\,dx = dt$

$\Rightarrow \quad dx = \dfrac{dt}{3}$

$\therefore \quad I = \dfrac{1}{3}\int_{-3}^{6} e^{\{t\}} dt$

$= \dfrac{1}{3}\int_{-3\times 1}^{6\times 1} e^{t-[t]} dt$

$= \dfrac{1}{3}(6+3)\int_0^1 e^{t-0} dt$

$= \dfrac{9}{3}\int_0^1 e^t dt = \dfrac{9}{3}[e^t]_0^1$

$= 3(e-1)$

(iii) Let $I = \int_0^{2000\pi} \dfrac{dx}{1+e^{\sin x}}$

$I = 1000\int_0^{2\pi} \dfrac{dx}{1+e^{\sin x}}$ ...(i)

[$\because$ period of $e^{\sin x}$ is $2\pi$]

$= 1000\int_0^{2\pi} \dfrac{dx}{1+e^{\sin(2\pi - x)}}$

$= 1000\int_0^{2\pi} \dfrac{dx}{1+e^{-\sin x}}$

$I = 1000\int_0^{2\pi} \dfrac{e^{\sin x}dx}{e^{\sin x}+1}$ ...(ii)

On adding Eqs. (i) and (ii), we get

$2I = 1000\int_0^{2\pi} \dfrac{1+e^{\sin x}}{1+e^{\sin x}} dx$

$\Rightarrow \quad 2I = 1000\int_0^{2\pi} dx$

$\Rightarrow \quad 2I = 1000\,[x]_0^{2\pi}$

$\Rightarrow \quad 2I = 1000 \times 2\pi = 2000\pi$

$\therefore \quad I = 1000\pi$

(iv) Let $I = \int_{-\pi}^{199\pi} \sqrt{\dfrac{1-\cos 2x}{2}}\, dx$

$= \int_{-\pi}^{199\pi} |\sin x|\, dx$

$= \int_{(-1)\pi}^{199\pi} |\sin x|\, dx$

$= (199+1)\int_0^{\pi} |\sin x|\, dx$ [$\because$ period of $|\sin x|$ is $\pi$]

$= 200\int_0^{\pi} \sin x\, dx = 200\,\{-[\cos x]_0^{\pi}\}$

$= 200\,\{-(-1-1)\} = 400$

**2.** (i) Given, $f(x) = \int_x^{x^2} \sqrt{\sin t}\, dt$

$f'(x) = \sqrt{\sin x^2}\cdot\dfrac{d}{dx}(x^2) - \sqrt{\sin x}\cdot\dfrac{d}{dx}(x)$

[by Leibnitz rule]

$= \sqrt{\sin x^2}\,(2x) - \sqrt{\sin x}$

(ii) We have, $\lim_{x\to 0} \dfrac{\int_0^{x^2} \cos t^2\, dt}{x\sin x}$

$= \lim_{x\to 0} \dfrac{\cos x^2 \cdot 2x}{x\cos x + \sin x}$

[by L' Hospital's rule followed by Leibnitz rule]

$= \lim_{x\to 0} \dfrac{2[x(-\sin x^2)2x + \cos x^2]}{-x\sin x + \cos x + \cos x}$ [by L' Hospital rule]

$= \lim_{x\to 0} \dfrac{-4x^2\sin x^2 + 2\cos x^2}{-x\sin x + 2\cos x}$

$= \dfrac{2}{2} = 1$

(iii) We have, $\lim_{x\to\infty} \dfrac{\left[\int_0^x e^{t^2} dt\right]^2}{\int_0^x e^{2t^2} dt}$

Since, $e^{t^2} > 0, e^{2t^2} > 0$ in $[0, t]$, where $t > 0$

$\int_0^x e^{t^2}\, dt$ and $\int_0^x e^{2t^2} dt \to \infty$ as $x \to \infty$

$\therefore \quad L = \lim_{x\to\infty} \dfrac{\left[\int_0^x e^{t^2} dt\right]^2}{\int_0^x e^{2t^2} dt}$ is the form of $\dfrac{\infty}{\infty}$.

Therefore, using L' Hospital rule, we get

$L = \lim_{x\to\infty} \dfrac{2e^{x^2}\cdot\int_0^x e^{t^2} dt}{e^{2x^2}}$

$= 2\lim_{x\to\infty} \dfrac{\int_0^x e^{t^2} dt}{e^{x^2}}$ $\left[\dfrac{\infty}{\infty}\text{ form}\right]$

$= 2\lim_{x\to\infty} \dfrac{e^{x^2}}{2xe^{x^2}}$

$= \lim_{x\to\infty} \dfrac{e^{x^2}}{xe^{x^2}}$

$= \lim_{x\to\infty} \left(\dfrac{1}{x}\right) = 0$

(iv) Given, $f(x) = e^{g(x)}$

and $\quad g(x) = \int_2^x \dfrac{t}{1+t^4} dt$

$g'(x) = \dfrac{x}{1+x^4}$ [by Leibnitz rule]

Now, $\quad f'(x) = e^{g(x)} g'(x)$

$= e^{g(x)}$

Then, $\quad f'(2) = e^{g(2)}\, g'(2)$

$\Rightarrow \quad f'(2) = e^0 \times \left(\dfrac{2}{1+(2)^4}\right)$

$= \dfrac{2}{1+16}$ $\left[\because g(2) = \int_2^2 \dfrac{t}{1+t^4} dt = 0\right]$

$= \dfrac{2}{17}$

**3.** (a) Given, $\int_0^{f(x)} t^2 dt = x\cos \pi x$

On differentiating both sides by using Newton-Leibnitz rule, we get

$$[f(x)]^2 f'(x) = \pi x(-\sin \pi x) + \cos \pi x$$

$$[f(x)]^2 f'(x) = -\pi x \sin \pi x + \cos \pi x$$

$$\Rightarrow \quad f'(x) = \frac{-\pi x \sin \pi x + \cos \pi x}{[f(x)]^2} \quad \ldots(i)$$

Again, $\int_0^{f(x)} t^2 dt = x\cos \pi x$

$$\Rightarrow \quad \left[\frac{t^3}{3}\right]_0^{f(x)} = x\cos \pi x$$

$$\Rightarrow \quad \frac{[f(x)]^3}{3} = x\cos \pi x$$

$$\Rightarrow \quad [f(x)]^3 = 3x\cos \pi x$$

$$\Rightarrow \quad f(x) = (3x\cos \pi x)^{1/3}$$

$$\Rightarrow \quad f(9) = (27\cos 9\pi)^{1/3} = 3 \quad \ldots(ii)$$

Now, $$f'(9) = \frac{-\pi 9 \sin 9\pi + \cos 9\pi}{[f(9)]^2}$$

$$= \frac{-\pi 9 \sin 9\pi + \cos 9\pi}{[f(9)]^2}$$

$$= \frac{-1}{(3)^2} \quad \text{[by Eq. (ii)]}$$

$$= -\frac{1}{9}$$

**4.** (a) We have, $\lim_{x \to 1} \int_4^{f(x)} \frac{2t}{x-1} dt$

$$= \lim_{x \to 1} \frac{2f(x) f'(x)}{1} = 2f(1) \cdot f'(1)$$

$$= 2 \times 4 f'(1)$$

$$= 8f'(1)$$

**5.** (b) We have, $\int_0^{n\pi + v} \sqrt{\frac{1+\cos 2x}{2}}\, dx$

$$= \int_0^{n\pi + v} \sqrt{\cos^2 x}\, dx$$

$$= \int_0^{n\pi + v} |\cos x|\, dx$$

$$= \int_0^{v} |\cos x|\, dx + \int_v^{v + n\pi} |\cos x|\, dx$$

$$= -\int_0^{v} \cos x\, dx + n\int_0^{\pi} |\cos x|\, dx$$

$$= [-\sin x]_0^v + n\left[\int_0^{\pi/2} \cos x\, dx + \int_{\pi/2}^{\pi} -\cos x\, dx\right]$$

$$= [-\sin x]_0^v + n\int_0^{\pi/2} \cos x\, dx - n\int_{\pi/2}^{\pi} \cos x\, dx$$

$$= [-\sin x]_0^v + n[\sin x]_0^{\pi/2} - n[\sin x]_{\pi/2}^{\pi}$$

$$= -\sin v + n + n = 2n - \sin v$$

**6.** (c) We have, $f(x) = \int_2^x \frac{dt}{\sqrt{1+t^4}}$

$$f'(x) = \frac{1}{\sqrt{1+x^4}}$$

Since, $g$ is inverse of $f$.

$$\therefore \quad g[f(x)] = x$$

$$\Rightarrow \quad g'[f(x)]\, f'(x) = 1$$

$$\Rightarrow \quad g'[f(x)] = \frac{1}{f'(x)}$$

$$\Rightarrow \quad g'[f(x)] = \sqrt{1+x^4}$$

At $x = 2$, $\quad f(x) = 0$

$$\Rightarrow \quad g'(0) = \sqrt{1+(2)^4} = \sqrt{17}$$

**7.** (c) We have, $\lim_{x \to \infty} \frac{d}{dx} \int_{\sqrt{3}}^{\sqrt{x}} \frac{r^3}{(r+1)(r-1)} dr$

$$= \lim_{x \to \infty} \left[\frac{x^{3/2}}{(\sqrt{x}+1)(\sqrt{x}-1)} \times \frac{1}{2\sqrt{x}}\right] \quad \text{[by Leibnitz rule]}$$

$$= \lim_{x \to \infty} \frac{x}{2(\sqrt{x}+1)(\sqrt{x}-1)}$$

$$= \frac{1}{2} \lim_{x \to \infty} \frac{1}{\left(1+\frac{1}{\sqrt{x}}\right)\left(1-\frac{1}{\sqrt{x}}\right)}$$

$$= \frac{1}{2}$$

**8.** (b) Given, $f(x) = \int_0^{\sin^2 x} \sin^{-1}\sqrt{t}\, dt + \int_0^{\cos^2 x} \cos^{-1}\sqrt{t}\, dt$

$$\Rightarrow f'(x) = \sin^{-1}\sqrt{\sin^2 x} \cdot 2\sin x\cos x + \cos^{-1}\sqrt{\cos^2 x}\, 2\cos x(-\sin x)$$

$$= 2x\sin x\cos x - 2x\sin x\cos x$$

$$= 0$$

$\Rightarrow$ $f(x)$ is a constant function and is independent of $x$.

So, let's select $x = \frac{\pi}{4}$ $\quad \left(\because x \in \left[0, \frac{\pi}{2}\right]\right)$

$$\therefore \quad f(x) = \int_0^{1/2} \sin^{-1}\sqrt{t}\, dt + \int_0^{1/2} \cos^{-1}\sqrt{t}\, dt$$

$$= \int_0^{1/2} (\sin^{-1}\sqrt{t} + \cos^{-1}\sqrt{t})\, dt$$

$$= \int_0^{1/2} \frac{\pi}{2} dt$$

$$= \frac{\pi}{2}[t]_0^{1/2} = \frac{\pi}{2} \times \frac{1}{2} = \frac{\pi}{4}$$

**9.** (c) Let $\quad I = \int_{\pi}^{10\pi} |\sin x|\, dx$

$$= (10-1)\int_0^{\pi} |\sin x|\, dx$$

$$\left[\because \int_{mT}^{nT} f(x)\, dx = (n-m)\int_0^T f(x)\, dx \text{ and period of } |\sin x| \text{ is } \pi\right]$$

$$= 9\int_0^{\pi} \sin x\, dx = -9[\cos x]_0^{\pi}$$

$$= -9(-1-1) = 18$$

**10.** (d) We have, $\lim_{x \to 0} \frac{\int_0^{x^2} \sec^2 t\, dt}{x\sin x}$

$$= \lim_{x \to 0} \frac{\sec^2 x^2\, 2x}{x\cos x + \sin x} \quad \left[\frac{0}{0} \text{ form}\right]$$

[by L' Hospital rule]

$$= 2\lim_{x \to 0} \frac{x\, 2\sec x^2\, 2x\sec x^2 \tan x^2 + \sec^2 x^2}{-x\sin x + \cos x + \cos x}$$

$$= 2\lim_{x \to 0} \frac{4x^2 \sec^2 x^2 \tan x^2 + \sec^2 x^2}{-x\sin x + 2\cos x}$$

$$= 2\lim_{x \to 0} \frac{\sec^2 x^2 (4x^2 \tan x^2 + 1)}{-x\sin x + 2\cos x}$$

$$= 2 \times \frac{1}{2} = 1$$

**11.** (d) We have, $\lim_{x \to 2} \int_6^{f(x)} \frac{4t^3}{x-2} dt$

$$= \lim_{x \to 2} \frac{\int_6^{f(x)} 4t^3 dt}{x-2} = \lim_{x \to 2} \frac{4[f(x)]^3 f'(x)}{1}$$

$$= 4\lim_{x \to 2} [f(x)]^3 f'(x) = 4[f(2)]^3 f'(2)$$

$$= 4(6)^3 \times \frac{1}{48} \quad \left[\because f(2) = 6, \; f'(2) = \frac{1}{48}\right]$$

$$= 18$$

**12.** (c) We have, $F(x) = \int_0^x f(t)\,dt$

$$\Rightarrow \quad F'(x) = f(x) \quad \ldots\text{(i) [Leibnitz rule]}$$

and $\quad F(x^2) = x^2(1+x) = x^2 + x^3 \quad$ [given]

$$\Rightarrow \quad F(x) = x + x^{3/2}$$

$$\Rightarrow \quad F'(x) = 1 + \frac{3}{2}x^{1/2}$$

$$\Rightarrow \quad f(x) = 1 + \frac{3}{2}x^{1/2} \quad \text{[from Eq. (i)]}$$

$$\Rightarrow \quad f(4) = 1 + \frac{3}{2} \times (4)^{1/2}$$

$$= 1 + \frac{3}{2} \times 2$$

$$\therefore \quad f(4) = 4$$

**13.** (a) Given, $f(x) = \int_1^x \sqrt{2 - t^2}\,dt$

$$f'(x) = \sqrt{2 - x^2} \quad \text{[by Leibnitz rule]}$$

Now, $\quad x^2 - f'(x) = 0$

$$\Rightarrow \quad x^2 - \sqrt{2 - x^2} = 0$$

$$\Rightarrow \quad x^2 = \sqrt{2 - x^2}$$

$$\Rightarrow \quad x^4 = 2 - x^2$$

$$\Rightarrow \quad x^4 + x^2 - 2 = 0$$

$$\Rightarrow \quad x^4 + 2x^2 - x^2 - 2 = 0$$

$$\Rightarrow \quad x^2(x^2 + 2) - 1(x^2 + 2) = 0$$

$$\Rightarrow \quad (x^2 + 2)(x^2 - 1) = 0$$

$$\Rightarrow \quad x^2 + 2 \neq 0$$

$$\Rightarrow \quad x^2 = 1$$

$$\therefore \quad x = \pm 1$$

**14.** (c) We have, $I_1 = \int_3^{3+3T} f(2x)\,dx$

Put $2x = t \Rightarrow dx = \dfrac{dt}{2}$

$$\therefore \quad I_1 = \frac{1}{2}\int_6^{6+6T} f(t)\,dt$$

$$= \frac{1}{2} \times 6 \int_0^T f(t)\,dt$$

$$= 3I \quad \left[\because I = \int_0^T f(x)\,dx\right]$$

**15.** (b) Given, $f(x) = \int_{x^2}^{x^2+1} e^{-t^2}\,dt$

On differentiating both sides by using Leibnitz rule, we get

$$f'(x) = e^{-(x^2+1)^2} \cdot 2x - e^{-(x^2)^2} \cdot 2x$$

$$f'(x) = 2xe^{-(x^4+2x^2+1)}(1 - e^{2x^2+1})$$

[where, $e^{2x^2+1} > 1, \forall\, x$ and $e^{-(x^4+2x^2+1)} > 0, \forall x$]

$$\therefore \quad f'(x) \geq 0$$

which shows $\quad 2x \leq 0$

$$\Rightarrow \quad x \leq 0$$

**16.** (a) Given, $\int_0^{t^2} xf(x)\,dx = \dfrac{2}{5}t^5$

Using Newton-Leibnitz formula differentiating both sides, we get

$$t^2\{f(t^2)\}\left\{\frac{d}{dt}(t^2)\right\} - 0 \cdot f(0)\left[\frac{d}{dt}(0)\right] = 2t^4$$

$$\Rightarrow \quad t^2 f(t^2)\,2t = 2t^4$$

$$\Rightarrow \quad f(t^2) = t$$

$$\therefore \quad f\left(\frac{4}{25}\right) = \frac{2}{5} \quad \left[\text{putting } t = \frac{2}{5}\right]$$

**17.** (c) We have,

$$\int_{\sin x}^{1} t^2 f(t)\,dt = 1 - \sin x$$

On differentiating both sides using Newton-Leibnitz formula, we get

$$-\sin^2 x\, f(\sin x)\cos x = -\cos x$$

$$\Rightarrow \quad f(\sin x) = \frac{1}{\sin^2 x}$$

$$\Rightarrow \quad f(\sin x) = \operatorname{cosec}^2 x$$

$f\left(\dfrac{1}{\sqrt{3}}\right)$ is obtained, when $\sin x = \dfrac{1}{\sqrt{3}}$

$$\Rightarrow \quad \operatorname{cosec} x = \sqrt{3}$$

i.e. $\quad f\left(\dfrac{1}{\sqrt{3}}\right) = (\sqrt{3})^2 = 3$

**18.** (a) We have,

$$\lim_{x \to \pi/4} \frac{\int_2^{\sec^2 x} f(t)\,dt}{x^2 - \dfrac{\pi^2}{16}}$$

$$= \lim_{x \to \pi/4} \frac{f(\sec^2 x)\,2\sec x \cdot \sec x \tan x}{2x}$$

[by L' Hospital's rule]

$$= \lim_{x \to \pi/4} \frac{2f(\sec^2 x)\sec^2 x \tan x}{2x}$$

$$= \frac{2f(2)}{\dfrac{\pi}{4}} = \frac{8}{\pi}f(2)$$

**19.** (a) Let $I = \lim_{t \to 0} \log\left[\dfrac{1}{t}\int_0^t (1 + a\sin bx)^{c/x}\,dx\right]$

$$= \log\left[\lim_{t \to 0} \frac{(1 + a\sin bt)^{c/t}}{1}\right]$$

Using L' Hospital's rule, we get

$$= \log\left[e^{\lim_{t \to 0} \frac{a(\sin bt)c \times b}{bt}}\right]$$

$$= \log(e^{abc})$$

$$= (abc)\log_e e$$

$$= abc$$

**20.** (c) Given, $\int_2^x f(t)\,dt = \dfrac{x^2}{2} + \int_x^2 t^2 f(t)\,dt$

On differentiating both sides using Newton-Leibnitz formula, we have

$$f(x) = x + [-x^2 f(x)]$$

$$\Rightarrow \quad f(x) = x - x^2 f(x)$$

$$\Rightarrow \quad (1 + x^2) f(x) = x$$

$$\Rightarrow \quad f(x) = \frac{x}{1 + x^2}$$

$$\Rightarrow \quad f(-x) = \frac{-x}{1 + x^2} = -f(x)$$

$\Rightarrow f(x)$ is odd function.

Now, $\int_{-\pi/4}^{\pi/4} \dfrac{f(x) + x^9 - x^3 + x + 1}{\cos^2 x}\,dx$

$$= \int_{-\pi/4}^{\pi/4} [f(x)\sec^2 x + (x^9 - x^3 + x)\sec^2 x + \sec^2 x]\,dx$$

$$= \int_{-\pi/4}^{\pi/4} f(x)\sec^2 x\,dx + \int_{-\pi/4}^{\pi/4} (x^9 - x^3 + x)\sec^2 x\,dx$$

$$+ \int_{-\pi/4}^{\pi/4} \sec^2 x\,dx$$

$$= 0 + 0 + \int_{-\pi/4}^{\pi/4} \sec^2 x dx$$

[$\because f(x)$ and $(x^9 - x^3 + x)$ are odd functions]

$$= [\tan x]_{-\pi/4}^{\pi/4}$$

$$= \tan\frac{\pi}{4} - \tan\left(\frac{-\pi}{4}\right)$$

$$= \tan\frac{\pi}{4} + \tan\frac{\pi}{4}$$

$$= 1 + 1 = 2$$

**21.** (c) Given,

$$y^2 + \int_0^x (\sin t + a^2 t^3 + bt)\, dt = \alpha, \alpha > 0$$

$$\Rightarrow \quad y^2 + \left[-\cos t + \frac{a^2 t^4}{4} + \frac{bt^2}{2}\right]_0^x = \alpha$$

$$\Rightarrow \quad y^2 + \left[-\cos x + \frac{a^2x^4}{4} + \frac{bx^2}{2} + 1\right] = \alpha$$

cuts $\quad y = f^{-1}(x) \quad$ ...(i)

$$\Rightarrow \quad y^2 + (1 - \cos x) + \frac{a^2x^4}{4} + \frac{bx^2}{2} > 0, \text{ as } \alpha > 0.$$

Since, $f(x)$ is linear.

$\therefore$ It must have atleast two solutions, when $b > 0$.

**22.** (c) Given,

$$\int_0^x \sqrt{1 - [f'(t)]^2}\, dt = \int_0^x f(t)\, dt, \ 0 \le x \le 1$$

On differentiating w.r.t. $x$ by using Leibnitz rule, we get

$$\sqrt{1 - [f'(x)]^2} = f(x)$$

$$\Rightarrow \quad f'(x) = \pm \sqrt{1 - [f(x)]^2}$$

$$\Rightarrow \quad \int \frac{f'(x)}{\sqrt{1 - [f(x)]^2}} dx = \pm \int dx$$

$$\Rightarrow \quad \sin^{-1}[f(x)] = \pm x + C$$

Put $x = 0 \Rightarrow \sin^{-1}[f(0)] = C$

$\Rightarrow \quad C = \sin^{-1}(0) = 0 \quad [\because f(0) = 0]$

$\therefore \quad f(x) = \pm \sin x$

But $\quad f(x) \ge 0, \forall x \in [0, 1]$

$\therefore \quad f(x) = \sin x$

As we know that

$$\sin x < x, \forall x > 0$$

$$\therefore \quad \sin\left(\frac{1}{2}\right) < \frac{1}{2} \text{ and } \sin\left(\frac{1}{3}\right) < \frac{1}{3}$$

$$f\left(\frac{1}{2}\right) < \frac{1}{2} \text{ and } f\left(\frac{1}{3}\right) < \frac{1}{3}$$

**23.** (b) We have,

$$\lim_{x \to 0} \frac{1}{x^3} \int_0^x \frac{t \log(1 + t)}{t^4 + 4} dt$$

Using L' Hospital's rule followed by Leibnitz rule, we have

$$= \lim_{x \to 0} \frac{\frac{x \log(1 + x)}{4 + x^4}}{3x^2}$$

$$= \lim_{x \to 0} \frac{\log(1 + x)}{3x} \cdot \frac{1}{4 + x^2}$$

$$= \frac{1}{12} \quad \left[\because \lim_{x \to 0} \frac{\log(1 + x)}{x} = 1\right]$$

**24.** (b) We have,

$$e^{-x} f(x) = 2t \int_0^x \sqrt{t^4 + 1}\, dt, \ x \in (-1, 1)$$

On differentiating w.r.t. $x$ using Leibnitz rule, we get

$$e^{-x}[f'(x) - f(x)] = \sqrt{x^4 + 1}$$

$$f'(x) = f(x) + \sqrt{x^4 + 1}\, e^x$$

$\because f^{-1}$ is inverse of $f$.

$\therefore \quad f^{-1}[f(x)] = x$

$\Rightarrow \quad (f^{-1})'[f(x)]\, f'(x) = 1$

$$\Rightarrow \quad (f^{-1})'[f(x)] = \frac{1}{f'(x)} = \frac{1}{f(x) + \sqrt{x^4 + 1}\, e^x}$$

At $x = 0$, $\quad f(x) = 2$

$$\therefore \quad (f^{-1})'(2) = \frac{1}{2 + 1} = \frac{1}{3}$$

**25.** (a) $\displaystyle\int_0^1 f(x)\, dx = \int_0^1 \frac{\sin x - x\cos x}{x^2} dx$

$$= \int_0^1 \frac{\sin x}{x^2} dx - \int_0^1 \frac{\cos x}{x} dx$$

$$= \left[\sin x\left(-\frac{1}{x}\right)\right]_0^1 + \int_0^1 \cos x\left(\frac{1}{x}\right) dx - \int_0^1 \frac{\cos x}{x} dx$$

$$= -\left[\frac{\sin x}{x}\right]_0^1$$

$$= 1 - \sin(1)$$

**26.** (b) $\displaystyle\lim_{t \to 0} \frac{\int_0^t f(x)\, dx}{t^2} = \lim_{t \to 0} \frac{\int_0^t \frac{\sin x - x\cos x}{x^2} dx}{t^2}$

Using L' Hospital's rule, we have

$$\lim_{t \to 0} \frac{\sin t - t\cos t}{t^2 \cdot 2t} = \lim_{t \to 0} \frac{\cos t\, (\tan t - t)}{2t^3}$$

$$= \frac{1}{2} \lim_{t \to 0} \frac{\cos^2 t - 1}{3t^2} = \frac{1}{6}$$

**27.** (d) $\displaystyle\lim_{x \to 0} \frac{\int_0^x \frac{t^2 dt}{(a + t^r)^{1/p}}}{bx - \sin x}$

Using L' Hospital's rule, we have

$$\lim_{x \to 0} \frac{\frac{x^2}{(a + x^r)^{1/p}}}{b - \cos x}$$

For existence of limit

$$\lim_{x \to 0} D^r \to 0$$

$\therefore \quad b - 1 = 0$

$\Rightarrow \quad b = 1$

**28.** (a) $\displaystyle I = \lim_{x \to 0} \frac{x^2}{(a + x^r)^{1/p}} \cdot \frac{x^2}{1 - \cos x} \cdot \frac{1}{x^2}$

$$= 2 \lim_{x \to 0} \frac{1}{(a + x^r)^{1/p}}$$

$$= \frac{2}{a^{1/p}}$$

If $p = 3$ and $I = 1$

$$\therefore \quad 1 = \frac{2}{a^{1/3}}$$

$\Rightarrow \quad a = 8$

**29.** (b) Again, $p = 2$ and $a = 9$, then

$$I = \frac{2}{(9)^{1/2}} = \frac{2}{3}$$

**30.** (4) Let $I = \int_0^{400\pi} \sqrt{1 - \cos 2x}\, dx$

$$= \int_0^{400\pi} \sqrt{2 \cdot \sin^2 x}\, dx$$

$$= 400\sqrt{2} \int_0^{400\pi} |\sin x|\, dx$$

$$= 400\sqrt{2} \int_0^{\pi} |\sin x|\, dx \qquad [\because \text{period of } |\sin x| \text{ is } \pi]$$

$$= 400\sqrt{2} \int_0^{\pi} \sin x dx$$

$$= -400\sqrt{2}\, [\cos x]_0^{\pi}$$

$$= -400\sqrt{2}\, (-1 - 1)$$

$$= 800\sqrt{2}$$

But $k \cdot 200\sqrt{2} = 800\sqrt{2}$

$\Rightarrow \qquad k = 4$

**31.** (5) Since, $x - [x]$ is periodic with period 1.

$\therefore e^{x - [x]}$ has period 1.

$$\therefore \qquad I = \int_0^{100} e^{x - [x]} dx$$

$$= \int_0^{100 \times 1} e^{x - [x]} dx$$

$$= 100 \int_0^1 e^{x - [x]} dx = 100 \int_0^1 e^{x - 0} dx$$

$$= 100\, [e^x]_0^1 = 100\, (e^1 - e^0)$$

$$= 100(e - 1)$$

But $\qquad k \cdot 20\, (e - 1) = 100\, (e - 1)$

$\Rightarrow \qquad k = 5$

**32.** (4) $f(x) = \begin{cases} x - [x], & \text{if } [x] \text{ is odd} \\ 1 + [x] - x, & \text{if } [x] \text{ is even} \end{cases}$

$f(x)$ and $\cos \pi x$ are both periodic with period 2 and both are even.

$$\therefore \qquad \int_{-10}^{10} f(x) \cos \pi x\, dx = 2 \int_0^{10} f(x) \cos \pi x\, dx$$

$$= 10 \int_0^2 f(x) \cos \pi x\, dx$$

Now, $\qquad \int_0^1 f(x) \cos \pi x\, dx = \int_0^1 (1 - x) \cos \pi x\, dx$

$$= -\int_0^1 4 \cos \pi x\, dx$$

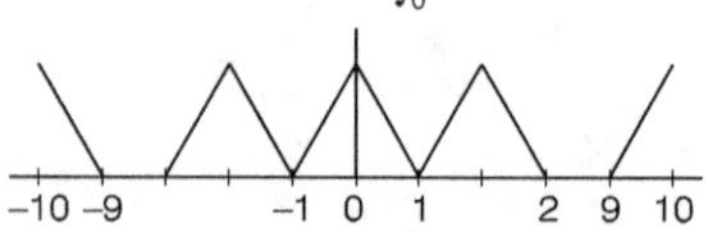


$$\int_1^2 f(x) \cos \pi x\, dx = \int_1^2 (x - 1) \cos \pi x\, dx$$

$$= -\int_0^1 4 \cos \pi u\, du$$

$$\therefore \qquad \int_{-10}^{10} f(x) \cos \pi x\, dx = -20 \int_0^1 4 \cos \pi u\, du$$

$$= \frac{40}{\pi^2}$$

$$\Rightarrow \qquad \frac{\pi^2}{10} \int_{-10}^{10} f(x) \cos \pi x\, dx = 4$$

**33.** (0) Let $I = \int_{-4}^{-5} e^{(x+5)^2} dx + 3 \int_{1/3}^{2/3} e^{9\left(x - \frac{2}{3}\right)^2} dx$

$$= \int_{-4}^{-5} e^{(x+5)^2} dx + 3 \int_{1/3}^{2/3} e^{(3x-2)^2} dx$$

$$= I_1 + I_2$$

Note that in both integrals $I_1$ and $I_2$, functions has same format, i.e. $e^t$.

Also, $e^{t^2}$ is non-integrable.

Now, in $I_1$, let $\qquad x + 5 = y$

and in $I_2$ $\qquad 3x - 2 = -t$

$$\Rightarrow \qquad I = \int_1^0 e^{y^2} dy + \int_1^0 e^t (-dt)$$

$$= 0$$

**34.** (2) Given,

$$y = \int_0^x f(t) \cdot \sin[2(x - t)]\, dt$$

$$= \int_0^x f(t)\, (\sin 2x \cos 2t - \sin 2t \cos 2x) dx$$

$$= \sin 2x \int_0^x f(t) \cos 2t\, dt - \cos 2x \int_0^x f(t) \sin 2t\, dt \qquad \ldots\text{(i)}$$

$$\Rightarrow \qquad \frac{dy}{dx} = 2\cos 2x \int_0^x f(t) \cos 2t\, dt + \sin 2x\, [f(x) \cos 2x]$$

$$+ 2\sin 2x \int_0^x f(t) \sin 2t\, dt - \cos 2x [f(x) \sin 2x]$$

$$= 2\cos 2x \int_0^x f(t) \cos 2t\, dt + 2\sin 2x \int_0^x f(t) \sin 2t\, dt \qquad \ldots\text{(ii)}$$

Again, differentiating Eq. (ii) w.r.t. $x$, we get

$$\Rightarrow \qquad \frac{d^2y}{dx^2} = -4\sin 2x \int_0^x f(t) \cos 2t\, dt + 2\cos 2x\, [f(x) \cos 2x]$$

$$+ 4\cos 2x \int_0^x f(t) \sin t\, dt + 2\sin 2x\, [f(x) \sin 2x]$$

$$= -4y + 2f(x)$$

$$\Rightarrow \qquad \frac{d^2y}{dx^2} + 4y = 2f(x)$$

$\therefore \qquad k = 2$

**35.** (2) Given, $\quad f(x) = \int_0^{x^2} \frac{t^2 - 5t + 4}{e^t + 2} dt$

$$\Rightarrow \qquad f'(x) = \frac{x^4 - 5x^2 + 4}{e^{x^2} + 2} \cdot 2x$$

$$f'(x) = 0$$

$$\Rightarrow \qquad x^5 - 5x^3 + 4x = 0$$

$$\Rightarrow \qquad x(x^4 - 5x^2 + 4) = 0$$

$$\Rightarrow \qquad x = 0 \text{ and } x^4 - 5x^2 + 4 = 0$$

$$\Rightarrow \qquad x = 0 \text{ and } x^4 - 4x^2 - x^2 + 4 = 0$$

$$\Rightarrow \qquad x^2\, (x^2 - 4) - 1\, (x^2 - 4) = 0$$

$$\Rightarrow \qquad (x^2 - 4)\, (x^2 - 1) = 0$$

$$\Rightarrow \qquad (x - 2)\, (x + 2)\, (x - 1)\, (x + 1) = 0$$

$$\Rightarrow \qquad x = \pm 2, \pm 1, 0$$

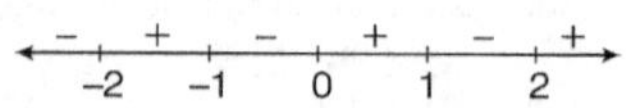


$\because f'(x)$ changes sign from positive to negative at $x = -1$ and $x = 1$.

$\therefore$ Points of local maxima is 2.

## DPP-5 **Estimation, Gamma Function and Limit as the Sum**

**1.** (i) Since, $1 + x^4 < (1 + x^2)^2$

$\therefore \quad \sqrt{1+x^4} < 1 + x^2$

$\Rightarrow \quad \dfrac{1}{\sqrt{1+x^4}} > \dfrac{1}{1+x^2}$ ...(i)

Also, $\quad \dfrac{1}{\sqrt{1+x^4}} \ge 1$ ...(ii)

Thus, $\quad \dfrac{1}{1+x^2} < \dfrac{1}{\sqrt{1+x^4}} \le 1$

$\Rightarrow \quad \int_0^1 \dfrac{dx}{1+x^2} < \int_0^1 \dfrac{dx}{\sqrt{1+x^4}} \le \int_0^1 1\,dx$

$\therefore \quad \dfrac{\pi}{4} < \int_0^1 \dfrac{dx}{\sqrt{1+x^4}} \le 1$ **Hence proved.**

(ii) For $0 \le I \le 1$, we have $e^0 \le e^{x^2} \le e^1$

$\therefore \quad e^0(1-0) \le \int_0^1 e^{x^2}dx \le e^1(1-0)$

$\Rightarrow \quad 1 \le \int_0^1 e^{x^2}\,dx \le e$ **Hence proved.**

(iii) For all $x \in [0, 1]$, we have

$1 - x^2 < 1 + x^6 < (1+x)^2$

$\Rightarrow \quad \sqrt{1-x^2} < \sqrt{1+x^6} < 1 + x$

$\Rightarrow \quad \dfrac{1}{1+x} < \dfrac{1}{\sqrt{1+x^6}} < \dfrac{1}{\sqrt{1-x^2}}$

$\Rightarrow \quad [\log(1+x)]_0^1 < \int_0^1 \dfrac{1}{\sqrt{1+x^6}}\,dx < [\sin^{-1}x]_0^1$

$\therefore \quad \log 2 < \int_0^1 \dfrac{dx}{\sqrt{1+x^6}} < \dfrac{\pi}{2}$ **Hence proved.**

(iv) Let $\quad f(x) = \dfrac{\sin x}{x}$

$\Rightarrow \quad f'(x) = \dfrac{x\cos x - \sin x}{x^2}$

$= \dfrac{\cos x(x - \tan x)}{x^2} < 0$ for $x \in \left(0, \dfrac{\pi}{2}\right)$

Thus, $f$ is decreasing function. Hence, maximum value of $f = f(0^+) = 1$

and minimum value of $f = f\left(\dfrac{\pi}{2}\right) = \dfrac{2}{\pi}$

Hence, $m\left(\dfrac{\pi}{2} - 0\right) < \int_0^{\pi/2} \dfrac{\sin x}{x}\,dx < M\left(\dfrac{\pi}{2} - 0\right)$

$\Rightarrow \quad \dfrac{2}{\pi} \times \dfrac{\pi}{2} < \int_0^{\pi/2} \dfrac{\sin x}{x}dx < \dfrac{\pi}{2}$

$1 < \int_0^{\pi/2} \dfrac{\sin x}{x}dx < \dfrac{\pi}{2}$ **Hence proved.**

**2.** (i) We have, $\Gamma(2) = \int_0^\infty e^{-x}x^{2-1}dx$

$= \lim_{b\to\infty} \int_0^b e^{-x}x\,dx$

$= \lim_{b\to\infty} [-xe^{-x} - e^{-x}]_0^b$

$= \lim_{b\to\infty} [(-be^{-b} - e^{-b}) - (0-1)]$

$= \lim_{b\to\infty} \left[\dfrac{-b}{e^b} - \dfrac{1}{e^b} + 1\right]$

$= \lim_{b\to\infty} \left[\dfrac{-(b+1)}{e^b} + 1\right]$

$= \lim_{b\to\infty} -\dfrac{1}{e^b} + 1$ [using L' Hospital's rule]

$= 1$

(ii) Let $I = \int_0^{\pi/2} \sin^4 x\cos^6 x\,dx$

$= \dfrac{\Gamma\left(\frac{4+1}{2}\right)\Gamma\left(\frac{6+1}{2}\right)}{2\Gamma\left(\frac{4+6+2}{2}\right)}$ $\left[\because \int_0^{\pi/2}\sin^m x\cos^n x\,dx = \dfrac{\Gamma\left(\frac{m+1}{2}\right)\Gamma\left(\frac{n+1}{2}\right)}{2\Gamma\left(\frac{m+n+2}{2}\right)}\right.$

$= \dfrac{\Gamma(5/2)\,\Gamma(7/2)}{2\Gamma(6)}$

$= \dfrac{\frac{3}{2}\cdot\frac{1}{2}\Gamma\left(\frac{1}{2}\right)\cdot\frac{5}{2}\cdot\frac{3}{2}\cdot\frac{1}{2}\cdot\Gamma\left(\frac{1}{2}\right)}{2\times5\times4\times3\times2\times1}$ $[\because \Gamma(n+1) = n\,\Gamma(n)]$

$= \dfrac{45\pi}{32\times240}$ $[\because \Gamma(1/2) = \sqrt{\pi}]$

$= \dfrac{3\pi}{512}$

(iii) Let $I = \int_0^1 x^5(1-x^2)^4dx$

Put $x^2 = t \Rightarrow 2x\,dx = dt \Rightarrow x\,dx = \dfrac{dt}{2}$

$\therefore \quad I = \dfrac{1}{2}\int_0^1 t^2(1-t)^4dt$

$= \dfrac{1}{2}\int_0^1 t^{3-1}(1-t)^{5-1}dt \left[\because B(m,n) = \int_0^1 x^{m-1}(1-x)^{n-1}dx\right]$

$= \dfrac{1}{2}B(3, 5)$

$= \dfrac{1}{2}\cdot\dfrac{\Gamma(3)\,\Gamma(5)}{\Gamma(8)}$ $\left[\because B(m,n) = \dfrac{\Gamma(m)\,\Gamma(n)}{\Gamma(m+n)}\right]$

$= \dfrac{1}{2}\cdot\dfrac{2\times1\times4\times3\times2\times1}{7\times6\times5\times4\times3\times2\times1} = \dfrac{1}{210}$

(iv) Let $I = \int_a^b (x-a)^m(x-b)^n\,dx$

Put $x = a\cos^2\theta + b\sin^2\theta$

$\Rightarrow \quad dx = 2(b-a)\sin\theta\cos\theta\,d\theta$

$\therefore I = \int_0^{\pi/2} (b-a)^{m+n}\cdot\sin^{2m}\theta\cdot\cos^{2n}\theta\cdot2(b-a)\sin\theta\cos\theta\,d\theta$

$= 2(b-a)^{m+n+1}\int_0^{\pi/2}\sin^{2m+1}\theta\cdot\cos^{2n+1}\theta\,d\theta$

$= 2(b-a)^{m+n+1}\cdot\dfrac{\Gamma\left(\frac{2m+1+1}{2}\right)\Gamma\left(\frac{2n+1+1}{2}\right)}{2\,\Gamma\left(\frac{2m+1+2n+1+2}{2}\right)}$

$= (b-a)^{m+n+1}\cdot\dfrac{\Gamma(m+1)\,\Gamma(n+1)}{\Gamma(m+n+2)}$

$= (b-a)^{m+n+1}\cdot\dfrac{(m)!\,(n)!}{(m+n+1)!}$

**3.** (i) Let $S = \lim_{n\to\infty}\left(\dfrac{n}{n^2+1^2} + \dfrac{n}{n^2+2^2} + \dfrac{n}{n^2+3^2} + \cdots + \dfrac{n}{n^2+n^2}\right)$

$= \lim_{n\to\infty}\sum_{r=1}^{n}\dfrac{n}{n^2+r^2}$

$= \dfrac{1}{n}\lim_{n\to\infty}\sum_{r=1}^{n}\dfrac{1}{1+\left(\frac{r}{n}\right)^2} = \int_0^1\dfrac{1}{1+x^2}dx$

$\left[\begin{array}{l}\lim_{n\to\infty}\left(\frac{r}{n}\right) \text{ for first term is } 0\\ \text{and } \lim_{n\to\infty}\left(\frac{r}{n}\right) \text{ for last term is } 1\\ \text{Replace } \sum_{r=0}^{n} \text{ by } \int \text{ and } \frac{r}{n} \text{ by } x \text{ and } \frac{1}{n} \text{ by } dx\end{array}\right]$

$= [\tan^{-1}x]_0^1 = \dfrac{\pi}{4}$

(ii) Let $S = \lim_{n\to\infty}\left\{\frac{1}{n}+\frac{1}{n+1}+\frac{1}{n+2}+\cdots+\frac{1}{n+2n}\right\}$

$$= \lim_{n\to\infty}\sum_{r=0}^{2n}\left(\frac{1}{n+r}\right)$$

$$= \lim_{n\to\infty}\left\{\frac{1}{n}\sum_{r=0}^{2n}\left(\frac{1}{1+\frac{r}{n}}\right)\right\}$$

Put $\frac{r}{n}=x$, then for lower limit,

$$\lim_{n\to\infty}\frac{r}{n}=\lim_{n\to\infty}\frac{0}{n}=0 \quad [\because r=0 \text{ for first term}]$$

for upper limit,

$$\lim_{n\to\infty}\frac{r}{n}=\lim_{n\to\infty}\frac{2n}{n}=2 \quad [\because r=2n \text{ for last term}]$$

$\therefore$ $$S=\lim_{n\to\infty}\frac{1}{n}\sum_{r=0}^{2n}\frac{1}{1+\frac{r}{n}}$$

$$=\int_0^2\frac{1}{1+x}dx=[\log(1+x)]_0^2$$

$$=\log 3-\log 1=\log 3$$

(iii) Let $S = \lim_{n\to\infty}\left(\frac{1^2}{1^3+n^3}+\frac{2^2}{2^3+n^3}+\frac{3^2}{3^3+n^3}+\cdots+\frac{n^2}{n^3+n^3}\right)$

$$=\lim_{n\to\infty}\sum_{r=1}^{n}\left(\frac{r^2}{r^3+n^3}\right)$$

$$=\lim_{n\to\infty}\left[\sum_{r=s}^{n}\frac{r^2}{n^3\left(1+\frac{r^3}{n^3}\right)}\right]$$

$$=\lim_{n\to\infty}\left[\sum_{r=1}^{n}\frac{1}{n}\cdot\frac{\frac{r^2}{n^2}}{\left(1+\frac{r^3}{n^3}\right)}\right]$$

$$=\lim_{n\to\infty}\left[\frac{1}{n}\sum_{r=1}^{n}\left(\frac{\left(\frac{r}{n}\right)^2}{1+\left(\frac{r}{n}\right)^3}\right)\right]$$

For lower limit,

$$\lim_{n\to\infty}\frac{r}{n}=\lim_{n\to\infty}\frac{1}{n}=0$$

For upper limit,

$$\lim_{n\to\infty}\frac{r}{n}=\lim_{n\to\infty}\frac{n}{n}=1$$

$\therefore$ $S=\int_0^1\left(\frac{x^2}{1+x^3}\right)dx$

Put $1+x^3=t \Rightarrow 3x^2\,dx=dt$

$\therefore$ $$S=\frac{1}{3}\int_1^2\frac{dt}{t}$$

$$=\frac{1}{3}[\log|t|]_1^2$$

$$=\frac{1}{3}[\log 2-\log 1]$$

$$=\frac{1}{3}\log 2$$

(iv) Let $L=\lim_{n\to\infty}\left(\frac{n!}{n}\right)^{1/n}$

$\Rightarrow$ $$\log L=\lim_{n\to\infty}\log\left(\frac{1\cdot2\cdot3\cdots n}{n^n}\right)^{1/n}$$

$\Rightarrow$ $$\log L=\lim_{n\to\infty}\log\left(\frac{1}{n}\cdot\frac{2}{n}\cdot\frac{3}{n}\cdots\frac{n}{n}\right)^{1/n}$$

$\Rightarrow$ $$\log L=\lim_{n\to\infty}\frac{1}{n}\sum_{r=1}^{n}\log\left(\frac{r}{n}\right)$$

$\Rightarrow$ $$\log L=\int_0^1\log x\,dx$$

$$=[x\log x-x]_0^1=-1$$

$\therefore$ $$L=e^{-1}=\frac{1}{e}$$

**4.** (c) Given, $I_1=\int_0^1 2^{x^2}dx$

$$I_2=\int_0^1 2^{x^3}dx$$

$$I_3=\int_1^2 2^{x^2}dx$$

$$I_4=\int_1^2 2^{x^3}dx$$

Since, $2^{x^3}<2^{x^2}$ for $0<x<1$ and $2^{x^3}>2^{x^2}$ for $x>1$.

$\therefore$ $$\int_0^1 2^{x^3}dx<\int_0^1 2^{x^2}dx$$

and $$\int_1^2 2^{x^3}dx>\int_1^2 2^{x^2}dx$$

$\Rightarrow$ $$I_2<I_1 \text{ and } I_4>I_3$$

**5.** (a) Given,

$$S_n=\frac{n}{(n+1)(n+2)}+\frac{n}{(n+2)(n+4)}+\frac{n}{(n+3)(n+6)}+\cdots+\frac{1}{6n^2}$$

$$=\sum_{r=1}^{n}\frac{n}{(n+r)(n+2r)}$$

$$=\sum_{r=1}^{n}\frac{n}{n^2\left(1+\frac{r}{n}\right)\left(1+\frac{2r}{n}\right)}$$

$$=\frac{1}{n}\sum_{r=1}^{n}\frac{1}{\left(1+\frac{r}{n}\right)\left(1+\frac{2r}{n}\right)}$$

Now, $$\lim_{n\to\infty}S_n=\lim_{n\to\infty}\left[\frac{1}{n}\sum_{r=1}^{n}\frac{1}{\left(1+\frac{r}{n}\right)\left(1+\frac{2r}{n}\right)}\right]$$

For upper limit,

$$\lim_{n\to\infty}\frac{r}{n}=\lim_{n\to\infty}\frac{n}{n}=1$$

For lower limit,

$$\lim_{n\to\infty}\frac{r}{n}=\lim_{n\to\infty}\frac{1}{n}=0$$

$$=\int_0^1\frac{1}{(1+x)(1+2x)}dx$$

Now, $$\frac{1}{(1+x)(1+2x)}=\frac{A}{1+x}+\frac{B}{1+2x}$$

$$1=A(1+2x)+B(1+x)$$

Put $$x=-1/2$$

$\Rightarrow$ $$1=\frac{B}{2}\Rightarrow B=2$$

Put $x=-1 \Rightarrow$ $$1=-A$$

$\Rightarrow$ $$A=-1$$

$\therefore$ $$\int_0^1\frac{dx}{(1+x)(1+2x)}=-\int_0^1\frac{dx}{1+x}+2\int_0^1\frac{dx}{1+2x}$$

$$=[-\log(1+x)]_0^1+[\log(1+2x)]_0^1$$

$$=-\log 2+(\log 3-0)$$

$$=-\log 2+\log 3=\log(3/2)$$

**6.** (b) We have, $\lim_{n\to\infty} \frac{1^p + 2^p + \ldots + n^p}{n^{p+1}}$

$$\Rightarrow \lim_{n\to\infty} \sum_{r=1}^{n} \frac{r^p}{n^{p+1}} = \lim_{n\to\infty} \sum_{r=1}^{n} \left(\frac{r}{n}\right)^p \cdot \frac{1}{n}$$

Put $\frac{r}{n} = x$

For lower limit, $\lim_{n\to\infty} \frac{r}{n} = \lim_{n\to\infty} \frac{1}{n} = 0$

For upper limit, $\lim_{n\to\infty} \frac{r}{n} = \lim_{n\to\infty} \frac{n}{n} = 1$

$$= \int_0^1 x^p dx = \left[\frac{x^{p+1}}{p+1}\right]_0^1$$

$$= \left[\frac{1}{p+1} - 0\right] = \frac{1}{p+1}$$

**7.** (b) We have, $\lim_{n\to\infty} \sum_{r=1}^{n} \frac{1}{n} e^{r/n}$

On replacing $\frac{r}{n} = x$

$\Rightarrow \frac{1}{n} = dx$

and $\lim_{n\to\infty} \sum_{r=1}^{n} = \int$

For lower limit,

$$\lim_{n\to\infty} \frac{r}{n} = \lim_{n\to\infty} \frac{1}{n} = 0$$

For upper limit,

$$\lim_{n\to\infty} \frac{r}{n} = \lim_{n\to\infty} \frac{n}{n} = 1$$

$\therefore \int_0^1 e^x dx = [e^x]_0^1 = (e - 1)$

**8.** (d) We have,

$$\lim_{n\to\infty} \left(\frac{1}{n^2} \sec^2 \frac{1}{n^2} + \frac{2}{n^2} \sec^2 \frac{4}{n^2} + \cdots + \frac{1}{n} \sec^2 1\right)$$

$$= \lim_{n\to\infty} \left(\frac{1}{n^2} \sec^2 \frac{1}{n^2} + \frac{2}{n^2} \sec^2 \frac{4}{n^2} + \cdots + \frac{1}{n} \sec^2 \frac{n^2}{n^2}\right)$$

$$= \lim_{n\to\infty} \sum_{r=1}^{n} \left(\frac{r}{n^2} \sec^2 \frac{r^2}{n^2}\right)$$

$$= \lim_{n\to\infty} \frac{1}{n} \sum_{r=1}^{n} \left[\left(\frac{r}{n}\right) \sec^2 \left(\frac{r}{n}\right)^2\right]$$

On replacing $\frac{r}{n} = x$

$\Rightarrow 1/n = dx$

$\lim_{n\to\infty} \sum_{r=1}^{n} = \int$

For lower limit, $\lim_{n\to\infty} \frac{r}{n} = \lim_{n\to\infty} \frac{1}{n} = 0$

For upper limit, $\lim_{n\to\infty} \frac{r}{n} = \lim_{n\to\infty} \frac{n}{n} = 1$

$\therefore \int_0^1 x \sec^2 x^2 dx$

Put $x^2 = t \Rightarrow 2x\,dx = dt$

$$\therefore \int_0^1 \sec^2 t \frac{dt}{2} = \frac{1}{2} \int_0^1 \sec^2 t\, dt$$

$$= \frac{1}{2} [\tan t]_0^1$$

$$= \frac{1}{2} [\tan(1) - \tan(0)]$$

$$= \frac{1}{2} \tan 1$$

**9.** (b) We have,

$\sin x < x$ for $0 < x < 1$

$\Rightarrow \frac{\sin x}{\sqrt{x}} < \frac{x}{\sqrt{x}}$ for $0 < x < 1$

$$\Rightarrow \int_0^1 \frac{\sin x}{\sqrt{x}} dx < \int_0^1 \sqrt{x} dx$$

$$\Rightarrow I < \frac{2}{3} [x^{3/2}]_0^1$$

$$\Rightarrow I < \frac{2}{3}$$

Also, $\cos x < 1$ for all $x \in (0,1)$

$\frac{\cos x}{\sqrt{x}} < \frac{1}{\sqrt{x}}$ for all $x \in (0,1)$

$$\Rightarrow \int_0^1 \frac{\cos x}{\sqrt{x}} dx < \int_0^1 \frac{1}{\sqrt{x}} dx$$

$$\Rightarrow J < 2 [x^{1/2}]_0^1$$

$$\therefore J < 2$$

**10.** (c) Given, $I(m, n) = \int_0^1 \underset{\text{I}}{t^m} \underset{\text{II}}{(1+t)^n} dt$

$$= \left[(1+t)^n \cdot \frac{t^{m+1}}{m+1}\right]_0^1 - \int_0^1 \left[n(1+t)^{n-1} \cdot \frac{t^{m+1}}{m+1}\right] dt$$

$$= \frac{2^n}{m+1} - \frac{n}{m+1} \cdot \int_0^1 (1+t)^{n-1} \cdot t^{m+1} dt$$

$$\therefore I(m, n) = \frac{2^n}{m+1} - \frac{n}{m+1} \cdot I(m+1, n-1)$$

**11.** (c) We have,

$$\lim_{n\to\infty} \sum_{r=1}^{4n} \frac{\sqrt{n}}{\sqrt{r}\,(3\sqrt{r} + 4\sqrt{n})^2}$$

$$\therefore T_r = \frac{1}{\sqrt{\frac{r}{n}}\, n \left(3\sqrt{\frac{r}{n}} + 4\right)^2}$$

$$S = \frac{1}{n} \sum_{r=1}^{4n} \frac{1}{\left(3\sqrt{\frac{r}{n}} + 4\right)^2 \sqrt{\frac{r}{n}}}$$

$$= \int_0^4 \frac{dx}{\sqrt{x}(3\sqrt{x} + 4)^2}$$

Put $3\sqrt{x} + 4 = t$

$$\Rightarrow \frac{3}{2} \cdot \frac{1}{\sqrt{x}} dx = dt$$

$$\therefore S = \frac{2}{3} \int_4^{10} \frac{dt}{t^2}$$

$$= \frac{2}{3} \left[\frac{1}{t}\right]_{10}^{4}$$

$$= \frac{2}{3} \left[\frac{1}{4} - \frac{1}{10}\right]$$

$$= \frac{2}{3} \cdot \frac{6}{40} = \frac{1}{10}$$

**12.** (c) To find $I = \left|\int_{10}^{19} \frac{\sin x}{1 + x^8} dx\right| \le \int_{10}^{19} \left|\frac{\sin x}{1 + x^8}\right| dx$ ...(i)

$$\left[\because \int_a^b f(x)dx = \int_a^b |f(x)|dx\right]$$

Since, $|\sin x| \le 1$ for $x \ge 10$

The inequality $\left|\frac{\sin x}{1 + x^8}\right| \le \frac{1}{|1 + x^8|}$ ...(ii)

Also, $10 \le x \le 19$

$\Rightarrow 1 + x^8 > 10^8$

$$\Rightarrow \quad \frac{1}{1+x^8} < \frac{1}{10^8}$$

or $$\frac{1}{|1+x^8|} < 10^{-8} \quad \ldots \text{(iii)}$$

From Eqs. (ii) and (iii), we get

$$\left|\frac{\sin x}{1+x^8}\right| < 10^{-8} \text{ is fulfilled.}$$

$$\left|\int_{10}^{19} \frac{\sin x}{1+x^8} dx\right| < \int_{10}^{19} 10^{-8}\, dx$$

$$\therefore \quad \int_{10}^{19} \frac{\sin x}{1+x^8} dx < (19-10)10^{-8} < 10^{-7}$$

**13.** (a) We have,

$$I = \int_0^1 \frac{x^4(1-x)^4}{1+x^2} dx$$

$$= \int_0^1 \frac{x^4}{1+x^2}(1-4x+6x^2-4x^3+x^4)\,dx$$

$$= \int_0^1 \frac{x^4-4x^5+6x^6-4x^7+x^8}{1+x^2} dx$$

$$= \int_0^1 \left(x^6-4x^5+5x^4-4x^2+4-\frac{4}{1+x^2}\right) dx$$

$$= \left[\frac{x^7}{7}-\frac{2x^6}{3}+x^5-\frac{4x^3}{3}+4x\right]_0^1 - \pi$$

$$= \frac{1}{7}-\frac{2}{3}+1-\frac{4}{3}+4-\pi = \frac{22}{7}-\pi$$

**14.** (a,d) Given, $S_n = \sum_{k=0}^{n} \frac{n}{n^2+kn+k^2}$

$$= \sum_{k=0}^{n} \frac{1}{n} \cdot \left(\frac{1}{1+\frac{k}{n}+\frac{k^2}{n^2}}\right) < \lim_{n\to\infty} \sum_{k=0}^{n} \frac{1}{n}\left(\frac{1}{1+\frac{k}{n}+\left(\frac{k}{n}\right)^2}\right)$$

$$= \int_0^1 \frac{1}{1+x+x^2} dx$$

$$= \left[\frac{2}{\sqrt{3}} \tan^{-1}\left\{\frac{2}{\sqrt{3}}\left(x+\frac{1}{2}\right)\right\}\right]_0^1$$

$$= \frac{2}{\sqrt{3}} \cdot \left(\frac{\pi}{3}-\frac{\pi}{6}\right) = \frac{\pi}{3\sqrt{3}}$$

i.e. $\quad S_n < \frac{\pi}{3\sqrt{3}}$

Similarly, $\quad T_n > \frac{\pi}{3\sqrt{3}}$

**15.** (b,d) Here,

$$\lim_{n\to\infty} \frac{(1^a+2^a+\cdots+n^a)}{(n+1)^{a-1}\{(na+1)+(na+2)+\cdots+(na+n)\}} = \frac{1}{60}$$

$$\Rightarrow \quad \lim_{n\to\infty} \frac{\sum_{r=1}^{n} r^a}{(n+1)^{a-1}\left[n^2a+\frac{n(n+1)}{2}\right]} = \frac{1}{60}$$

$$\Rightarrow \quad \lim_{n\to\infty} \frac{2\sum_{r=1}^{n}\left(\frac{r}{n}\right)^a}{\left(1+\frac{1}{n}\right)^{a-1}(2na+n+1)} = \frac{1}{60}$$

$$\Rightarrow \lim_{n\to\infty} \frac{1}{n}\left[2\sum_{r=1}^{n}\left(\frac{r}{n}\right)^a\right] \cdot \lim_{n\to\infty} \frac{1}{\left(1+\frac{1}{n}\right)^{a-1}\left(2a+1+\frac{1}{n}\right)} = \frac{1}{60}$$

$$\Rightarrow \quad 2\int_0^1 (x^a)dx \cdot \frac{1}{(2a+1)} = \frac{1}{60}$$

$$\Rightarrow \quad \frac{2\cdot[x^{a+1}]_0^1}{(2a+1)(a+1)} = \frac{1}{60}$$

$$\Rightarrow \quad \frac{2}{(2a+1)(a+1)} = \frac{1}{60}$$

$$\because \quad \frac{2}{(2a+1)(a+1)} = \frac{1}{60}$$

$$\Rightarrow \quad (2a+1)(a+1) = 120$$

$$\Rightarrow \quad 2a^2+3a+1-120 = 0$$

$$\Rightarrow \quad 2a^2+3a-119 = 0$$

$$\Rightarrow \quad (2a+17)(a-7) = 0$$

$$\therefore \quad a = 7, -\frac{17}{2}$$

**16.** (d) Given, $S_n = \int_0^{\pi/2} \frac{\sin(2n-1)x}{\sin x} dx$

Then, $\quad S_{n+1} = \int_0^{\pi/2} \frac{\sin(2n+1)x}{\sin x} dx$

$$\therefore S_{n+1} - S_n = \int_0^{\pi/2} \frac{\{\sin(2n+1)x - \sin(2n-1)x\}}{\sin x} dx$$

$$= 2\int_0^{\pi/2} \frac{\sin x \cdot \cos(2n)x}{\sin x} dx$$

$$= 2\int_0^{\pi/2} \cos(2n)x\,dx$$

$$= \frac{2\cdot[\sin(2nx)]_0^{\pi/2}}{2n} = \frac{1}{n}\cdot(0) = 0$$

**17.** (b) Given, $V_n = \int_0^{\pi/2} \frac{\sin^2 nx}{\sin^2 x} dx$

Then, $\quad V_{n+1} = \int_0^{\pi/2} \frac{\sin^2(n+1)x}{\sin x} dx$

$$\therefore \quad V_{n+1} - V_n = \int_0^{\pi/2} \frac{\sin^2(n+1)x - \sin^2 nx}{\sin^2 x} dx$$

$$= \int_0^{\pi/2} \frac{\sin(2n+1)x \cdot \sin x}{\sin^2 x} dx$$

$$\therefore \quad V_{n+1} - V_n = S_{n+1}$$

**Explanations for Q. Nos. (18-19)**

$$\because I_{n+2} - I_{n+1} = \int_0^{\pi} \left[\frac{1-\cos(n+2)x}{1-\cos x} - \frac{1-\cos(n+1)x}{1-\cos x}\right] dx$$

$$= \int_0^{\pi} \frac{\cos(n+1)x - \cos(n+2)x}{1-\cos x} dx$$

$$= \int_0^{\pi} \frac{2\sin\left(n+\frac{3}{2}\right)x \cdot \sin\frac{x}{2}}{2\sin^2\frac{x}{2}} dx$$

$$= \int_0^{\pi} \frac{\sin\left(n+\frac{3}{2}\right)x}{\sin\frac{x}{2}} dx \quad \ldots \text{(i)}$$

$$\therefore \quad I_{n+1} - I_n = \int_0^{\pi} \frac{\sin\left(n+\frac{1}{2}\right)x}{\sin\left(\frac{x}{2}\right)} dx \quad \ldots \text{(ii)}$$

From Eqs. (i) and (ii), we get

$$(I_{n+2} - I_{n+1}) - (I_{n+1} - I_n)$$

$$= \int_0^{\pi/2} \frac{\sin\left(n + \frac{3}{2}\right)x - \sin\left(n + \frac{1}{2}\right)x}{\sin\frac{x}{2}}\,dx$$

$$= 2\int_0^{\pi/2} \frac{\sin\frac{x}{2}\cdot\cos(n+1)x}{\sin\frac{x}{2}}\,dx$$

$$= \frac{2[\sin(n+1)x]_0^{\pi/2}}{n+1}$$

$$= \frac{2}{n+1}\cdot(0) = 0$$

$\therefore \quad I_{n+2} + I_n = 2I_{n+1}$

$\Rightarrow \quad I_{n+2} - I_{n+1} = I_{n+1} - I_n = I_n - I_{n+1} = \dots = I_1 - I_0$

$\therefore \quad I_{n+2} - I_{n+1} = \pi$

$$\left\{\because I_1 = \int_0^{\pi} \frac{1-\cos x}{1-\cos x}\,dx = \pi \text{ and } I_0 = 0\right\}$$

$\therefore \quad I_n = I_1 + (n-1)\pi$

$= \pi + (n-1)\pi$

$= n\pi$

Hence, $\int_0^{\pi} \frac{1-\cos nx}{1-\cos x}\cdot dx = n\pi$

Now, $\int_0^{\pi/2} \frac{\sin^2 nx}{\sin^2 x}\,dx = \int_0^{\pi/2} \frac{1-\cos 2nx}{1-\cos 2x}\,dx$

$$= \frac{1}{2}\int_0^{\pi} \frac{1-\cos n\theta}{1-\cos\theta}\,d\theta,$$

where $2x = \theta \Rightarrow 2dx = d\theta$

$$= \frac{1}{2}(n\pi)$$

**18.** (b) **19.** (b)

**Explanations for Q. Nos. (20-21)**

Given,

$$u_n = \int_0^{\pi/2} \cos^n x\cdot\cos nx\,dx$$

$$= \left[\frac{\cos^n x\sin nx}{n}\right]_0^{\pi/2} - \int_0^{\pi/2} n\cos^{n-1}x\cdot(-\sin x)\cdot\frac{\sin(nx)}{n}\,dx$$

$$= 0 + \int_0^{\pi/2} \cos^{n-1}x\cdot\sin x\cdot\sin(nx)\,dx$$

$$= \int_0^{\pi/2} \cos^{n-1}x\cdot[\cos(n-1)x - \cos x\cdot\cos nx]\,dx$$

$$[\because \cos(n-1)x = \cos(nx)\cos x + \sin(nx)\cdot\sin x]$$

$$= \int_0^{\pi/2} \cos^{n-1}x\cdot\cos(n-1)x - \int_0^{\pi/2}\cos^n x\cdot\cos nx\,dx$$

$\Rightarrow \quad u_n = u_{n-1} - u_n$

$\Rightarrow \quad 2u_n = u_{n-1}$

$\therefore \quad \frac{u_n}{u_{n-1}} = \frac{1}{2}$, for all $n = 2, 3, \dots$

Now, $u_n = u_1\left(\frac{1}{2}\right)^{n-1} \quad \left\{\because u_1 = \int_0^{\pi/2}\cos^2 x\,dx = \frac{\pi}{4}\right\}$

$\Rightarrow \quad u_n = \frac{\pi}{4}\cdot\left(\frac{1}{2}\right)^{n-1}$

$\therefore \quad u_n = \frac{\pi}{2^{n+1}}$

**20.** (c) **21.** (b)

**22.** (1) Let $I(\alpha) = \int_0^1 \frac{x^\alpha - 1}{\log x}\,dx$

Then, $\frac{d[I(\alpha)]}{d\alpha} = \int_0^1 \frac{x^\alpha\cdot\log x}{\log x}\,dx = \int_0^1 x^\alpha\,dx$

$$= \left[\frac{x^{\alpha+1}}{\alpha+1}\right]_0^1 = \frac{1}{\alpha+1}$$

$\Rightarrow \quad \frac{d[I(\alpha)]}{d\alpha} = \frac{1}{\alpha+1}$

On integrating w.r.t. $\alpha$, we get

$I(\alpha) = \log|\alpha + 1| + C \quad \dots$(i)

If $\alpha = 0$, then $I(0) = \int_0^1 \frac{x^0 - 1}{\log x}\,dx = 0$

On putting $\alpha = 0$ in Eq. (i), we get

$0 = \log(1) + C \Rightarrow C = 0$

$\therefore \quad I(\alpha) = \frac{1}{\alpha+1} \Rightarrow k = 1$

**23.** (3) We have,

$$\lim_{n\to\infty}\frac{\pi}{6n}\left[\sec^2\left(\frac{\pi}{6n}\right) + \sec^2\left(2\cdot\frac{\pi}{6n}\right) + \dots + \sec^2\left((n-1)\frac{\pi}{6n}\right) + \frac{4}{3}\right]$$

$$= \lim_{n\to\infty}\frac{\pi}{6}\cdot\frac{1}{n}\cdot\sum_{r=0}^{n}\sec^2\left(\frac{r\cdot\pi}{6n}\right)$$

$$= \frac{\pi}{6}\cdot\int_0^1 \sec^2\left(\frac{\pi}{6}x\right)dx = \frac{\pi}{6}\cdot\left[\frac{\tan\left(\frac{\pi x}{6}\right)}{\frac{\pi}{6}}\right]_0^1$$

$$= \tan\frac{\pi}{6} - 0 = \frac{1}{\sqrt{3}} = \frac{1}{\sqrt{k}}$$

$\therefore \quad k = 3$

**24.** (8) We have,

$$\lim_{\delta x\to 0}\delta x\,(x_1 + x_2 + x_3 + \dots + x_n)$$

$$= \lim_{n\to\infty}\frac{4}{n}\left[\frac{4}{n} + \frac{8}{n} + \frac{12}{n} + \dots + 4\cdot\frac{n}{n}\right] \quad \left[\because \delta x = \frac{4}{n}\right]$$

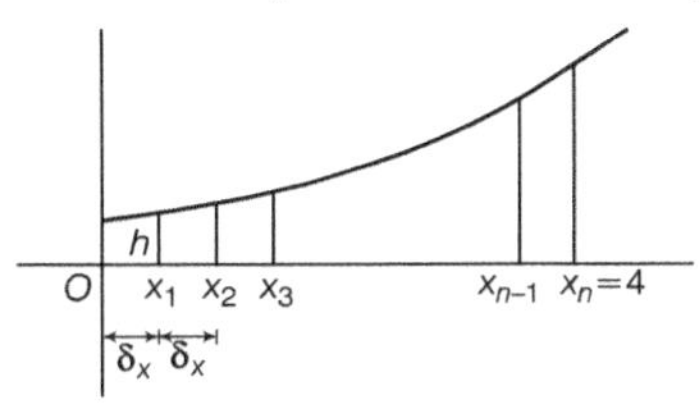


$$= \lim_{n\to\infty}\frac{16}{n^2}(1 + 2 + 3 + \dots + n)$$

$$= \lim_{n\to\infty}\frac{16}{n^2}\cdot\frac{n(n+1)}{2} = 8$$

**25.** (2) Let $I = \int_0^\infty x^{2n}\cdot x\cdot e^{-x^2}\,dx$

Put $x^2 = t \Rightarrow 2x\,dx = dt$

$\therefore \quad I = \frac{1}{2}\int_0^\infty t^n\cdot e^{-t}dt = \frac{1}{2}\int_0^\infty e^{-t}\cdot t^{(n+1)-1}dt$

$= \frac{1}{2}\Gamma(n+1) = \frac{n!}{2} \quad [\text{as } \Gamma(n+1) = n!, n \in N]$

$\therefore \quad k = 2$

## Revisal Problems (JEE Main)

**1.** (b) Let $I = \int_0^a \frac{g(x)}{f(x) + f(a-x)} dx$ ...(i)

$\Rightarrow \quad I = \int_0^a \frac{g(a-x)}{f(a-x) + f(x)} dx$ ...(ii)

On adding Eqs. (i) and (ii), we get

$\therefore \quad 2I = \int_0^a \left[ \frac{g(x)}{f(x) + f(a-x)} + \frac{g(a-x)}{f(a-x) + f(x)} \right] dx$

$\Rightarrow \quad I = 0$, if $g(x) = -g(a-x)$

**2.** (a) We have, $\int_1^2 \{a^2 + (4-4a)x + 4x^3\} dx \le 12$

$\Rightarrow \quad [a^2x + (2-2a)x^2 + x^4]_1^2 \le 12$

$\Rightarrow \quad (2a^2 + 8 - 8a + 16) - (a^2 + 2 - 2a + 1) \le 12$

$\Rightarrow \quad a^2 - 6a + 21 \le 12$

$\Rightarrow \quad a^2 - 6a + 9 \le 0$

$\Rightarrow \quad (a-3)^2 \le 0$

$\Rightarrow \quad a = 3$

**3.** (d) We have, $\int_{-a}^a \{f(x) + f(-x)\} \{g(x) + g(-x)\} dx$

Integral is an odd function.

$\therefore \int_{-a}^a \{f(x) + f(-x)\} \{g(x) + g(-x)\} dx = 0$

**4.** (d) We have, $\int_0^1 \frac{\sin t}{1+t} dt = \alpha$

Let $\quad I = \int_{4\pi-2}^{4\pi} \frac{\sin t/2}{4\pi + 2 - t} dt$

Put $t/2 = x \Rightarrow dt = 2dx$

Now, $\quad I = \int_{2\pi-1}^{2\pi} \frac{2\sin x}{4\pi + 2 - 2x} dx = \int_{2\pi-1}^{2\pi} \frac{\sin x}{2\pi + 1 - x} dx$

$\Rightarrow \quad I = -\int_{2\pi-1}^{2\pi} \frac{\sin(2\pi - x)}{2\pi + 1 - x} dx$

Let $\quad 2\pi - x = z \Rightarrow dx = -dz$

$\therefore \quad I = \int_1^0 \frac{\sin z}{z+1} dz$

$\Rightarrow \quad I = -\int_0^1 \frac{\sin t}{t+1} dt = -\alpha$

$\therefore \quad I = -\alpha$

**5.** (b) Let $I_1 = \int_{\sin^2 t}^{1+\cos^2 t} x f\{x(2-x)\} dx$

$\Rightarrow \quad I_1 = \int_{\sin^2 t}^{1+\cos^2 t} (1 + \cos^2 t + \sin^2 t - x) f(2-x)(x) dx$

$\Rightarrow \quad I_1 = \int_{\sin^2 t}^{1+\cos^2 t} (2-x) f\{x(2-x)\} dx$

$\Rightarrow \quad I_1 = \int_{\sin^2 t}^{1+\cos^2 t} 2f\{x(2-x)\} dx - \int_{\sin^2 t}^{1+\cos^2 t} x f\{x(2-x)\} dx$

$\Rightarrow \quad I_1 = 2I_2 - I_1$

$\Rightarrow \quad 2I_1 = 2I_2 \Rightarrow \frac{I_1}{I_2} = 1$

**6.** (b) Let $\quad I = \int_0^a \frac{dx}{1 + e^{f(x)}}$

$= \int_0^a \frac{dx}{1 + e^{f(a-x)}}$

$= \int_0^a \frac{dx}{1 + e^{-f(x)}} \quad [\because f(a-x) + f(x) = 0]$

$\Rightarrow \quad I = \int_0^a \frac{e^{f(x)}}{1 + e^{f(x)}} dx$

$\Rightarrow \quad 2I = \int_0^a dx \Rightarrow I = \frac{a}{2}$

**7.** (a) We have,

$\int_0^a f(x)\, dx = \lambda,$

$\int_0^a f(2a - x) dx = \mu$

Now, $\quad \int_0^{2a} f(x)\, dx = \int_0^a f(x)\, dx + \int_0^a f(2a-x)\, dx$

$\Rightarrow \quad \int_0^{2a} f(x) dx = \lambda + \mu$

**8.** (c) We have, $\frac{d\,F(x)}{dx} = \frac{e^{\sin x}}{x}$

$\Rightarrow \quad F(x) = \int \frac{e^{\sin x}}{x} dx$

Now, $\int_1^4 \frac{2e^{\sin x^2}}{x} dx = F(k) - F(1)$

Put $x^2 = t \Rightarrow 2x dx = dt$

When $x = 1$, then $\quad t = 1$

and when $x = 4$, then $\quad t = 16$

$\therefore \quad \int_1^{16} \frac{e^{\sin t}}{t} dt = [F(t)]_1^{16}$

$= F(16) - F(1)$

$\therefore \quad k = 16$

**9.** (b) Let $\quad I = \int_{-2}^2 \frac{\sin^2 x}{\left[\frac{x}{\pi}\right] + \frac{1}{2}} dx$

$= \int_{-2}^0 \frac{\sin^2 x}{-1 + \frac{1}{2}} dx + \int_0^2 \frac{\sin^2 x}{0 + \frac{1}{2}} dx$

$= \int_{-2}^0 -2\sin^2 x dx + \int_0^2 2\sin^2 x dx$

$= -\int_0^2 2\sin^2 x dx + \int_0^2 \sin^2 x dx$

$= 0$

**10.** (c) We have,

$I = \int_{-\pi/2}^{\pi/2} \left( \left[\frac{x}{\pi}\right] + \frac{1}{2} \right) dx$

Let $\quad f(x) = \left[\frac{x}{\pi}\right] + \frac{1}{2}$

$\Rightarrow \quad f(-x) = \left[\frac{-x}{\pi}\right] + \frac{1}{2}$

$\Rightarrow \quad f(-x) = -1 - \left[\frac{x}{\pi}\right] + \frac{1}{2} \quad (\because [-x] = 1 - [x], x \notin z)$

$\Rightarrow \quad f(x) = -\left( \left[\frac{x}{\pi}\right] + \frac{1}{2} \right) = -f(x)$

$\therefore f(x)$ is an odd function.

Hence, $I = \int_{-\pi/2}^{\pi/2} \left[\frac{x}{\pi}\right] + \frac{1}{2} dx = 0$

**11.** (d) Let $\quad I = \int_{-\pi/2}^{\pi/2} \frac{1}{e^{\sin x} + 1} dx$

$= \int_{-\pi/2}^{\pi/2} \frac{1}{e^{-\sin x} + 1} dx$

$\Rightarrow \quad I = \int_{-\pi/2}^{\pi/2} \frac{e^{\sin x}}{1 + e^{\sin x}} dx$

$\Rightarrow \quad 2I = \int_{-\pi/2}^{\pi/2} \frac{1 + e^{\sin x}}{1 + e^{\sin x}} dx = \int_{-\pi/2}^{\pi/2} dx$

$\Rightarrow \quad 2I = [x]_{-\pi/2}^{\pi/2} = \pi$

$\therefore \quad I = \pi/2$

**12.** (a) We have, $I_1 = \int_0^{\pi/2} \cos\theta f(\sin\theta + \cos^2\theta)d\theta$

and $I_2 = \int_0^{\pi/2} \sin 2\theta f(\sin\theta + \cos^2\theta)d\theta$

$\Rightarrow \quad I_1 - I_2 = \int_0^{\pi/2} (\cos\theta - \sin 2\theta) f(\sin\theta + \cos^2\theta)d\theta$

Put $\sin\theta + \cos^2\theta = x$

$\Rightarrow \quad (\cos\theta - \sin 2\theta)d\theta = dx$

$\therefore \quad I_1 - I_2 = \int_1^1 f(x)dx = 0$

$\Rightarrow \quad I_1 = I_2$

**13.** (d) Given, $I_n = \int_{1/n}^{\frac{an-1}{n}} \frac{\sqrt{x}}{\sqrt{a-x}+\sqrt{x}}dx$

$$I_n = \int_{1/n}^{\frac{an-1}{n}} \frac{\sqrt{\frac{an-1}{n}+\frac{1}{n}-x}}{\sqrt{a-\left(\frac{an-1}{n}+\frac{1}{n}-x\right)}+\sqrt{\frac{an-1}{n}+\frac{1}{n}-x}}dx$$

$$\Rightarrow \quad I_n = \int_{1/n}^{\frac{an-1}{n}} \frac{\sqrt{a-x}}{\sqrt{x}+\sqrt{a-x}}dx$$

$$\Rightarrow \quad 2I_n = \int_{1/n}^{\frac{an-1}{n}} \frac{\sqrt{x}+\sqrt{a-x}}{\sqrt{x}+\sqrt{a-x}}dx$$

$$\Rightarrow \quad 2I_n = \int_{1/n}^{\frac{an-1}{n}} dx = [x]_{1/n}^{\frac{an-1}{n}}$$

$$\Rightarrow \quad 2I_n = \frac{an-1}{n} - \frac{1}{n} = \frac{an-2}{n}$$

$$\Rightarrow \quad I_n = \frac{an-2}{2n}$$

Similarly, $I_{n-1} = \dfrac{a(n-1)-2}{2(n-1)}$

$$\Rightarrow \quad I_n - nI_{n-1} = \frac{an-2}{2n} - \frac{na(n-1)-2}{2(n-1)}$$

$$= \frac{a(n^2-n) - 2n + 2 - (n^2-n)an + 2n}{2n(n-1)}$$

$$= \frac{a(n^2-n)(n+1)+2}{2n(n-1)}$$

$$= \frac{na(n^2-1)+2}{2n(n-1)}$$

**14.** (d) Let $I_0 = \dfrac{\int_0^n [x]dx}{\int_0^n \{x\}dx}$

Let $I_1 = \int_0^n [x]\,dx$

$$= \int_0^{10} 0dx + \int_1^2 dx + \int_2^3 2dx + \cdots + \int_{n-1}^n (n-1)dx$$

$$= 0 + 1 + 2 + 3 + \cdots + (n-1) = \frac{n(n-1)}{2}$$

and $I_2 = \int_0^n \{x\}\,dx$

$$= \int_0^n x - [x]\,dx$$

$$= \int_0^n x dx - \int_0^n [x]dx$$

$$= \left[\frac{x^2}{2}\right]_0^n - \frac{n(n-1)}{2}$$

$$= \frac{n^2}{2} - \frac{n^2}{2} + \frac{n}{2} = \frac{n}{2}$$

$$\therefore \quad \frac{\int_0^n [x]dx}{\int_0^n \{x\}\,dx} = \frac{\frac{n(n-1)}{2}}{\frac{n}{2}} = (n-1)$$

**15.** (b) Given, $f(x) = \max(x + |x|, x - [x])$

The graph of $f(x)$ is

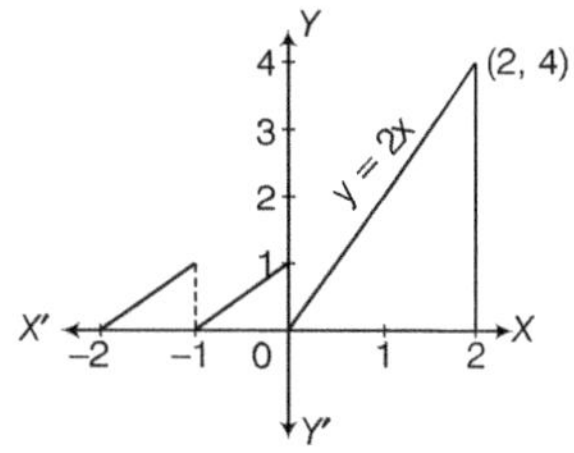


$$\therefore \quad \int_{-2}^2 f(x)\,dx = \frac{1}{2} \times 2 + \frac{1}{2} \times 2 \times 4$$

$$= 1 + 4 = 5$$

**16.** (b) We have, $\int_0^1 \frac{e^t}{t+1}dt = a$

Let $I_1 = \int_0^1 \frac{e^{1-t}}{2-t}dt = a$

and $I_2 = \int_{b-1}^b \frac{e^{-t}}{t-b-1}dt$

Put $t = b - 1 + x$

$\Rightarrow \quad dt = dx$

When $t = b - 1$, then $x = 0$

and when $t = b$, then $x = 1$

$$\therefore \quad I_2 = \int_0^1 \frac{e^{-(b-1+x)}}{b-1+x-b-1}dx$$

$$= \int_0^1 \frac{e^{-b} \cdot e^{1-x}}{x-2}dx$$

$$= -e^{-b}\int_0^1 \frac{e^{1-x}}{2-x}dx$$

$$= -e^{-b}\int_0^1 \frac{e^{1-t}}{2-t}dt$$

$$= -e^{-b} \cdot a = -ae^{-b}$$

**17.** (c) We know that $f(x)$ is periodic with period $T$.

Then, $\int_{a+T}^{b+T} f(x)\,dx = \int_a^b f(x)\,dx$

**18.** (a) We have,

$$\int_0^x f(t)\,dt = x + \int_x^1 t\,f(t)\,dt$$

$\Rightarrow \quad f(x) = 1 + 0 - x\,f(x)$ [using Leibnitz's rule]

$$\Rightarrow \quad f(x) = \frac{1}{1+x}$$

$$\therefore \quad f(1) = \frac{1}{2}$$

**19.** (b) Given, $I_n = \int_1^e (\log x)^n dx$

$$\Rightarrow \quad I_n = [(\log x)^n \cdot x]_1^e - \int_1^e \frac{n(\log x)^{n-1}}{x} \cdot x dx$$

$$\Rightarrow \quad I_n = e - nI_{n-1}$$

$\therefore I_n + nI_{n-1} = e$

**20.** (c) We have, $\left|\int_a^b f(x)\,dx\right| = \int_a^b |f(x)|\,dx$

We know that, if $\left|\int_a^b f(x)dx\right| = \int_a^b |f(x)|\,dx$, then

$f(x)$ have same sign.

$\therefore f(x)$ is above the $X$-axis or below the $X$-axis.

Hence, $f(x)$ has no root in $(a,b)$.

**21.** (d) We have, $\int_0^{100\pi} \sum_{r=1}^{10} \tan rx\, dx$

$= \int_0^{100\pi} \tan x dx + \int_0^{100\pi} \tan 2x\, dx + \cdots + \int_0^{100\pi} \tan 10x\, dx$

$= 100\int_0^{\pi} \tan x\, dx + 200\int_0^{\pi/2} \tan 2x dx + \cdots + 1000\int_0^{\pi/10} \tan 10x dx$

[$\because \tan x$ is periodic]

$= 100\left[\int_0^{\pi/2} \tan x dx + \int_0^{\pi/2} \tan(\pi - x) dx\right]$

$+ 200\left[\int_0^{\pi/4} \tan 2x\, dx + \int_0^{\pi/4} \tan 2\left(\frac{\pi}{2} - x\right) dx\right]$

$\cdots + 1000\left[\int_0^{\pi/20} \tan 10x\, dx + \int_0^{\pi/20} \tan 10\left(\frac{\pi}{10} - x\right) dx\right]$

$= 100\left[\int_0^{\pi/2} \tan x\, dx - \int_0^{\pi/2} \tan x\, dx\right]$

$+ 200\left[\int_0^{\pi/4} \tan 2x\, dx - \int_0^{\pi/4} \tan 2x dx\right]$

$+ 1000\left[\int_0^{\pi/20} \tan 10x dx - \int_0^{\pi/20} \tan 10x\, dx\right]$

$= 0$

**22.** (a) Given, $I_n = \int_0^{\pi/4} \tan^n x dx$

$\Rightarrow \quad I_n = \int_0^{\pi/4} \tan^{n-2} x (\sec^2 x - 1) dx$

$\Rightarrow \quad I_n = \int_0^{\pi/4} \tan^{n-2} x \sec^2 x\, dx - \int_0^{\pi/4} \tan^{n-2} x dx$

$\Rightarrow \quad I_n = \left[\frac{\tan^{n-1} x}{x - 1}\right]_0^{\pi/4} - I_{n-2}$

$\Rightarrow \quad I_n + I_{n-2} = \frac{1}{n-1}$

$\Rightarrow \quad n - 1 = \frac{1}{I_n + I_{n-2}}$

$\therefore \quad \frac{1}{I_2 + I_4} = 4 - 1 = 3$

$\Rightarrow \quad \frac{1}{I_3 + I_5} = 5 - 1 = 4$

$\Rightarrow \quad \frac{1}{I_4 + I_6} = 6 - 1 = 5$

Hence, $\frac{1}{I_2 + I_4}, \frac{1}{I_3 + I_5}, \frac{1}{I_4 + I_6}$ are in AP.

**23.** (b) Let $I = \int_1^{e^6} \left[\frac{\log x}{3}\right] dx$

$= \int_1^{e^3} 0 dx + \int_{e^3}^{e^6} dx = e^6 - e^3$

**24.** (a) $f(x)$ is an increasing function in $[a, b]$, then

$(b - a) f(a) < \int_a^b f(x)\, dx < (b - a) f(b)$

$\Rightarrow \quad f(a) < \frac{1}{b-a}\int_a^b f(x)\, dx < f(b)$

Hence, minimum value is $f(a)$ and maximum value is $f(b)$.

**25.** (c) We have, $f(x + y) = f(x) + f(y)$

Put $x = 0, y = 0$

$\Rightarrow \quad f(0 + 0) = f(0) + f(0)$

$\Rightarrow \quad f(0) = 2 f(0)$

$\Rightarrow \quad f(0) = 0$

$\therefore \quad \int_{-2}^{2} f(x)\, dx = \int_0^2 f(x)\, dx + \int_0^2 f(-x)\, dx$

$= \int_0^2 f(x - x)\, dx$

$= \int_0^2 f(0)\, dx = 0$

**26.** (d) We have, $f(x) = \begin{cases} |x - [x]|, & \text{when } [x] \text{ is odd} \\ |x - [x] - 1|, & \text{when } [x] \text{ is even} \end{cases}$

$f(x) = \begin{cases} |\{x\}|, & [x] \text{ is odd} \\ |\{x\} - 1|, & [x] \text{ is even} \end{cases}$

$f(x) = \begin{cases} \{x\}, & [x] \text{ is odd} \\ 1 - \{x\} & [x] \text{ is even} \end{cases}$

Graph of $f(x)$ is

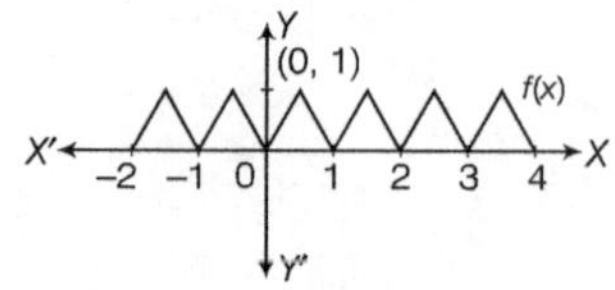


$\therefore \quad \int_{-2}^{4} f(x)\, dx = \frac{1}{2} \times 6 = 3$

**27.** (b) We have,

$\int_{-1}^{x} f(t)\, dt + f'''(3) \int_x^0 f(t)\, dt = \int_1^x t^3 dt - f'(1) \int_0^x t^2 dt + f''(2) \int_x^3 t dt$

On differentiating w. r. t. $x$, we have

$f(x) - f'''(3) f(x) = x^3 - x^2 f'(1) - f''(2)\, x$

$\Rightarrow \quad f'(x)\,[1 - f'''(3)] = 3x^2 - 2x f'(1) - f''(2)$

$\therefore \quad f'(4)\,[1 - f'''(3)] = 48 - 8f'(1) - f''(2)$

**28.** (a) Let $I = \int_{-1}^{1} [x^2 + \{x\}] dx$

$= \int_{-1}^{1} (x^2 + x - [x])\, dx \qquad (\because \{x\} = x - [x])$

$= \int_{-1}^{0} [x^2 + x + 1]\, dx + \int_0^1 [x^2 + x]\, dx$

[$\because [x] = -1, -1 \le x < 0, [x] = 0, 0 \le x \le 1$]

$= \int_0^1 [x^2 + x]\, dx \qquad [\because [x^2 + x + 1] = 0; -1 \le x < 0]$

$= \int_0^{\frac{\sqrt{5}-1}{2}} [x^2 + x]\, dx + \int_{\frac{\sqrt{5}-1}{2}}^{1} [x^2 + x]\, dx$

$= \int_0^{\frac{\sqrt{5}-1}{2}} 0 dx + \int_{\frac{\sqrt{5}-1}{2}}^{1} dx = 1 - \left(\frac{\sqrt{5}-1}{2}\right) = \frac{3 - \sqrt{5}}{2}$

**29.** (a) We have, $I = \int_0^1 \frac{dx}{1 + x^{\pi/2}}$

$0 \le x \le 1 \Rightarrow x^2 < x^{\pi/2} < x$

$\Rightarrow \quad 1 + x^2 < 1 + x^{\pi/2} < 1 + x$

$\Rightarrow \quad \frac{1}{1 + x} < \frac{1}{1 + x^{\pi/2}} < \frac{1}{1 + x^2}$

$\Rightarrow \quad \int_0^1 \frac{1}{1 + x} dx < \int_0^1 \frac{dx}{1 + x^{\pi/2}} < \int_0^1 \frac{1}{1 + x^2} dx$

$\Rightarrow \quad [\log(1 + x)]_0^1 < I < [\tan^{-1} x]_0^1$

$\Rightarrow \quad \log 2 < I < \pi/4$

**30.** (a) We have, $f(x) = \int_0^x \{f(t)\}^{-1} dt$

$\Rightarrow \quad f'(x) = \{f(x)\}^{-1}$

$\Rightarrow \quad f(x) f'(x) = 1$

$\Rightarrow \quad \int f(x) f'(x)\, dx = \int dx$

$\Rightarrow \quad \frac{\{f(x)\}^2}{2} = x + C$

$\Rightarrow \quad \frac{\{f(1)\}^2}{2} = 1 + C \; [\because f(1) = \int_0^1 \{f(t)\}^{-1} dt = \sqrt{2}]$

$\Rightarrow \quad \frac{2}{2} = 1 + C \Rightarrow C = 0$

$\therefore \quad \frac{\{f(x)\}^2}{2} = x \Rightarrow f(x) = \sqrt{2x}$

**31.** (c) Given, $I_n = \int_0^1 x^n e^{-x} dx$

$\Rightarrow \quad I_n = [x^n \int e^{-x} dx]_0^1 - \int_0^1 [nx^{n-1} \int e^{-x} dx]\, dx$

$\Rightarrow \quad I_n = [-x^n e^{-x}]_0^1 + nI_{n-1}$

$\therefore \quad I_n - nI_{n-1} = -\frac{1}{e}$

**32.** (b) We have, $\lim_{n \to \infty} \frac{1}{n} \cdot \sum_{r=1}^{2n} \frac{r}{\sqrt{n^2 + r^2}}$

$$= \lim_{n \to \infty} \frac{1}{n} \sum_{r=1}^{2n} \frac{r/n}{\sqrt{1 + \left(\frac{r}{n}\right)^2}} = \int_0^2 \frac{x}{\sqrt{1 + x^2}} dx$$

$$= [\sqrt{1 + x^2}]_0^2 = \sqrt{5} - 1 = \sqrt{5} - 1$$

**33.** (a) We have, $\phi(x) = \int_1^x e^{-t^2/2}(1 - t^2)\, dt$

$\Rightarrow \quad \phi'(x) = e^{-x^2/2}(1 - x^2)$

$\Rightarrow \quad \phi'(x) = 0$

$\therefore \quad e^{-x^2/2}(1 - x^2) = 0 \Rightarrow \quad x = \pm 1$

Hence, extremum of $\phi(x)$ is $\{-1, 1\}$.

**34.** (c) Given, $I_n = \int_0^{\pi/2} \frac{\cos^2 nx}{\sin x} dx$

$\Rightarrow \quad I_n - I_{n-1} = \int_0^{\pi/2} \left[\frac{\cos^2 nx - \cos^2(n-1)}{\sin x}\right] dx$

$\Rightarrow \quad I_n - I_{n-1} = \int_0^{\pi/2} \left[\frac{\sin^2(n-1)x - \sin^2 nx}{\sin x}\right] dx$

$\Rightarrow \quad I_n - I_{n-1} = \int_0^{\pi/2} \frac{\sin(n - 1 + n)\, x \sin(n - 1 - n)\, x}{\sin x} dx$

$\Rightarrow \quad I_n - I_{n-1} = \int_0^{\pi/2} -\sin(2n - 1)\, x dx$

$\Rightarrow \quad I_n - I_{n-1} = \left[\frac{\cos(2n - 1)x}{2n - 1}\right]_0^{\pi/2}$

$\Rightarrow \quad I_n - I_{n-1} = \frac{1}{2n - 1}$

$\therefore \quad I_2 - I_1 = \frac{1}{5}, I_3 - I_2 = \frac{1}{7}, I_4 - I_3 = \frac{1}{9}$

Hence, $\frac{1}{5}, \frac{1}{7}, \frac{1}{9}$ are in HP.

$\therefore \quad I_2 - I_1, I_3 - I_2, I_4 - I_3$ are in HP.

**35.** (b) We have, $\int_n^{n+1} f(x)\, dx = n^3$

$\therefore \quad \int_{-3}^3 f(x)\, dx = \int_{-3}^{-2} f(x)\, dx + \int_{-2}^{-1} f(x)\, dx + \int_{-1}^0 f(x)\, dx$

$\quad + \int_0^1 f(x)\, dx + \int_1^2 f(x)\, dx + \int_2^3 f(x)\, dx$

$= -27 - 8 - 1 + 0 + 1 + 8 = -27$

**36.** (b) We have, $I = \int_0^1 \frac{x}{8 + x^3} dx$

$0 \le x \le 1$

$\Rightarrow \quad 0 < x^3 < 1$

$\Rightarrow \quad 8 < 8 + x^3 < 9$

$\Rightarrow \quad \frac{1}{9} < \frac{1}{8 + x^3} < \frac{1}{8}$

$\Rightarrow \quad 0 < \frac{x}{8 + x^3} < \frac{1}{9}$

$\Rightarrow \quad \int_0^1 0\, dx < \int_0^1 \frac{x}{8 + x^3} dx < \int_0^1 \frac{1}{9} dx \quad \Rightarrow 0 < I < \frac{1}{9}$

$\therefore$ $I$ lies between $\left(0, \frac{1}{9}\right)$.

**37.** (c) Given, $\int_0^x t\, f(t)\, dt = \sin x - x\cos x - \frac{x^2}{2}$

$\Rightarrow \quad xf(x) = \cos x + x\sin x - \cos x - x$

$\Rightarrow \quad xf(x) = x(\sin x - 1)$

$\Rightarrow \quad f(x) = \sin x - 1$

$\Rightarrow \quad f\left(\frac{\pi}{6}\right) = \sin\frac{\pi}{6} - 1 = \frac{1}{2} - 1 = -\frac{1}{2}$

**38.** (a) Given, $\int_0^x \frac{bt\cos 4t - a\sin 4t}{t^2} dt = \frac{a\sin 4x}{x}$

$\Rightarrow \quad \frac{bx\cos 4x - a\sin 4x}{x^2} = \frac{4ax\cos 4x - a\sin 4x}{x^2}$

Equating the coefficient of $\sin 4x$ and $\cos 4x$, we get

$b = 4a$

$\therefore \quad a = \frac{1}{4}$ and $b = 1$

**39.** (a) Given, $\int_0^{t^2} xf(x)\, dx = \frac{2}{5} t^5$

$t^2 f(t^2) 2t = \frac{2 \times 5t^4}{5}$

$\Rightarrow \quad 2t^3 f(t^2) = 2t^4 \quad \Rightarrow \quad f(t^2) = t$

$\Rightarrow \quad f\left\{\left(\frac{2}{5}\right)^2\right\} = \frac{2}{5}$

$\therefore \quad f\left(\frac{4}{25}\right) = \frac{2}{5}$

**40.** (d) Let $I = \int_{-2}^2 |\,[x]\,|\, dx$

$= \int_{-2}^{-1} 2\, dx + \int_{-1}^0 1\, dx + \int_0^1 0\, dx + \int_1^2 1\, dx$

$= 2[x]_{-2}^{-1} + [x]_{-1}^0 + 0 + [x]_1^2$

$= 2(-1 + 2) + (0 + 1) + (2 - 1)$

$= 2 + 1 + 1 = 4$

**41.** (c) We have, $\phi(x) = \int_{5\pi/4}^x (3\sin t + 4\cos t)\, dt$

$\Rightarrow \quad \phi'(x) = 3\sin x + 4\cos x$

$\phi'(x) < 0, \forall\ x \in \left[\frac{5\pi}{4}, \frac{4\pi}{3}\right]$

$\Rightarrow \phi$ is decreasing function on $\left[\frac{5\pi}{4}, \frac{4\pi}{3}\right]$.

$\therefore \phi\left(\frac{4\pi}{3}\right)$ is minimum value of $\phi(x)$.

$\therefore \phi\left(\frac{4\pi}{3}\right) = \int_{5\pi/4}^{4\pi/3} (3\sin t + 4\cos t)\, dt$

$\Rightarrow \phi\left(\frac{4\pi}{3}\right) = [-3\cos t + 4\sin t]_{5\pi/4}^{4\pi/3} = \frac{3}{2} - 2\sqrt{3} + \frac{1}{\sqrt{2}}$

$\therefore$ Minimum value of $\phi(x)$ at $4\pi/3$ is

$-2\sqrt{3} + \frac{3}{2} + \frac{1}{\sqrt{2}}$

**42.** (d) Given, $f\left(\frac{1}{x}\right) + x^2 f(x) = 0$

$\Rightarrow \quad f(x) = \frac{-1}{x^2} f\left(\frac{1}{x}\right)$

Now, $\int_{\sin\theta}^{\operatorname{cosec}\theta} f(x)\, dx = \int_{\sin\theta}^{\operatorname{cosec}\theta} \frac{-1}{x^2} f\left(\frac{1}{x}\right) dx$

$= \int_{\operatorname{cosec}\theta}^{\sin\theta} f(t)\, dt$

$\Rightarrow \quad \int_{\sin\theta}^{\operatorname{cosec}\theta} f(x)\, dx = -\int_{\sin\theta}^{\operatorname{cosec}\theta} f(x)\, dx$

$\Rightarrow \quad 2\int_{\sin\theta}^{\operatorname{cosec}\theta} f(x)\, dx = 0$

$\Rightarrow \quad \int_{\sin\theta}^{\operatorname{cosec}\theta} f(x)\, dx = 0$

**43.** (c) We have, $f\left(\frac{1}{x}\right) + x^2 f(x) = 0$

Now, $I = \int_{1/x}^{x} f(t) dt$

$\Rightarrow \quad I = \int_{1/x}^{x} \frac{-1}{t^2} f\left(\frac{1}{t}\right) dt$

Put $\frac{1}{t} = y \Rightarrow \frac{-1}{t^2} dt = dy$

$\therefore \quad I = \int_{x}^{1/x} f(y) dy$

$\Rightarrow \quad I = -\int_{1/x}^{x} f(t) dt$

$\Rightarrow \quad I = -I$

$\Rightarrow \quad 2I = 0 \Rightarrow I = 0$

**44.** (a) Given, $f(x) = \int_1^x (t^2 - 3t + 2) dt$

$\Rightarrow \quad f'(x) = x^3 - 3x^2 + 2x$

$\Rightarrow \quad f'(x) = x(x-2)(x-3)$

$f(x)$ is increasing in $x \in (1, 2)$ and decreasing in $(2, 3)$.

$\therefore$ Range of $f(x)$ in $x \in [1, 3]$ is

$f(2) = \int_1^2 t(t^2 - 3t + 2) dt$ and $f(3) = \int_1^3 t(t^2 - 3t + 2) dt$

$\Rightarrow \quad f(2) = \left[\frac{t^4}{4} - t^3 + t^2\right]_1^2$ and $f(3) = \left[\frac{t^4}{4} - t^3 + t^2\right]_1^3$

$\Rightarrow \quad f(2) = (4 - 8 + 4) - \left(\frac{1}{4} - 1 + 1\right)$

and $f(3) = \left(\frac{81}{4} - 27 + 9\right) - \left(\frac{1}{4} - 1 + 1\right)$

$\Rightarrow \quad f(2) = \frac{-1}{4}$ and $f(3) = 2$

$\therefore$ Range of $f(x)$ is $[-1/4, 2]$.

**45.** (c) Given, $f(x) = \frac{e^x + 1}{e^x - 1}$

and $\int_0^1 x^3 \frac{e^x + 1}{e^x - 1} dx = \alpha$

Let $F(x) = x^3 f(x)$

$\Rightarrow \quad F(x) = x^3\left(\frac{e^x + 1}{e^x - 1}\right)$

$\Rightarrow \quad F(-x) = -x^3\left(\frac{e^{-x} + 1}{e^{-x} - 1}\right)$

$= -x^3\left(\frac{1 + e^x}{1 - e^x}\right)$

$= x^3\left(\frac{e^x + 1}{e^x - 1}\right) = F(x)$

$\therefore F(x)$ is an even function.

Hence, $\int_{-1}^{1} t^3 f(t) dt = 2\int_0^1 t^3 f(t) dt$

$= 2\int_0^1 x^3\left(\frac{e^x + 1}{e^x - 1}\right) dx = 2\alpha$

**46.** (d) Let $I = \int_{\pi/6}^{\pi/3} \frac{dx}{1 + \sqrt{\tan x}}$

$\Rightarrow \quad I = \int_{\pi/6}^{\pi/3} \frac{\sqrt{\cos x}}{\sqrt{\cos x} + \sqrt{\sin x}} dx$

$\Rightarrow \quad I = \int_{\pi/6}^{\pi/3} \frac{\sqrt{\cos\left(\frac{\pi}{3} + \frac{\pi}{6} - x\right)}}{\sqrt{\cos\left(\frac{\pi}{3} + \frac{\pi}{6} - x\right)} + \sqrt{\sin\left(\frac{\pi}{3} + \frac{\pi}{6} - x\right)}} dx$

$\Rightarrow \quad I = \int_{\pi/6}^{\pi/3} \frac{\sqrt{\sin x}}{\sqrt{\sin x} + \sqrt{\cos x}} dx$

$\Rightarrow \quad 2I = \int_{\pi/6}^{\pi/3} \frac{\sqrt{\cos x} + \sqrt{\sin x}}{\sqrt{\sin x} + \sqrt{\cos x}} dx$

$\Rightarrow \quad 2I = \int_{\pi/6}^{\pi/3} dx = [x]_{\pi/6}^{\pi/3} = \frac{\pi}{3} - \frac{\pi}{6} = \frac{\pi}{6}$

$\Rightarrow \quad I = \frac{\pi}{12}$

Hence, Statement I is incorrect and Statement II is correct.

**47.** (c) Let $\int_0^1 f(t) dt = k$ ...(i)

So, $f(x) = xk + 1$

Now, putting $f(t) = kt + 1$ in Eq. (i)

$\int_0^1 (kt + 1) dt = k$

$\Rightarrow \quad \frac{k}{2} + 1 = k \Rightarrow k = 2$

$\therefore \quad f(x) = 2x + 1$

Also, $\int_0^3 f(x) dx = \int_0^3 (2x + 1) dx$

$= [x^2 + x]_0^3 = 12$

**48.** (d) We have, $I = \int_{-\pi/4}^{\pi/4} \frac{dx}{1 - \sin x}$

$= \int_{-\pi/4}^{\pi/4} \frac{1 + \sin x}{\cos^2 x} dx$

$= \int_{-\pi/4}^{\pi/4} (\sec^2 x + \sec x \tan x) dx$

$= [\tan x + \sec x]_{-\pi/4}^{\pi/4}$

$= (1 + \sqrt{2}) - (-1 + \sqrt{2}) = 2$

Hence, Statement I is incorrect and Statement II is correct.

**49.** (c) We have, $f(x) = \int_0^x \sqrt{1 + t^2}\, dt$

$\Rightarrow \quad f'(x) = \sqrt{1 + x^2}$

$\because f'(x)$ is an even function.

$\therefore f(x)$ is an odd function.

e.g. $\quad f'(x) = x \cdot \sin x,$

then $\quad f(x) = -x\cos x + \sin x + C$

$f(-x) = -\sin x + x\cos x + C$

$\therefore \quad f(x) + f(-x) = \text{Constant}$ [need not to be 0]

**50.** (a) We have,

$f(x) = \sin^3 x$

and $\quad p(x) = x^2 + x + 1$

Now, $\quad I = \int_0^{2\pi} p(x) f''(x) dx$

$\Rightarrow \quad I = [p(x) f'(x)]_0^{2\pi} - \int p'(x) f'(x) dx$

$\Rightarrow \quad I = [(x^2 + x + 1)(3\sin x^2 \cos x)]_0^{2\pi}$

$- [p'(x) f(x) - \int -p''f(x) f(x) dx]_0^{2\pi}$

$\Rightarrow \quad I = 0 - [(2x + 1)\sin^3 x]_0^{2\pi} + [(p''(x) \int f(x) dx]_0^{2\pi} - \int_0^{2\pi} f(x) dx$

$\Rightarrow \quad I = 0 + 2x \times 0 - 0$

$\Rightarrow \quad I = 0$

$\therefore \quad \int_0^{2\pi} \sin^3 x\, dx = \int_0^{\pi} \sin^3 x dx + \int_0^{\pi} \sin^3 (2\pi - x) dx$

$= \int_0^{\pi} \sin^3 x dx - \int_0^{\pi} \sin^3 x dx = 0$

Hence, Statement I is correct and Statement II is correct explanation of Statement I.

## Revisal Problems (JEE Advanced)

**1.** (d) We have, $\int_a^0 (3^{-2x} - 2\cdot 3^{-x})dx \ge 0$

$$\left[-\frac{1}{2}\frac{3^{-2x}}{\log 3} + 2\frac{3^{-x}}{\log 3}\right]_a^0 \ge 0$$

$$\Rightarrow \left(-\frac{1}{2\log 3} + \frac{2}{\log 3}\right) - \left(-\frac{1}{2}\frac{3^{-2a}}{\log 3} + 2\frac{3^{-a}}{\log 3}\right) \ge 0$$

$$\Rightarrow \frac{3}{2\log 3} - \frac{1}{\log 3}\left(2\cdot 3^{-a} - \frac{1}{2}\cdot 3^{-2a}\right) \ge 0$$

$$\Rightarrow \frac{3}{2\log 3} - \frac{1}{2\log 3}3^{-2a}(4\cdot 3^a - 1) \ge 0$$

$$\Rightarrow 3 - 3^{-2a}(4\cdot 3^a - 1) \ge 0$$

$$\Rightarrow 3^{-2a}(4\cdot 3^a - 1) \le 3$$

or $4\cdot 3^a - 1 \le 3\cdot 3^{2a}$ [put $3^a = t$]

$$\Rightarrow 3t^2 - 4t + 1 \ge 0$$

$$\Rightarrow (t-1)(3t-1) \ge 0 \Rightarrow t \le \frac{1}{3} \text{ or } t \ge 1$$

$$\Rightarrow 3^a \le 3^{-1} \text{ or } 3^a \ge 1 = 3^0$$

$$a \le -1 \text{ or } a \ge 0$$

$$\Rightarrow -\infty < a \le -1 \text{ or } 0 \le a < \infty$$

$$\therefore a \in (-\infty, -1] \cup [0, \infty)$$

**2.** (d) We have, $\sqrt{5x - 6 - x^2} + \frac{\pi}{2}\int_0^x dz > x\int_0^\pi \sin^2 x dx$

$$\Rightarrow \sqrt{5x - 6 - x^2} + \frac{\pi}{2}x > \frac{\pi}{2}x$$

$$\Rightarrow \sqrt{5x - 6 - x^2} > 0$$

$$\Rightarrow 5x - 6 - x^2 > 0$$

$$\Rightarrow x^2 - 5x + 6 < 0$$

$$\Rightarrow (x-2)(x-3) < 0 \Rightarrow 2 < x < 3$$

$$\therefore x \in (2, 3)$$

**3.** (a) We have, $\int_{\pi/4}^{5\pi/4}(|\cos t|\sin t + |\sin t|\cos t)dt$

Break the limit $\left[\frac{\pi}{4}, \frac{5\pi}{4}\right]$ into $\left[\frac{\pi}{4}, \frac{\pi}{2}\right], \left[\frac{\pi}{2}, \pi\right]$ and $\left[\pi, \frac{5\pi}{4}\right]$, we get

$$\int_{\pi/4}^{\pi/2}(\cos t \sin t + \sin t \cos t)dt + \int_{\pi/2}^{\pi}(-\cos t \sin t + \sin t \cos t)dt + \int_{\pi}^{5\pi/4}(-\cos t \sin t - \sin t \cos t)dt$$

$$= \int_{\pi/4}^{\pi/2} 2\sin t \cos t\, dt + \int_{\pi}^{5\pi/4} -2\sin t \cos t\, dt$$

$$= \int_{\pi/4}^{\pi/2}\sin 2t - \int_{\pi}^{5\pi/4} -\sin 2t\, dt$$

$$= \left[-\frac{1}{2}\cos 2t\right]_{\pi/4}^{\pi/2} + \left[\frac{1}{2}\cos 2t\right]_{\pi}^{5\pi/4}$$

$$= -\frac{1}{2}[-1-0] + \frac{1}{2}[0-1] = \frac{1}{2} - \frac{1}{2} = 0$$

**4.** (a) Let $I = \int_{\pi/6}^{\pi/4}\frac{1 + \cot x}{e^x \sin x}dx$

$$\Rightarrow I = \int_{\pi/6}^{\pi/4} e^{-x}[\text{cosec } x + \text{cosec } x \cdot \cot x]dx$$

$$\Rightarrow I = -\int_{-\pi/6}^{-\pi/4} e^x[-\text{cosec } x + \text{cosec } x \cdot \cot x]dx$$

$$\because I = [e^x(-\text{cosec } x)]_{-\pi/4}^{-\pi/6}$$

$$= -\left[e^{-\pi/6}\text{cosec}\left(-\frac{\pi}{6}\right) - e^{-\pi/4}\text{cosec}\left(-\frac{\pi}{4}\right)\right]$$

$$= -[-2e^{-\pi/6} + e^{-\pi/4}\sqrt{2}] = 2e^{-\pi/6} - \sqrt{2}e^{-\pi/4}$$

But it is $ae^{-\pi/6} + be^{-\pi/4}$

On comparing, we get $a = 2$ and $b = -\sqrt{2}$

Hence, $a + b = 2 - \sqrt{2}$

**5.** (c) Given, $U_n = \int_0^1 x^n(2-x)^n dx$

and $V_n = \int_0^1 x^n(1-x)^n dx$

In $U_n$, put $x = 2t \Rightarrow dx = 2dt$

$$\therefore U_n = \int_0^{1/2} 2^n \cdot t^n \cdot (2-2t)^n \cdot 2\, dt \quad \text{...(i)}$$

Now, $V_n = 2\int_0^{1/2} x^n \cdot (1-x)^n dx$ ...(ii)

From Eqs. (i) and (ii), we get

$$U_n = 2^{2n}\cdot V_n$$

**6.** (b) Given, $S(x) = \int_{x^2}^{x^3}\log t\, dt$

$$S'(x) = \log x^3 \cdot 3x^2 - \log x^2 \cdot 2x$$

and $H(x) = \frac{S'(x)}{x} = 3x\log x^3 - 2\log x^2$

$$\therefore H(x) = 9x\log x - 4\log x = \log x(9x - 4)$$

Clearly, its domain $x > 0$ and $H(x)$ is derivable and continuous in the given interval.

**7.** (a) We have, $f(x) = \frac{\sin x}{x}$

Let $I = \int_0^{\pi/2} f(x)\cdot f\left(\frac{\pi}{2} - x\right)dx$

$$= \int_0^{\pi/2}\frac{\sin x}{x}\cdot\frac{\cos x}{\frac{\pi}{2} - x}dx = \int_0^{\pi/2}\frac{\sin 2x}{x(\pi - 2x)}dx$$

Put $2x = t$

$$\Rightarrow dx = \frac{dt}{2}$$

$$\therefore I = \frac{1}{2}\int_0^\pi \frac{\sin t}{\frac{t}{2}(\pi - t)}dt = \int_0^\pi \frac{\sin t}{t(\pi - t)}dt$$

$$= \frac{1}{\pi}\int_0^\pi\left[\frac{\sin t}{t} + \frac{\sin t}{\pi - t}\right]dt$$

$$= \frac{1}{\pi}\left[\int_0^\pi \frac{\sin t}{t}dt + \int_0^\pi \frac{\sin t}{\pi - t}dt\right]$$

$$= \frac{1}{\pi}\left[\int_0^\pi \frac{\sin t}{t}dt + \int_\pi^0 \frac{\sin(\pi - \theta)}{\theta}(-d\theta)\right]$$

$\left[\begin{array}{l}\text{put } \pi - t = \theta \\ \Rightarrow dt = -d\theta\end{array}\right]$

$$= \frac{1}{\pi}\left[\int_0^\pi \frac{\sin t}{t}dt + \int_0^\pi \frac{\sin\theta}{\theta}d\theta\right]$$

$$= \frac{2}{\pi}\int_0^\pi f(x)dx$$

**8.** (a) Given, $\int_0^{f(x)} t^2 dt = x\cos \pi x$

On differentiating w.r.t. $x$ both sides, we get

$$[f(x)]^2 f'(x) = \pi x \sin \pi x + \cos \pi x$$

On putting $x = 9$, we get

$$[f(9)]^2 f'(9) = -1 \quad \text{...(i)}$$

Also, $\left[\frac{t^3}{3}\right]_0^{f(x)} = x\cos \pi x$

$$\Rightarrow \frac{[f(x)]^3}{3} - 0 = x\cos \pi x$$

$$\Rightarrow [f(9)]^3 = 3\cdot 9\cos 9\pi = -27$$

$\therefore f(9) = -3$, putting in Eq. (i), we get

$$(-3)^2 f'(9) = -1$$

$$\Rightarrow f'(9) = -\frac{1}{9}$$

**9.** (a) Let $f(a)=\int_a^{a^2}\frac{dx}{x+\sqrt{x}}$

$\therefore \quad f'(a)=\frac{1}{a^2+a}\times 2a-\frac{1}{a+\sqrt{a}}$

$\Rightarrow \quad \frac{2}{a+1}-\frac{1}{a+\sqrt{a}}=0$

$\Rightarrow \quad 2a+2\sqrt{a}-a-1=0$

$\Rightarrow \quad 2\sqrt{a}+a-1=0$

$\Rightarrow \quad 2\sqrt{a}=-a+1$

$\Rightarrow \quad 4a=a^2+1-2a$

$\Rightarrow \quad a^2-2a-4a+1=0$

$\Rightarrow \quad a^2-6a+1=0$

$\therefore \quad a=\frac{6\pm\sqrt{36-4}}{2}=\frac{6\pm 4\sqrt{2}}{2}$

$\Rightarrow \quad a=3\pm 2\sqrt{2}=(\sqrt{2}\pm 1)^2$

and $\quad f''(a)=\frac{2}{2\sqrt{a}}+1$

It is smallest, when $\quad \sqrt{a}=\tan\pi/8$

$\therefore \quad a=\tan^2\pi/8$

**10.** (d) $f(x)=1+x+\int_1^x(\log^2 t+2\log t)dt$

$\Rightarrow \quad f'(x)=1+\log^2 x+2\log x=0$

$\Rightarrow \quad (\log x+1)^2=0$

$\Rightarrow \quad \log x=-1$

$\Rightarrow \quad x=e^{-1}$

$\therefore \quad f(e^{-1})=1+e^{-1}+\int_0^{e^{-1}}[1+2(-1)]dt$

$=1+e^{-1}-\int_0^{e^{-1}}-1dt$

$=1+e^{-1}+[t]_0^{e^{-1}}$

$=1+e^{-1}+e^{-1}=1+2e^{-1}$

**11.** (b) We have, $\lim_{\lambda\to 0}\left[\int_0^1(1+x)^\lambda dx\right]^{\frac{1}{\lambda}}$

First take $\int_0^1(1+x)^\lambda dx=\left[\frac{(1+x)^{\lambda+1}}{\lambda+1}\right]_0^1$

$=\frac{2^{\lambda+1}}{\lambda+1}-\frac{1}{\lambda+1}=\frac{2^{\lambda+1}-1}{\lambda+1}$

Now, $\quad \lim_{\lambda\to 0}\left[\frac{2^{\lambda+1}-1}{\lambda+1}\right]^{\frac{1}{\lambda}}$ $\quad [1^\infty$ form$]$

$\therefore \quad =e^{\lim_{\lambda\to 0}\frac{2^{\lambda+1}-1-\lambda-1}{\lambda+1}\times\frac{1}{\lambda}}$

$=e^{\lim_{\lambda\to 0}\frac{2^{\lambda+1}-\lambda-2}{\lambda^2+\lambda}}$ $\quad \left[\frac{0}{0}\text{ form}\right]$

$=e^{\lim_{\lambda\to 0}\frac{2^{\lambda+1}\log 2-1}{2\lambda+1}}$

$=e^{2\log 2-1}=e^{2\log 2}\cdot e^{-1}$

$=4e^{-1}=\frac{4}{e}$

**12.** (a) Let $I=\int_0^1 e^{e^x}(1+xe^x)dx$

Put $e^x=t$

$\Rightarrow x=\log t \Rightarrow dx=\frac{dt}{t}$

$\therefore \quad I=\int_1^e e^t(1+t\log t)\frac{dt}{t}=\int_1^e e^t\left(\log t+\frac{1}{t}\right)dt$

$=[e^t\cdot\log t]_1^e=e^e-0=e^e$

**13.** (b) Let $f(x)=\int_{-1}^1\frac{dx}{(1+e^x)(1+x^2)}$ ...(i)

Put $x=-t \Rightarrow \quad dx=-dt$

Now, $\quad I=f(-x)=-\int_1^{-1}\frac{dt}{(1+e^{-t})(1+t^2)}$

$=-\int_1^{-1}\frac{e^t}{(e^t+1)(1+t^2)}dt$

$\Rightarrow \quad I=\int_{-1}^1\frac{e^t}{(1+e^t)(1+t^2)}dt$ ...(ii)

On adding Eqs. (i) and (ii), we get

$2I=\int_{-1}^1\frac{e^t+1}{(e^t+1)(t^2+1)}dt=\int_{-1}^1\frac{dt}{t^2+1}$

$\Rightarrow \quad 2I=2\int_0^1\frac{dt}{t^2+1}=2[\tan^{-1}t]_0^1$

$\Rightarrow \quad I=\tan^{-1}1=\frac{\pi}{4}$

**14.** (b) We have, $\quad \int_0^2(t-\log_2 a)\,dt=\log_2\left(\frac{4}{a^2}\right)$

$\Rightarrow \quad \left[\frac{t^2}{2}-\log_2 a\cdot t\right]_0^2=\log_2\left(\frac{4}{a^2}\right)$

$\Rightarrow \quad 2-2\log_2 a=2\log_2\frac{2}{a}$

$\Rightarrow \quad 2-2\log_2 a=2-2\log_2 a$, which is equal.

But domain of $\log_2 a$ is $a>0$.

$\therefore$ It exists, when $a\in R^+$.

**15.** (d) We have, $f''(x)=-\frac{1}{x^2}-\pi^2\sin\pi x$

$\Rightarrow \quad f'(x)=-\frac{x^{-1}}{-1}+\pi^2\frac{\cos\pi x}{\pi}+C_1$

$\Rightarrow \quad f'(x)=\frac{1}{x}+\pi\cos\pi x+C_1$

$\Rightarrow \quad f(x)=\log x+\pi\frac{\sin\pi x}{\pi}+C_1x+C_2$

$\Rightarrow \quad f(x)=\log x+\sin\pi x+C_1x+C_2$

$\Rightarrow \quad f(1)=C_1+C_2=0$ ...(i)

$\Rightarrow \quad f'(2)=\frac{1}{2}+\pi\cos 2\pi+C_1$

$\Rightarrow \quad \pi+\frac{1}{2}=\frac{1}{2}+\pi+C_1$ $\quad \left[\text{given, } f'(2)=\pi+\frac{1}{2}\right]$

$\Rightarrow \quad C_1=0$

On putting $C_1=0$ in Eq. (i), we get

$0+C_2=0$

$\Rightarrow \quad C_2=0$

$\therefore \quad f(x)=\log x+\sin\pi x$

$\Rightarrow \quad f\left(\frac{1}{2}\right)=\log\frac{1}{2}-\sin\frac{\pi}{2}=-\log 2+1$

**16.** (c) We have, $\quad I=\int_0^1\frac{\tan^{-1}x}{x}dx$

Put $x=\tan\theta\Rightarrow dx=\sec^2\theta\,d\theta$

$\therefore \quad I=\int_0^{\pi/4}\frac{\theta\sec^2\theta}{\tan\theta}d\theta=\int_0^{\pi/4}\frac{\theta}{\sin\theta\cos\theta}d\theta$

$=\int_0^{\pi/4}\frac{2\theta}{\sin 2\theta}d\theta$

Put $2\theta=t \Rightarrow \quad d\theta=\frac{dt}{2}$

$\therefore \quad I=\frac{1}{2}\int_0^{\pi/2}\frac{t}{\sin t}dt=\frac{1}{2}\int_0^{\pi/2}\frac{x}{\sin x}dx$

$\left[\text{by property } \int_a^b f(x)dx=\int_a^b f(t)dt\right]$

**17.** (a) We have, $I(a) = \int_0^{\pi}\left(\frac{x}{a} + a\sin x\right)^2 dx$

$$\Rightarrow \quad I(a) = \int_0^{\pi}\left(\frac{x^2}{a^2} + a^2\sin^2 x + 2x\sin x\right)dx$$

$$\Rightarrow \quad I(a) = \int_0^{\pi}\left(\frac{x^2}{a^2} + \frac{a^2(1-\cos 2x)}{2} + 2x\sin x\right)dx$$

$$= \left[\frac{x^3}{3a^2} + \frac{a^2}{2}\left(x - \frac{1}{2}\sin 2x\right) + 2(-x\cos x + \sin x)\right]_0^{\pi}$$

$$= \frac{\pi^3}{3a^2} + \frac{a^2}{2}(\pi) + 2(-\pi\cos\pi + 0)$$

$$\Rightarrow \quad I(a) = \frac{\pi^3}{3a^2} + \frac{\pi a^2}{2} + 2\pi$$

$$I'(a) = \frac{\pi^3}{3}\left(\frac{-2}{a^3}\right) + \pi a = 0$$

$$\Rightarrow \quad -\frac{2}{3}\frac{\pi^3}{a^3} = -\pi a$$

$$\Rightarrow \quad \frac{2}{3}\pi^2 = a^4$$

$$\Rightarrow \quad a = \pm\sqrt{\pi\sqrt{\frac{2}{3}}}$$

$$\Rightarrow \quad I''(a) = -\frac{2}{3}\pi^3\frac{(-3)}{a^4} + \pi = \frac{2\pi^3}{a^4} + \pi$$

$\therefore$ $I''(a)$ is positive for $a = \sqrt{\pi\sqrt{\frac{2}{3}}}$, which is minima.

**18.** (d) $A = \int_{-1}^{1} f(x)dx = \int_{-1}^{1}(ax^2 + bx + c)dx$

$$= 2\int_0^1 (ax^2 + c)dx = 2\left[\frac{ax^3}{3} + cx\right]_0^1$$

$$= 2\left[\frac{a}{3} + c\right] = \frac{2}{3}[a + 3c]$$

Now, we will check all four options.

(a) $A = f(-1) + f(1) = a - b + c + a + b + c$
$= 2(a + c)$ [incorrect]

(b) $A = f\left(-\frac{1}{2}\right) + f\left(\frac{1}{2}\right)$

$= \frac{a}{4} - \frac{b}{2} + c + \frac{a}{4} + \frac{b}{2} + c = \frac{a}{2} + 2c$ [incorrect]

(c) $A = \frac{1}{2}[f(-1) + 2f(0) + f(1)]$

$= \frac{1}{2}[a - b + c + 2c + a + b + c]$

$= \frac{1}{2}[2a + 4c]$ [incorrect]

(d) $A = \frac{1}{3}[f(-1) + 4f(0) + f(1)]$

$= \frac{1}{3}[a - b + c + 4c + a + b + c]$

$= \frac{1}{3}[2a + 6c] = \frac{2}{3}(a + 3c)$ [correct]

**19.** (c) Let $I = \int_0^{\infty}\frac{x^2 + ax + 1}{1 + x^4}\cdot\tan^{-1}\frac{1}{x}dx$ ...(i)

Put $x = \frac{1}{t} \Rightarrow dx = -\frac{dt}{t^2}$

$$\Rightarrow \quad I = \int_{\infty}^{0}\frac{\frac{1}{t^2} + \frac{a}{t} + 1}{1 + \frac{1}{t^4}}\cdot\tan^{-1}t\left(-\frac{1}{t^2}\right)dt$$

$$\Rightarrow \quad I = -\int_{\infty}^{0}\frac{t^2 + at + 1}{1 + t^4}\tan^{-1}t\,dt \qquad \text{...(ii)}$$

On adding Eqs. (i) and (ii), we get

$$2I = \int_0^{\infty}\frac{x^2 + ax + 1}{1 + x^4}\cdot\frac{\pi}{2}dx$$

$$\therefore \quad I = \frac{\pi}{4}\left[\int_0^{\infty}\frac{x^2 + 1}{1 + x^4}dx + a\int_0^{\infty}\frac{x}{1 + x^4}dx\right]$$

$$= \frac{\pi}{4}\left[\underset{\text{I}}{\int_0^{\infty}\frac{1 + \frac{1}{x^2}}{x^2 + \frac{1}{x^2}}dx} + a\underset{\text{II}}{\int_0^{\infty}\frac{x}{1 + (x^2)^2}dx}\right]$$

Put $x - \frac{1}{x} = t$ in Ist integral and $x^2 = u$ in IInd integral, we get

$$I = \frac{\pi}{4}\left[\int_0^{\infty}\frac{dt}{t^2 + 2} + \frac{a}{2}\int_0^{\infty}\frac{du}{1 + u^2}\right]$$

$$= \frac{\pi}{4}\left[\frac{1}{\sqrt{2}}\tan^{-1}\left(\frac{t}{\sqrt{2}}\right) + \frac{a}{2}\tan^{-1}u\right]_0^{\infty}$$

$$= \frac{\pi}{4}\left[\frac{1}{\sqrt{2}}\tan^{-1}\left(\frac{x - \frac{1}{x}}{\sqrt{2}}\right) + \frac{a}{2}\tan^{-1}x^2\right]_0^{\infty}$$

$$= \frac{\pi}{4}\left[\frac{1}{\sqrt{2}}\cdot\frac{\pi}{2} + \frac{a}{2}\cdot\frac{\pi}{2}\right] = \frac{\pi^2}{8}\left(\frac{1}{\sqrt{2}} + \frac{a}{2}\right)$$

$$= \lim_{a\to\infty}\frac{\pi^2}{8}\frac{1}{a}\left(\frac{1}{\sqrt{2}} + \frac{a}{2}\right) = \frac{\pi^2}{k}$$

$$\Rightarrow \quad \lim_{a\to\infty}\frac{\pi^2}{8}\left(\frac{2 + a\sqrt{2}}{2\sqrt{2}a}\right) = \frac{\pi^2}{k}$$

$$\Rightarrow \quad \frac{\pi^2}{8}\times\frac{\sqrt{2}}{2\sqrt{2}} = \frac{\pi^2}{k} \Rightarrow \frac{1}{16} = \frac{1}{k}$$

Hence, $k = 16$

**20.** (d) We have, $I = \int_{\pi/2}^{3\pi/2} e^{-x/2}\frac{\sqrt{1 - \sin x}}{1 + \cos x}dx$

$$= \int_{\pi/2}^{3\pi/2} e^{-x/2}\left(\frac{\cos\frac{x}{2} - \sin\frac{x}{2}}{2\cos^2\frac{x}{2}}\right)dx$$

$$= \frac{1}{2}\int_{\pi/2}^{3\pi/2} e^{-x/2}\left[\sec\frac{x}{2} - \sec\frac{x}{2}\tan\frac{x}{2}\right]dx$$

Put $x = -2t \Rightarrow dx = -2dt$

$$\therefore \quad I = \frac{1}{2}(-2)\int_{-\pi/4}^{-3\pi/4} e^t[\sec t + \sec t\tan t]dt$$

$$= \int_{-3\pi/4}^{-\pi/4} e^t[\sec t + \sec t\tan t]dt$$

$$= [e^t\sec t]_{-3\pi/4}^{-\pi/4} = e^{-\pi/4}\sqrt{2} + e^{-3\pi/4}\sqrt{2}$$

$$= \sqrt{2}[e^{-\pi/4} + e^{-3\pi/4}]$$

**21.** (a) Given, $\int_1^x t\log t\,dt = \frac{1}{4}$

$$\Rightarrow \quad \left[\log t\cdot\frac{t^2}{2} - \int\frac{1}{t}\cdot\frac{t^2}{2}dt\right]_1^x = \frac{1}{4}$$

$$\Rightarrow \quad \left[\frac{t^2}{2}\log t - \frac{1}{2}\cdot\frac{t^2}{2}\right]_1^x = \frac{1}{4}$$

$$\Rightarrow \quad \left(\frac{x^2}{2}\log x - \frac{x^2}{4}\right) - \left(-\frac{1}{4}\right) = \frac{1}{4}$$

$$\Rightarrow \quad \frac{x^2}{2}\left(\log x - \frac{1}{2}\right) = 0$$

$$\Rightarrow \quad x = 0 \text{ or } \log x = \frac{1}{2}$$

$$\Rightarrow \quad x = e^{\frac{1}{2}} = \sqrt{e}$$

$x = 0$ is unadmissible because $x > 1$.

**22.** (c) We have, $I = \int_0^{\pi/2} \sin(|2x - \alpha|)dx;\ \alpha \in [0, \pi]$

Put $2x - \alpha = t$

$\Rightarrow dx = \frac{dt}{2}$

$\therefore I = \frac{1}{2}\int_0^{\pi-\alpha} \sin|t|\,dt;\ \pi - \alpha > 0$

$= \frac{1}{2}\int_0^{\pi-\alpha} \sin t\, dt$

$= -\frac{1}{2}[\cos t]_0^{\pi-\alpha}$

$= -\frac{1}{2}[\cos(\pi - \alpha) - \cos 0]$

$= -\frac{1}{2}[-\cos\alpha - 1] = \frac{1+\cos\alpha}{2}$

**23.** (b) We have,

$\int_{-1/2}^{1/2}[\sin^{-1}(3x - 4x^3) - \cos^{-1}(4x^3 - 3x)]dx$

$= \int_{-1/2}^{1/2}[3\sin^{-1}x - (2\pi - 3\cos^{-1}x)]dx$

$= \int_{-1/2}^{1/2}[3(\sin^{-1}x + \cos^{-1}x) - 2\pi]dx$

$= \int_{-1/2}^{1/2}\left[3\cdot\frac{\pi}{2} - 2\pi\right]dx$

$= -\frac{\pi}{2}\int_{-1/2}^{1/2}dx = -\frac{\pi}{2}\times 2\int_0^{1/2}dx$

$= -\pi[x]_0^{1/2} = -\frac{\pi}{2}$

**24.** (c) We have, $f(x) = \int_2^x \frac{dt}{\sqrt{1+t^4}}$

$\therefore f'(x) = \frac{1}{\sqrt{1+x^4}}$

and $gof(x) = x$

Now, on differentiating w.r.t. $x$, we get

$g'of(x)\cdot f'(x) = 1$

On putting $x = 0$, we get

$g'of(0) = \frac{1}{f'(0)} = \sqrt{17}$

$\therefore f(0) = 0$

Hence, $g'(0) = \sqrt{17}$

**25.** (a) We have, $a_n = \int_0^{\pi/2}(1 - \sin t)^n \sin 2t\, dt$

$a_n = \int_0^{\pi/2}(1 - \sin t)^n 2\sin t \cos t\, dt$

Put $1 - \sin t = u$

$\Rightarrow \cos t\, dt = -du$

$\therefore a_n = -\int_1^0 2u^n(1-u)\,du = 2\int_0^1(u^n - u^{n+1})du$

$= 2\left[\frac{u^{n+1}}{n+1} - \frac{u^{n+2}}{n+2}\right]_0^1 = 2\left[\frac{1}{n+1} - \frac{1}{n+2}\right]$

Therefore, $\frac{a_n}{n} = 2\left[\frac{1}{n(n+1)} - \frac{1}{n(n+2)}\right]$

$\therefore \lim_{n\to\infty}\sum_{n=1}^{n}\frac{a_n}{n} = 2\left[\Sigma\left(\frac{1}{n} - \frac{1}{n+1}\right) - \frac{1}{2}\Sigma\left(\frac{1}{n} - \frac{1}{n+2}\right)\right]$

$= 2\sum_{n=1}^{n}\left(\frac{1}{n} - \frac{1}{n+1}\right) - \sum_{n=1}^{n}\left(\frac{1}{n} - \frac{1}{n+2}\right)$

$= 2(1) - \left[\left(1 - \frac{1}{3}\right) + \left(\frac{1}{2} - \frac{1}{4}\right) + \left(\frac{1}{3} - \frac{1}{5}\right) + \cdots\right]$

$= 2 - \frac{3}{2} = \frac{1}{2}$

**26.** (a) We have,

$I = \int_0^{3\pi/4}[(1+x)\sin x + (1-x)\cos x]dx$

$= \int_0^{3\pi/4}[(\sin x + \cos x) + x(\sin x - \cos x)]dx$

$= \int_0^{3\pi/4} x(\sin x - \cos x)dx + \int_0^{3\pi/4}(\sin x + \cos x)dx$

$= x[(-\cos x - \sin x)]_0^{3\pi/4} - \int_0^{3\pi/4}(-\cos x - \sin x)\,dx + \int_0^{3\pi/4}(\sin x + \cos x)dx$

$= -\frac{3\pi}{4}\left(\cos\frac{3\pi}{4} + \sin\frac{3\pi}{4}\right) + 2\int_0^{3\pi/4}(\cos x + \sin x)dx$

$= -\frac{3\pi}{4}\left(-\frac{1}{\sqrt{2}} + \frac{1}{\sqrt{2}}\right) - 0 + 2[\sin x - \cos x]_0^{3\pi/4}$

$= 0 + 2\left[\left(\sin\frac{3\pi}{4} - \cos\frac{3\pi}{4}\right) - (0-1)\right] = 2\left[\frac{1}{\sqrt{2}} + \frac{1}{\sqrt{2}} + 1\right]$

$= 2\left(1 + \frac{2}{\sqrt{2}}\right) = 2(1+\sqrt{2}) = 2\tan\frac{3\pi}{8}$

**27.** (d) We have, $\lim_{n\to\infty} n^2\cdot C_n$ [it is $0\cdot\infty$ form]

$= \lim_{n\to\infty}\frac{C_n}{\frac{1}{n^2}}$ $\left[\frac{0}{0}\text{ form}\right]$

$= \lim_{n\to\infty}\frac{C'_n}{-\frac{2}{n^3}}$ ...(i)

Now, we have $C_n = \int_{1/(n+1)}^{1/n}\frac{\tan^{-1}nx}{\sin^{-1}nx}dx$

$\therefore C'_n = \frac{\tan^{-1}1}{\sin^{-1}1}\cdot\left(-\frac{1}{n^2}\right) - \frac{\tan^{-1}\left(\frac{n}{n+1}\right)}{\sin^{-1}\left(\frac{n}{n+1}\right)}\cdot\left(-\frac{1}{(n+1)^2}\right)$

$\Rightarrow C'_n = -\frac{1}{2n^2} + \frac{1}{(n+1)^2}\cdot\frac{\tan^{-1}\left(\frac{n}{n+1}\right)}{\sin^{-1}\left(\frac{n}{n+1}\right)}$

On putting Eq. (i), we get

$= \lim_{n\to\infty}\frac{-\frac{1}{2n^2} + \frac{1}{(n+1)^2}\cdot\frac{\tan^{-1}\left(\frac{n}{n+1}\right)}{\sin^{-1}\left(\frac{n}{n+1}\right)}}{-\frac{2}{n^3}}$

$= \lim_{n\to\infty}\frac{-\frac{1}{2n^2} + \frac{1}{2(n+1)^2}}{-\frac{2}{n^3}}$

$= \lim_{n\to\infty}\frac{\frac{1}{2}\left[\frac{1}{(n+1)^2} - \frac{1}{n^2}\right]}{-\frac{2}{n^3}}$

$= \lim_{n\to\infty}\frac{\frac{1}{2}\left[\frac{n^2 - n^2 - 1 - 2n}{n^2(n+1)^2}\right]}{-\frac{2}{n^3}}$

$= \lim_{n\to\infty}\frac{2n+1}{2n^2(n+1)^2}\times\frac{n^3}{2}$

$= \frac{1}{4}\lim_{n\to\infty}\frac{2n^4 + n^3}{n^2(n+1)^2} = \frac{1}{4}\times 2 = \frac{1}{2}$

**28.** (a) We have, $g(x) = \int_2^x \frac{t}{1+t^4} dt$

Put $t^2 = u$

$\Rightarrow \quad t\,dt = \frac{1}{2} du$

$\therefore \quad g(x) = \frac{1}{2}\int_4^{x^2} \frac{du}{1+u^2} = \frac{1}{2}[\tan^{-1} u]_4^{x^2}$

$\Rightarrow \quad g(x) = \frac{1}{2}[\tan^{-1} x^2 - \tan^{-1} 4]$

$\because \quad g(2) = 0$

We have, $\quad f(x) = e^{g(x)}$

$f'(x) = e^{g(x)} \cdot g'(x)$ ...(i)

and $\quad g'(x) = \frac{x}{1+x^4}$

$\therefore \quad f'(x) = e^{g(x)} \cdot \frac{x}{1+x^4}$

$\Rightarrow \quad f'(2) = e^{g(2)} \cdot \frac{2}{17} = e^0 \cdot \frac{2}{17}$

$\Rightarrow \quad f'(2) = \frac{2}{17}$

**29.** (b) We have,

$$f(x) = \sin x + \int_0^x f'(t)(2\sin t - \sin^2 t)dt$$

On differentiating both sides, we get

$$f'(x) = \cos x + f'(x)(2\sin x - \sin^2 x)$$

$\Rightarrow \quad f'(x)[1 - 2\sin x + \sin^2 x] = \cos x$

$\Rightarrow \quad f'(x) = \frac{\cos x}{(\sin x - 1)^2}$

So, $\quad f(x) = \int \frac{\cos x}{(\sin x - 1)^2} dx$

Put $\sin x - 1 = t$

$\Rightarrow \quad \cos x\,dx = dt$

$\therefore \quad f(x) = \int \frac{dt}{t^2} = -\frac{1}{t} + C$

$\Rightarrow \quad f(x) = -\frac{1}{\sin x - 1} + C$

On putting $x = 0$, then

$f(0) = 1 + C \Rightarrow 0 = 1 + C$

Hence, $\quad C = -1$

$\therefore \quad f(x) = \frac{1}{1-\sin x} - 1 = \frac{1 - 1 + \sin x}{1 - \sin x}$

$= \frac{\sin x}{1 - \sin x}$

**30.** (b) We have, $\int_{-1}^1 (px+q)(x^{2n+1} + a_n x + b_n)dx = 0$

$\Rightarrow \quad \int_{-1}^1 [px^{2n+2} + pa_n x^2 + b_n px + qx^{2n+1} + a_n qx + b_n q]dx = 0$

$\Rightarrow \quad \int_{-1}^1 px^{2n+2} + p\cdot a_n x^2 + x(b_n p + a_n q) + qx^{2n+1} + b_n q\,]dx = 0$

Since, term having odd exponent have the integral zero.

$\therefore \int_{-1}^1 (px^{2n+2} + a_n px^2 + qb_n)dx$

$= 2\int_0^1 (px^{2n+2} + a_n px^2 + qb_n)dx$

$= 2\left[p\frac{x^{2n+3}}{2n+3} + a_n \cdot p\frac{x^3}{3} + b_n q \cdot x\right]_0^1$

$= 2\left[\frac{p}{2n+3} + \frac{a_n p}{3} + b_n q\right] = 0$

Hence, $b_n = 0$ and $a_n = -\frac{3}{2n+3}$

**31.** (c) We have, $f(t) = \int_1^{t^2} \frac{\sqrt{1+u^4}}{u} du$

$\Rightarrow \quad f'(t) = \frac{\sqrt{1+t^8}}{t^2} \times 2t = \frac{2\sqrt{1+t^8}}{t}$

and $\quad F(x) = \int_1^x f(t)dt$

$\Rightarrow \quad F'(x) = f(x)$

$\Rightarrow \quad F''(x) = f'(x)$

$\therefore \quad F''(2) = f'(2) = \frac{2\sqrt{1+2^8}}{2}$

$= \sqrt{1+256} = \sqrt{257}$

**32.** (c) We have, $f(x) = \int_{-1}^x e^{t^2} dt$

and $\quad h(x) = f[1 + g(x)]$; $g(x) < 0$ for $x > 0$

Now, $\quad h(x) = \int_{-1}^{1+g(x)} e^{t^2} dt = f[1 + g(x)]$

$\Rightarrow \quad h'(x) = e^{\{1+g(x)\}^2} \cdot g'(x)$

$\Rightarrow \quad h'(1) = e$

$\Rightarrow \quad e^{\{1+g(1)\}^2} \cdot g'(1) = e$

$[1 + g(1)]^2 = 1$

$1 + g(1) = \pm 1$

$\therefore \quad g(1) = 0$ [not possible]

or $\quad g(1) = -2$

**33.** (c) Let $I = \int_3^7 f^{-1}(x)dx$

Put $f^{-1}(x) = t$

$\Rightarrow \quad x = f(t)$

$\Rightarrow \quad dx = f'(t)\,dt$

$\therefore \quad I = \int_{f^{-1}(3)}^{f^{-1}(7)} t\,f'(t)dt$

Given, $f(2) = 3$, then $f^{-1}(3) = 2$

and if $f(5) = 7$, then $f^{-1}(7) = 5$

$\therefore \quad I = \int_2^5 t\,f'(t)dt$

$= [t \cdot f(t)]_2^5 - \int_2^5 f(t)\,dt$

$= 5f(5) - 2f(2) - 17$

$= 35 - 6 - 17 = 35 - 23$

$= 12$

**34.** (c) We have, $f(x) = \int_0^{g(x)} \frac{dt}{\sqrt{1+t^2}}$

$\Rightarrow \quad f'(x) = \frac{1}{\sqrt{1+[g(x)]^2}} \times g'(x)$

$\Rightarrow \quad f'\left(\frac{\pi}{2}\right) = \frac{g'\left(\frac{\pi}{2}\right)}{\sqrt{1+g^2\left(\frac{\pi}{2}\right)}}$; $g\left(\frac{\pi}{2}\right) = 0$

$= g'\left(\frac{\pi}{2}\right)$

$\Rightarrow \quad g(x) = \int_0^{\cos x} (1 + \sin t^2)dt$

$\Rightarrow \quad g'(x) = [1 + \sin(\cos^2 x)](-\sin x)$

$\Rightarrow \quad g'\left(\frac{\pi}{2}\right) = -1$

Hence, $\quad f'\left(\frac{\pi}{2}\right) = -1$

and $\quad h'(0^+) = -1$

$\therefore \quad f'\left(\frac{\pi}{2}\right) = h'(0^+)$

**35.** (d) Let $I = \lim_{t \to 0} \int_0^{2\pi} \left| \dfrac{\sin(x+t) - \sin x}{t} \right| dx$

$$= \int_0^{2\pi} \lim_{t \to 0} \left| \dfrac{2\cos\left(x + \frac{t}{2}\right) \cdot \sin\frac{t}{2}}{t} \right| dx$$

$$= \int_0^{2\pi} |\cos x|\, dx = 4$$

**36.** (a) We have, $\lim_{x \to \infty} x^3 \int_{-1/x}^{1/x} \dfrac{\log(1+t^2)}{1+e^t} dt$

On putting $x = \dfrac{1}{y}$

$$L = \lim_{y \to 0} \frac{1}{y^3} \int_{-y}^{y} \frac{\log(1+t^2)}{1+e^t} dt$$

Take $I = \int_{-y}^{y} \dfrac{\log(1+t^2)}{1+e^t} dt$ ...(i)

Put $t = -\theta \Rightarrow dt = -d\theta$

$\therefore$ $I = -\int_{y}^{-y} \dfrac{\log(1+\theta^2)}{1+e^{-\theta}} d\theta$

$\Rightarrow$ $I = \int_{-y}^{y} \dfrac{e^{\theta}\log(1+\theta^2)}{e^{\theta}+1} d\theta$ ...(ii)

On adding Eqs. (i) and (ii), we get

$$2I = \int_{-y}^{y} \frac{\log(1+\theta^2)}{e^{\theta}+1}(e^{\theta}+1) d\theta$$

$\Rightarrow$ $2I = 2\int_0^y \log(1+\theta^2) d\theta$

$\Rightarrow$ $I = \int_0^y \log(1+\theta^2) d\theta$

$$= [\log(1+\theta^2)\,\theta]_0^y - \int_0^y \frac{1}{1+\theta^2} 2\theta \cdot \theta\, d\theta$$

$$= y\log(1+y^2) - 2\int_0^y \left(1 - \frac{1}{\theta^2+1}\right) d\theta$$

$$= y\log(1+y^2) - 2[\theta - \tan^{-1}\theta]_0^y$$

$$= y\log(1+y^2) - 2[y - \tan^{-1} y]$$

$$= \lim_{y \to 0} \frac{y\log(1+y^2) - 2y + 2\tan^{-1} y}{y^3}$$

$$= \lim_{y \to 0} \frac{y \cdot \frac{1}{1+y^2} 2y + \log(1+y^2) - 2 + \frac{2}{1+y^2}}{3y^2}$$

$$= \lim_{y \to 0} \frac{\frac{2y^2+2}{y^2+1} + \log(1+y^2) - 2}{3y^2}$$

$$= \lim_{y \to 0} \frac{\log(1+y^2)}{3y^2} = \frac{1}{3}$$

**37.** (b) Given, $x = \int_0^y \dfrac{dt}{\sqrt{1+t^2}}$

$\Rightarrow$ $x = \left[\log t + \sqrt{1+t^2}\right]_0^y$

$\Rightarrow$ $x = \log(y + \sqrt{1+y^2})$

$\Rightarrow$ $y + \sqrt{1+y^2} = e^x$

$\Rightarrow$ $\sqrt{1+y^2} = e^x - y$

$\Rightarrow$ $1 + y^2 = e^{2x} + y^2 - 2ye^x$

$\Rightarrow$ $2ye^x = e^{2x} - 1$

$\Rightarrow$ $y = \dfrac{1}{2}(e^x - e^{-x}) \Rightarrow \dfrac{dy}{dx} = \dfrac{1}{2}(e^x + e^{-x})$

$\therefore$ $\dfrac{d^2y}{dx^2} = \dfrac{1}{2}(e^x - e^{-x}) = y$

**38.** (a) Let $I = \int_1^a x \cdot a^{-[\log_a x]} dx$

Put $\log_a x = t \Rightarrow x = a^t$

$\Rightarrow$ $dx = a^t \cdot \log a\, dt$

$\therefore$ $I = \int_0^1 a^t \cdot a^{-[t]} \cdot a^t \log a\, dt$

$$= \log a \int_0^1 a^{t-[t]} \cdot a^t\, dt$$

In $0 < t < 1; t - [t] = t$

$\therefore$ $\log a \int_0^1 a^{2t}\, dt = \left[\log a \cdot \dfrac{a^{2t}}{2\log a}\right]_0^1$

$\Rightarrow$ $\dfrac{1}{2}[a^2 - 1] = \dfrac{e-1}{2}$

$\Rightarrow$ $a^2 = e \Rightarrow a = \sqrt{e}$

**39.** (c) We have, $\lim_{n \to \infty} \sum_{k=1}^{n} \dfrac{n}{n^2 + k^2x^2}$

$$= \lim_{n \to \infty} \frac{1}{n} \sum_{k=1}^{n} \frac{1}{1 + \frac{k^2}{n^2}x^2}$$

$$= \int_0^1 \frac{1}{1+\theta^2x^2} d\theta$$

$$= \frac{1}{x}[\tan^{-1}\theta x]_0^1$$

$$= \frac{1}{x}[\tan^{-1} x] = \frac{\tan^{-1} x}{x}$$

**40.** (d) We have, $\int_0^a f(x)dx + \int_0^b f^{-1}(x)dx$ ...(i)

Now, $\int_0^b f^{-1}(x)dx$

Put $f^{-1}(x) = t \Rightarrow x = f(t)$

$\Rightarrow$ $dx = f'(t)dt$

$$\int_0^{b} f^{-1}(x)dx = \int_0^a t\, f'(t)\,dt = [t \cdot f(t)]_0^a - \int_0^a f(t)\,dt$$

$$= af(a) - \int_0^a f(t)dt$$

On putting in Eq. (i), we get

$$\int_0^a f(x)dx + af(a) - \int_0^a f(t)dt = af(a) = ab$$

**41.** (b) We have, $\int_{19}^{37} (\{x\}^2 + 3\sin 2\pi x)dx$

Since, given integrand is periodic function with period 1.

$\therefore (37 - 19)\int_0^1 [(\{x\})^2 + 3\sin 2\pi x]dx$

$$= 18\int_0^1 (x^2 + 3\sin 2\pi x)dx$$

$$= 18\left[\frac{x^3}{3} - \frac{3}{2\pi}\cos 2\pi x\right]_0^1$$

$$= 18\left[\left(\frac{1}{3} - \frac{3}{2\pi}\right) + \frac{3}{2\pi}\right]$$

$$= \frac{18}{3} = 6$$

**42.** (b) Let $I = \int_0^1 \dfrac{1}{1+x} \cdot n \cdot x^{n-1} dx$

$$= \left[\frac{1}{1+x} \cdot x^n\right]_0^1 + \int_0^1 \frac{x^n}{(1+x)^2} dx$$

$$= \frac{1}{2} + \int_0^1 \frac{x^n}{(1+x)^2} dx = \frac{1}{2} + 0$$

when $\int_0^1 \dfrac{x^n}{(1+x)^2} dx = 0$, as $n \to \infty$

$$= \frac{1}{2}$$

**43.** (a) We have,

$$\int_{-1/\sqrt{3}}^{1/\sqrt{3}} \frac{x^4}{1-x^4} \cos^{-1}\left(\frac{2x}{1+x^2}\right) dx$$

$$= \int_{-1/\sqrt{3}}^{1/\sqrt{3}} \frac{x^4}{1-x^4}\left(\frac{\pi}{2} - \sin^{-1}\frac{2x}{1+x^2}\right) dx$$

$$= \frac{\pi}{2}\underbrace{\int_{-1/\sqrt{3}}^{1/\sqrt{3}} \frac{x^4}{1-x^4} dx}_{\text{I}} - \underbrace{\int_{-1/\sqrt{3}}^{1/\sqrt{3}} \frac{x^4}{1-x^4} \sin^{-1}\left(\frac{2x}{1+x^4}\right) dx}_{\text{II}}$$

$$= \frac{\pi}{2}\int_{-1/\sqrt{3}}^{1/\sqrt{3}} \frac{x^4}{1-x^4} dx - 0$$

(in second integration, the integrand is odd function)

$$= \frac{\pi}{2}\cdot 2\int_0^{1/\sqrt{3}} \frac{x^4}{1-x^4} dx$$

$$= k\int_0^{1/\sqrt{3}} \frac{x^4}{1-x^4} dx$$

Hence, $k = \pi$

**44.** (a) We have, $g(x) = f^{-1}(x)$

Now, $\int_2^{10} g(y)dy = \int_2^{10} f^{-1}(y)dy$

Put $f^{-1}(y) = t$

$\Rightarrow$ $y = f(t)$

$\Rightarrow$ $dy = f'(t)dt$

$\therefore$ $\int_2^{10} g(y)dy = \int_{f^{-1}(2)}^{f^{-1}(10)} t \cdot f'(t)dt$

$= \int_1^5 t\, f'(t)dt$

$= [t\, f(t)]_1^5 - \int_1^5 f(t)dt$

Hence, $\int_1^5 f(t)dx + \int_2^{10} g(y)dy = 5f(5) - f(1)$

$= 5\times 10 - 2$

$= 48$

**45.** (c) As $f(x)$ is differentiable and monotonic.

$\Rightarrow$ $f^{-1}(x)$ exists

Now, check option (c),

Put $f^{-1}(x) = t$

$\Rightarrow$ $x = f(t)$

$\Rightarrow$ $dx = f'(t)\,dt$

$\therefore$ $\int_{f(a)}^{f(b)} 2x\{b - f^{-1}(x)\}dx = \int_a^b 2f(t)(b-t)f'(t)\,dt$

$= \int_a^b (b-t)\{2f(t)f'(t)\}dt$

$= [(b-t)\{f(t)\}^2]_a^b + \int_a^b \{f(t)\}^2 dt$

$= -(b-a)\{f(t)\}^2 + \int_a^b \{f(t)\}^2 dt$

$= \int_a^b [\{f(t)\}^2 - \{f(a)\}^2]dt$

$= \int_a^b [(f^2(x) - f^2(a)]dx$

**46.** (c) Given, $I = \int_0^{1/2} \frac{dx}{\sqrt{1-x^{2n}}}$, since $n \geq 1$ and $0 \leq x \leq \frac{1}{2}$

$\therefore$ $\sqrt{1-x^2} \leq \sqrt{1-x^{2n}} \leq 1$

$\Rightarrow$ $\frac{1}{\sqrt{1-x^2}} \geq \frac{1}{\sqrt{1-x^{2n}}} \geq 1$

$\Rightarrow$ $\int_0^{1/2} 1\,dx \leq I \leq \int_0^{1/2} \frac{dx}{\sqrt{1-x^2}}$

$\therefore$ $\frac{1}{2} \leq I \leq \frac{\pi}{6}$

**47.** (c) We have, $f(4-x) = f(4+x)$

On replacing $x$ by $4-x$, we get

$f(4-4+x) = f(4+4-x)$

$f(x) = f(8-x) = f(x+8)$

Thus, $f(x)$ is periodic function with period 8.

Now, $\int_0^{200} f(x)\,dx = \int_0^{8\times 25} f(x)dx$

$= 25\int_0^8 f(x)dx = 25\times 5 = 125$

**48.** (d) Given, $f[x + f(y)] = f(x) + y, \forall x, y \in R$

On putting $g = 0$,

$f[x + f(0)] = f(x)$

$\Rightarrow$ $f(x+1) = f(x)$

Hence, $f(x)$ is periodic function with period 1.

We have, $\int_0^{10} f(10-x)\,dx$

Put $10 - x = t \Rightarrow$ $dx = -dt$

$\therefore$ $\int_0^{10} f(10-x)dx - \int_{10}^0 f(t)\,dt = \int_0^{10} f(t)\,dt$

$= 10\int_0^1 f(t)\,dt = 10\int_0^1 f(x)\,dx$

**49.** (d) Let $I = \int_0^\infty \frac{\log_e x}{x^2+a^2}dx$

Put $x = a\tan\theta$

$\Rightarrow$ $dx = a\sec^2\theta d\theta$

$\therefore$ $I = \int_0^{\pi/2} \frac{\log(a\tan\theta)}{a^2\sec^2\theta}\times a\sec^2\theta\, d\theta$

$= \frac{1}{a}\int_0^{\pi/2} (\log a + \log\tan\theta)\,d\theta$

$= \frac{1}{a}\log a\cdot\frac{\pi}{2} + \frac{1}{a}\int_0^{\pi/2} \log\tan\theta\, d\theta$

$= \frac{\pi\log a}{2a} + 0 = \frac{\pi\log a}{2a}$

**50.** (b) Let $I = \int_{-2008}^{2008} \frac{f'(x) + f'(-x)}{(2008)^x + 1}dx$ ...(i)

Put $x = -\theta$ $\Rightarrow$ $dx = -d\theta$

$\therefore$ $I = -\int_{2008}^{-2008} \frac{f'(-\theta) + f'(\theta)}{(2008)^{-\theta} + 1}d\theta$

$= \int_{-2008}^{2008} \frac{(2008)^\theta[f'(\theta) + f'(-\theta)]}{1 + (2008)^\theta}d\theta$ ...(ii)

On adding Eqs. (i) and (ii), we get

$2I = \int_{-2008}^{2008} [f'(\theta) + f'(-\theta)]d\theta$

$\Rightarrow$ $2I = 2\int_0^{2008} [f'(\theta) + f'(-\theta)]d\theta$

$\therefore$ $I = [f(\theta) - f(-\theta)]_0^{2008}$

$= f(2008) - f(-2008)$

**51.** (a,c) We have, $\int_1^{xy} f(t)dt = y\int_1^x f(t)dt + x\int_1^y f(t)dt$

On differentiating w.r.t. $x$, treating $y$ constant, we get

$f(xy)\cdot y = y\,f(x) + \int_1^y f(t)dt$

On putting $x = 1$, we get

$f(y)y = yf(1) + \int_1^y f(t)dt$

$\Rightarrow$ $yf(y) = 3y + \int_1^y f(t)dt$

On differentiating w.r.t. $y$, we get

$yf'(y) + f(y) = 3 + f(y)$

$\Rightarrow$ $f'(y) = \frac{3}{y}$

$\Rightarrow$ $f(y) = 3\log y + C$

On puttting $y = 1$, we get

$$f(1) = 3 = C$$

Now, $f(y) = 3\log y + 3 = 3(\log y + 1)$

$\therefore$ $f(y) = 3\log(ey)$

Hence, $f(x) = 3\log ex$

**52.** (a,b) Here, $S_n < \int_2^3 \frac{x}{1+x^2}dx < T_n$ for all $n \in N$.

$$\Rightarrow \quad S_n < \frac{1}{2}\int_5^{10}\frac{dt}{t} < T_n$$

$$\Rightarrow \quad S_n < \frac{1}{2}[\log(t)]_5^{10} < T_n$$

$$\therefore \quad S_n < \frac{1}{2}\log 2 < T_n$$

**53.** (b,c) Since, $x \in \left[0, \frac{\pi}{2}\right]$

$$\Rightarrow \quad 1 \le 1 + \sin^3 x \le 2$$

$$\Rightarrow \quad \frac{1}{\sqrt{2}} \le \frac{1}{\sqrt{1+\sin^3 x}} \le 1$$

$$\Rightarrow \quad \int_0^{\pi/2}\frac{1}{\sqrt{2}}dx \le \int_0^{\pi/2}\frac{dx}{\sqrt{1+\sin^3 x}} \le \int_0^{\pi/2}dx$$

$$\therefore \quad \frac{\pi}{2\sqrt{2}} \le I \le \frac{\pi}{2}$$

**54.** (a,b,c) We have,

$$\int_0^\infty e^{x^2}dx = \frac{\sqrt{\pi}}{2}$$

Now, check options.

(a) $\int_0^\infty e^{-2x^2}dx$

Put $\sqrt{2}x = t \Rightarrow dx = \frac{dt}{\sqrt{2}}$

$$\therefore \quad \int_0^\infty e^{-2x^2}dx = \frac{1}{\sqrt{2}}\int_0^\infty e^{-t^2}dt$$

$$= \frac{1}{\sqrt{2}} \times \frac{\sqrt{\pi}}{2} = \frac{\sqrt{\pi}}{2\sqrt{2}}$$

(b) $\int_0^\infty xe^{-x^2}dx$

Put $x^2 = t \Rightarrow x\,dx = \frac{1}{2}dt$

$$\therefore \quad \int_0^\infty xe^{-x^2}dx = \frac{1}{2}\int_0^\infty e^{-t}dt = \frac{1}{2}[-e^{-t}]_0^\infty$$

$$= -\frac{1}{2}[e^{-\infty} - e^0] = -\frac{1}{2}[0-1] = \frac{1}{2}$$

(c) $\int_0^\infty x^2 e^{-x^2}dx$

Put $x^2 = t \Rightarrow 2x\,dx = dt$

$$\Rightarrow \quad dx = \frac{dt}{2\sqrt{t}}$$

$$\therefore \quad \int_0^\infty x^2 e^{-x^2}dx = \int_0^\infty \frac{t\cdot e^{-t}}{2\sqrt{t}}dt$$

$$= \frac{1}{2}\int_1^\infty \sqrt{t}\cdot e^t dt = \frac{1}{2}\int_0^\infty \sqrt{t}\,e^{-t}dt$$

$$= \frac{1}{2}\left[[(-\sqrt{t}\,e^{-t})]_0^\infty + \int_0^\infty \frac{1}{2\sqrt{t}}e^{-t}dt\right]$$

$$= \frac{1}{2}\left[0 + \frac{1}{2}\int_0^\infty t^{\frac{-1}{2}}e^{-t}dt\right] = \frac{1}{4}\int_0^\infty t^{-\frac{1}{2}}\cdot e^{-t}dt$$

We know that, $\int_0^\infty e^{-x}x^{n-1}dx = \Gamma n$

$$\Rightarrow \quad \frac{1}{4}\int_0^\infty t^{\frac{1}{2}-1}e^{-t}dt = \frac{1}{4}\Gamma\frac{1}{2} = \frac{1}{4}\sqrt{\pi} = \frac{\sqrt{\pi}}{4}$$

**55.** (b,d) We have, $(6x+1)x = 1 + \left[\cos\frac{\pi}{4}\right]$

$$\Rightarrow \quad 6x^2 + x - 1 = 0$$

$$\Rightarrow \quad (2x+1)(3x-1) = 0$$

$$\Rightarrow \quad \alpha = -\frac{1}{2} \text{ and } \beta = \frac{1}{3}$$

Now, $\int_0^{2\alpha}\sin\left[\frac{\pi[x]}{2}\right]dx + \int_0^{3\beta}\cos(\pi[x])dx$

$$= \int_0^{-1}\sin\left(\frac{\pi[x]}{2}\right) + \int_0^1 \cos(\pi[x])\,dx$$

$$= \int_0^1 \cos(\pi[x])dx - \int_{-1}^0 \sin\left(\frac{\pi[x]}{2}\right)dx$$

$$= \int_0^1 \cos 0\,dx - \int_{-1}^0 \sin\left(-\frac{\pi}{2}\right)dx$$

$$= [x]_0^1 + [x]_{-1}^0$$

$$= 1 + 0 - (-1)$$

$$= 1 + 1 = 2$$

Now, $\alpha + \beta \ne 2$

$\therefore$ $2\alpha + 9\beta = -1 + 3 = 2$

**56.** (c,d) Given, $f(x) = \int_1^x \frac{\log t}{1+t}dt$

Here, $f(x) + f\left(\frac{1}{x}\right) = \frac{\log^2 x}{2} = 2$

$$\Rightarrow \quad \log_e x = \pm 2$$

$$\therefore \quad x = e^{\pm 2}$$

**57.** (c,d) $f(x) = \int_{-1}^1 (1-|t|)\cos xt\,dt$

$$= 2\int_0^1 (1-|t|)\cos xt\,dt$$

$$= 2\left[\left\{(1-t)\frac{\sin xt}{x}\right\}_0^1 - \int_0^1 (-1)\frac{\sin xt}{x}dt\right]$$

$$= 2\left[0 + \frac{-\cos xt}{x^2}\right]_0^1 = \frac{2(1-\cos x)}{x^2}$$

$$= -\frac{2}{x^2}[\cos x - 1]$$

$$\therefore \lim_{x\to 0} f(x) = \lim_{x\to 0}\frac{2\left(2\sin^2\frac{x}{2}\right)}{x^2} = \lim_{x\to 0}\frac{\sin^2\frac{x}{2}}{\frac{x^2}{4}} = 1$$

$$\Rightarrow \quad f(0) = \int_{-1}^1 (1-|t|)dt$$

$$= 2\int_0^1 (1-t)\,dt = 2\left[t - \frac{t^2}{2}\right]_0^1$$

$$= 2\left(1 - \frac{1}{2}\right) = 2 \times \frac{1}{2} = 1$$

Hence, $f(x)$ is continuous at $x = 0$.

**58.** (b,d) Let $f(x) = x^2 + ax^2 + bx^3$

where, $a = \int_{-1}^1 t\,f(t)\,dt$ and $b = \int_{-1}^1 f(t)\,dt$

Now, $a = \int_{-1}^1 t[(a+1)t^2 + bt^3]dt$

$$a = 2b\int_0^1 t^4\,dt = \frac{2b}{5} \qquad \text{...(i)}$$

Again, $b = \int_{-1}^1 f(t)dt = \int_{-1}^1 [(a+1)t^2 + bt^3]dt$

$$= 2\int_0^1 (a+1)t^2 dt$$

$$\Rightarrow \quad b = \frac{2(a+1)}{3} \qquad \text{...(ii)}$$

From Eqs. (i) and (ii),

$$\frac{5a}{2}=\frac{2(a+1)}{3}$$

$$\left(\frac{5}{2}-\frac{2}{3}\right)a=\frac{2}{3}$$

$$\Rightarrow \quad \frac{11a}{6}=\frac{2}{3}$$

$$\Rightarrow \quad a=\frac{4}{11} \text{ and } b=\frac{10}{11}$$

Hence, $\int_{-1}^{1} t\, f(t)dt=\frac{4}{11}$ and $\int_{-1}^{1} f(t)\,dt=\frac{10}{11}$

Now, $f(x)=(a+1)x^2+bx^3$

$\Rightarrow \quad f(1)=(a+1)+b$

$\Rightarrow \quad f(-1)=a+1-b$

Hence, $f(1)+f(-1)=2(a+1)=\frac{30}{11}$

and $f(1)-f(-1)=2b=\frac{20}{11}$

**59.** (a,b,d) We have,

$$f[f(x)]=-x+1$$

On replacing $x$ by $f(x)$, we get

$$f[f\{f(x)\}]=-f(x)+1$$

$\Rightarrow \quad f(1-x)=-f(x)+1$

$\Rightarrow \quad f(x)+f(1-x)=1 \quad \ldots(\text{i})$

Now, $J=\int_0^1 f(x)dx=\int_0^1 f(1-x)dx$

$\Rightarrow \quad 2J=\int_0^1 [f(x)+f(1-x)]dx$

$\Rightarrow \quad 2J=\int_0^1 dx=1$

$\therefore \quad J=\frac{1}{2}$

Let $K=\int_0^{\pi/2}\frac{\sin x}{(\sin x+\cos x)^3}dx \quad \ldots(\text{ii})$

$K=\int_0^{\pi/2}\frac{\cos x}{(\cos x+\sin x)^3}dx \quad \ldots(\text{iii})$

From Eqs. (ii) and (iii), we get

$$2K=\int_0^{\pi/2}\frac{\sin x+\cos x}{(\sin x+\cos x)^3}dx$$

$$\Rightarrow \quad 2K=\int_0^{\pi/2}\frac{1}{(\sin x+\cos x)^2}dx$$

$$=\int_0^{\pi/2}\frac{1}{1+2\sin x\cos x}dx$$

$$=\int_0^{\pi/2}\frac{\sec^2 x}{1+\tan^2 x+2\tan x}dx$$

Put $1+\tan x=t \Rightarrow \sec^2 x\,dx=dt$

$$\Rightarrow \quad 2K=\int_1^{\infty}\frac{dt}{t^2}=\left[-\frac{1}{t}\right]_1^{\infty}=-[0-1]$$

$\Rightarrow \quad 2K=1$

Then, $K=\frac{1}{2}=J$

**60.** (c,d) We have, $[f'(x)]^2+[g'(x)]^2=1$

and $f(x)+\int_0^x g(t)dt=\sin x(\cos x-\sin x)$

On differentiating both sides, we get

$$f'(x)+g(x)=\cos 2x-\sin 2x \quad \ldots(\text{i})$$

On squaring both sides, we get

$$[f'(x)]^2+[g(x)]^2+2f'(x)\cdot g(x)=1-\sin 4x$$

$\Rightarrow \quad 2f'(x)\cdot g(x)=-\sin 4x$

$g(x)=-\frac{\sin 4x}{2f'(x)}$, substituting in Eq. (i)

$$f'(x)-\frac{\sin 4x}{2f'(x)}=\cos 2x-\sin 2x$$

On putting $f'(x)=t$

$$2t^2-2(\cos 2x-\sin 2x)t-\sin 4x=0$$

$$\Rightarrow \quad t=\frac{2(\cos 2x-\sin 2x)\pm\sqrt{4(1-\sin 4x)+8\sin 4x}}{4}$$

$$\Rightarrow \quad 2t=\cos 2x-\sin 2x\pm\sqrt{1+4\sin 4x}$$

On taking positive sign, we get

$$2t=\cos 2x-\sin 2x+\sqrt{1+4\sin 4x}$$

$$=\cos 2x-\sin 2x+\cos 2x+\sin 2x$$

$\Rightarrow \quad t=\cos 2x$

On taking negative sign, we get

$$t=-\sin 2x$$

Since, $f'(x)=\cos 2x$ or $f'(x)=-\sin 2x$

$f(x)=\frac{1}{2}\sin 2x+C_1$

or $f(x)=\frac{\cos 2x}{2}+C_2$

$f(0)=0 \Rightarrow C_1=0$ and $C_2=\frac{-1}{2}$

$f(x)=\frac{1}{2}\sin 2x$ or $f(x)=\frac{\cos 2x-1}{2}$

If $f'(x)=\cos 2x$, then $g(x)=-\sin 2x$

If $f(x)=\frac{\cos 2x-1}{2}$, then $g(x)=\cos 2x$

$\therefore \quad f(x)=-\sin^2 x$ and $g(x)=\cos 2x$

**61.** (a)

(i) We have,

$$f(x)=\int_0^{g(x)}\frac{dt}{\sqrt{1+t^3}}$$

$$f'(x)=\frac{1}{\sqrt{1+[g(x)]^3}}\times g'(x)$$

Also, $g(x)=\int_0^{\cos x}(1+\sin t^2)dt$

$\Rightarrow \quad g'(x)=[1+\sin(\cos^2 x)](-\sin x)$

$\Rightarrow \quad g'\left(\frac{\pi}{2}\right)=-1$

$$\Rightarrow \quad f'\left(\frac{\pi}{2}\right)=\frac{1}{\sqrt{1+\left[g\left(\frac{\pi}{2}\right)\right]^3}}\cdot g'\left(\frac{\pi}{2}\right) \qquad \left[\because g\left(\frac{\pi}{2}\right)=0\right]$$

$\therefore \quad f'\left(\frac{\pi}{2}\right)=-1$

(ii) $\int_0^x f(t)\,dt=[f(x)]^2$

On differentiating w.r.t. $x$, we get

$$f(x)=2f(x)\cdot f'(x)$$

$\Rightarrow \quad f'(x)=\frac{1}{2}$

$\Rightarrow \quad f(x)=\frac{x}{2}+C$

On putting $x=0$, we get

$$f(0)=C=0$$

Hence, $f(x)=\frac{x}{2}$

$\Rightarrow \quad f(2)=\frac{2}{2}=1$

(iii) We have, $\int_a^b (2 + x - x^2)dx$

Since, the given integrand is parabola. This parabola contains maximum area, when it enclosed between −1 and 2.

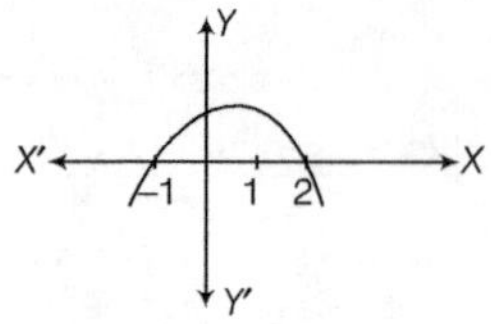

$$\int_a^b (2 + x - x^2)dx = \int_{-1}^2 (2 + x - x^2)dx$$

Hence, $a = -1$ and $b = 2$

$\therefore \quad a + b = -1 + 2 = 1$

(iv) $\lim_{x \to 0} \left(\frac{\sin 2x}{x^3} + a + \frac{b}{x^2}\right) = 0$

$\Rightarrow \quad \lim_{x \to 0} \frac{\sin 2x + ax^3 + bx}{x^3} = 0,$

it is $\frac{0}{0}$ form.

$\Rightarrow \quad \lim_{x \to 0} \frac{\cos 2x \cdot 2 + 3ax^2 + b}{3x^2} = 0$

It is $\frac{0}{0}$ form; $2 + b = 0 \Rightarrow b = -2$

$\Rightarrow \quad \lim_{x \to 0} -\frac{4 \sin 2x + 6ax}{6x} = 0$

$\Rightarrow \quad \lim_{x \to 0} -\frac{4 \sin 2x}{3 \cdot 2x} + \frac{6a}{6} = 0$

$\Rightarrow \quad -\frac{4}{3} + a = 0 \Rightarrow a = \frac{4}{3}$

$\Rightarrow \quad 3a = 4$

$\therefore \quad 3a + b = 4 - 2 = 2$

Hence, (i) → (s), (ii) → (r), (iii) → (r), (iv) → (q)

**62.** (a)

(i) Let $I = \int_0^1 [1 + (2008)\, x^{2008}] e^{x^{2008}} dx$

$$= \int_0^1 \left[e^{x^{2008}} + 2008 \cdot x^{2008} \cdot e^{x^{2008}}\right] dx$$

$$= \int_0^1 [f(x) + xf'(x)]dx, \text{ where } f(x) = e^{x^{2008}}$$

$$= \left[xe^{x^{2008}}\right]_0^1 = e - 0 = e$$

(ii) Let $I = I_1 + I_2$

$$I_2 = \int_1^{1/e} \sqrt{-\log x}\, dx$$

Put $-\log x = t^2 \Rightarrow x = e^{-t^2}$

$\Rightarrow \quad dx = e^{-t^2}(-2t)dt$

$\therefore \quad I_2 = -2\int_0^1 t \cdot te^{-t^2} dt$

$$= -2\int_0^1 t^2 \cdot e^{-t^2} dt$$

$$= \int_0^1 t \left[-2te^{-t^2}\right] dt$$

I II

$$= \left[t \cdot e^{-t^2}\right]_0^1 - \int_0^1 e^{-t^2} dt = \frac{1}{e} - \int_0^1 e^{-t^2} dt$$

$\therefore \quad I_2 + I_1 = \frac{1}{e} = e^{-1}$

(iii) Let $I = \lim_{n \to \infty} \left[\left(\frac{1}{n}\right)^1 \left(\frac{2}{n}\right)^2 \left(\frac{3}{n}\right)^3 \cdots \left(\frac{n}{n}\right)^n\right]^{\frac{1}{n^2}}$

$$I = \lim_{n \to \infty} \prod_{r=1}^{n} \left(\frac{r}{n}\right)^{\frac{r}{n^2}}$$

On taking log both sides, we get

$$\log I = \lim_{n \to \infty} \sum_{r=1}^{n} \frac{r}{n^2} \log\left(\frac{r}{n}\right) = \frac{1}{n} \lim_{n \to \infty} \sum_{r=1}^{n} \frac{r}{n} \log\left(\frac{r}{n}\right)$$

$\therefore \quad \log I = \int_0^1 x \log x\, dx = \log x \cdot \frac{x^2}{2} - \int \frac{1}{x} \cdot \frac{x^2}{2} dx$

$$= \left[\frac{x^2 \log x}{2} - \frac{x^2}{4}\right]_0^1$$

$\Rightarrow \quad \log I = -\frac{1}{4}$

$\therefore \quad I = e^{-\frac{1}{4}}$

Hence, (i) → (s), (ii) → (p), (iii) → (q)

**63.** (a) We have, $f(x) = \int_0^1 e^{x+t} f(t) dt$

$\Rightarrow \quad f(x) = e^x \int_0^1 e^t f(t) dt \quad$ or $\quad f(x) = A \cdot e^x$

where, $A = \int_0^1 e^t\, f(t) dt$

$$f(t) = Ae^t$$

where, $A = \int_0^1 e^t f(t)\, dt$

$$A = \int_0^1 e^t \cdot A \cdot e^t dt = A \int_0^1 e^{2t} dt$$

which implies either $A = 0$ or $\int_0^1 e^{2t}\, dt = 0$

But $\quad \int_0^1 e^{2t} dt \neq 0$

Hence, $f(x) = 0 \Rightarrow f(1) = 0$

**64.** (a) Given, $g(x) = e^x \int_0^1 e^t g(t)\, dt + x$

$\Rightarrow \quad g(x) = Be^x + x$

or $g(t) = Be^t + t$, where $g(t) = Be^t + t$

$$B = \int_0^1 e^t (Be^t + t)dt$$

$\Rightarrow \quad B = B\int_0^1 e^{2t} dt + \int_0^1 e^t \cdot t\, dt$

But $\int_0^1 e^{2t} dt = \frac{1}{2}(e^2 - 1)$ and $\int_0^1 t\, e^t dt = 1$

$\therefore \quad B = \frac{B}{2}(e^2 - 1) + 1$

$\Rightarrow \quad 2B = B(e^2 - 1) + 2$

$\Rightarrow \quad 3B = Be^2 + 2 \Rightarrow B = \frac{2}{3 - e^2}$

$\Rightarrow \quad g(x) = \left(\frac{2}{3 - e^2}\right)e^x + x$

$\Rightarrow \quad g(0) = \frac{2}{3 - e^2}$ and $f(0) = 0$

$\therefore \quad g(0) - f(0) = \frac{2}{3 - e^2}$

**65.** (b) $g(2) = \frac{2e^2}{3 - e^2} + 2 = \frac{6}{3 - e^2}$

$\therefore \quad \frac{g(0)}{g(2)} = \frac{2}{3 - e^2} \times \frac{3 - e^2}{6} = \frac{1}{3}$

**66.** (c) $D(t) = \frac{|0 + 0 - 1|}{\sqrt{a^2e^{2t} + b^2e^{-2t}}} = \frac{1}{\sqrt{a^2e^{2t} + b^2e^{-2t}}}$

$$I = \int_0^1 \frac{dt}{[D(t)]^2} = \int_0^1 (a^2e^{2t} + b^2e^{-2t})dt$$

$$= \left[a^2 \frac{1}{2} e^{2t} - \frac{b^2}{2} e^{-2t}\right]_0^1 = \frac{1}{2}[(a^2e^2 - b^2e^{-2}) - (a^2 - b^2)]$$

$$= \frac{1}{2}[a^2(e^2 - 1) - b^2(e^{-2} - 1)] = \frac{1}{2}\left[a^2(e^2 - 1) - b^2 \frac{(1 - e^2)}{e^2}\right]$$

$$= \frac{1}{2}(e^2 - 1)\left[a^2 + \frac{b^2}{e^2}\right] = \frac{1}{2}(e^2 - 1)\left(a^2 + \frac{b^2}{e^2}\right)$$

**67.** (d) We have, $I = \dfrac{e^2-1}{2}\left(a^2 + \dfrac{b^2}{e^2}\right)$

We have, $ab = 1 \Rightarrow b = \dfrac{1}{a}$

$\therefore$ $I = \dfrac{e^2-1}{2}\left(a^2 + \dfrac{1}{a^2e^2}\right)$

$= \dfrac{e^2-1}{2}\left[\left(a - \dfrac{1}{ae}\right)^2 + \dfrac{2}{e}\right]$

$I$ is minimum, if $a = \dfrac{1}{ae}$.

$\therefore$ $a^2 = \dfrac{1}{e} \Rightarrow \dfrac{1}{b^2} = \dfrac{1}{e}$

$\Rightarrow$ $b = \sqrt{e}$

**68.** (b) $I$ is minimum, if $a - \dfrac{1}{ea} = 0$

Hence, $I_{min} = \dfrac{e^2-1}{2} \times \dfrac{2}{e} = \dfrac{e^2-1}{e} = e - \dfrac{1}{e}$

**69.** (4) Since, $I$ is independent of $x$, so $\dfrac{dI}{dx} = 0$.

$\Rightarrow$ $f(xy)\cdot y - f(x) = 0$

$\Rightarrow$ $f(xy)\cdot y = f(x)$

On putting, $y = \dfrac{2}{x} \Rightarrow f(2)\left(\dfrac{2}{x}\right) = f(x)$

$\Rightarrow$ $f(x) = \dfrac{4}{x}$ $\quad [\because f(2) = 2]$

$\therefore$ $\int_1^e \dfrac{4}{x}dx = 4\log e = 4$

**70.** (4) $I = \int_{k\pi}^{(k+1)\pi} \dfrac{|\sin 2x|}{|\sin x| + |\cos x|}dx$

Given integrand is periodic with period $\dfrac{\pi}{2}$.

$\therefore$ $I = \int_0^{\pi} \dfrac{|\sin 2x|}{|\sin x| + |\cos x|}dx$

$= 2\int_0^{\pi/2} \dfrac{|\sin 2x|}{|\sin x| + |\cos x|}dx$

$\Rightarrow$ $I = 2\cdot 2\int_0^{\pi/4} \dfrac{|\sin 2x|}{|\sin x| + |\cos x|}dx$

$= 4\int_0^{\pi/4} \dfrac{|\sin 2x|}{|\sin x| + |\cos x|}dx$

$\Rightarrow$ $I = 4\int_0^{\pi/4} \dfrac{\sin 2x}{\sin x + \cos x}dx$ $\quad$ ...(i)

and $J = \int_0^{\pi/4} \dfrac{1}{\sin x + \cos x}dx$ $\quad$ ...(ii)

$\Rightarrow$ $I + 4J = 4\int_0^{\pi/4} \dfrac{1 + \sin 2x}{\sin x + \cos x}dx$

$= 4\int_0^{\pi/4} (\sin x + \cos x)dx$

$= 4[-\cos x + \sin x]_0^{\pi/4}$

$= 4[0 - (-1)] = 4$

**71.** (1) We have, $f(x + 13) + f(x + 630) = 0$

On replacing $x$ by $x - 13$, we get

$f(x) + f(x + 617) = 0$ $\quad$ ...(i)

Again, on replacing $x$ by $x + 617$, we get

$f(x + 617) + f(x + 1234) = 0$ $\quad$ ...(ii)

On subtracting Eq. (i) from Eq. (ii), we get

$f(x + 1234) = f(x)$

$\because f(x)$ is periodic with period $T = 1234$.

Now, $\sum_{p=1}^{777} \int_p^{p+T} f(x)dx = \lambda \int_0^T f(x)\,dx$

$\Rightarrow$ $\sum_{p=1}^{777} \int_0^T f(x)dx = \lambda \int_0^T f(x)\,dx$

$\Rightarrow$ $\int_0^T f(x)\,dx \times 777 = \lambda \int_0^T f(x)\,dx$

$\therefore$ $\lambda = 777$

Then, $\dfrac{\lambda + T}{2011} = \dfrac{1234 + 777}{2011} = \dfrac{2011}{2011} = 1$

**72.** (5) We have, $f(2 - x) = f(2 + x)$

On replacing $x$ by $2 - x$, we get

$f(2 - 2 + x) = f(2 + 2 - x)$

$f(x) = f(4 - x) = f(x + 4)$

Since, $f(x)$ is periodic with period 4.

We have, $\int_0^2 f(x)dx = 5$

$\Rightarrow$ $\dfrac{1}{25}\int_0^{50} f(x)dx = \dfrac{1}{25} \times \dfrac{25}{2}\int_0^4 f(x)dx$

$= \dfrac{1}{2}\int_0^4 f(x)dx = \int_0^2 f(x)dx$

$= 5$

**73.** (3) We have, $\int_1^{xy} f(t)\,dt = y\int_1^x f(t)dt + x\int_1^y f(t)\,dt$

On differentiating w.r.t. $x$, treating $y$ as constant, we get

$f(xy)\cdot y = y\,f(x) + \int_1^y f(t)\,dt$

On putting $x = 1$, we get

$f(y)\cdot y = y\,f(1) + \int_1^y f(t)\,dt$

$\Rightarrow$ $f(y)\cdot y = 3y + \int_1^y f(t)\,dt$

Now, on differentiating w.r.t. $y$, we get

$yf'(y) + f(y) = 3 + f(y)$

$\Rightarrow$ $f'(y) = \dfrac{3}{y} = \dfrac{\lambda}{y}$

Hence, $\lambda = 3$

**74.** (1) We have, $f(x) = \int_a^x f[(x)]^{-2}dx$

$f'(x) = [f(x)]^{-2}\cdot 1$

$\therefore$ $[f(x)]^2 f'(x) = 1$

**75.** (1) Let $I = \int_0^x \left(x - [x] - \dfrac{1}{2}\right)dx$

$= \int_0^{[x]+\{x\}} \left(x - [x] - \dfrac{1}{2}\right)dx$

$= \int_0^{[x]} \left(\{x\} - \dfrac{1}{2}\right)dx + \int_{[x]}^{[x]+\{x\}} \left(\{x\} - \dfrac{1}{2}\right)dx$

$= [x]\int_0^1 \left(\{x\} - \dfrac{1}{2}\right)dx + \int_0^{\{x\}} \left(\{x\} - \dfrac{1}{2}\right)dx$

$= [x]\int_0^1 \left(x - \dfrac{1}{2}\right)dx + \int_0^{\{x\}} \left(x - \dfrac{1}{2}\right)dx$

$= [x]\left(\dfrac{x^2}{2} - \dfrac{x}{2}\right)_0^1 + \left(\dfrac{x^2}{2} - \dfrac{1}{2}x\right)_0^{\{x\}}$

$= 0 + \dfrac{\{x\}(\{x\} - 1)}{2} = \dfrac{\{x\}(\{x\} - 1)}{2}$

On comparing,

$\dfrac{f(x)\cdot g(x)}{2} = \dfrac{\{x\}(\{x\} - 1)}{2}$

$\therefore$ $|f(x) - g(x)| = 1$

## JEE Main & AIEEE Archive

**1.** (c) Apply the property, $\int_a^b f(x)dx = \int_a^b f(a+b-x)dx$ and then add.

Let $I = \int_2^4 \frac{\log x^2}{\log x^2 + \log(36-12x+x^2)}dx$

$= \int_2^4 \frac{2\log x}{2\log x + \log(6-x)^2}dx = \int_2^4 \frac{2\log x}{2[\log x + \log(6-x)]}dx$

$\Rightarrow \quad I = \int_2^4 \frac{\log x}{[\log x + \log(6-x)]}dx \quad \ldots(i)$

$\Rightarrow \quad I = \int_2^4 \frac{\log(6-x)}{\log(6-x)+\log x}dx \quad \ldots(ii)$

$\left[\because \int_a^b f(x)dx = \int_a^b f(a+b-x)dx\right]$

On adding Eqs. (i) and (ii), we get

$$2I = \int_2^4 \frac{\log x + \log(6-x)}{\log x + \log(6-x)}dx$$

$\Rightarrow \quad 2I = \int_2^4 dx = [x]_2^4 \Rightarrow 2I = 2 \Rightarrow I = 1$

**2.** (d) $\int_0^\pi \sqrt{\left(1-2\sin\frac{x}{2}\right)^2}dx = \int_0^\pi \left|1-2\sin\frac{x}{2}\right|dx$

$= \int_0^{\pi/3}\left(1-2\sin\frac{x}{2}\right)dx - \int_{\pi/3}^{\pi}\left(1-2\sin\frac{x}{2}\right)dx$

$\left[\because |x-a| = \begin{cases} x-a, & x \ge a \\ -(x-a), & x < a\end{cases}\right]$

$= \left[x + 4\cos\frac{x}{2}\right]_0^{\pi/3} - \left[x + 4\cos\frac{x}{2}\right]_{\pi/3}^{\pi}$

$= 4\sqrt{3} - 4 - \frac{\pi}{3}$

**3.** (d) Let $I = \int_{\pi/6}^{\pi/3} \frac{dx}{1+\sqrt{\tan x}} \quad \ldots(i)$

$\therefore \quad I = \int_{\pi/6}^{\pi/3} \frac{dx}{1+\sqrt{\tan\left(\frac{\pi}{2}-x\right)}} = \int_{\pi/6}^{\pi/3}\frac{dx}{1+\sqrt{\cot x}}$

$\Rightarrow \quad I = \int_{\pi/6}^{\pi/3} \frac{\sqrt{\tan x}}{1+\sqrt{\tan x}}dx \quad \ldots(ii)$

On adding Eqs. (i) and (ii), we get

$2I = \int_{\pi/6}^{\pi/3} dx \Rightarrow 2I = [x]_{\pi/6}^{\pi/3}$

$\therefore \quad I = \frac{1}{2}\left[\frac{\pi}{3} - \frac{\pi}{6}\right] = \frac{\pi}{12}$

Hence, Statement I is incorrect.

But $\int_a^b f(x)\,dx = \int_a^b f(a+b-x)\,dx$ is a correct statement by property of definite integrals.

**4.** (b,c) Given, integral $g(x) = \int_0^x \cos 4t\, dt$

To find, $g(x+\pi)$ in terms of $g(x)$ and $g(\pi)$.

$$g(x) = \int_0^x \cos 4t\, dt$$

$\Rightarrow \quad g(x+\pi) = \int_{t=0}^{t=x+\pi} \cos 4t\, dt$

$= \int_0^x \cos 4t\, dt + \int_x^{x+\pi}\cos 4t\, dt$

$= g(x) + I_1 \quad$ [say]

where, $I_1 = \int_x^{x+\pi} \cos 4t\, dt$

$= \int_0^\pi \cos 4t\, dt \quad$ [using definite integral property]

$= g(\pi)$

Also, $I_1 = \left(\frac{\sin 4t}{4}\right)_0^\pi = \left(\frac{\sin 4\pi}{4} - \frac{\sin 0}{4}\right) = 0$

i.e. the value of $I_1$ is zero.

Since, $g(\pi) = I_1 = 0$, so value of $g(x+\pi)$ does not depend on $g(\pi)$ and we can add or subtract $g(\pi)$ to or from $g(x)$.

Hence, $\quad g(x+\pi) = g(x) + g(\pi)$

or $\quad g(x+\pi) = g(x) - g(\pi)$

**5.** (d) Let $I = \int_0^1 \frac{8\log(1+x)}{(1+x^2)}dx$

Put $x = \tan\theta \Rightarrow dx = \sec^2\theta\, d\theta$

When $x = 0 \Rightarrow \tan\theta = 0$

$\therefore \quad \theta = 0$

When $\quad x = 1 = \tan\theta$

$\Rightarrow \quad \theta = \frac{\pi}{4}$

$\therefore \quad I = \int_0^{\pi/4} \frac{8\log(1+\tan\theta)}{1+\tan^2\theta}\cdot\sec^2\theta d\theta$

$I = 8\int_0^{\pi/4} \log(1+\tan\theta)\, d\theta \quad \ldots(i)$

Using $\int_0^a f(x)\,dx = \int_0^a f(a-x)\,dx$

$\therefore \quad I = 8\int_0^{\pi/4}\log\left\{1+\tan\left(\frac{\pi}{4}-\theta\right)\right\}d\theta$

$= 8\int_0^{\pi/4}\log\left\{1+\frac{1-\tan\theta}{1+\tan\theta}\right\}d\theta$

$= 8\int_0^{\pi/4}\log\left\{\frac{2}{1+\tan\theta}\right\}d\theta \quad \ldots(ii)$

On adding Eqs. (i) and (ii), we get

$$2I = 8\int_0^{\pi/4}\left[\log\{1+\tan\theta\} + \log\left\{\frac{2}{1+\tan\theta}\right\}\right]d\theta$$

$\Rightarrow \quad I = 4\int_0^{\pi/4}\log 2\, d\theta = 4\cdot\log 2[\theta]_0^{\pi/4}$

$= 4\log 2\cdot\left(\frac{\pi}{4} - 0\right) = \pi\log 2$

**6.** (d) Here, $\int_0^{1.5} x[x^2]\,dx$

$I = \int_0^1 x\cdot 0\, dx + \int_1^{\sqrt{2}} x\cdot 1\, dx + \int_{\sqrt{2}}^{1.5} x\cdot 2\, dx$

$= 0 + \left[\frac{x^2}{2}\right]_1^{\sqrt{2}} + [x^2]_{\sqrt{2}}^{1.5}$

$= \frac{1}{2}\{2-1\} + \{(1.5)^2 - 2\}$

$= \frac{1}{2} + 2.25 - 2 = \frac{1}{2} + \frac{1}{4} = \frac{3}{4}$

**7.** (b) We have, $\quad p'(x) = p'(1-x),\ \forall\ x \in [0,1],$

$p(0) = 1,\ p(1) = 41 \quad$ [given]

On integrating, we get

$$\int p'(x)\,dx = \int p'(1-x)\,dx$$

$\Rightarrow \quad p(x) = -p(1-x) + C$

Put $x = 1$, $\quad p(1) = -p(0) + C$

$\Rightarrow \quad 41 = -1 + C$

$\Rightarrow \quad C = 42$

$\therefore \quad p(x) + p(1-x) = 42 \quad \ldots(i)$

Now, $I = \int_0^1 p(x)dx$

$\Rightarrow$ $I = \int_0^1 p(1-x)dx \left[\because \int_0^a f(x)\,dx = \int_0^a f(a-x)\,dx\right]$

$\Rightarrow$ $2I = \int_0^1 [p(x) + p(1-x)]dx$ [on adding]

$= \int_0^1 42\,dx = 42$ [by Eq. (i)]

$\therefore$ $I = 21$

**8.** (d) Let $I = \int_0^{\pi} [\cot x]\,dx$ ...(i)

$\Rightarrow$ $I = \int_0^{\pi} [\cot(\pi - x)]\,dx$

$= \int_0^{\pi} [-\cot x]\,dx$ ...(ii)

On adding Eqs. (i) and (ii), we get

$$2I = \int_0^{\pi} [\cot x]\,dx + \int_0^{\pi} [-\cot x]\,dx$$

$$= \int_0^{\pi} (-1)\,dx \qquad \left[\because [x] + [-x] = \begin{cases} -1, \text{if } x \notin Z \\ 0, \text{if } x \in Z \end{cases}\right]$$

$$= [-x]_0^{\pi}$$

$$= -\pi$$

$\therefore$ $I = -\dfrac{\pi}{2}$

**9.** (c) Since, $I = \int_0^1 \dfrac{\sin x}{\sqrt{x}}\,dx < \int_0^1 \dfrac{x}{\sqrt{x}}\,dx,$

because in $x \in (0, 1)$, $x > \sin x$

$\Rightarrow$ $I < \int_0^1 \sqrt{x}\,dx = \dfrac{2}{3}[x^{3/2}]_0^1$

$\Rightarrow$ $I < \dfrac{2}{3}$

and $J = \int_0^1 \dfrac{\cos x}{\sqrt{x}}\,dx < \int_0^1 x^{-\frac{1}{2}}\,dx$

$= 2[x^{1/2}]_0^1 = 2$

$\therefore$ $J < 2$

**10.** (a) Since, $f(x) = \int_1^x \dfrac{\log t}{1+t}\,dt$

and $F(e) = f(e) + f\left(\dfrac{1}{e}\right)$

$\Rightarrow$ $F(e) = \int_1^e \dfrac{\log t}{1+t}\,dt + \int_1^{1/e} \dfrac{\log t}{1+t}\,dt$

Put $t = \dfrac{1}{t}$ in second integration,

$$\therefore\ F(e) = \int_1^e \frac{\log t}{1+t}\,dt + \int_1^e \frac{-\log t}{1+\frac{1}{t}}\,d\left(\frac{1}{t}\right)$$

$$= \int_1^e \frac{\log t}{1+t}\,dt - \int_1^e \frac{t\log t}{(1+t)} \times \left(\frac{-dt}{t^2}\right)$$

$$= \int_1^e \frac{\log t}{1+t}\,dt + \int_1^e \frac{\log t}{t(1+t)}\,dt$$

$$= \int_1^e \frac{\log t}{1+t}\,dt + \int_1^e \frac{\log t}{t}\,dt - \int_1^e \frac{\log t}{(1+t)}\,dt$$

$$= \int_1^e \frac{\log t}{t}\,dt \qquad \left[\because \frac{1}{t(1+t)} = \frac{1}{t} - \frac{1}{t+1}\right]$$

$$= \left[\frac{(\log t)^2}{2}\right]_1^e = \frac{1}{2}[(\log e)^2 - (\log 1)^2]$$

$$= \frac{1}{2}$$

**11.** (a) Given, $\int_{\sqrt{2}}^{x} \dfrac{dt}{t\sqrt{t^2-1}} = \dfrac{\pi}{2}$

$\Rightarrow$ $[\sec^{-1} t]_{\sqrt{2}}^{x} = \dfrac{\pi}{2} \Rightarrow \sec^{-1} x - \sec^{-1}\sqrt{2} = \dfrac{\pi}{2}$

$\Rightarrow$ $\sec^{-1} x = \dfrac{\pi}{2} + \dfrac{\pi}{4} = \dfrac{3\pi}{4}$

$\therefore$ $x = \sec\dfrac{3\pi}{4} = -\sqrt{2}$

**12.** (a) Let $I = \int_3^6 \dfrac{\sqrt{x}}{\sqrt{9-x} + \sqrt{x}}\,dx$ ...(i)

$= \int_3^6 \dfrac{\sqrt{9-x}}{\sqrt{9-9+x} + \sqrt{9-x}}\,dx$

$\Rightarrow$ $I = \int_3^6 \dfrac{\sqrt{9-x}}{\sqrt{x} + \sqrt{9-x}}\,dx$ ...(ii)

On adding Eqs. (i) and (ii), we get

$$2I = \int_3^6 \frac{\sqrt{x} + \sqrt{9-x}}{\sqrt{x} + \sqrt{9-x}}\,dx = [x]_3^6 = 6 - 3 = 3$$

$\therefore$ $I = \dfrac{3}{2}$

**13.** (b) Let $I = \int_{-3\pi/2}^{-\pi/2} [(x+\pi)^3 + \cos^2(x + 3\pi)]\,dx$

$\Rightarrow$ $I = \int_{-3\pi/2}^{-\pi/2} [(x+\pi)^3 + \cos^2 x]\,dx$ ...(i)

$$\Rightarrow\ I = \int_{-3\pi/2}^{-\pi/2} \left[\left(-\frac{\pi}{2} - \frac{3\pi}{2} - x - \pi\right)^3 + \cos^2\left(-\frac{\pi}{2} - \frac{3\pi}{2} - x\right)\right]dx$$

$$\left[\because \int_a^b f(x)\,dx = \int_a^b f(a+b-x)\,dx\right]$$

$\Rightarrow$ $I = \int_{-3\pi/2}^{-\pi/2} [-(x+\pi)^3 + \cos^2 x]\,dx$ ...(ii)

On adding Eqs. (i) and (ii), we get

$$2I = \int_{-3\pi/2}^{-\pi/2} 2\cos^2 x\,dx$$

$$= \int_{-3\pi/2}^{-\pi/2} (1 + \cos 2x)\,dx$$

$$= \left[x + \frac{\sin 2x}{2}\right]_{-3\pi/2}^{-\pi/2}$$

$$= \left[-\frac{\pi}{2} + \frac{\sin(-\pi)}{2} - \left(-\frac{3\pi}{2} + \frac{\sin(-3\pi)}{2}\right)\right]$$

$$= -\frac{\pi}{2} + \frac{3\pi}{2} = \pi$$

$\therefore$ $I = \dfrac{\pi}{2}$

**14.** (c) Let $I = \int_0^{\pi} x f(\sin x)\,dx$ ...(i)

$\Rightarrow$ $I = \int_0^{\pi} (\pi - x) f[\sin(\pi - x)]\,dx$

$\Rightarrow$ $I = \int_0^{\pi} (\pi - x) f(\sin x)\,dx$ ...(ii)

On adding Eqs. (i) and (ii), we get

$$2I = \int_0^{\pi} \pi f(\sin x)\,dx$$

$\Rightarrow$ $I = \dfrac{\pi}{2}\int_0^{\pi} f(\sin x)\,dx$ ...(iii)

$$\left[\because \int_0^{2a} f(x)\,dx = \begin{cases} 2\int_0^a f(x)\,dx, \text{if } f(2a-x) = f(x) \\ 0 \quad, \text{if } f(2a-x) = -f(x) \end{cases}\right]$$

$\Rightarrow$ $I = \pi\int_0^{\pi/2} f(\sin x)\,dx$

Put $\frac{\pi}{2} - x = t$

$\Rightarrow x = \frac{\pi}{2} - t$

Put $dx = -dt$ in Eq. (iii), we get

$$I = \frac{\pi}{2}\int_{-\pi/2}^{\pi/2} f(\cos t)\,dt$$

$$= \frac{\pi}{2}\int_{-\pi/2}^{\pi/2} f(\cos x)\,dx$$

$$= \pi\int_0^{\pi/2} f(\cos x)\,dx$$

[$\because f(\cos x)$ is an even function.]

**15.** (a) Since, $\int_1^a [x]f'(x)\,dx = \int_1^2 f'(x)\,dx + \int_2^3 2f'(x)\,dx + \ldots + \int_{[a]}^a [a]f'(x)\,dx$

$$= [f(x)]_1^2 + 2\,[f(x)]_2^3 + \ldots + [a][f(x)]_{[a]}^a$$

$$= f(2) - f(1) + 2f(3) - 2f(2) + \ldots + [a]f(a) - [a]f([a])$$

$$= [a]f(a) - \{f(1) + f(2) + \ldots + f([a])\}$$

**16.** (a) Let $A = \lim_{n\to\infty}\left[\frac{1}{n^2}\sec^2\frac{1}{n^2} + \frac{2}{n^2}\sec^2\frac{4}{n^2} + \ldots + \frac{n}{n^2}\sec^2 1\right]$

$$= \lim_{n\to\infty}\frac{1}{n}\left[\frac{1}{n}\sec^2\left(\frac{1}{n}\right)^2 + \frac{2}{n}\sec^2\left(\frac{2}{n}\right)^2 + \ldots + \frac{n}{n}\sec^2\left(\frac{n}{n}\right)^2\right]$$

$$= \lim_{n\to\infty}\frac{1}{n}\sum_{r=1}^{n}\left(\frac{r}{n}\right)\sec^2\left(\frac{r}{n}\right)^2$$

$\Rightarrow A = \int_0^1 x\sec^2(x^2)\,dx$

Put $x^2 = t \Rightarrow x\,dx = \frac{dt}{2}$

$\therefore A = \frac{1}{2}\int_0^1 \sec^2 t\,dt$

$$= \frac{1}{2}[\tan t]_0^1 = \frac{1}{2}\tan 1$$

**17.** (a) $\lim_{x\to 2}\int_6^{f(x)}\frac{4t^3}{x-2}\,dt = \lim_{x\to 2}\frac{\int_6^{f(x)} 4t^3\,dt}{(x-2)}$ $\left[\frac{0}{0}\text{form}\right]$

[by Leibnitz's rule]

$$= \lim_{x\to 2}\frac{4\{f(x)\}^3}{1}f'(x) = 4\{f(2)\}^3 f'(2)$$

$$= 4\times(6)^3\times\frac{1}{48} \quad \left[\because f(2) = 6 \text{ and } f'(2) = \frac{1}{48}\right]$$

$= 18$

**18.** (c) Given that,

$$I_1 = \int_0^1 2^{x^2}\,dx$$

$$I_2 = \int_0^1 2^{x^3}\,dx,$$

$$I_3 = \int_1^2 2^{x^2}\,dx$$

and $I_4 = \int_1^2 2^{x^3}\,dx$

Since, $2^{x^3} < 2^{x^2}$ for $0 < x < 1$

and $2^{x^3} > 2^{x^2}$ for $x > 1$

$\therefore \int_0^1 2^{x^3}\,dx < \int_0^1 2^{x^2}\,dx$

and $\int_1^2 2^{x^3}\,dx > \int_1^2 2^{x^2}\,dx$

$\Rightarrow I_2 < I_1$ and $I_4 > I_3$

**19.** (c) Let $I = \int_{-\pi}^{\pi}\frac{\cos^2 x}{1+a^x}\,dx,\ a > 0$ ...(i)

Put $x = -x$, we get

$$I = \int_{-\pi}^{\pi}\frac{\cos^2 x}{1+a^{-x}}\,dx \quad \text{...(ii)}$$

On adding Eqs. (i) and (ii), we get

$$2I = \int_{-\pi}^{\pi}\frac{(1+a^x)\cos^2 x}{(1+a^x)}\,dx$$

$$= \int_{-\pi}^{\pi}\cos^2 x\,dx$$

$$= \int_{-\pi}^{\pi}\left(\frac{\cos 2x + 1}{2}\right)dx$$

$$= \frac{1}{2}\left[\left(\frac{\sin 2x}{2} + x\right)\right]_{-\pi}^{\pi}$$

$$= \frac{1}{2}\left[\frac{\sin 2\pi}{2} + \pi - \left(\frac{\sin(-2\pi)}{2} - \pi\right)\right]$$

$\Rightarrow 2I = \frac{1}{2}(\pi + \pi)$

$\therefore I = \frac{\pi}{2}$

**20.** (a) We have, $\int_{-2}^{3}|1-x^2|\,dx$

$$= \int_{-2}^{-1}(x^2-1)\,dx + \int_{-1}^{1}(1-x^2)\,dx + \int_1^3(x^2-1)\,dx$$

$$= \left[\frac{x^3}{3} - x\right]_{-2}^{-1} + \left[x - \frac{x^3}{3}\right]_{-1}^{1} + \left[\frac{x^3}{3} - x\right]_1^3$$

$$= \left(-\frac{1}{3} + 1 + \frac{8}{3} - 2\right) + \left(1 - \frac{1}{3} + 1 - \frac{1}{3}\right) + \left(9 - 3 - \frac{1}{3} + 1\right)$$

$$= \frac{4}{3} + \frac{4}{3} + \frac{20}{3} = \frac{28}{3}$$

**21.** (c) We have, $\int_0^{\pi/2}\frac{(\sin x + \cos x)^2}{\sqrt{1+\sin 2x}}\,dx$

$$= \int_0^{\pi/2}\frac{(\sin x + \cos x)^2}{\sqrt{(\sin x + \cos x)^2}}\,dx$$

$$= \int_0^{\pi/2}(\sin x + \cos x)\,dx$$

$$= [-\cos x + \sin x]_0^{\pi/2}$$

$$= -\cos\frac{\pi}{2} + \sin\frac{\pi}{2} + \cos 0 - \sin 0$$

$$= -0 + 1 + 1 - 0 = 2$$

**22.** (b) Let $I = \int_0^{\pi} x f(\sin x)\,dx$ ...(i)

$\therefore I = \int_0^{\pi}(\pi - x)f[\sin(\pi - x)]\,dx$

$\Rightarrow I = \int_0^{\pi}(\pi - x)f(\sin x)\,dx$ ...(ii)

On adding Eqs. (i) and (ii), we get

$$2I = \int_0^{\pi}(x + \pi - x)\,f(\sin x)\,dx$$

$\Rightarrow 2I = \pi\int_0^{\pi} f(\sin x)\,dx$

$\Rightarrow 2I = 2\pi\int_0^{\pi/2} f(\sin x)\,dx$

$\Rightarrow I = \pi\int_0^{\pi/2} f(\sin x)\,dx$

$\Rightarrow A\int_0^{\pi/2} f(\sin x)\,dx = \pi\int_0^{\pi/2} f(\sin x)\,dx \quad \left[\because I = A\int_0^{\pi/2} f(\sin x)\,dx\right]$

$\therefore A = \pi$

**23.** (a) Given that, $f(x) = \dfrac{e^x}{1+e^x}$

$\therefore \quad f(a) = \dfrac{e^a}{1+e^a}$ ...(i)

and $f(-a) = \dfrac{e^{-a}}{1+e^{-a}} = \dfrac{1}{1+e^a}$ ...(ii)

On adding Eqs. (i) and (ii), we get

$$f(a) + f(-a) = 1$$

$\Rightarrow \quad f(a) = 1 - f(-a)$

Let $f(-a) = t \Rightarrow f(a) = 1 - t$

Now, $I_1 = \int_t^{1-t} xg\,[x(1-x)]\,dx$ ...(iii)

$\left[\because I = \int_a^b f(x)\,dx = \int_a^b f(a+b-x)\,dx\right]$

$\Rightarrow \quad I_1 = \int_t^{1-t} (1-x)g[x(1-x)]\,dx$ ...(iv)

On adding Eqs. (iii) and (iv), we get

$2I_1 = \int_t^{1-t} g[x(1-x)]\,dx = I_2$ [given]

$\therefore \quad \dfrac{I_2}{I_1} = 2$

**24.** (b) Now, $\lim\limits_{n\to\infty} \sum\limits_{r=1}^{n} \dfrac{1}{n} e^{r/n} = \int_0^1 e^x\,dx = [e^x]_0^1 = e - 1$

**25.** (b) Given that, $f(y) = e^y$, $g(y) = y$

and $F(t) = \int_0^t f(t-y)g(y)\,dy$

$= \int_0^t e^{t-y} \cdot y\,dy = e^t \int_0^t e^{-y} y\,dy$

$= e^t \left[(-ye^{-y})_0^t - \int_0^t 1(-e^{-y})\,dy\right]$

$= e^t[(-te^{-t} - 0) - (e^{-y})_0^t]$

$= e^t[-te^{-t} - e^{-t} + 1]$

$\therefore \quad F(t) = e^t - (1+t)$

**26.** (b) Let $I = \int_a^b xf(x)\,dx$ ...(i)

$\Rightarrow \quad I = \int_a^b (a+b-x)\,f(a+b-x)\,dx$

$\Rightarrow \quad I = \int_a^b (a+b-x)f(x)\,dx$ $[\because f(a+b-x) = f(x)]$

$\Rightarrow \quad I = (a+b)\int_a^b f(x)\,dx - \int_a^b x\,f(x)\,dx$

$\Rightarrow \quad I = (a+b)\int_a^b f(x)\,dx - I$ [from Eq. (i)]

$\Rightarrow \quad 2I = (a+b)\int_a^b f(x)\,dx$

$\therefore \quad I = \left(\dfrac{a+b}{2}\right)\int_a^b f(x)\,dx$

**27.** (c) $\lim\limits_{x\to 0} \dfrac{\int_0^{x^2} \sec^2 t\,dt}{x\sin x}$ $\left[\dfrac{0}{0}\text{ form}\right]$

Using L'Hospital's rule,

$= \lim\limits_{x\to 0} \dfrac{\sec^2 x^2 \cdot 2x}{x\cos x + \sin x}$

$\left[\because \dfrac{d}{dx}\int_0^{x^2} \sec^2 t\,dt = \sec^2 x^2 \cdot 2x \text{ (Leibnitz's rule)}\right]$

Again, using L'Hospital's rule, we get

$= \lim\limits_{x\to 0} \dfrac{2x \cdot 2\sec^2 x^2 \cdot \tan x^2 \cdot 2x + 2\sec^2 x^2}{-x\sin x + \cos x + \cos x}$

$= \dfrac{0 + 2\sec^2 0}{0 + 2\cos 0} = 1$

**28.** (c) Given, $I = \int_0^1 x(1-x)^n\,dx$

Put $1 - x = z \Rightarrow -dx = dz$

$\therefore \quad I = \int_1^0 (1-z)z^n(-dz)$

$= \int_0^1 (1-z)\,z^n\,dz$ $\left[\because \int_a^b f(x)\,dx = -\int_b^a f(x)\,dx\right]$

$= \int_0^1 (z^n - z^{n+1})\,dz$

$= \left[\dfrac{z^{n+1}}{n+1} - \dfrac{z^{n+2}}{n+2}\right]_0^1$

$= \dfrac{1}{n+1} - \dfrac{1}{n+2}$

**29.** (d) $\lim\limits_{n\to\infty} \dfrac{\left[\begin{matrix}1 + 2^4 + 3^4 \\ + \ldots + n^4\end{matrix}\right]}{n^5} - \lim\limits_{n\to\infty} \dfrac{\left[\begin{matrix}1 + 2^3 + 3^3 \\ + \ldots + n^3\end{matrix}\right]}{n^5}$

$= \lim\limits_{n\to\infty} \dfrac{1}{n}\sum\limits_{r=1}^{n}\left(\dfrac{r}{n}\right)^4 - \lim\limits_{n\to\infty}\dfrac{1}{n} \times \lim\limits_{n\to\infty}\dfrac{1}{n}\sum\limits_{r=1}^{n}\left(\dfrac{r}{n}\right)^3$

$= \int_0^1 x^4\,dx - \lim\limits_{n\to\infty}\dfrac{1}{n} \times \int_0^1 x^3\,dx$

$= \left[\dfrac{x^5}{5}\right]_0^1 - 0 = \dfrac{1}{5}$

**30.** (d) Given, $\dfrac{d}{dx}F(x) = \dfrac{e^{\sin x}}{x}$, $x > 0$

On integrating both sides, we get

$F(x) = \int \dfrac{e^{\sin x}}{x}\,dx$ ...(i)

Also, $\int_1^4 \dfrac{3}{x} e^{\sin x^3}\,dx = \int_1^4 \dfrac{3x^2}{x^3} \cdot e^{\sin x^3}\,dx$

$= F(k) - F(1)$

Let $x^3 = z$

$\Rightarrow \quad 3x^2\,dx = dz$

$\therefore \quad \int_1^{64} \dfrac{e^{\sin z}}{z}\,dz = F(k) - F(1)$

$\Rightarrow \quad [F(z)]_1^{64} = F(k) - F(1)$ [from Eq.(i)]

$\Rightarrow \quad F(64) - F(1) = F(k) - F(1)$

$\therefore \quad k = 64$

**31.** (c) Given, $f'(x) = f(x)$

and $f(0) = 1$

Let $f(x) = e^x$ ...(i)

Also, $f(x) + g(x) = x^2$

$\Rightarrow \quad g(x) = x^2 - e^x$ ...(ii)

Now, $\int_0^1 f(x)g(x)\,dx = \int_0^1 e^x(x^2 - e^x)\,dx$ [from Eqs. (i) and (ii)]

$= \int_0^1 x^2e^x\,dx - \int_0^1 e^{2x}\,dx$

$= [x^2e^x - \int 2xe^x\,dx]_0^1 - \dfrac{1}{2}[e^{2x}]_0^1$

$= [x^2e^x - 2xe^x + 2e^x]_0^1 - \dfrac{1}{2}(e^2 - 1)$

$= [(x^2 - 2x + 2)e^x]_0^1 - \dfrac{1}{2}e^2 + \dfrac{1}{2}$

$= [(1 - 2 + 2)e^1 - (0 - 0 + 2)e^0] - \dfrac{1}{2}e^2 + \dfrac{1}{2}$

$= e - 2 - \dfrac{e^2}{2} + \dfrac{1}{2}$

$= e - \dfrac{e^2}{2} - \dfrac{3}{2}$

**32.** (a) Since, $|\sin x|$ is a periodic function with period $\pi$.

$$\therefore \int_0^{10\pi} |\sin x|\, dx = 10 \int_0^{\pi} |\sin x|\, dx$$
$$= 10 \int_0^{\pi} \sin x\, dx = 10[-\cos x]_0^{\pi}$$
$$= 10[-\cos \pi + \cos 0]$$
$$= 10[1 + 1] = 20$$

**33.** (b) Since, $I_n = \int_0^{\pi/4} \tan^n x\, dx$

$$\therefore \quad I_{n+2} = \int_0^{\pi/4} \tan^{n+2} x\, dx$$

Now, $I_n + I_{n+2} = \int_0^{\pi/4} \tan^n x\, dx + \int_0^{\pi/4} \tan^{n+2} x\, dx$

$$= \int_0^{\pi/4} \tan^n x (1 + \tan^2 x)\, dx$$
$$= \int_0^{\pi/4} \sec^2 x \tan^n x\, dx$$

Put $\tan x = t$

$\Rightarrow \quad \sec^2 x\, dx = dt$

$$\therefore \quad I_n + I_{n+2} = \int_0^1 t^n\, dt$$
$$= \left[\frac{t^{n+1}}{n+1}\right]_0^1 = \frac{1}{n+1}$$
$$\Rightarrow \lim_{n \to \infty} n[I_n + I_{n+2}] = \lim_{n \to \infty} \frac{n}{n+1}$$
$$= \lim_{n \to \infty} \frac{1}{1 + \frac{1}{n}} = 1$$

**34.** (d) We have, $\int_0^2 [x^2]\, dx$

$$= \int_0^1 [x^2]\, dx + \int_1^{\sqrt{2}} [x^2]\, dx + \int_{\sqrt{2}}^{\sqrt{3}} [x^2]\, dx + \int_{\sqrt{3}}^{2} [x^2]\, dx$$
$$= \int_0^1 0\, dx + \int_1^{\sqrt{2}} 1\, dx + \int_{\sqrt{2}}^{\sqrt{3}} 2\, dx + \int_{\sqrt{3}}^{2} 3\, dx$$
$$= [x]_1^{\sqrt{2}} + [2x]_{\sqrt{2}}^{\sqrt{3}} + [3x]_{\sqrt{3}}^{2}$$
$$= \sqrt{2} - 1 + 2\sqrt{3} - 2\sqrt{2} + 6 - 3\sqrt{3}$$
$$= 5 - \sqrt{3} - \sqrt{2}$$

**35.** (b) Let $I = \int_{-\pi}^{\pi} \frac{2x(1 + \sin x)}{1 + \cos^2 x}\, dx$

$$= \int_{-\pi}^{\pi} \frac{2x}{1 + \cos^2 x}\, dx + \int_{-\pi}^{\pi} \frac{2x \sin x}{1 + \cos^2 x}\, dx$$
$$\Rightarrow \quad I = 0 + 4 \int_0^{\pi} \frac{x \sin x}{1 + \cos^2 x}\, dx \quad \ldots(i)$$

$\left[\because \frac{2x}{1 + \cos^2 x}\right.$ is an odd function

and $\frac{2x \sin x}{1 + \cos^2 x}$ is an even function$\left.\right]$

$$\Rightarrow \quad I = 4 \int_0^{\pi} \frac{(\pi - x) \sin (\pi - x)}{1 + \cos^2 (\pi - x)}\, dx$$
$$\Rightarrow \quad I = 4 \int_0^{\pi} \frac{\pi \sin x}{1 + \cos^2 x}\, dx - 4 \int_0^{\pi} \frac{x \sin x}{1 + \cos^2 x}\, dx$$
$$\Rightarrow \quad I = 4\pi \int_0^{\pi} \frac{\sin x}{1 + \cos^2 x}\, dx - I \quad \text{[from Eq. (i)]}$$
$$\Rightarrow \quad I = 2\pi \int_0^{\pi} \frac{\sin x}{1 + \cos^2 x}\, dx$$

Put $\cos x = t \Rightarrow -\sin x\, dx = dt$

$$\therefore \quad I = -2\pi \int_1^{-1} \frac{1}{1 + t^2}\, dt = 2\pi[\tan^{-1} t]_{-1}^{1}$$
$$= 2\pi \left[\frac{\pi}{4} + \frac{\pi}{4}\right] = \pi^2$$

**36.** (a) Let $I = \int_0^{\pi/2} \frac{\sqrt{\sin x}}{\sqrt{\cos x} + \sqrt{\sin x}}\, dx \quad \ldots(i)$

$$I = \int_0^{\pi/2} \frac{\sqrt{\sin\left(\frac{\pi}{2} - x\right)}}{\sqrt{\cos\left(\frac{\pi}{2} - x\right)} + \sqrt{\sin\left(\frac{\pi}{2} - x\right)}}\, dx$$
$$I = \int_0^{\pi/2} \frac{\sqrt{\cos x}}{\sqrt{\sin x} + \sqrt{\cos x}}\, dx \quad \ldots(ii)$$

On adding Eqs. (i) and (ii), we get

$$2I = \int_0^{\pi/2} 1\, dx = [x]_0^{\pi/2}$$
$$\therefore \quad I = \frac{\pi}{4}$$

## JEE Advanced & IIT JEE Archive

**1.** (0) Here, $f(x) = \begin{cases} [x], & x \le 2 \\ 0, & x > 2 \end{cases}$

$$\therefore I = \int_{-1}^{2} \frac{x f(x^2)}{2 + f(x+1)}\, dx$$
$$= \int_{-1}^{0} \frac{x f(x^2)}{2 + f(x+1)}\, dx + \int_0^1 \frac{x f(x^2)}{2 + f(x+1)}\, dx + \int_1^{\sqrt{2}} \frac{x f(x^2)}{2 + f(x+1)}\, dx$$
$$+ \int_{\sqrt{2}}^{\sqrt{3}} \frac{x f(x^2)}{2 + f(x+1)}\, dx + \int_{\sqrt{3}}^{2} \frac{x f(x^2)}{2 + f(x+1)}\, dx$$
$$= \int_{-1}^{0} 0\, dx + \int_0^1 0\, dx + \int_1^{\sqrt{2}} \frac{x \cdot 1}{2 + 0}\, dx + \int_{\sqrt{2}}^{\sqrt{3}} 0\, dx + \int_{\sqrt{3}}^{2} 0\, dx$$

$$\left[\begin{aligned} \because -1 < x < 0 &\Rightarrow 0 < x^2 < 1 \Rightarrow [x^2] = 0 \\ 0 < x < 1 &\Rightarrow 0 < x^2 < 1 \Rightarrow [x^2] = 0 \end{aligned}\right]$$

$$1 < x < \sqrt{2} \Rightarrow \begin{cases} 1 < x^2 < 2 \Rightarrow [x^2] = 1 \\ 2 < x + 1 < 1 + \sqrt{2} \Rightarrow f(x+1) = 0 \end{cases}$$

$$\sqrt{2} < x < \sqrt{3} \Rightarrow 2 < x^2 < 3 \Rightarrow f(x^2) = 0$$

and $\sqrt{3} < x < 2 \Rightarrow 3 < x^2 < 4 \Rightarrow f(x^2) = 0$

$$\Rightarrow \quad I = \int_1^{\sqrt{2}} \frac{x}{2}\, dx = \left[\frac{x^2}{4}\right]_1^{\sqrt{2}}$$
$$= \frac{1}{4}(2 - 1) = \frac{1}{4}$$
$$\therefore \quad 4I = 1 \Rightarrow 4I - 1 = 0$$

**2.** (9) Here, $\alpha = \int_0^1 e^{(9x + 3\tan^{-1} x)} \left(\frac{12 + 9x^2}{1 + x^2}\right) dx$

Put $9x + 3\tan^{-1} x = t$

$$\Rightarrow \quad \left(9 + \frac{3}{1 + x^2}\right) dx = dt$$
$$\therefore \quad \alpha = \int_0^{9 + 3\pi/4} e^t\, dt = [e^t]_0^{9 + 3\pi/4} = e^{9 + 3\pi/4} - 1$$
$$\Rightarrow \quad \log_e |1 + \alpha| = 9 + \frac{3\pi}{4}$$
$$\Rightarrow \log_e |\alpha + 1| - \frac{3\pi}{4} = 9$$

**3.** (7) Here, $\lim_{x\to 1}\frac{F(x)}{G(x)}=\frac{1}{14}$

$\Rightarrow \quad \lim_{x\to 1}\frac{F'(x)}{G'(x)}=\frac{1}{14}$ [using L' Hospital's rule]...(i)

As $\quad F(x)=\int_{-1}^{x} f(t)dt \Rightarrow F'(x)=f(x)$ ...(ii)

and $\quad G(x)=\int_{-1}^{x} t\,|f\{f(t)\}|dt$

$\Rightarrow \quad G'(x)=x|f\{f(x)\}|$ ...(iii)

$\therefore \quad \lim_{x\to 1}\frac{F(x)}{G(x)}=\lim_{x\to 1}\frac{F'(x)}{G'(x)}=\lim_{x\to 1}\frac{f(x)}{x|f\{f(x)\}|}$

$=\frac{f(1)}{1|f\{f(1)\}|}=\frac{1/2}{|f(1/2)|}$ ...(iv)

Given, $\quad \lim_{x\to 1}\frac{F(x)}{G(x)}=\frac{1}{14}$

$\therefore \quad \frac{\frac{1}{2}}{\left|f\left(\frac{1}{2}\right)\right|}=\frac{1}{14} \Rightarrow \left|f\left(\frac{1}{2}\right)\right|=7$

**4.** (a,b) Here, $f(x)=7\tan^8 x+7\tan^6 x-3\tan^4 x-3\tan^2 x$

for all $x\in\left(\frac{-\pi}{2},\frac{\pi}{2}\right)$.

$\therefore \quad f(x)=7\tan^6 x\sec^2 x-3\tan^2 x\sec^2 x$

$=(7\tan^6 x-3\tan^2 x)\sec^2 x$

Now, $\int_0^{\pi/4} x f(x)dx=\int_0^{\pi/4} \underset{\text{I}}{x}\,\underset{\text{II}}{(7\tan^6 x-3\tan^2 x)\sec^2 x}\,dx$

$=[x(\tan^7 x-\tan^3 x)]_0^{\pi/4}-\int_0^{\pi/4} 1\,(\tan^7 x-\tan^3 x)\,dx$

$=0-\int_0^{\pi/4}\tan^3 x(\tan^4 x-1)dx$

$=-\int_0^{\pi/4}\tan^3 x\,(\tan^2 x-1)\sec^2 x\,dx$

Put $\tan x=t \Rightarrow \sec^2 x\,dx=dt$

$\therefore \quad \int_0^{\pi/4} x f(x)dx=-\int_0^1 t^3(t^2-1)dt$

$=\int_0^1 (t^3-t^5)dt=\left[\frac{t^4}{4}-\frac{t^6}{6}\right]_0^1=\frac{1}{4}-\frac{1}{6}=\frac{1}{12}$

Also, $\int_0^{\pi/4} f(x)dx=\int_0^{\pi/4}(7\tan^6 x-3\tan^2 x)\sec^2 x dx$

$=\int_0^1 (7t^6-3t^2)dt=[t^7-t^3]_0^1=0$

**5.** (d) Here, $\quad f'(x)=\frac{192\,x^3}{2+\sin^4 \pi x}$

$\therefore \quad \frac{192x^3}{3}\le f'(x)\le\frac{192x^3}{2}$

On integrating between the limits $\frac{1}{2}$ to $x$, we get

$$\int_{1/2}^{x}\frac{192x^3}{3}dx\le\int_{1/2}^{x} f'(x)dx\le\int_{1/2}^{x}\frac{192x^3}{2}dx$$

$\Rightarrow \quad \frac{192}{12}\left(x^4-\frac{1}{16}\right)\le f(x)-f(0)\le 24x^4-\frac{3}{2}$

$\Rightarrow \quad 16x^4-1\le f(x)\le 24x^4-\frac{3}{2}$

Again, integrating between the limits $\frac{1}{2}$ to 1, we get

$$\int_{1/2}^{1}(16x^4-1)dx\le\int_{1/2}^{1} f(x)dx\le\int_{1/2}^{1}\left(24x^4-\frac{3}{2}\right)dx$$

$\Rightarrow \quad \left[\frac{16x^5}{5}-x\right]_{1/2}^{1}\le\int_{1/2}^{1} f(x)dx\le\left[\frac{24x^5}{5}-\frac{3}{2}x\right]_{1/2}^{1}$

$\Rightarrow \quad \left(\frac{11}{5}+\frac{2}{5}\right)\le\int_{1/2}^{1} f(x)dx\le\left(\frac{33}{10}+\frac{6}{10}\right)$

$\Rightarrow \quad 2.6\le\int_{1/2}^{1} f(x)\,dx\le 3.9$

So, from the options, possible values of $m$ and $M$ are 1 and 12.

**6.** (a,c) Let $I_1=\int_0^{4\pi} e^t(\sin^6 at+\cos^6 at)dt$

$=\int_0^{\pi} e^t(\sin^6 at+\cos^6 at)dt+\int_{\pi}^{2\pi} e^t(\sin^6 at+\cos^6 at)dt$

$+\int_{2\pi}^{3\pi} e^t(\sin^6 at+\cos^6 at)dt+\int_{3\pi}^{4\pi} e^t(\sin^6 at+\cos^6 at)dt$

$\therefore \quad I_1=I_2+I_3+I_4+I_5$ ...(i)

Now, $\quad I_3=\int_{\pi}^{2\pi} e^t(\sin^6 at+\cos^6 at)dt$

Put $\quad t=\pi+t$

$\Rightarrow \quad dt=dt$

$\therefore \quad I_3=\int_0^{\pi} e^{\pi+t}\cdot(\sin^6 at+\cos^6 at)dt$

$=e^{\pi}\cdot I_2$ ...(ii)

Now, $\quad I_4=\int_{2\pi}^{3\pi} e^t(\sin^6 at+\cos^6 at)dt$

Put $t=2\pi+t \Rightarrow dt=dt$

$\therefore \quad I_4=\int_0^{\pi} e^{t+2\pi}(\sin^6 at+\cos^6 at)dt=e^{2\pi}\cdot I_2$ ...(iii)

and $\quad I_5=\int_{3\pi}^{4\pi} e^t(\sin^6 at+\cos^6 at)dt$

Put $t=3\pi+t$

$\therefore \quad I_5=\int_0^{\pi} e^{3\pi+t}(\sin^6 at+\cos^6 at)dt=e^{3\pi}\cdot I_2$ ...(iv)

From Eqs. (i), (ii), (iii) and (iv), we get

$I_1=I_2+e^{\pi}\cdot I_2+e^{2\pi}\cdot I_2+e^{3\pi}\cdot I_2=(1+e^{\pi}+e^{2\pi}+e^{3\pi})I_2$

$\therefore \quad L=\frac{\int_0^{4\pi} e^t(\sin^6 at+\cos^6 at)dt}{\int_0^{\pi} e^t(\sin^6 at+\cos^6 at)dt}=(1+e^{\pi}+e^{2\pi}+e^{3\pi})$

$=\frac{1\cdot(e^{4\pi}-1)}{e^{\pi}-1}$ for $a\in R$

**7.** (a,b,c) According to the given data,

$F(x)<0,\ \forall x\in(1,3)$

We have, $f(x)=xF(x)$

$\Rightarrow \quad f'(x)=F(x)+xF'(x)$ ...(i)

$\Rightarrow \quad f'(1)=F(1)+F'(1)<0$ [given, $F(1)=0$ and $F'(x)<0$]

Also, $\quad f(2)=2F(2)<0$ [using $F(x)<0, \forall x\in(1,3)$]

Now, $\quad f'(x)=F(x)+xF'(x)<0$ [using $F(x)<0,\ \forall\ x\in(1,3)$]

$\Rightarrow \quad f'(x)<0$

**8.** (c,d) Given, $\quad \int_1^3 x^2F'(x)dx=-12$

$\Rightarrow \quad [x^2F(x)]_1^3-\int_1^3 2x\cdot F(x)dx=-12$

$\Rightarrow \quad 9F(3)-F(1)-2\int_1^3 f(x)dx=-12$ [$\because xF(x)=f(x)$]

$\Rightarrow \quad -36-0-2\int_1^3 f(x)dx=-12$

$\therefore \quad \int_1^3 f(x)dx=-12$

and $\quad \int_1^3 x^3F''(x)dx=40$

$\Rightarrow \quad [x^3F'(x)]_1^3-\int_1^3 3x^2F'(x)dx=40$

$\Rightarrow \quad [x^2\cdot(xF'(x))]_1^3-3\times(-12)=40$

$\Rightarrow \quad \{x^2\cdot[f'(x)-F(x)]\}_1^3=4$

$\Rightarrow \quad 9[f'(3)-F(3)]-[f'(1)-F(1)]=4$

$\Rightarrow \quad 9[f'(3)+4]-[f'(1)-0]=4$

$\Rightarrow \quad 9f'(3)-f'(1)=-32$

**9.** (a) **Plan** This type of question can be done using appropriate substitution.

Given, $I = \int_{\pi/4}^{\pi/2} (2 \operatorname{cosec} x)^{17}\, dx$

$$= \int_{\pi/4}^{\pi/2} \frac{2^{17}(\operatorname{cosec} x)^{16} \operatorname{cosec} x (\operatorname{cosec} x + \cot x)}{(\operatorname{cosec} x + \cot x)} dx$$

Let $\quad \operatorname{cosec} x + \cot x = t$

$\Rightarrow \quad (-\operatorname{cosec} x \cdot \cot x - \operatorname{cosec}^2 x)\, dx = dt$

and $\quad \operatorname{cosec} x - \cot x = 1/t$

$\Rightarrow \quad 2 \operatorname{cosec} x = t + \frac{1}{t}$

$$\therefore \quad I = -\int_{\sqrt{2}+1}^{1} 2^{17} \left( \frac{t + \frac{1}{t}}{2} \right)^{16} \frac{dt}{t}$$

Let $t = e^u \Rightarrow dt = e^u du$.

When $t = 1, e^u = 1 \Rightarrow u = 0$

and when $t = \sqrt{2} + 1, e^u = \sqrt{2} + 1$

$\Rightarrow \quad u = \ln(\sqrt{2} + 1)$

$$\therefore \quad I = -\int_{\ln(\sqrt{2}+1)}^{0} 2(e^u + e^{-u})^{16} \frac{e^u}{e^u} du$$

$$= 2\int_0^{\ln(\sqrt{2}+1)} (e^u + e^{-u})^{16} du$$

**10.** (b) **Plan** Newton-Leibnitz's formula

$$\frac{d}{dx}\left[\int_{\phi(x)}^{\psi(x)} f(t)\, dt\right] = f\{\psi(x)\}\left\{\frac{d}{dx}\psi(x)\right\} - f\{\phi(x)\}\left\{\frac{d}{dx}\phi(x)\right\}$$

Given, $F(x) = \int_0^{x^2} f(\sqrt{t})\, dt$

$\therefore \quad F'(x) = 2x\, f(x)$

Also, $\quad F'(x) = f'(x)$

$\Rightarrow \quad 2x\, f(x) = f'(x) \Rightarrow \frac{f'(x)}{f(x)} = 2x$

$\Rightarrow \quad \int \frac{f'(x)}{f(x)} dx = \int 2x\, dx$

$\Rightarrow \quad \ln f(x) = x^2 + C \Rightarrow f(x) = e^{x^2 + C}$

$\Rightarrow \quad f(x) = K e^{x^2} \quad [K = e^C]$

Now, $\quad f(0) = 1$

$\therefore \quad 1 = K$

Hence, $\quad f(x) = e^{x^2}$

$\Rightarrow \quad F(2) = \int_0^4 e^t dt = [e^t]_0^4 = e^4 - 1$

**11.** (2) **Plan** Integration by parts

$$\int f(x)\, g(x)\, dx = f(x) \int g(x)\, dx - \int \left[\frac{d}{dx}[f(x)] \int g(x)\, dx\right] dx$$

Given, $\quad I = \int_0^1 \underset{\text{I}}{4x^3} \underset{\text{II}}{\frac{d^2}{dx^2}(1 - x^2)^5}\, dx$

$$= \left[4x^3 \frac{d}{dx}(1 - x^2)^5\right]_0^1 - \int_0^1 12x^2 \frac{d}{dx}(1 - x^2)^5\, dx$$

$$= \left[4x^3 \times 5(1 - x^2)^4\ (-2x)\right]_0^1$$

$$- 12\left[[x^2(1 - x^2)^5]_0^1 - \int_0^1 2x(1 - x^2)^5\, dx\right]$$

$$= 0 - 0 - 12(0 - 0) + 12\int_0^1 2x(1 - x^2)^5\, dx$$

$$= 12 \times \left[-\frac{(1 - x^2)^6}{6}\right]_0^1$$

$$= 12\left[0 + \frac{1}{6}\right] = 2$$

**12.** (b) Let $I = \int_{-\pi/2}^{\pi/2} \left(x^2 + \log\left(\frac{\pi - x}{\pi + x}\right)\right) \cos x\, dx$

As, $\int_{-a}^{a} f(x)\, dx = 0$, when $f(-x) = -f(x)$

$$\therefore \quad I = \int_{-\pi/2}^{\pi/2} x^2 \cos x\, dx + 0 = 2\int_0^{\pi/2} (x^2 \cos x)\, dx$$

$$= 2\{(x^2 \sin x)_0^{\pi/2} - \int_0^{\pi/2} 2x \cdot \sin x\, dx\}$$

$$= 2\left[\frac{\pi^2}{4} - 2\{(-x \cdot \cos x)_0^{\pi/2} - \int_0^{\pi/2} 1 \cdot (-\cos x)\, dx\}\right]$$

$$= 2\left[\frac{\pi^2}{4} - 2(\sin x)_0^{\pi/2}\right] = 2\left[\frac{\pi^2}{4} - 2\right] = \left(\frac{\pi^2}{2} - 4\right)$$

**13.** (a) Put $\quad x^2 = t \Rightarrow x\,dx = dt/2$

$$\therefore \quad I = \int_{\log 2}^{\log 3} \frac{\sin t \cdot \frac{dt}{2}}{\sin t + \sin(\log 6 - t)} \quad \ldots \text{(i)}$$

Using, $\int_a^b f(x)\, dx = \int_a^b f(a + b - x)\, dx$

$$I = \frac{1}{2}\int_{\log 2}^{\log 3} \frac{\sin(\log 2 + \log 3 - t)}{\sin(\log 2 + \log 3 - t) + \sin(\log 6 - (\log 2 + \log 3 - t))} dt$$

$$I = \frac{1}{2}\int_{\log 2}^{\log 3} \frac{\sin(\log 6 - t)}{\sin(\log 6 - t) + \sin(t)} dt$$

$$\therefore \quad I = \frac{1}{2}\int_{\log 2}^{\log 3} \frac{\sin(\log 6 - t)}{\sin(\log 6 - t) + \sin t} dt \quad \ldots \text{(ii)}$$

On adding Eqs. (i) and (ii), we get

$$2I = \frac{1}{2}\int_{\log 2}^{\log 3} \frac{\sin t + \sin(\log 6 - t)}{\sin(\log 6 - t) + \sin t} dt$$

$$\therefore \quad 2I = \frac{1}{2}[t]_{\log 2}^{\log 3} = \frac{1}{2}(\log 3 - \log 2)$$

$$I = \frac{1}{4}\log\left(\frac{3}{2}\right)$$

**14.** (b,d) **Concept Involved**

Converting infinite series into definite integral

i.e. $\lim\limits_{n \to \infty} \frac{h(n)}{n}$

$$= \lim_{n \to \infty} \frac{1}{n} \sum_{r = g(n)}^{h(n)} f\left(\frac{r}{n}\right) = \int f(x) dx$$

$$= \lim_{n \to \infty} \frac{g(n)}{n}$$

where, $\frac{r}{n}$ is replaced with $x$.

$\Sigma$ is replaced with integral.

Here, $\lim\limits_{n \to \infty} \frac{1^a + 2^a + \ldots + n^a}{(n + 1)^{a-1}\{(na + 1) + (na + 2) + \ldots + (na + n)\}} = \frac{1}{60}$

$$\Rightarrow \quad \lim_{n \to \infty} \frac{\sum_{r=1}^{n} r^a}{(n + 1)^{a-1} \cdot \left[n^2 a + \frac{n(n + 1)}{2}\right]} = \frac{1}{60}$$

$$\Rightarrow \quad \lim_{n \to \infty} \frac{2\sum_{r=1}^{n} \left(\frac{r}{n}\right)^a}{\left(1 + \frac{1}{n}\right)^{a-1} \cdot (2na + n + 1)} = \frac{1}{60}$$

$$\Rightarrow \quad \lim_{n \to \infty} \frac{1}{n}\left[2\sum_{r=1}^{n} \left(\frac{r}{n}\right)^a\right] \cdot \lim_{n \to \infty} \frac{1}{\left(1 + \frac{1}{n}\right)^{a-1} \cdot \left(2a + 1 + \frac{1}{n}\right)} = \frac{1}{60}$$

$$\Rightarrow \quad 2\int_0^1 (x^a)\, dx \cdot \frac{1}{1 \cdot (2a + 1)} = \frac{1}{60}$$

$$\Rightarrow \quad \frac{2\cdot[x^{a+1}]_0^1}{(2a+1)\cdot(a+1)} = \frac{1}{60}$$

$$\Rightarrow \quad \frac{2}{(2a+1)\cdot(a+1)} = \frac{1}{60}$$

$$\therefore \quad \frac{2}{(2a+1)(a+1)} = \frac{1}{60}$$

$$\Rightarrow \quad (2a+1)(a+1) = 120$$

$$\Rightarrow \quad 2a^2 + 3a + 1 - 120 = 0$$

$$\Rightarrow \quad 2a^2 + 3a - 119 = 0$$

$$\Rightarrow \quad (2a+17)(a-7) = 0$$

$$\Rightarrow \quad a = 7, \frac{-17}{2}$$

**15.** (a) Let $I = \int_0^1 \frac{x^4(1-x)^4}{1+x^2}dx$

$$= \int_0^1 \frac{(x^4-1)(1-x)^4 + (1-x)^4}{(1+x^2)}dx$$

$$= \int_0^1 (x^2-1)(1-x)^4 dx + \int_0^1 \frac{(1+x^2-2x)^2}{(1+x^2)}dx$$

$$= \int_0^1 \left\{(x^2-1)(1-x)^4 + (1+x^2) - 4x + \frac{4x^2}{(1+x^2)}\right\}dx$$

$$= \int_0^1 \left((x^2-1)(1-x)^4 + (1+x^2) - 4x + 4 - \frac{4}{1-x^2}\right)dx$$

$$= \int_0^1 \left(x^6 - 4x^5 + 5x^4 - 4x^2 + 4 - \frac{4}{1+x^2}\right)dx$$

$$= \left[\frac{x^7}{7} - \frac{4x^6}{6} + \frac{5x^5}{5} - \frac{4x^3}{3} + 4x - 4\tan^{-1}x\right]_0^1$$

$$= \frac{1}{7} - \frac{4}{6} + \frac{5}{5} - \frac{4}{3} + 4 - 4\left(\frac{\pi}{4} - 0\right)$$

$$= \frac{22}{7} - \pi$$

**16.** (4) Given, $f(x) = \begin{cases} x - [x], & \text{if } [x] \text{ is odd} \\ 1 + [x] - x, & \text{if } [x] \text{ is even} \end{cases}$

$f(x)$ and $\cos \pi x$ both are periodic with period 2 and both are even.

$$\therefore \quad \int_{-10}^{10} f(x)\cos \pi x\, dx = 2\int_0^{10} f(x)\cos \pi x dx$$

$$= 10\int_0^2 f(x)\cos \pi x\, dx$$

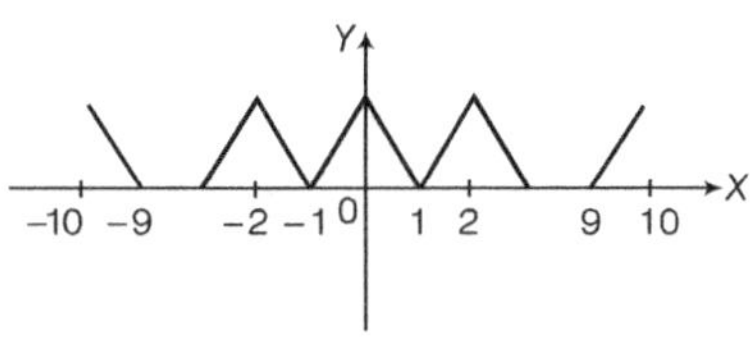


Now, $\int_0^1 f(x)\cos \pi x\, dx = \int_0^1 (1-x)\cos \pi x\, dx = -\int_0^1 u\cos \pi u\, du$

and $\int_1^2 f(x)\cos \pi x\, dx = \int_1^2 (x-1)\cos \pi x dx$

$$= -\int_0^1 u\cos \pi u\, du$$

$$\therefore \quad \int_{-10}^{10} f(x)\cos \pi x dx = -20\int_0^1 u\cos \pi u\, du = \frac{40}{\pi^2}$$

$$\Rightarrow \quad \frac{\pi^2}{10}\int_{-10}^{10} f(x)\cos \pi x\, dx = 4$$

**17.** (a,b,c) Given, $I_n = \int_{-\pi}^{\pi} \frac{\sin nx}{(1+\pi^x)\sin x}dx$ ...(i)

Using $\int_a^b f(x)\,dx = \int_a^b f(b+a-x)\,dx$

We get, $I_n = \int_{-\pi}^{\pi} \frac{\pi^x \sin nx}{(1+\pi^x)\sin x}dx$ ...(ii)

On adding Eqs. (i) and (ii), we have

$$2I_n = \int_{-\pi}^{\pi} \frac{\sin nx}{\sin x}dx$$

$$= 2\int_0^{\pi} \frac{\sin nx}{\sin x}dx$$

$\left[\because f(x) = \frac{\sin nx}{\sin x} \text{ is an even function.}\right]$

$$\Rightarrow \quad I_n = \int_0^{\pi} \frac{\sin nx}{\sin x}dx$$

Now, $I_{n+2} - I_n = \int_0^{\pi} \frac{\sin(n+2)x - \sin nx}{\sin x}dx$

$$= \int_0^{\pi} \frac{2\cos(n+1)x\cdot\sin x}{\sin x}dx$$

$$= 2\int_0^{\pi} \cos(n+1)x\, dx$$

$$= 2\left[\frac{\sin(n+1)x}{(n+1)}\right]_0^{\pi} = 0$$

$\therefore \quad I_{n+2} = I_n$ ...(iii)

Since, $I_n = \int_0^{\pi} \frac{\sin nx}{\sin x}dx$

$\Rightarrow \quad I_1 = \pi$ and $I_2 = 0$

$\therefore$ From Eq. (iii),

$$I_1 = I_3 = I_5 = \ldots = \pi$$

and $I_2 = I_4 = I_6 = \ldots = 0$

$$\Rightarrow \quad \sum_{m=1}^{10} I_{2m+1} = 10\pi$$

and $\sum_{m=1}^{10} I_{2m} = 0$

**18.** (c) Given, $\int_0^x \sqrt{1-[f'(t)]^2}\,dt = \int_0^x f(t)dt, 0 \le x \le 1$

On differentiating both sides w.r.t. $x$ by using Leibnitz rule, we get

$$\sqrt{1-[f'(x)]^2} = f(x)$$

$$\Rightarrow \quad f'(x) = \pm\sqrt{1-[f(x)]^2}$$

$$\Rightarrow \quad \int \frac{f'(x)}{\sqrt{1-[f(x)]^2}}dx = \pm\int dx$$

$$\Rightarrow \quad \sin^{-1}[f(x)] = \pm x + C$$

Put $x = 0 \Rightarrow \sin^{-1}[f(0)] = C$

$\Rightarrow \quad C = \sin^{-1}(0) = 0$ $\quad [\because f(0) = 0]$

$\therefore \quad f(x) = \pm \sin x$

But $\quad f(x) \ge 0, \forall\ x \in [0, 1]$

$\therefore \quad f(x) = \sin x$

As we know that,

$$\sin x < x, \forall\ x > 0$$

$$\therefore \quad \sin\left(\frac{1}{2}\right) < \frac{1}{2} \text{ and } \sin\left(\frac{1}{3}\right) < \frac{1}{3}$$

$$\Rightarrow \quad f\left(\frac{1}{2}\right) < \frac{1}{2} \text{ and } f\left(\frac{1}{3}\right) < \frac{1}{3}$$

**19.** (a,d) Given, $S_n = \sum_{k=0}^{n} \frac{n}{n^2 + kn + k^2}$

$$= \sum_{k=0}^{n} \frac{1}{n} \cdot \left( \frac{1}{1 + \frac{k}{n} + \frac{k^2}{n^2}} \right) < \lim_{n \to \infty} \sum_{k=0}^{n} \frac{1}{n} \left( \frac{1}{1 + \frac{k}{n} + \left(\frac{k}{n}\right)^2} \right)$$

$$= \int_0^1 \frac{1}{1 + x + x^2} dx$$

$$= \left[ \frac{2}{\sqrt{3}} \tan^{-1} \left\{ \frac{2}{\sqrt{3}} \left( x + \frac{1}{2} \right) \right\} \right]_0^1$$

$$= \frac{2}{\sqrt{3}} \cdot \left( \frac{\pi}{3} - \frac{\pi}{6} \right) = \frac{\pi}{3\sqrt{3}}, \text{ i.e. } S_n < \frac{\pi}{3\sqrt{3}}$$

Similarly, $T_n > \frac{\pi}{3\sqrt{3}}$

**20.** (a)

(i) Let $I = \int_{-1}^{1} \frac{dx}{1 + x^2}$

Put $x = \tan\theta \Rightarrow dx = \sec^2\theta \, d\theta$

$\therefore \quad I = 2\int_0^{\pi/4} d\theta = \frac{\pi}{2}$

(ii) Let $I = \int_0^1 \frac{dx}{\sqrt{1 - x^2}}$

Put $x = \sin\theta \Rightarrow dx = \cos\theta \, d\theta$

$\therefore \quad I = \int_0^{\pi/2} 1 d\theta = \frac{\pi}{2}$

(iii) $\int_2^3 \frac{dx}{1 - x^2} = \frac{1}{2}\left[ \log\left(\frac{1 + x}{1 - x}\right) \right]_2^3 = \frac{1}{2}\left[ \log\left(\frac{4}{-2}\right) - \log\left(\frac{3}{-1}\right) \right]$

$$= \frac{1}{2}\left[ \log\left(\frac{2}{3}\right) \right]$$

(iv) $\int_1^2 \frac{dx}{x\sqrt{x^2 - 1}} = [\sec^{-1} x]_1^2 = \frac{\pi}{3} - 0 = \frac{\pi}{3}$

Hence, (i) → (s), (ii) → (s), (iii) → (p), (iv) → (r)

**21.** (b) Here, $f(x) = (1 - x)^2 \cdot \sin^2 x + x^2 \geq 0, \forall x$.

and $g(x) = \int_1^x \left( \frac{2(t - 1)}{t + 1} - \log t \right) f(t) \, dt$

$\Rightarrow \quad g'(x) = \left\{ \frac{2(x - 1)}{(x + 1)} - \log x \right\} \cdot \underbrace{f(x)}_{+\text{ve}}$ ...(i)

For $g'(x)$ to be increasing or decreasing.

Let $\phi(x) = \frac{2(x - 1)}{(x + 1)} - \log x$

$$\phi'(x) = \frac{4}{(x + 1)^2} - \frac{1}{x} = \frac{-(x - 1)^2}{x(x + 1)^2}$$

$\phi'(x) < 0$, for $x > 1$

$\Rightarrow \quad \phi(x) < \phi(1) \Rightarrow \phi(x) < 0$ ...(ii)

From Eqs. (i) and (ii), we get

$g'(x) < 0$ for $x \in (1, \infty)$

$\therefore g(x)$ is decreasing for $x \in (1, \infty)$.

**22.** (c) Here, $f(x) + 2x = (1 - x)^2 \cdot \sin^2 x + x^2 + 2x$ ...(i)

where, P : $f(x) + 2x = 2(1 + x^2)$ ...(ii)

$\therefore \quad 2(1 + x^2) = (1 - x)^2 \sin^2 x + x^2 + 2x$

$\Rightarrow \quad (1 - x)^2 \sin^2 x = x^2 - 2x + 2$

$\Rightarrow \quad (1 - x)^2 \sin^2 x = (1 - x)^2 + 1 \Rightarrow (1 - x)^2 \cos^2 x = -1$

which is never possible.

$\therefore$ P is incorrect.

Again, let $h(x) = 2f(x) + 1 - 2x(1 + x)$

where, $h(0) = 2f(0) + 1 - 0 = 1$

$h(1) = 2f(1) + 1 - 4 = -3$

As, $h(0)\,h(1) < 0$

$\Rightarrow h(x)$ must have a solution.

$\therefore$ Q is correct.

**23.** (c) $\int_0^{\pi/2} \sin x \, dx = \frac{\frac{\pi}{2} - 0}{4} \left[ \sin 0 + \sin\left(\frac{\pi}{2}\right) + 2\sin\left(\frac{0 + \frac{\pi}{2}}{2}\right) \right]$

$$= \frac{\pi}{8}(1 + \sqrt{2})$$

**24.** (a) $F'(c) = (b - a) f'(c) + f(a) - f(b)$

$F''(c) = f''(c)(b - a) < 0 \Rightarrow F'(c) = 0$

$\Rightarrow \quad f'(c) = \frac{f(b) - f(a)}{b - a}$

**25.** (b) Given, $\lim_{t \to a} \frac{\int_a^t f(x)dx - \frac{(t - a)}{2}\{f(t) + f(a)\}}{(t - a)^3} = 0$

Using L'Hospital's rule

$$\lim_{h \to 0} \frac{\int_a^{a+h} f(x)\,dx - \frac{h}{2}\{f(a + h) + f(a)\}}{h^3} = 0$$

$$\Rightarrow \quad \lim_{h \to 0} \frac{f(a + h) - \frac{1}{2}\{f(a + h) + f(a)\} - \frac{h}{2}\{f'(a + h)\}}{3h^2} = 0$$

Again, using L' Hospital's rule

$$\lim_{h \to 0} \frac{f'(a+h) - \frac{1}{2}f'(a+h) - \frac{1}{2}f'(a+h) - \frac{h}{2}f''(a+h)}{6h} = 0$$

$$\Rightarrow \quad \lim_{h \to 0} \frac{-\frac{h}{2}f''(a + h)}{6h} = 0$$

$\Rightarrow \quad f''(a) = 0, \forall a \in R$

$\Rightarrow f(x)$ must have maximum degree 1.

**26.** (c) Let $I = \int_{-2}^{0} [x^3 + 3x^2 + 3x + 3 + (x + 1)\cos(x + 1)]\, dx$

$$= \int_{-2}^{0} [(x + 1)^3 + 2 + (x + 1)\cos(x + 1)]\, dx$$

Put $x + 1 = t$

$\Rightarrow \quad dx = dt$

$\therefore \quad I = \int_{-1}^{1} (t^3 + 2 + t\cos t)\, dt$

$$= \int_{-1}^{1} t^3\, dt + 2\int_{-1}^{1} dt + \int_{-1}^{1} t\cos t\, dt$$

$$= 0 + 2 \cdot 2[x]_0^1 + 0$$

[since, $t^3$ and $t\cos t$ are odd functions.]

$\therefore \quad I = 4$

**27.** (a) Since, $\int_{\sin x}^{1} t^2 f(t)\, dt = 1 - \sin x$, thus to find $f(x)$.

On differentiating both sides using Newton-Leibnitz formula

i.e. $\frac{d}{dx} \int_{\sin x}^{1} t^2 f(t)\, dt = \frac{d}{dx}(1 - \sin x)$

$\Rightarrow \{1^2 f(1)\} \cdot (0) - (\sin^2 x) \cdot f(\sin x) \cdot \cos x = -\cos x$

$\Rightarrow \quad f(\sin x) = \frac{1}{\sin^2 x}$

For $f\left(\frac{1}{\sqrt{3}}\right)$ is obtained, when $\sin x = \frac{1}{\sqrt{3}}$

i.e. $f\left(\frac{1}{\sqrt{3}}\right) = (\sqrt{3})^2 = 3$

# 3. Differential Equations

## DPP-1 Order, Degree and Formation of Differential Equations

**1.** (i) We have, $\sqrt{\frac{d^2y}{dx^2}} = \sqrt[3]{\frac{dy}{dx} + 3}$

$$\left(\frac{d^2y}{dx^2}\right)^{1/2} = \left(\frac{dy}{dx} + 3\right)^{1/3}$$

$$\Rightarrow \quad \left(\frac{d^2y}{dx^2}\right)^{3} = \left(\frac{dy}{dx} + 3\right)^{2}$$

Hence, order = 2 and degree = 3.

(ii) We have, $\frac{d^2y}{dx^2} = \sin\left(\frac{dy}{dx}\right)$

Hence, order = 2 and degree = not defined.

(iii) We have, $\frac{dy}{dx} = \sqrt{3x + 5}$

$$\Rightarrow \quad \left(\frac{dy}{dx}\right)^2 = 3x + 5$$

Hence, order = 1 and degree = 2.

(iv) We have, $\left[1 + \left(\frac{dy}{dx}\right)^2\right]^{1/2} = x^2 + y$

$$\Rightarrow \quad \left[1 + \left(\frac{dy}{dx}\right)^2\right] = (x^2 + y)^2$$

Hence, order = 1 and degree = 2.

(v) We have, $\left(1 + \frac{dy}{dx}\right)^{1/2} = \frac{d^2y}{dx^2}$

$$\Rightarrow \quad 1 + \frac{dy}{dx} = \left(\frac{d^2y}{dx^2}\right)^2$$

Hence, order = 2 and degree = 2.

(vi) We have, $\frac{dy}{dx} = \sin y$

Hence, order = 1 and degree = 1.

**2.** (i) Equation of the circle touching the $X$-axis at origin is

$$x^2 + (y - a)^2 = a^2$$

$$\Rightarrow \quad x^2 + y^2 - 2ay = 0 \quad \ldots\text{(i)}$$

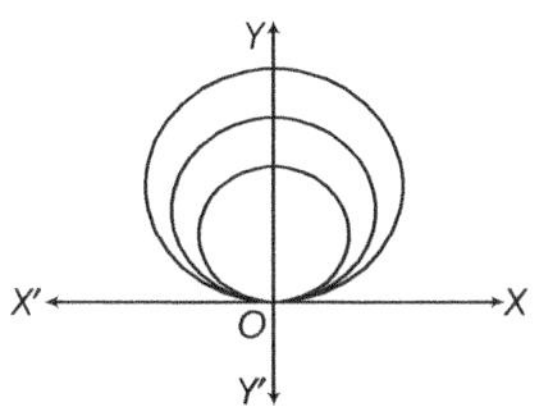


On differen tiating Eq. (i) w.r.t. $x$, we get

$$2x + 2yy' - 2ay' = 0$$

$$\Rightarrow \quad x + yy' = ay'$$

$$\Rightarrow \quad \frac{x + yy'}{y'} = a \quad \ldots\text{(ii)}$$

On putting the value of $a$ in Eq. (i), we get

$$x^2 + y^2 - 2\left(\frac{x + yy'}{y'}\right)y = 0$$

$$\Rightarrow \quad x^2y' + y^2y' - 2xy - 2y^2y' = 0$$

$$\Rightarrow \quad (x^2 - y^2)\, y' = 2xy$$

(ii) Equation of the circle touching the $Y$-axis at origin is

$$(x - a)^2 + y^2 = a^2$$

$$\Rightarrow \quad x^2 + y^2 - 2ax = 0 \quad \ldots\text{(i)}$$

On differentiating Eq. (i) w.r.t. $x$, we get

$$2x + 2yy' - 2a = 0$$

$$\Rightarrow \quad x + yy' = a \quad \ldots\text{(ii)}$$

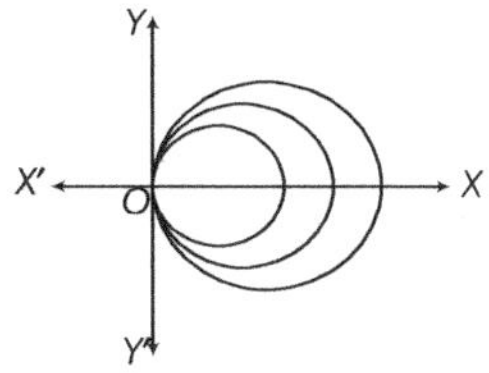


On putting the value of $a$ in Eq. (i), we get

$$x^2 + y^2 - 2(x + yy')\,x = 0$$

$$\Rightarrow \quad x^2 + y^2 - 2x^2 - 2xyy' = 0$$

$$\Rightarrow \quad y^2 - x^2 = 2xyy'$$

**3.** Equation of the circle whose centre $(h, k)$ and radius $r$, is

$$(x - h)^2 + (y - k)^2 = r^2 \quad \ldots\text{(i)}$$

On differentiating Eq. (i) w.r.t. $x$, we get

$$x - h + (y - k)\frac{dy}{dx} = 0 \quad \ldots\text{(ii)}$$

Again, differentiating w.r.t. $x$, we get

$$1 + (y - k)\frac{d^2y}{dx^2} + \left(\frac{dy}{dx}\right)^2 = 0$$

$$\Rightarrow \quad (y - k) = \frac{-\left[1 + \left(\frac{dy}{dx}\right)^2\right]}{\frac{d^2y}{dx^2}}$$

On putting the value of $(y - k)$ in Eq. (ii), we get

$$(x - h) - \frac{\left[1 + \left(\frac{dy}{dx}\right)^2\right]\frac{dy}{dx}}{\frac{d^2y}{dx^2}} = 0$$

$$\Rightarrow \quad x - h = \frac{\left[1 + \left(\frac{dy}{dx}\right)^2\right]\frac{dy}{dx}}{\frac{d^2y}{dx^2}}$$

On putting the values of $(x - h)$ and $(y - k)$ in Eq. (i), we get

$$\left[\frac{\left\{1 + \left(\frac{dy}{dx}\right)^2\right\}\frac{dy}{dx}}{\frac{d^2y}{dx^2}}\right]^2 + \left[\frac{1 + \left(\frac{dy}{dx}\right)^2}{\frac{d^2y}{dx^2}}\right]^2 = r^2$$

$$\Rightarrow \quad \frac{\left[1 + \left(\frac{dy}{dx}\right)^2\right]^2\left(\frac{dy}{dx}\right)^2}{\left(\frac{d^2y}{dx^2}\right)^2} + \frac{\left[1 + \left(\frac{dy}{dx}\right)^2\right]^2}{\left(\frac{d^2y}{dx^2}\right)^2} = r^2$$

$$\Rightarrow \quad \left[1 + \left(\frac{dy}{dx}\right)^2\right]^2\left(\frac{dy}{dx}\right)^2 + \left[1 + \left(\frac{dy}{dx}\right)^2\right]^2 = r^2\left[\frac{d^2y}{dx^2}\right]^2$$

$$\Rightarrow \quad \left[1 + \left(\frac{dy}{dx}\right)^2\right]^2\left[1 + \left(\frac{dy}{dx}\right)^2\right] = r^2\left[\frac{d^2y}{dx^2}\right]^2$$

$$\therefore \quad \left[1 + \left(\frac{dy}{dx}\right)^2\right]^3 = r^2\left[\frac{d^2y}{dx^2}\right]^2$$

**4.** The equation of circle in the first quadrant which touches the coordinate axes is

$$(x - a)^2 + (y - a)^2 = a^2 \quad \ldots(\text{i})$$

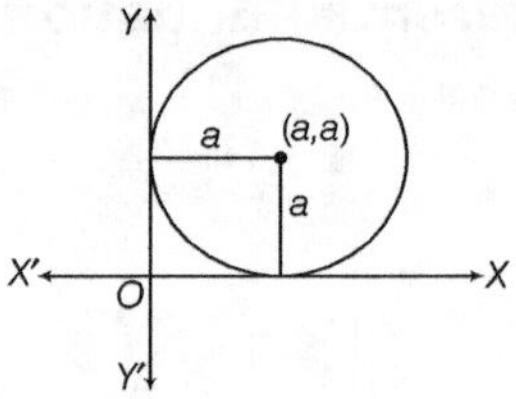


On differentiating Eq. (i) w.r.t. $x$, we get

$$(x - a) + (y - a)\,y' = 0$$

$$\Rightarrow \quad x - a + yy' - ay' = 0$$

$$\Rightarrow \quad x + yy' = a(1 + y')$$

$$\Rightarrow \quad a = \frac{x + yy'}{1 + y'}$$

On putting the value of $a$ in Eq. (i), we get

$$\left[x - \left(\frac{x + yy'}{1 + y'}\right)\right]^2 + \left[y - \left(\frac{x + yy'}{1 + y'}\right)\right]^2 = \left(\frac{x + yy'}{1 + y'}\right)^2$$

$$\Rightarrow (x + xy' - x - yy')^2 + (y + yy' - x - yy')^2 = (x + yy')^2$$

$$\Rightarrow \quad [(x - y)y']^2 + (y - x)^2 = (x + yy')^2$$

$$\Rightarrow \quad (x - y)^2 (1 + y'^2) = (x + yy')^2$$

$$\therefore \quad (x - y)^2\left[1 + \left(\frac{dy}{dx}\right)^2\right] = \left[x + y\frac{dy}{dx}\right]^2$$

**5.** (i) The equation of straight lines, which are non-horizontal, is

$$x = my + c$$

$$\therefore \quad \frac{dx}{dy} = m \quad \text{and} \quad \frac{d^2x}{dy^2} = 0$$

(ii) The equation of straight lines, which are non-vertical, is

$$y = mx + c$$

$$\therefore \quad \frac{dy}{dx} = m \quad \text{and} \quad \frac{d^2y}{dx^2} = 0$$

**6.** The equation of conic whose axes coincide with coordinate axes, is

$$\frac{x^2}{a^2} \pm \frac{y^2}{b^2} = 1 \quad \ldots(\text{i})$$

On differentiating Eq. (i) w.r.t. $x$, we get

$$\frac{x}{a^2} \pm \frac{y}{b^2}y' = 0$$

$$\therefore \quad yy' = \pm \frac{b^2}{a^2}x \quad \ldots(\text{ii})$$

and

$$yy'' + y'^2 = \pm \frac{b^2}{a^2} \quad \ldots(\text{iii})$$

On putting the value of $\pm \frac{b^2}{a^2}$ in Eq. (iii), we get

$$yy' = x(yy'' + y'^2)$$

$$\Rightarrow \quad y\frac{dy}{dx} = x\left[y\frac{d^2y}{dx^2} + \left(\frac{dy}{dx}\right)^2\right]$$

**7.** The equation of all conic whose centre lies at origin, is

$$ax^2 + 2hxy + by^2 = 1 \quad \ldots(\text{i})$$

On differentiating Eq. (i) w.r.t. $x$, we get

$$ax + hxy' + hy + byy' = 0$$

On multiplying both sides by $x$, we get

$$ax^2 + hx^2y' + hxy + bxyy' = 0 \quad \ldots(\text{ii})$$

On subtracting Eq. (ii) from Eq. (i), we get

$$h(xy - x^2y') + b(y^2 - xyy') = 1$$

$$\Rightarrow \quad (hx + by)\,y - xy'\,(hx + by) = 1$$

$$\Rightarrow \quad (hx + by)\,(y - xy') = 1$$

$$\Rightarrow \quad hx + by = \frac{1}{y - xy'} \quad \ldots(\text{iii})$$

Again, differentiating w.r.t. $x$, we get

$$h + by' = -\frac{(y' - xy'' - y')}{(y - xy')^2} \quad \ldots(\text{iv})$$

From Eqs. (iii) and (iv), we get

$$b(y - xy') = \frac{y - xy' - x^2y''}{(y - xy')^2}$$

$$\Rightarrow \quad b = \frac{y - xy' - x^2y''}{(y - xy')^3}$$

Again, differentiating w.r.t. $x$, we get

$$0 = \frac{(y' - y' - 3xy'' - x^2y''')}{(y - xy')^3} + \frac{3(y - xy' - x^2y'')\,xy''}{(y - xy')^4}$$

$$\Rightarrow \quad (3xy'' + x^2y''')\,(y - xy') = 3xy''(y - xy' - x^2y'')$$

**8.** Let the equation of line be $y = mx + c$.

Here, $p$ is the distance from origin to the line $y = mx + c$.

$$\therefore \quad p = \frac{c}{\sqrt{1 + m^2}}$$

$$\Rightarrow \quad c = p\sqrt{1 + m^2}$$

So, the equation of line which are at a constant distance $p$ from origin, is

$$y = mx + p\sqrt{1 + m^2} \quad \ldots(\text{i})$$

$$\Rightarrow \quad \frac{dy}{dx} = m$$

On putting the value of $m$ in Eq. (i), we get

$$y = x\frac{dy}{dx} + p\sqrt{1 + \left(\frac{dy}{dx}\right)^2}$$

$$\Rightarrow \quad \left(y - x\frac{dy}{dx}\right)^2 = p^2\left[1 + \left(\frac{dy}{dx}\right)^2\right]$$

**9.** The equation of all parabola having their axes of symmetry coincident with the axes of $x$, is

$$y^2 = 4a(x - h) \quad \ldots(\text{i})$$

On differentiating Eq. (i) w.r.t. $x$, we get

$$2yy' = 4a$$

Again, differentiating w.r.t. $x$, we get

$$yy'' + y'^2 = 0$$

So, the required differential equation is

$$y\frac{d^2y}{dx^2} + \left(\frac{dy}{dx}\right)^2 = 0.$$

**10.** The equation of parabola whose axis of symmetric is parallel to $X$-axis, is

$$x = ay^2 + by + c$$

Now,

$$\frac{dy}{dx} = 2ay + b$$

$$\Rightarrow \quad \frac{d^2x}{dy^2} = 2a$$

$$\Rightarrow \quad \frac{d^3x}{dy^3} = 0$$

**11.** (d) We have, $\left(1 + 3\frac{dy}{dx}\right)^{2/3} = 4\frac{d^3y}{dx^3}$

$$\Rightarrow \quad \left(1 + 3\frac{dy}{dx}\right)^2 = \left(4\frac{d^3y}{dx^3}\right)^3$$

Hence, order = 3 and degree = 3.

**12.** (c) We have,

$$\sqrt[3]{\frac{dy}{dx}} - 4\frac{d^2y}{dx^2} - 7x = 0$$

$$\Rightarrow \quad \sqrt[3]{\frac{dy}{dx}} = 4\frac{d^2y}{dx^2} + 7x$$

$$\Rightarrow \quad \frac{dy}{dx} = \left(4\frac{d^2y}{dx^2} + 7x\right)^3$$

Hence, order $a = 2$ and degree $b = 3$

$\therefore \quad a + b = 2 + 3 = 5$

**13.** (c) We have, $y = (c_1 + c_2)\cos(x + c_3) - c_4 e^{x + c_5}$

$$y = (c_1 + c_2)\cos(x + c_3) - (c_4 e^{c_5} \cdot e^x)$$

$$y = A\cos(x + c_3) - Be^x \quad \left[\because c_1 + c_2 = A \text{ and } c_4 e^{c_5} = B\right]$$

The above equation have three arbitrary constants $A$, $B$ and $c_3$. Hence, the order of differential equation is 3.

**14.** (b) We have, $y = a + bx + ce^{-x}$

$$\therefore \quad \frac{dy}{dx} = b - ce^{-x}$$

$$\Rightarrow \quad \frac{d^2y}{dx^2} = ce^{-x}$$

$$\Rightarrow \quad \frac{d^3y}{dx^3} = -ce^{-x}$$

Now, $\frac{d^3y}{dx^3} + \frac{d^2y}{dx^2} = 0$

$\Rightarrow \quad y''' + y'' = 0$

**15.** (d) We have, $y = e^{cx}$

$\Rightarrow \quad \log y = cx \quad \ldots(i)$

On differentiating Eq. (i) w.r.t. $x$, we get

$$\frac{1}{y}y' = c$$

On putting the value of $c$ in Eq. (i), we get

$$\log y = \frac{y'x}{y}$$

$$\Rightarrow \quad y\log y = xy'$$

**16.** (c) We have, $x^2 + y^2 - 2ay = 0 \quad \ldots(i)$

On differentiating Eq. (i) w.r.t. $x$, we get

$$x + yy' - ay' = 0$$

$$\Rightarrow \quad a = \frac{x + yy'}{y'}$$

On putting the value of $a$ in Eq. (i), we get

$$x^2 + y^2 - 2y\left(\frac{x + yy'}{y'}\right) = 0$$

$$\Rightarrow \quad x^2y' + y^2y' - 2xy - 2y^2y' = 0$$

$$\Rightarrow \quad (x^2 - y^2)y' = 2xy$$

**17.** (c) We have, $y^2 = 2c(x + \sqrt{c}) \quad \ldots(i)$

$\Rightarrow \quad 2yy' = 2c$

$\Rightarrow \quad yy' = c$

On putting the value of $c$ in Eq. (i), we get

$$y^2 = 2yy'[x + (yy')^{1/2}]$$

$$\Rightarrow \quad y^2 = 2xyy' + 2(yy')^{3/2}$$

$$\Rightarrow \quad y^2 - 2xyy' = 2(yy')^{3/2}$$

$$\Rightarrow \quad (y^2 - 2xy')^2 = 4(yy')^3$$

Hence, order = 1 and degree = 3.

**18.** (c) We have, $Ax^2 + By^2 = 1$

$\Rightarrow \quad 2Ax + 2Byy' = 0 \quad \ldots(i)$

$\Rightarrow \quad A + B(yy'' + y'^2) = 0 \quad \ldots(ii)$

On solving Eqs. (i) and (ii), we get

$$x(yy'' + y'^2) = yy'$$

Hence, order = 2 and degree = 1

**19.** (c) The equation of family of circle with the fixed radius 5 units and centre on the line $y = 2$, is

$$(x - a)^2 + (y - 2)^2 = 25 \quad \ldots(i)$$

On differentiating Eq.(i) w.r.t. $x$, we get

$$2(x - a) + 2(y - 2)y' = 0$$

$$\Rightarrow \quad (x - a) = -(y - 2)y'$$

On putting the value of $(x - a)$ in Eq. (i), we get

$$(y - 2)^2 y'^2 + (y - 2)^2 = 25$$

$$\Rightarrow \quad (y - 2)^2 y'^2 = 25 - (y - 2)^2$$

**20.** (c) We have, $D^3y - 3D^2y - 4Dy + 12y = 0$

$\Rightarrow \quad (D^3 - 3D^2 - 4D + 12)y = 0$

$\Rightarrow \quad D^3 - 3D^2 - 4D + 12 = 0$

Let $m$ be the root of equation, then

$$m^3 - 3m^2 - 4m + 12 = 0$$

$\Rightarrow \quad m = \pm 2, 3$

$\because \quad m \in N$

$\therefore \quad m = \{2, 3\}$

$\therefore$ number of solutions is 2.

**21.** (c) We have, $m^3 - 7m + 6 = 0$

Let the roots of the given equation be $m_1$, $m_2$ and $m_3$.

then, $m_1 + m_2 + m_3 = 0$

and $m_1m_2 + m_2m_3 + m_1m_3 = -7$

Hence, the sum of product of two roots at a time is –7.

**22.** (a) We have, $y = c_1e^{m_1x} + c_2e^{m_2x} + c_3e^{m_3x}$

where, $c_1, c_2, c_3$ are arbitrary constants and $m_1, m_2, m_3$ are roots of $m^3 - 7m + 6 = 0$.

So, the differential equation of curve is

$$D^3y - 7Dy + 6y = 0$$

$$\Rightarrow \quad y_3 - 7y_1 + 6y = 0$$

**23.** (c) Let $(h, k)$ be the vertex of parabola, whose latusrectum is $4a$.

Hence, the equation of parabola parallel to $X$-axis is

$$(y - k)^2 = 4a(x - h)$$

**24.** (d) We have, $(y - k)^2 = 4a(x - h) \quad \ldots(i)$

On differentiating Eq. (i) w.r.t. $x$, we get

$$2(y - k)y' = 4a \quad \ldots(ii)$$

Again, differentiating w.r.t. $x$, we get

$$(y - k)y'' + y'^2 = 0$$

$$\Rightarrow \quad (y - k) = \frac{-y'^2}{y''}$$

On putting the value of $(y - k)$ in Eq. (ii), we get

$$\frac{-y'^3}{y''} = 2a$$

$$\Rightarrow \quad y'^3 + 2ay'' = 0$$

Hence, order = 2 and degree = 1.

**25.** (a) The differential equation of given parabola is

$$y_1^3 + 2ay_2 = 0.$$

**26.** (3) We have, $y = Ae^{3x} + Be^{5x}$

$\therefore \quad \frac{dy}{dx} = 3Ae^{3x} + 5Be^{5x}$

$\Rightarrow \quad \frac{dy}{dx} = 3(y - Be^{5x}) + 5Be^{5x}$

$\Rightarrow \quad \frac{dy}{dx} = 3y + 2Be^{5x}$

$\Rightarrow \quad \frac{d^2y}{dx^2} = \frac{3dy}{dx} + 10Be^{5}$

$\Rightarrow \quad \frac{d^2y}{dx^2} = \frac{3dy}{dx} + 5\left(\frac{dy}{dx} - 3y\right)$

$\Rightarrow \quad \frac{d^2y}{dx^2} = \frac{8dy}{dx} - 15y$

Hence, order (*a*)= 2 and degree (*b*) = 1

$\therefore \quad a + b = 2 + 1 = 3$

**27.** (1) We have, $\frac{x^2}{a^2} + \frac{y^2}{a^2 + \lambda} = 1$

On differentiating both sides w.r.t. *x*, we get

$$\frac{x}{a^2} + \frac{yy'}{a^2 + \lambda} = 0$$

$\Rightarrow \quad yy' = \frac{-a^2 + \lambda}{a^2} x \quad \text{...(i)}$

and $\quad yy'' + y'^2 = \frac{-(a^2 + \lambda)}{a^2} \quad \text{...(ii)}$

On putting the value of $\frac{-(a^2 + \lambda)}{a^2}$ in Eq. (i), we get

$$yy' = (yy'' + y^2)\,x$$

Hence, degree = 1.

**28.** (1) We have,

$$xy = ae^x + be^{-x} + x^2$$

$\Rightarrow \quad xy_1 + y = ae^x - be^{-x} + 2x$

$\Rightarrow \quad xy_2 + y_1 + y_1 = ae^x + be^x + 2$

$\Rightarrow \quad xy_2 + 2y_1 = xy - x^2 + 2$

$\Rightarrow \quad xy_2 + 2y_1 - xy + x^2 - 2 = 0$

Hence, $a = 2$ and $b = -1$

$\therefore \quad a + b = 2 - 1 = 1$

**29.** (1) The equation of line which are equally inclined to coordinate axes, is

$$y = \pm x$$

$\Rightarrow \quad \frac{dy}{dx} = \pm 1$

Hence, $\quad |\lambda| = |\pm 1| = 1$

**30.** (5) We have, $\quad x = x(t)$

It given that, $\quad \frac{dx}{dt} = 5\%$ of $x$

Hence, $\lambda = 5$

## DPP-2 Equation in Variable Separable Form and Reducible to Variable Separation Form

**1.** (d) Given, $xy\frac{dy}{dx} = \frac{1 + y^2}{1 + x^2}(1 + x + x^2)$

$\Rightarrow \quad \frac{y}{1 + y^2}dy = \frac{1 + x + x^2}{x(1 + x^2)}$

$\Rightarrow \quad \frac{y}{1 + y^2}dy = \left(\frac{1}{x} + \frac{1}{1 + x^2}\right)dx$

On integrating both sides, we get

$$\int \frac{y}{1 + y^2}dy = \int \left(\frac{1}{x} + \frac{1}{1 + x^2}\right)dx$$

$\Rightarrow \quad \frac{1}{2}\log(1 + y^2) = \log x + \tan^{-1} x + C$

$\Rightarrow \quad \log\sqrt{1 + y^2} - \log x = \tan^{-1} x + C$

$\Rightarrow \quad \log \frac{\sqrt{1 + y^2}}{x} = \tan^{-1} x + C$

$\Rightarrow \quad \frac{\sqrt{1 + y^2}}{x} = Ce^{\tan^{-1} x}$

$\Rightarrow \quad \sqrt{1 + y^2} = Cxe^{\tan^{-1} x}$

**2.** (a) Given, $(x^3 - y^2x^3)\frac{dy}{dx} + y^3 + x^2y^3 = 0$

$\Rightarrow \quad x^3(1 - y^2)\,dy = -y^3(1 + x^2)dx$

$\Rightarrow \quad \frac{y^2 - 1}{y^3}dy = \frac{(1 + x^2)}{x^3}dx$

$\Rightarrow \quad \left(\frac{1}{y} - \frac{1}{y^3}\right)dy = \left(\frac{1}{x^3} + \frac{1}{x}\right)dx$

On integrating both sides, we get

$$\int \left(\frac{1}{y} - \frac{1}{y^3}\right)dy = \int \left(\frac{1}{x^3} + \frac{1}{x}\right)dx$$

$\Rightarrow \quad \log y + \frac{1}{2y^2} = -\frac{1}{2x^2} + \log x + C$

$\Rightarrow \quad \log\left(\frac{x}{y}\right) = \left(\frac{1}{2}\right)\left(\frac{1}{x^2} + \frac{1}{y^2}\right) + C$

**3.** (b) We have,

$$\sin^{-1}\left(\frac{dy}{dx}\right) = x + y$$

$\Rightarrow \quad \frac{dy}{dx} = \sin(x + y)$

Put $x + y = v \Rightarrow 1 + \frac{dy}{dx} = \frac{dv}{dx}$

$\therefore \quad \frac{dv}{dx} - 1 = \sin v$

$\Rightarrow \quad \frac{dv}{1 + \sin v} = dx$

On integrating both sides, we get

$$\int \frac{1 - \sin v}{\cos^2 v}dv = \int dx$$

$\Rightarrow \quad \int (\sec^2 v - \sec v \tan v)\,dv = \int dx$

$\Rightarrow \quad \tan v - \sec v = x + C$

$\Rightarrow \quad \tan(x + y) - \sec(x + y) = x + C$

$\Rightarrow \quad x = \tan(x + y) - \sec(x + y) + C$

**4.** (c) Given, $\frac{dy}{dx} = \cos(x + y)$

Put $x + y = v \Rightarrow \frac{dy}{dx} = \frac{dv}{dx} - 1$

$\Rightarrow \quad \frac{dv}{dx} - 1 = \cos v$

$\Rightarrow \quad \frac{dv}{1 + \cos v} = dx$

On integrating both sides, we get

$$\int \frac{dv}{1 + \cos v} = \int dx$$

$\Rightarrow \quad \frac{1}{2}\int \sec^2 \frac{v}{2} dv = \int dx$

$\Rightarrow \quad \tan\frac{v}{2} = x + C$

$\Rightarrow \quad \tan\left(\frac{x + y}{2}\right) = x + C$

**5.** (a) Given, $\quad y\log y + xy' = 0$

$\Rightarrow \quad y\log y + x\frac{dy}{dx} = 0$

On integrating both sides, we get

$$\int \frac{dy}{y\log y} = -\int \frac{dx}{x}$$

$\Rightarrow \quad \log(\log y) = -\log x + C$

$\Rightarrow \quad \log\log y + \log x = C$

$\Rightarrow \quad x(\log y) = C$

It is given that, $\quad y(1) = e$

$\therefore \quad 1 \cdot \log e = C \Rightarrow C = 1$

Hence, $\quad x(\log y) = 1$

**6.** (c) Here, $\frac{dy}{dx} = y + 3 > 0$ and $y(0) = 2$

On integrating, we get

$$\int \frac{dy}{y + 3} = \int dx$$

$\Rightarrow \quad \log|y + 3| = x + C$

Since, $\quad y(0) = 2$

$\Rightarrow \quad \log|2 + 3| = 0 + C$

$\therefore \quad C = \log_e 5$

$\Rightarrow \quad \log|y + 3| = x + \log_e 5$

when, $x = \log_e 2$

$\Rightarrow \quad \log|y + 3| = \log_e 2 + \log_e 5 = \log_e 10$

$\Rightarrow \quad y + 3 = 10$

$\therefore \quad y = 7$

**7.** (a) We have, $\frac{2 + \sin x}{y + 1} \cdot \frac{dy}{dx} = -\cos x$

$\Rightarrow \quad \int \frac{dy}{y + 1} = -\int \frac{\cos x}{2 + \sin x} dx$

$\Rightarrow \quad \log(y + 1) = -\log(2 + \sin x) + \log C$

$\Rightarrow \quad \log(y + 1) + \log(2 + \sin x) = \log C$

$\Rightarrow \quad (y + 1)(2 + \sin x) = C$

It is given that, $y(0) = 1$

Put $x = 0$ and $y = 1$, then

$$2(2 + \sin 0) = C \Rightarrow C = 4$$

So, the equation is

$$(y + 1)(2 + \sin x) = 4$$

Now, put $x = \pi/2$

$\therefore \quad (y + 1)\left(2 + \sin\frac{\pi}{2}\right) = 4$

$\Rightarrow \quad (y + 1)(3) = 4 \Rightarrow y = \frac{1}{3}$

**8.** (d) We have, $\quad \frac{dy}{dx} = \frac{\sqrt{1 - y^2}}{y}$

On integrating both sides, we get

$$\int \frac{y}{\sqrt{1 - y^2}} dy = \int dx$$

$\Rightarrow \quad -\sqrt{1 - y^2} = x + C$

On squaring both sides, we get

$$(x + c)^2 + y^2 = 1$$

Which is the equation of circle whose fixed radius is 1 and centre is variable on $X$-axis.

**9.** (b) Given, $\quad \frac{dy}{dx} = \frac{1 - x}{y}$

On integrating both sides, we get

$$\int y\,dy = \int (1 - x)\,dx$$

$\Rightarrow \quad \frac{y^2}{2} = x - \frac{x^2}{2} + C$

$\Rightarrow \quad x^2 + y^2 - 2x = C$

$\Rightarrow \quad (x - 1)^2 + y^2 = C + 1$

Which represents equation of the circle.

**10.** (b) We have, $Dy = 100 - y \Rightarrow \frac{dy}{dx} = 100 - y$

On integrating both sides, we get

$$\int \frac{dy}{100 - y} = \int dx$$

$\Rightarrow \quad -\log(100 - y) = x + C$

$\Rightarrow \quad 100 - y = Ce^{-x}$

$\Rightarrow \quad y = 100 - Ce^{-x}$

$\Rightarrow \quad 50 = 100 - Ce^0 \quad [\because y(0) = 50]$

$\Rightarrow \quad C = 50$

$\therefore \quad y = 100 - 50e^{-x}$

The graph of curve $y = 100 - 50e^{-x}$ is

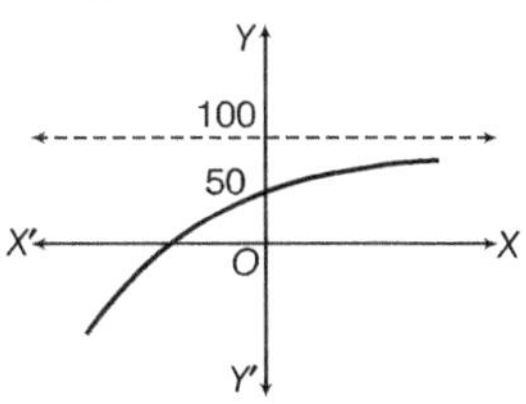


**11.** (b) We have, $\quad \frac{dp(t)}{dt} = 0.5\,p(t) - 450$

$\Rightarrow \quad \frac{dp(t)}{dt} = \frac{1}{2}p(t) - 450$

$\Rightarrow \quad \frac{dp(t)}{dt} = \frac{p(t) - 900}{2}$

$\Rightarrow \quad \int \frac{dp(t)}{p(t) - 900} = \int \frac{dt}{2}$

$\Rightarrow \quad \log[p(t) - 900] = t/2 + C$

$\Rightarrow \quad p(t) - 900 = Ce^{t/2}$

Put $p(0) = 850$, then

$\Rightarrow \quad 850 - 900 = Ce^0 \Rightarrow C = -50$

$\therefore \quad p(t) = 900 - 50e^{t/2}$

Put $p(t) = 0$, then

$$0 = 900 - 50e^{t/2}$$

$\Rightarrow \quad 50e^{t/2} = 900 \quad \Rightarrow e^{t/2} = 18$

$\Rightarrow \quad t/2 = \log 18$

$\Rightarrow \quad t = 2\log 18$

12. (c) Given, $\frac{dP}{dx} = 100 - 12\sqrt{x}$

$\Rightarrow \quad dP = (100 - 12\sqrt{x})\,dx$

On integrating both sides, we get

$$\int dP = \int (100 - 12\sqrt{x})\,dx$$

$\Rightarrow \quad P = 100x - 8(x)^{3/2} + C$

When $x = 0$, $P = 2000$, then

$$C = 2000$$

Now, when $x = 25$, then

$$P = 100\,(25) - 8(25)^{3/2} + 2000$$
$$= 2500 - 8(125) + 2000 = 3500$$

13. (d) We have, $f'(x) < 2f(x) \Rightarrow \frac{f'(x)}{f(x)} < 2$

On integrating both sides, we get

$$\log f(x) < 2x + C$$

$\Rightarrow \quad f(x) < e^{2x + C}$

$\Rightarrow \quad f(x) < e^{2x - 1} \quad \left[\because f\left(\frac{1}{2}\right) = 1\right]$

Hence, $f(x) > 0$

$\therefore \quad 0 < f(x) < e^{2x - 1}$

$\Rightarrow \quad 0 < \int_{1/2}^{1} f(x)\,dx < \int_{1/2}^{1} e^{2x - 1}\,dx$

$\Rightarrow \quad 0 < \int_{1/2}^{1} f(x)\,dx < \left[\frac{e^{2x - 1}}{2}\right]_{1/2}^{1}$

$\Rightarrow \quad 0 < \int_{1/2}^{1} f(x)\,dx < \frac{e - 1}{2}$

14. (a,c) Given, $\frac{dy}{dx} = \frac{ax + h}{by + k}$

$\Rightarrow \quad (by + k)\,dy = (ax + h)dx$

On integrating both sides, we get

$$\int (by + k)\,dy = \int (ax + h)dx$$

$\Rightarrow \quad \frac{by^2}{2} + ky = \frac{ax^2}{2} + hx + C$

$\Rightarrow \quad by^2 + 2ky = ax^2 + 2hx + C$

It represents parabola of either $a = 0, b \neq 0$ or $a \neq 0, b = 0$.

15. (a,c) The equation of normal at $(x, y)$ is

$$(X - x) + (Y - y)\frac{dy}{dx} = 0$$

$\Rightarrow \quad \frac{X}{x + y\frac{dy}{dx}} + \frac{Y}{\frac{x + y\,dy/dx}{dy/dx}} = 1$

$\Rightarrow \quad OA = x + y\frac{dy}{dx}$ and $OB = \frac{x + y\frac{dy}{dx}}{dy/dx}$

Also, $\frac{1}{OA} + \frac{1}{OB} = 1$

$\Rightarrow \quad 1 + \frac{dy}{dx} = x + y\frac{dy}{dx}$

$\Rightarrow \quad (y - 1)\,dy/dx + (x - 1) = 0$

On integrating both sides, we get

$$(y - 1)^2 + (x - 1)^2 = C$$

Since, the curve passes through the point (5, 4) and $C = 25$.

Hence, the curve is $(x - 1)^2 + (y - 1)^2 = (5)^2$.

$\therefore$ Centre of circle = (1, 1) and radius = 5.

16. (a,b,d) Given, $\frac{dy}{dx} + y\cos x = \cos x$

$\Rightarrow \quad \frac{dy}{dx} = \cos x\,(y - 1)$

$\Rightarrow \quad \frac{dy}{y - 1} = \cos x dx$

On integrating both sides, we get

$$\int \frac{dy}{y - 1} = \int \cos x dx$$

$\Rightarrow \quad \log(y - 1) = \sin x + C$

$\Rightarrow \quad y - 1 = Ce^{\sin x}$

Since, it passes through the point (0, 1).

$\therefore \quad 1 - 1 = Ce^{0}$

$\Rightarrow \quad C = 0$

$\Rightarrow \quad y = 1$

$\therefore \quad f(x) = 1$

So, it is constant function and it is continuous and differentiable, $\forall x \in R$.

Hence, every constant function is periodic.

17. (b) Given, $(1 + y^2)\,dx - xydy = 0$

$\Rightarrow \quad \frac{y}{1 + y^2}\,dy = \frac{dx}{x}$

On integrating both sides, we get

$$\int \frac{y}{1 + y^2} dy = \int \frac{dx}{x}$$

$\Rightarrow \quad \frac{1}{2}\log(1 + y^2) = \log xC$

$\Rightarrow \quad 1 + y^2 = x^2C^2$

Since, it passes through the point (1, 0).

$$C^2 = 1$$

Hence, the equation of the curve is

$$x^2 - y^2 = 1.$$

18. (a) Given, $\frac{dy}{dx} + \frac{1 + y^2}{\sqrt{1 - x^2}} = 0$

$\Rightarrow \quad \frac{dy}{y^2 + 1} = \frac{-dx}{\sqrt{1 - x^2}}$

$\Rightarrow \quad \int \frac{dy}{y^2 + 1} = -\int \frac{dx}{\sqrt{1 - x^2}}$

$\Rightarrow \quad \tan^{-1} y = -\sin^{-1} x + C$

$\therefore \quad \tan^{-1} y + \sin^{-1} x = C$

19. (a) Here, $PQ$ is constant length of $k$ i.e. $k$ is length of normal.

We know that, length of normal $= y\sqrt{1 + \left(\frac{dy}{dx}\right)^2}$

$\therefore \quad k = y\sqrt{1 + \left(\frac{dy}{dx}\right)^2}$

$\Rightarrow \quad 1 + \left(\frac{dy}{dx}\right)^2 = \frac{k^2}{y^2}$

$\Rightarrow \quad \left(\frac{dy}{dx}\right)^2 = \frac{k^2 - y^2}{y^2}$

$\therefore \quad y\frac{dy}{dx} = \pm\sqrt{k^2 - y^2}$

**20.** (c) We have, $$y\frac{dy}{dx} = \pm\sqrt{k^2 - y^2}$$

$$\Rightarrow \quad \frac{y}{\sqrt{k^2 - y^2}}dy = \pm\, dx$$

On integrating both sides, we get

$$\int \frac{y}{\sqrt{k^2 - y^2}}\,dy = \pm\int dx$$

$$\Rightarrow \quad -\sqrt{k^2 - y^2} = \pm\, x + C$$

$$\Rightarrow \quad k^2 - y^2 = (x \pm C)^2$$

Since, it passes through the point $(0, k)$. $\therefore\ C = 0$

Hence, the equation of curve is $x^2 + y^2 = k^2$.

**21.** (b) The equation of curve is

$$x^2 + y^2 = k^2$$

On differentating both sides w.r.t. $x$, we get

$$2x + 2y\frac{dy}{dx} = 0$$

$$\Rightarrow \quad \frac{dy}{dx} = \frac{-x}{y}$$

$$\Rightarrow \quad \left(\frac{dy}{dx}\right)_{(1,\,1)} = -1$$

Hence, the equation of tangent passing the point (0, 1) is

$$y - 1 = \frac{dy}{dx}(x - 0)$$

$$\Rightarrow \quad y - 1 = -(x) \Rightarrow x + y = 1$$

**22.** (2) Given, $\dfrac{dy}{dx} - x\tan(y - x) = 1$

Put $y - x = v \Rightarrow \dfrac{dy}{dx} - 1 = \dfrac{dv}{dx}$

$$\Rightarrow \quad \frac{dv}{dx} + 1 - x\tan v = 1 \Rightarrow \frac{dv}{dx} = x\tan v$$

$$\Rightarrow \quad \frac{dv}{\tan v} = x\,dx$$

On integrating both sides, we get

$$\int \cot v\,dv = \int x\,dx$$

$$\Rightarrow \quad \log \sin v = \frac{x^2}{2} + C$$

$$\Rightarrow \quad \log \sin(y - x) = \frac{x^2}{2} + C$$

$$\therefore \quad k = 2$$

**23.** (1) Given, $\dfrac{dy}{dx} = (x + y - 1) + \dfrac{x + y}{\log(x + y)}$

Put $x + y = v \Rightarrow \dfrac{dy}{dx} = \dfrac{dv}{dx} - 1$

$$\therefore \quad \frac{dv}{dx} - 1 = (v - 1) + \frac{v}{\log v}$$

$$\Rightarrow \quad \frac{dv}{dx} = \frac{v\log v + v}{\log v}$$

$$\Rightarrow \quad \frac{\log v}{v(\log v + 1)}dv = dx$$

On integrating both sides, we get

$$\int \frac{\log v}{v(\log v + 1)}dv = \int dx$$

$$\Rightarrow \int\left[\frac{1}{v} + \frac{1}{v(\log v + 1)}\right]dv = \int dx$$

$$\Rightarrow \quad \log v - \log(\log v + 1) = x + C$$

$$\Rightarrow \log(x + y) - \log[\log(x + y) + 1] = x + C$$

$$\therefore \quad k = 1$$

**24.** (1) We have, $\dfrac{dy}{dx} = \cos(x + y) + \sin(x + y)$

Put $x + y = v \Rightarrow 1 + \dfrac{dy}{dx} = \dfrac{dv}{dx}$

$$\therefore \quad \frac{dv}{dx} - 1 = \cos v + \sin v$$

$$\Rightarrow \quad \frac{dv}{dx} = 1 + \cos v + \sin v$$

$$\Rightarrow \quad \frac{dv}{1 + \sin v + \cos v} = dx$$

On integrating both sides, we get

$$\int \frac{dv}{1 + \dfrac{2\tan\frac{v}{2}}{1 + \tan^2\frac{v}{2}} + \dfrac{1 - \tan^2\frac{v}{2}}{1 + \tan^2\frac{v}{2}}} = \int dx$$

$$\Rightarrow \quad \int \frac{\sec^2\frac{v}{2}}{1 + \tan^2\frac{v}{2} + 2\tan\frac{v}{2} + 1 - \tan^2\frac{v}{2}}dv = \int dx$$

$$\Rightarrow \quad \int \frac{\sec^2 v/2}{2\left(1 + \tan\frac{v}{2}\right)}dv = \int dx$$

$$\Rightarrow \quad \log\left|1 + \tan\frac{v}{2}\right| = x + C$$

$$\Rightarrow \quad \log\left[1 + \tan\left(\frac{x + y}{2}\right)\right] = x + C$$

Hence, $$f\left(\frac{x + y}{2}\right) = \tan\left(\frac{x + y}{2}\right)$$

$$\therefore \quad f(x) = \tan x$$

Now, $$\lim_{x \to 0}\frac{f(x)}{x} = \lim_{x \to 0}\frac{\tan x}{x} = 1$$

**25.** (9) Let equation of tangent to the curve $y = f(x)$ at $(x, y)$ is

$$Y - y = \frac{dy}{dx}(X - x)$$

Since, $y$-intercept of tangent is $-\dfrac{xdy}{dx} + y$.

It is given that, $$-x\frac{dy}{dx} + y = x^3$$

$$\Rightarrow \quad \frac{dy}{dx} - \frac{y}{x} = -x^2$$

$$\Rightarrow \quad \frac{xdy - ydx}{x^2} = -x\,dx$$

On integrating both sides, we get

$$\int d\left(\frac{y}{x}\right) = -\int x\,dx$$

$$\Rightarrow \quad \frac{y}{x} = -\frac{x^2}{2} + C$$

$$\Rightarrow \quad y = -\frac{x^3}{2} + Cx$$

Since, it passes through the point (1, 1).

$$\therefore \quad 2 = -1 + 2C$$

$$\Rightarrow \quad C = 3/2$$

When $x = -3$, then

$$y = \frac{-(-3)^3}{2} + \left(\frac{-9}{2}\right) = \frac{27}{2} - \frac{9}{2} = \frac{18}{2} = 9$$

## DPP-3 Homogeneous Differential Equations and Equation Reducible to Homogeneous Form

**1.** (a) It is given that slope of any point $P(x, y)$ is

$$\frac{y}{x} - \cos^2\left(\frac{y}{x}\right)$$

$$\therefore \quad \frac{dy}{dx} = \frac{y}{x} - \cos^2\left(\frac{y}{x}\right)$$

Put $y = vx \Rightarrow \frac{dy}{dx} = v + x\frac{dv}{dx}$

$$\therefore \quad v + x\frac{dv}{dx} = v - \cos^2 v$$

$$\Rightarrow \quad \frac{dv}{\cos^2 v} = -\frac{dx}{x}$$

On integrating both sides, we get

$$\int \frac{dv}{\cos^2 v} = -\int \frac{dx}{x}$$

$$\Rightarrow \quad \int \sec^2 v \, dv = -\int \frac{dx}{x}$$

$$\Rightarrow \quad \tan v = -\log x + \log C$$

$$\Rightarrow \quad \tan\frac{y}{x} = \log\frac{C}{x}$$

$$\Rightarrow \quad y = x\tan^{-1}\log\frac{C}{x}$$

Since, this curve passes through the point $(1, \pi/4)$.

$$\therefore \quad \frac{\pi}{4} = 1\tan^{-1}\log C \Rightarrow C = e$$

Hence, the equation of curve is

$$y = x\tan^{-1}\log(e/x).$$

**2.** (c) Given, $x\frac{dy}{dx} = y(\log y - \log x + 1)$

$$\Rightarrow \quad \frac{dy}{dx} = \frac{y}{x}\left(\log\frac{y}{x} + 1\right)$$

Put $y = vx \Rightarrow \frac{dy}{dx} = v + x\frac{dv}{dx}$

$$\therefore \quad v + x\frac{dv}{dx} = v(\log v + 1) \Rightarrow x\frac{dv}{dx} = v\log v$$

$$\Rightarrow \quad \frac{dv}{v\log v} = \frac{dx}{x} \Rightarrow \int\frac{dv}{v\log v} = \int\frac{dx}{x}$$

$$\Rightarrow \quad \log(\log v) = \log x + \log C$$

$$\Rightarrow \quad \log v = Cx$$

$$\Rightarrow \quad \log\left(\frac{y}{x}\right) = Cx \qquad [\because v = y/x]$$

**3.** (c) Equation of circle passing through the origin and having centre at the $X$-axis, is

$$(x-a)^2 + y^2 = a^2 \Rightarrow x^2 + y^2 = 2ax \qquad \ldots(i)$$

On differentiating both sides w.r.t. $x$, we get

$$2x + 2y\frac{dy}{dx} = 2a \qquad \ldots(ii)$$

From Eqs. (i) and (ii), we get

$$x^2 + y^2 = 2x^2 + 2xy\,dy/dx$$

$$\Rightarrow \quad y^2 = x^2 + 2xy\,dy/dx$$

**4.** (d) We have, $\frac{dy}{dx} = \frac{x+y}{x}$

$$\Rightarrow \quad \frac{dy}{dx} = 1 + \frac{y}{x}$$

Put $y = vx \Rightarrow \frac{dy}{dx} = v + x\frac{dv}{dx}$

$$\therefore \quad v + x\frac{dv}{dx} = 1 + v \Rightarrow dv = \frac{dx}{x}$$

On integrating both sides, we get

$$\int dv = \int\frac{dx}{x}$$

$$\Rightarrow \quad v = \log x + C$$

$$\Rightarrow \quad y = x\log x + Cx$$

When $x = 1$ and $y = 1$, then $C = 1$

$$\therefore \quad y = x\log x + x$$

**5.** (c) Given, $(x^2 + y^2)\,dy = xy\,dx$

$$\Rightarrow \quad \frac{dy}{dx} = \frac{xy}{x^2 + y^2}$$

Put $y = vx \Rightarrow \frac{dy}{dx} = v + x\frac{dv}{dx}$

$$\therefore \quad v + x\frac{dv}{dx} = \frac{v}{1+v^2} \Rightarrow x\frac{dv}{dx} = \frac{v}{1+v^2} - v$$

$$\Rightarrow \quad x\frac{dv}{dx} = \frac{v - v - v^3}{1+v^2}$$

$$\Rightarrow \quad \frac{1+v^2}{v^3}dv = -\frac{dx}{x}$$

On integrating both sides, we get

$$\int\frac{1+v^2}{v^3}dv = -\int\frac{dx}{x}$$

$$\Rightarrow \quad \int\left(\frac{1}{v^3} + \frac{1}{v}\right)dv = -\int\frac{dx}{x}$$

$$\Rightarrow \quad \frac{-1}{2v^2} + \log v = -\log x + C$$

$$\Rightarrow \quad \frac{-x^2}{2y^2} + \log\frac{y}{x} = -\log x + C$$

$$\Rightarrow \quad \log\frac{y}{x} + \log x = C + \frac{x^2}{2y^2} \Rightarrow \log y = C + \frac{x^2}{2y^2}$$

When $x = 1$, $y = 1$, then $C = -\frac{1}{2}$

$$\therefore \quad \log y = -\frac{1}{2} + \frac{x^2}{2y^2}$$

When $x = x_0$, then $y = e$

$$\therefore \quad \log e = -\frac{1}{2} + \frac{x_0^2}{2e^2} \Rightarrow 1 + \frac{1}{2} = \frac{x_0^2}{2e^2}$$

$$\Rightarrow \quad 3e^2 = x_0^2$$

$$\Rightarrow \quad x_0 = \sqrt{3}e$$

**6.** (a) Slope of the curve at point $P(x, y)$ is $\frac{y}{x} + \sec\left(\frac{y}{x}\right)$.

$$\therefore \quad \frac{dy}{dx} = \frac{y}{x} + \sec\left(\frac{y}{x}\right)$$

$$\Rightarrow \quad v + x\frac{dv}{dx} = v + \sec v \qquad \left[\because y = vx \Rightarrow \frac{dy}{dx} = v + x\frac{dv}{dx}\right]$$

$$x\frac{dv}{dx} = \sec v$$

On integrating both sides, we get

$$\Rightarrow \quad \int\frac{dv}{\sec v} = \int\frac{dx}{x} \Rightarrow \int\cos v\,dv = \int\frac{dx}{x}$$

$$\Rightarrow \quad \sin v = \log x + C$$

$$\Rightarrow \quad \sin y/x = \log x + C$$

When $x = 1$, $y = \pi/6$, then $C = \frac{1}{2}$

$\therefore$ Equation of the curve is

$$\sin\left(\frac{y}{x}\right) = \log x + \frac{1}{2}.$$

**7.** (a) Given, $(3xy + y^2)dx + (x^2 + xy)dy = 0$

$$\Rightarrow \quad \frac{dy}{dx} = -\frac{(3xy + y^2)}{x^2 + xy}$$

$$\Rightarrow \quad v + x\frac{dv}{dx} = -\frac{(3v + v^2)}{1 + v} \quad \left[\because y = vx \Rightarrow \frac{dy}{dx} = v + x\frac{dv}{dx}\right]$$

$$\Rightarrow \quad x\frac{dv}{dx} = \frac{-3v - v^2}{1 + v} - v$$

$$\Rightarrow \quad x\frac{dv}{dx} = \frac{-3v - v^2 - v - v^2}{1 + v}$$

$$\Rightarrow \quad x\frac{dv}{dx} = \frac{-4v - 2v^2}{1 + v}$$

$$\Rightarrow \quad \frac{1 + v}{2v + v^2}dv = -\frac{2dx}{x}$$

On integrating both sides, we get

$$\int \frac{1 + v}{v(2 + v)}dv = -2\int \frac{dx}{x}$$

$$\Rightarrow \quad \int \frac{v + 1}{v^2 + 2v}dv = -2\int \frac{dx}{x}$$

$$\Rightarrow \quad \frac{1}{2}\log(v^2 + 2v) = -2\log x + \log C$$

$$\Rightarrow \quad \frac{1}{2}\log\left(\frac{y^2}{x^2} + \frac{2y}{x}\right) + 2\log x = \log C$$

$$\Rightarrow \quad \log\left[\frac{(y^2 + 2xy)}{x^2} \times x^4\right] = 2\log C$$

$$\Rightarrow \quad x^2(y^2 + 2xy) = C$$

**8.** (c) We have,

$$y\{x\cos(y/x) + y\sin(y/x)\}dx - x\{y\sin(y/x) - x\cos(y/x)\}dy = 0$$

$$\Rightarrow \quad \frac{dy}{dx} = \frac{y\{x\cos(y/x) + y\sin(y/x)\}}{x\{y\sin(y/x) - x\cos(y/x)\}}$$

$$\Rightarrow \quad v + x\frac{dv}{dx} = \frac{v(\cos v + v\sin v)}{v\sin v - \cos v} \quad \left[\because y = vx \Rightarrow v + x\frac{dv}{dx} = \frac{dy}{dx}\right]$$

$$\Rightarrow \quad v + x\frac{dv}{dx} = \frac{v\cos v + v^2\sin v}{v\sin v - \cos v}$$

$$\Rightarrow \quad x\frac{dv}{dx} = \frac{v\cos v + v^2\sin v}{v\sin v - \cos v} - v$$

$$\Rightarrow \quad x\frac{dv}{dx} = \frac{v\cos v + v^2\sin v - v^2\sin v + v\cos v}{v\sin v - \cos v}$$

$$\Rightarrow \quad x\frac{dv}{dx} = \frac{2v\cos v}{v\sin v - \cos v}$$

$$\Rightarrow \quad \frac{v\sin v - \cos v}{v\cos v}dv = \frac{2dx}{x}$$

On integrating both sides, we get

$$\int\left(\tan v - \frac{1}{v}\right)dv = 2\int \frac{dx}{x}$$

$$\Rightarrow \quad \log\sec v - \log v = 2\log x + \log C$$

$$\Rightarrow \quad \log\frac{\sec v}{v} - \log x^2 = \log C$$

$$\Rightarrow \quad \frac{\sec v}{x^2 v} = C$$

$$\Rightarrow \quad xy\cos(y/x) = C$$

**9.** (c) We have,

$$y\cos\left(\frac{y}{x}\right)(x\,dy - y\,dx) + x\sin\left(\frac{y}{x}\right)(x\,dy + y\,dx) = 0$$

$$\Rightarrow \quad \frac{dy}{dx} = \frac{y^2\cos(y/x) - xy\sin(y/x)}{xy\cos(y/x) + x^2\sin(y/x)}$$

$$\Rightarrow \quad v + x\frac{dv}{dx} = \frac{v^2\cos v - v\sin v}{v\cos v + \sin v} \quad \left[\because y = vx \Rightarrow \frac{dy}{dx} = v + x\frac{dv}{dx}\right]$$

$$\Rightarrow \quad x\frac{dv}{dx} = \frac{v^2\cos v - v\sin v}{v\cos v + \sin v} - v$$

$$\Rightarrow \quad x\frac{dv}{dx} = \frac{v^2\cos v - v\sin v - v^2\cos v - v\sin v}{v\cos v + \sin v}$$

$$\Rightarrow \quad x\frac{dv}{dx} = -\frac{2v\sin v}{v\cos v + \sin v}$$

$$\Rightarrow \quad \frac{v\cos v + \sin v}{v\sin v}dv = -\frac{2dx}{x}$$

On integrating both sides, we get

$$\int\left(\cot v + \frac{1}{v}\right)dv = -2\int \frac{dx}{x}$$

$$\Rightarrow \quad \log\sin v + \log v = -2\log x + \log C$$

$$\Rightarrow \quad \log v\sin v \cdot x^2 = \log C$$

$$\Rightarrow \quad \frac{y}{x}\sin\left(\frac{y}{x}\right)\cdot x^2 = C$$

$$\Rightarrow \quad xy\sin(y/x) = C$$

**10.** (b) We have, $\dfrac{dy}{dx} = \dfrac{x^2 + y^2 - 2x - 2y + 2}{xy - x - y + 1}$

$$\Rightarrow \quad \frac{dy}{dx} = \frac{(x^2 - 2x + 1) + (y^2 - 2y + 1)}{x(y - 1) - 1(y - 1)}$$

$$\Rightarrow \quad \frac{dy}{dx} = \frac{(x - 1)^2 + (y - 1)^2}{(x - 1)(y - 1)}$$

Put $(y - 1) = v(x - 1)$

$$\therefore \quad \frac{dy}{dx} = v + (x - 1)\frac{dv}{dx}$$

$$\Rightarrow \quad v + (x - 1)\frac{dv}{dx} = \frac{1 + v^2}{v}$$

$$\Rightarrow \quad (x - 1)\frac{dv}{dx} = \frac{1 + v^2 - v^2}{v} \Rightarrow v\,dv = \frac{dx}{x - 1}$$

On integrating both sides, we get

$$\int v\,dv = \int \frac{dx}{x - 1}$$

$$\Rightarrow \quad \frac{v^2}{2} = \log|x - 1| + C$$

$$\Rightarrow \quad \frac{(y - 1)^2}{2(x - 1)^2} = \log|x - 1| + C$$

$$\Rightarrow \quad \frac{(y - 1)^2}{(x - 1)^2} = 2\log|x - 1| + C$$

**11.** (a) We have,

$$xy\cdot\log\left(\frac{x}{y}\right)dx + \left\{y^2 - x^2\log\left(\frac{x}{y}\right)\right\}dy = 0$$

$$\Rightarrow \quad \frac{dx}{dy} = \frac{-\{y^2 - x^2\log(x/y)\}}{xy\log(x/y)}$$

$$\Rightarrow \quad v + y\frac{dv}{dy} = -\frac{(1 - v^2\log v)}{v\log v} \quad \left[\because x = vy \Rightarrow \frac{dx}{dy} = v + y\frac{dv}{dy}\right]$$

$$\Rightarrow \quad y\frac{dv}{dy} = \frac{-1 + v^2\log v - v^2\log v}{v\log v}$$

$$\Rightarrow \quad v\log v\,dv = -\frac{dy}{y}$$

On integrating both sides, we get

$$\int v\log v\,dv = \int -\frac{dy}{y}$$

$\Rightarrow \quad \log v \int v\,dv - \int \frac{d \log v}{dv} \cdot \int v dv \cdot dv = -\log y + \log C$

$\Rightarrow \quad \frac{v^2}{2} \log v - \int \frac{1}{v} \cdot \frac{v^2}{2} dv = -\log y + \log C$

$\Rightarrow \quad \frac{v^2}{2} \log v - \frac{v^2}{4} = -\log y + \log C$

$\Rightarrow \quad 2v^2 \log v - v^2 = 4 \log C/x$

$\Rightarrow \quad 2x^2 \log\left(\frac{x}{y}\right) - x^2 = 4y^2 \log\left(\frac{C}{x}\right)$

**12.** (b) Given, $(1 + e^{x/y})dx + e^{x/y}(1 - x/y)dy = 0$

$\Rightarrow \quad \frac{dx}{dy} = \frac{-e^{x/y}(1 - x/y)}{1 + e^{x/y}}$

$\Rightarrow \quad v + y\frac{dv}{dy} = -\frac{e^v(1-v)}{1+e^v} \quad \left[\because x = vy \Rightarrow \frac{dx}{dy} = v + y\frac{dv}{dy}\right]$

$\Rightarrow \quad y\frac{dv}{dy} = \frac{-e^v + ve^v - v - ve^v}{1+e^v}$

$\Rightarrow \quad y\frac{dv}{dy} = \frac{-e^v - v}{1 + e^v}$

$\Rightarrow \quad \frac{1+e^v}{v+e^v} dv = -\frac{dy}{y}$

On integrating both sides, we get

$\Rightarrow \quad \int \frac{1+e^v}{v+e^v} dv = \int -\frac{dy}{y}$

$\Rightarrow \quad \log(v + e^v) = \log C/y$

$\Rightarrow \quad v + e^v = C/y$

$\Rightarrow \quad e^{x/y} + x/y = C/y$

**13.** (d) We have, $\quad \frac{dy}{dx} = \frac{2x - y + 3}{x + 2y + 4} \quad \ldots(i)$

The given equation is non-homogeneous.

On solving $2x - y + 3 = 0$ and $x + 2y + 4 = 0$, we get

$x = -2$ and $y = -1$

Put $x = -2 + u$ and $y = -1 + v$

$\Rightarrow \quad dx = du$ and $dy = dv$

$\Rightarrow \quad \frac{dy}{dx} = \frac{dv}{du}$

Eq. (i) reduces to

$\frac{dv}{dx} = \frac{2(-2+u) - (-1+v) + 3}{(-2+u) + 2(-1+v) + 4}$

$\Rightarrow \quad \frac{dv}{du} = \frac{2u - v}{u + 2v}$

$\Rightarrow \quad t + u\frac{dt}{du} = \frac{2-t}{1+2t} \quad \left[\because v = ut \Rightarrow \frac{dv}{du} = t + u\frac{dt}{du}\right]$

$\Rightarrow \quad u\frac{dt}{du} = \frac{2 - t - t - 2t^2}{1 + 2t}$

On integrating both sides, we get

$\frac{1+2t}{t^2+t+1} dt = -2\frac{du}{u} \quad \Rightarrow \quad \int \frac{2t+1}{t^2+t-1} dt = -2\int \frac{du}{u}$

$\Rightarrow \quad \log(t^2 + t - 1) = \log \frac{C}{u^2} \quad \Rightarrow \quad t^2 + t - 1 = \frac{C}{u^2}$

$\Rightarrow \quad \frac{v^2}{u^2} + \frac{v}{u} - 1 = \frac{C}{u^2} \quad \left[\because t = \frac{v}{u}\right]$

$\Rightarrow \quad v^2 + uv - u^2 = C$

$\Rightarrow \quad (y+1)^2 + (x+2)(y+1) - (x+2)^2 = C$

$[\because v = y + 1, u = x + 2]$

$\Rightarrow \quad y^2 + 2y + 1 + xy + x + 2y + 2 - x^2 - 4x - 4 = C$

$\Rightarrow \quad y^2 - x^2 + xy + 4y - 3x - 1 = C$

**14.** (c) We have, $\quad \frac{dy}{dx} = \frac{x + y + 1}{x - y - 1} \quad \ldots(i)$

On solving $x + y + 1 = 0$ and $x - y - 1 = 0$, we get

$x = 0, y = -1$

Now, put $x = u$ and $y = v - 1$

$\Rightarrow \quad \frac{dy}{dx} = \frac{dv}{du}$

Eq. (i) reduces to

$\frac{dv}{du} = \frac{u + v}{u - v}$

$\Rightarrow \quad t + u\frac{dt}{du} = \frac{1+t}{1-t} \quad \left[\because v = ut \Rightarrow \frac{dv}{du} = t + u\frac{dt}{du}\right]$

$\Rightarrow \quad u\frac{dt}{du} = \frac{1 + t - t + t^2}{1 - t}$

$\Rightarrow \quad \frac{1-t}{1+t^2} dt = \frac{du}{u}$

On integrating both sides, we get

$\int \left(\frac{1}{1+t^2} - \frac{t}{1+t^2}\right) dt = \int \frac{du}{u}$

$\Rightarrow \quad \tan^{-1} t - \frac{1}{2} \log|1 + t^2| = \log uC$

$\Rightarrow \quad 2\tan^{-1} t = 2\log uC + \log(1 + t^2)$

$\Rightarrow \quad 2\tan^{-1} t = \log u^2 C(1 + t^2)$

$\Rightarrow \quad Cu^2(1 + t^2) = e^{2\tan^{-1} t}$

$\Rightarrow \quad Cu^2\left(1 + \frac{v^2}{u^2}\right) = e^{2\tan^{-1}\frac{v}{u}} \quad \left[\because t = \frac{v}{u}\right]$

$\Rightarrow \quad C(u^2 + v^2) = e^{2\tan^{-1}\frac{v}{u}}$

$\Rightarrow \quad C[(x^2 + (y+1)^2] = e^{2\tan^{-1}\frac{y+1}{x}} \quad [\because v = y + 1, u = x]$

$\Rightarrow \quad C\sqrt{x^2 + (y+1)^2} = e^{\tan^{-1}\left(\frac{y+1}{x}\right)}$

**15.** (c) Given, $(2x^2 + 3y^2 - 7)x \cdot dx - (3x^2 + 2y^2 - 8)y \cdot dy = 0$

$\Rightarrow \quad \frac{dy}{dx} = \frac{(2x^2 + 3y^2 - 7)x}{(3x^2 + 2y^2 - 8)y}$

On solving $2x^2 + 3y^2 - 7 = 0$ and $3x^2 + 2y^2 - 8 = 0$, we get

$x^2 = 2$ and $y^2 = 1$

Now, put $x^2 = u + 2$ and $y^2 = v + 1$

$\Rightarrow \quad 2x dx = du$ and $2y dy = dv$

$\Rightarrow \quad \frac{ydy}{xdx} = \frac{dv}{du}$

Eq. (i) reduces to $\quad \frac{dv}{du} = \frac{2u + 3v}{3u + 2v}$

$\Rightarrow \quad t + u\frac{dt}{du} = \frac{2 + 3t}{3 + 2t} \quad \left[\because v = tu \Rightarrow \frac{dv}{du} = t + u\frac{dt}{du}\right]$

$\Rightarrow \quad u\frac{dt}{du} = \frac{2 + 3t - 3t - 2t^2}{3 + 2t}$

$\Rightarrow \quad \frac{3+2t}{1-t^2} dt = \frac{2du}{v}$

On integrating both sides, we get

$\int \left(\frac{5/2}{1-t} + \frac{1/2}{1+t}\right) dt = 2\int \frac{du}{u}$

$\Rightarrow \quad -\frac{5}{2}\log|1-t| + \frac{1}{2}\log|1+t| = 2\log u + \log C$

$\Rightarrow \quad \frac{1}{2}\log \frac{1+t}{(1-t)^5} = \log u^2 C$

$\Rightarrow \quad \log \frac{1+t}{(1-t)^5} = \log u^4 C$

$\Rightarrow \quad \dfrac{1+t}{(1-t)^5} = u^4 C$

$\Rightarrow \quad \dfrac{u+v}{(u-v)^5} = C \qquad \left[\because t = \dfrac{v}{u}\right]$

$\Rightarrow \quad u + v = C(u-v)^5$

$\Rightarrow \quad x^2 + y^2 - 3 = C(x^2 - y^2 - 1)^5$

$[\because u = x^2 - 2, v = y^2 - 1]$

**16.** (a) We have,

$$\frac{dy}{dx} + \frac{\cos x(3\cos y - 7\sin x - 3)}{\sin y(3\sin x - 7\cos y + 7)} = 0$$

$$\Rightarrow \quad \frac{dy}{dx} = \frac{-\cos x(3\cos y - 7\sin x - 3)}{\sin y(3\sin x - 7\cos y + 7)}$$

On solving $3\cos y - 7\sin x - 3 = 0$ and $3\sin x - 7\cos y + 7 = 0$, we get

$$\sin x = 0, \cos y = 1$$

Now, put $\sin x = u$ and $\cos y = v + 1$

$\Rightarrow \quad \cos x\, dx = du$ and $-\sin y dy = dv$

$$\Rightarrow \quad \frac{dy}{dx} = \frac{-\cos x}{\sin y}\frac{dv}{du}$$

Eq. (i) reduces to $\quad \dfrac{dv}{du} = \dfrac{3v - 7u}{3u - 7v}$

$\Rightarrow \quad t + u\dfrac{dt}{du} = \dfrac{3t - 7}{3 - 7t} \qquad \left[\because v = ut \Rightarrow \dfrac{dv}{du} = t + u\dfrac{dt}{du}\right]$

$$\Rightarrow \quad u\frac{dt}{du} = \frac{3t - 7 - 3t + 7t^2}{3 - 7t}$$

$$\Rightarrow \quad \frac{3-7t}{t^2 - 1}dt = 7\frac{du}{u}$$

On integrating both sides, we get

$$\int\left(\frac{-5}{t+1} - \frac{2}{t-1}\right)dt = 7\int\frac{du}{u}$$

$\Rightarrow \quad -5\log|t+1| - 2\log|t-1| = 7\log u + C$

$\Rightarrow \quad 5\log|t+1| + 2\log|t-1| = -7\log u + C$

$\Rightarrow \quad \log(t+1)^5(t-1)^2 = \log C/u^7$

$\Rightarrow \quad (t+1)^5(t-1)^2 = \dfrac{C}{u^7}$

$\Rightarrow \quad \dfrac{(u+v)^5(v-u)^2}{u^7} = \dfrac{C}{u^7} \qquad \left[\because t = \dfrac{v}{u}\right]$

$\Rightarrow \quad (u+v)^5(v-u)^2 = C$

$\Rightarrow (\sin x + \cos y - 1)^5(\cos y - \sin x - 1)^2 = C$

**17.** (c) Given, $\quad \dfrac{dy}{dx} = \dfrac{6x - 4y + 3}{3x - 2y + 1} \qquad \ldots$(i)

$$\Rightarrow \quad \frac{dy}{dx} = \frac{2(3x-2y) + 3}{3x - 2y + 1}$$

Put $3x - 2y = v$

$$\Rightarrow \quad 3 - 2\frac{dy}{dx} = \frac{dv}{dx} \Rightarrow \frac{dy}{dx} = \frac{1}{2}\left(3 - \frac{dv}{dx}\right)$$

Eq. (i) reduces to

$$\frac{1}{2}\left(3 - \frac{dv}{dx}\right) = \frac{2v+3}{v+1}$$

$$\frac{dv}{dx} = 3 - \left(\frac{4v+6}{v+1}\right)$$

$$\Rightarrow \quad \frac{dv}{dx} = \frac{3v + 3 - 4v - 6}{v+1}$$

$$\Rightarrow \quad \frac{dv}{dx} = \frac{-v-3}{v+1}$$

$$\Rightarrow \quad \frac{v+1}{v+3}dv = -dx$$

On integrating both sides, we get

$$\int\left(\frac{v+3}{v+3} - \frac{2}{v+3}\right)dv = -\int dx$$

$\Rightarrow \quad v - 2\log|v+3| = -x + C$

$\Rightarrow \quad 3x - 2y + x = 2\log|3x - 2y + 3| + C$

$\Rightarrow \quad 4x - 2y = 2\log|3x - 2y + 3| + C$

**18.** (a) We have,

$$(2x + 4y + 3)\frac{dy}{dx} = 2y + x + 1$$

$$\frac{dy}{dx} = \frac{2y + x + 1}{2(x+2y) + 3} \qquad \ldots\text{(i)}$$

Put $x + 2y = v$

$$\Rightarrow \quad 1 + 2\frac{dy}{dx} = \frac{dv}{dx}$$

Eq. (i) reduces to

$$\frac{1}{2}\left(\frac{dv}{dx} - 1\right) = \frac{v+1}{2v+3}$$

$$\Rightarrow \quad \frac{dv}{dx} = \frac{2v+2}{2v+3} + 1$$

$$\Rightarrow \quad \frac{dv}{dx} = \frac{4v+5}{2v+3}$$

On integrating both sides, we get

$$\int\frac{2v+3}{4v+5}dv = \int dx$$

$$\Rightarrow \quad \frac{1}{2}\int\left[\frac{4v+5}{4v+5} + \frac{1}{4v+5}\right]dv = \int dx$$

$$\frac{1}{2}v + \frac{1}{8}\log|4v+5| = x + C$$

$\Rightarrow \quad 4v + \log|4v+5| = 8x + C$

$\Rightarrow \quad 4x + 8y + \log|4x + 8y + 5| = 8x + C \qquad [\because v = x + 2y]$

$\therefore \quad \log|4x + 8y + 5| = 4x - 8y + C$

**19.** (c) We have, $\quad \dfrac{dy}{dx} = \dfrac{y}{x} - \cot\left(\dfrac{y}{x}\right)\cos\left(\dfrac{y}{x}\right)$

$\Rightarrow \quad v + x\dfrac{dv}{dx} = v - \cot v\cos v \left[\because y = vx \Rightarrow \dfrac{dy}{dx} = v + x\dfrac{dv}{dx}\right]$

$$\Rightarrow \quad x\frac{dv}{dx} = -\frac{\cos^2 v}{\sin v}$$

$$\Rightarrow \quad \frac{\sin v}{\cos^2 v}dv = -\frac{dx}{x}$$

On integrating both sides, we get

$$\int\sec v\tan v\, dv = -\int\frac{dx}{x}$$

$\Rightarrow \quad \sec v = -\log x + C$

$\Rightarrow \quad \sec\left(\dfrac{y}{x}\right) = \log\dfrac{1}{x} + C$

Hence, $\quad f(x) = \log|1/x|$

**20.** (a) Now, $\displaystyle\lim_{x\to 0}\frac{f(1+x)}{x} = \lim_{x\to 0}\frac{\log\frac{1}{1+x}}{x} \qquad \left[\because f(x) = \log\left(\frac{1}{x}\right)\right]$

$$= \lim_{x\to 0}\frac{-\log(1+x)}{x} = -1$$

**21.** (0) We have, $\quad f(x, y) = \sin\left(\dfrac{x}{y}\right)$

$$= f(\lambda x, \lambda y) = \sin\left(\frac{\lambda x}{\lambda y}\right)$$

$$= f(\lambda x, \lambda y) = \lambda^0\sin\left(\frac{x}{y}\right)$$

$$= \lambda^0 f(x, y)$$

Hence, the degree of homogenity of $f(x, y)$ is 0.

**22.** (0) Given, $(x + y)dx + (3x + 3y - 4)dy = 0$

$$\Rightarrow \quad \frac{dy}{dx} = \frac{-(x+y)}{3(x+y)-4} \quad \ldots(i)$$

Put $x + y = v$

$$\Rightarrow \quad 1 + \frac{dy}{dx} = \frac{dv}{dx}$$

$$\Rightarrow \quad \frac{dy}{dx} = \frac{dv}{dx} - 1$$

Eq. (i) reduces to

$$\frac{dv}{dx} - 1 = \frac{-v}{3v-4} \quad \Rightarrow \quad \frac{dv}{dx} = \frac{-v}{3v-4} + 1$$

$$\Rightarrow \quad \frac{dv}{dx} = \frac{-v+3v-4}{3v-4}$$

$$\Rightarrow \quad \frac{dv}{dx} = \frac{2(v-2)}{3v-4}$$

On integrating both sides, we get

$$\int \frac{3v-4}{v-2} dv = 2\int dx$$

$$\Rightarrow \quad \int \left(3 + \frac{2}{v-2}\right) dv = 2\int dx$$

$$\Rightarrow \quad 3v + 2\log|v-2| = 2x + C_1$$

$$\Rightarrow \quad 3x + 3y + 2\log|x+y-2| = 2x + C_1$$

$$\Rightarrow \quad x + 3y + 2\log|x+y-2| = C_1$$

Therefore, $a = -2, b = 1, c = 1$

$\therefore \quad a + b + c = -2 + 1 + 1 = 0$

**23.** (2) We have,

$$(2x + 2y + 3)dy = (x - 2y + 1)dx$$

$$\Rightarrow \quad \frac{dy}{dx} = \frac{x-2y+1}{2x+2y+3}$$

On solving $x - 2y + 1 = 0$ and $2x + 2y + 3 = 0$, we get

$$x = -\frac{4}{3}, y = -\frac{1}{6}$$

Now, put $x = u - \frac{4}{3}$ and $y = v - \frac{1}{6}$

$$\Rightarrow \quad dx = du \text{ and } dy = dv$$

$$\therefore \quad \frac{dy}{dx} = \frac{dv}{du}$$

Eq. (i) reduces to $\frac{dv}{dx} = \frac{u-2v}{2u+2v}$

$$\Rightarrow \quad t + u\frac{dt}{du} = \frac{1-2t}{2+2t} \quad \left[\because v = tu \Rightarrow \frac{dv}{du} = t + u\frac{dt}{du}\right]$$

$$\Rightarrow \quad u\frac{dt}{du} = \frac{1-2t-2t-2t^2}{2(1+t)}$$

On integrating both sides, we get

$$\int \frac{1+t}{2t^2+4t-1} dt = \frac{-1}{2}\int \frac{du}{u}$$

$$\Rightarrow \quad \frac{1}{4}\log|2t^2 + 4t - 1| = -\frac{1}{2}\log u + C$$

$$\Rightarrow \quad (2t^2 + 4t - 1)(u^2) = C = \left(2\frac{v^2}{u^2} + \frac{4v}{u} - 1\right)(u^2) = C \quad \left[\therefore t = \frac{v}{u}\right]$$

$$\Rightarrow \quad (2v^2 + 4uv - u^2) = C$$

$$\Rightarrow \quad 2\left(y + \frac{1}{6}\right)^2 + 4\left(x + \frac{4}{3}\right)\left(y + \frac{1}{6}\right) - \left(x + \frac{4}{3}\right)^2 = C \quad \left[\because u = x + \frac{4}{3}, v = y + \frac{1}{6}\right]$$

$$\Rightarrow \quad 2\left(y^2 + \frac{1}{3}y + \frac{1}{36}\right) + 4\left(xy + \frac{4}{3}y + \frac{1}{6}x + \frac{4}{18}\right) - \left(x^2 + \frac{8}{3}x + \frac{16}{9}\right) = C_1$$

$$\Rightarrow 2y^2 + \frac{2}{3}y + \frac{1}{18} + 4xy + \frac{16}{3}y + \frac{2}{3}x + \frac{16}{18} - x^2 - \frac{8}{3}x - \frac{16}{9} = C_1$$

$$\Rightarrow 2y^2 + 6y + 4xy = x^2 + 2x + \frac{16}{9} - \frac{16}{18} - \frac{1}{18} + C_1$$

$$\Rightarrow 2y^2 + 6y + 4xy = x^2 + 2x + C \quad \left[\because C = \frac{16}{9} - \frac{16}{18} - \frac{1}{8} + C_1\right]$$

$$\Rightarrow \quad y^2 + 2xy + 3y = \frac{x^2}{2} + x + C$$

Hence, $a = 2$

**24.** (2) We have,

$$\frac{dy}{dx} = \frac{\sin y + x}{\sin 2y - x\cos y}$$

$$\Rightarrow \quad \frac{dy}{dx} = \frac{\sin y + x}{2\sin y\cos y - x\cos y}$$

$$\Rightarrow \quad \frac{dy}{dx} = \frac{\sin y + x}{\cos y(2\sin y - x)}$$

$$\Rightarrow \quad 2\sin y\cos y\,dy - x\cos y\,dy = \sin y\,dx + x\,dx$$

$$\Rightarrow \quad 2\sin y\cos y\,dy = \sin y\,dx + x\cos y\,dy + x\,dx$$

On integrating both sides, we get

$$\sin^2 y = x\sin y + \frac{x^2}{2} + C$$

Hence, $a = 2$

**25.** (1) We have,

$$(1 + 2e^{x/y})dx + 2e^{x/y}(1 - x/y)dy = 0$$

$$\Rightarrow \quad \frac{dx}{dy} = \frac{-2e^{x/y}(1 - x/y)}{1 + 2e^{x/y}}$$

$$\Rightarrow \quad v + y\frac{dv}{dy} = -\frac{2e^v(1-v)}{1+2e^v} \quad \left[\because x = vy \Rightarrow \frac{dx}{dy} = v + y\frac{dv}{dy}\right]$$

$$\Rightarrow \quad y\frac{dv}{dy} = \frac{-2e^v + 2ve^v - v - 2ve^v}{1 + 2e^v}$$

$$\Rightarrow \quad y\frac{dv}{dy} = \frac{-(v + 2e^v)}{1 + 2e^v}$$

$$\Rightarrow \quad \frac{1 + 2e^v}{v + 2e^v} dv = -\frac{dy}{y}$$

On integrating both sides, we get

$$\int \frac{1+2e^v}{v+2e^v} dv = -\int \frac{dy}{y}$$

$$\Rightarrow \quad \log|v + 2e^v| = \log C/y$$

$$\Rightarrow \quad v + 2e^v = C/y$$

$$\Rightarrow \quad \frac{x}{y} + 2e^{x/y} = \frac{C}{y}$$

$$\Rightarrow \quad x + 2ye^{x/y} = C$$

Hence, $f(x) = e^x$

$$\therefore \quad \lim_{x\to 0} \frac{f(x)-1}{x} = \lim_{x\to 0} \frac{e^x - 1}{x} = 1$$

# DPP-4 Linear Differential Equations and Bernoulli's Equation

**1.** (d) We have, $(1+t)\frac{dy}{dt} - ty = 1$

$$\Rightarrow \quad \frac{dy}{dt} - \left(\frac{t}{1+t}\right)y = \frac{1}{1+t}$$

which is a linear differential equation of the form

$$\frac{dy}{dt} + Py = Q(t)$$

where, $P = \frac{-t}{1+t}$ and $Q = \frac{1}{1+t}$

Now, $\int P\,dt = \int \frac{-t}{1+t}dt = -\int \frac{t+1}{t+1}dt + \int \frac{1}{1+t}dt$

$$= -t + \log(1+t)$$

$\therefore$ IF $= e^{\int P\,dt} = e^{-t+\log(1+t)} = e^{-t}\cdot e^{\log(1+t)} = e^{-t}(1+t)$

So, the general solution is

$$y\cdot \text{IF} = \int Q\cdot \text{IF}\, dt$$

$$\Rightarrow \quad ye^{-t}(1+t) = \int \frac{1}{1+t}\cdot e^{-t}(1+t)\,dt$$

$$\Rightarrow \quad ye^{-t}(1+t) = -e^{-t} + C$$

When $t = 0$, $y = -1$, then $C = 0$

$$\therefore \quad y = \frac{-1}{1+t}$$

When $t = 1$, then $y = \frac{-1}{2}$

**2.** (a) It is given that, $\frac{d(px)}{dx} = p(x)$ for all $x \geq 1$

$$\Rightarrow \quad \frac{d\,p(x)}{dx} - p(x) > 0$$

On multiplying both sides by $e^{-x}$, we get

$$e^{-x}\frac{dp(x)}{dx} - e^{-x}p(x) > 0$$

$$\Rightarrow \quad d\,[p(x)e^{-x}] > 0$$

Hence, $p(x)e^{-x}$ is increasing, $\forall\, x \in [1, \infty]$.

Now, $\quad p(1) = 0$

$\therefore$ $p(x) > 0$ for all $x > 1$.

**3.** (a) We have, $x^2\frac{dy}{dx}\cos\left(\frac{1}{x}\right) - y\sin\left(\frac{1}{x}\right) = -1$

$$\Rightarrow \quad \frac{dy}{dx} - \frac{y}{x^2}\tan\left(\frac{1}{x}\right) = \frac{-\sec(1/x)}{x^2}$$

which is a linear differential equation.

where, $P = \frac{-\tan(1/x)}{x^2}$ and $Q = \frac{-\sec(1/x)}{x^2}$

Now, $\quad \text{IF} = e^{\int \frac{-\tan(1/x)}{x^2}dx} = e^{\log\sec\left(\frac{1}{x}\right)} = \sec\left(\frac{1}{x}\right)$

So, the solution is

$$y\cdot\sec\frac{1}{x} = \int \frac{-\sec^2\left(\frac{1}{x}\right)}{x^2}dx$$

$$\Rightarrow \quad y\sec\frac{1}{x} = \tan\frac{1}{x} + C$$

$$\Rightarrow \quad y = \sin\left(\frac{1}{x}\right) + C\cos\left(\frac{1}{x}\right)$$

When $x \to \infty$ and $y \to -1$, then $C = -1$

Hence, the required solution is

$$y = \sin\frac{1}{x} - \cos\frac{1}{x}$$

**4.** (c) Given, $1 + y^2 + (x - e^{\tan^{-1}y})\frac{dy}{dx} = 0$

$$\therefore \quad (x - e^{\tan^{-1}y})\frac{dy}{dx} = -(1+y^2)$$

$$\Rightarrow \quad \frac{dx}{dy} = \frac{-(x - e^{\tan^{-1}y})}{1+y^2}$$

$$\Rightarrow \quad \frac{dx}{dy} + \frac{x}{1+y^2} = \frac{e^{\tan^{-1}y}}{1+y^2}$$

Now, $\quad \text{IF} = e^{\int \frac{1}{1+y^2}dy} = e^{\tan^{-1}y}$

Hence, the solution of given differential equation is

$$xe^{\tan^{-1}y} = \int \frac{e^{2\tan^{-1}y}}{1+y^2}dy + C$$

$$\Rightarrow \quad xe^{\tan^{-1}y} = \frac{1}{2}e^{2\tan^{-1}y} + C$$

$$\Rightarrow \quad 2xe^{\tan^{-1}y} = e^{2\tan^{-1}y} + C$$

**5.** (b) Given, $y\,dx + (x + x^2y)\,dy = 0$

$$\Rightarrow \quad \frac{dy}{dx} = \frac{-y}{x + x^2y} \quad \Rightarrow \quad \frac{dx}{dy} = \frac{-(x + x^2y)}{y}$$

$$\Rightarrow \quad \frac{dx}{dy} + \frac{x}{y} = -x^2 \quad \Rightarrow \quad \frac{-1}{x^2}\cdot\frac{dx}{dy} - \frac{1}{xy} = 1$$

Put $\frac{1}{x} = z \quad \Rightarrow \quad -\frac{1}{x^2}\cdot\frac{dx}{dy} = \frac{dz}{dy}$

$$\Rightarrow \quad \frac{dz}{dy} - \frac{z}{y} = 1$$

which is a linear differential equation of the form

$$\frac{dz}{dy} + P(y)\,z = Q(y)$$

where, $P = \frac{-1}{y}$ and $Q = 1$

$$\therefore \quad \text{IF} = e^{\int \frac{-1}{y}dy} = e^{-\log y} = \frac{1}{y}$$

Hence, the solution of given differential equation is

$$\frac{z}{y} = \int \frac{1}{y}dy + C$$

$$\Rightarrow \quad \frac{z}{y} = \log y + C$$

$$\Rightarrow \quad \frac{1}{xy} = \log y + C \qquad \left[\because z = \frac{1}{x}\right]$$

$$\Rightarrow \quad \frac{-1}{xy} + \log y = C$$

**6.** (c) We have, $(2x - 10y^3)\frac{dy}{dx} + y = 0$

$$\Rightarrow \quad \frac{dy}{dx} = \frac{-y}{2x - 10y^3}$$

$$\Rightarrow \quad \frac{dx}{dy} = \frac{10y^3 - 2x}{y}$$

$$\Rightarrow \quad \frac{dx}{dy} + \frac{2x}{y} = 10y^2$$

$$\text{IF} = e^{\int \frac{2}{y}dy} = e^{2\log y} = y^2$$

Hence, the solution of given differential equation is

$$xy^2 = \int 10y^4\,dy + C$$

$$\Rightarrow \quad xy^2 = 2y^5 + C$$

**7.** (a) We have, $y' + y\phi'(x) - \phi(x)\phi'(x) = 0$

$$\Rightarrow \quad \frac{dy}{dx} + \phi'(x)\,y = \phi(x)\,\phi'(x)$$

which is a linear differential equation,

where $P = \phi'(x)$ and $Q = \phi(x)\,\phi'(x)$

$$\therefore \quad \text{IF} = e^{\int \phi'(x)dx} = e^{\phi(x)}$$

Hence, the solution of given differential equation is

$$y \cdot e^{\phi(x)} = \int e^{\phi(x)} \cdot \phi(x)\,\phi'(x)\,dx$$

$$\Rightarrow \quad y \cdot e^{\phi(x)} = e^{\phi(x)}[\phi(x) - 1] + C$$

$$\Rightarrow \quad y = Ce^{-\phi(x)} + \phi(x) - 1$$

**8.** (d) We have, $y^2dx + \left(x - \frac{1}{y}\right)dy = 0$

$$\Rightarrow \quad \frac{dx}{dy} + \frac{x}{y^2} = \frac{1}{y^3}$$

$$\therefore \quad \text{IF} = e^{\int \frac{1}{y^2}dy} = e^{-\frac{1}{y}}$$

Hence, the solution of given differential equation is

$$xe^{-1/y} = \int \frac{e^{-1/y}}{y^3}\,dy$$

$$\Rightarrow \quad xe^{-1/y} = -\int e^t \cdot t\,dt \quad \left[\because \text{put } \frac{-1}{y} = t \Rightarrow \frac{1}{y^2}dy = dt\right]$$

$$\Rightarrow \quad xe^{-1/y} = -e^t(t-1) + C$$

$$\Rightarrow \quad xe^{-1/y} = e^{-1/y} + \frac{e^{-1/y}}{y} + C$$

When $x = 1$ and $y = 1$, then $C = \frac{-1}{e}$

$$\therefore \quad x = 1 + \frac{1}{y} - \frac{e^{1/y}}{e}$$

**9.** (a) We have, $\lim_{t \to x} \frac{t^2 f(x) - x^2 f(t)}{t - x} = 1$

$$\Rightarrow \quad \lim_{t \to x} \frac{2t\,f(x) - x^2 f'(t)}{1} = 1$$

[applying L' Hospital's rule]

$$\Rightarrow \quad 2xf(x) - x^2f'(x) = 1$$

$$\Rightarrow \quad x^2f'(x) = 2xf(x) - 1$$

$$\Rightarrow \quad f'(x) - \frac{2f(x)}{x} = \frac{-1}{x^2}$$

which is a form of linear differential equation.

where, $P = \frac{-2}{x}$ and $Q = \frac{-1}{x^2}$

$$\therefore \quad \text{IF} = e^{\int -2/x\,dx} = e^{-2\log x} = \frac{1}{x^2}$$

Hence, the solution of given differential equation is

$$f(x) \cdot \frac{1}{x^2} = \int \frac{-1}{x^4}dx$$

$$\Rightarrow \quad f(x) \cdot \frac{1}{x^2} = \frac{1}{3x^3} + C$$

$$\Rightarrow \quad f(x) = \frac{1}{3x} + Cx^2$$

$$\Rightarrow \quad f(1) = \frac{1}{3} + C = 1 \quad \Rightarrow \quad C = \frac{2}{3}$$

$$\therefore \quad f(x) = \frac{1}{3x} + \frac{2x^2}{3}$$

**10.** (a) We have, $y' - y\tan x = 2x\sec x$

$$\Rightarrow \quad \frac{dy}{dx} - y\tan x = 2x\sec x$$

which is a form of linear differential equation.

where, $P = -\tan x$ and $Q = 2x\sec x$

$$\therefore \quad \text{IF} = e^{\int -\tan x\,dx} = e^{\log\cos x} = \cos x$$

So, the solution of given differential equation is

$$y\cos x = \int 2x\sec x\cos x\,dx + C$$

$$y\cos x = x^2 + C$$

When $x = 0$ and $y = 0$, then $C = 0$

$$\therefore \quad y = x^2\sec x$$

$$\Rightarrow \quad y\left(\frac{\pi}{4}\right) = \left(\frac{\pi}{4}\right)^2 \sec\frac{\pi}{4} = \frac{\pi^2}{16} \times \sqrt{2} = \frac{\pi^2}{8\sqrt{2}}$$

**11.** (a,b) We have, $y' + y = \cos x - \sin x$

$$\Rightarrow \quad \frac{dy}{dx} + y = \cos x - \sin x$$

$$\therefore \quad \text{IF} = e^{\int dx} = e^x$$

Hence, the solution of given differential equation is

$$y \cdot e^x = \int e^x(\cos x - \sin x)\,dx + C$$

$$\Rightarrow \quad y \cdot e^x = e^x\cos x + C$$

$$\Rightarrow \quad y = \cos x + Ce^{-x}$$

When $x = \frac{\pi}{2}$ and $y = 0$, then $C = 0$

$$\therefore \quad y(0) = 1,\ y\left(\frac{\pi}{4}\right) = \frac{1}{\sqrt{2}} \text{ and } y\left(\frac{\pi}{3}\right) = \frac{1}{2}$$

**12.** (a,c) We have, $y' + y\tan x = 2x + x^2\tan x$

$$\Rightarrow \quad \frac{dy}{dx} + y\tan x = 2x + x^2\tan x$$

$$\therefore \quad \text{IF} = e^{\int \tan x\,dx} = e^{\log\sec x} = \sec x$$

Hence, the solution of given differential equation is

$$y\sec x = \int (2x + x^2\tan x)\sec x\,dx + C$$

$$\Rightarrow \quad y\sec x = \int (2x\sec x + x^2\sec x\tan x)\,dx + C$$

$$\Rightarrow \quad y\sec x = \sec x\int 2x\,dx - \int \frac{d}{dx}\sec x \int 2x\,dx \cdot dx + \int x^2\sec x\tan x\,dx + C$$

$$\Rightarrow \quad y\sec x = x^2\sec x - \int x^2\sec x\tan x\,dx + \int x^2\sec x\tan x\,dx + C$$

$$\Rightarrow \quad y\sec x = x^2\sec x + C$$

When $x = 0$ and $y = 0$, then $C = 0$

$$\therefore \quad y = x^2$$

Now, $y\left(\frac{\pi}{4}\right) = \left(\frac{\pi}{4}\right)^2 = \frac{\pi^2}{16}$

and $y\left(\frac{\pi}{3}\right) = \left(\frac{\pi}{3}\right)^2 = \frac{\pi^2}{9}$

**13.** (b,c) We have, $y' + \frac{y}{x} = \cos x + \frac{\sin x}{x}$

$$\Rightarrow \quad \frac{dy}{dx} + \frac{y}{x} = \cos x + \frac{\sin x}{x}$$

$$\text{IF} = e^{\int \frac{1}{x}dx} = e^{\log x} = x$$

Hence, the solution of given differential equation is

$$y \cdot x = \int \left(\cos x + \frac{\sin x}{x}\right)x\,dx + C$$

$$\Rightarrow \quad y \cdot x = \int (x\cos x + \sin x)\,dx + C$$

$$\Rightarrow \quad y \cdot x = x\sin x + C$$

$$\Rightarrow \quad y = \sin x + C/x$$

When $x = 0$ and $y = 0$, then $C = 0$

$$\therefore \quad y = \sin x$$

Now, $y\left(\frac{\pi}{2}\right) = 1$ and $y\left(\frac{\pi}{4}\right) = \frac{1}{\sqrt{2}}$

**14.** (a,c) We have, $y - (x + 2y^2)\,y' = 0$

$$\Rightarrow \quad y - (x + 2y^2)\frac{dy}{dx} = 0 \Rightarrow \frac{dx}{dy} = \frac{x + 2y^2}{y}$$

$$\Rightarrow \quad \frac{dx}{dy} - \frac{x}{y} = 2y$$

$$\therefore \quad \text{IF} = e^{\int \frac{-1}{y} dy} = e^{-\log y} = \frac{1}{y}$$

Hence, the solution of given differential equation is

$$\frac{x}{y} = \int \frac{2y}{y}\,dy + C \Rightarrow \frac{x}{y} = 2y + C$$

$$\Rightarrow \quad x = 2y^2 + Cy$$

When $x = 0$, $y = 1$, then $C = -2$

$$\therefore \quad x = 2y^2 - 2y \Rightarrow x = 2y(y - 1)$$

When $y = 2$, then $x = 4$

**15.** (a) Tangent drawn at any point $(x, y)$ is

$$Y - y = \frac{dy}{dx}(X - x)$$

When $Y = 0$, then $X = x - y\frac{dy}{dx}$

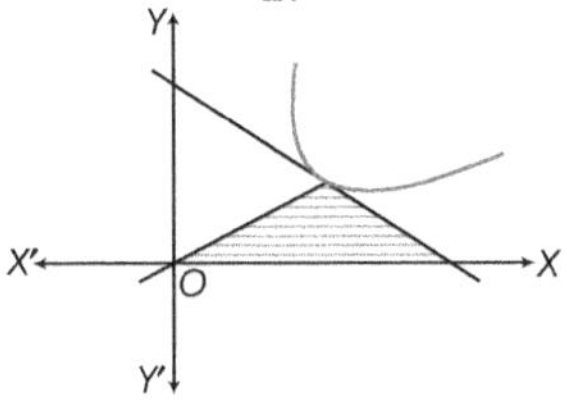


Area of triangle $= k^2$ [given]

i.e $\quad \left|\frac{1}{2} \cdot XY\right| = k^2$

$$\Rightarrow \quad xy - y^2\frac{dx}{dy} = \pm 2k^2$$

$$\Rightarrow \quad xdy - ydx = \pm \frac{2k^2}{y}$$

**16.** (c) We have, $xdy - ydx = \pm 2k^2/y$

$$\Rightarrow \quad \frac{dx}{dy} - \frac{x}{y} = \mp \frac{2k^2}{y^2}$$

$$\therefore \quad \text{IF} = e^{\int \frac{-1}{y} dy} = e^{-\log y} = \frac{1}{y}$$

Hence, the solution of given differential equation is

$$\frac{x}{y} = \int \mp \frac{2k^2}{y^3}\,dy + C$$

$$\Rightarrow \quad \frac{x}{y} = \mp \frac{k^2}{y^2} + C \Rightarrow x = \mp \frac{k^2}{y} + Cy$$

When $x = 0$ and $y = 1$, then $C = \pm k^2$

$\therefore$ Equation of the curve is $x = \mp \frac{k^2}{y} \pm k^2 y$

$$\Rightarrow \quad xy = \pm k^2(1 - y^2)$$

**17.** (b) Length of subnormal $= y\frac{dy}{dx}$ at $P(x, y)$

$$\therefore \quad OP^2 = x^2 + y^2$$

It is given that, $y\frac{dy}{dx} \propto OP^2$

$$\Rightarrow \quad y\frac{dy}{dx} = \lambda(x^2 + y^2) \quad [\because \lambda \text{ is a constant}]$$

$$\Rightarrow \quad y\frac{dy}{dx} - \lambda y^2 = \lambda x^2$$

**18.** (a) We have, $y\frac{dy}{dx} - \lambda y^2 = \lambda x^2$ ...(i)

Put $y^2 = z \Rightarrow 2y\frac{dy}{dx} = \frac{dz}{dx}$

Eq. (i) reduces to $\frac{1}{2}\frac{dz}{dx} - \lambda z = \lambda x^2$

$$\Rightarrow \quad \frac{dz}{dx} - 2\lambda z = 2\lambda x^2$$

which is a form of linear differential equation.

where, $P = -2\lambda$ and $Q = 2\lambda x^2$

$$\therefore \quad \text{IF} = e^{\int -2\lambda dx} = e^{-2\lambda x}$$

Hence, the solution of given differential equation is

$$z \cdot e^{-2\lambda x} = \int 2\lambda x^2 e^{-2\lambda x} dx$$

$$\Rightarrow \quad ze^{-2\lambda} = 2\lambda\left[x^2 \int e^{-2\lambda x} - \int \frac{dx^2}{dx} \int e^{-2\lambda x} dx \cdot dx\right] + C$$

$$\Rightarrow \quad ze^{-2\lambda x} = 2\lambda\left[\frac{x^2 e^{-2\lambda x}}{-2\lambda} + \int \frac{2x\,e^{-2\lambda x} dx}{2\lambda}\right] + C$$

$$\Rightarrow \quad ze^{-2\lambda x} = -x^2 e^{-2\lambda x} + 2\int xe^{-2\lambda x} dx + C$$

$$\Rightarrow \quad ze^{-2\lambda x} = dx - x^2 e^{-2\lambda x} dx + 2\left[x\int e^{-2\lambda x} dx - \int \left\{\frac{dy}{dx}\int e^{-2\lambda x} \cdot dx\right\} dx\right] + C$$

$$\Rightarrow \quad ze^{-2\lambda x} = -x^2 e^{-2\lambda x} + 2\left[\frac{xe^{-2\lambda x}}{-2\lambda} + \int \frac{e^{-2\lambda x}}{2\lambda} dx\right] + C$$

$$\Rightarrow \quad ze^{-2\lambda x} = -x^2 e^{-2\lambda x} - \frac{xe^{-2\lambda x}}{\lambda} - \frac{e^{-2\lambda x}}{2\lambda^2} + C$$

$$\Rightarrow \quad z = -x^2 - \frac{x}{\lambda} - \frac{1}{2\lambda^2} + Ce^{2\lambda x}$$

$$\Rightarrow \quad y^2 = -x^2 - \frac{x}{\lambda} - \frac{1}{2\lambda^2} + Ce^{2\lambda x} \quad [\because z = y^2]$$

**19.** (6) Given, $6\int_1^x f(x)\,dx = 3xf(x) - x^3$

On differentiating both sides w.r.t. $x$, we get

$$6f(x) = 3xf'(x) + 3f(x) - 3x^2$$

$$\Rightarrow \quad 2f(x) = xf'(x) + f(x) - x^2 \Rightarrow xf'(x) - f(x) = x^2$$

$$\Rightarrow \quad f'(x) - \frac{f(x)}{x} = x \Rightarrow \frac{dy}{dx} - \frac{y}{x} = x \quad \left[\because y = f(x) \Rightarrow \frac{dy}{dx} = f'(x)\right]$$

which is a form of linear differential equation.

where, $P = \frac{-1}{x}$ and $Q = x$

$$\therefore \quad \text{IF} = e^{\int \frac{-1}{x} dx} = e^{-\log x} = \frac{1}{x}$$

Hence, the solution of given differential equation is

$$\frac{y}{x} = \int \frac{x}{x} dx + C$$

$$\frac{y}{x} = x + C \Rightarrow y = x^2 + Cx$$

When $x = 1$ and $y = 2$, then $C = 1$

$$\therefore \quad y = x^2 + x$$

When $x = 2$, then $\quad y = 6$

**20.** (0) Given, $y'(x) + y(x)\,g'(x) = g(x)\,g'(x)$

which is a form of linear differential equation.

where, $P = g'(x)$ and $Q = g(x)\,g'(x)$

$$\therefore \quad \text{IF} = e^{\int g'(x) dx} = e^{g(x)}$$

Hence, the solution of given differential equation is

$$y(x) \cdot e^{g(x)} = \int e^{g(x)} \cdot g(x)[g'(x)\,dx] + C$$

$$\Rightarrow \quad y(x)\,e^{g(x)} = e^{g(x)}[g(x) - 1] + C$$

We have, $y(0) = g(0) = g(2) = 0$, then $C = 1$

$$\Rightarrow \quad y(2)\,e^{g(2)} = e^{g(2)}\,[g(2) - 1] + 1 \Rightarrow y(2) = 0$$

## DPP-5 Solving by Inspection Method and Orthogonal Trajectory

**1.** (c) We have,

$$x\,dy + y\,dx + 2x^3dx = 0$$

$$\Rightarrow \quad d(xy) + 2x^3dx = 0$$

On integrating both sides, we get

$$\int d(xy) + \int 2x^3dx = 0$$

$$\therefore \quad xy + \frac{x^4}{2} = C$$

**2.** (a) Given, $x\,dx + y\,dy = a(x^2 + y^2)\,dy$

$$\Rightarrow \quad \frac{x\,dx + y\,dy}{x^2 + y^2} = a\,dy$$

$$\Rightarrow \quad \frac{1}{2}d\log(x^2 + y^2) = a\,dy$$

On integrating both sides, we get

$$\int d[\log(x^2 + y^2)] = 2\int a\,dy$$

$$\therefore \quad \log(x^2 + y^2) = 2ay + C$$

**3.** (c) We have,

$$(1 + xy)\,x\,dy + (1 - xy)\,y\,dx = 0$$

$$\Rightarrow \quad x\,dy + x^2y\,dy + y\,dx - xy^2dx = 0$$

$$\Rightarrow \quad x\,dy + y\,dx + x^2y\,dy - xy^2dx = 0$$

On dividing by $x^2y^2$, we get

$$\frac{x\,dy + y\,dx}{x^2y^2} + \frac{x\,dy - y\,dx}{xy} = 0$$

On integrating both sides, we get

$$\int d\left(-\frac{1}{xy}\right) + \int d\left[\log\left(\frac{y}{x}\right)\right] = C$$

$$\therefore \quad -\frac{1}{xy} + \log\left(\frac{y}{x}\right) = C$$

**4.** (a) We have,

$$y\,dx - x\,dy + (1 + x^2)\,dx + x^2\sin y\,dy = 0$$

On dividing by $x^2$, we get

$$\frac{y\,dx - x\,dy}{x^2} + \left(\frac{1}{x^2} + 1\right)dx + \sin y\,dy = 0$$

$$-\left(\frac{x\,dy - y\,dx}{x^2}\right) + \left(\frac{1}{x^2} + 1\right)dx + \sin y\,dy = 0$$

On integrating both sides, we get

$$-\int d\left(\frac{y}{x}\right) + \int\left(\frac{1}{x^2} + 1\right)dx + \int \sin y\,dy = C$$

$$\Rightarrow \quad -\frac{y}{x} + x - \frac{1}{x} - \cos y = C$$

$$\therefore \quad \frac{y}{x} + \frac{1}{x} - x + \cos y + C = 0$$

**5.** (b) Given, $(1 + x\sqrt{x^2 + y^2})\,dx + (-1 + \sqrt{x^2 + y^2})\,y\,dy = 0$

$$dx + x\sqrt{x^2 + y^2}\,dx - y\,dy + y\sqrt{x^2 + y^2}\,dy = 0$$

$$\Rightarrow \quad dx - y\,dy + x\sqrt{x^2 + y^2}\,dx + y\sqrt{x^2 + y^2}\,dy = 0$$

$$\Rightarrow \quad dx - y\,dy + \frac{1}{3}d(x^2 + y^2)^{3/2} = 0$$

On integrating both sides, we get

$$\int dx - \int y\,dy + \int \frac{1}{3}d(x^2 + y^2)^{3/2} = 0$$

$$\Rightarrow \quad x - \frac{y^2}{2} + \frac{1}{3}(x^2 + y^2)^{3/2} = C$$

$$\therefore \quad 2x - y^2 + \frac{2}{3}(x^2 + y^2)^{3/2} = C$$

**6.** (a) We have, $\dfrac{x\,dy}{x^2 + y^2} = \left(\dfrac{y}{x^2 + y^2} - 1\right)dx$

$$\Rightarrow \quad \frac{x\,dy}{x^2 + y^2} - \frac{y\,dx}{x^2 + y^2} = -dx$$

$$\Rightarrow \quad \frac{x\,dy - y\,dx}{x^2 + y^2} = -dx \Rightarrow d\tan^{-1}\left(\frac{y}{x}\right) = -dx$$

$$\Rightarrow \quad \int d\tan^{-1}\left(\frac{y}{x}\right) = \int -dx$$

$$\therefore \quad \tan^{-1}\left(\frac{y}{x}\right) = -x + C$$

**7.** (d) Given, $ye^{x/y}dx = (xe^{x/y} + y^2\sin y)\,dy$

$$\Rightarrow \quad ye^{x/y}dx - xe^{x/y}dy = y^2\sin y\,dy$$

$$\Rightarrow \quad e^{x/y}\left(\frac{y\,dx - x\,dy}{y^2}\right) = \sin y\,dy$$

$$\Rightarrow \quad \int de^{x/y} = \int \sin y\,dy$$

$$\therefore \quad e^{x/y} = -\cos y + C$$

**8.** (c) We have,

$$\{y(1 + x^{-1}) + \sin y\}\,dx + \{x + \log x + x\cos y\}\,dy = 0$$

$$\Rightarrow y\,dx + \frac{y}{x}dx + \sin y\,dx + x\,dy + \log x\,dy + x\cos y\,dy = 0$$

$$\Rightarrow y\,dx + x\,dy + \frac{y}{x}dx + \log x\,dy + \sin y\,dx + x\cos y\,dy = 0$$

$$\Rightarrow \quad d(xy) + d(y\log x) + d(x\sin y) = 0$$

$$\Rightarrow \quad \int d(xy) + \int d(y\log x) + \int d(x\sin y) = 0$$

$$\therefore \quad xy + y\log x + x\sin y = C$$

**9.** (c) We have,

$$x\cos\left(\frac{y}{x}\right)\cdot\{y\,dx + x\,dy\} = y\sin\left(\frac{y}{x}\right)\{x\,dy - y\,dx\}$$

$$\Rightarrow xy\cos\left(\frac{y}{x}\right)dx + x^2\cos\frac{y}{x}dy = xy\sin\left(\frac{y}{x}\right)dy - y^2\sin\left(\frac{y}{x}\right)dx$$

$$\Rightarrow x^2\cos\frac{y}{x}dy - xy\sin\left(\frac{y}{x}\right)dy = -\left\{y^2\sin\left(\frac{y}{x}\right) + xy\cos\left(\frac{y}{x}\right)\right\}dx$$

$$\Rightarrow \quad \frac{dy}{dx} = \frac{-\left[y^2\sin\left(\frac{y}{x}\right) + xy\cos\left(\frac{y}{x}\right)\right]}{x^2\cos\left(\frac{y}{x}\right) - xy\sin\left(\frac{y}{x}\right)}$$

$$\Rightarrow v + x\frac{dv}{dx} = \frac{-(v^2\sin v + v\cos v)}{\cos v - v\sin v}\left[\because y = vx \Rightarrow \frac{dy}{dx} = v + x\frac{dv}{dx}\right]$$

$$\Rightarrow \quad x\frac{dv}{dx} = \frac{-v^2\sin v - v\cos v - v\cos v + v^2\sin v}{\cos v - v\sin V}$$

$$\Rightarrow \quad x\frac{dv}{dx} = \frac{-2v\cos v}{\cos v - v\sin v}$$

$$\Rightarrow \quad \frac{\cos v - v\sin v}{v\cos v}dv = \frac{-2\,dx}{x}$$

$$\Rightarrow \quad \int\left(\frac{1}{v} - \tan v\right)dV = -2\int\frac{dx}{x}$$

$$\Rightarrow \quad \log v - \log\sec v = -2\log x + \log C$$

$$\Rightarrow \quad \log v + \log x^2 = \log C + \log\sec v$$

$$\Rightarrow \quad vx^2 = C\sec v$$

$$\Rightarrow \quad \frac{y}{x}\cdot x^2 = C\sec\left(\frac{y}{x}\right) \qquad \left[\because v = \frac{y}{x}\right]$$

$$\therefore \quad xy = C\sec\left(\frac{y}{x}\right)$$

**10.** (a) We have, $x \sin\left(\frac{y}{x}\right) dy = \left\{ y \sin\left(\frac{y}{x}\right) - x \right\} dx$

$$\Rightarrow \quad x \sin\left(\frac{y}{x}\right) dy - y \sin\left(\frac{y}{x}\right) dx = -x\,dx$$

$$\Rightarrow \quad \sin\left(\frac{y}{x}\right)(x\,dy - y\,dx) = -x\,dx$$

$$\Rightarrow \quad \sin\left(\frac{y}{x}\right)\frac{(x\,dy - y\,dx)}{x^2} = -\frac{1}{x}dx$$

$$\Rightarrow \quad \sin\left(\frac{y}{x}\right) d\left(\frac{y}{x}\right) = -\frac{1}{x}dx$$

$$\Rightarrow \quad \int \sin\left(\frac{y}{x}\right) d\left(\frac{y}{x}\right) = -\int \frac{dx}{x}$$

$$\Rightarrow \quad -\cos\left(\frac{y}{x}\right) = -\log x + C$$

$$\therefore \quad \log x - \cos\left(\frac{y}{x}\right) = C$$

**11.** (c) We have, $\quad y = cx^2 \quad$ ...(i)

On differentiating w.r.t. $x$, we get

$$\frac{dy}{dx} = 2cx \quad \text{...(ii)}$$

From Eqs. (i) and (ii), we get

$$y = \frac{1}{2} x \frac{dy}{dx}$$

$$\Rightarrow \quad 2y = x\frac{dy}{dx} \quad \text{...(iii)}$$

Now, to obtain orthogonal trajectory replace $\frac{dy}{dx}$ by $\left(-\frac{dx}{dy}\right)$ in Eq. (iii), we get

$$\Rightarrow \quad 2y = -x\frac{dx}{dy}$$

$$\Rightarrow \quad 2y\,dy = -x\,dx$$

On integrating both sides, we get

$$\int 2y\,dy = -\int x\,dx$$

$$\Rightarrow \quad y^2 = -\frac{x^2}{2} + C_1$$

$$\therefore \quad x^2 + 2y^2 = 2C_1$$

**12.** (d) We have, $\quad x^2 + y^2 - 2cx = 0 \quad$ ...(i)

On differentiating w.r.t. $x$ both sides, we get

$$2x + 2y\frac{dy}{dx} - 2c = 0$$

$$\Rightarrow \quad c = x + y\frac{dy}{dx}$$

On putting the value of $c$ in Eq. (i), we get

$$x^2 + y^2 - 2\left(x + y\frac{dy}{dx}\right)x = 0$$

$$\Rightarrow \quad y^2 - x^2 = 2xy\frac{dy}{dx}$$

For orthogonal trajectories, we replace $\frac{dy}{dx}$ by $\left(-\frac{dx}{dy}\right)$, we get

$$y^2 - x^2 = -2xy\frac{dx}{dy} \quad \Rightarrow \quad \frac{dy}{dx} = \frac{2xy}{x^2 - y^2}$$

$$\Rightarrow \quad v + x\frac{dv}{dx} = \frac{2v}{1 - v^2} \quad \left[\because y = vx \Rightarrow \frac{dy}{dx} = v + x\frac{dv}{dx}\right]$$

$$\Rightarrow \quad x\frac{dv}{dx} = \frac{2v - v + v^3}{1 - v^2}$$

On integrating both sides, we get

$$\int \frac{1 - v^2}{v + v^3} dv = \int \frac{dx}{x}$$

$$\Rightarrow \quad \int \left(\frac{1}{v} - \frac{2v}{1 + v^2}\right) dv = \int \frac{dx}{x}$$

$$\Rightarrow \quad \log v - \log(1 + v^2) = \log x + \log C$$

$$\Rightarrow \quad \frac{v}{1 + v^2} = xC$$

$$\Rightarrow \quad \frac{yx}{x^2 + y^2} = Cx$$

$$\therefore \quad x^2 + y^2 + Cy = 0$$

**13.** (a) We have, $\quad \frac{x\,dx + y\,dy}{x\,dy - y\,dx} = \sqrt{\frac{1 - x^2 - y^2}{x^2 + y^2}}$

$$\Rightarrow \quad \frac{x\,dx + y\,dy}{\sqrt{1 - (x^2 + y^2)}} = \frac{x\,dy - y\,dx}{\sqrt{x^2 + y^2}}$$

$$\Rightarrow \quad \frac{x\,dx + y\,dy}{(\sqrt{x^2 + y^2})\sqrt{1 - (x^2 + y^2)}} = \frac{x\,dy - y\,dx}{x^2 + y^2}$$

$$\Rightarrow \quad d(\sin^{-1}\sqrt{x^2 + y^2}) = d\tan^{-1}\left(\frac{y}{x}\right)$$

$$\Rightarrow \quad \int d(\sin^{-1}\sqrt{x^2 + y^2}) = \int d\tan^{-1}\left(\frac{y}{x}\right)$$

$$\Rightarrow \quad \sin^{-1}\sqrt{x^2 + y^2} = \tan^{-1}\left(\frac{y}{x}\right) + C$$

When $x = 1$, $y = 0$, then $\quad C = \frac{\pi}{2}$

$$\therefore \sin^{-1}\sqrt{x^2 + y^2} = \tan^{-1}\left(\frac{y}{x}\right) + \frac{\pi}{2},$$

when $x = 0$, $y = 0$.

**14.** (a) Given, $\frac{\sqrt{x^2 - y^2}}{\sqrt{1 + \left(\sqrt{x^2 - y^2}\right)^2}} \cdot (x\,dx - y\,dy) = x\,dy - y\,dx$

$$\Rightarrow \quad \frac{1}{2}\frac{\sqrt{x^2 - y^2}}{\sqrt{1 + (x^2 - y^2)^2}} \cdot d(x^2 - y^2) = x(dy - dx) - y(dx - dy) + x\,dx - y\,dy$$

$$\Rightarrow \quad \frac{1}{2\sqrt{x^2 - y^2} \cdot \sqrt{1 + (\sqrt{x^2 - y^2})^2}} \cdot d(x^2 - y^2) = \frac{-(x + y)(dx - dy)}{x^2 - y^2} + \frac{1}{2}\frac{d(x^2 - y^2)}{(x^2 - y^2)}$$

$$\Rightarrow \quad \frac{1}{\sqrt{1 - \left(\sqrt{x^2 - y^2}\right)^2}} \cdot d(\sqrt{x^2 - y^2}) = -\frac{d(x - y)}{x - y} + \frac{1}{2} \cdot \frac{d(x^2 - y^2)}{x^2 - y^2}$$

On integrating, we get

$$\log\left\{\sqrt{x^2 - y^2} + \sqrt{1 + x^2 - y^2}\right\} = -\log(x - y) + \frac{1}{2}\log(x^2 - y^2) + \log C$$

$$\Rightarrow \quad \sqrt{x^2 - y^2} + \sqrt{1 + x^2 - y^2} = \left(\frac{\sqrt{x^2 - y^2}}{x - y}\right) \cdot C$$

i.e. $\quad \sqrt{x^2 - y^2} + \sqrt{1 + x^2 - y^2} = \sqrt{\frac{x + y}{x - y}} \cdot C$

**15.** (c,d) Equation of tangent to the curve $y = f(x)$ at $(x, y)$ is

$$Y - y = \frac{dy}{dx}(X - x)$$

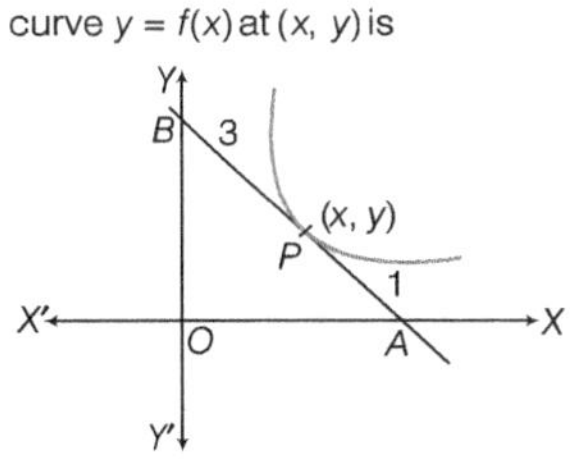


Thus, cuts the $X$-axis at $A$ and $Y$-axis at $B$.

$$\therefore \quad A\left(\frac{x\frac{dy}{dx} - y}{\frac{dy}{dx}}, 0\right)$$

and $B\left(0, -x\frac{dy}{dx} + y\right)$

Now, it is given that $BP : AP = 3:1$

$$x = \frac{\frac{3\left(x\frac{dy}{dx} - y\right)}{dy/dx} + 1 \times 0}{3+1}$$

$$\Rightarrow \quad 4x\frac{dy}{dx} = 3x\frac{dy}{dx} - 3y \Rightarrow x\frac{dy}{dx} + 3y = 0$$

$$\Rightarrow \quad \int \frac{dy}{y} = \int \frac{-3\,dx}{x}$$

$$\Rightarrow \quad \log y = -3\log x + \log C$$

$$\Rightarrow \quad yx^3 = C$$

Curve passes through (1, 1).

$\therefore \quad C = 1$

Equation of curve $= x^3y = 1$

The curve is also passes through $\left(2, \frac{1}{8}\right)$.

**16.** (c) It is given that, $\frac{dV}{dt} \propto V$ for each reservoir.

For reservoir $A$, $\quad \frac{dV_A}{dt} \propto V_A$

$$\Rightarrow \quad \frac{dV_A}{dt} = -k_1V_A \quad [\because k_1 \text{ is proportionality constant}]$$

$$\Rightarrow \quad \int \frac{dV_A}{V_A} = -\int k_1\,dt$$

$$\Rightarrow \quad \log V_A = -k_1t + C_1$$

$$\Rightarrow \quad V_A = C_1e^{-k_1t} \quad \text{...(i)}$$

Similarly for $B$, $\quad V_B = C_2e^{-k_2t}$ ...(ii)

Now, at $t = 0$, $\quad V_A = 2V_B$

From Eqs. (i) and (ii), we get

$$\frac{C_1}{C_2} = 2$$

Also, at $t = 1$, $\quad V_A = \frac{3}{2}V_B$

$$\Rightarrow \quad e^{-(k_1-k_2)}\cdot\frac{C_1}{C_2} = \frac{3}{2} \Rightarrow e^{-(k_1-k_2)} = \frac{3}{4}$$

$$\Rightarrow \quad k_1 - k_2 = -\log\left(\frac{3}{4}\right)$$

If $\quad t = \frac{1}{2}, \frac{V_A}{V_B} = \frac{C_1}{C_2}e^{\frac{1}{2}(k_1-k_2)} \Rightarrow \frac{V_A}{V_B} = 2\,(e^{k_1-k_2})^{1/2}$

$$\Rightarrow \quad \frac{V_A}{\sqrt{3}} = 2 \times \left(\frac{3}{4}\right)^{1/2} \Rightarrow V_A = \sqrt{3}\,V_B$$

$$\therefore \quad k = \sqrt{3}$$

**17.** (a,d) Let at $t = t_0$, $V_A = V_B$

Then, $\quad C_1e^{-k_1t_0} = C_2e^{-k_2t_0}$

$$\Rightarrow \quad \frac{C_1}{C_2}e^{-t_0(k_1-k_2)} = 1 \quad \Rightarrow e^{-t_0(k_1-k_2)} = \frac{1}{2}$$

$$\Rightarrow \quad (e^{-(k_1-k_2)})^{t_0} = \frac{1}{2} \quad \Rightarrow \left(\frac{3}{4}\right)^{t_0} = \frac{1}{2}$$

$$\Rightarrow \quad t_0\log\left(\frac{3}{4}\right) = -\log 2$$

$$\Rightarrow \quad t_0 = \frac{-\log 2}{\log\frac{3}{4}}$$

$$\Rightarrow \quad t_0 = \frac{-\log 2}{-\log\frac{4}{3}} = \log_{4/3}2$$

$$\Rightarrow \quad t_0 = \frac{1}{\log_2 4/3} = \frac{1}{\log_2 4 - \log_2 3} = \frac{1}{\log 2 - \log_2 3}$$

**18.** (b) $\because \frac{V_A}{V_B} = \frac{C_1}{C_2}e^{-(k_1-k_2)t}$

Now, $\quad f(t) = \frac{C_1}{C_2}e^{-(k_1-k_2)t}$

$$\Rightarrow \quad f'(t) = -2(k_1-k_2)e^{-(k_1-k_2)t}$$

$$\Rightarrow \quad f'(t) = 2\log\frac{3}{4}\cdot e^{-(k_1-k_2)t} < 0$$

$\therefore f(t)$ is decreasing.

### Explanations for Q. Nos. (19-21)

The equation of tangents with equal abscissa are

$$Y - f(x) = f'(x)(X - x)$$

and $\quad Y - g(x) = g'(x)(X - x)$

These tangents intersect at $Y$-axis.

$$\therefore \quad f(x) - x\,f'(x) = g(x) - x\,g'(x)$$

$$\Rightarrow \quad f(x) - g(x) = x\,[f'(x) - g'(x)]$$

$$\Rightarrow \quad \frac{f'(x) - g'(x)}{f(x) - g(x)} = \frac{1}{x}$$

On integrating both sides, we get

$$\log|f(x) - g(x)| = \log x$$

$$\Rightarrow \quad f(x) - g(x) = \pm C_1x \quad \text{...(i)}$$

Now, equation of normal with equal abscissa $x$ are

$$Y - f(x) = -\frac{1}{f'(x)}(X - x)$$

and $\quad Y - g(x) = -\frac{1}{g'(x)}(X - x)$

Also, these normal intersect on the $X$-axis.

$$\Rightarrow \quad x + f(x)\,f'(x) = x + g(x)\,g'(x)$$

$$\Rightarrow \quad f(x)\,f'(x) = g(x)\,g'(x)$$

Again on integrating, we get

$$f^2(x) - g^2(x) = C_2$$

$$\Rightarrow \quad [f(x) + g(x)]\,[f(x) - g(x)] = C_2$$

$$\Rightarrow \quad f(x) + g(x) = \pm\frac{C_2}{C_1x} \quad \text{[by Eq. (i)] ...(ii)}$$

From Eqs. (i) and (ii), we get

$$2f(x) = \pm\left(\frac{\lambda_2}{x} + C_1x\right), 2g(x) = \pm\left(\frac{\lambda_2}{x} - C_1x\right)$$

We have, $\quad f(1) = 1$ and $g(2) = 3$

Hence, $\quad f(x) = \frac{2}{x} - x$ and $g(x) = \frac{2}{x} + x$

Now, $\int_1^2 [g(x) - f(x)]\,dx = \int_1^2\left(\frac{2}{x} + x - \frac{2}{x} + x\right)dx$

$$= \int_1^2 2x\,dx = [x^2]_1^2 = 4 - 1 = 3$$

**19.** (a) **20.** (b) **21.** (b)

**22.** (c) Given, $f''(x) - 2f'(x) + f(x) \ge e^x$

$$\Rightarrow \quad f''(x)e^{-x} - 2f'(x)e^{-x} + f(x)e^{-x} \ge 1$$

$$\Rightarrow f''(x)e^{-x} - f'(x)e^{-x} - f'(x)e^{-x} + f(x)e^{-x} \ge 1$$

$$\Rightarrow \quad \frac{d[f'(x)e^{-x}]}{dx} - \frac{d[f(x)e^{-x}]}{dx} \ge 1$$

$$\Rightarrow \quad \frac{d^2}{dx}[e^{-x}f(x)] \ge 1, \forall x \in [0, 1]$$

$\therefore \quad \phi(x) = e^{-x}f(x)$ is concave up

$$f(0) = f(1) = 0 \quad \Rightarrow \phi(0) = 0 = \phi(1)$$

Here, $\phi'(x) < 0, x \in \left(0, \frac{1}{4}\right)$ and $\phi'(x) > 0, x \in \left(\frac{1}{4}, 1\right)$

$$\Rightarrow \quad e^{-x}f'(x) - e^{-x}f(x) < 0, x \in \left(0, \frac{1}{4}\right)$$

$$\Rightarrow \quad f'(x) < f(x), x \in \left(0, \frac{1}{4}\right)$$

**23.** (d) Here, $\phi(x) < 0 \Rightarrow e^{-x}f(x) < 0$

$\therefore \quad f(x) < 0$

Hence, $-\infty < f(x) < 0$

**24.** (2) Given, $\int_a^x t\, y(t)\, dt = x^2 + y(x)$

On differentiating w.r.t. $x$ both sides, we get

$$xy(x) = 2x + y'(x)$$

$$\Rightarrow \quad xy = 2x + \frac{dy}{dx} \quad \left[\because y = y(x) \Rightarrow \frac{dy}{dx} = y'(x)\right]$$

$$\Rightarrow \quad \frac{dy}{dx} - xy = -2x$$

$$\therefore \quad \text{IF} = e^{\int -x\,dx} = e^{-x^2/2}$$

The solution of differential equation is

$$y \cdot e^{-x^2/2} = \int -2x \cdot e^{-x^2/2} dx$$

$$\Rightarrow \quad ye^{-x^2/2} = -\int 2e^{-x^2/2}\, d\left(\frac{x^2}{2}\right) + C$$

$$\Rightarrow \quad ye^{-x^2/2} = 2e^{-x^2/2} + C \Rightarrow y = 2 + Ce^{x^2/2}$$

If $x = a \Rightarrow \quad a^2 + y = 0 \Rightarrow y = -a^2$

Hence, $-a^2 = 2 + Ce^{a^2/2} \Rightarrow C = -(2 + a^2)e^{-a^2/2}$

$$\therefore \quad y = 2 - (2 + a^2)e^{(x^2 - a^2)/2} \Rightarrow \lambda = 2$$

**25.** (1) Given, $\int_0^1 f(tx)\, dt = n \cdot f(x)$ ...(i)

Put $tx = u \Rightarrow \quad \frac{dt}{du} = \frac{1}{x}$

Form Eq. (i) reduces to

$$\frac{1}{x}\int_0^x f(u)\, du = n\, f(x)$$

$$\Rightarrow \quad \int_\theta^x f(u)\, du = nx\, f(x)$$

On differentiating w.r.t. $x$ both sides, we get

$$f(x) = n\, f(x) + nx\, f'(x)$$

$$\Rightarrow \quad nx\, f'(x) + (n-1)\, f(x) = 0$$

$$\Rightarrow \quad \frac{f'(x)}{f(x)} = \frac{1-n}{nx}$$

On integrating both sides, we get

$$\log f(x) = \frac{1-n}{n} \log x + \log C$$

$$\Rightarrow \quad \log f(x) = \log x^{\frac{1-n}{n}} \cdot C$$

$$\Rightarrow \quad f(x) = C\, x^{\frac{1-n}{n}}$$

Hence, $\quad \lambda = 1 - n$ and $\mu = n$

$$\Rightarrow \quad \lambda + \mu = 1 - n + n = 1$$

## Revisal Problems (JEE Main)

**1.** (b) Equation of parabola whose axis of symmetry is parallel to X-axis is

$$x = ay^2 + by + c$$

Since, this equation has three arbitrary constants.

$\therefore$ Order of differential equation is 3.

**2.** (d) Here, $\frac{d^2y}{dx^2} + 3\left(\frac{dy}{dx}\right)^2 = x^2 \log\left(\frac{d^2y}{dx^2}\right)$

Hence, this is not form in polynomial in all the differential coefficient.

$\therefore$ Degree of equation does not exist.

**3.** (a) Given, $\sqrt{1-x^2} + \sqrt{1-y^2} = a(x-y)$

Put $x = \sin A$ and $y = \sin B$

$$\sqrt{1-\sin^2 A} + \sqrt{1-\sin^2 B} = a(\sin A - \sin B)$$

$$\Rightarrow \quad \cos A + \cos B = a(\sin A - \sin B)$$

$$\Rightarrow 2\cos\left(\frac{A+B}{2}\right)\cos\left(\frac{A-B}{2}\right) = 2a \sin\left(\frac{A-B}{2}\right)\cos\left(\frac{A+B}{2}\right)$$

$$\Rightarrow \quad \cot\left(\frac{A-B}{2}\right) = a$$

$$\Rightarrow \quad A - B = 2\cot^{-1} a$$

$$\Rightarrow \quad \sin^{-1} x - \sin^{-1} y = 2\cot^{-1} a$$

On differentiating both sides w.r.t. $x$, we get

$$\frac{1}{\sqrt{1-x^2}} - \frac{1}{\sqrt{1-y^2}}\frac{dy}{dx} = 0$$

$$\Rightarrow \quad \frac{dy}{dx} = \frac{\sqrt{1-y^2}}{\sqrt{1-x^2}}$$

$\therefore$ Degree = 1

**4.** (b) Here, $\left(\frac{d^3y}{dx^3}\right)^{2/3} + 4 - 3\frac{d^2y}{dx^2} + 5\frac{dy}{dx} = 0$

$$\Rightarrow \quad \left(\frac{d^3y}{dx^3}\right)^{2/3} = 3\frac{d^2y}{dx^2} - 5\frac{dy}{dx} - 4$$

$$\Rightarrow \quad \left(\frac{d^3y}{dx^3}\right)^2 = \left(3\frac{d^2y}{dx^2} - 5\frac{dy}{dx} - 4\right)^3$$

$\therefore$ Degree = 2

**5.** (a) Here,

$$\sqrt{1+x^2} + \sqrt{1+y^2} = \lambda\left[x\sqrt{1+y^2} - y\sqrt{1+x^2}\right]$$

Put $x = \tan A$, $y = \tan B$, then

$$\sqrt{1+\tan^2 A} + \sqrt{1+\tan^2 B} = \lambda\,[\tan A\sqrt{1+\tan^2 B} - \tan B\sqrt{1+\tan^2 A}]$$

$$\Rightarrow \quad \sec A + \sec B = \lambda\,(\tan A \sec B - \tan B \sec A)$$

$$\Rightarrow \quad \cos A + \cos B = \lambda\,(\sin A - \sin B)$$

$$\Rightarrow 2\cos\left(\frac{A+B}{2}\right)\cos\left(\frac{A-B}{2}\right) = \lambda\left[2\sin\left(\frac{A-B}{2}\right)\cos\left(\frac{A+B}{2}\right)\right]$$

$$\Rightarrow \quad A - B = 2\cot^{-1}\lambda$$

$$\Rightarrow \quad \tan^{-1} x - \tan^{-1} y = 2\cot^{-1}\lambda$$

On differentiating both sides w.r.t. $x$, we get

$$\frac{1}{1+x^2} - \frac{1}{1+y^2}\frac{dy}{dx} = 0$$

$$\Rightarrow \quad \frac{dy}{dx} = \frac{1+y^2}{1+x^2}$$

$\therefore$ Degree = 1

**6.** (b) We have,

$$y = a(x + a)^2 \quad \text{...(i)}$$

On differentiating w.r.t. $x$, we get

$$y' = 2a(x + a)$$

On multiplying by $(x + a)$, we get

$$y'(x + a) = 2a(x + a)^2$$

$$y'(x + a) = 2y \quad \text{[from Eq. (i)]}$$

$$\Rightarrow \quad x + a = \frac{2y}{y'}$$

$$\Rightarrow \quad a = \frac{2y}{y'} - x$$

On substituting the value of $a$ in Eq. (i), we get

$$y = \left(\frac{2y - xy'}{y'}\right)\left(\frac{2y}{y'}\right)^2$$

$$\Rightarrow \quad y'^3 y = 4y^2(2y - xy')$$

$$\Rightarrow \quad y'^3 = 8y^2 - 4xyy'$$

$\therefore$ Degree = 3

**7.** (a) Here, $x^2 = 4y \quad \text{...(i)}$

$$\Rightarrow \quad y' = \frac{x}{2}$$

$$\Rightarrow \quad x = 2y'$$

On substituting the value of $x$ in Eq. (i), we get

$$(2y')^2 = 4y$$

$$\Rightarrow \quad y'^2 - y = 0$$

$\therefore$ Order = 1 and degree = 2.

**8.** (b) Equation of parabola whose axis is $X$-axis, is

$$y^2 = 4a(x - h)$$

On differentiating w.r.t. $x$, we get

$$2yy' = 4a$$

Again differentiating w.r.t. $x$, we get

$$yy'' + y'^2 = 0$$

$\therefore$ Degree = 1 and order = 2.

**9.** (c) We have, $\left(\frac{dy}{dx}\right)^2 - x\frac{dy}{dx} + y = 0 \quad \text{...(i)}$

On differentiating w.r.t. $x$, we get

$$2\frac{dy}{dx}\cdot\frac{d^2y}{dx^2} - x\frac{d^2y}{dx^2} - \frac{dy}{dx} + \frac{dy}{dx} = 0$$

$$\Rightarrow \quad \frac{d^2y}{dx^2}\left(2\frac{dy}{dx} - x\right) = 0$$

$$\Rightarrow \quad \frac{d^2y}{dx^2} = 0 \text{ or } 2\frac{dy}{dx} - x = 0$$

$$\Rightarrow \quad \frac{dy}{dx} = K \text{ or } \frac{dy}{dx} = \frac{x}{2}$$

On putting the value of $\frac{dy}{dx} = K$ and $\frac{x}{2}$ in Eq. (i), we get

$$K^2 - Kx + y = 0$$

and $$\frac{x^2}{4} - \frac{x^2}{2} + y = 0$$

or $$K^2 - Kx + y \text{ and } 4y = x^2$$

For $K = 2$, $$y = 2x - 4 \text{ and } y = \frac{x^2}{4}$$

**10.** (a) We have,

$$y^2 = 2c(x + \sqrt{c}) \quad \text{...(i)}$$

On differentiating w.r.t. $x$, we get

$$2yy' = 2c$$

$$\Rightarrow \quad yy' = c$$

On putting the value of $C$ in Eq. (i), we get

$$y^2 = 2yy'[x + (yy')^{1/2}]$$

$$\Rightarrow \quad y = 2xy' + 2y'(yy')^{1/2}$$

$$\Rightarrow \quad y - 2xy' = 2\sqrt{y}\,(y')^{3/2}$$

On squaring both sides, we get

$$(y - 2xy')^2 = 4yy'^3$$

$\therefore$ Order = 1 and degree = 3.

**11.** (d) We know that, length of normal of curve = $y\sqrt{1 + \left(\frac{dy}{dx}\right)^2}$

Since, length of normal is $C$.

$$\therefore \quad C = y\sqrt{1 + \left(\frac{dy}{dx}\right)^2}$$

$$\Rightarrow \quad C^2 = y^2\left[1 + \left(\frac{dy}{dx}\right)^2\right]$$

$\therefore$ Degree = 2

**12.** (b) Length of subnormal = $y\frac{dy}{dx}$

It is given that, length of subnormal is varies as the square of ordinate.

$$\therefore \quad y\frac{dy}{dx} = \lambda y^2$$

$$\Rightarrow \quad \frac{dy}{y} = \lambda\, dx$$

On integrating both sides, we get

$$\log y = \lambda x + C_1$$

$$\Rightarrow \quad y = e^{\lambda x + C_1}$$

$$\Rightarrow \quad y = C e^{\lambda x}$$

**13.** (d) Given, $x\,dy - y\,dx = 0$

$$\Rightarrow \quad \frac{dy}{y} = \frac{dx}{x}$$

On integrating both sides, we get

$$\log y = \log x + \log C$$

$$\Rightarrow \quad y = Cx$$

This equation represents straight lines passing through the origin.

**14.** (b) We have, $x = 1 + \frac{dy}{dx} + \frac{1}{2!}\left(\frac{dy}{dx}\right)^2 + \frac{1}{3!}\left(\frac{dy}{dx}\right)^3 + \cdots \infty$

$$\Rightarrow \quad x = e^{dy/dx}$$

$$\Rightarrow \quad \log x = \frac{dy}{dx}$$

$\therefore$ Degree = 1

**15.** (b) We have, $a^{n-1}y = x^n \quad \text{...(i)}$

On differentiating w.r.t. $x$, we get

$$a^{n-1}\frac{dy}{dx} = nx^{n-1} \quad \text{...(ii)}$$

From Eqs. (i) and (ii), we get

$$x\frac{dy}{dx} = ny \quad \text{...(iii)}$$

For orthogonal trajectories, replace $\frac{dy}{dx}$ by $\left(-\frac{dx}{dy}\right)$.

$$\therefore \quad -x\frac{dx}{dy} = ny$$

$$\Rightarrow \quad ny\,dy + x\,dx = 0$$

On integrating both sides, we get

$$\frac{ny^2}{2} + \frac{x^2}{2} = C$$

$$\Rightarrow \quad ny^2 + x^2 = C$$

**16.** (a) We have, $x^2 + y^2 - 2ay = 0$ ...(i)

On differentiating w.r.t. $x$, we get

$$x + y\frac{dy}{dx} - a\frac{dy}{dx} = 0$$

$$\Rightarrow \quad a = \frac{x + y\frac{dy}{dx}}{\frac{dy}{dx}}$$

On putting the value of $a$ in Eq. (i), we get

$$x^2 + y^2 - \frac{2y\left(x + y\frac{dy}{dx}\right)}{\frac{dy}{dx}} = 0$$

$$\Rightarrow \quad x^2\frac{dy}{dx} + y^2\frac{dy}{dx} - 2xy - 2y^2\frac{dy}{dx} = 0$$

$$\Rightarrow \quad (x^2 - y^2)\frac{dy}{dx} = 2xy$$

For orthogonal trajectories, $\frac{dy}{dx}$ is replaced by $\left(-\frac{dx}{dy}\right)$.

$$\Rightarrow \quad -(x^2 - y^2)\frac{dx}{dy} = 2xy$$

$$\Rightarrow \quad \frac{dy}{dx} = -\frac{(x^2 - y^2)}{2xy}$$

$$\Rightarrow \quad v + x\frac{dv}{dx} = -\frac{(1 - v^2)}{2v} \quad \left[\because y = vx \Rightarrow \frac{dy}{dx} = v + x\frac{dv}{dx}\right]$$

$$\Rightarrow \quad x\frac{dv}{dx} = \frac{-1 + v^2 - 2v^2}{2v}$$

$$\Rightarrow \quad \frac{2v}{v^2 + 1}dv = -\frac{dx}{x}$$

On integrating both sides, we get

$$\log(v^2 + 1) = -\log x + \log k$$

$$\Rightarrow \quad (v^2 + 1)x = k$$

$$\Rightarrow \quad (x^2 + y^2) = kx \quad [\because v = y/x]$$

$$\Rightarrow \quad x^2 + y^2 - kx = 0$$

**17.** (d) Equation of tangent of any curve at $P(x, y)$ is

$$Y - y = \frac{dy}{dx}(X - x)$$

Coordinates of $A\left(\frac{x\frac{dy}{dx} - y}{\frac{dy}{dx}}, 0\right)$ and $B\left(0, -x\frac{dy}{dx} + y\right)$.

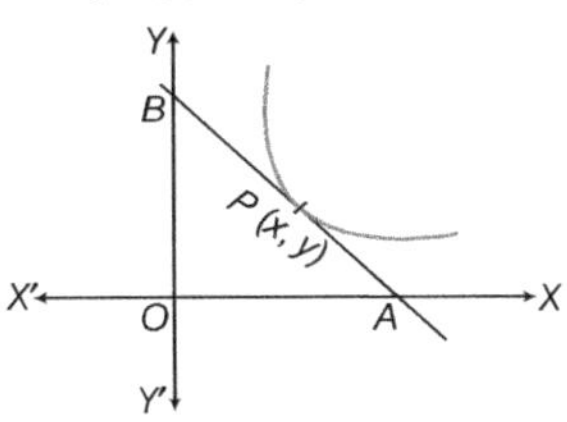


It is given that, $P$ is mid-point of $AB$.

$$\therefore \quad x = \frac{x\frac{dy}{dx} - y}{2\frac{dy}{dx}}$$

$$\Rightarrow \quad 2x\frac{dy}{dx} = x\frac{dy}{dx} - y$$

$$\Rightarrow \quad x\frac{dy}{dx} = -y$$

$$\Rightarrow \quad \frac{dy}{y} = -\frac{dx}{x}$$

On integrating both sides, we get

$$\log y = -\log x + \log C$$

$$\Rightarrow \quad xy = C$$

which represents the equation of a hyperbola.

**18.** (d) It is given that, length of subnormal is twice of abscissa.

$$\therefore \quad y\frac{dy}{dx} = 2x$$

$$\Rightarrow \quad y\,dy = 2x\,dx$$

On integrating both sides, we get

$$\frac{y^2}{2} = x^2 + C$$

$$\Rightarrow \quad y^2 - 2x^2 = C$$

which represents the equation of a hyperbola.

**19.** (c) We have,

$$\frac{x}{x^2 + y^2}dy = \left(\frac{y}{x^2 + y^2} - 1\right)dx$$

$$\Rightarrow \quad x\,dy = [y - (x^2 + y^2)]\,dx$$

$$\Rightarrow \quad x\,dy - y\,dx = -(x^2 + y^2)\,dx$$

$$\Rightarrow \quad \frac{x\,dy - y\,dx}{x^2 + y^2} = -dx$$

$$\Rightarrow \quad \frac{\frac{x\,dy - y\,dx}{x^2}}{1 + \frac{y^2}{x^2}} = -dx$$

$$\Rightarrow \quad d\tan^{-1}\left(\frac{y}{x}\right) = -dx$$

On integrating both sides, we get

$$\tan^{-1}\left(\frac{y}{x}\right) = -x + C$$

$$\Rightarrow \quad \frac{y}{x} = \tan(C - x)$$

$$\Rightarrow \quad y = x\tan(C - x)$$

**20.** (a) We have, $dy = \{y\,\phi'(x) - \phi(x)\,\phi'(x)\}\,dx = 0$

$$\frac{dy}{dx} + y\,\phi'(x) = \phi(x)\,\phi'(x)$$

This is a form of linear differential equation,

where $P = \phi'(x)$ and $Q = \phi(x)\,\phi'(x)$

$$\therefore \quad \text{IF} = e^{\int \phi'(x)dx} = e^{\phi(x)}$$

The solution is

$$y \cdot e^{\phi(x)} = \int \phi(x) \cdot \phi'(x)\,e^{\phi(x)}\,dx$$

$$\Rightarrow \quad y\,e^{\phi(x)} = e^{\phi(x)}[\phi(x) - 1] + C$$

$$\therefore \quad y = \{\phi(x) - 1\} + C\,e^{-\phi(x)}$$

**21.** (a) We have, $\frac{dy}{dx} = \frac{x^2 + y^2 + 1}{2xy}$

$$\Rightarrow \quad 2xy\,dy - y^2dx = (x^2 + 1)\,dx$$

$$\Rightarrow \quad \frac{2xy\,dy - y^2dx}{x^2} = \frac{(x^2 + 1)\,dx}{x^2}$$

$$\Rightarrow \quad d\left(\frac{y^2}{x}\right) = \left(1 + \frac{1}{x^2}\right)dx$$

On integrating both sides, we get

$$\frac{y^2}{x} = x - \frac{1}{x} + C$$

$$\Rightarrow \quad y^2 = x^2 + Cx - 1$$

When $x = 1$, $y = 1$ and $C = 1$

$$\therefore \quad y^2 - x^2 - x + 1 = 0$$

which represents the equation of a hyperbola.

**22.** (b) Given, $x^3 \frac{dy}{dx} + 4x^2 \tan y = e^x \sec y$

On dividing by $x^3 \sec y$ both sides, we get

$$\frac{1}{\sec y}\frac{dy}{dx} + \frac{4 \tan y}{x \sec y} = \frac{e^x}{x^3}$$

$$\Rightarrow \cos y \frac{dy}{dx} + \frac{4 \sin y}{x} = e^x \cdot x^{-3} \quad \text{...(i)}$$

Let $\sin y = Z$

Then, $\cos y \frac{dy}{dx} = \frac{dZ}{dx}$

Eq. (i) reduces to $\frac{dZ}{dx} + \frac{4Z}{x} = e^x x^{-3}$

This is a form of linear differential equation,

where $P = \frac{4}{x}, Q = x^{-3}e^x.$

$$\therefore \quad \text{IF} = e^{\int \frac{4}{x}dx} = e^{4 \log x} = x^4$$

The solution of differential equation is

$$Zx^4 = \int e^x x^{-3} \cdot x^4 dx + C$$

$$\Rightarrow Zx^4 = \int x e^x dx + C$$

$$\Rightarrow Zx^4 = x e^x - e^x + C$$

$$\Rightarrow Z = e^x(x-1)x^{-4} + Cx^{-4}$$

$$\Rightarrow \sin y = e^x(x-1)x^{-4} + Cx^{-4} \quad [\because Z = \sin y]$$

When $x = 1$, $y = 0$ and $C = 0$

$$\therefore \quad \sin y = e^x(x-1)x^{-4}$$

**23.** (a) We have,

$$y(xy + 2x^2y^2)dx + x(xy - x^2y^2)dy = 0$$

$$\Rightarrow (xy^2 + 2x^2y^3)dx + (x^2y - x^3y^2)dy = 0$$

$$\Rightarrow xy^2dx + x^2y\,dy = x^3y^2dy - 2x^2y^3dx$$

$$\Rightarrow xy(y\,dx + x\,dy) = x^2y^2(x\,dy - 2y\,dx)$$

$$\Rightarrow \frac{y\,dx + x\,dy}{x^2y^2} = \frac{x\,dy - 2y\,dx}{xy}$$

$$\Rightarrow d\left(-\frac{1}{xy}\right) = \frac{dy}{y} - \frac{2}{x}dx$$

On integrating both sides, we get

$$-\frac{1}{xy} = \log y - 2\log x + C$$

$$\Rightarrow 2\log|x| - \log|y| - \frac{1}{xy} = C$$

**24.** (c) We have,

$$x = 1 + xy\frac{dy}{dx} + \frac{(xy)^2}{2!}\left(\frac{dy}{dx}\right)^2 + \frac{(xy)^3}{3!}\left(\frac{dy}{dx}\right)^3 + \cdots$$

$$\Rightarrow x = e^{xy\,dy/dx}$$

$$\Rightarrow \log x = xy\frac{dy}{dx}$$

$$\Rightarrow y\,dy = \frac{\log x\,dx}{x}$$

On integrating both sides, we get

$$\frac{y^2}{2} = \frac{(\log x)^2}{2} + C$$

$$\Rightarrow y = \pm\sqrt{(\log x)^2 + 2C}$$

**25.** (d) We have, $\left(\frac{x+y-1}{x+y-2}\right)\frac{dy}{dx} = \left(\frac{x+y+1}{x+y+2}\right)$

$$\Rightarrow \frac{dy}{dx} = \frac{(x+y+1)(x+y-2)}{(x+y+2)(x+y-1)} \quad \text{...(i)}$$

Let $x + y = z \Rightarrow 1 + \frac{dy}{dx} = \frac{dz}{dx}$

From Eq. (i) reduces to

$$\frac{dz}{dx} = \frac{(z+1)(z-2)}{(z+2)(z-1)} + 1$$

$$\Rightarrow \frac{dz}{dx} = \frac{z^2 - z - 2 + z^2 + z - 2}{z^2 + z - 2}$$

$$\Rightarrow \frac{dz}{dx} = \frac{2(z^2-2)}{z^2+z-2}$$

$$\Rightarrow \frac{z^2+z-2}{z^2-2}dz = 2\,dx$$

$$\Rightarrow \int\left(\frac{z^2-2}{z^2-2} + \frac{z}{z^2-2}\right)dz = 2\int dx$$

$$\Rightarrow z + \frac{1}{2}\log(z^2-2) = 2x + C$$

$$\Rightarrow 2z + \log|z^2 - 2| = 4x + 2C$$

$$\Rightarrow 2(x+y) + \log|(x+y)^2 - 2| = 4x + 2C$$

$$\Rightarrow \log|(x+y)^2 - 2| = 2(x+y) + 2C$$

When $x = 1$, $y = 1$, then $C = \frac{1}{2}\log 2 - 2$

$$\because \quad \log|(x+y)^2 - 2| = 2(x+y) + \log 2 - 4$$

$$\Rightarrow \log\left|\frac{(x+y)^2 - 2}{2}\right| = 2(x+y) - 4$$

**26.** (b) We have, $\frac{dy}{dx} = \frac{ax+3}{2y+f}$

$$\Rightarrow (2y+f)dy = (ax+3)dx$$

On integrating both sides, we get

$$y^2 + fy = \frac{ax^2}{2} + 3x + C$$

$$\Rightarrow -ax^2 + 2y^2 + 2fy - 6x - C = 0$$

This equation represents circles, if $a = -2$.

**27.** (c) We have, $\frac{dy}{dx} = \frac{x^2+y^2}{x^2-y^2}$

Slope at the point (1, 0).

i.e. $\left(\frac{dy}{dx}\right)_{(1,0)} = \frac{1^2+0^2}{1^2-0^2} = 1$

**28.** (a) Given, $xy = 2$

$$\Rightarrow y = \frac{2}{x} \Rightarrow \left(\frac{dy}{dx}\right)_1 = -\frac{2}{x^2}$$

Let the slope of curve is $\left(\frac{dy}{dx}\right)_2$.

It is given that, these are cut orthogonally.

$$\therefore \left(\frac{dy}{dx}\right)_2 \times \left(\frac{dy}{dx}\right)_1 = -1$$

$$\Rightarrow \left(\frac{dy}{dx}\right)_2 = \frac{x^2}{2}$$

$$\Rightarrow y = \frac{x^3}{6} + C$$

**29.** (b) We have,

$$y(x^2y + e^x)dx = e^x dy$$

$$\Rightarrow x^2y^2dx + ye^xdx = e^xdy$$

$$\Rightarrow x^2y^2dx = e^xdy - ye^xdx$$

$$\Rightarrow x^2dx = \frac{e^xdy - ye^xdx}{y^2}$$

$$\Rightarrow x^2dx = -d\left(\frac{e^x}{y}\right)$$

On integrating both sides, we get

$$\int x^2 dx = -\int d\left(\frac{e^x}{y}\right)$$

$$\Rightarrow \quad \frac{x^3}{3} = -\frac{e^x}{y} + C$$

$$\Rightarrow \quad yx^3 + 3e^x = 3Cy$$

**30.** (a) Given, $(1+t)\frac{dy}{dt} - ty = 1$

$$\Rightarrow \quad \frac{dy}{dt} - \left(\frac{t}{1+t}\right)y = \frac{1}{1+t}$$

$$\therefore \quad \text{IF} = e^{\int \frac{-t}{1+t}dt} = e^{\int \left[-\left(\frac{t+1}{t+1}\right) + \frac{1}{t+1}\right]dt}$$

$$= e^{-t + \log(t+1)} = e^{-t}(1+t)$$

The solution of equation is

$$ye^{-t}(1+t) = \int \frac{e^{-t}(1+t)}{1+t}dt$$

$$\Rightarrow \quad ye^{-t}(1+t) = -e^t + C = y = -\frac{1}{1+t} + Ce^t$$

when $t = 0$, $y = -1$ and $C = 0$

$$y = \frac{-1}{1+t}$$

$$\Rightarrow \quad y(1) = -\frac{1}{2}$$

**31.** (b) Asymptotes of the rectangular hyperbola $x^2 - y^2 = a^2$ is $xy = 0$.

On differentiating w.r.t. $x$, we get

$$x\frac{dy}{dx} + y = 0$$

$$\Rightarrow \quad x\frac{dy}{dx} = -y$$

**32.** (c) Equation of parabola having vertex at the origin and the $Y$-axis as the axis of symmetry is

$$x^2 = 4ay \quad \text{...(i)}$$

On differentiating w.r.t. $x$, we get

$$2x = 4a\frac{dy}{dx} \quad \text{...(ii)}$$

From Eqs. (i) and (ii), we get

$$x\frac{dy}{dx} - 2y = 0$$

**33.** (c) Equation of ellipse having major and minor axes along $X$ and $Y$-axes, respectively and minor axis is equal to half of major axes, is

$$\frac{x^2}{a^2} + \frac{y^2}{(a/2)^2} = 1$$

$$\Rightarrow \quad x^2 + 4y^2 = a^2$$

On differentiating w.r.t. $x$, we get

$$2x + 8yy' = 0$$

$$\Rightarrow \quad x + 4yy' = 0$$

**34.** (b) Since, $\sin x$ is an integrating factor of

$$\frac{dy}{dx} + Py = Q$$

$$\therefore \quad \sin x = e^{\int P dx}$$

$$\Rightarrow \quad \log \sin x = \int P dx$$

On differentiating w.r.t. $x$, we get

$$\cot x = P$$

$$\therefore \quad P = \cot x$$

**35.** (c) Since, slope of line joining to point $P(x, y)$ and origin is $\frac{y}{x}$.

Slope of tangent at $P$ is $\frac{dy}{dx}$.

It is given that,

$$\frac{dy}{dx} = \frac{2y}{x}$$

$$\Rightarrow \quad \frac{dy}{y} = 2\frac{dx}{x}$$

$$\Rightarrow \quad \int \frac{dy}{y} = 2\int \frac{dx}{x}$$

$$\Rightarrow \quad \log y = 2\log x + \log C$$

$$\Rightarrow \quad y = x^2 C$$

which represents the equation of a parabola.

**36.** (a) Equation of circle which touches $X$-axis is

$$(x-h)^2 + (y-k)^2 = k^2$$

Since, this equation have two arbitrary constants $h$ and $k$.

$\therefore$ Order is 2.

Thus, Statement I is correct and also Statement II is correct and correct explanation of Statement I.

**37.** (d) We have,

$$y = A + \log Bx$$

$$\Rightarrow \quad y = A + \log B + \log x$$

$$y' = \frac{1}{x}$$

$\therefore$ Order = 1

So, Statement I is incorrect and Statement II is correct.

**38.** (a) We know that, each straight line passing through the centre of circle is an orthogonal trajectory of the family of circle.

$\therefore$ Orthogonal trajectory of circle $(x-a)^2 + (y-b)^2 = r^2$ is

$$y - b = m(x - a)$$

$$\Rightarrow \quad y = mx - am + b$$

$\therefore$ Statement I is correct and Statement II is also correct and correct explanation of Statement I.

**39.** (d) Since, $\sin x\frac{d^2y}{dx^2} + \cos x\frac{dy}{dx} + \tan x = 0$ is not linear differential equation.

$\therefore$ Statement I is incorrect and Statement II is correct.

**40.** (c) We have,

$$(xy-1)\frac{dy}{dx} + y^2 = 0$$

$$\Rightarrow \quad \frac{dy}{dx} = \frac{-y^2}{xy-1}$$

$$\Rightarrow \quad \frac{dx}{dy} = \frac{xy-1}{-y^2}$$

$$\Rightarrow \quad \frac{dx}{dy} + \frac{x}{y} = \frac{1}{y^2}$$

It is a form of linear differential equation

$$\frac{dx}{dy} + Px = Q,$$

where $P = \frac{1}{y}$ and $Q = \frac{1}{y^2}$

$$\therefore \quad \text{IF} = e^{\int \frac{1}{y}dy} = e^{\log y} = y$$

Solution is $x \cdot y = \int \frac{y}{y^2}dy + C$

$$xy = \log y + C$$

Hence, Statement I is correct and Statement II is incorrect.

# Revisal Problems (JEE Advanced)

**1.** (c) We have, $y(x + y^3)\,dx = x(y^3 - x)dy$

$$\Rightarrow \quad xydx + y^4dx = xy^3dy - x^2dy$$

$$\Rightarrow \quad xydx + x^2dy = xy^3dy - y^4dx$$

$$\Rightarrow \quad x(ydx + xdy) = x^2y^3\left[\frac{1}{x}dy - \frac{y}{x^2}dx\right]$$

$$\Rightarrow \quad \frac{dxy}{x^2y^2} = \frac{y}{x}\left[\frac{1}{x}dy - \frac{y}{x^2}dx\right]$$

$$\frac{d(xy)}{(xy)^2} = \frac{y}{x}d\left(\frac{y}{x}\right)$$

On integrating both sides, we get

$$\Rightarrow \quad -\frac{1}{xy} = \frac{1}{2}\left(\frac{y}{x}\right)^2 + C$$

It is passes through $(4, -2)$.

$$\Rightarrow \quad \frac{1}{8} = \frac{1}{2}\left(\frac{-1}{2}\right)^2 + C \Rightarrow C = 0$$

$$\Rightarrow \quad \frac{-1}{xy} = \frac{y^2}{2x^2} \Rightarrow y^3 = -2x$$

$$\therefore \quad f(x) = (-2x)^{1/3}$$

The second equation is given as

$$Y = \int_{1/8}^{\sin^2 x} \sin^{-1}\sqrt{t}\,dt + \int_{1/8}^{\cos^2 x} \cos^{-1}\sqrt{t}\,dt$$

On differentiating w.r.t. $x$, we get

$$\frac{dy}{dx} = \sin^{-1}\sqrt{\sin^2 x}\cdot 2\sin x\cos x - \cos^{-1}\sqrt{\cos^2 x}\cdot 2\sin x\cos x$$

$$= 2x\sin x\cos x - 2x\sin x\cos x$$

$$\Rightarrow \quad \frac{dy}{dx} = 0 \Rightarrow y = C$$

Here, put $\sin x = \cos x = \dfrac{1}{\sqrt{2}}$

$$\Rightarrow \quad y = \int_{1/8}^{1/2} \sin^{-1}\sqrt{t}\,dt + \int_{1/8}^{1/2} \cos^{-1}\sqrt{t}\,dt$$

$$\therefore \quad y = \int_{1/8}^{1/2} (\sin^{-1}\sqrt{t} + \cos^{-1}\sqrt{t})\,dt = \left[\frac{\pi}{2}t\right]_{1/8}^{1/2} = \frac{3\pi}{16}$$

Now, area between, $y = (-2x)^{1/3}$ and $y = 3\pi/16$ is

At $y = \dfrac{3\pi}{16}$; $\quad x = \dfrac{-1}{2}\left(\dfrac{3\pi}{16}\right)^3$

$$\therefore \quad \text{Area} = \int_{-\frac{1}{2}\left(\frac{3\pi}{16}\right)^3}^{0} \left[\frac{3\pi}{16} + (2x)^{1/3}\right]dx$$

$$= \left[\frac{3\pi}{16}x + \frac{3}{4}\cdot 2^{1/3}(x)^{4/3}\right]_{-\frac{1}{2}\left(\frac{3\pi}{16}\right)^3}^{0}$$

$$= \frac{1}{8}\left(\frac{3\pi}{16}\right)^4$$

**2.** (b) Given, $\displaystyle\lim_{t \to a} \frac{\int_a^t f(x)\,dx - \left(\frac{t-a}{2}\right)\{f(t) + f(a)\}}{(t-a)^3} = 3a$

$$\Rightarrow \quad \lim_{t \to a} \frac{f(t) - \left(\frac{t-a}{2}\right)(f'(t)) - \frac{1}{2}[f(t) + f(a)]}{3(t-a)^2} = 3a$$

$$\Rightarrow \quad \lim_{t \to a} \frac{\frac{1}{2}[f(t) - f(a)] - \frac{1}{2}f'(t)(t-a)}{3(t-a)^2} = 3a$$

$$\Rightarrow \quad \lim_{t \to a} \frac{f'(t) - f''(t)(t-a) - f'(t)}{12(t-a)} = 3a$$

$$\therefore \quad \lim_{t \to a} -f''(t) = 36a$$

$$\Rightarrow \quad f''(a) = -36a$$

On integrating both sides, we get

$$f'(a) = -18a^2 + C$$

$$\Rightarrow \quad f'(1) = -18 + C$$

$$\Rightarrow \quad 42 = -18 + C \Rightarrow C = 60$$

$$\therefore \quad f'(a) = -18a^2 + 60$$

Again, integrating both sides, we get

$$f(a) = -6a^3 + 60a + C$$

$$\Rightarrow \quad f(1) = -6 + 60 + C$$

$$\Rightarrow \quad -6 = -6 + 60 + C$$

$$\Rightarrow \quad C = -60$$

$$\therefore \quad f(0) = -60$$

**3.** (c) At the point of maxima

$$f'(x) = 0 \text{ and } f''(x) < 0.$$

$$\Rightarrow \quad f''(x) = -9 + x^2 + f^2(x) < 0$$

$$\Rightarrow \quad x^2 + f^2(x) < 9$$

$\Rightarrow P[x_1, f(x)]$ lies inside the circle $x^2 + y^2 = 9$.

$\therefore$ No tangent is possible.

**4.** (d) Given, $f(x)\cdot\sin 2x - \cos x + (1 + \sin^2 x)\,f'(x) = 0$

$$(1 + \sin^2 x)\,f'(x) + f(x)\sin 2x = \cos x$$

$$\Rightarrow \quad \frac{dy}{dx} + \frac{\sin 2x\; y}{1 + \sin^2 x} = \frac{\cos x}{1 + \sin^2 x} \qquad \left[\because y = f(x) \Rightarrow \frac{dy}{dx} = f'(x)\right]$$

$$\therefore \quad \text{IF} = e^{\int \frac{\sin 2x}{1 + \sin^2 x}dx} = e^{\log(1 + \sin^2 x)} = 1 + \sin^2 x$$

Solution of the equation is

$$y\cdot(1 + \sin^2 x) = \int \frac{\cos x}{1 + \sin^2 x}(1 + \sin^2 x)\,dx + C$$

$$\Rightarrow \quad y(1 + \sin^2 x) = \int \cos dx + C$$

$$\Rightarrow \quad y(1 + \sin^2 x) = \sin x + C$$

When, $x = 0$ and $y = 0$, then $C = 0$

Hence, $\quad y(1 + \sin^2 x) = \sin x$

$$\Rightarrow \quad y = \frac{\sin x}{1 + \sin^2 x}$$

$$\therefore \quad y\left(\frac{\pi}{6}\right) = \frac{\sin(\pi/6)}{1 + \sin^2 \pi/6} = \frac{1/2}{1 + 1/4} = \frac{2}{5}$$

**5.** (c) It is given that, $\dfrac{dy}{dt} = -k\sqrt{y}$

$$\Rightarrow \quad \frac{dy}{\sqrt{y}} = -k\,dt$$

On integrating both sides, we get

$$2\sqrt{y} = -kt + C$$

when, $t = 0$ and $y = 4$, then $C = 4$

So, $\quad 2\sqrt{y} = -\dfrac{1}{15}t + 4 \qquad [\because k = 1/15]$

when $y = 0$, then $\quad t = 60$

So, $\quad t = 60$ min

**6.** (a) We have, $Y = Ax + Be^{2x}$ $\quad$ ...(i)

On differentiating w.r.t. $x$, we get

$$\frac{dy}{dx} = A + 2Be^{2x}$$

From Eq. (i), we have

$$\frac{dy}{dx} = A + 2(y - Ax)$$

$$\Rightarrow \quad \frac{dy}{dx} = A + 2y - 2Ax \qquad \text{...(ii)}$$

Again, differentiating w.r.t. $x$, we get

$$\frac{d^2y}{dx^2} = \frac{2dy}{dx} - 2A$$

From Eq. (ii), $$\frac{d^2y}{dx^2} = \frac{2dy}{dx} - 2\frac{\left(\frac{dy}{dx} - 2y\right)}{1-2x}$$

$$\Rightarrow \quad (1-2x)\left[\frac{d^2y}{dx^2} - \frac{2dy}{dx}\right] + 2\left(\frac{dy}{dx} - 2y\right) = 0$$

$\therefore$ $k = 2$ and $l = -2$

**7.** (a) Let the equation of tangent of curve at $P(x, y)$ is

$$Y - y = \frac{dy}{dx}(X - x)$$

$$x\text{-intercept} = \frac{xdy/dx - y}{dy/dx}$$

It is given that, $y =$ $x$-intercept

$$\therefore \quad y = \frac{xdy/dx - y}{dy/dx}$$

$$\Rightarrow \quad y\frac{dy}{dx} = x\frac{dy}{dx} - y$$

$$\Rightarrow \quad \frac{dy}{dx} = \frac{y}{x-y}$$

$$\Rightarrow \quad v + x\frac{dv}{dx} = \frac{v}{1-v} \qquad \left[\because y = vx \Rightarrow \frac{dy}{dx} = v + \frac{xdv}{dx}\right]$$

$$\Rightarrow \quad x\frac{dv}{dx} = \frac{v - v + v^2}{1-v}$$

$$\Rightarrow \quad \frac{1-v}{v^2}dv = \frac{dx}{x}$$

On integrating both sides, we get

$$\int\left(\frac{1}{v^2} - \frac{1}{v}\right)dv = \int\frac{dx}{x}$$

$$\Rightarrow \quad \frac{-1}{v} - \log v = \log xC$$

$$\log Cx + \log v = -\frac{1}{v}$$

$$\Rightarrow \quad Cx\,v = e^{-1/v}$$

$$\Rightarrow \quad Cy = e^{-x/y} \qquad [\because y = vx]$$

$$\Rightarrow \quad ye^{x/y} = C$$

when, $x = 1$, $y = 1$ then, $C = e$

Hence, $ye^{x/y} = e$

**8.** (a) Given, $f'(x)\sin x + f(x)\cos x = 1$

$$\Rightarrow \quad d\,f(x)\sin x = 1$$

On integrating both sides, we get

$$f(x)\sin x = x + C$$

$$\Rightarrow \quad f(x) = \frac{x+C}{\sin x}$$

$f(x)$ being bounded when $x \to 0, C = 0$

$$f(x) = \frac{x}{\sin x}$$

Now, $\frac{x}{\sin x}$ is increasing function from 0 to $\frac{\pi}{2}$.

$\therefore$ Minimum value of $f(x) = 1$ at $x = 0$

and maximum value of $f(x) = \frac{\pi}{2}$ at $x = \frac{\pi}{2}$

$$\therefore \quad 1\left(\frac{\pi}{2} - 0\right) < \int_0^{\pi/2} f(x)\,dx < \frac{\pi}{2}\left(\frac{\pi}{2} - 0\right)$$

$$\Rightarrow \quad \frac{\pi}{2} < \int_0^{\pi/2} f(x)\,dx < \frac{\pi^2}{4}$$

$$\Rightarrow \quad \frac{\pi}{2} < I < \frac{\pi^2}{4}$$

**9.** (c) It is given that,

Area of shaded region $= y^3$

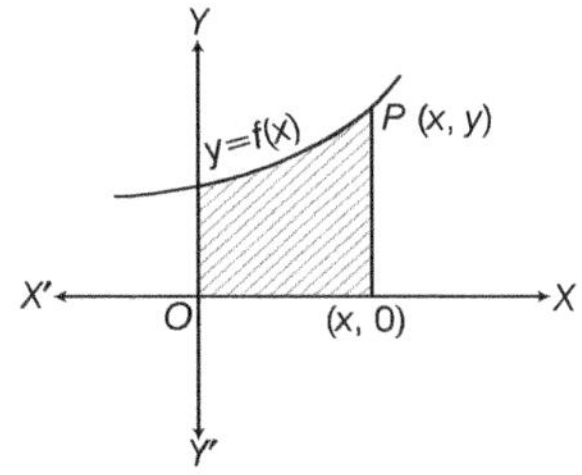


$$\Rightarrow \quad \int_0^x y\,dx = y^3$$

On differentiating w.r.t. $x$, we get

$$y = 3y^2\frac{dy}{dx}$$

$$\Rightarrow \quad 3y\,dy = dx$$

On integrating, both sides, we get

$$\Rightarrow \quad \frac{3y^2}{2} = x + C$$

$\therefore$ $3y^2 = 2x + 2C$, which represents a parabola.

**10.** (c) Given, $y = u^m$

On differentiating w.r.t. $x$, we get

$$\frac{dy}{dx} = mu^{m-1}\frac{dy}{dx}$$

On substituting the values of $y$ and $\frac{dy}{dx}$ in

$$2x^4y\frac{dy}{dx} + y^4 = 4x^6$$

$$\therefore \quad 2x^4u^m \cdot mu^{m-1}\frac{du}{dx} + u^{4m} = 4x^6$$

$$\Rightarrow \quad \frac{du}{dx} = \frac{4x^6 - u^{4m}}{2mx^4 \cdot u^{2m-1}}$$

This equation is homogeneous,

if $4m = 6 \Rightarrow m = 3/2$

or $4 + 2m - 1 = 6$

$\therefore$ $m = 3/2$

**11.** (a) Since, $y = \sin kt$

On differentiating.w.r.t. $x$, we get

$$y' = k\cos kt$$

Again, differentiating.w.r.t. $x$, we get

$$y'' = -k^2\sin kt$$

$$\Rightarrow \quad y'' + k^2y = 0$$

It is given that, $y'' + 9y = 0$

$$\Rightarrow \quad k^2 = 9$$

Hence, $k = \pm 3$

Now, Case II

Since, $y = e^{kt}$

$$\Rightarrow \quad y' = ke^{kt}$$

$$\therefore \quad y'' = k^2e^{kt}$$

It is given that, $y'' + y' - 6y = 0$

$$\Rightarrow \quad k^2e^{kt} + ke^{kt} - 6e^{kt} = 0$$

$$\Rightarrow \quad k^2 + k - 6 = 0$$

$$\Rightarrow \quad (k+3)(k-2) = 0$$

$$\therefore \quad k = -3 \text{ and } k = 2$$

Hence, for $k = -3$ both cases are correct.

**12.** (c) Let moisture content in wet porous substance be $M$.
Now, it is given that

$$\frac{dM}{dt} = -kM$$

$$\Rightarrow \quad \frac{dM}{M} = -k\,dt$$

On integrating both sides, we get

$$\log M = -kt + C$$

$$\therefore \quad M = Ce^{-kt}$$

When $t = 0$ and $M = M_0$,

then $\quad C = M_0$

Hence, $\quad M = M_0 e^{-kt}$

when $t = \frac{1}{2}$ and $M = \frac{M_0}{2}$

Then, $\quad k = \log 2$

$$\therefore \quad M = M_0 e^{-t\log 2}$$

When $\quad M = \frac{M_0}{1000}$

Then, $\quad t = \log_2 1000$

$$\Rightarrow \quad t = \frac{\log_{10} 1000}{\log_{10} 2} = \frac{3}{0.3010} = 10 \text{ h (approx.)}$$

**13.** (a) Equation of normal of curve at point $P(x, y)$ is

$$Y - y = -\frac{dx}{dy}(X - x)$$

It passes through (1, 0).

$$\therefore \quad a - y = -\frac{dx}{dy}(1 - x)$$

$$\Rightarrow \quad y\,dy = (1 - x)\,dx$$

On integrating both sides, we get

$$\int y\,dy = \int (1 - x)\,dx$$

$$\Rightarrow \quad \frac{y^2}{2} = x - \frac{x^2}{2} + C$$

This curve is passes through the origin.

$$\therefore \quad C = 0$$

Hence, equation of curve is

$$x^2 + y^2 - 2x = 0$$

Intersection point of curve $x^2 + y^2 - 2x = 0$ and parabola $y^2 = 4x$ is (0, 0).

∴ Slope of tangent of parabola

i.e. $\quad \frac{dy}{dx} = \frac{2}{y}$

$$\Rightarrow \quad \left(\frac{dy}{dx}\right)_{(0,\,0)} = \infty$$

∴ Equation of common tangent is $x = 0$.

**14.** (b) We have,

$$(x + 1) f'(x) - 2(x^2 + x) f(x) = \frac{ex^2}{x + 1}$$

$$\Rightarrow \quad f'(x) - \frac{2x(x + 1)}{x + 1} f(x) = \frac{e^{x^2}}{(x + 1)^2}$$

$$\Rightarrow \quad \frac{dy}{dx} - 2xy = \frac{e^{x^2}}{(x + 1)^2} \qquad \left[\because y = f(x) \Rightarrow \frac{dy}{dx} = f'(x)\right]$$

$$\therefore \quad \text{IF} = e^{\int -2x\,dx} = e^{-x^2}$$

Solution of given differential equation is

$$y \cdot e^{-x^2} = \int \frac{e^{x^2}}{(x + 1)^2} \cdot e^{-x^2} dx + C$$

$$\Rightarrow \quad ye^{-x^2} = \int \frac{dx}{(x + 1)^2} + C$$

$$\Rightarrow \quad ye^{-x^2} = \frac{-1}{x + 1} + C$$

$$\Rightarrow \quad y = \frac{-e^{x^2}}{x + 1} + Ce^{x^2}$$

When $x = 0$, $y = 5$, then $C = 6$

Hence, $\quad y = \frac{-e^{x^2}}{x + 1} + 6e^{x^2}$

$$\Rightarrow \quad y = e^{x^2}\left[\frac{-1 + 6x + 6}{x + 1}\right]$$

$$\Rightarrow \quad y = \left(\frac{6x + 5}{x + 1}\right)e^{x^2}$$

**15.** (a) Let ordinate = $PM$

and projection of ordinate on normal = $PN$

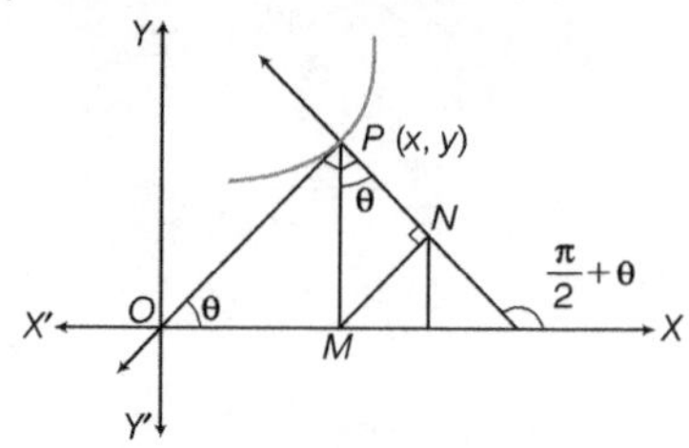


$$\therefore \quad PN = PM\cos\theta = a$$

$$\Rightarrow \quad y\cos\theta = a$$

$$\Rightarrow \quad y = a\sec\theta$$

$$\Rightarrow \quad y^2 = a^2(1 + \tan^2\theta)$$

$$\Rightarrow \quad \tan^2\theta = \frac{y^2}{a^2} - 1$$

$$\Rightarrow \quad \tan\theta = \sqrt{\frac{y^2 - a^2}{a^2}}$$

$$\Rightarrow \quad \frac{dy}{dx} = \frac{\sqrt{y^2 - a^2}}{a} \qquad [\because \tan\theta = dy/dx]$$

$$\Rightarrow \quad \frac{dy}{\sqrt{y^2 - a^2}} = \frac{dx}{a}$$

On integrating both sides, we get

$$a\log\left|y + \sqrt{y^2 - a^2}\right| = x + C$$

**16.** (c) We have, $\frac{dy}{dx} + y\tan x = x\tan x + 1$

$$\text{IF} = e^{\int \tan x\,dx} = e^{\log\sec x} = \sec x$$

Solution of given differential equation is

$$y\sec x = \int (x\tan x + 1)\sec x\,dx + C$$

$$\Rightarrow \quad y\sec x = \int (x\sec x\tan x + \sec x)\,dx + C$$

$$\Rightarrow \quad y\sec x = x\sec x - \int \sec x\,dx + \int \sec x\,dx + C$$

$$\Rightarrow \quad y\sec x = x\sec x + C$$

$$\Rightarrow \quad y = x + C\cos x$$

This curve intersect at $Y$-axis at $(0, c)$.

$$\Rightarrow \quad \frac{dy}{dx} = 1 - c\sin x$$

$$\therefore \quad \left(\frac{dy}{dx}\right)_{(0,\,c)} = 1$$

Hence, slope is constant for all values of $y$.

∴ Tangents are parallel.

**17.** (a) Given, $y = z^{\alpha}$

On differentiating w.r.t. $x$, we get

$$\frac{dy}{dx} = \alpha z^{\alpha - 1}\frac{dz}{dx}$$

On substituting the value of $y$ and $\frac{dy}{dx}$ in

$$(x^2y^2 - 1)\frac{dy}{dx} + 2xy^3 = 0$$

$$\Rightarrow (x^2z^{2\alpha} - 1)\left(\alpha z^{\alpha - 1}\frac{dz}{dx}\right) + 2x\,z^{3\alpha} = 0$$

$$\Rightarrow \frac{dz}{dx} = \frac{-2x\,z^{3\alpha}}{\alpha x^2\,z^{3\alpha - 1} - \alpha z^{\alpha - 1}}$$

For homogeneous,

$$3\alpha + 1 = \alpha - 1$$

$$\therefore \quad \alpha = -1$$

**18.** (a) We have, $\int_0^x t\,y(t)\,dt = x^2y(x)$

On differentiating w.r.t. $x$, we get

$$xy(x) = x^2y'(x) + 2xy(x)$$

$$\Rightarrow x^2y'(x) + xy(x) = 0$$

$$\Rightarrow xy'(x) = -y(x)$$

$$\Rightarrow \frac{dy}{dx} = \frac{-y}{x} \quad \left[\because y = y(x) \Rightarrow \frac{dy}{dx} = y'(x)\right]$$

On integrating both sides, we get

$$\int \frac{dy}{y} = -\int \frac{dx}{x}$$

$$\Rightarrow \log y = -\log x + \log C$$

$$\Rightarrow xy = C$$

It passes through (2, 3).

$$\therefore \quad 2 \cdot 3 = C \Rightarrow C = 6$$

Hence, $\quad xy = 6$

**19.** (a,b,d) Equation of normal at any point $P(x, y)$ is

$$Y - y = \frac{-dx}{dy}(X - x)$$

Perpendicular distance from origin,

$$\frac{x + y\frac{dy}{dx}}{\sqrt{1 + \left(\frac{dy}{dx}\right)^2}} = y$$

$$\Rightarrow \left(x + y\frac{dy}{dx}\right)^2 = y^2\left[1 + \left(\frac{dy}{dx}\right)^2\right]$$

$$\Rightarrow x^2 + 2xy\frac{dy}{dx} + y^2\left(\frac{dy}{dx}\right)^2 = y^2 + y^2\left(\frac{dy}{dx}\right)^2$$

$$\Rightarrow \frac{dy}{dx} = \frac{y^2 - x^2}{2xy} \text{ [homogeneous form]}$$

$$\Rightarrow 2xy\frac{dy}{dx} + x^2 = y^2$$

$$\Rightarrow 2y\frac{dy}{dx} + x = \frac{y^2}{x}$$

$$\Rightarrow 2y\frac{dy}{dx} - \frac{y^2}{x} = -x$$

Put $y^2 = z \quad \Rightarrow \quad 2y\frac{dy}{dx} = \frac{dz}{dx}$

$\therefore \quad \frac{dz}{dx} - \frac{z}{x} = -x \Rightarrow$ Linear differential equation form

$$\text{IF} = e^{\int \frac{-1}{x}dx} = e^{-\log x} = \frac{1}{x}$$

Solution of the equation is

$$\frac{z}{x} = \int \frac{-x}{x}\,dx + C$$

$$\Rightarrow \frac{z}{x} = -x + C$$

$$\Rightarrow z = -x^2 + Cx$$

$$\Rightarrow y^2 = -x^2 + Cx$$

$$\Rightarrow x^2 + y^2 - Cx = 0$$

which is the family of circle touching the $Y$-axis at origin.

**20.** (c,d) We have,

$$f^2(x) + 4f'(x)f(x) + [f'(x)]^2 = 0$$

$$\Rightarrow y^2 + \frac{4dy}{dx}\cdot y + \left(\frac{dy}{dx}\right)^2 = 0 \quad \left[\because y = f(x) \Rightarrow \frac{dy}{dx} = f'(x) \text{ and } y^2 = f^2(x)\right]$$

$$\Rightarrow \left(\frac{dy}{dx}\right)^2 + 4y\frac{dy}{dx} + y^2 = 0$$

$$\Rightarrow \frac{dy}{dx} = \frac{-4y \pm \sqrt{16y^2 - 4y^2}}{2}$$

$$\Rightarrow \frac{dy}{dx} = \frac{-4y \pm 2\sqrt{3}y}{2}$$

$$\Rightarrow \frac{dy}{dx} = -2y \pm \sqrt{3}y$$

$$\Rightarrow \frac{dy}{y} = (-2 \pm \sqrt{3})dx$$

On integrating both sides, we get

$$\log y = (-2 \pm \sqrt{3})\,x + C_1$$

$$\Rightarrow y = Ce^{(-2 \pm \sqrt{3})x}$$

**21.** (a,b,c) We have,

$$\frac{dy}{dx}\sin x - y\cos x + \frac{\sin^2 x}{x^2} = 0$$

$$\Rightarrow \frac{dy}{dx} - y\frac{\cos x}{\sin x} = \frac{-\sin x}{x^2}$$

$$\therefore \quad \text{IF} = e^{\int \frac{-\cos x}{\sin x}dx} = e^{-\log \sin x} = \frac{1}{\sin x}$$

Solution of the equation is

$$\frac{y}{\sin x} = \int \frac{-\sin x}{x^2} \times \frac{1}{\sin x}\,dx + C$$

$$\Rightarrow \frac{y}{\sin x} = -\int \frac{1}{x^2}dx + C$$

$$\Rightarrow \frac{y}{\sin x} = \frac{1}{x} + C$$

$$\Rightarrow y = \frac{\sin x}{x} + C\sin x$$

When $x \to \infty$, $y \to 0$, then $C = 0$.

$$\therefore \quad f(x) = \frac{\sin x}{x}$$

(a) $\lim_{x \to 0} \frac{\sin x}{x} = 1$

(b) Let $I = \int_0^{\pi/2} f(x)\,dx = \int_0^{\pi/2} \frac{\sin x}{x}dx$

Since, $\frac{\sin x}{x}$ is decreasing function.

So, $\quad x > 0$

$$\Rightarrow f(x) < f(0)$$

$$\Rightarrow \int_0^{\pi/2} f(x)\,dx < \frac{\pi}{2} \text{ and } x < \frac{\pi}{2}$$

$$\Rightarrow \int_0^{\pi/2} f(x)\,dx > 1$$

**22.** (a,b,c) We have,

$$f(x) = \int_0^x f(t)\cos t - \cos(t - x)\,dt$$

$$\Rightarrow \quad f(x) = \int_0^x f(t)\cos t - \int_0^x \cot(t - x)\,dt$$

$$\Rightarrow \quad f(x) = \int_0^x f(t)\cos t - \int_0^x \cos(-t)\,dt \quad \left[\because \int_0^a f(x)dx = \int_0^a f(a-x)dx\right]$$

On differentiating both sides, we get

$$f'(x) = f(x)\cos x - \cos x$$

$$\Rightarrow f'(x) - f(x)\cos x = -\cos x$$

$$\Rightarrow \quad \frac{dy}{dx} - y\cos x = -\cos x \qquad \left[\because y = f(x) \Rightarrow \frac{dy}{dx} = f'(x)\right]$$

$$\text{IF} = e^{\int -\cos x\,dx} = e^{-\sin x}$$

Solution of the equation is

$$y \cdot e^{-\sin x} = -\int e^{-\sin x}\cos x\,dx + C$$

$$\Rightarrow \quad y \cdot e^{-\sin x} = e^{-\sin x} + C$$

$$\Rightarrow \quad y = Ce^{\sin x} + 1$$

If $x = 0$, $y = 0$, then $C = -1$

$$\therefore \quad f(x) = -e^{\sin x} + 1$$

Minimum value of $f(x)$ is $1 - e$ at $x = \pi/2$ and maximum value of $f(x)$ is $1 - e^{-1}$ at $x = -\pi/2$.

Now, $\quad f(x) = 1 - e^{\sin x}$

$$\Rightarrow \quad f(x) = -\cos x e^{\sin x}$$

$$\Rightarrow \quad f'(0) = -1$$

$$\Rightarrow \quad f''(x) = -[\cos^2 x e^{\sin x} - \sin x e^{\sin x}]$$

$$\therefore \quad f''\left(\frac{\pi}{2}\right) = e$$

**23.** (a,b,d) Given, $\frac{dy}{dx} + y = f(x)$

$$\text{IF} = e^{\int dx} = e^x$$

$$\Rightarrow \quad ye^x = \int e^x f(x)\,dx + C$$

Now, if $0 \le x \le 2$, then

$$\Rightarrow \quad ye^x = \int e^x e^{-x}\,dx + C$$

$$\Rightarrow \quad ye^x = x + C$$

When, $x = 0$, $y = 1$, then $C = 1$

$$\therefore \quad y = xe^{-x} + e^{-x}$$

$$\Rightarrow \quad y(1) = e^{-1} + e^{-1} = 2e^{-1}$$

$$\Rightarrow \quad \frac{dy}{dx} = -xe^{-x} + e^{-x} - e^{-x} = -xe^{-x}$$

Then, $\quad \left(\frac{dy}{dx}\right)_{x=1} = -e^{-1}$

If $x > 2$ then, $\quad ye^x = \int e^x \cdot e^{-2}dx + C$

$$\Rightarrow \quad ye^x = e^{x-2} + C$$

$$\Rightarrow \quad y = e^{-2} + Ce^{-x}$$

As $y$ is continuous,

$$\lim_{x \to 2} xe^{-x} + e^{-x} = \lim_{x \to 2}(e^{-2} + Ce^{-x})$$

$$\Rightarrow \quad 3e^{-2} = (C + 1)e^{-2}$$

$$\Rightarrow \quad C = 2$$

For $x > 2$, $\quad y = e^{-2} + 2e^{-x}$

when $x = 3$, $\quad y = e^{-2} + 2e^{-3} \Rightarrow y' = -2e^{-x}$

$$\therefore \quad y'(3) = -2e^{-3}$$

**24.** (b) Given, $f(x) = (e^{-x} + e^x)\cos x - 2x - \int_0^x (x - t)f'(t)\,dt$

$$f(0) = (1 + 1) - 0 - 0 = 2$$

$$\therefore \quad f(0) = 2$$

Now, $f(x) = (e^{-x} + e^x)\cos x - 2x - x\int_0^x f'(t)\,dt + \int_0^x t f'(t)\,dt$

$$\Rightarrow \quad f(x) = (e^{-x} + e^x)\cos x - 2x - x[f(x)]_0^x + \int_0^x t f'(t)\,dt$$

$$\Rightarrow \quad f(x) = (e^{-x} + e^x)\cos x - 2x - xf(x) + xf(0) + \int_0^x t f'(t)dt$$

$$\Rightarrow \quad f(x) = (e^{-x} + e^x)\cos x - 2x - xf(x) + 2x + \int_0^x t f'(t)dt$$

$$\Rightarrow \quad f(x) = (e^{-x} + e^x)\cos x - xf(x) + \int_0^x t f'(t)dt$$

On differentiating w.r.t. $x$, we get

$$f'(x) = (e^{-x} + e^x)(-\sin x) + \cos x(-e^{-x} + e^x) - xf'(x) - f(x) + xf'(x)$$

$$\Rightarrow \quad f'(x) = e^x(\cos x - \sin x) - e^{-x}(\cos x + \sin x) - f(x)$$

$$\Rightarrow \quad f'(x) + f(x) = e^x(\cos x - \sin x) - e^{-x}(\cos x + \sin x)$$

$$\Rightarrow \quad \frac{dy}{dx} + y = e^x(\cos x - \sin x) - e^{-x}(\cos x + \sin x)$$

**25.** (d) $f'(x) + f(x) = e^x(\cos x - \sin x) - e^{-x}(\cos x + \sin x)$

$$\Rightarrow f''(x) + f'(x) = e^x[\cos x - \sin x - \sin x - \cos x) - e^{-x}[-\cos x - \sin x - \sin x + \cos x]$$

$$\Rightarrow f''(x) + f'(x) = -2e^x\sin x + 2e^{-x}\sin x$$

$$\Rightarrow f''(0) + f'(0) = -2 + 2 = 0$$

**26.** (a) $f'(x) + f(x) = e^x(\cos x - \sin x) - e^{-x}(\cos x + \sin x)$

$\therefore \quad \text{IF} = e^{\int dx} = e^x$

Solution of the equation is

$$f(x) \cdot e^x = \int e^{2x}(\cos x - \sin x) - (\cos x + \sin x)\,dx$$

$$\Rightarrow \quad f(x)e^x = \int e^{2x}(\cos x - \sin x)\,dx - (\sin x - \cos) + C$$

Let $\quad I = \int e^{2x}(\cos x - \sin x)dx$

$$= \int e^{2x}\cos x\,dx - \int e^{2x}\sin x dx$$

$$= \frac{1}{5}e^{2x}[\sin x + 2\cos x] - \frac{e^{2x}}{5}[2\sin x - \cos x]$$

$$= e^{2x}\left[\frac{3}{5}\cos x - \frac{1}{5}\sin x\right]$$

$$\therefore \quad f(x)e^x = e^{2x}\left[\frac{3}{5}\cos x - \frac{1}{5}\sin x\right] - \sin x + \cos x + \frac{2}{5}$$

$$\Rightarrow \quad f(x) = \frac{e^x}{5}[3\cos x - \sin x] - e^{-x}(\sin x - \cos x) + \frac{2}{5}e^{-x}$$

**27.** (a) Given, $(1 + x^2)\frac{dy}{dx} + 2xy = 4x^2$

$$\Rightarrow \quad \frac{dy}{dx} + \frac{2xy}{1 + x^2} = \frac{4x^2}{1 + x^2}$$

$$\therefore \quad \text{IF} = e^{\int \frac{2x}{1+x^2}dx} = e^{\log|1 + x^2|} = 1 + x^2$$

Solution of the equation is

$$y(1 + x^2) = \int \frac{4x^2}{1 + x^2} \times (1 + x^2)\,dx$$

$$y(1 + x^2) = \int 4x^2 dx$$

$$y(1 + x^2) = \frac{4x^3}{3} + C$$

when, $x = 0$, $y = 0$, then $C = 0$

$$\therefore \quad y = \frac{4x^3}{3(1 + x^2)}$$

$$\Rightarrow \quad \frac{dy}{dx} = \frac{4}{3}\left[\frac{(1 + x^2)(3x^2) - x^3(2x)}{(1 + x^2)^2}\right] \Rightarrow \frac{dy}{dx} = \frac{4}{3}\left[\frac{3x^2 + x^4}{(1 + x^2)^2}\right]$$

$$\frac{dy}{dx} > 0, \ \forall\ x \ne 0$$

$\frac{dy}{dx} = 0$ at $x = 0$ and it does not change sign.

$\Rightarrow \quad x = 0$ is the point of inflection.

$\therefore \quad y = f(x)$ is increasing, $\forall\ x \in R$.

**28.** (c) We have, $y = \frac{4}{3}\left(\frac{x^3}{1+x^2}\right)$

Since, $x \to \infty, y \to \infty$ and $x \to -\infty, y \to -\infty$

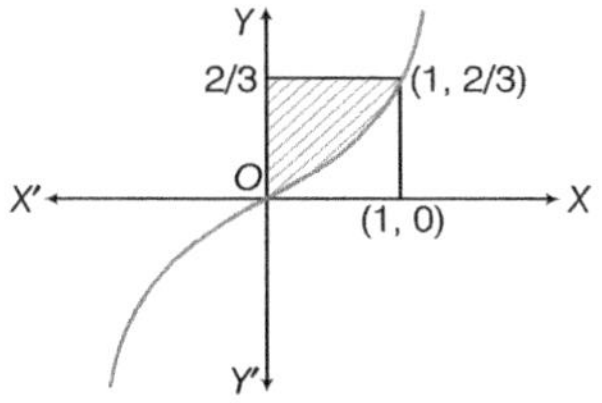


Area enclosed by $y = f^{-1}(x)$,

$X$-axis and ordinate at $x = 2/3$.

$$A = \frac{2}{3} - \frac{4}{3}\int_0^1 \frac{x^3}{1+x^2}dx$$

Put $1 + x^2 = t \Rightarrow 2x dx = dt$

$$\Rightarrow \quad A = \frac{2}{3} - \frac{2}{3}\int_1^2 \frac{t-1}{t}dt$$

$$\Rightarrow \quad A = \frac{2}{3} - \frac{2}{3}[t - \log t]_1^2$$

$$\therefore \quad A = \frac{2}{3}[1 - 2 + \log 2 + 1] = \frac{2}{3}\log_e 2$$

**29.** (c) $f(x)$ is not transcidental function.

**30.** (4) Given, $y' + \frac{y}{x} = \frac{1}{(1+\log xy)^2}$

$$\Rightarrow \quad \frac{dy}{dx} + \frac{y}{x} = \frac{1}{(1+\log xy)^2} \quad \text{...(i)}$$

Let $xy = v$

$$\Rightarrow \quad x\frac{dy}{dx} + y = \frac{dv}{dx} \Rightarrow \frac{dy}{dx} + \frac{y}{x} = \frac{1}{x}\frac{dv}{dx}$$

Eq. (i) reduces to

$$\frac{1}{x}\frac{dv}{dx} = \frac{1}{(1+\log v)^2}$$

$$\Rightarrow \quad (1 + \log v)^2 dv = x dx$$

On integrating, we get

$$\int (1 + 2\log v + (\log v)^2]dv = \int x\,dx$$

$$\Rightarrow \quad v + \int (\log v)^2 dv + 2\int \log v\,dv = \frac{x^2}{2} + C$$

$$\Rightarrow v + v(\log v)^2 - \int \frac{2\log v}{v}\cdot v dv + 2\int \log v dv = \frac{x^2}{2} + C$$

$$\Rightarrow \quad v + v(\log v)^2 = \frac{x^2}{2} + C$$

$$\Rightarrow \quad v[1 + (\log v)]^2 = \frac{x^2}{2} + C$$

$$\Rightarrow \quad xy[1 + (\log xy)^2] = \frac{x^2}{2} + C \quad [\because v = xy]$$

$$\therefore \quad \lambda = 2, \mu = 2 \Rightarrow \lambda + \mu = 4$$

**31.** (5) Given, $y'(x^2 - 2x^3y^3) = (3x^2y^4 + 2xy)$

$$\Rightarrow \quad dy(x^2 - 2x^3y^3) = (3x^2y^4 + 2xy)dx$$

$$\Rightarrow \quad x^2dy - 2x^3y^3dy - 3x^2y^4dx - 2xy\,dx = 0$$

$$\Rightarrow \quad 3x^2y^4dx + 2x^3y^3dy + 2xydx - x^2dy = 0$$

$$\Rightarrow \quad 3x^2y^2dx + 2x^3ydy + \frac{2xydx - x^2dy}{y^2} = 0$$

$$\Rightarrow \quad d(x^3y^2) + d\left(\frac{x^2}{y}\right) = 0$$

On integrating both sides, we get

$$x^3y^2 + \frac{x^2}{y} = C$$

$$\therefore \quad \lambda = 3, \mu = 2$$

$$\Rightarrow \quad \lambda + \mu = 3 + 2 = 5$$

**32.** (2) Given, $y(2xy + 1)dx + x(1 + 2xy + x^2y^2)dy = 0$

$$\Rightarrow \quad 2xy^2dx + ydx + xdy + 2x^2ydy + x^3y^2dy = 0$$

$$\Rightarrow \quad 2xy^2dx + ydx + xdy + 2x^2ydy = -x^3y^2dy$$

$$\Rightarrow \quad \frac{(2xy+1)(ydx + xdy)}{x^3y^3} = \frac{-1}{y}dy$$

$$\Rightarrow \quad \left(2 + \frac{1}{xy}\right)\left(\frac{ydx + xdy}{x^2y^2}\right) = \frac{-1}{y}dy$$

$$\Rightarrow \quad -\int\left(2 + \frac{1}{xy}\right)d\left(\frac{1}{xy}\right) = \frac{-1}{y}dy$$

On integrating both sides, we get

$$\frac{-2}{xy} - \frac{1}{2(xy)^2} = -\log y + C$$

$$\Rightarrow \quad \frac{-2}{xy} - \frac{1}{2(xy)^2} + \log y = C$$

Hence, $\lambda = 2$

**33.** (4) Given, $\cos\left(\frac{x}{y}\right)(ydx - xdy) = xy^3(xdy + ydx)$

$$\cos\left(\frac{x}{y}\right)\left(\frac{ydx - xdy}{y^2}\right) = xy(xdy + ydx)$$

$$\Rightarrow \quad \cos\left(\frac{x}{y}\right)d\left(\frac{x}{y}\right) = xy d(xy)$$

On integrating, we get

$$\sin\left(\frac{x}{y}\right) = \frac{(xy)^2}{2} + C$$

$$\therefore \quad \lambda = 2, \mu = 2$$

Hence, $\lambda + \mu = 2 + 2 = 4$

**34.** (1) Given, $\left\{y\left(1 + \frac{1}{x}\right) + \cos y\right\}dx + \{x + \log x - x\sin y\}dy = 0$

$$\Rightarrow ydx + \frac{y}{x}dx + \cos y dx + xdy + \log x dy - x\sin y dy = 0$$

$$\Rightarrow (ydx + xdy) + \frac{y}{x}dx + \log x dy + \cos y dx - x\sin y dy = 0$$

$$\Rightarrow \quad d\,xy + d(y\log x) + d(x\cos y) = C$$

On integrating both sides, we get

$$xy + y\log x + x\cos y = C$$

$$\therefore \quad \lambda = 1$$

**35.** (0) We have,

$$xy' - y = mx^2\sqrt{x^2 - y^2}$$

$$\Rightarrow \quad xdy - ydx = mx^2\sqrt{x^2 - y^2}\,dx$$

$$\Rightarrow \quad \frac{xdy - ydx}{\sqrt{x^2 - y^2}} = mx^2dx$$

$$\Rightarrow \quad \frac{xdy - ydx}{x\sqrt{1 - \frac{y^2}{x^2}}} = mx^2dx$$

$$\Rightarrow \quad \frac{\frac{xdy - ydx}{x^2}}{\sqrt{1 - y^2/x^2}} = mxdx$$

$$\Rightarrow \quad d\sin^{-1}\left(\frac{y}{x}\right) = mxdx$$

On integrating both sides, we get

$$\sin^{-1}\frac{y}{x} = \frac{mx^2}{2} + C$$

$$\therefore \quad \lambda = 2, \mu = 2$$

$$\Rightarrow \quad \lambda - \mu = 2 - 2 = 0$$

# JEE Main & AIEEE Archive

**1.** (c) Given differential equation is

$$(x\log x)\frac{dy}{dx} + y = 2x\log x \qquad [\because x \ge 1]$$

$$\Rightarrow \quad \frac{dy}{dx} + \frac{y}{x\log x} = 2$$

Which is a linear differential equation.

$$\therefore \quad \text{IF} = e^{\int \frac{1}{x\log x}dx} = e^{\log(\log x)} = \log x$$

Now, the solution of given differential equation is given by

$$y\cdot\log x = \int \log x\cdot 2dx$$

$$\Rightarrow \quad y\cdot\log x = 2\int \log x\, dx$$

$$\Rightarrow \quad y\cdot\log x = 2\,[x\log x - x] + C$$

At $x = 1$, $\quad C = 2$

$$\Rightarrow \quad y\cdot\log x = 2\,[x\log x - x] + 2$$

At $x = e$, $\Rightarrow y = 2(e - e) + 2 \Rightarrow y = 2$

**2.** (a) Given, $\frac{dp}{dt} - \frac{1}{2}p(t) = -200$

$$\therefore \quad \text{IF} = e^{\int -\left(\frac{1}{2}\right)dt} = e^{-t/2}$$

Hence, the solution of given differential equation is

$$p(t)\cdot e^{-t/2} = \int -200\cdot e^{-t/2}dt$$

$$p(t)\cdot e^{-t/2} = 400\,e^{-t/2} + K$$

$$\because \quad p(0) = 100$$

$$\Rightarrow \quad K = -300$$

$$\therefore \quad p(t)e^{-t/2} = 400\,e^{-t/2} - 300$$

$$\Rightarrow \quad p(t) = 400 - 300\,e^{t/2}$$

**3.** (c) Given, $\frac{dP}{dx} = (100 - 12\sqrt{x})$

$$\Rightarrow \quad dP = (100 - 12\sqrt{x})\,dx$$

On integrating both sides, we get

$$\int dP = \int (100 - 12\sqrt{x})\,dx$$

$$\Rightarrow \quad P = 100x - 8x^{3/2} + C$$

When $x = 0$, then $\quad P = 2000 \Rightarrow C = 2000$

Now, when $x = 25$, then

$$P = 100 \times 25 - 8 \times (25)^{3/2} + 2000$$

$$= 2500 - 8 \times 125 + 2000$$

$$= 4500 - 1000 = 3500$$

**4.** (a) Given,

(i) The population of mouse at time $t$ satisfies the differential equation

$$p'(t) = \frac{dp(t)}{dt} = 0.5p(t) - 450$$

(ii) Population of mouse at time

$$t = 0 \text{ is } p(0) = 850$$

**To find** The time at which the population of the mouse will become zero i.e. to find the value of $t$ at which $p(t) = 0$.

Let's solve the differential equation first.

$$p'(t) = \frac{dp(t)}{dt} = 0.5p(t) - 450$$

$$\Rightarrow \quad \frac{2dp(t)}{p(t) - 900} = dt$$

On integrating both sides, we get

$$\int \frac{2dp(t)}{p(t) - 900} = \int dt$$

$$\Rightarrow \quad 2\log|\,p(t) - 900| = t + C$$

where, $C$ is the constant of integration.

To find the value of $C$, let's substitute $t = 0$.

$$\Rightarrow \quad 2\log|\,p(0) - 900\,| = 0 + C$$

$$\Rightarrow \quad C = 2\log|850 - 900| \qquad [\because p(0) = 850]$$

$$\therefore \quad C = 2\log 50$$

Now, substituting the value of $C$ back in the solution, we get

$$2\log|\,p(t) - 900\,| = t + 2\log 50$$

Here, since we want to find the value of $t$ at which $p(t) = 0$, hence substituting $p(t) = 0$, we get

$$2\log|\,0 - 900\,| = t + 2\log 50$$

$$\Rightarrow \quad t = 2\log\left|\frac{900}{50}\right|$$

$$\therefore \quad t = 2\log 18$$

**5.** (d) Here, $\frac{dy}{dx} = y + 3 > 0$ and $y(0) = 2$

On integrating both sides, we get

$$\int \frac{dy}{y+3} = \int dx$$

$$\Rightarrow \quad \log|y + 3| = x + C$$

Since, $\quad y(0) = 2$

$$\Rightarrow \quad \log|2 + 3| = 0 + C$$

$$\therefore \quad C = \log_e 5$$

$$\Rightarrow \quad \log|y + 3| = x + \log_e 5$$

when, $\quad x = \log_e 2$

$$\Rightarrow \quad \log|y + 3| = \log_e 2 + \log_e 5$$

$$= \log_e 10$$

$$\Rightarrow \quad y + 3 = 10$$

$$\therefore \quad y = 7$$

**6.** (a) Given, $\frac{d\{V(t)\}}{dt} = -k(T - t)$

$$\therefore \quad d\{V(t)\} = -k(T - t)\,dt$$

When $t = 0$, $V(t) = I$ $\qquad \ldots$(i)

$$\Rightarrow \quad \int_0^T d\{V(t)\} = \int_0^T -k(T - t)\,dt$$

$$\Rightarrow \quad V(T) - V(0) = k\left[\frac{(t - T)^2}{2}\right]_0^T$$

$$\Rightarrow \quad V(T) - I = \frac{k}{2}[(T - T)^2 - (0 - T)^2]$$

$$\therefore \quad V(T) = I - \frac{k}{2}T^2$$

**7.** (d) Here, $\frac{dx}{dy} + \frac{1}{y^2}\cdot x = \frac{1}{y^3}$ $\qquad$ [linear differential equation in $x$]

$$\therefore \quad \text{IF} = e^{\int \frac{1}{y^2}dy} = e^{-\frac{1}{y}}$$

Complete solution is

$$x\cdot e^{-1/y} = \int \frac{1}{y^3}\cdot e^{-1/y}\,dy$$

$$\Rightarrow \quad x\cdot e^{-1/y} = \int \frac{1}{y}\cdot\frac{1}{y^2}\cdot e^{-1/y}\,dy$$

Put $-\frac{1}{y} = t \Rightarrow \frac{1}{y^2}dy = dt$

$$\Rightarrow \quad xe^{-1/y} = \int -t\cdot e^t dt$$

$$\Rightarrow \quad xe^{-1/y} = -\{t\cdot e^t - \int 1\cdot e^t\,dt\} + C$$

$$\Rightarrow \quad xe^{-1/y} = -te^t + e^t + C$$

$$\Rightarrow \quad xe^{-1/y} = \frac{1}{y}\cdot e^{-1/y} + e^{-1/y} + C$$

$$\because \quad y(1) = 1$$

$\therefore \quad e^{-1} = e^{-1} + e^{-1} + C \Rightarrow C = -\frac{1}{e}$

$\therefore \quad xe^{-1/y} = \frac{1}{y} \cdot e^{-1/y} + e^{-1/y} - \frac{1}{e}$

$\Rightarrow \quad x = \frac{1}{y} + 1 - \frac{1}{e} \cdot e^{1/y}$

**8.** (a) Since, $\cos x\, dy = y \sin x\, dx - y^2 dx$

$\Rightarrow \quad \frac{1}{y^2}\frac{dy}{dx} - \frac{1}{y}\tan x = -\sec x$

which is the Bernaulli's form.

Put $-\frac{1}{y} = z \Rightarrow \frac{1}{y^2}\frac{dy}{dx} = \frac{dz}{dx}$

$\Rightarrow \quad \frac{dz}{dx} + (\tan x) z = -\sec x$

This is a linear differential equation. Therefore,

$$\text{IF} = e^{\int \tan x dx} = e^{\log \sec x} = \sec x$$

Hence, the solution is

$$z \cdot (\sec x) = \int -\sec x \cdot \sec x\, dx + C_1 = -\int \sec^2 x\, dx + C_1$$

$\Rightarrow \quad -\frac{1}{y}\sec x = -\tan x + C_1$

$\Rightarrow \quad \sec x = y(\tan x + C) \quad$ [put $C = -C_1$]

**9.** (d) Given, $y = c_1 e^{c_2 x} \Rightarrow y' = c_2 c_1 e^{c_2 x}$

$\Rightarrow \quad y' = c_2 y \quad \ldots$(i)

$\Rightarrow \quad y'' = c_2 y' \quad \ldots$(ii)

$\Rightarrow \quad y'' = \frac{(y')^2}{y} \quad \left[\text{from Eq. (i)}, c_2 = \frac{y'}{y}\right]$

$\Rightarrow \quad yy'' = (y')^2$

**10.** (d) The equation of family of circles with centre on $y = 2$ and radius 5 is

$(x - \alpha)^2 + (y - 2)^2 = 5^2 \quad \ldots$(i)

$\Rightarrow \quad x^2 + \alpha^2 - 2\alpha x + y^2 + 4 - 4y = 25$

On differentiating w.r.t. $x$, we get

$$2x - 2\alpha + 2y\frac{dy}{dx} - 4\frac{dy}{dx} = 0$$

$\Rightarrow \quad \alpha = x + \frac{dy}{dx}(y - 2)$

On putting the value of $\alpha$ in Eq. (i), we get

$$\left[x - x - \frac{dy}{dx}(y - 2)\right]^2 + (y - 2)^2 = 5^2$$

$\Rightarrow \quad \left(\frac{dy}{dx}\right)^2 (y - 2)^2 = 25 - (y - 2)^2$

i.e. $\quad y'^2 (y - 2)^2 = 25 - (y - 2)^2$

**11.** (a) Given equation can be rewritten as

$$\frac{dy}{dx} - \frac{1}{x} \cdot y = 1$$

Now, $\text{IF} = e^{-\int \frac{1}{x} dx} = e^{-\log x} = \frac{1}{x}$

$\therefore$ Required solution,

$$y\left(\frac{1}{x}\right) = \int \frac{1}{x} dx = \log x + C$$

$\because \quad y(1) = 1$

$\Rightarrow \quad 1 = \log 1 + C$

$\Rightarrow \quad C = 1$

$\therefore \quad y = x \log x + x$

**12.** (c) General equation of all such circles, which pass through the origin and whose centre lie on $X$-axis, is

$x^2 + y^2 + 2gx = 0 \quad \ldots$(i)

On differentiating w.r.t. $x$, we get

$$2x + 2y\frac{dy}{dx} + 2g = 0$$

$\Rightarrow \quad 2g = -\left(2x + 2y\frac{dy}{dx}\right)$

On putting the value of $2g$ in Eq. (i), we get

$$x^2 + y^2 + \left(-2x - 2y\frac{dy}{dx}\right)x = 0$$

$\Rightarrow \quad x^2 + y^2 - 2x^2 - 2xy\frac{dy}{dx} = 0$

$\Rightarrow \quad y^2 = x^2 + 2xy\frac{dy}{dx}$

which is the required equation.

**13.** (c) The given equation is

$$Ax^2 + By^2 = 1$$

On differentiating w.r.t. $x$, we get

$2Ax + 2By\frac{dy}{dx} = 0 \quad \ldots$(i)

Again, differentiating Eq. (i) w.r.t. $x$, we get

$2A + 2B\left\{\left(\frac{dy}{dx}\right)^2 + y\frac{d^2y}{dx^2}\right\} = 0 \quad \ldots$(ii)

Eliminating $A$ and $B$ from Eqs. (i) and (ii), we get

$$y\frac{d^2y}{dx^2} + \left(\frac{dy}{dx}\right)^2 - \frac{y}{x} \cdot \frac{dy}{dx} = 0$$

This is the required differential equation, whose order is two and degree is one.

**14.** (b) Given equation of family of curves is

$y^2 = 2c(x + \sqrt{c}) \quad \ldots$(i)

On differentiating Eq. (i) w.r.t. $x$, we get

$$2yy_1 = 2c$$

$\Rightarrow \quad c = yy_1$

On putting the value of $c$ in Eq. (i), we get

$$y^2 = 2yy_1(x + \sqrt{yy_1})$$

$\therefore \quad (y^2 - 2yy_1 x)^2 = 4(yy_1)^3$

Hence, the degree and order of above equation are three and one, respectively.

**15.** (b) Given,

$$x\frac{dy}{dx} = y(\log y - \log x + 1)$$

$\therefore \quad \frac{dy}{dx} = \left(\frac{y}{x}\right)\left[\log\left(\frac{y}{x}\right) + 1\right] \quad \ldots$(i)

which is the homogeneous equation.

Put $y = tx \Rightarrow \frac{dy}{dx} = t + x\frac{dt}{dx}$

$\therefore \quad t + x\frac{dt}{dx} = t \log t + t \quad$ [from Eq. (i)]

$\Rightarrow \quad t \log t\, dx = x\, dt$

$\Rightarrow \quad \int \frac{dt}{t \log t} = \int \frac{dx}{x}$

On integrating, we get

$$\log t = \log x + \log C$$

$\Rightarrow \quad \log\left(\frac{y}{x}\right) = Cx$

**16.** (c) The equation of the family of curves is

$$x^2 + y^2 - 2ay = 0 \quad \ldots(i)$$

On differentiating w.r.t. $x$, we get

$$2x + 2yy' - 2ay' = 0$$

$$\Rightarrow \quad \frac{2x + 2yy'}{y'} = 2a \quad \ldots(ii)$$

$$\Rightarrow \quad 2a = \frac{x^2 + y^2}{y} \quad \text{[from Eq. (i)]}$$

On putting the value of $2a$ in Eq. (ii), we get

$$\frac{2x + 2yy'}{y'} = \frac{x^2 + y^2}{y}$$

$$\Rightarrow \quad 2xy + 2y^2y' = x^2y' + y^2y'$$

$$\therefore \quad (x^2 - y^2)y' = 2xy$$

**17.** (b) Given that, $y\,dx + (x + x^2y)\,dy = 0$

$$\Rightarrow \quad \frac{y\,dx + x\,dy}{x^2y^2} = -\frac{1}{y}dy$$

On integrating both sides, we get

$$\int d\left(-\frac{1}{xy}\right) = -\int \frac{1}{y}dy$$

$$\Rightarrow \quad -\frac{1}{xy} = -\log y + C$$

$$\Rightarrow \quad -\frac{1}{xy} + \log y = C$$

**18.** (b) General equation of parabola, whose axis is $X$-axis, is

$$y^2 = 4a(x + h)$$

On differentiating w.r.t. $x$, we get

$$2y\frac{dy}{dx} = 4a$$

$$\Rightarrow \quad y\frac{dy}{dx} = 2a$$

Again differentiating w.r.t. $x$, we get

$$y\left(\frac{dy}{dx}\right)^2 + y\frac{d^2y}{dx^2} = 0$$

This is a differential equation whose degree and order are one and two, respectively.

**19.** (b) Given differential equation can be rewritten as

$$(1 + y^2)\frac{dx}{dy} + x = e^{\tan^{-1}y} \quad \Rightarrow \quad \frac{dx}{dy} + \frac{1}{1 + y^2}x = \frac{e^{\tan^{-1}y}}{1 + y^2}$$

$$\therefore \quad \text{IF} = e^{\int P\,dy} = e^{\int \frac{1}{1+y^2}dy} = e^{\tan^{-1}y}$$

$$\therefore \quad \text{Required solution, } xe^{\tan^{-1}y} = \int \frac{e^{2\tan^{-1}y}}{1 + y^2}dy + C_1$$

Put $t = \tan^{-1}y \Rightarrow dt = \dfrac{dy}{1 + y^2}$

$$\therefore \quad xe^{\tan^{-1}y} = \int e^{2t}\,dt + C_1 \quad \Rightarrow \quad xe^{\tan^{-1}y} = \frac{1}{2}e^{2t} + C_1$$

$$\Rightarrow \quad xe^{\tan^{-1}y} = \frac{1}{2}e^{2\tan^{-1}y} + C_1$$

$$\Rightarrow \quad 2xe^{\tan^{-1}y} = e^{2\tan^{-1}y} + C \quad [\text{put } 2C_1 = C]$$

**20.** (c) The given differential equation can be rewritten as

$$\left(1 + 3\frac{dy}{dx}\right)^2 = 4\left(\frac{d^3y}{dx^3}\right)^3$$

This shows that the order and degree of given equation are 3 and 3, respectively.

**21.** (b) $\because \dfrac{d^2y}{dx^2} = e^{-2x}$

On integrating both sides, we get

$$\frac{dy}{dx} = \frac{e^{-2x}}{-2} + C$$

Again integrating, we get

$$y = \frac{e^{-2x}}{4} + Cx + D$$

**22.** (a) The general equation of all non-vertical lines in a plane is $ax + by = 1$, where $b \neq 0$.

On differentiating, we get

$$a + b\frac{dy}{dx} = 0$$

Again differentiating, we get

$$b\frac{d^2y}{dx^2} = 0 \quad \Rightarrow \quad \frac{d^2y}{dx^2} = 0 \quad [\because b \neq 0]$$

## JEE Advanced & IIT JEE Archive

**1.** (3) Since, $F'(a) + 2$ is the area bounded by $x = 0$, $y = 0$, $y = f(x)$ and $x = a$.

$$\therefore \quad \int_0^a f(x)dx = F'(a) + 2$$

Using Newton-Leibnitz formula, $f(a) = F''(a)$

and $\quad f(0) = F''(0) \quad \ldots(i)$

Given, $\quad F(x) = \int_x^{x^2 + \pi/6} 2\cos^2 t\,dt$

On differentiating, $F'(x) = 2\cos^2\left(x^2 + \dfrac{\pi}{6}\right)\cdot 2x - 2\cos^2 x \cdot 1$

Again differentiating,

$$F''(x) = 4\left\{\cos^2\left(x^2 + \frac{\pi}{6}\right) - 2x\cos\left(x^2 + \frac{\pi}{6}\right)\sin\left(x^2 + \frac{\pi}{6}\right)2x\right\} + \{4\cos x \cdot \sin x\}$$

$$= 4\left\{\cos^2\left(x^2 + \frac{\pi}{6}\right) - 4x^2\cos\left(x^2 + \frac{\pi}{6}\right)\sin\left(x^2 + \frac{\pi}{6}\right)\right\} + 2\sin 2x$$

$$\therefore \quad F''(0) = 4\left\{\cos^2\left(\frac{\pi}{6}\right)\right\} = 3$$

$$\therefore \quad f(0) = 3$$

**2.** (a,c) Here, $\quad (1 + e^x)y' + ye^x = 1$

$$\Rightarrow \quad \frac{dy}{dx} + e^x \cdot \frac{dy}{dx} + ye^x = 1$$

$$\Rightarrow \quad dy + e^x dy + ye^x\,dx = dx$$

$$\Rightarrow \quad dy + d(e^x y) = dx$$

On integrating both sides, we get

$$y + e^x y = x + C$$

Given, $\quad y(0) = 2$

$$\Rightarrow \quad 2 + e^0 \cdot 2 = 0 + C$$

$$\Rightarrow \quad C = 4$$

$$\therefore \quad y(1 + e^x) = x + 4$$

$$\Rightarrow \quad y = \frac{x + 4}{1 + e^x}$$

Now at $x = -4$, $\quad y = \dfrac{-4 + 4}{1 + e^{-4}} = 0$

$$\therefore \quad y(-4) = 0 \quad \ldots(i)$$

For critical points, $\dfrac{dy}{dx} = 0$

i.e. $\dfrac{dy}{dx} = \dfrac{(1+e^x)\cdot 1 - (x+4)e^x}{(1+e^x)^2} = 0$

$\Rightarrow \quad e^x (x+3) - 1 = 0$

or $\quad e^{-x} = (x+3)$

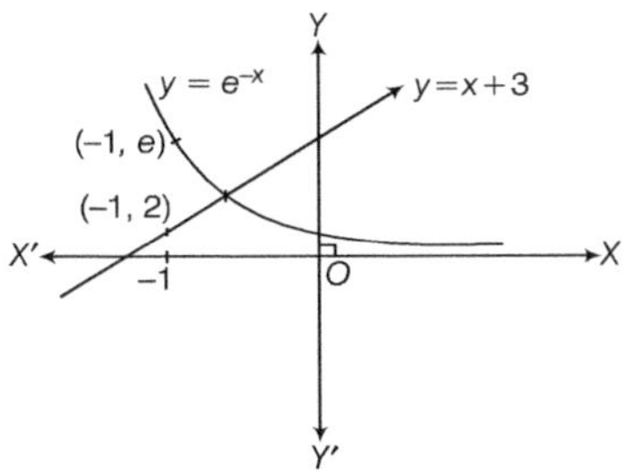


Clearly, the intersection point lies between (– 1, 0).

$\therefore$ $y(x)$ has a critical point in the interval (– 1, 0).

**3.** (b,c) Since, centre lies on $y = x$.

$\therefore$ Equation of circle is

$$x^2 + y^2 - 2ax - 2ay + c = 0$$

On differentiating, we get

$$2x + 2yy' - 2a - 2ay' = 0$$

$$\Rightarrow \quad x + yy' - a - ay' = 0$$

$$\Rightarrow \quad a = \frac{x + yy'}{1 + y'}$$

Again differentiating, we get

$$0 = \frac{(1+y')[1 + yy'' + (y')^2] - (x + yy')\cdot(y'')}{(1+y')^2}$$

$$\Rightarrow (1+y')[1 + (y')^2 + yy''] - (x + yy')(y'') = 0$$

$$\Rightarrow \quad 1 + y'[(y')^2 + y' + 1] + y''(y - x) = 0$$

On comparing with $Py'' + Qy' + 1 = 0$, we get

$$P = y - x$$

and $\quad Q = (y')^2 + y' + 1$

**4.** (b) **Plan**

(i) Solution of the differential equation $\dfrac{dy}{dx} + Py = Q$ is

$$y\cdot(\text{IF}) = \int Q\cdot(\text{IF})\,dx + C$$

where, IF $= e^{\int P\,dx}$.

(ii) $\int_{-a}^{a} f(x)\,dx = 2\int_0^a f(x)\,dx$, if $f(-x) = f(x)$

Given differential equation is

$$\frac{dy}{dx} + \frac{x}{x^2 - 1}\,y = \frac{x^4 + 2x}{\sqrt{1 - x^2}}$$

This is a linear differential equation.

$$\therefore \quad \text{IF} = e^{\int \frac{x}{x^2-1}dx} = e^{\frac{1}{2}\ln|x^2-1|} = \sqrt{1 - x^2}$$

Solution is

$$y\sqrt{1-x^2} = \int \frac{x(x^3+2)}{\sqrt{1-x^2}}\cdot\sqrt{1-x^2}\,dx$$

or $\quad y\sqrt{1-x^2} = \int (x^4 + 2x)\,dx = \dfrac{x^5}{5} + x^2 + C$

$$f(0) = 0 \Rightarrow C = 0$$

$$\Rightarrow \quad f(x)\sqrt{1-x^2} = \frac{x^5}{5} + x^2$$

Now, $\displaystyle\int_{-\sqrt{3}/2}^{\sqrt{3}/2} f(x)\,dx = \int_{-\sqrt{3}/2}^{\sqrt{3}/2} \frac{x^2}{\sqrt{1-x^2}}\,dx$ [using property]

$$= 2\int_0^{\sqrt{3}/2} \frac{x^2}{\sqrt{1-x^2}}\,dx$$

$$= 2\int_0^{\pi/3} \frac{\sin^2\theta}{\cos\theta}\cos\theta\,d\theta \qquad [\text{taking } x = \sin\theta]$$

$$= 2\int_0^{\pi/3} \sin^2\theta\,d\theta = \int_0^{\pi/3} (1 - \cos 2\theta)\,d\theta$$

$$= \left(\theta - \frac{\sin 2\theta}{2}\right)_0^{\pi/3} = \frac{\pi}{3} - \frac{\sin 2\pi/3}{2}$$

$$= \frac{\pi}{3} - \frac{\sqrt{3}}{4}$$

**5.** (d) **Concept Involved** When ever we have linear differential equation containing inequality we should always check for increasing or decreasing

i.e. for $\quad \dfrac{dy}{dx} + Py < 0 \Rightarrow \dfrac{dy}{dx} + Py > 0$

Multiply by integrating factor i.e. $e^{\int P dx}$ and convert into total differential equation.

Here, $f'(x) < 2f(x)$, multiplying by $e^{-\int 2dx}$

$$f'(x)\cdot e^{-2x} - 2e^{-2x}f(x) < 0 \Rightarrow \frac{d}{dx}(f(x)\cdot e^{-2x}) < 0$$

$\therefore \quad \phi(x) = f(x)e^{-2x}$ is decreasing for $x \in \left[\dfrac{1}{2}, 1\right]$.

Thus, when $x > \dfrac{1}{2}$

$$\phi(x) < \phi\left(\frac{1}{2}\right)$$

$$\Rightarrow \quad e^{-2x}f(x) < e^{-1}\cdot f\left(\frac{1}{2}\right)$$

$$\Rightarrow \quad f(x) < e^{2x-1}\cdot 1, \text{ given } f\left(\frac{1}{2}\right) = 1$$

$$\Rightarrow \quad 0 < \int_{1/2}^{1} f(x)dx < \int_{1/2}^{1} e^{2x-1}dx$$

$$\Rightarrow \quad 0 < \int_{1/2}^{1} f(x)dx < \left[\frac{e^{2x-1}}{2}\right]_{1/2}^{1}$$

$$\Rightarrow \quad 0 < \int_{1/2}^{1} f(x)dx < \frac{e-1}{2}$$

**6.** (a) **Concept Involved**

Solving of homogeneous differential equation i.e. substitute

$$\frac{y}{x} = v$$

$$\therefore \quad y = vx \Rightarrow \frac{dy}{dx} = v + x\frac{dv}{dx}$$

Here, slope of the curve at $(x, y)$ is

$$\frac{dy}{dx} = \frac{y}{x} + \sec\left(\frac{y}{x}\right)$$

Put $\dfrac{y}{x} = v$

$$\therefore \quad v + x\frac{dv}{dx} = v + \sec(v)$$

$$\Rightarrow \quad x\frac{dv}{dx} = \sec(v) \Rightarrow \int \frac{dv}{\sec v} = \int \frac{dx}{x}$$

$$\Rightarrow \quad \int \cos v\,dv = \int \frac{dx}{x} \Rightarrow \sin v = \log x + \log C$$

$$\Rightarrow \quad \sin\left(\frac{y}{x}\right) = \log(Cx)$$

As it passes through $\left(1, \frac{\pi}{6}\right) \Rightarrow \sin\left(\frac{\pi}{6}\right) = \log C$

$$\Rightarrow \quad \log C = \frac{1}{2}$$

$$\therefore \quad \sin\left(\frac{y}{x}\right) = \log x + \frac{1}{2}$$

**7.** (b) Here, $f''(x) - 2f'(x) + f(x) \geq e^x$

$$\Rightarrow f''(x)e^{-x} - f'(x)e^{-x} - f'(x)e^{-x} + f(x)e^{-x} \geq 0$$

$$\Rightarrow \frac{d}{dx}[f'(x)e^{-x}] - \frac{d}{dx}[f(x)e^{-x}] \geq 1$$

$$\Rightarrow \frac{d}{dx}[f'(x)e^{-x} - f(x)e^{-x}] \geq 1$$

$$\Rightarrow \frac{d^2}{dx^2}[e^{-x}f(x)] \geq 1 \text{ for all } x \in [0, 1]$$

$\therefore$ $\phi(x) = e^{-x}f(x)$ is concave $n\phi$.

$$f(0) = f(1) = 0 \quad \Rightarrow \phi(0) = 0 = \phi(1)$$

$$\Rightarrow \quad \phi(x) < 0$$

$$\Rightarrow \quad e^{-x}f(x) < 0$$

$$\therefore \quad f(x) < 0$$

**8.** (c) Here, $\phi'(x) < 0,\ x \in \left(0, \frac{1}{4}\right)$

and $\quad \phi'(x) > 0,\ x \in \left(\frac{1}{4}, 1\right)$.

$$\Rightarrow e^{-x}f'(x) - e^{-x}f(x) < 0, x \in \left(0, \frac{1}{4}\right)$$

$$\Rightarrow f'(x) < f(x),\ 0 < x < \frac{1}{4}.$$

**9.** (a,d) **Concept Involved** Linear differential equation under one variable.

$$\frac{dy}{dx} + Py = Q; \quad \text{IF} = e^{\int P dx}$$

$\therefore$ Solution is, $y(\text{IF}) = \int Q \cdot (\text{IF}) dx + C$

$$y' - y\tan x = 2x\sec x \text{ and } y(0) = 0$$

$$\Rightarrow \frac{dy}{dx} - y\tan x = 2x\sec x$$

$$\therefore \quad \text{IF} = \int e^{-\tan x}dx = e^{\log|\cos x|} = \cos x$$

Solution is $y \cdot \cos x = \int 2x\sec x \cdot \cos x dx + C$

$$\Rightarrow \quad y \cdot \cos x = x^2 + C,$$

As $\quad y(0) = 0 \Rightarrow C = 0$

$$\therefore \quad y = x^2 \sec x$$

Now, $\quad y\left(\frac{\pi}{4}\right) = \frac{\pi^2}{8\sqrt{2}}$

$$\Rightarrow \quad y'\left(\frac{\pi}{4}\right) = \frac{\pi}{\sqrt{2}} + \frac{\pi^2}{8\sqrt{2}}$$

$$y\left(\frac{\pi}{3}\right) = \frac{2\pi^2}{9}$$

$$\Rightarrow \quad y'\left(\frac{\pi}{3}\right) = \frac{4\pi}{3} + \frac{2\pi^2}{3\sqrt{3}}$$

**10.** (0) Given, $\frac{dy}{dx} + y \cdot g'(x) = g(x)\, g'(x)$

$$\therefore \quad \text{IF} = e^{\int g'(x)dx} = e^{g(x)}$$

Solution is $\quad y(e^{g(x)}) = \int g(x) \cdot g'(x) \cdot e^{g(x)} dx + C$

Put $g(x) = t$, $\quad g'(x)dx = dt$

$$y(e^{g(x)}) = \int t \cdot e^t\, dt + C$$

$$= t \cdot e^t - \int 1 \cdot e^t\, dt + C$$

$$= t \cdot e^t - e^t + C$$

$$\Rightarrow \quad y e^{g(x)} = (g(x) - 1)e^{g(x)} + C \quad \ldots\text{(i)}$$

Given, $\quad y(0) = 0,\ g(0) = g(2) = 0$

$\therefore$ Eq. (i) becomes,

$$y(0) \cdot e^{g(0)} = (g(0) - 1) \cdot e^{g(0)} + C$$

$$\Rightarrow \quad 0 = (-1) \cdot 1 + C$$

$$\Rightarrow \quad C = 1$$

$$\therefore \quad y(x) \cdot e^{g(x)} = (g(x) - 1)e^{g(x)} + 1$$

$$\Rightarrow \quad y(2) \cdot e^{g(2)} = (g(2) - 1)e^{g(2)} + 1, \quad \text{where } g(2) = 0$$

$$\Rightarrow \quad y(2) \cdot 1 = (-1) \cdot 1 + 1$$

$$\therefore \quad y(2) = 0$$

**11.** (2) Given, $\quad g\{f(x)\} = x$

$$\Rightarrow \quad g'\{f(x)\}f'(x) = 1 \quad \ldots\text{(i)}$$

If $f(x) = 1 \Rightarrow x = 0,\ f(0) = 1$

Substitute $x = 0$ in Eq. (i), we get

$$g'(1) = \frac{1}{f'(0)}$$

$$\Rightarrow \quad g'(1) = 2$$

$$\left[\because f'(x) = 3x^2 + \frac{1}{2}e^{x/2} \Rightarrow f'(0) = \frac{1}{2}\right]$$

**Alternate Method**

Given, $\quad f(x) = x^3 + e^{x/2}$

$$\Rightarrow \quad f'(x) = 3x^2 + \frac{1}{2}e^{x/2}$$

For $x = 0, f(0) = 1$, $\quad f'(0) = \frac{1}{2}$

and $\quad g(x) = f^{-1}(x)$

On replacing $x$ by $f(x)$, we have

$$g[f(x)] = x$$

$$\Rightarrow \quad g'[f(x)] \cdot f'(x) = 1$$

Put $x = 0$, we get

$$g'(1) = \frac{1}{f'(0)} = 2$$

**12.** (0) From given integral equation, $f(0) = 0$.

Also, differentiating the given integral equation w.r.t. $x$

$$f'(x) = f(x)$$

If $f(x) \neq 0 \Rightarrow \quad \frac{f'(x)}{f(x)} = 1$

$$\Rightarrow \quad \log f(x) = x + C$$

$$\Rightarrow \quad f(x) = e^C e^x$$

$$\because \quad f(0) = 0 \Rightarrow e^C = 0, \text{ a contradiction.}$$

$$\therefore \quad f(x) = 0,\ \forall\ x \in R$$

$$\Rightarrow \quad f(\ln 5) = 0$$

**Alternate Method**

Given, $\quad f(x) = \int_0^x f(t)\, dt$

$$\Rightarrow \quad f(0) = 0$$

and $\quad f'(x) = f(x)$

If $f(x) \neq 0$

$$\Rightarrow \quad \frac{f'(x)}{f(x)} = 1$$

$$\Rightarrow \quad \ln f(x) = x + C$$

$$\Rightarrow \quad f(x) = e^C \cdot e^x$$

$$\because \quad f(0) = 0$$

$\Rightarrow e^C = 0$, a contradiction.

$$\therefore \quad f(x) = 0, \forall\ x \in R$$

$$\Rightarrow \quad f(\ln 5) = 0$$

**13.** (c) Given, $\dfrac{dy}{dx} = \dfrac{y\sqrt{y^2-1}}{x\sqrt{x^2-1}}$

On integrating both sides, we get

$$\int \frac{dy}{y\sqrt{y^2-1}} = \int \frac{dx}{x\sqrt{x^2-1}}$$

$\Rightarrow \quad \sec^{-1} y = \sec^{-1} x + C$

At $x = 2$, $y = \dfrac{2}{\sqrt{3}}$;

$$\frac{\pi}{6} = \frac{\pi}{3} + C \Rightarrow C = -\frac{\pi}{6}$$

Now, $\quad y = \sec\left(\sec^{-1} x - \dfrac{\pi}{6}\right)$

$$= \cos\left[\cos^{-1}\frac{1}{x} - \cos^{-1}\frac{\sqrt{3}}{2}\right]$$

$$= \cos\left[\cos^{-1}\left(\frac{\sqrt{3}}{2x} + \sqrt{1-\frac{1}{x^2}}\sqrt{1-\frac{3}{4}}\right)\right]$$

$\therefore \quad y = \dfrac{\sqrt{3}}{2x} + \dfrac{1}{2}\sqrt{1-\dfrac{1}{x^2}}$

Hence, Statement I is correct but Statement II is incorrect.

**14.** (a) Given, $\displaystyle\lim_{t \to x} \frac{t^2 f(x) - x^2 f(t)}{t - x} = 1$

$\Rightarrow \quad x^2 f'(x) - 2x f(x) + 1 = 0$

$\Rightarrow \quad \dfrac{x^2 f'(x) - 2x f(x)}{(x^2)^2} + \dfrac{1}{x^4} = 0$

$\Rightarrow \quad \dfrac{d}{dx}\left[\dfrac{f(x)}{x^2}\right] = -\dfrac{1}{x^4}$

On integrating both sides, we get

$$f(x) = Cx^2 + \frac{1}{3x}$$

Also, $f(1) = 1$, $\quad C = \dfrac{2}{3}$

Hence, $\quad f(x) = \dfrac{2}{3}x^2 + \dfrac{1}{3x}$

**15.** (c) Given, $\dfrac{dy}{dx} = \dfrac{\sqrt{1-y^2}}{y} \Rightarrow \displaystyle\int \frac{y}{\sqrt{1-y^2}}dy = \int dx$

$\Rightarrow \quad -\sqrt{1-y^2} = x + C$

$\Rightarrow \quad (x+C)^2 + y^2 = 1$

where, centre $(-C, 0)$ and radius $= 1$.

**16.** (a,c) Since, $BP : AP = 3 : 1$. Then, equation of tangent is

$$Y - y = f'(x)(X - x)$$

The intercept on the coordinate axes are

$$A\left[x - \frac{y}{f'(x)}, 0\right] \text{ and } B[0, y - x f'(x)]$$

Since, $P$ is internally intercepts a line $AB$.

$\therefore \quad x = \dfrac{3\left(x - \dfrac{y}{f'(x)}\right) + 1 \times 0}{3+1}$

$\Rightarrow \quad \dfrac{dy}{dx} = \dfrac{y}{-3x} \Rightarrow \dfrac{dy}{y} = -\dfrac{1}{3x}dx$

On integrating both sides, we get

$$xy^3 = C$$

Since, curve passes through $(1, 1)$, then $C = 1$.

$\therefore \quad xy^3 = 1$

$\therefore$ At $x = \dfrac{1}{8} \quad \Rightarrow y = 2$.

**17.** Since, the length of tangent $= \left| y\sqrt{1+\left(\dfrac{dx}{dy}\right)^2}\right| = 1$

$\Rightarrow \quad y^2\left[1 + \left(\dfrac{dx}{dy}\right)^2\right] = 1$

$\therefore \quad \dfrac{dy}{dx} = \pm \dfrac{y}{\sqrt{1-y^2}}$

$\Rightarrow \quad \displaystyle\int \frac{\sqrt{1-y^2}}{y}dy = \pm \int x\,dx$

$\Rightarrow \quad \displaystyle\int \frac{\sqrt{1-y^2}}{y}dy = \pm x + C$

Put $y = \sin\theta \Rightarrow dy = \cos\theta\,d\theta$

$\therefore \quad \displaystyle\int \frac{\cos\theta}{\sin\theta}\cdot\cos\theta\,d\theta = \pm x + C$

$\Rightarrow \quad \displaystyle\int \frac{\cos^2\theta}{\sin^2\theta}\cdot\sin\theta\,d\theta = \pm x + C$

Again, put $\cos\theta = t \Rightarrow -\sin\theta\,d\theta = dt$

$\therefore \quad \displaystyle -\int \frac{t^2}{1-t^2}dt = \pm x + C$

$\Rightarrow \quad \displaystyle\int \left(1 - \frac{1}{1-t^2}\right)dt = \pm x + C$

$\Rightarrow \quad t - \log\left|\dfrac{1+t}{1-t}\right| = \pm x + C$

$\Rightarrow \quad \sqrt{1-y^2} - \log\left|\dfrac{1+\sqrt{1-y^2}}{1-\sqrt{1-y^2}}\right| = \pm x + C$

**18.** (a) Given, $x\,dy = y(dx + y\,dy)$, $y > 0$

$\Rightarrow \quad x\,dy - y\,dx = y^2 dy$

$\Rightarrow \quad \dfrac{x\,dy - y\,dx}{y^2} = dy$

$\Rightarrow \quad d\left(\dfrac{x}{y}\right) = -dy$

On integrating both sides, we get

$$\frac{x}{y} = -y + C \quad \ldots(i)$$

$\because \quad y(1) = 1 \Rightarrow x = 1, y = 1$

$\therefore \quad C = 2$

From Eq. (i), $\quad \dfrac{x}{y} + y = 2$

Again, for $x = -3$

$\Rightarrow \quad -3 + y^2 = 2y$

$\Rightarrow \quad y^2 - 2y - 3 = 0$

$\Rightarrow \quad (y+1)(y-3) = 0$

As $y > 0$ take $y = 3$, neglecting $y = -1$.

# 4. Area

## DPP-1 Area Under the Curves

**1.** (i) Graph of $y = \sec^2 x$ is given by

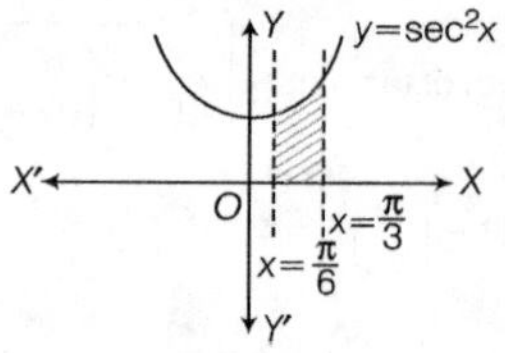


$\therefore$ Area $= \int_{\pi/6}^{\pi/3} \sec^2 x dx = [\tan x]_{\pi/6}^{\pi/3} = \sqrt{3} - \frac{1}{\sqrt{3}} = \frac{2}{\sqrt{3}}$ sq unit

(ii) Given, curve $y = \sin 2x$

The graph of $\sin 2x$ could be plotted as

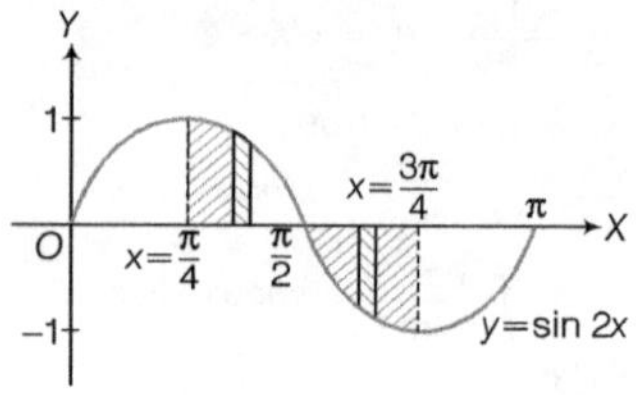


$\therefore$ Required area $= \int_{\pi/4}^{\pi/2} \sin 2x\, dx - \int_{\pi/2}^{3\pi/4} \sin 2x\, dx$

$= \left[\frac{\cos 2x}{2}\right]_{\pi/4}^{\pi/2} - \left[\frac{\cos 2x}{2}\right]_{\pi/2}^{3\pi/4}$

$= \left[-\frac{1}{2} - 0\right] - \left[0 + \frac{1}{2}\right]$

$= -1$

$= 1$ sq unit [$\because$ area cannot be negative]

(iii) Given equation of curve

$$y = 4x^2$$

$\Rightarrow \quad x^2 = \frac{1}{4} y \quad \ldots(i)$

Clearly, it represents a parabola opening upward and vertex at the origin and axis of symmetry along the positive direction of $Y$-axis as shown in figure given below.

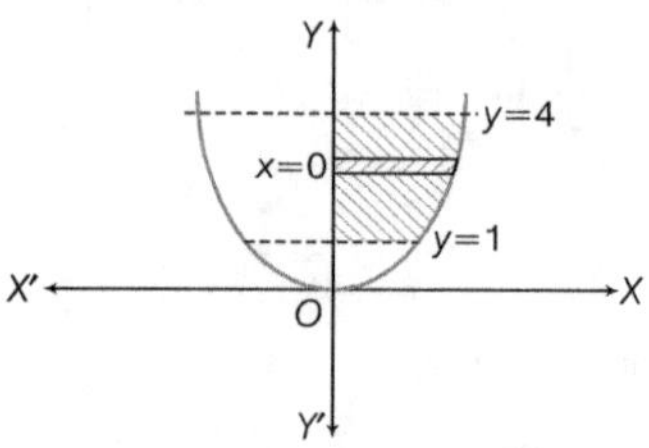


Clearly, $y = 1$ and $y = 4$ represent a straight line parallel to $X$-axis and $x = 0$ is $Y$-axis.

$\therefore$ Required area $= \int_1^4 x dy = \int_1^4 \frac{\sqrt{y}}{2} dy$

$= \frac{1}{2} \int_1^4 y^{1/2} dy = \frac{1}{2} \times \frac{2}{3} [y^{3/2}]_1^4$

$= \frac{1}{3} [(4)^{3/2} - (1)^{3/2}]$

$= \frac{1}{3} [8 - 1]$

$= \frac{7}{3}$ sq units

(iv) Given, $\quad y = x^2 + 2$

$\Rightarrow \quad x^2 = y - 2 \quad \ldots(i)$

Clearly, Eq. (i) represents a parabola opening upward and vertex at (0, 2) as shown in figure given below.

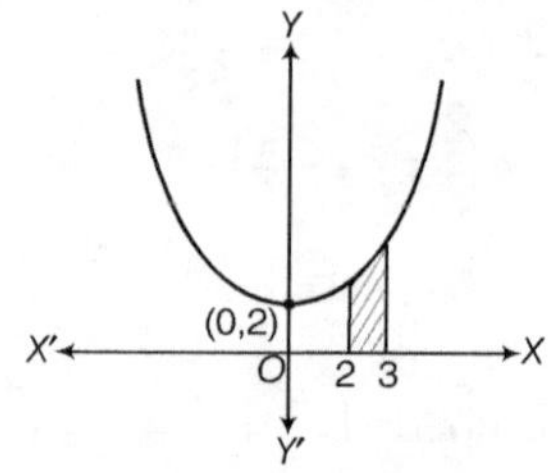


Clearly, $x = 2$, $x = 3$ is a line parallel to $Y$-axis.

$\therefore$ Required area $= \int_2^3 y\, dx = \int_2^3 (x^2 + 2) dx$

$= \left[\frac{x^3}{3} + 2x\right]_2^3 = (9 + 6) - \left(\frac{8}{3} + 4\right)$

$= \frac{25}{3}$ sq units

(v) Given, $\quad y = \sqrt{1 - x^2}$

and $\quad y^2 = 1 - x^2$

$\Rightarrow \quad x^2 + y^2 = 1$

Clearly, it represents a circle with centre at (0, 0) and radius 1 unit as shown in figure.

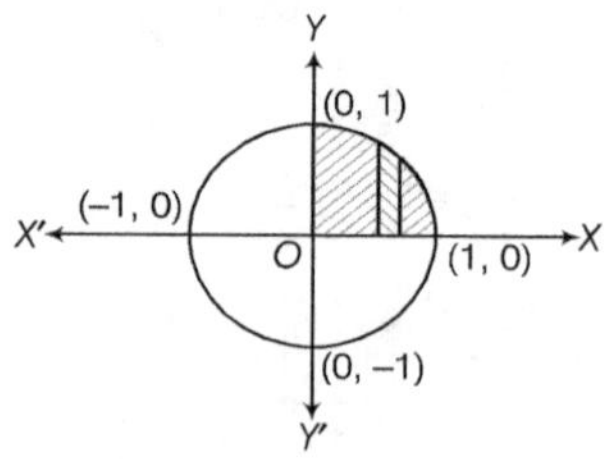


$\therefore$ Required area $= \int_0^1 y\, dx = \int_0^1 \sqrt{1 - x^2}\, dx$

$= \left[\frac{x}{2}\sqrt{1 - x^2} + \frac{1}{2} \sin^{-1} x\right]_0^1$

$= \frac{\pi}{4}$ sq unit

(vi) Given, $\quad y = x|x|$

i.e. $\quad y = \begin{cases} -x^2, & x < 0 \\ x^2, & x \geq 0 \end{cases}$

The graph of the curve is given below.

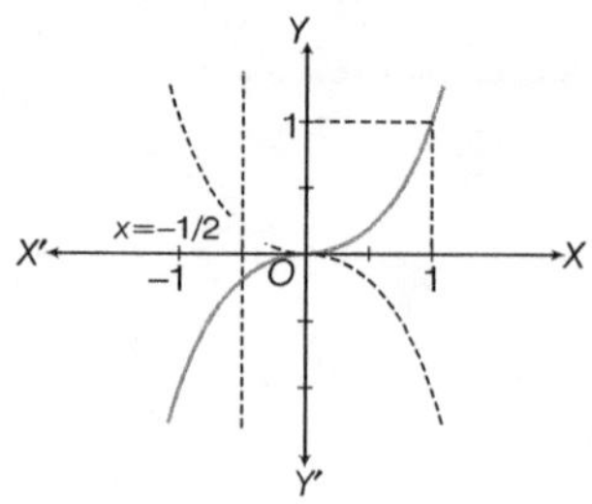

$\therefore$ Required area $= \int_{-1/2}^{0} y\,dx + \int_{0}^{1} y\,dx$

$$= \int_{-1/2}^{0} (-x^2)\,dx + \int_{0}^{1} x^2\,dx$$

$$= -\left[\frac{x^3}{3}\right]_{-1/2}^{0} + \left[\frac{x^3}{3}\right]_{0}^{1}$$

$$= -\frac{1}{24} + \frac{1}{3}$$

$$= \frac{7}{24} \text{ sq unit}$$

**2.** (i) Given curves $\quad x = y^2 \quad$ ...(i)

and $\quad x = 3 - 2y^2 \quad$ ...(ii)

To find points of intersection of curves (i) and (ii), we solve Eqs. (i) and (ii) simultaneously, then

$$y^2 = 3 - 2y^2$$

$$\Rightarrow \quad 3y^2 = 3$$

$$\Rightarrow \quad y^2 = 1$$

$$\Rightarrow \quad y = \pm 1$$

$\therefore$ Points of intersection are (1, 1) and (1, – 1).

Clearly, $y^2 = x$ represents a parabola with vertex (0, 0) and $y^2 = -\frac{1}{2}(x - 3)$ also represents a parabola with vertex (3, 0) as shown in figure.

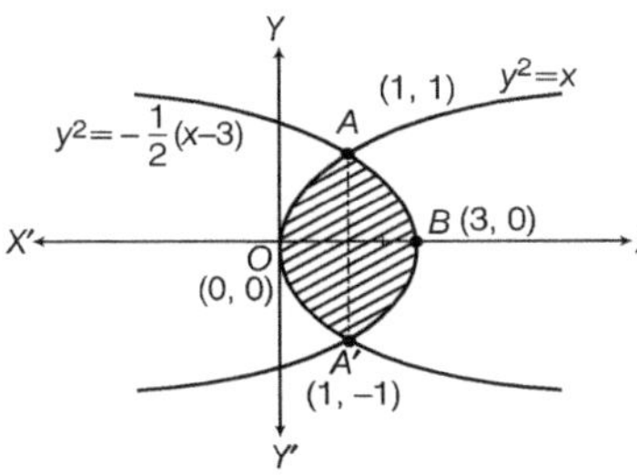


$\therefore$ Required area

$= \text{Area } (OABA'O)$

$= 2 \times \text{Area } (OABO)$

$$= 2\left[\int_0^1 y\,dx \text{ (for } y^2 = x) + \int_1^3 y\,dx \left(\text{for } y^2 = \frac{3-x}{2}\right)\right]$$

$$= 2\left[\int_0^1 \sqrt{x}\,dx + \int_1^3 \sqrt{\frac{3-x}{2}}\,dx\right]$$

$$= 2\left[\frac{2}{3}[x^{3/2}]_0^1 - \frac{1}{\sqrt{2}}[(3-x)^{3/2}]_1^3 \times \frac{2}{3}\right]$$

$$= 2\left[\frac{2}{3} - \frac{\sqrt{2}}{3}(-2\sqrt{2})\right]$$

$$= 2\left[\frac{2}{3} + \frac{4}{3}\right]$$

$= 4$ sq units

(ii) Given curves are $\quad y = \sqrt{x} \quad$ ...(i)

and $\quad y = x \quad$ ...(ii)

To find the point of intersection, we solve Eqs. (i) and (ii) simultaneously, then

$$x = \sqrt{x}$$

$$\Rightarrow \quad x^2 = x$$

$$\Rightarrow \quad x(x - 1) = 0$$

$$\Rightarrow \quad x = 0, 1$$

$\therefore$ Points of intersection are (0, 0) and (1, 1).

Clearly, $y = \sqrt{x}$ represents a parabola with vertex (0, 0) and $y = x$ represents a straight line passing through the origin, as shown in figure below.

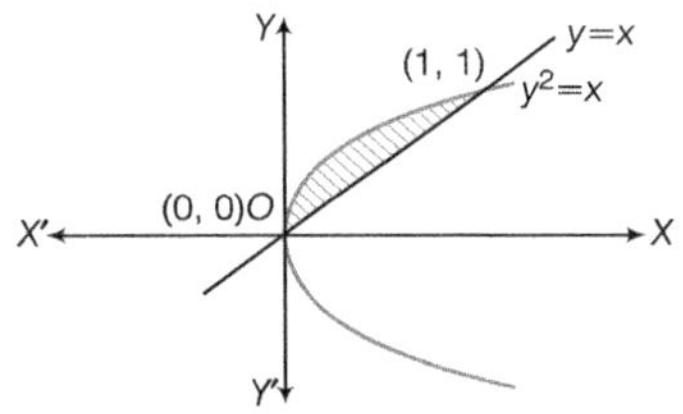


$\therefore$ Required area $= \int_0^1 [y \text{ (for } y^2 = x) - y\,dx \text{ (for } y = x)]\,dx$

$$= \int_0^1 (\sqrt{x} - x)\,dx$$

$$= \left[\frac{2}{3}x^{3/2} - \frac{x^2}{2}\right]_0^1$$

$$= \left(\frac{2}{3} - \frac{1}{2}\right)$$

$$= \frac{1}{6} \text{ sq unit}$$

(iii) Since, the inverse of a logarithmic functions is an exponential function and *vice-versa* and these two curves are on the opposite side of the line $y = x$. Thus, $y = 2^x$ and $y = \log x$ do not intersect. Their graphs are shown below. The shaded region in figure shows the area bounded by the given curves. Let us slice this region into vertical strips as shown in figure.

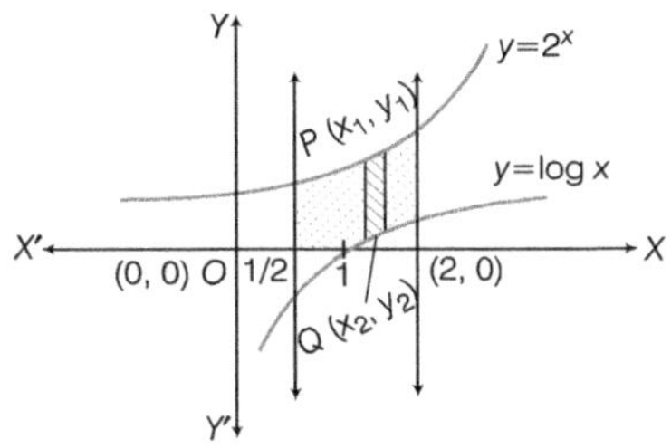


For the approximating rectangle shown in figure, we have

length $= (y_1 - y_2)$, width $= \Delta x$

Area $= (y_1 - y_2)\,dx$

As the approximating rectangle can move horizontally between $x = \frac{1}{2}$ and $x = 2$.

So, required area is given by

$$= \int_{1/2}^{2} (y_1 - y_2)\,dx = \int_{1/2}^{2} (2^x - \log x)\,dx$$

$$= \left[\frac{2^x}{\log 2} - x\log x + x\right]_{1/2}^{2}$$

$$= \left[\left\{\frac{4}{\log 2} - 2\log 2 + 2\right\} - \left\{\frac{\sqrt{2}}{\log 2} + \frac{1}{2}\log 2 + \frac{1}{2}\right\}\right]$$

$$= \left(\frac{4 - \sqrt{2}}{\log 2} - \frac{5}{2}\log 2 + \frac{3}{2}\right) \text{ sq units}$$

(iv) The equation $y = 2x - x^2$ represents a parabola opening downwards having vertex at (1, 1) and crossing $X$-axis at (0, 0) and (2, 0).

The equation $2^x$ represents an exponential curve as shown in figure below. Lines $x = 0$ and $x = 2$ shown in the following figure. The shaded region in the figure represents area bounded by these curves.

We slice the shaded region into vertical strips. The length osf each such strips is $(y_1 - y_2)$ and width $dx$.

$\therefore$ Area $= (y_1 - y_2)\,dx$

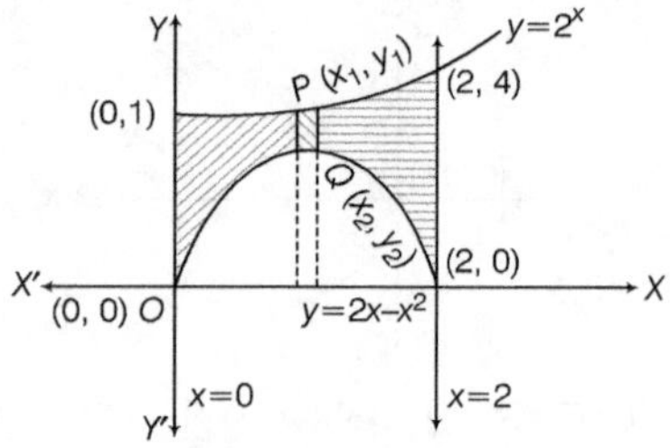


$\therefore$ Required area $= \int_0^2 (y_1 - y_2)\,dx = \int_0^2 (2^x - 2x + x^2)\,dx$

$\left[\begin{array}{l} P(x, y_1) \text{ and } Q(x, y_2) \text{ lies on as } y = 2^x \text{ and } y = 2x - x^2 \\ \text{respectively, where} \\ y_1 = 2^x \text{ and } y_2 = 2x - x^2 \end{array}\right]$

$$= \left[\frac{2^x}{\log 2} - x^2 + \frac{x^3}{3}\right]_0^2$$

$$= \frac{4}{\log 2} - 4 + \frac{8}{3} - \frac{1}{\log 2}$$

$$= \left(\frac{3}{\log 2} - \frac{4}{3}\right) \text{sq unit}$$

(v) Given, $y = (x-1)(x-2)(x-3)$

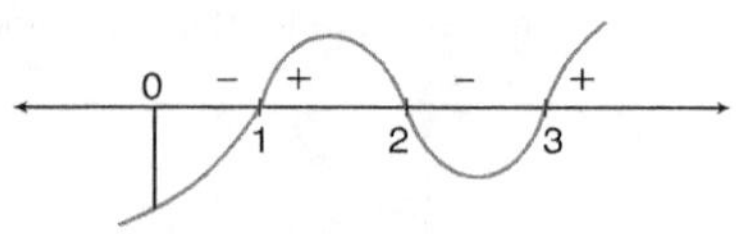


$\therefore$ Required area $= \int_0^3 y\,dx$

$$= -\int_0^1 y\,dx + \int_1^2 y\,dx - \int_2^3 y\,dx$$

$$= -\int_0^1 (x-1)(x-2)(x-3)\,dx + \int_1^2 (x-1)(x-2)(x-3)\,dx - \int_2^3 (x-1)(x-2)(x-3)\,dx$$

$$= -\int_0^1 (x^3 - 6x^2 + 11x - 6)\,dx + \int_1^2 (x^3 - 6x^2 + 11x - 6)\,dx - \int_2^3 (x^3 - 6x^2 + 11x - 6)\,dx$$

$$= -\left[\frac{x^4}{4} - \frac{6x^3}{3} + \frac{11x^2}{2} - 6x\right]_0^1 + \left[\frac{x^4}{4} - \frac{6x^3}{3} + \frac{11x^2}{2} - 6x\right]_1^2 - \left[\frac{x^4}{4} - \frac{6x^3}{3} + \frac{11x^2}{2} - 6x\right]_2^3$$

$$= -\left[\frac{1}{4} - 2 + \frac{11}{2} - 6\right] + \left[(4 - 16 + 22 - 12) - \left(\frac{1}{4} - 2 + \frac{11}{2} - 6\right)\right] - \left[\left(\frac{81}{4} - 54 + \frac{99}{2} - 18\right) - (4 - 16 + 22 - 12)\right]$$

$$= -\left(-\frac{9}{4}\right) + \left[(-2) - \left(-\frac{9}{4}\right)\right] - \left[\left(\frac{9}{4}\right) - (-2)\right]$$

$$= \frac{9}{4} + \left[-2 + \frac{9}{4}\right] + \left[\frac{9}{4} - 2\right]$$

$$= \frac{11}{4} \text{ sq units}$$

(vi) Given curves are

$$x = -2y^2 \quad \text{...(i)}$$

and $$x = 1 - 3y^2 \quad \text{...(ii)}$$

To find points of intersection, we solve Eqs. (i) and (ii) simultaneously, we get

$$-2y^2 = 1 - 3y^2 \Rightarrow y^2 = 1$$

$\Rightarrow$ $y = \pm 1$

$\therefore$ Points of intersection are $(-2, 1)$ and $(-2, -1)$.

Clearly, curve $x = -2y^2$ represents left hand parabola with vertex (0, 0)

and curve $x = 1 - 3y^2$ represents left hand parabola with vertex (1, 0) as shown in the figure.

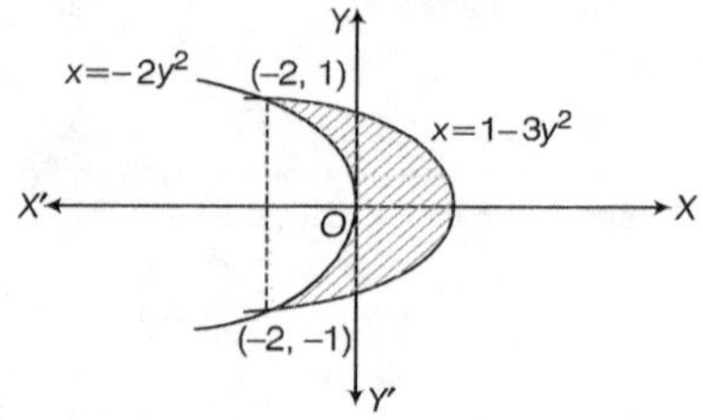


$\therefore$ Required area

$$= 2\left[\int_0^1 x \cdot dy \text{ (for } x = 1 - 3y^2) - \int_0^1 x\,dy \text{ (for } x = -2y^2)\right]$$

$$= 2\left[\int_0^1 (1 - 3y^2)\,dy - \int_0^1 (-2y^2)\,dy\right]$$

$$= 2\left[\int_0^1 (1 - 3y^2 + 2y^2)\,dy\right]$$

$$= 2\left[\int_0^1 (1 - y^2)\,dy\right] = 2\left[y - \frac{y^3}{3}\right]_0^1$$

$$= 2\left[\left(1 - \frac{1}{3}\right) - 0\right] = \frac{4}{3} \text{ sq units}$$

**3.** (d) We have,

$$\int_1^b f(x)\,dx = \sqrt{b^2 + 1} - \sqrt{2}$$

On differentiating w.r.t. $b$, we get

$$f(b) = \frac{b}{\sqrt{b^2 + 1}}$$

$\Rightarrow$ $$f(x) = \frac{x}{\sqrt{x^2 + 1}}$$

**4.** (a) Here, area of rectangle $= (2a\cos\theta)(2b\sin\theta)$

$= (2ab\sin 2\theta)$

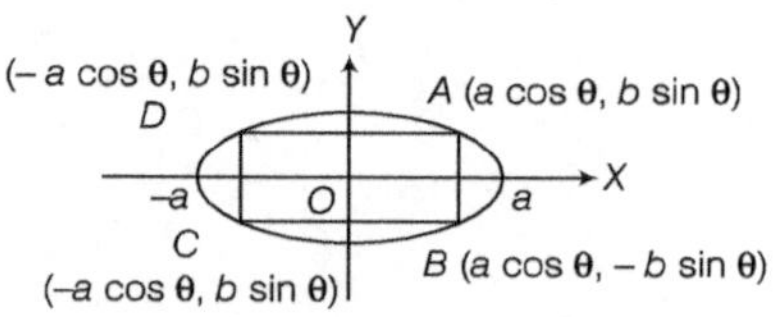


which is maximum, when $\sin 2\theta = 1$.

$\therefore$ $(\text{Area})_{max} = 2ab$

**5.** (a) The area enclosed by $y = \log(x + e)$ and the coordinate axes is shaded in the figure given below.

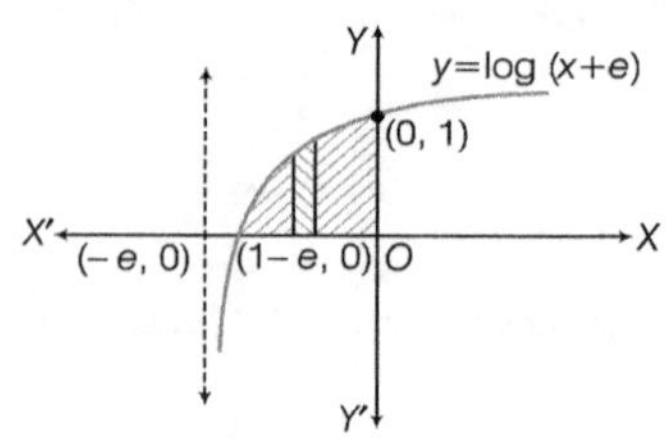

$\therefore$ Required area $= \int_{1-e}^{0} y\,dx$

$$= \int_{1-e}^{0} \log(x+e)\,dx$$

$$= [x\log(x+e)]_{1-e}^{0} - \int_{1-e}^{0} \frac{x}{x+e}\,dx$$

$$= [x\log(x+e)]_{1-e}^{0} - [x - e\log(x+e)]_{1-e}^{0}$$

$$= 0 - [(0-e) - (1-e) + 0]$$

$$= 1 \text{ sq unit}$$

**6.** (d) Given curves

$$y = |x-1| \quad \text{...(i)}$$

and $\quad y = 3 - |x| \quad$ ...(ii)

The graph of the curves is given below

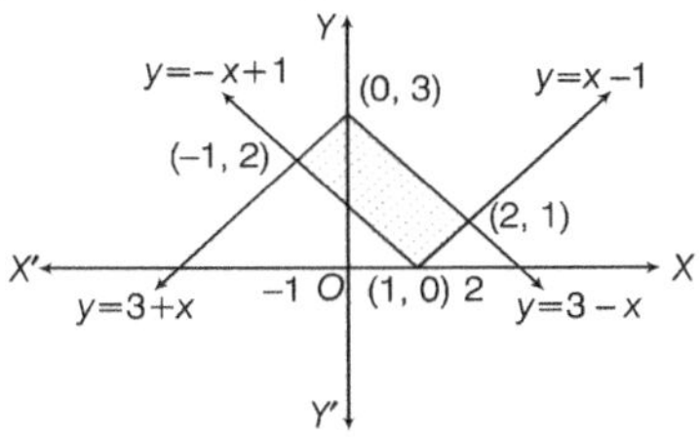


$\therefore$ The required area

$$= \int_{-1}^{0} \{(3+x) - (-x+1)\} + \int_{0}^{1} \{(3-x) - (-x+1)\}\,dx + \int_{1}^{2} \{(3-x) - (x-1)\}\,dx$$

$$= \int_{-1}^{0} (2+2x)\,dx + \int_{0}^{1} 2\,dx + \int_{1}^{2} (4-2x)\,dx$$

$$= [2x + x^2]_{-1}^{0} + [2x]_0^1 + [4x - x^2]_1^2$$

$$= 4 \text{ sq units}$$

**7.** (a) Let $A$ be the required area, then

$$A = \int_1^3 |x-2|\,dx$$

$$= \int_1^2 -(x-2)\,dx + \int_2^3 (x-2)\,dx$$

$$= \left[2x - \frac{x^2}{2}\right]_1^2 + \left[\frac{x^2}{2} - 2x\right]_2^3$$

$$= \frac{1}{2} + \frac{1}{2} = 1 \text{ sq unit}$$

**8.** (d) Given curves are

$$y^2 = 4x \quad \text{...(i)}$$

and $\quad x^2 = 4y \quad$ ...(ii)

Clearly, $y^2 = 4x$ and $x^2 = 4y$ intersect at (4, 4).

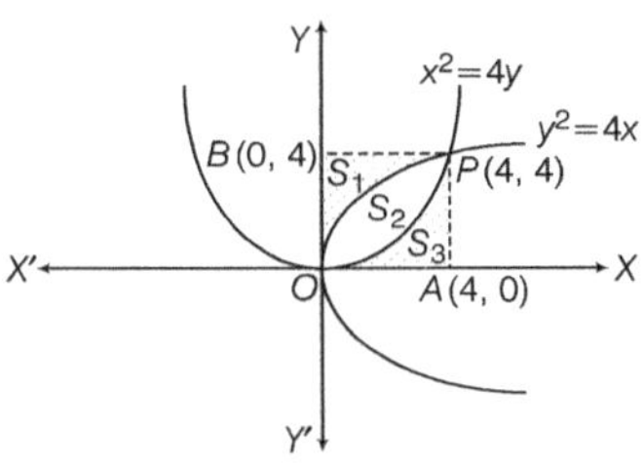


$\therefore \quad S_2 + S_3 = \int_0^4 \sqrt{4x}\,dx$

$\Rightarrow \quad S_2 + S_3 = 2\int_0^4 \sqrt{x}\,dx = \frac{4}{3}[x^{3/2}]_0^4 = \frac{32}{3}$ sq units

and $\quad S_3 = \int_0^4 \frac{x^2}{4}\,dy = \frac{1}{4}\left[\frac{x^3}{3}\right]_0^4 = \frac{16}{3}$ sq units

Thus, we have

$$S_2 + S_3 = \frac{32}{3} \text{ and } S_3 = \frac{16}{3}$$

$\Rightarrow \quad S_2 = \frac{16}{3}$ sq units

$\because \quad S_1 + S_2 + S_3 =$ Area of square $OAPB$

$= 4 \times 4 = 16$ sq units

$\therefore \quad S_1 = \frac{16}{3}$ sq units $\quad \left[\because S_2 = S_3 = \frac{16}{3} \text{ sq units}\right]$

Hence, $S_1 : S_2 : S_3 = 1 : 1 : 1$

**9.** (c) Let $A$ be the area enclosed between the given curves

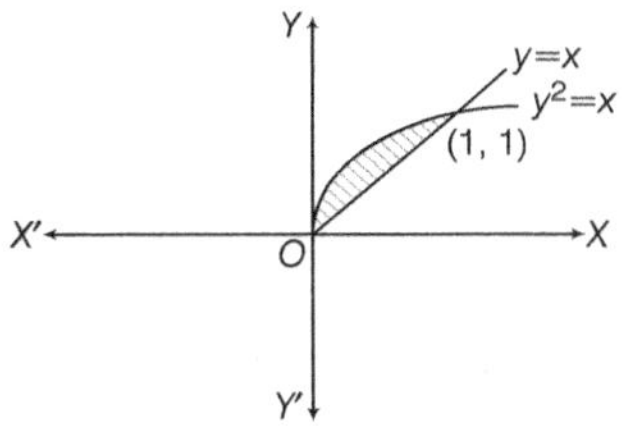


then, $\quad A = \int_0^1 (\sqrt{x} - x)\,dx$

$$= \left[\frac{2}{3}x^{3/2} - \frac{x^2}{2}\right]_0^1 = \frac{1}{6} \text{ sq unit}$$

**10.** (a) Given equation of parabola is

$$y^2 - 4y - x + 5 = 0$$

The equation of tangent at (2, 3) is

$$3y - 2(y+3) - \frac{(x+2)}{2} + 5 = 0$$

$\Rightarrow \quad 2y - x - 4 = 0$

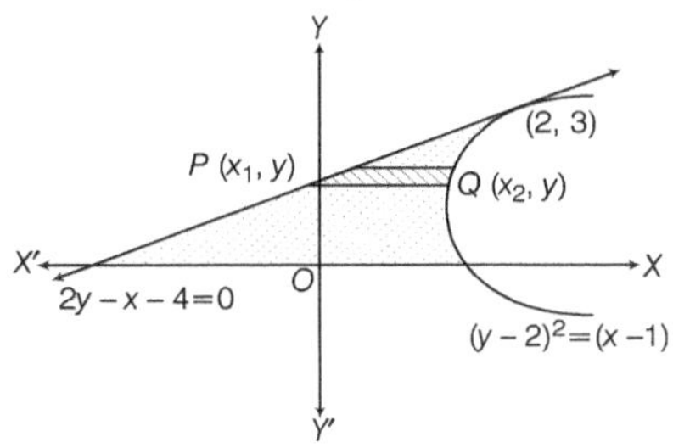


$\therefore$ Required area is given by

$$= \int_0^3 (x_2 - x_1)\,dx$$

$$= \int_0^3 [\{(y-2)^2 + 1\} - \{2y - 4\}]\,dy$$

$$= \int_0^3 (y^2 - 6y + 9)\,dy = \int_0^3 (3-y)^2\,dy$$

$$= -\left[\frac{(3-y)^3}{3}\right]_0^3$$

$$= 9 \text{ sq units}$$

**11.** (d) We have,

$$\int_{\pi/4}^{\beta} f(x)\,dx = \beta\sin\beta + \frac{\pi}{4}\cos\beta + \sqrt{2}\,\beta$$

On differentiating both sides w.r.t. $\beta$, we get

$$f(\beta) = \beta\cos\beta + \sin\beta - \frac{\pi}{4}\sin\beta + \sqrt{2}$$

$\therefore \quad f\left(\frac{\pi}{2}\right) = 1 - \frac{\pi}{4} + \sqrt{2}$

**12.** (a) Same as Q. 2 part (vi) of this DPP.

**13.** (a) Given curves are

$$y = \cos x \qquad \ldots(i)$$

and

$$y = \sin x \qquad \ldots(ii)$$

The graph of curves (i) and (ii) is shown below.

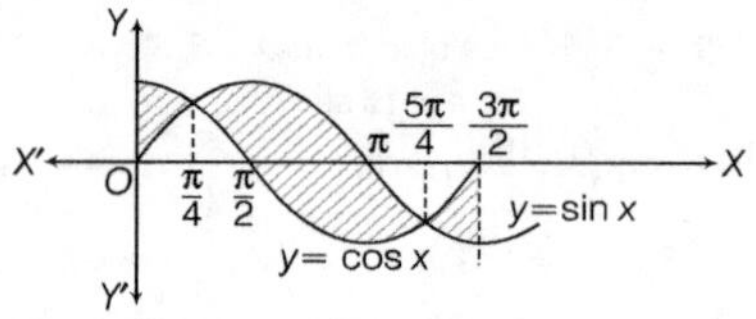


$\therefore$ Required area

$$= \int_0^{\pi/4} (\cos x - \sin x)\, dx + \int_{\pi/4}^{5\pi/4} (\sin x - \cos x)\, dx + \int_{5\pi/4}^{3\pi/2} (\cos x - \sin x)\, dx$$

$$= [\sin x + \cos x]_0^{\pi/4} + [-\cos x - \sin x]_{\pi/4}^{5\pi/4} + [\sin x + \cos x]_{5\pi/4}^{3\pi/2}$$

$$= (4\sqrt{2} - 2) \text{ sq units}$$

**14.** (a) Given curves are

$$y = \sqrt{x} \Rightarrow y^2 = x \qquad \ldots(i)$$

and

$$2y - x + 3 = 0 \qquad \ldots(ii)$$

Clearly, $y^2 = x$, represents a parabola having vertex (0, 0) and $2y - x + 3 = 0$ represents a straight line.

To find point of intersection, solving Eqs. (i) and (ii), we get

$$\left(\frac{x-3}{2}\right)^2 = x$$

$$\Rightarrow (x-3)^2 = 4x$$

$$\Rightarrow x^2 + 9 - 6x = 4x$$

$$\Rightarrow x^2 - 10x + 9 = 0$$

$$\Rightarrow x^2 - 9x - x + 9 = 0$$

$$\Rightarrow x(x-9) - 1(x-9) = 0$$

$$\Rightarrow (x-9)(x-1) = 0$$

$$\Rightarrow x = 1, 9$$

When $x = 1$, $y = \pm 1$

When $x = 9$, $y = \pm 3$

$\therefore$ Points of intersection are $(1, -1)$ and $(9, 3)$.

The graph of curve (i) and (ii) is shown below.

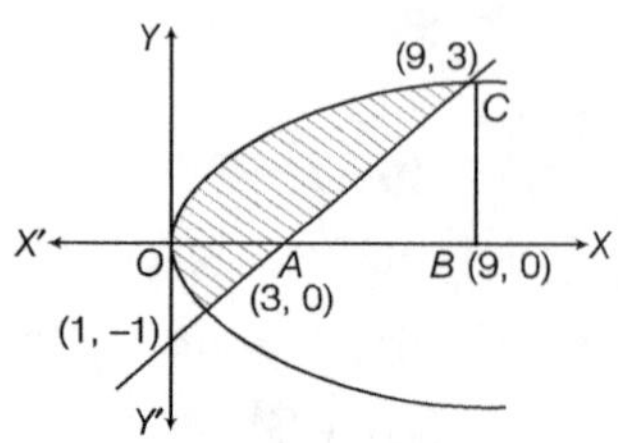


$\therefore$ Required area $= \int_0^9 y\,dx$ (for $y^2 = x$) $- \int_3^9 y\,dx$ $\left(\text{for } y = \frac{x-3}{2}\right)$

$$= \int_0^9 \sqrt{x}\, dx - \int_3^9 \left(\frac{x-3}{2}\right) dx$$

$$= \frac{2}{3}[x^{3/2}]_0^9 - \frac{1}{2}\left[\frac{x^2}{2} - 3x\right]_3^9$$

$$= \frac{2}{3}(27) - \frac{1}{2}\left(\frac{81}{2} - 27 - \frac{9}{2} + 9\right)$$

$$= 18 - \frac{1}{2}\left(\frac{72}{2} - 18\right) = 18 - \frac{1}{2}\left(\frac{36}{2}\right)$$

$$= 18 - 9 = 9 \text{ sq units}$$

**15.** (c) Given curve,

$$f(x) = x^2 + bx - b$$

$$\Rightarrow f'(x) = 2x + b$$

$\therefore$ Slope of tangent $m = f'(x)$ at (1, 1) is

$$m = 2 + b$$

Equation of tangent

$$y - 1 = (2 + b)(x - 1)$$

$$\Rightarrow y - 1 = (2 + b)x - (2 + b)$$

$$\Rightarrow (2 + b)x - y + 1 - 2 - b = 0$$

$$\Rightarrow (2 + b)x - y - 1 - b = 0$$

$$\Rightarrow \frac{x}{\frac{1+b}{2+b}} + \frac{y}{-(1+b)} = 1 \qquad \ldots(i)$$

The graph of Eqs. (i) is as follows

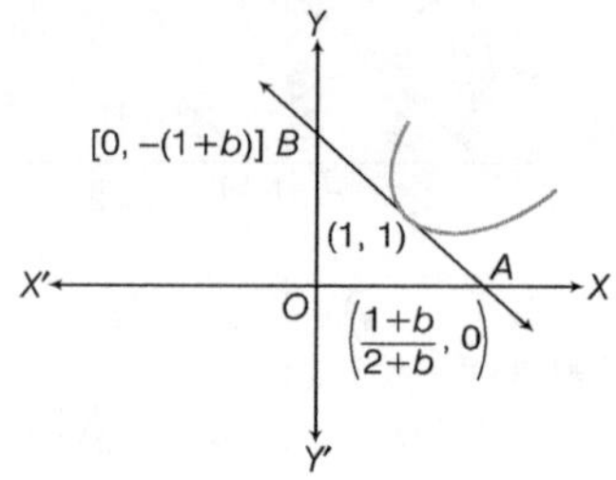


$\because$ Area of $\Delta OAB = 2$ sq units

$$\Rightarrow \frac{1}{2} \times \frac{(1+b)}{2+b} \times \{-(1+b)\} = 2$$

$$\Rightarrow \frac{-(1+b)^2}{2+b} = 4$$

$$\Rightarrow -(1 + b^2 + 2b) = 8 + 4b$$

$$\Rightarrow b^2 + 6b + 9 = 0$$

$$\Rightarrow (b+3)^2 = 0$$

$$\therefore b = -3$$

**16.** (a) The region bounded by given curves on the right side of $x = 1$ is shown.

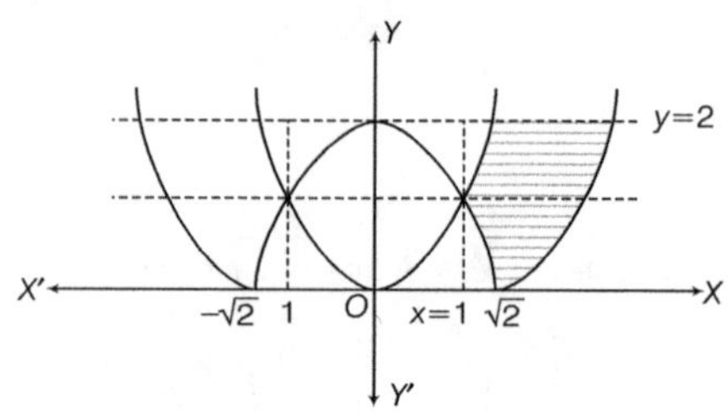


Thus, the required area

$$= \int_1^{\sqrt{2}} \{x^2 - (2 - x^2)\}\, dx + \int_{\sqrt{2}}^2 \{4 - x^2\}\, dx$$

$$= \int_1^{\sqrt{2}} (2x^2 - 2)\, dx + \int_{\sqrt{2}}^2 (4 - x^2)\, dx$$

$$= \left[2\frac{x^3}{3} - 2x\right]_1^{\sqrt{2}} + \left[4x - \frac{x^3}{3}\right]_{\sqrt{2}}^2$$

$$= \left(\frac{4\sqrt{2}}{3} - 2\sqrt{2}\right) - \left(\frac{2}{3} - 2\right) + \left(8 - \frac{8}{3}\right) - \left(4\sqrt{2} - \frac{2\sqrt{2}}{3}\right)$$

$$= \left(-4\sqrt{2} + \frac{20}{3}\right) = \left(\frac{20 - 12\sqrt{2}}{3}\right) \text{ sq units}$$

**17.** (b) Let $f(x) = y = \sin x + \cos x$, when $0 \le x \le \frac{\pi}{2}$

and $g(x) = y = |\cos x - \sin x|$

$$= \begin{cases} \cos x - \sin x, \text{ for } 0 \le x \le \frac{\pi}{4} \\ \sin x - \cos x, \text{ for } \frac{\pi}{4} \le x \le \frac{\pi}{2} \end{cases}$$

The graph of these two functions could be plotted as

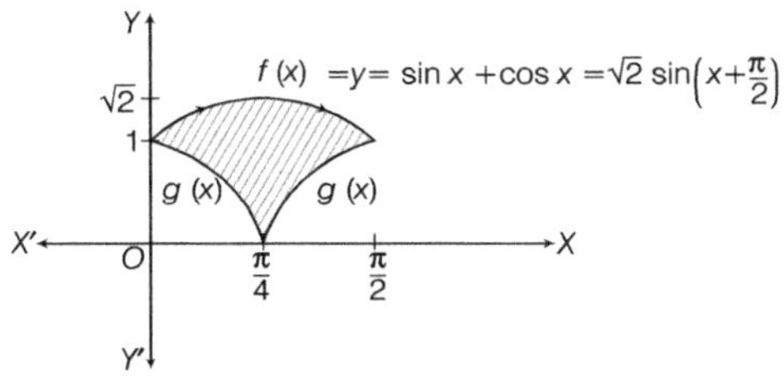


$\therefore$ Area bounded $= \int_0^{\pi/4} \{(\sin x + \cos x) - (\cos x - \sin x)\}\, dx$

$$+ \int_{\pi/4}^{\pi/2} \{(\sin x + \cos x) - (\sin x - \cos x)\}\, dx$$

$$= \int_0^{\pi/4} 2\sin x\, dx + \int_{\pi/4}^{\pi/2} 2\cos x\, dx$$

$$= -2\,[\cos x]_0^{\pi/4} + 2\,[\sin x]_{\pi/4}^{\pi/2}$$

$$= 4 - 2\sqrt{2}$$

$$= 2\sqrt{2}\,(\sqrt{2} - 1) \text{ sq units}$$

**18.** (a) Given curves are $y = ax^2$ ...(i)

and $x = ay^2$ ...(ii)

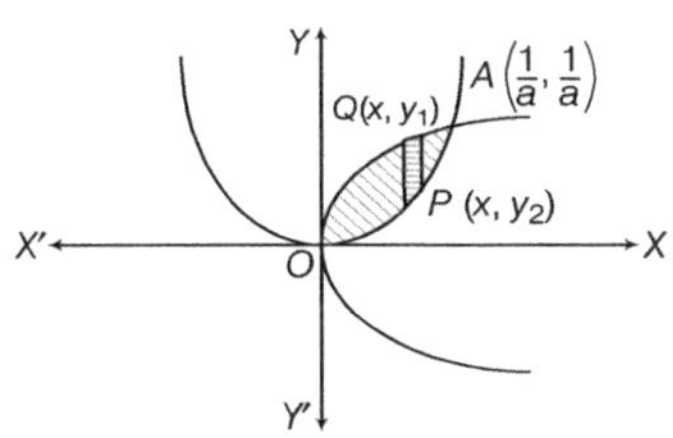


We have, $\int_0^{1/a} (y_2 - y_1)\, dx = 1$

$\Rightarrow$ $\int_0^{1/a} \left(\sqrt{\frac{x}{a}} - ax^2\right) dx = 1$

$\Rightarrow$ $\left[\frac{2}{3\sqrt{a}}\, x^{3/2} - \frac{ax^3}{3}\right]_0^{1/a} = 1$

$\Rightarrow$ $\frac{2}{3\sqrt{a}} \times \frac{1}{a^{3/2}} - \frac{a}{3} \times \frac{1}{a^3} = 1$

$\Rightarrow$ $\frac{2}{3a^2} - \frac{1}{3a^2} = 1$

$\Rightarrow$ $\frac{1}{3a^2} = 1 \quad \Rightarrow \quad a^2 = \frac{1}{3}$

$\Rightarrow$ $a = \frac{1}{\sqrt{3}}$ $[\because a > 0]$

**19.** (c) Given curves are

$$y = \log_e x \text{ and } y = (\log_e x)^2$$

On solving these two curves simultaneously, we have

$$\log_e x = (\log_e x)^2$$

$\Rightarrow$ $\log_e x = 0, 1$

$\Rightarrow$ $x = 1$ and $x = e$

Also, for $1 < x < e$, $0 < \log_e x < 1$

$\Rightarrow$ $\log_e x > (\log_e x)^2$

For $x > e$, $\log_e x < (\log_e x)^2$

$\Rightarrow$ $y = (\log_e x)^2 > 0$ for all $x > 0$

and when $x \to 0$, $(\log_e x)^2 \to \infty$

Using the above information, we can plot the graph of the functions.

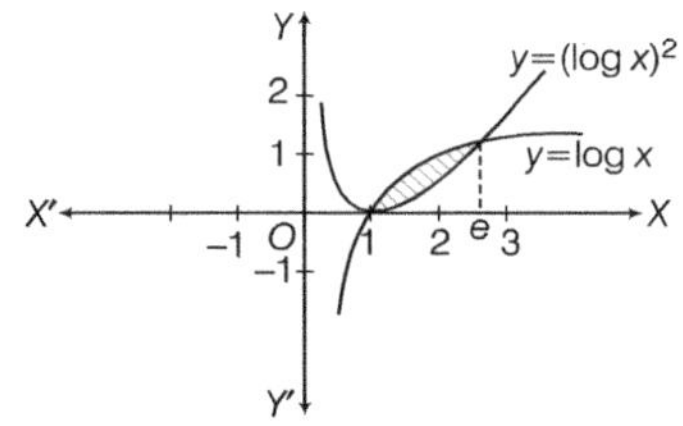


Then, required area

$$= \int_1^e [\log x - (\log x)^2]\, dx$$

$$= \int_1^e \log x\, dx - \int_1^e (\log x)^2\, dx$$

$$= [x \log x - x]_1^e - [x(\log x)^2]_1^e + \int_1^e \frac{2\log x}{x}\, x\, dx$$

$$= 1 - e + 2\,[x \log x - x]_1^e$$

$$= (3 - e) \text{ sq unit}$$

**20.** (b) Given, $y = -\sqrt{-x}$

$\Rightarrow$ $y^2 = -x$

and $x = -\sqrt{-y}$

$\Rightarrow$ $x^2 = -y$

where, $x$ and $y$ both are negative.

We know that area enclosed between the curves $y^2 = 4ax$ and $x^2 = 4by$ is $\frac{16ab}{3}$ sq units.

$\therefore$ $A = \frac{16ab}{3}$

$$= \frac{16}{3} \times \frac{1}{4} \times \frac{1}{4} \qquad \left[\because a = b = \frac{1}{4}\right]$$

$$= \frac{1}{3} \text{ sq unit}$$

**21.** (b) Given curves $y = x^2$ ...(i)

and $y^2 = |x|$ ...(ii)

The graph of the curve (i) and (ii) shown below.

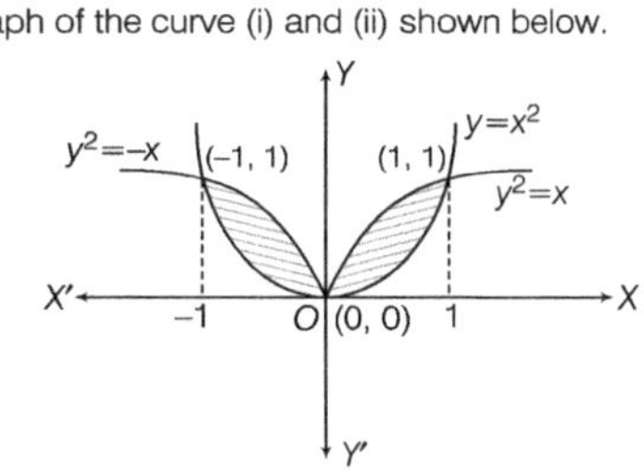


Clearly, both the curves intersect at (−1, 1), (0, 0) and (1, 1).

$\therefore$ Required area $= 2\int_0^1 [y \text{ (for } y^2 = |x|) - y \text{ (for } y = x^2)]\, dx$

$$= 2\int_0^1 [\sqrt{|x|} - x^2]\, dx$$

$$= 2\left[\frac{2}{3}|x|^{3/2} - \frac{x^3}{3}\right]_0^1$$

$$= 2\left[\frac{2}{3} - \frac{1}{3}\right] = \frac{2}{3} \text{ sq unit}$$

**22.** (b) Given equation of curve

$$y = x(1 - \log_e x)$$

$\therefore$ Required area $= \int_{1/e}^{e} x(1 - \log x)\,dx$

$$= \int_{1/e}^{e} (x - x\log x)\,dx$$

$$= \int_{1/e}^{e} x\,dx - \int_{1/e}^{e} x\log x\,dx$$

$$= \left[\frac{x^2}{2}\right]_{1/e}^{e} - \left[\frac{x^2}{2}\left(\log x - \frac{1}{2}\right)\right]_{1/e}^{e}$$

$$= \left[\frac{x^2}{2}\left(\frac{3}{2} - \log_e x\right)\right]_{1/e}^{e} = \frac{e^2}{4} - \frac{1}{2e^2}\left(\frac{5}{2}\right)$$

$$= \frac{e^2}{4} - \frac{5}{4e^2} = \left(\frac{e^2 - 5e^{-2}}{4}\right) \text{sq units}$$

**23.** (a,b,c,d) Required area $= \int_{2n\pi}^{2n\pi + 2\pi} \sin x\,dx$

$= \int_0^{2\pi} \sin x\,dx$ [option a]

$= 2\int_0^{\pi} \sin x\,dx$ [option b]

$\left[\because \int_0^{2a} f(x)\,dx = 2\int_0^{a} f(x)\,dx, \text{ if } f(2a - x) = f(x)\right]$

$= 4\int_0^{\pi/2} \sin x\,dx$ [option c]

$= 4[\cos x]_0^{\pi/2}$

$= 4(0 - 1) = -4$

$= 4$ sq units [$\because$ area cannot be negative] [option d]

**24.** (b,d) The two curves meet at $mx = x - x^2$ or $x^2 = x(1 - m)$.

$\therefore$ $x = 0, 1 - m$

Now, $A = \int_0^{1-m} (x - x^2 - mx)\,dx$

$$= \left[(1 - m)\frac{x^2}{2} - \frac{x^3}{3}\right]_0^{1-m} = \frac{9}{2}, \text{ if } m < 1$$

$\Rightarrow (1 - m)^3\left[\frac{1}{2} - \frac{1}{3}\right] = \frac{9}{2}$

$\Rightarrow (1 - m)^3 = 27$

$\Rightarrow m = -2$

But if $(m > 1)$, then $(1 - m)$ is negative.

Now, $\left[(1 - m)\frac{x^2}{2} - \frac{x^3}{3}\right]_{1-m}^{0} = \frac{9}{2}$

$\Rightarrow -(1 - m)^3\left(\frac{1}{2} - \frac{1}{3}\right) = \frac{9}{2}$

$\Rightarrow -(1 - m)^3 = 27$

$\therefore m = 4$

**25.** (a,b) Given, $C_1 : y = \frac{1}{1 + x^2}$ ...(i)

and $C_2 : y = \frac{x^2}{2}$ ...(ii)

The graph of curves $C_1$ and $C_2$ is as follows

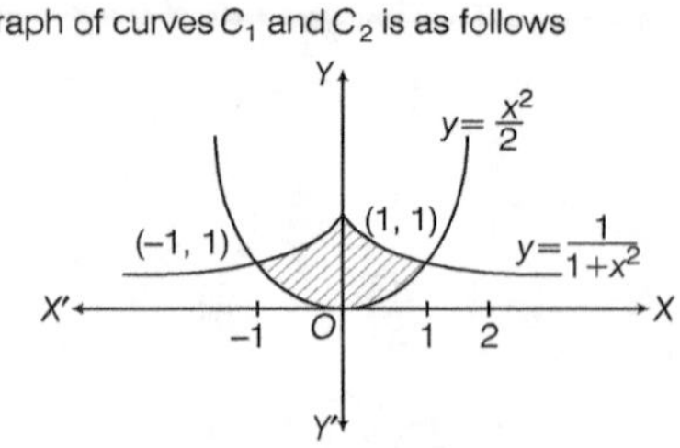


Area between $C_1$ and $y = 0$

$$= 2\int_0^{\infty} y\,dx$$

$$= 2\int_0^{\infty} \frac{1}{1 + x^2}\,dx$$

$$= 2[\tan^{-1} x]_0^{\infty}$$

$$= 2 \cdot \frac{\pi}{2} = \pi \text{ sq units}$$

The points of intersection of curves $C_1$ and $C_2$ is (1, 1) and (–1, 1).

$\therefore$ Required area between $C_1$ and $C_2$

$$= 2\int_0^{1}\left(\frac{1}{1 + x^2} - \frac{x^2}{2}\right)dx$$

$$= 2\left[\int_0^{1} \frac{1}{1 + x^2}\,dx - \int_0^{1} \frac{x^2}{2}\,dx\right]$$

$$= 2\left[[\tan^{-1} x]_0^1 - \frac{1}{2}\left[\frac{x^3}{3}\right]_0^1\right]$$

$$= 2\left[\frac{\pi}{4} - \frac{1}{6}\right] = \left(\frac{\pi}{2} - \frac{1}{3}\right) \text{sq units}$$

**26.** (a,d) $(|x|, |y|) \le 1 \Rightarrow |x| \le 1$ and $|y| \le 1$

which represent square bounded by $x = \pm 1$ and $y = \pm 1$.

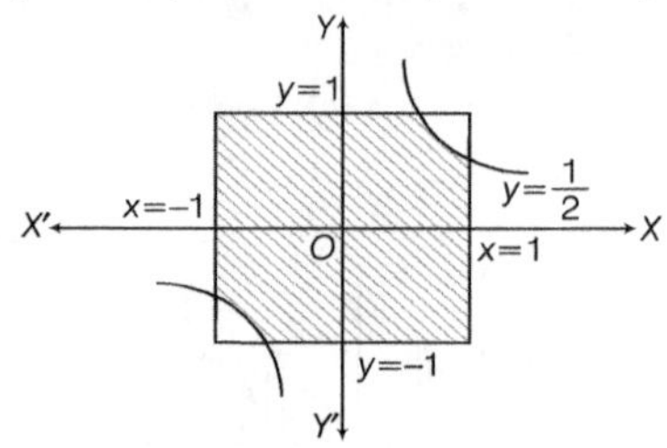


Now, shaded area $= 2\int_{1/2}^{1}\left(1 - \frac{1}{2x}\right)dx$

$$= 2\left[x - \frac{1}{2}\log x\right]_{1/2}^{1}$$

$$= 2\left[(1 - 0) - \left(\frac{1}{2} - \frac{1}{2}\log\frac{1}{2}\right)\right]$$

$$= (1 - \log 2)$$

$\therefore$ Required area $= 4 - (1 - \log 2)$

$= (3 + \log 2)$ sq units

Clearly, which is less than $4\log 3$.

**27.** (c) Given, $f(x) = |x|$, $g(x) = |x - 1|$ and $h(x) = |x + 1|$

Graph for $f(x)$, $g(x)$ and $h(x)$ is shown below.

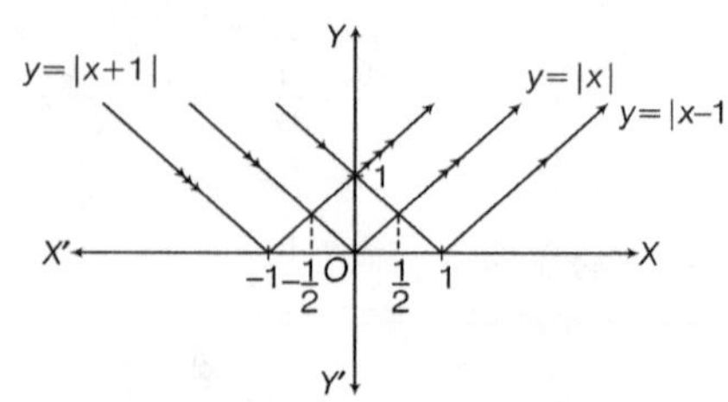


(i) $\text{Min}\{f(x), g(x)\} = \begin{cases} -x, & x \le 0 \\ x, & 0 \le x \le \frac{1}{2} \\ 1 - x, & \frac{1}{2} \le x \le 1 \\ x - 1, & x \ge 1 \end{cases}$

$\therefore$ Required area $= \int_0^{1/2} x\,dx + \int_{1/2}^{1} (1-x)\,dx$

$$= \left[\frac{x^2}{2}\right]_0^{1/2} + \left[x - \frac{x^2}{2}\right]_{1/2}^{1}$$

$$= \frac{1}{8} + \left(1 - \frac{1}{2} - \frac{1}{2} + \frac{1}{8}\right) = \frac{1}{4}$$

(ii) $\text{Min}\{f(x), h(x)\} = \begin{cases} -(x+1), & x \le -1 \\ x+1, & -1 \le x \le -\frac{1}{2} \\ -x, & -\frac{1}{2} \le x \le 0 \\ x, & 0 \le x \le \frac{1}{2} \end{cases}$

$\therefore$ Required area $= \int_{-1}^{-1/2} (x+1)\,dx + \int_{-1/2}^{0} (-x)\,dx$

$$= \left[\frac{x^2}{2} + x\right]_{-1}^{-1/2} - \left[\frac{x^2}{2}\right]_{-1/2}^{0}$$

$$= \left(\frac{1}{8} - \frac{1}{2} - \frac{1}{2} + 1\right) - \left(0 - \frac{1}{8}\right)$$

$$= \frac{1}{4} \text{ sq unit}$$

(iii) $\text{Min}\{f(x), g(x), h(x)\} = \begin{cases} -(x+1), & x \le -1 \\ (x+1), & -1 \le x \le -\frac{1}{2} \\ (-x), & -\frac{1}{2} \le x \le 0 \\ x, & 0 \le x \le \frac{1}{2} \\ 1-x, & \frac{1}{2} \le x \le 1 \\ x-1, & x \ge 1 \end{cases}$

$\therefore$ Required area

$$= \int_{-1}^{-1/2} (x+1)\,dx + \int_{-1/2}^{0} (-x)\,dx + \int_0^{1/2} x\,dx + \int_{1/2}^{1} (1-x)\,dx$$

$$= \left[\frac{x^2}{2} + x\right]_{-1}^{-1/2} - \left[\frac{x^2}{2}\right]_{-1/2}^{0} + \left[\frac{x^2}{2}\right]_0^{1/2} + \left[x - \frac{x^2}{2}\right]_{1/2}^{1}$$

$$= \left\{\frac{1}{8} - \frac{1}{2} - \frac{1}{2} + 1\right\} - \left\{0 - \frac{1}{8}\right\} + \left\{\frac{1}{8} - 0\right\} + \left\{1 - \frac{1}{2} - \frac{1}{2} + \frac{1}{8}\right\}$$

$$= \frac{1}{8} - \frac{1}{2} - \frac{1}{2} + 1 + \frac{1}{8} + \frac{1}{8} + 1 - \frac{1}{2} - \frac{1}{2} + \frac{1}{8}$$

$$= \frac{4}{8} = \frac{1}{2} \text{ sq unit}$$

(iv) The graph of $\min\{f(x), g(x), h(x)\}$ and $y = \frac{1}{2}$ could be plotted as

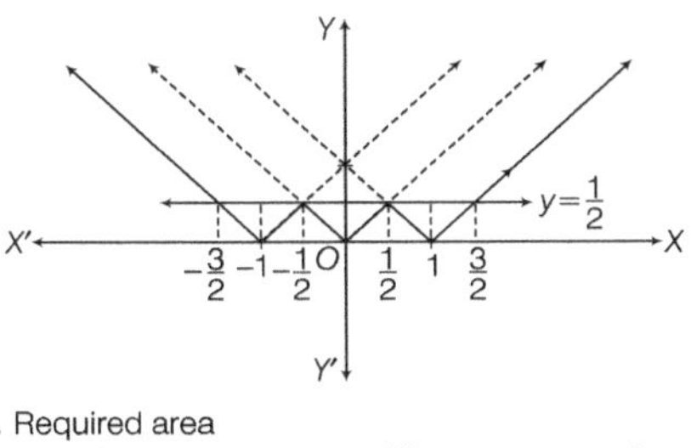


$\therefore$ Required area

$$= \int_{-3/2}^{-1} -(x+1)\,dx + \int_{-1}^{-1/2} (x+1)\,dx + \int_{-1/2}^{0} (-x)\,dx + \int_0^{1/2} x\,dx + \int_{1/2}^{1} (1-x)\,dx + \int_1^{3/2} (x-1)$$

$$= -\left[\frac{x^2}{2} - x\right]_{-3/2}^{-1} + \left[\frac{x^2}{2} + x\right]_{-1}^{-1/2} - \left[\frac{x^2}{2}\right]_{-1/2}^{0} + \left[\frac{x^2}{2}\right]_0^{1/2} + \left[x - \frac{x^2}{2}\right]_{1/2}^{1} + \left[\frac{x^2}{2} - x\right]_1^{3/2}$$

$$= -\left\{\frac{1}{2} - 1 - \frac{9}{8} + \frac{3}{2}\right\} + \left\{\frac{1}{8} - \frac{1}{2} - \frac{1}{2} + 1\right\} - \left\{-\frac{1}{8}\right\} + \left\{\frac{1}{8}\right\} + \left\{1 - \frac{1}{2} - \frac{1}{2} + \frac{1}{8}\right\} + \left\{\frac{9}{8} - \frac{3}{2} - \frac{1}{2} + 1\right\}$$

$$= \frac{1}{8} + \frac{1}{2} + \frac{1}{8} = \frac{3}{4} \text{ sq unit}$$

Hence, (i) $\rightarrow$ (q), (ii) $\rightarrow$ (q), (iii) $\rightarrow$ (r), (iv) $\rightarrow$ (s)

**Explanations for Q. Nos. (28–30)**

Here, $C_1 : |x + y| \le , C_2 : |x - y| \le 1$

$C_3 : |x| \le 1/2$

$C_4 = |y| \le 1/2$

$C_5 : x^2 + y^2 = \frac{1}{3}$, is shown as

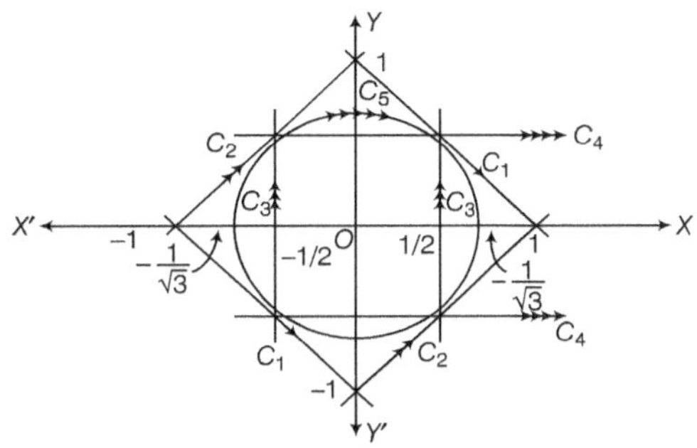


**28.** (c) Area bounded by $C_1, C_2$, which does not contain $C_5$

= Area of square − Area of circle

$$= (\sqrt{2})^2 - \pi\left(\frac{1}{\sqrt{3}}\right)^2 = 2 - \frac{\pi}{3}$$

**29.** (d) Area of curve $C_5$ which does not contain $C_3$ and $C_4$ is shown as

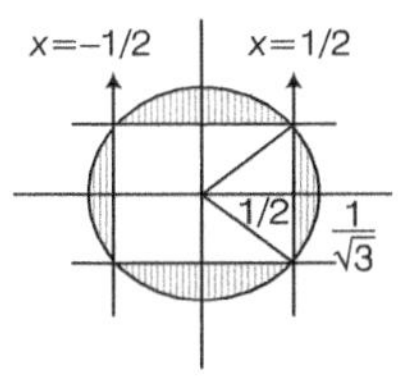


$$\therefore \quad \text{Area} = 4\left[2\int_{1/2}^{1/\sqrt{3}} \sqrt{\frac{1}{3} - x^2}\,dx\right]$$

$$= 8\left[\frac{\pi}{36} - \frac{1}{8\sqrt{3}}\right] = \left(\frac{2\pi}{9} - \frac{1}{\sqrt{3}}\right)$$

**30.** (a) Area bounded by $C_1$ and $C_2$ but not bounded by $C_3$ and $C_4$ is

$$= (\sqrt{2})^2 - (1)^2 = 1$$

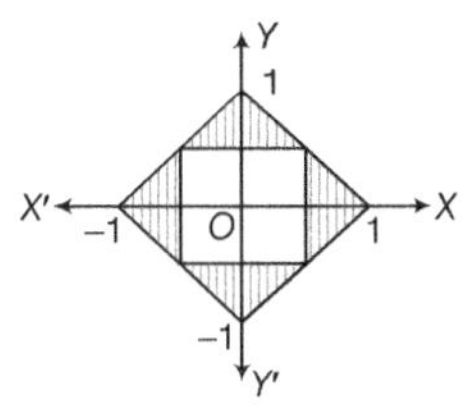

# DPP-2 Curve Sketching and Its Applications in Finding the Area

**1.** (b) Given curve $(y - x)^2 = x^3$

$$y - x = \pm x\sqrt{x}$$

$\Rightarrow$ $$y = x + x\sqrt{x} \quad \text{...(i)}$$

and $$y = x - x\sqrt{x} \quad \text{...(ii)}$$

which could be drawn as shown in figure.

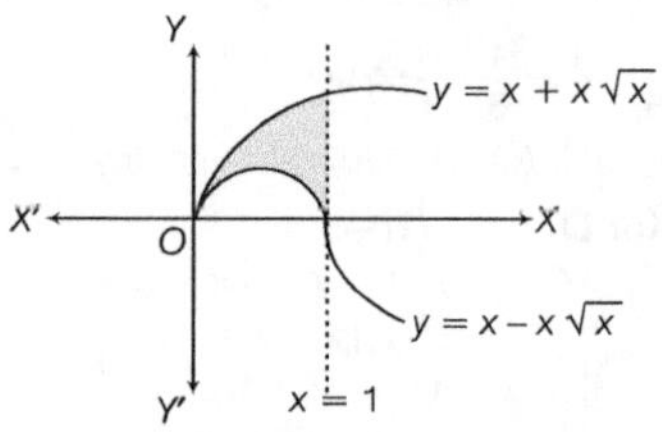


Hence, the required area

$$= \int_0^1 \{(x + x\sqrt{x}) - (x - x\sqrt{x})\}\, dx$$

$$= \int_0^1 (2x\sqrt{x})\, dx = 2\int_0^1 x^{3/2}\, dx$$

$$= 2 \times \frac{2}{5}[x^{5/2}]_0^1 = \frac{4}{5} \text{ sq unit}$$

**2.** (c) Given curve $5y^2 + 6xy + 2x^2 - 1 = 0$

$$y = \frac{-6x \pm \sqrt{36x^2 - 20(2x^2 - 1)}}{10}$$

$\Rightarrow$ $$y = \frac{-3x \pm \sqrt{5 - x^2}}{5} \quad \text{...(i)}$$

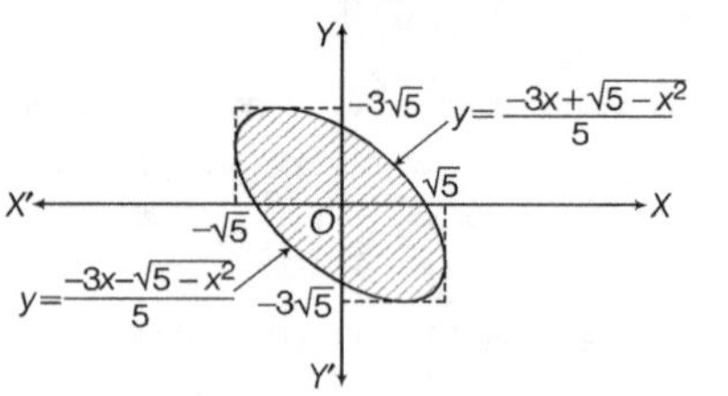


Since, $y$ is real, RHS of Eq. (i) is also real.

$\Rightarrow$ $$-\sqrt{5} \le x \le \sqrt{5}$$

If $x = -\sqrt{5}$, $\quad y = 3\sqrt{5}$

If $x = \sqrt{5}$, $\quad y = -3\sqrt{5}$

If $x = 0$, $\quad y = \pm \frac{1}{\sqrt{5}}$

If $y = 0$, $\quad x = \pm \frac{1}{\sqrt{2}}$

$\therefore$ Required area, $$A = \int_{-\sqrt{5}}^{\sqrt{5}} \left(\frac{-3x + \sqrt{5 - x^2}}{5} - \frac{-3x - \sqrt{5 - x^2}}{5}\right) dx$$

$$= \int_{-\sqrt{5}}^{\sqrt{5}} \left(\frac{-3x + \sqrt{5 - x^2} + 3x + \sqrt{5 - x^2}}{5}\right) dx$$

$$= \frac{2}{5}\int_{-\sqrt{5}}^{\sqrt{5}} \sqrt{5 - x^2}\, dx = \frac{4}{5}\int_0^{\sqrt{5}} \sqrt{5 - x^2}\, dx$$

Put $x = \sqrt{5}\sin\theta \Rightarrow dx = \sqrt{5}\cos\theta\, d\theta$

When $x = 0$, $\quad \theta = 0$

$x = \sqrt{5}$, $\quad \theta = \frac{\pi}{2}$

$\therefore$ $$A = \frac{4}{5}\int_0^{\pi/2} \sqrt{5 - 5\sin^2\theta}\, \sqrt{5}\cos\theta\, d\theta$$

$$= 4\int_0^{\pi/2} \cos^2\theta\, d\theta = 4 \times \frac{1}{2} \cdot \frac{\pi}{2} = \pi \text{ sq units}$$

**3.** (c) Let $x_1, x_2$ be the roots of equation

$$x^2 + 2x - 3 = mx + 1$$

$\Rightarrow$ $$x^2 + (2 - m)x - 4 = 0$$

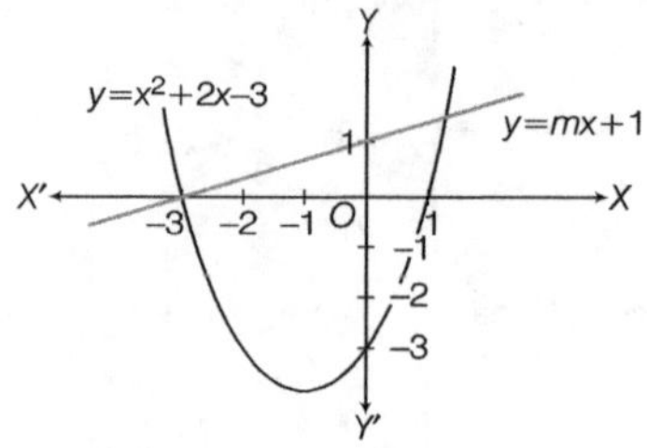


Now, $x_1 + x_2 = m - 2$

and $x_1 \cdot x_2 = -4$

$\therefore$ Required area $$(A) = \int_{x_1}^{x_2} [(mx + 1) - (x^2 + 2x - 3)]\, dx$$

$$= \left[(m - 2)\frac{x^2}{2} - \frac{x^3}{3} + 4x\right]_{x_1}^{x_2}$$

$$= (m - 2)\left(\frac{x_2^2 - x_1^2}{2}\right) - \frac{1}{3}(x_2^3 - x_1^3) + 4(x_2 - x_1)$$

$$= (x_2 - x_1)\left[\frac{(m - 2)^2}{2} - \frac{1}{3}\{(x_2 + x_1)^2 - x_1x_2\} + 4\right]$$

$$= \sqrt{(x_2 + x_1)^2 - 4x_1x_2}\left[\frac{(m - 2)^2}{2} - \frac{1}{3}\{(m - 2)^2 + 4\} + 4\right]$$

$$= \frac{\sqrt{(m - 2)^2 + 16}}{6}\left[\frac{1}{6}(m - 2)^2 + \frac{16}{3}\right]$$

$$= \frac{[(m - 2)^2 + 16]^{3/2}}{6}$$

$\Rightarrow A$ is minimum, if $m = 2$.

Hence, $A_{min} = \frac{32}{3}$ sq units

**4.** (b) Given, $$y = 2x^2 \quad \text{...(i)}$$

and $$y = x^4 - 2x^2 \quad \text{...(ii)}$$

Clearly, the points of intersection of curves (i) and (ii) are (0,0), (2,8) and (–2,8).

The graph curves (i) and (ii) could be plotted

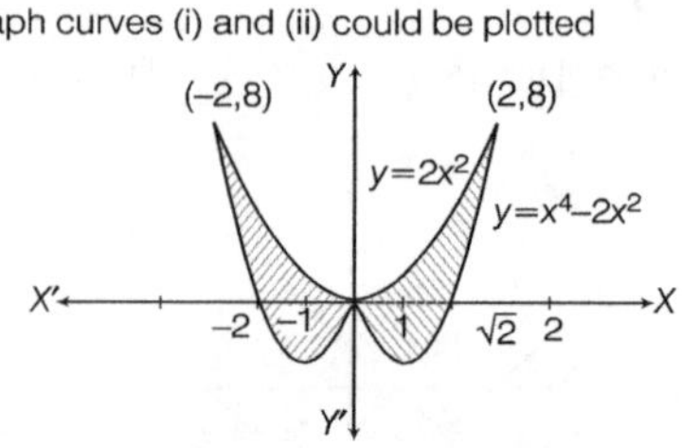


$\therefore$ Required area $$= 2\left[\int_0^2 (2x^2 - x^4 + 2x^2)\, dx\right]$$

$$= 2\left[\frac{2x^3}{3} - \frac{x^5}{5} + \frac{2x^3}{3}\right]_0^2 = 2\left[\frac{4}{3}x^3 - \frac{x^5}{5}\right]_0^2$$

$$= 2\left(\frac{32}{3} - \frac{32}{5}\right) = 2\left[\frac{64}{15}\right]$$

$$= \frac{128}{15} \text{ sq units}$$

**5.** (a) As we know,

$$[-\sin^2 x] = 0 \text{ or } -1$$

But $\sec^{-1}(0)$ is not defined.

$\Rightarrow \quad \sec^{-1}[-\sin^2 x] = \sec^{-1}(-1) = \pi$

Thus, to find area bounded between

$$y = x^2 \text{ and } y = \pi$$

i.e. when $x^2 = \pi$ or $(x = -\sqrt{\pi}$ to $x = \sqrt{\pi})$

$\therefore \quad \text{Area} = \int_{-\sqrt{\pi}}^{\sqrt{\pi}} (\pi - x^2)\,dx = \left[\pi x - \frac{x^3}{3}\right]_{-\sqrt{\pi}}^{\sqrt{\pi}}$

$$= \pi(\sqrt{\pi} + \sqrt{\pi}) - \frac{1}{3}(\pi\sqrt{\pi} + \pi\sqrt{\pi})$$

$$= \frac{4\pi}{3}\sqrt{\pi} \text{ sq units}$$

**6.** (c) Given, $\int_0^4 f(x)\,dx - \int_0^4 g(x)\,dx = 10$

$\Rightarrow \quad (A_1 + A_3 + A_4) - (A_2 + A_3 + A_4) = 10$

$$A_1 - A_2 = 10 \quad \text{...(i)}$$

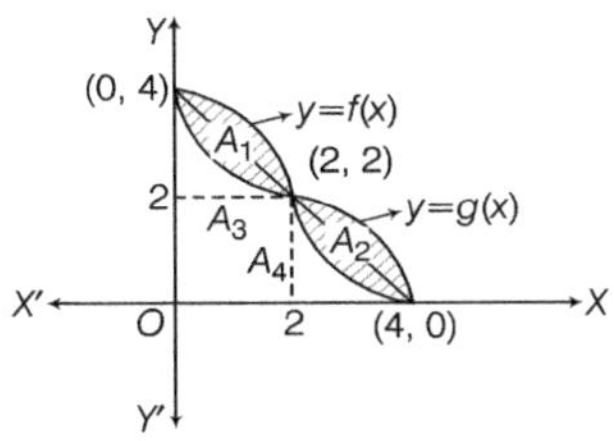


and $\quad \int_2^4 g(x)\,dx - \int_2^4 f(x)\,dx = 5$

$$(A_2 + A_4) - A_4 = 5$$

$$A_2 = 5 \quad \text{...(ii)}$$

On adding Eqs. (i) and (ii), we get

$$A_1 = 15 \text{ sq units}$$

**7.** (d) Given, $y = xe^{-x}$, $xy = 0$, $x = c$

$\Rightarrow \quad y' = e^{-x} - xe^{-x}$

$= (1 - x)e^{-x}$, increasing for $x < 1$

Now, $\quad y'' = -e^{x} - [e^{-x} - xe^{-x}]$

$$= e^{-x}[-1 - 1 + x]$$

$$= (x - 2)e^{-x}$$

For point of inflection

$$y'' = 0$$

$\Rightarrow \quad x = 2$

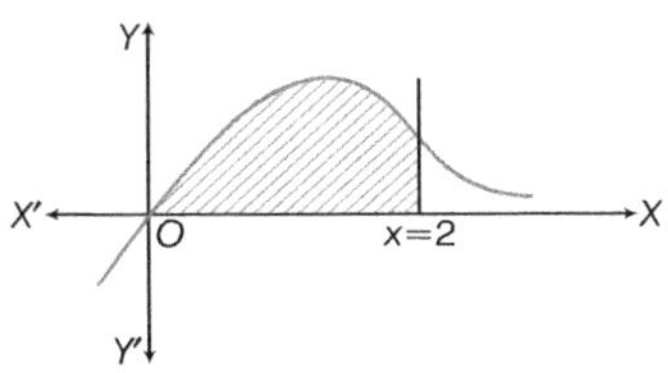


$\therefore \quad \text{Area} = \int_0^2 xe^{-x}\,dx$

$$= [-xe^{-x}]_0^2 + \int_0^2 e^{-x}\,dx$$

$$= [-2e^{-2}] - [e^{-x}]_0^2$$

$$= -2e^{-2} - (e^{-2} - 1)$$

$$= 1 - e^{-2} - 2e^{-2} = 1 - 3e^{-2}$$

**8.** (d) Given, $\quad C_1 : y = e^x \quad \text{...(i)}$

and $\quad C_2 : y = e^{a-x} \quad \text{...(ii)}$

On solving Eqs. (i) and (ii), we get

$$e^{2x} = e^a$$

$\Rightarrow \quad x = \frac{a}{2}$

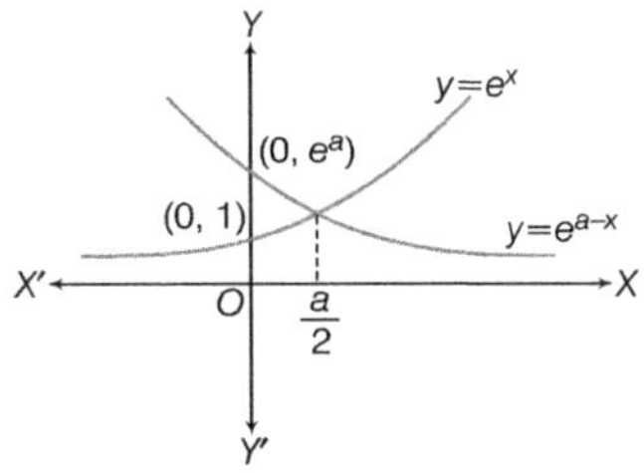


Now, $\quad S = \int_0^{a/2} (e^a \cdot e^{-x} - e^x)\,dx$

$$= [-(e^a \cdot e^{-x} + e^x)]_0^{a/2}$$

$$= (e^a + 1) - (e^{a/2} + e^{a/2})$$

$$= e^a - 2e^{a/2} + 1$$

$$= (e^{a/2} - 1)^2$$

Then, $\quad \frac{S}{a^2} = \left(\frac{e^{a/2} - 1}{a}\right)^2 = \frac{1}{4}\left(\frac{e^{a/2} - 1}{a/2}\right)^2$

$\therefore \quad \lim_{a \to 0} \frac{S}{a^2} = \frac{1}{4}$

**9.** (d) Given, $\quad y = \log^2 x - 1$

$\Rightarrow \quad y'' = \frac{2\log x}{x} = 0$

$\Rightarrow \quad x = 1$

If $x > 1$, $y$ increases and $0 < x < 1$, $y$ is decreases.

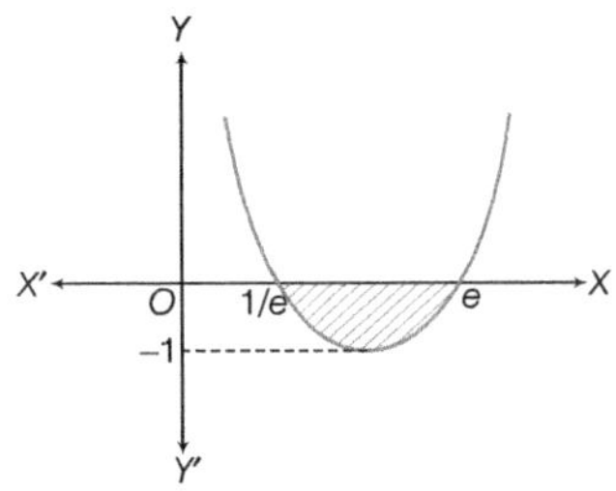


$\therefore$ Required area $= \left|\int_{1/e}^{e} (\log^2 x - 1)\,dx\right|$

$$= \left|\int_{1/e}^{e} \log^2 x\,dx - \int_{1/e}^{e} dx\right|$$

$$= \left|[x\log^2 x]_{1/e}^{e} - 2\int_{1/e}^{e} \frac{\log x}{x} \cdot x\,dx - \left(e - \frac{1}{e}\right)\right|$$

$$= \left|\left(e - \frac{1}{e}\right) - 2\int_{1/e}^{e} \left(\frac{\log x}{x}\right) x\,dx - \left(e - \frac{1}{e}\right)\right|$$

$$= \left|-2\left[[x\log x]_{1/e}^{e} - \int_{1/e}^{e} dx\right]\right|$$

$$= \left|-2\left[\left(e + \frac{1}{e}\right) - \left(e - \frac{1}{e}\right)\right]\right|$$

$$= \left|-\frac{4}{e}\right| = \frac{4}{e}$$

**10.** (d) Given curves are $x = y^2 - 1$ and $x = |y| \sqrt{1 - y^2}$

The graph of curves could be plotted as

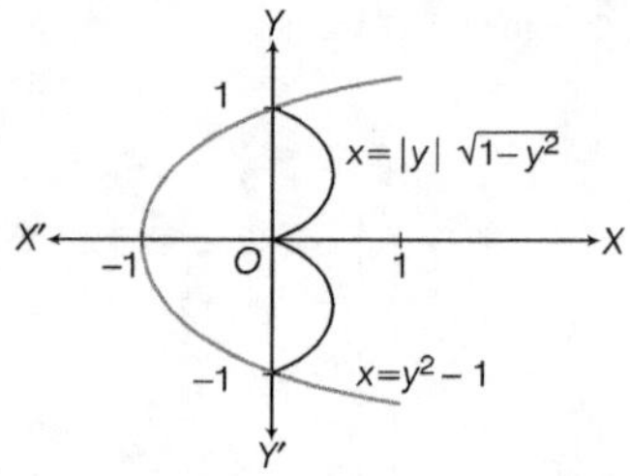


$$\therefore \text{Required area} = 2 \int_0^1 [y\sqrt{1-y^2} - (y^2 - 1)]\, dy$$

$$= 2\left[\int_0^1 y\sqrt{1-y^2}\, dy - \int_0^1 (y^2 - 1)\, dy\right]$$

$$= 2\left[\int_0^1 y\sqrt{1-y^2}\, dy - \left(\frac{y^3}{3} - y\right)_0^1\right]$$

$$= 2\left[-\frac{1}{2}\int_1^0 \sqrt{t}\, dt - \left(\frac{y^3}{3} - y\right)_0^1\right], \text{ where } t = 1 - y^2$$

$$= 2\left[\frac{1}{2}\int_0^1 \sqrt{t}\, dt - \left(\frac{1}{3} - 1\right)\right]$$

$$= 2\left[\frac{1}{2} \times \frac{2}{3}[t^{3/2}]_0^1 + \frac{2}{3}\right] = 2\left[\frac{1}{3} + \frac{2}{3}\right]$$

$= 2$ sq units

**11.** (c) Let $\dfrac{p}{\sqrt{p^2+q^2}} \cdot x + \dfrac{q}{\sqrt{p^2+q^2}} \cdot y = u$ and

$$\frac{q}{\sqrt{p^2+q^2}} \cdot x - \frac{p}{\sqrt{p^2+q^2}} \cdot y = v$$

Then, the axis get rotated through an angle $\theta$, where

$$\cos\theta = \frac{p}{\sqrt{p^2+q^2}} \text{ and } \sin\theta = \frac{q}{\sqrt{p^2+q^2}}$$

Now, the equation of given curve becomes $|u| + |v| = a$.

Then, the area bounded $= 2a^2$

$\therefore$ Statement I is correct.

The equation of curve is

$|\alpha x + \beta y| + |\beta x - \alpha y| = a$, which is equivalent to

$$\left|\frac{\alpha}{\sqrt{\alpha^2+\beta^2}} \cdot x + \frac{\beta}{\sqrt{\alpha^2+\beta^2}} \cdot y\right| + \left|\frac{\beta}{\sqrt{\alpha^2+\beta^2}} \cdot x - \frac{\alpha}{\sqrt{\alpha^2+\beta^2}} \cdot y\right| = \frac{\alpha}{\sqrt{\alpha^2+\beta^2}}$$

$$\text{Area bounded} = \frac{2a^2}{\alpha^2+\beta^2}$$

$\therefore$ Statement II is incorrect.

**12.** (b) Given curve $\quad y = x^2 - 4x + 3$

$$y = (x-2)^2 - 1$$

$\Rightarrow \quad y + 1 = (x-2)^2 \quad \ldots(\text{i})$

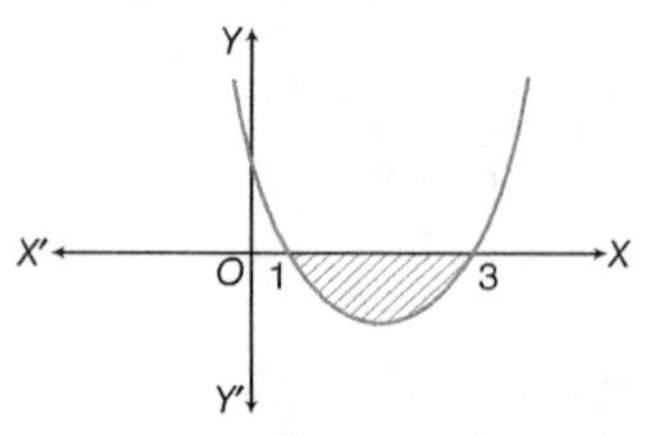


$$\therefore \text{Required area} = \left|\int_1^3 y\, dx\right| = \left|\int_1^3 (x^2 - 4x + 3)\, dx\right|$$

$$= \left|\left[\frac{x^3}{3} - \frac{4x^2}{2} + 3x\right]_1^3\right|$$

$$= \left|(9 - 18 + 9) - \left(\frac{1}{3} - 2 + 3\right)\right|$$

$$= \left|-\frac{4}{3}\right| = \frac{4}{3} \text{ sq units}$$

Hence, Statement I is correct.

Clearly, Statement II is correct but not the correct explanation of Statement I.

**Explanations for Q. Nos. (13-15)**

Given, $\quad y = \dfrac{1}{x^2} \quad \ldots(\text{i})$

and $\quad y = \dfrac{1}{4(x-1)} \quad \ldots(\text{ii})$

On solving Eqs. (i) and (ii), we get

$$x^2 = 4(x-1)$$

$\Rightarrow \quad x^2 - 4x + 4 = 0$

$\Rightarrow \quad (x-2)^2 = 0$

Therefore, curves touch each other.

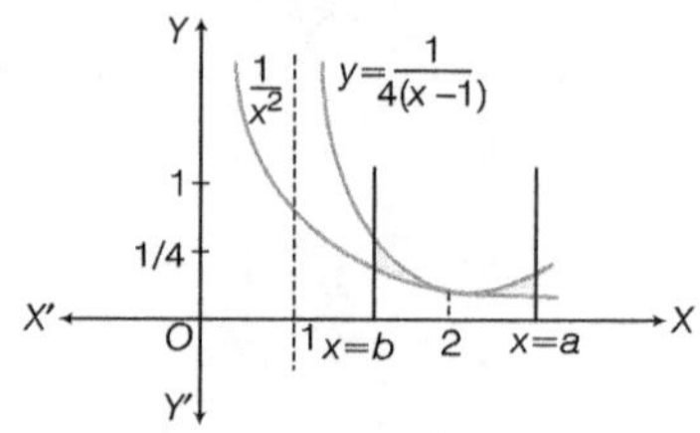


$$\therefore \quad A = \int_2^a \left[\frac{1}{4(x-1)} - \frac{1}{x^2}\right] dx = \frac{1}{a}$$

$$\Rightarrow \quad \left[\frac{1}{4}\log(x-1) + \frac{1}{x}\right]_2^a = \frac{1}{a}$$

$$\Rightarrow \quad \frac{1}{4}\log(a-1) + \frac{1}{a} - \frac{1}{2} = \frac{1}{a}$$

$\Rightarrow \quad \log(a-1) = 2$

$\Rightarrow \quad a = e^2 + 1$

Also, $\quad 1 - \dfrac{1}{b} = \displaystyle\int_b^2 \left[\frac{1}{4(x-1)} - \frac{1}{x^2}\right] dx$

$\Rightarrow \quad b = 1 + e^{-2}$

**13.** (a) $\log_e a - \log_e b = \log_e \dfrac{a}{b} = \log_e \left[\dfrac{1+e^2}{1+e^{-2}}\right] = \log e^2 = 2$

**14.** (d) Given, $A = \begin{bmatrix} \log(a-1) & 0 \\ 0 & \log(b-1) \end{bmatrix}$

$$= \begin{bmatrix} \log(1+e^2-1) & 0 \\ 0 & \log(1+e^{-2}-1) \end{bmatrix}$$

$$= \begin{bmatrix} \log e^2 & 0 \\ 0 & \log e^{-2} \end{bmatrix}$$

$$\Rightarrow \quad A = \begin{bmatrix} 2 & 0 \\ 0 & -2 \end{bmatrix} = [-4 - 0] = [-4]$$

$\Rightarrow \quad |A| = -4$

Now, $\quad A^{-1} = \dfrac{1}{-4}\begin{bmatrix} -2 & 0 \\ 0 & 2 \end{bmatrix} = \dfrac{1}{4}\begin{bmatrix} 2 & 0 \\ 0 & -2 \end{bmatrix}$

$\therefore \quad A^{-1} = \dfrac{A}{4}$

**15.** (c) Given, $z = \log(a-1) + i\log(b-1)$

$$= \log(1 + e^2 - 1) + i\log(1 + e^{-2} - 1)$$
$$= \log e^2 + i\log(e^{-2}) = 2 + (-2)i$$
$$z = 2 - 2i$$

$\therefore \quad \arg(z) = -\left(\tan^{-1}\left|-\frac{2}{2}\right|\right) = -(\tan^{-1}|-1|) = -\frac{\pi}{4}$

**16.** (b) Given, $y^3 - 3y + x = 0$

On differentiating w.r.t. x, we get

$$3y^2y' - 3y' + 1 = 0$$

$\Rightarrow \quad y' = \frac{1}{3(1-y^2)} \quad \text{...(i)}$

$\Rightarrow \quad y'(-10\sqrt{2}) = \frac{1}{3[1-(2\sqrt{2})^2]} = \frac{1}{3(1-8)} = -\frac{1}{21}$

Again differentiating Eq. (i), we get

$$y'' = \frac{6yy'^2}{3(1-y^2)}$$

$\therefore \quad y''(-10\sqrt{2}) = \frac{6\cdot 2\sqrt{2}\left(-\frac{1}{21}\right)^2}{3(1-8)} = -\frac{4\sqrt{2}}{7^3 3^2}$

**17.** (a) Required area $= \int_a^b f(x)\,dx = [x\,f(x)]_a^b - \int_a^b x\,f'(x)\,dx$

$$= b\,f(b) - a\,f(a) + \int_a^b \frac{x}{3[f^2(x) - 1]}\,dx$$

**18.** (d) $y' = \frac{1}{3[1 - f^2(x)]}$ which is an even function.

$\therefore \quad \int_{-1}^{1} g'(x)\,dx = g(1) - g(-1)$

Since, $\quad y^3 - 3y + x = 0 \quad \text{...(i)}$

and $\quad y = g(x)$

$[g(x)]^3 - 3[g(x)] + x = 0 \quad$ [by Eq. (i)]

At $x = 1$, $\quad [g(1)]^3 - 3g(1) + 1 = 0 \quad \text{...(ii)}$

At $x = -1$, $\quad [g(-1)]^3 - 3g(-1) - 1 = 0 \quad \text{...(iii)}$

On adding Eqs. (ii) and (iii), we get

$$[g(1)]^3 + [g(-1)]^3 - 3\{g(1) + g(-1)\} = 0$$

$\Rightarrow \{g(1) + g(-1)\}\{g^2(1) + g^2(-1) - g(1)\,g(-1) - 3\} = 0$

$\Rightarrow g(1) + g(-1) = 0, \; g(1) = -g(-1)$

$\therefore \quad \int_{-1}^{1} g'(x) = g(1) - g(-1) = g(1) - [-g(-1)] = 2g(1)$

**19.** (c) $\because f'(x) = 2 + 6x + 12x^2 > 0, \forall x \in R$

$\therefore f(x)$ is strictly increasing in $R$.

$\because \quad f(0) = 1, f(-1) = -2, f(-1/2) = 1/4$ and $f(-3/4) = -1/2$

$\therefore f(x) = 0$ has only one real root $\in (-3/4, -1/2)$.

**20.** (a) Let the real root be $-\alpha$.

$\Rightarrow \quad t = |S| = \alpha$

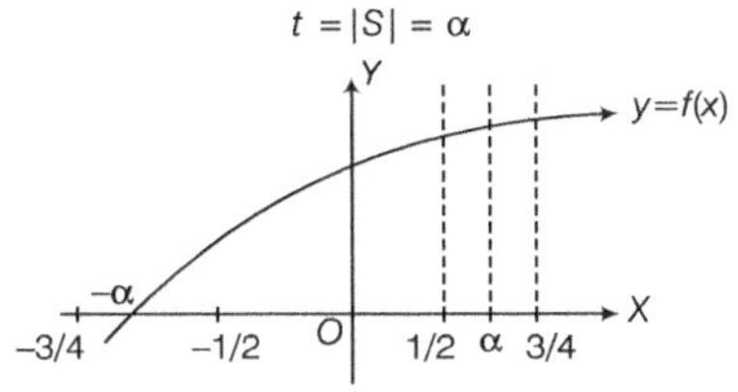


Required area, $A = \int_0^\alpha f(x)\,dx$ and

$$\int_0^{1/2} f(x)\,dx < A < \int_0^{3/4} f(x)\,dx$$

$\Rightarrow \quad [x + x^2 + x^3 + x^4]_0^{1/2} < A < [x + x^2 + x^3 + x^4]_0^{3/4} < [4x]_0^{3/4}$

$\Rightarrow \quad 15/16 < A < 3$

**21.** (b) $f'(x) = 2(6x^2 + 3x + 1)$

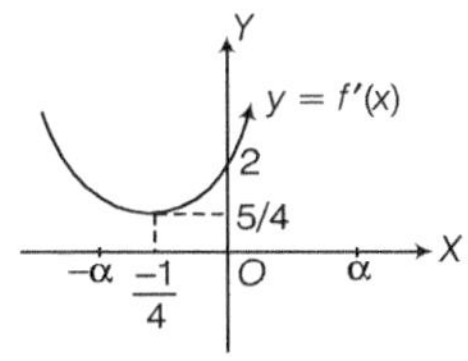


$\Rightarrow \quad f'(x)$ is decreasing in $(-\alpha, -1/4)$ and increasing in $(-1/4, \alpha)$.

or $\quad f'(x)$ is decreasing in $(-t, -1/4)$ and Increasing in $(-1/4, t)$.

**22.** (7) We have, $f(x) = \max\{x^2, (1-x)^2, 2x(1-x)\}$

Graphically it could be shown as

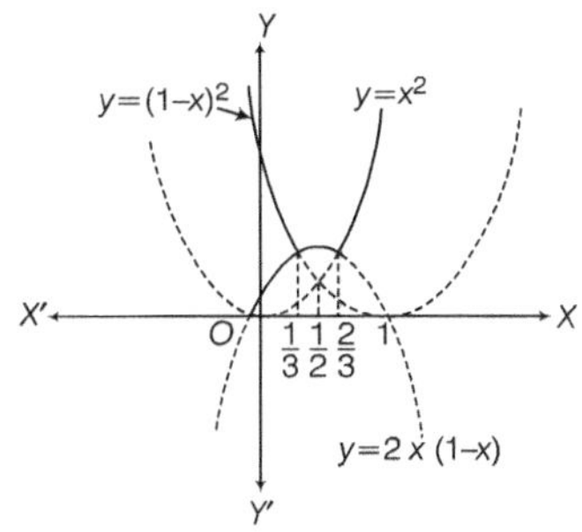


From the figure, it is clear that maximum graph i.e. maximum of the given curves are considered and other are neglected.

$\therefore$ For $x \in \left[0, \frac{1}{3}\right]$, $x^2 < 2x(1-x) \le (1-x)^2$

For $x \in \left[\frac{1}{3}, \frac{1}{2}\right]$, $\quad x^2 \le (1-x)^2 \le 2x(1-x)$

For $x \in \left[\frac{1}{2}, \frac{2}{3}\right]$, $\quad (1-x)^2 \le x^2 \le 2x(1-x)$

For $x \in \left[\frac{2}{3}, 1\right]$, $\quad (1-x)^2 \le 2x(1-x) \le x^2$

Hence, $f(x)$ can be written as

$$f(x) = \begin{cases} (1-x)^2, & \text{for } 0 \le x \le \frac{1}{3} \\ 2x(1-x), & \text{for } \frac{1}{3} \le x \le \frac{2}{3} \\ x^2, & \text{for } \frac{2}{3} \le x \le 1 \end{cases}$$

Hence, the area bounded by the curve $y = f(x)$; X-axis and the line $x = 0$, $x = 1$ is given by

$$= \int_0^{1/3} (1-x)^2\,dx + \int_{1/3}^{2/3} 2x(1-x)\,dx + \int_{2/3}^{1} x^2\,dx$$
$$= \int_0^{1/3} (1 + x^2 - 2x)\,dx + \int_{1/3}^{2/3} (2x - 2x^2)\,dx + \int_{2/3}^{1} x^2\,dx$$
$$= \left[x + \frac{x^3}{3} - x^2\right]_0^{1/3} + \left[x^2 - \frac{2x^3}{3}\right]_{1/3}^{2/3} + \left[\frac{x^3}{3}\right]_{2/3}^{1}$$
$$= \left(\frac{1}{3} + \frac{1}{81} - \frac{1}{9}\right) + \left(\frac{4}{9} - \frac{16}{81} - \frac{1}{9} + \frac{2}{81}\right) + \left(\frac{1}{3} - \frac{8}{81}\right)$$
$$= \frac{17}{27} \text{ sq unit}$$

But area is given $\frac{m}{n}$.

$\Rightarrow \quad \frac{m}{n} = \frac{17}{27}$

$\therefore \quad 2m - n = (2 \times 17 - 27) = 7$

**23.** (4) Given curves are

$$y = \cos x - \cos^2 x \quad \text{...(i)}$$

and

$$y = x^2 \left(x^2 - \frac{\pi^2}{4}\right) \quad \text{...(ii)}$$

The graph of these two curve could be plotted as

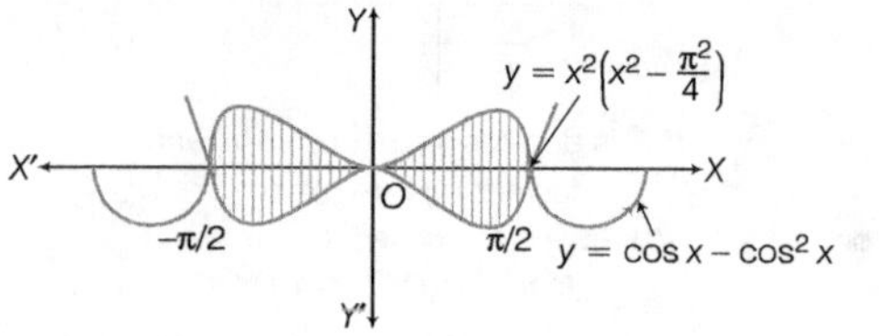


$\therefore$ Required area

$$= 2\int_0^{\pi/2} \left[\cos x - \cos^2 x - \left\{x^2\left(x^2 - \frac{\pi^2}{4}\right)\right\}\right] dx$$

$$= 2\int_0^{\pi/2} \left(\cos x - \cos^2 x - x^4 + \frac{\pi^2}{4} x^2\right) dx$$

$$= 2\left[\sin x - \frac{x}{2} - \frac{\sin 2x}{4} - \frac{x^5}{5} + \frac{\pi^2}{12} x^3\right]_0^{\pi/2}$$

$$= \left(2 - \frac{\pi}{2} + \frac{\pi^5}{120}\right) \text{sq units}$$

But given area is $\left(a - \frac{\pi}{b} + \frac{\pi^5}{120}\right)$ sq units

$\Rightarrow \quad a = 2$ and $b = 2$

$\therefore \quad a + b = 4$

**24.** (6) We see that $f$ is continuous everywhere except possibly at integral points. Let $n$ be an integer, then

$$f(n^-) = n - 1 + (1)^2 = n$$

and $\quad f(n^+) = n + (0)^2 = n$

$\therefore \quad f(n) = n$

$\Rightarrow f$ is everywhere continuous, also $f(x) \to \pm\infty$

as $\quad x \to \pm\infty$

$\Rightarrow$ Range of $f \in R$, thus onto

and $f(x) = n + (x - n)^2$, for all $x \in [n, n + 1]$, where $n \in I$.

$\therefore$ $f$ is monotonically increasing.

$\Rightarrow f$ is one-one, thus invertible.

Obviously for $x \in (0, 1)$, $f(x) = x^2$ and $f^{-1}(x) = \sqrt{x}$

and for $x \in \left[1, \frac{3}{2}\right]$, $f(x) = 1 + (x - 1)^2$

and $\quad f^{-1}(x) = 1 + \sqrt{x - 1}$

$$\Rightarrow \quad f(x) = \begin{cases} x^2, & 0 \le x < 1 \\ 1 + (x - 1)^2, & 1 \le x \le \frac{3}{2} \end{cases}$$

$\therefore$ Required area $= \int_0^{3/2} \{f^{-1}(x) - f(x)\}\, dx$

$$= \int_0^1 (\sqrt{x} - x^2)\, dx + \int_1^{3/2} [(1 + \sqrt{x - 1}) - \{1 + (x - 1)^2\}] dx$$

$$= \left[\frac{2}{3} x^{3/2} - \frac{x^3}{3}\right]_0^1 + \left[\frac{2}{3}(x - 1)^{3/2} - \frac{(x - 1)^3}{3}\right]_1^{3/2}$$

$$= \left(\frac{2}{3} - \frac{1}{3}\right) + \left\{\frac{2}{3}\left(\frac{1}{2}\right)^{3/2} - \frac{(1/2)^3}{3}\right\} = \frac{1}{3} + \frac{1}{3\sqrt{2}} - \frac{1}{24}$$

$$= \frac{8\sqrt{2} + 8 - \sqrt{2}}{24\sqrt{2}} = \frac{7\sqrt{2} + 8}{24\sqrt{2}} \text{ sq units}$$

But given area is $\frac{a + 7\sqrt{2}}{b\sqrt{c}}$.

$\Rightarrow \quad a = 8, b = 24$ and $c = 2$

$\therefore \quad b - 2a - c = 24 - 16 - 2 = 6$

**25.** (8) Take $\quad x = y = 1$

Now, $\quad y = \frac{1}{x}$

$$\Rightarrow \quad 0 = f\left(x \cdot \frac{1}{x}\right)$$

$$= f(x) + f\left(\frac{1}{x}\right)$$

$$\Rightarrow \quad f\left(\frac{1}{x}\right) = -f(x)$$

$$\therefore \quad f\left(\frac{x}{y}\right) = f(x) + f\left(\frac{1}{y}\right)$$

$$= f(x) - f(y) \quad \text{...(i)}$$

Now, $\quad f'(x) = \lim_{h \to 0} \frac{f(x + h) - f(x)}{h}$

$$= \lim_{h \to 0} \frac{f\left(\frac{x + h}{x}\right)}{h} \quad \text{[by Eq. (i)]}$$

$$= \lim_{h \to 0} \frac{f\left(1 + \frac{h}{x}\right) - f(1)}{\frac{h}{x} \cdot x} = \frac{f'(1)}{x} \quad [\because f(1) = 0]$$

$$f'(x) = \frac{2}{x}$$

$\Rightarrow \quad f(x) = 2 \log x + C$

$[\because f(1) = 0 \Rightarrow C = 0]$

$\Rightarrow \quad f(x) = 2 \log x$

Thus, $f(x) = 2 \log x$ and $y = |x^3 - 6x^2 + 11x - 6|$ could be plotted as

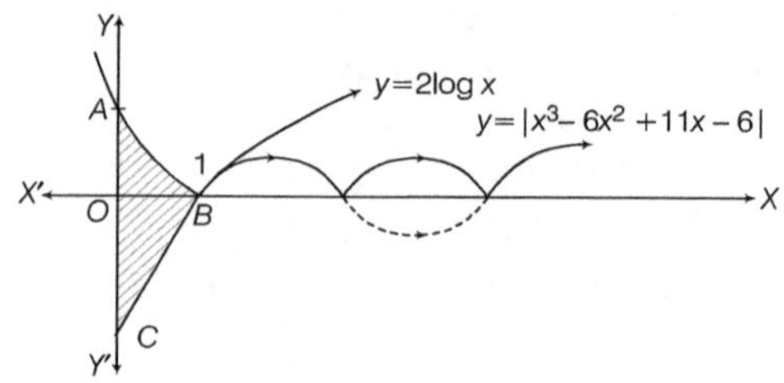


Hence, required area = Area $OABCO$

$$= \int_0^1 (x^3 - 6x^2 + 11x - 6)\, dx + \int_{-\infty}^0 e^{y/2} dy$$

$$= \left[\frac{x^4}{4} - \frac{6x^3}{3} + \frac{11x^2}{2} - 6x\right]_0^1 + 2[e^{y/2}]_{-\infty}^0$$

$$= \left(\frac{1}{4} - 2 + \frac{11}{2} - 6\right) - (0) + 2(e^0 - e^{-\infty})$$

$$= \frac{1}{4} + \frac{11}{2} - 8 + 2$$

$$= -\frac{1}{4} \text{ sq unit} = \frac{1}{4} \text{ sq unit} \quad [\because \text{area cannot be negative}]$$

But given area is $\frac{2}{\lambda}$.

$$\therefore \quad \frac{2}{\lambda} = \frac{1}{4}$$

$$\Rightarrow \quad \lambda = 8$$

# Revisal Problems (JEE Main)

**1.** (c) We have,

$$0 \le \frac{x^2}{64} < 1, \text{ if } -8 < x < 8$$

$$\Rightarrow \quad 2 \le \frac{x^2}{64} + 2 < 3, \text{ if } |x| < 8$$

$$\Rightarrow \quad y = \left[\frac{x^2}{64} + 2\right] = 2, \text{ if } |x| < 8$$

The graph of given curves is as shown in figure.

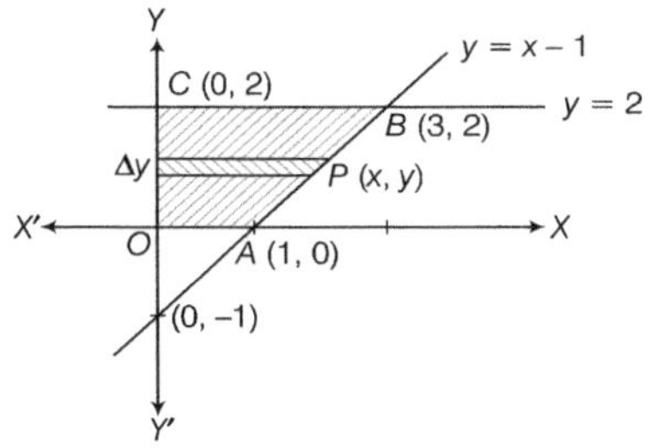


$\therefore$ Required area = Area of the shaded region

$$= \int_0^2 x\,dy$$

$$= \int_0^2 (y+1)\,dy$$

$$= \frac{1}{2}[(y+1)^2]_0^2$$

$$= \frac{9}{2} - \frac{1}{2} = 4 \text{ sq units}$$

**Alternate Method**

$\therefore$ Required area = Area of trapezium $OABC$

$$= \frac{1}{2}(OA + BC) \times OC$$

$$= \frac{1}{2}(1+3) \times 2$$

$$= 4 \text{ sq units}$$

**2.** (d) Given curves are

$$y = x e^{|x|} \quad \text{...(i)}$$

and $\quad |x| = 1 \quad$ ...(ii)

The graph of the given curves could be plotted as

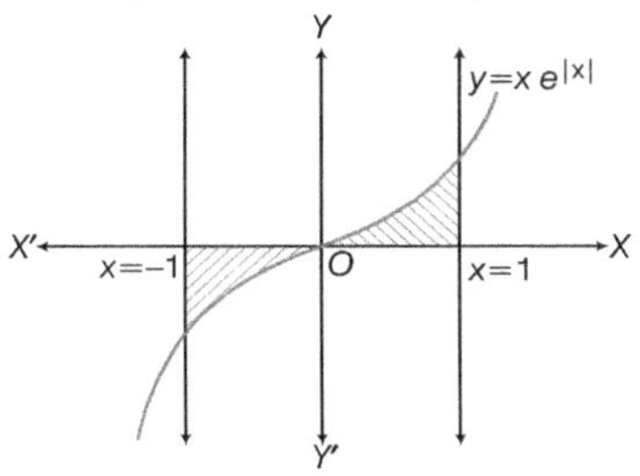


$\therefore$ Required area = 2 (Area in the first quadrant)

$$= 2\int_0^1 y\,dx$$

$$= 2\int_0^1 x e^{|x|} dx$$

$$= 2\int_0^1 x \cdot e^x dx$$

$$= 2\,[(x-1)e^x]_0^1$$

$$= 2\,[0 - (-1)]$$

$$= 2 \text{ sq units}$$

**3.** (b) Given curves are $\quad y = \log x \quad$ ...(i)

$y = \log|x|, \quad$ ...(ii)

$y = |\log x| \quad$ ...(iii)

and $\quad y = |\log|x|| \quad$ ...(iv)

The graph of the given curves could be plotted as

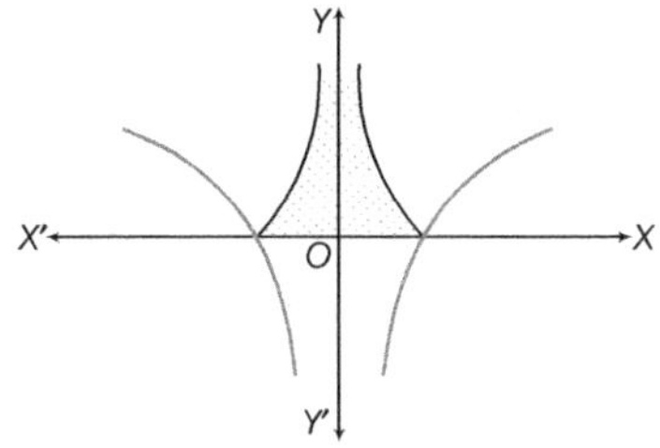


$\therefore$ Required area = Area of the shaded region

= 2 (Area in first quadrant)

$$= 2\int_0^1 |\log x|\,dx$$

$$= -2\int_0^1 \log x\,dx$$

$$= -2\,[x(\log x - 1)]_0^1 = 2 \text{ sq units}$$

**4.** (c) It is given that $\int_0^a f(x)\,dx = 1 + \frac{a^2}{2}\sin a$

On differentiating w.r.t. $a$, we get

$$f(a) = a\sin a + \frac{a^2}{2}\cos a$$

**5.** (c) We have, $|x - 1| \le 2 \Leftrightarrow -1 \le x \le 3$

The graph of $|x - 1| \le 2$ and $x^2 - y^2 = 1$ could be plotted as

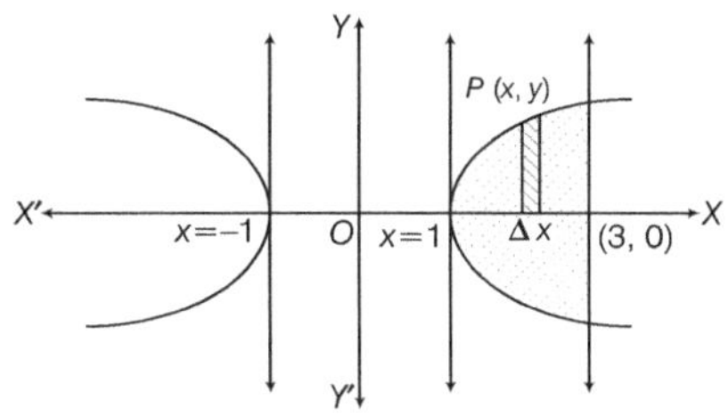


$\therefore$ Required area = Area of the shaded region

$$= 2\int_1^3 \sqrt{x^2 - 1}\,dx$$

$$= 2\left[x\sqrt{x^2-1} - \log\left(x + \sqrt{x^2 - 1}\right)\right]_1^3$$

$$= \{3\sqrt{8} - \log(3 + \sqrt{8})\}$$

$$= 6\sqrt{2} - \log|3 + 2\sqrt{2}|$$

**6.** (d) Clearly, $f : [0, 2\pi] \to [0, 2\pi]$ given by

$f(x) = x + \sin x$ is a bijection. So, its inverse exists, the graph of $f^{-1}(x)$ is the mirror image of the graph of $f(x)$ in the line $y = x$.

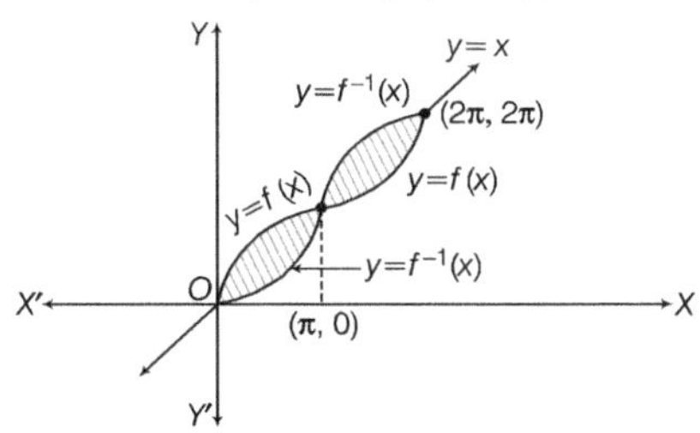

$\therefore$ Required area = 4 (Area of one loop)

$$= 4\left[\int_0^{\pi} (x + \sin x)\,dx - \int_0^{\pi} x\,dx\right]$$
$$= 4\int_0^{\pi} \sin x\,dx$$
$$= -4\,[\cos x]_0^{\pi}$$
$$= -4\,[\cos \pi - \cos 0]$$
$$= 8 \text{ sq units}$$

**7.** (b) Clearly, $f(x)$ is everywhere continuous and differentiable except at $x = 2$. The sign of $f(x)$ for different values of $x$ are shown in figure given below.

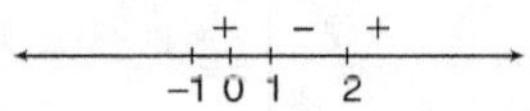


A rough sketch of the curve $y = f(x)$ is as shown below.

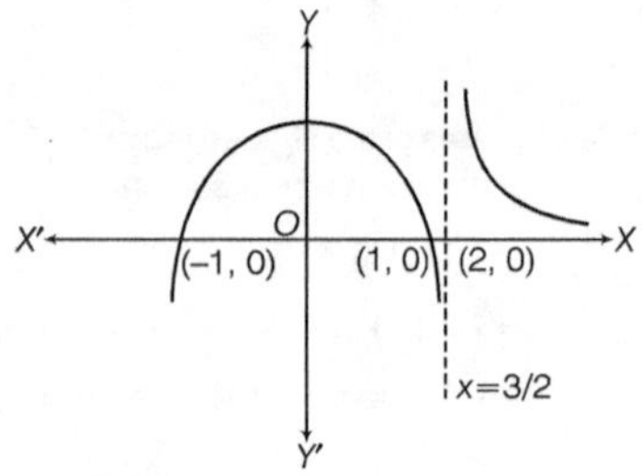


$\therefore$ Required area $= \int_0^{3/2} f(x)\,dx$

$$= \int_0^1 f(x)\,dx + \left|\int_1^{3/2} f(x)\,dx\right|$$
$$= \int_0^1 \frac{(x-1)(x+1)}{x-2}\,dx - \int_1^{3/2} \frac{(x-1)(x+1)}{(x-2)}\,dx$$
$$= \int_0^1 \left(x + 2 + \frac{3}{x-2}\right)dx - \int_1^{3/2} \left(x + 2 + \frac{3}{x-2}\right)dx$$
$$= \left[\frac{(x+2)^2}{2} + 3\log|x-2|\right]_0^1 - \left[\frac{(x+2)^2}{2} + 3\log|x-2|\right]_1^{3/2}$$
$$= \left[\frac{9}{2} - 2 - 3\log 2\right] - \left[\left(\frac{49}{8} + 3\log\frac{1}{2}\right) - \frac{9}{2}\right]$$
$$= \frac{7}{8} \text{ sq unit}$$

**8.** (b) We have, $f(x) = \max\left\{\sin x, \cos x, \frac{1}{2}\right\}$

$$= \begin{cases} \cos x, & 0 \le x \le \frac{\pi}{4} \\ \sin x, & \frac{\pi}{4} \le x \le \frac{5\pi}{6} \\ \frac{1}{2}, & \frac{5\pi}{6} \le x \le \frac{5\pi}{3} \end{cases}$$

$\therefore$ Required area $= \int_0^{5\pi/3} f(x)\,dx$

$$= \int_0^{\pi/4} \cos x\,dx + \int_{\pi/4}^{5\pi/6} \sin x\,dx + \int_{5\pi/6}^{5\pi/3} \frac{1}{2}\,dx$$
$$= [\sin x]_0^{\pi/4} - [\cos x]_{\pi/4}^{5\pi/6} + \frac{1}{2}[x]_{5\pi/6}^{5\pi/3}$$
$$= \frac{1}{\sqrt{2}} - \left(-\frac{\sqrt{3}}{2} - \frac{1}{\sqrt{2}}\right) + \frac{1}{2}\left(\frac{5\pi}{3} - \frac{5\pi}{6}\right)$$
$$= \frac{1}{\sqrt{2}} + \frac{\sqrt{3}}{2} + \frac{1}{\sqrt{2}} + \frac{5\pi}{12}$$
$$= \frac{\sqrt{3}}{2} + \sqrt{2} + \frac{5\pi}{12}$$

**9.** (d) Clearly, $y = f(x)$ passes through the points (2, 0) and (0, 1).

$\therefore$ $0 = f(2)$ and $1 = f(0)$

Also, $\int_0^2 f(x)\,dx = \frac{3}{4}$ [given]

$\Rightarrow$ $\int_0^2 x f'(x)\,dx = [x f(x)]_0^2 - \int_0^2 f(x)\,dx$

$\Rightarrow$ $\int_0^2 x f'(x) = [2 f(2) - 0 f(0)] - \frac{3}{4}$

$$= 2 \times 0 - 0 \times 1 - \frac{3}{4}$$
$$= -\frac{3}{4}$$

**10.** (c) Given, $R = \{(x, y) : |x| \le |y| \text{ and } x^2 + y^2 \le 1\}$

The given region could be plotted as

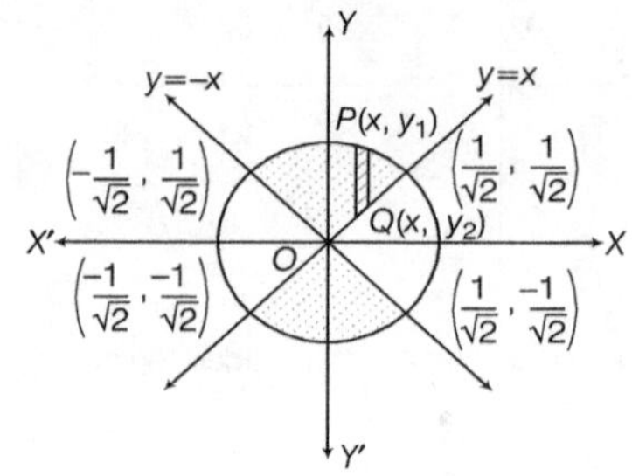


$\therefore$ Required area = 4 (Area of the shaded region in first quadrant)

$$= 4\int_0^{1/\sqrt{2}} (y_1 - y_2)\,dx$$
$$= 4\int_0^{1/\sqrt{2}} (\sqrt{1 - x^2} - x)\,dx$$
$$= 4\left[\frac{1}{2}x\sqrt{1 - x^2} + \frac{1}{2}\sin^{-1} x - \frac{x^2}{2}\right]_0^{1/\sqrt{2}}$$
$$= 4\left[\frac{1}{2\sqrt{2}} \times \frac{1}{\sqrt{2}} + \frac{1}{2} \times \frac{\pi}{4} - \frac{1}{4}\right]$$
$$= \frac{\pi}{2} \text{ sq units}$$

**11.** (a) Given, $\begin{bmatrix} 4a^2 & 4a & 1 \\ 4b^2 & 4b & 1 \\ 4c^2 & 4c & 1 \end{bmatrix} \begin{bmatrix} f(-1) \\ f(1) \\ f(2) \end{bmatrix} = \begin{bmatrix} 3a^2 + 3a \\ 3b^3 + 3b \\ 3c^2 + 3c \end{bmatrix}$

$\Rightarrow$ $4a^2 f(-1) + 4a\,f(1) + f(2) = 3a^2 + 3a$ ...(i)

$\Rightarrow$ $4b^2 f(-1) + 4b\,f(1) + f(2) = 3b^2 + 3b$ ...(ii)

$\Rightarrow$ $4c^2 f(-1) + 4c\,f(1) + f(2) = 3c^2 + 3c$ ...(iii)

On comparing the coefficients of $a^2$, $a$ and constant terms on both sides, we get

$$f(-1) = \frac{3}{4} = f(1)$$

and $f(2) = 0$ ...(iv)

Let $f(x) = Ax^2 + Bx + C$ ...(v)

From Eqs. (iv) and (v), we get

$$A = -\frac{1}{4}, B = 0, C = 1$$

$\therefore$ $f(x) = -\frac{1}{4}x^2 + 1$

Let $B\left(t, 1 - \frac{t^2}{4}\right)$ be any point on the parabola

$$f(x) = y = -\frac{x^2}{4} + 1$$

As, $AB$ chord subtend right angle at $V$.

$$\Rightarrow \quad \left(-\frac{1}{2}\right) \times \left(\frac{\frac{t^2}{4}}{-t}\right) = -1$$

$$\Rightarrow \quad t = -8$$

$$\Rightarrow \quad B(-8, -15)$$

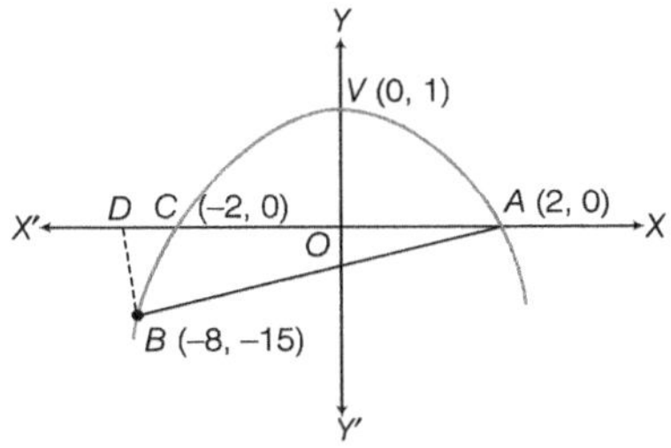


$\therefore$ Required area = Area ($BCVAB$)

$$= 2 \times \int_0^2 \left(1 - \frac{x^2}{4}\right) dx + \frac{1}{2} \times 10 \times 15 - \int_{-8}^{-2} \left(1 - \frac{x^2}{4}\right) dx$$

$$= \frac{125}{3} \text{ sq units}$$

**12.** (b) Clearly, $\quad R_1 = \int_0^b (1 - x)^2 \, dx$

and $\quad R_2 = \int_b^1 (1 - x)^2 \, dx$

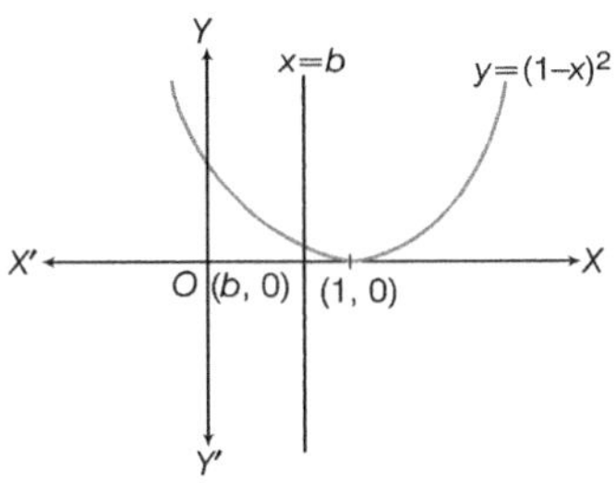


$$\Rightarrow \quad R_1 = \left[\frac{(x-1)^3}{3}\right]_0^b$$

$$R_2 = \left[\frac{(x-1)^3}{3}\right]_b^1$$

$$\Rightarrow \quad R_1 = \frac{(b-1)^3}{3} + \frac{1}{3}$$

and $\quad R_2 = -\frac{(b-1)^3}{3}$

$$\because \quad R_1 - R_2 = \frac{1}{4}$$

$$\Rightarrow \quad \frac{2}{3}(b-1)^3 + \frac{1}{3} = \frac{1}{4}$$

$$\Rightarrow \quad \frac{2}{3}(b-1)^3 = -\frac{1}{12}$$

$$\Rightarrow \quad (b-1)^3 = -\frac{1}{8}$$

$$\Rightarrow \quad b - 1 = -\frac{1}{2}$$

$$\therefore \quad b = \frac{1}{2}$$

**13.** (c) We have, $\quad R_1 = \int_{-1}^2 x f(x) \, dx$

$$\Rightarrow \quad R_1 = \int_{-1}^2 (1-x) f(1-x) \, dx$$

$$\left[\text{using } \int_a^b f(x) \, dx = \int_a^b f(a + b - x) \, dx\right]$$

$$\Rightarrow \quad R_1 = \int_{-1}^2 (1-x) f(x) \, dx \qquad [\because f(1-x) = f(x)]$$

$$\Rightarrow \quad R_1 = \int_{-1}^2 f(x) \, dx - R_1$$

$$\Rightarrow \quad 2R_1 = \int_{-1}^2 f(x) \, dx$$

It is given that

$$R_2 = \int_{-1}^2 f(x) \, dx.$$

$$\therefore \quad 2R_1 = R_2$$

**14.** (b) $y = Ax^2 + Bx + C$ passes through (1, 3).

$$\Rightarrow \quad 3 = A + B + C \qquad \text{...(i)}$$

Tangent at (2, 0) is

$$y = -4x + 8$$

$$\therefore \quad 4A + 2B + C = 0 \qquad \text{...(ii)}$$

and $\quad dy/dx = 2Ax + B$

passing through (2, 0) has slope –4.

$$\therefore \quad 4A + B = -4 \qquad \text{...(iii)}$$

From Eqs. (i) and (ii), we get

$$3A + B = -3 \qquad \text{...(iv)}$$

$$\Rightarrow \quad A = -1, B = 0 \text{ and } C = 4$$

$$\Rightarrow \quad f(x) = -x^2 + 4, \text{ shown as}$$

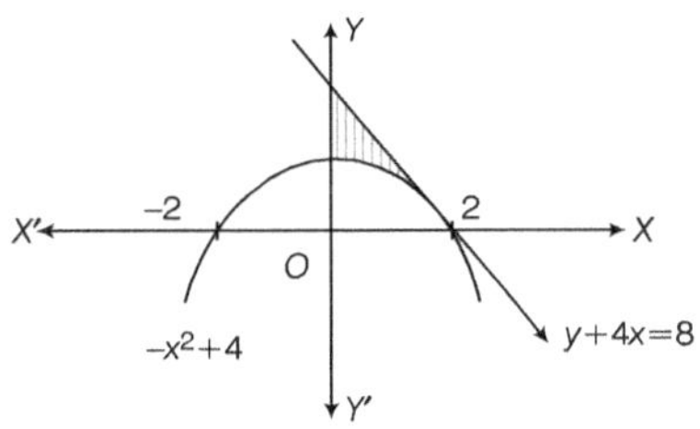


$$\therefore \quad \text{Area} = \int_0^2 [(8 - 4x) - (-x^2 + 4)] \, dx$$

$$= \int_0^2 (x^2 - 4x + 4) dx$$

$$= \int_0^2 (x-2)^2 dx$$

$$= \left[\frac{(x-2)^3}{3}\right]_0^2$$

$$= \frac{8}{3}$$

**15.** (b) Given curves are $y = x^2$ and $y = \sqrt{|x|}$.

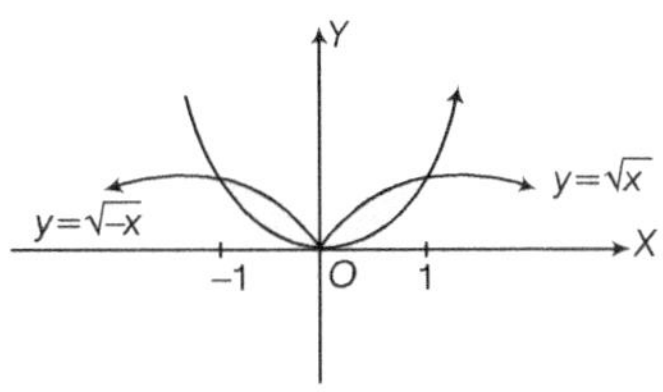


$\therefore$ Required area $= 2 \int_0^2 (\sqrt{x} - x^2) dx = \frac{2}{3}$ sq unit

**16.** (a) Equation of normal at (1, 1) is

$$\frac{y-1}{x-1} = -\frac{1}{3}$$

$$\Rightarrow \quad 3y - 3 = -x + 1$$

$$\Rightarrow \quad x + 3y = 4 \qquad \ldots\text{(i)}$$

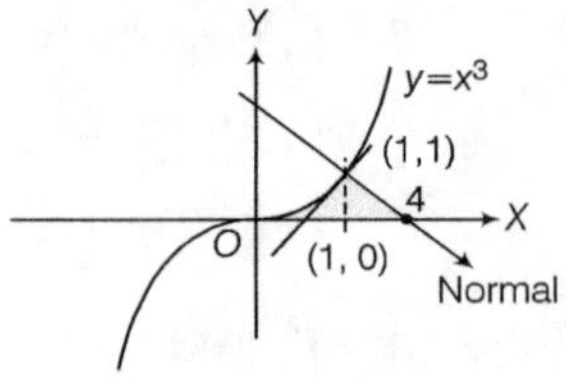


$$\therefore \quad \text{Area} = \int_0^1 x^3 dx + \frac{1}{2}(1)(3)$$

$$= \left[\frac{x^4}{4}\right]_0^1 + \frac{3}{2}$$

$$= \frac{1}{4} + \frac{3}{2}$$

$$= \frac{1+6}{4} = \frac{7}{4} \text{ sq units}$$

**17.** (b) We have, $f(x) = \sin x \qquad x \in \left(0, \frac{\pi}{2}\right]$

$$f(x) + f(\pi - x) = 2$$

$$\Rightarrow \quad f(x) = 2 - f(\pi - x) = 2 - \sin(\pi - x)$$

$$= 2 - \sin x, \qquad \text{where } x \in \left(\frac{\pi}{2}, \pi\right]$$

$$\Rightarrow \quad f(x) = f(2\pi - x)$$

$$= 2 - \sin(2\pi - x), \qquad \text{where } x \in \left(\pi, \frac{3\pi}{2}\right]$$

$$f(x) = f(2\pi - x) = -\sin x, \quad \text{where } x \in \left(\frac{3\pi}{2}, 2\pi\right]$$

i.e.

$$f(x) = \begin{cases} \sin x, & \text{for } x \in \left(0, \frac{\pi}{2}\right] \\ 2 - \sin x, & \text{for } x \in \left(\frac{\pi}{2}, \pi\right] \\ 2 + \sin x, & \text{for } x \in \left(\pi, \frac{3\pi}{2}\right] \\ -\sin x, & \text{for } x \in \left(\frac{3\pi}{2}, 2\pi\right] \end{cases}$$

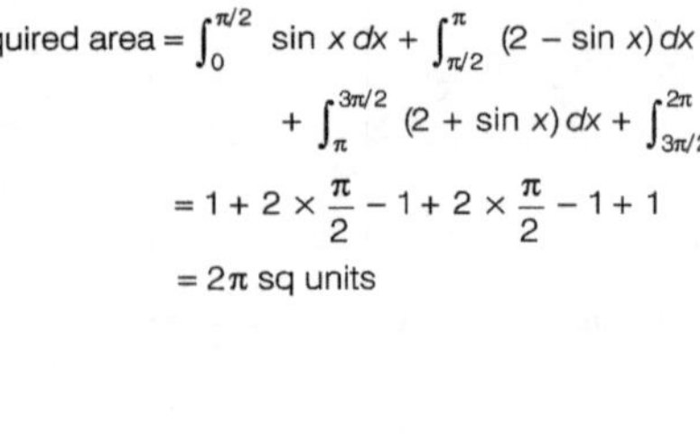


$$\therefore \text{ Required area} = \int_0^{\pi/2} \sin x \, dx + \int_{\pi/2}^{\pi} (2 - \sin x) \, dx$$

$$+ \int_{\pi}^{3\pi/2} (2 + \sin x) \, dx + \int_{3\pi/2}^{2\pi} (-\sin x) \, dx$$

$$= 1 + 2 \times \frac{\pi}{2} - 1 + 2 \times \frac{\pi}{2} - 1 + 1$$

$$= 2\pi \text{ sq units}$$

**18.** (d) $\because \sin^2 x \le \sin x, \forall x \in (0, \pi)$

Therefore, area of $y = \sin^2 x$ will be lesser from area of $y = \sin x$.

Thus, Statement I is incorrect.

$\therefore$ Statement II is obviously correct.

**19.** (c) Let the line $y = kx + 2$ cuts $y = x^2 - 3$ at $x = \alpha$ and $x = \beta$.

$\therefore$ Area bounded by the curves

$$= \int_\alpha^\beta (y_1 - y_2) \, dx$$

$$= \int_\alpha^\beta \{(kx + 2) - (x^2 - 3)\} \, dx$$

$$f(x) = \frac{(k^2 + 20)^{3/2}}{6}$$

which clearly shows that Statement II is incorrect but $f(k)$ is least, when $k = 0$.

**20.** (c) Given, $\qquad y = x - \lambda x^2$

and $\qquad y = \frac{x^2}{\lambda}, (\lambda > 0)$

On eliminating $y$ from $y = \frac{x^2}{\lambda}$ and $y = x - \lambda x^2$, we get

$$x^2 = \lambda x - \lambda^2 x^2$$

$$\Rightarrow \quad x = 0$$

Thus, the area enclosed between the parabolas is shown below.

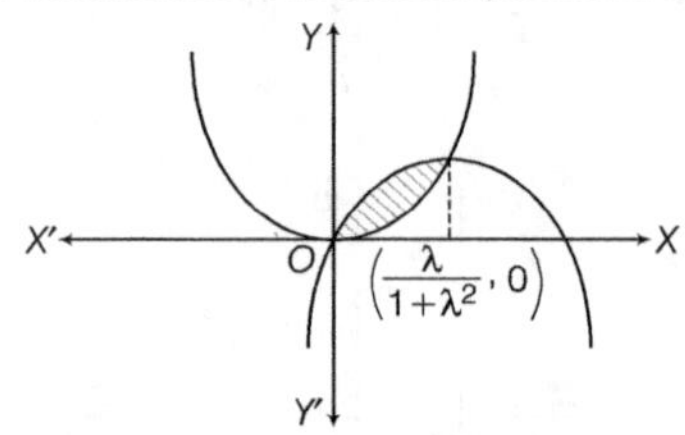


$$\therefore \quad \text{Area, } A = \int_0^{\frac{\lambda}{1+\lambda^2}} \left(x - \lambda x^2 - \frac{x^2}{\lambda}\right) dx$$

$$= \int_0^{\frac{\lambda}{1+\lambda^2}} \left\{x - x^2\left(\frac{1+\lambda^2}{\lambda}\right)\right\} dx$$

$$= \left[\frac{x^2}{2} - \frac{x^3}{3} \cdot \frac{1+\lambda^2}{\lambda}\right]_0^{\frac{\lambda}{1+\lambda^2}} = \frac{1}{6} \frac{\lambda^2}{(1+\lambda^2)^2}$$

For maximum value of $A$,

$$\frac{dA}{d\lambda} = 0$$

But $\qquad \frac{dA}{d\lambda} = \frac{1}{6} \frac{(1+\lambda^2)^2 2\lambda - \lambda^2 \cdot 2(1+\lambda^2) \cdot 2\lambda}{(1+\lambda^2)^4}$

$$= \frac{1}{3} \frac{\lambda(1-\lambda^2)}{(1+\lambda^2)^3}$$

Hence, $\frac{dA}{d\lambda} = 0$ gives $\lambda = -1, 0, 1$.

Since, $\lambda > 0$, therefore we consider only $\lambda = 1$.

Sign scheme for $\frac{dA}{d\lambda}$ around $\lambda = 1$ is shown below.

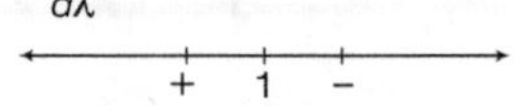


From the sign scheme, it is clear that $A$ is maximum.

Hence, Statement I is correct and Statement II is incorrect.

# Revisal Problems (JEE Advanced)

**1.** (d) If $[\,|x|\,] = 1$ and $[\,|y|\,] = 0$

Then, $\quad 1 \le |x| < 2,\ 0 \le |y| < 1$

$\Rightarrow \quad x \in (-2, -1] \cup [1, 2),\ y \in (-1, 1)$

If $[\,|x|\,] = 0,\ [\,|y|\,] = 1$

Then, $\quad x \in (-1, 1)$ and $y \in (-2, -1] \cup (1, 2]$

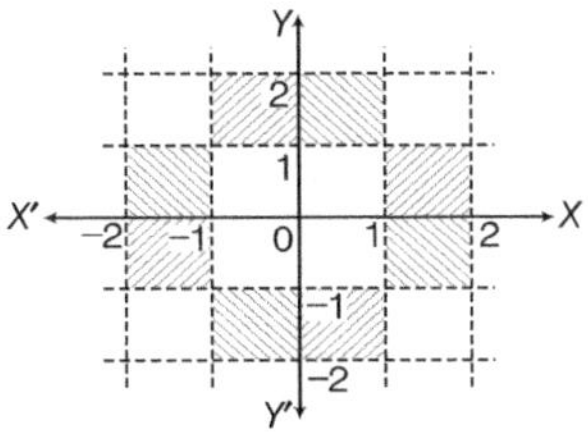


Hence, required area $= 4(2-1)[1-(-1)] = 8$ units

**2.** (b) $A'(x) = 0 \Rightarrow A(x) = c$

$\therefore \quad x^2 - 2x + 3 \ge c \quad \Rightarrow c \le 2$

$\therefore \quad \text{Area} = \int_0^5 cx\,dx = \dfrac{25c}{2} \le 25$

**3.** (d) Given, $y = \sin^{-1}|\sin x| = \begin{cases} x, & 0 \le x \le \dfrac{\pi}{2} \\ \pi - x, & \dfrac{\pi}{2} \le x < \pi \\ x - \pi, & \pi \le x < \dfrac{3\pi}{2} \\ 2\pi - x, & \dfrac{3\pi}{2} \le x < 2\pi \end{cases}$

and $\quad y = (\sin^{-1}|\sin x|)^2 = \begin{cases} x^2, & 0 \le x \le \dfrac{\pi}{2} \\ (\pi - x)^2, & \dfrac{\pi}{2} \le x < \pi \\ (x - \pi)^2, & \pi \le x < \dfrac{3\pi}{2} \\ (2\pi - x)^2, & \dfrac{3\pi}{2} \le x < 2\pi \end{cases}$

The graph of the given curve could be plotted as

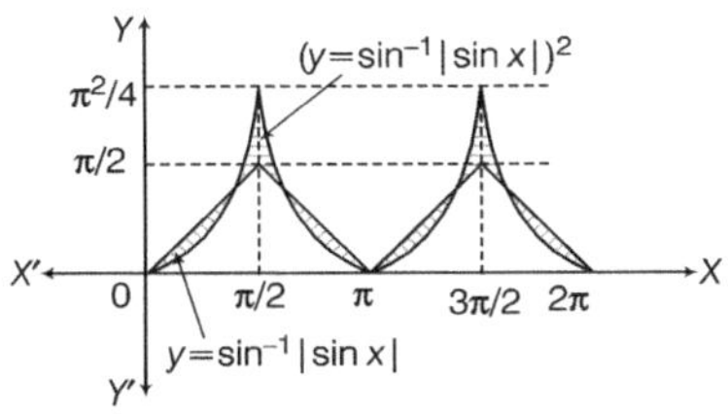


Thus, the required area is shaded region in the figure.

$\therefore$ Area $A = 4\int_0^1 (x - x^2)\,dx + 4\int_1^{\pi/2} (x^2 - x)\,dx$

$= \dfrac{4}{3} + \pi^2 \left[\dfrac{\pi - 3}{6}\right]$ sq units

**4.** (b) Given region is $R : \{(x, y) : x^2 + y^2 \le 5,\ \big||x| - |y|\big| > 1$

This could be plotted as,

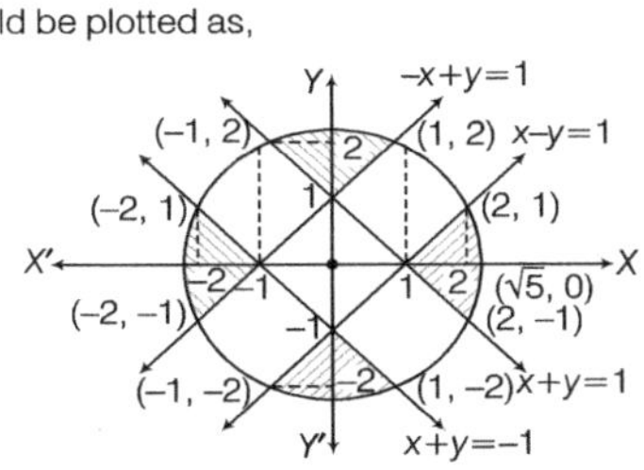


$\therefore$ Required area $= 4\left[2\left\{\int_1^2 (x-1)\,dx + \int_2^{\sqrt{5}} \sqrt{5 - x^2}\,dx\right\}\right]$

$= 8\left[\left[\dfrac{x^2}{2} - x\right]_1^2 + \left\{\dfrac{x}{2}\sqrt{5 - x^2} + \dfrac{5}{2}\sin^{-1}\dfrac{x}{\sqrt{5}}\right\}_2^{\sqrt{5}}\right]$

$= 8\left[\dfrac{1}{2} + \left\{\dfrac{5\pi}{4} - 1 - \dfrac{5}{2}\sin^{-1}\dfrac{2}{\sqrt{5}}\right\}\right]$

$= 8\left[\dfrac{1}{2} + \dfrac{5\pi}{4} - 1 - \dfrac{5}{2}\sin^{-1}\dfrac{2}{\sqrt{5}}\right]$

$= -4 + 10\pi - 20\sin^{-1}\dfrac{2}{\sqrt{5}}$

$= 20\cos^{-1}\left(\dfrac{2}{\sqrt{5}}\right) - 4 = 20\tan^{-1}\left(\dfrac{1}{2}\right) - 4$

$= 5\left[\pi - \tan^{-1}\left(\dfrac{24}{7}\right)\right] - 4$

**5.** (a) $\min\{|x|, |y|\} = 1$

$\max\{|x|, |y|\} = 2$

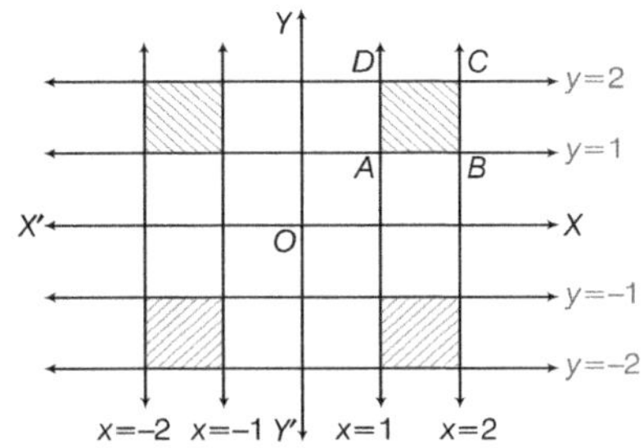


$\therefore$ Required area $= 4 \times$ Area of rectangle $(ABCD)$

$= 4 \times (1) = 4$ sq units

**6.** (a) The shaded area is the required region

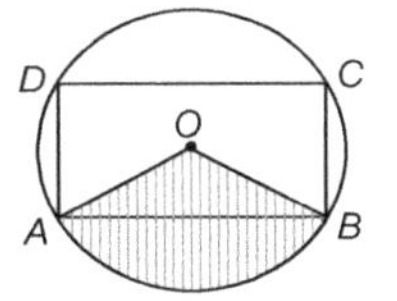


$= \dfrac{\pi r^2}{4} = 4\pi$ sq units

**7.** (b) Given, $ay^2 = x^2(a - x) \Rightarrow y = \pm x\sqrt{\dfrac{a - x}{a}}$

For the tracing of the curve, we must have $x \le a$.

For $0 < x \le a$, $y > 0$ and for $x < 0$, $y < 0$.

Also, $\quad y = 0 \Rightarrow x = 0, a$

Curve is symmetrical about $X$-axis,

when $\quad x \to -\infty,\ y \to -\infty$

Also, it can be verified that $y$ has only one point of maxima for $0 < x < a$.

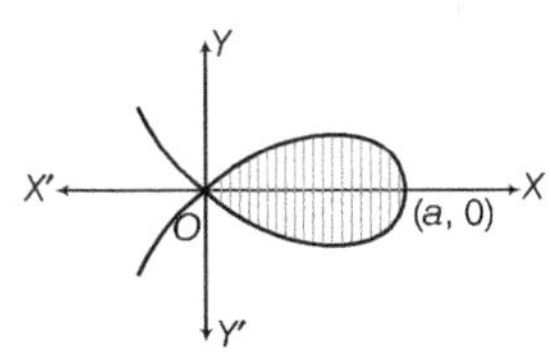


$\therefore$ Required area $= 2\int_0^a x\sqrt{\dfrac{a - x}{a}}\,dx$

Put $\sqrt{\frac{a-x}{a}} = t \Rightarrow 1 - \frac{x}{a} = t^2$

$$\Rightarrow \quad x = a(1-t^2)$$

$$\Rightarrow \quad dx = -a\,2t\,dt$$

$$\therefore \quad A = 2\int_1^0 a(1-t^2)(t)(-2at)\,dt$$

$$= 4a^2\int_0^1 (t^2 - t^4)\,dt$$

$$= 4a^2\left[\frac{t^3}{3} - \frac{t^5}{5}\right]_0^1$$

$$= 4a^2\left[\frac{1}{3} - \frac{1}{5}\right] = \frac{8a^2}{15} \text{ sq. unit}$$

**8.** (b) Given, $f(x) = \begin{cases} \sqrt{\{x\}}, & x \notin Z \\ 1, & x \in Z \end{cases}$ and $g(x) = \{x\}^2$

where both $f(x)$ and $g(x)$ are periodic with period 1 shown as

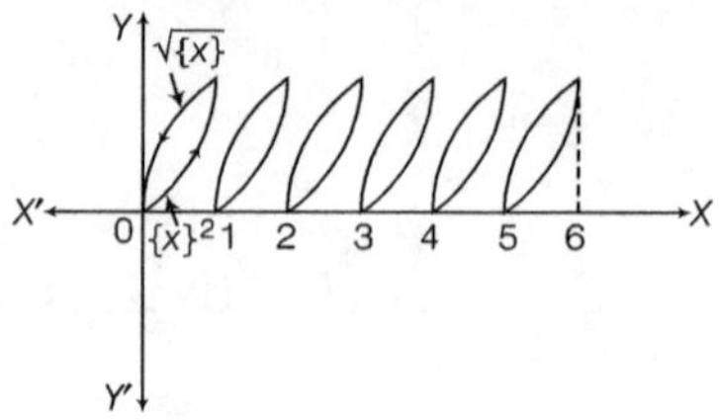


Thus, required area $= 6\int_0^1 (\sqrt{\{x\}} - \{x\}^2)\,dx = 6\int_0^1 [(x)^{1/2} - x^2]\,dx$

$$= 6\left[\frac{2}{3}x^{3/2} - \frac{x^3}{3}\right]_0^1 = 6\left(\frac{2}{3} - \frac{1}{3}\right) = \frac{6}{3}$$

$$= 2 \text{ sq units}$$

**9.** (b) It is easy to see that $ABC$ is an equilateral triangle with side of length 4.

$BD$ and $CE$ are angle bisectors of $\angle B$ and $\angle C$ respectively.

Any point inside the $\Delta AEC$ is nearer to $AC$ than $BC$ and any point inside the $\Delta\, BDA$ is nearer to $AB$ than $BC$.

So, any point inside the quadrilateral $AEGC$ will satisfy the given condition.

Hence, the shaded region is the required region whose area is to be found, shown in adjoining figure.

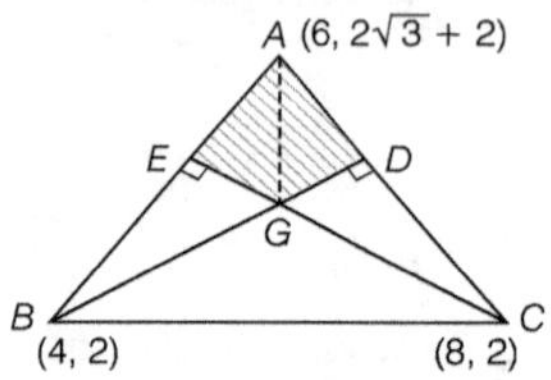


Thus, required area = 2(Area of $\Delta EAG$)

$$= 2 \times \frac{1}{2} \times AE \times EG = \frac{1}{2}AB \times \frac{1}{3}CE$$

$$= \frac{1}{6} \times 4 \times \sqrt{4^2 - 2^2} = \frac{4\sqrt{3}}{3} \text{ sq units}$$

**10.** (a) Given, $y = x^2$ ...(i)

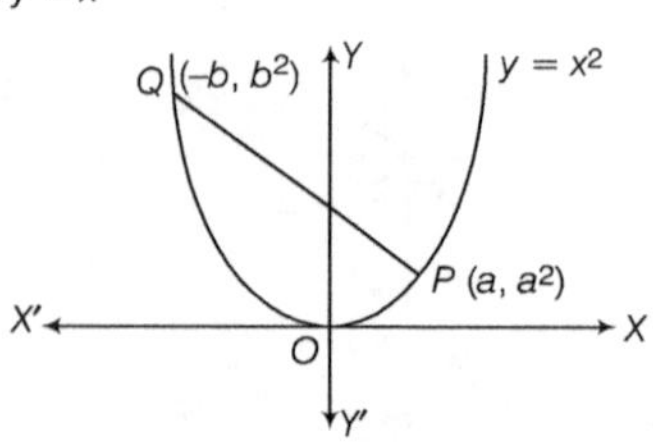


Slope of $PQ = \dfrac{a^2 - b^2}{a+b} = a - b$

Equation of $PQ$ is $y - a^2 = \dfrac{a^2 - b^2}{a+b}(x - a)$

$$\Rightarrow \quad y - a^2 = (a-b)(x-a)$$

$$\Rightarrow \quad y = a^2 + x(a-b) - a^2 + ab$$

$$\Rightarrow \quad y = (a-b)x + ab$$

$$\therefore \quad S_1 = \int_{-b}^{a} [(a-b)x + ab - x^2]\,dx$$

$$= \left[(a-b)\frac{x^2}{2} + abx - \frac{x^3}{3}\right]_{-b}^{a}$$

$$= \frac{(a+b)^3}{6} \quad \text{...(i)}$$

Also,
$$S_2 = \frac{1}{2}\begin{vmatrix} a & a^2 & 1 \\ -b & b^2 & 1 \\ 0 & 0 & 1 \end{vmatrix}$$

$$= \frac{1}{2}[ab^2 + a^2b]$$

$$= \frac{1}{2}ab(a+b) \quad \text{...(ii)}$$

Now,
$$\frac{S_1}{S_2} = \frac{(a+b)^3}{6} \cdot \frac{2}{ab(a+b)} = \frac{(a+b)^2}{3ab}$$

$$= \frac{1}{3}\left[\frac{a}{b} + \frac{b}{a} + 2\right]$$

$$\therefore \quad \left[\frac{S_1}{S_2}\right]_{\min} = \frac{4}{3}$$

**11.** (c) Given, $y = f(x)$ is function such that

$$f''(0) = 0 \quad \text{...(i)}$$

$$f''(x) = f'(x) \quad \text{...(ii)}$$

and
$$f'(0) = 1 \quad \text{...(iii)}$$

From Eq. (ii), we have

$$\frac{f''(x)}{f'(x)} = 1$$

On integrating both sides, we have

$$\log f'(x) = x + C$$

$$\because \quad f'(0) = 1$$

$$\therefore \quad C = 0$$

Now, $\quad f'(x) = e^x$

$$\Rightarrow \quad f(x) = e^x + K$$

$$\because \quad f(0) = 0$$

$$\therefore \quad K = -1$$

Then, $\quad f(x) = e^x - 1$

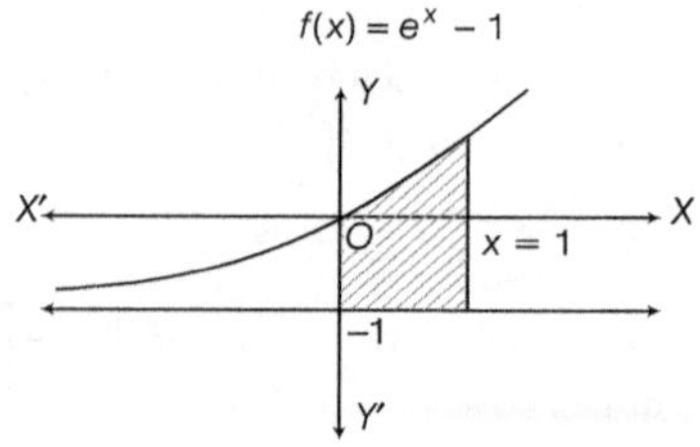


$\therefore$ Required area $= \int_0^1 (e^x - 1 + 1)\,dx = [e^x]_0^1 = (e - 1)$

**12.** (a) $y = ax^2 + bx + c$

When $x = 1$ and $y = 2$, then

$$2 = a + b + c \quad \text{...(i)}$$

When $x = 0$ and $y = 0$, then

$$\Rightarrow \quad c = 0$$

$$\therefore \quad a + b = 2 \quad \text{...(ii)}$$

Now, $\qquad \dfrac{dy}{dx} = [2ax + b]$

$\therefore \qquad \left(\dfrac{dy}{dx}\right)_{(0,\,0)} = b = 1 \qquad \text{...(iii)}$

[$\because$ slope of $y = x$ is 1]

From Eqs. (ii) and (iii), we have

$$a = 1 \text{ and } b = 1$$

Hence, the curve is $y = x^2 + x$.

The graph of the curve could be as shown.

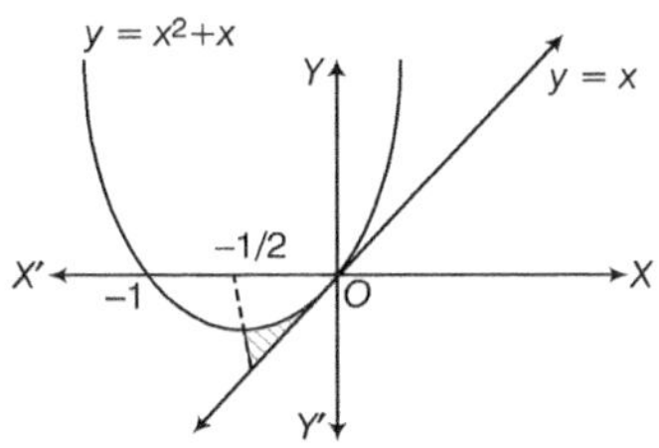


$\therefore$ Required area $= \displaystyle\int_{-1/2}^{0} (x^2 + x - x)\,dx$

$$= \int_{-1/2}^{0} x^2\,dx = \frac{1}{24} \text{ sq unit}$$

**13.** (a) Given differential equation is

$$\frac{dy}{dx} - y = \cos x - \sin x$$

which is linear.

$$\text{IF} = e^{-x}$$

$\therefore \qquad ye^{-x} = \displaystyle\int e^{-x}(\cos x - \sin x)dx$

Put $-x = t \Rightarrow -dx = dt$

$\therefore \qquad ye^{-x} = -\displaystyle\int e^{t}(\cos t + \sin t)\,dt$

$\Rightarrow \qquad ye^{-x} = -e^{t}\sin t + C$

$\Rightarrow \qquad y \cdot e^{-x} = e^{-x}\sin x + C$

Since, $y$ is bounded, when $x \to \infty \Rightarrow C = 0$

$\therefore \qquad y = \sin x$

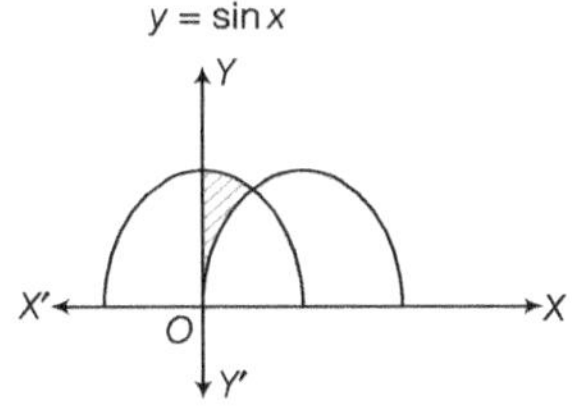


Hence, area $= \displaystyle\int_0^{\pi/4} (\cos x - \sin x)\,dx = \sqrt{2} - 1$

**14.** (a,c) $\qquad$ Area $(T) = \dfrac{c \cdot c^2}{2} = \dfrac{c^3}{2}$

Area $\qquad (R) = \dfrac{c^3}{2} - \displaystyle\int_0^c x^2\,dx$

$$= \frac{c^3}{2} - \frac{c^3}{3} = \frac{c^3}{6}$$

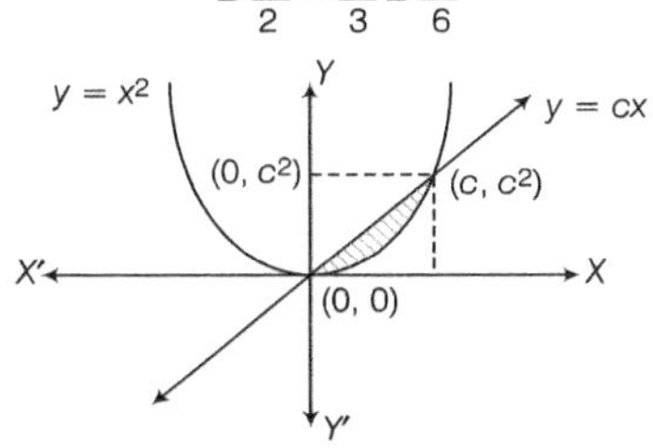


$\therefore \qquad \displaystyle\lim_{c \to 0^+} \frac{\text{area } (T)}{\text{area } (R)} = \lim_{c \to 0^+} \frac{c^3}{2} \cdot \frac{6}{c^3} = 3$

**15.** (a,c,d) $y = f(x) = \dfrac{x^2 - 1}{x^2 + 1} = 1 - \dfrac{2}{x^2 + 1}$

$$f'(x) = \frac{4x}{(x^2 + 1)^2},\ x > 0$$

$f(x)$ is increasing and $x < 0$, $f(x)$, is decreasing.

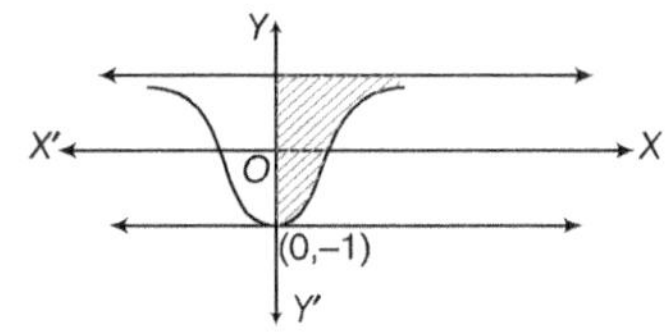


Range is $[-1, 1] \Rightarrow$ into

Minimum value occurs at $x = 0$ and $f(0) = -1$

$\therefore \qquad A = 2\displaystyle\int_0^{\infty}\left(1 - \frac{x^2 - 1}{x^2 + 1}\right)dx = 4\int_0^{\infty}\frac{dx}{x^2 + 1}$

$$= 4\,[\tan^{-1} x]_0^{\infty} = 4 \times \frac{\pi}{2} = 2\pi$$

**16.** (a,d) Given, $f(x) = \begin{cases} \cos x, & 0 \le x < \dfrac{\pi}{2} \\ \left(\dfrac{\pi}{2} - x\right)^2, & \dfrac{\pi}{2} \le x < \pi \end{cases}$

and $f$ is periodic with period $\pi$.

The graph of $f(x)$ could be plotted as

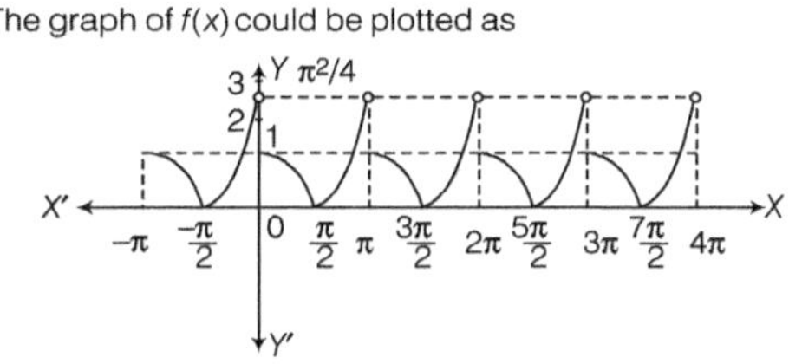


From the graph, the range of function is $\left[0, \dfrac{\pi^2}{4}\right]$.

It is discontinuous at $x = n\pi$, $n \in I$.

It is not differentiable at $x = \dfrac{n\pi}{2}$, $n \in I$

Area bounded by $y = f(x)$ and the X-axis from $-n\pi$ to $n\pi$ for $n \in N$.

$\therefore \qquad$ Area $= 2n \displaystyle\int_0^{\pi} f(x)\,dx = 2n\left[\int_0^{\pi/2} \cos x\,dx + \int_{\pi/2}^{\pi}\left(\frac{\pi}{2} - x\right)^2 dx\right]$

$$= 2n\left(1 + \frac{\pi^3}{24}\right)$$

**17.** (b,c,d) The graph of $f(x) = (x - 1)^2(x - 2) + 1$ on $[0, 2]$ is shown as

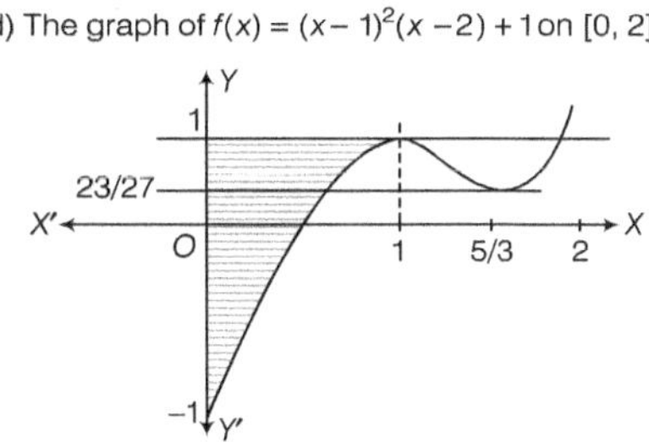


Here, $f(1) = f(2) = 1$

Range $\in [-1, 1]$.

Turning points are $(1, 1)$ and $(5/3, 23/27)$.

The value of $p$ for which $f(x) = p$ has 3 solutions, lies in the interval $(23/27, 1)$.

$\therefore \qquad$ Area $= \displaystyle\int_0^1 (1 - [(x - 1)^2 (x - 2) + 1])dx = \frac{7}{12}$

**18.** (b,c,d) Given, $f(x) = 2x - x^2$ ...(i)

and $g(x) = x^n$ ...(ii)

On solving Eqs. (i) and (ii), we get

$$2x - x^2 = x^n$$

$\Rightarrow$ $x = 0$ and $x = 1$

$\therefore$ $$\text{Area}(A) = \int_0^1 (2x - x^2 - x^n)\,dx$$

$$= \left[x^2 - \frac{x^3}{3} - \frac{x^{n+1}}{n+1}\right]_0^1$$

$$= 1 - \frac{1}{3} - \frac{1}{n+1} = \frac{2}{3} - \frac{1}{n+1}$$

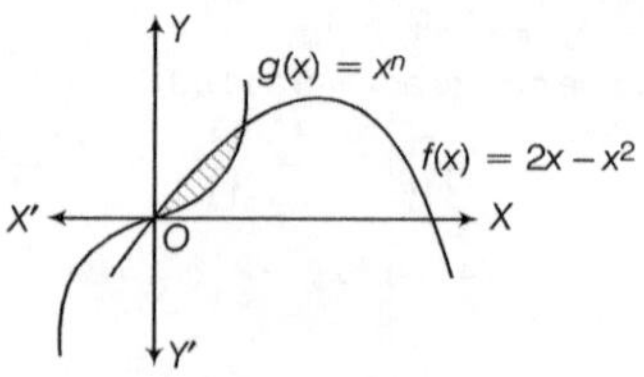


According to the question,

$$\frac{2}{3} - \frac{1}{n+1} = \frac{1}{2}$$

$\Rightarrow$ $$\frac{2}{3} - \frac{1}{2} = \frac{1}{n+1}$$

$\Rightarrow$ $$\frac{4-3}{6} = \frac{1}{n+1}$$

$\Rightarrow$ $n + 1 = 6$

$\Rightarrow$ $n = 5$

Thus, $n$ is divisor of 15, 20 and 30.

**19.** (b,c) Area of the region bounded by $y = x^2 - x$ and $y = mx$ is

$$S_m = -\int_0^{m+1} x[x - (m+1)]dx = \frac{(m+1)^3}{6}$$

$\therefore$ $$\int_\alpha^\beta (x - \alpha)(\alpha - \beta)\,dx = \frac{-(\beta - \alpha)^3}{6}, (\alpha < \beta)$$

Area of the region is $S_m - S_n = \dfrac{37}{6}$

$\Rightarrow$ $(m+1)^3 - (n+1)^3 = 37$

Let $(m+1) = a$ and $(n+1) = b$

$\therefore$ $(a - b)(a^2 + ab + b^2) = 37$, since 37 is prime.

$\Rightarrow$ $a - b = 1$ and $a^2 + ab + b^2 = 37$

On solving above equations, we get

$$a = 4, b = 3$$

$\therefore$ $m = 3, n = 2$

**20.** (a,c) Given, $|x - a_i| + |y| = b_i$, where $i \in N$

where $$a_{i+1} = a_i + \frac{3}{2} b_i$$

and $$b_{i+1} = \frac{b_i}{2}$$

$\Rightarrow$ $a_1 = 0, b_1 = 32$

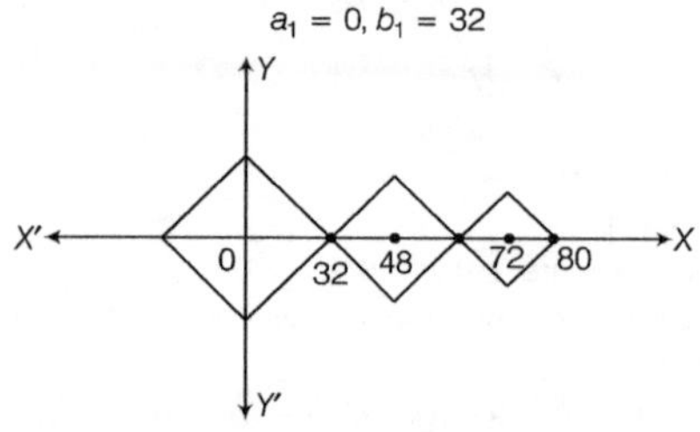


Now, $$a_2 = a_1 + \frac{3}{2} b_1 = 48,$$

and $$b_2 = \frac{b_1}{2} = 16$$

$\Rightarrow$ $$a_3 = 48 + \frac{3}{2} \times 16 = 72$$

and $$b_3 = \frac{16}{2} = 8$$

So, the three loops from $i = 1$ to $i = 3$ are alike.

Now, area of the loop (square) $= \frac{1}{2}$ (Diagonal)$^2$

$$A_i = \frac{1}{2}(2b_i)^2 = 2(b_i)^2$$

So, $$\frac{A_i + 1}{A_i} = \frac{2(b_i)^2 + 1}{2(b_i)^2} = \frac{1}{4}$$

So, the areas form a geometric progression.

$\therefore$ The sum of the GP upto infinite terms is

$$\lim_{n \to \infty} \sum_{i=1}^{n} A_i = A_1 \cdot \frac{1}{1 - r}$$

$$= 2(32)^2 \times \frac{1}{1 - \frac{1}{4}}$$

$$= 2 \times (32)^2 \times \frac{4}{3}$$

$$= \frac{8}{3}(32)^2 \text{ sq units}$$

Also, $$A_3 = 2(b_3)^2$$

$$= 2 \times (8)^2 = 128$$

**21.** (c) $f(x) = (x - 1)(x^2 - 7x + 13)$ for $f(x)$ to be prime atleast one of the factors must be prime.

Therefore, $x - 1 = 1$

$\Rightarrow$ $x = 2$ or $x^2 - 7x + 13 = 1$

$\Rightarrow$ $x^2 - 7x + 12 = 0$

$\Rightarrow$ $x = 3$ or $4$

$\Rightarrow$ $x = 2, 3, 4$

**22.** (b) $f(x)$ is many-one as it increases and decreases, also range of $f(x) \in R$

$\Rightarrow$ Many-one and onto.

**23.** (a) Required area $= \left|\int_0^1 f(x)\,dx\right|$

$$= -\int_0^1 (x^3 - 8x^2 + 20x - 13)dx$$

$$= -\left[\frac{x^4}{4} - \frac{8}{3}x^3 + 10x^2 - 13x\right]_0^1$$

$$= \frac{65}{12} \text{ sq units}$$

**24.** (d) Given, $f(x) = 3 + \displaystyle\int_0^x \frac{1 + \sin t}{t^2 + 2}\,dt$

$\Rightarrow$ $$f'(x) = 0 + \left(\frac{1 + \sin x}{x^2 + 2}\right)$$ [by Leibnitz's rule]

$\Rightarrow$ $$f'(x) = \frac{1 + \sin x}{x^2 + 2}$$

$\Rightarrow$ $$f'(0) = \frac{1}{2} = P'(0)$$ [given]

Now, $$f''(x) = \frac{(x^2 + 2)\cos x - (1 + \sin x)\,2x}{(x^2 + 2)^2}$$

$\Rightarrow$ $$f''(x) = \frac{(x^2 + 2)\cos x - 2x - 2x\sin x}{(x^2 + 2)^2}$$

$\Rightarrow \quad f''(0) = \frac{2}{4} = \frac{1}{2} = P''(0)$ [given]

and $\quad f(0) = 3 = P(0)$ [given]

Let $\quad p(x) = ax^2 + bx + c$

$P(0) = c = 3$

Now, $\quad p'(x) = 2ax + b$

$\Rightarrow \quad p'(0) = b = \frac{1}{2}$

and $\quad p''(x) = 2a = \frac{1}{2}$

$\Rightarrow \quad a = \frac{1}{4}$

Hence, $\quad p(x) = \frac{1}{4}x^2 + \frac{1}{2}x + 3$

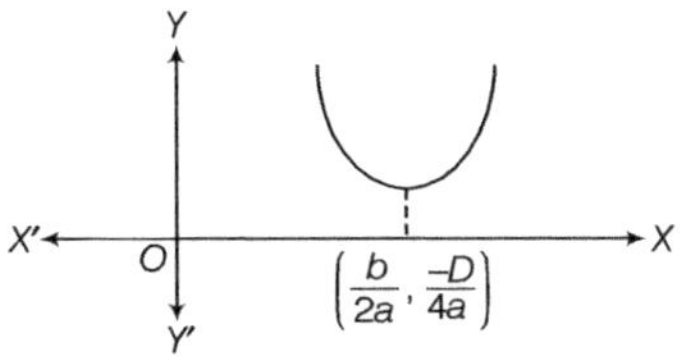


$\therefore$ Least value of $p(x)$ is

$$\frac{-D}{4a} = \frac{-\left(-\frac{11}{4}\right)}{4 \times \frac{1}{4}} = \frac{11}{4}$$

**25.** (a) $p(-2) = 1 - 1 + 3 = 3$

$\therefore$ Slope of tangent at $(-2, 3)$ to the curve $P(x)$ is

$$P'(x) = \frac{1}{2}x + \frac{1}{2}$$

Hence, $\quad \text{slope} = -\frac{1}{2}$

**26.** (d) Equation of tangent at $(-2, 3)$ to the curve $P(x)$ is

$$y - 3 = -\frac{1}{2}(x + 2)$$

$\Rightarrow \quad y - 3 = -\frac{1}{2}x - 1$

$\Rightarrow \quad y = -\frac{1}{2}x + 2$

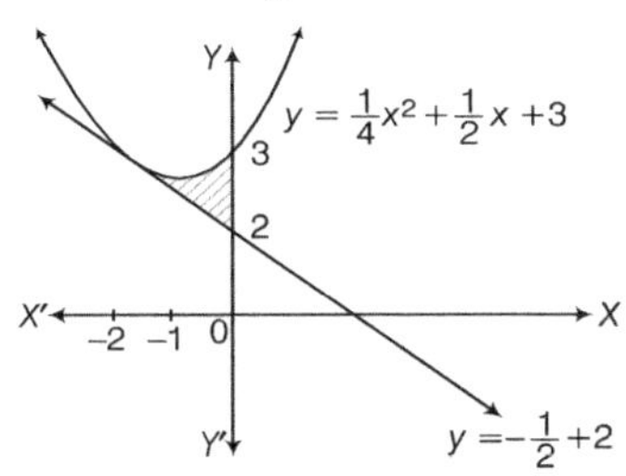


$\therefore \quad \text{Required area} = \int_{-2}^{0}\left[\left(\frac{x^2}{4} + \frac{x}{2} + 3\right) + \frac{x}{2} - 2\right]dx$

$= \int_{-2}^{0}\left(\frac{x^2}{4} + x + 1\right)dx$

$= \left[\frac{x^3}{12} + \frac{x^2}{2} + x\right]_{-2}^{0}$

$= -\left(-\frac{8}{12} + 2 - 2\right)$

$= \frac{8}{12} = \frac{2}{3}$

**27.** (d) Given, $f[x + f(y) + xf(y)] = y + f(x) + yf(x), \forall\ x, y \in A$

Keep $y$ constant and differentiating the expression w.r.t. $x$, we get

$$f'[x + f(y) + x(y)]\,[1 + f(y)] = f'(x)\,(1 + y) \quad \ldots\text{(i)}$$

Similarly, differentiate the expression w.r.t. $y$ and keep $x$ constant, we get

$$f'[x + f(y) + xf(y)]\,[(f'(y)\,(1 + x)] = 1 + f(x) \quad \ldots\text{(ii)}$$

On dividing Eq. (i) by Eq. (ii), we get

$$\frac{1 + f(y)}{f'(y)\,(1 + x)} = \frac{f'(x)\,(1 + y)}{[1 + f(x)]}$$

$\Rightarrow \quad \frac{1 + f(y)}{f'(y)\,(1 + y)} = \frac{f'(x)\,(1 + x)}{[1 + f(x)]} = C$

$\Rightarrow \quad f'(y) = \frac{1}{C}\left[\frac{1 + f(y)}{1 + y}\right]$

and $\quad f'(x) = C \cdot \frac{1 + f(x)}{1 + x}$

$\Rightarrow \quad C = \frac{1}{C}$

$\Rightarrow \quad C = \pm 1$

$\therefore \quad \frac{f'(x)}{1 + f(x)} = \pm \frac{1}{1 + x}$

$\Rightarrow \quad f(x) = \lambda_1 (1 + x)^{\pm 1} - 1$

On replacing $x, y \to 0$ in the Eq. (i), we get

$$f[f(0)] = f(0)$$

Now, $\quad f(x) = f(\lambda_1 - 1) = \lambda_1 \cdot \lambda_1^{\pm 1} - 1$

Since, $\quad f[f(0)] = f(0)$

$$\lambda_1 \cdot \lambda_1^{\pm 1} - 1 = \lambda_1 - 1$$

$\Rightarrow \quad \lambda_1^{1 \pm 1} = \lambda_1$

By taking positive sign, we get $\lambda_1 = 0, 1$

$$f(x) = -1 \text{ or } f(x) = x$$

By taking negative sign, we get

$$\lambda_1 = 1$$

$\Rightarrow \quad f(x) = \frac{1}{1 + x} - 1 = \frac{-x}{1 + x}$

**28.** (a) Required area $= \int_0^2\left(-\frac{x}{1 + x} - 1\right)dx = \log_e 2$

**29.** (d) At $x = 0$, $f(0) = 0$

Neither maxima nor minima nor a point of inflection.

**30.** (a) The graph of $f(|x|)$ is

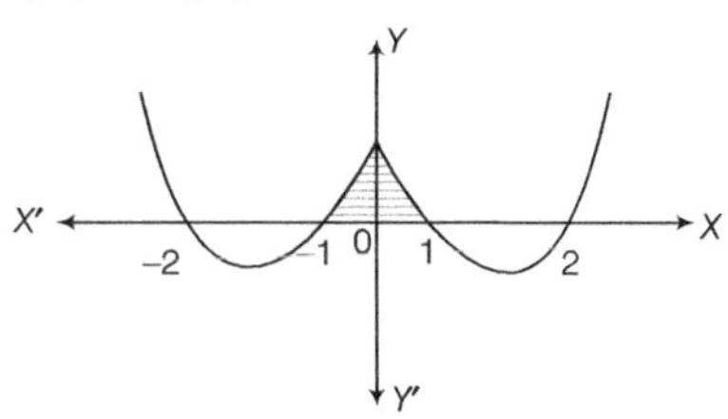


$\therefore$ Required area $= 2\int_0^1 f(x)\,dx + 2\left|\int_1^2 f(x)dx\right|$

$= 2\int_0^1 (x^2 - 3x + 2)\,dx + 2\left|\int_1^2 (x^2 - 3x + 2)dx\right|$

$= 2\left(\frac{1}{3} - \frac{3}{2} + 2\right) + \frac{1}{3}$

$= \frac{5}{3} + \frac{1}{3}$

$= 2$ sq units

**31.** (c) $\because 1 \le |x| \le 2$

$\therefore \quad x \in (-2, -1] \cup [1, 2]$

The graph of $|f(|x|)|$ is

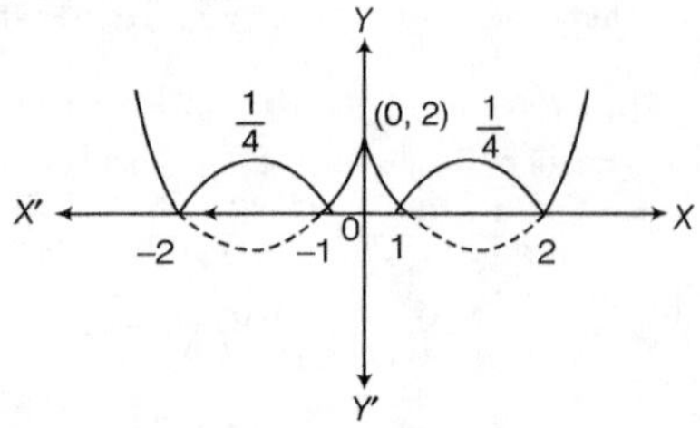


Hence, required area $= 2\left|\int_1^2 f(x)\,dx\right|$

$$= 2\left|\int_1^2 (x^2 - 3x + 2)\,dx\right|$$

$$= 2\left|\left\{\frac{x^3}{3} - \frac{3x^2}{2} + 2x\right\}_1^2\right|$$

$$= 2\left|\left\{\left(\frac{8}{3} - \frac{12}{2} + 4\right) - \left(\frac{1}{3} - \frac{3}{2} + 2\right)\right\}\right|$$

$$= 2\left|\frac{2}{3} - \frac{5}{6}\right| = \frac{1}{3} \text{ sq unit}$$

**32.** (d) From the figure in Q. 31

The line $y = \frac{1}{4}$ intersects the curve $|f(|x|)|$ at 6 points.

$\therefore$ For six real roots of $|f(|x|)| = a$

$\Rightarrow \quad a = \frac{1}{4}$

**33.** (2) $\quad f(x) = x^2 + \int_0^x e^{-t} f(x - t)\,dt$

$$= x^2 + \int_0^x e^{-(x-t)} \cdot f(t)\,dt$$

$$= x^2 + e^{-x} \cdot \int_0^x e^t f(t)\,dt$$

$$f'(x) = 2x + e^{-x}[e^x f(x)] - e^{-x} \cdot \int_0^x e^t f(t)\,dt$$

$$= 2x + f(x) - [f(x) - x^2]$$

$$f'(x) = x^2 + 2x$$

$\Rightarrow \quad f(x) = \frac{x^3}{3} + x^2 + C$, as $f(0) = 0$

$\therefore \quad f(x) = \frac{x^3}{3} + x^2$

Now, $f(1) + f(2) + \cdots + f(a)$

$$= \frac{1}{3}(1^3 + 2^3 + \cdots + 9^3) + (1^2 + 2^2 + \cdots + 9^2)$$

$$= 960 = 2(480)$$

$\therefore \quad \lambda = 2$

**34.** (5) $A = \int_0^1 x^6 (1 - x)^7 dx = \int_0^1 x^{7-1}(1 - x)^{8-1} dx$

$$= B(7, 8) \quad \left[\because B(m, n) = \int_0^1 x^{m-1}(1 - x)^{n-1} dx\right]$$

$$= \frac{\Gamma 7\, \Gamma 8}{\Gamma(7 + 8)} \quad \left[\because B(m, n) = \frac{\Gamma m\, \Gamma n}{\Gamma(m + n)}\right]$$

$\therefore \quad A = \frac{6!\,7!}{14!} \quad [\because \Gamma(n + 1) = n!]$

$\Rightarrow \quad A^{-1} = \frac{14!}{6! \cdot 7!} = 2^3 \cdot 3 \cdot 7 \cdot 11 \cdot 13$

$\therefore$ The number of prime factors is 5.

**35.** (4) Given curves are

$$x^2 = 4y \quad \ldots \text{(i)}$$

and $\quad y = \frac{8}{x^2 + 4} \quad \ldots$(ii)

For point of intersection on solving Eqs. (i) and (ii), we get

$$y = \frac{8}{4y + 4} \Rightarrow y = \frac{2}{y + 1}$$

$\Rightarrow \quad y^2 + y = 2$

$\Rightarrow \quad y^2 + y - 2 = 0$

$\Rightarrow \quad y^2 + 2y - y - 2 = 0$

$\Rightarrow \quad y(y + 2) - 1(y + 2) = 0$

$\Rightarrow \quad (y + 2)(y - 1) = 0$

$\Rightarrow \quad y = 1, -2$

But $y = -2$ is not possible.

When $y = 1$, then $x = \pm 2$

$\therefore$ Points of intersection are $(2, 1)$ and $(-2, 1)$.

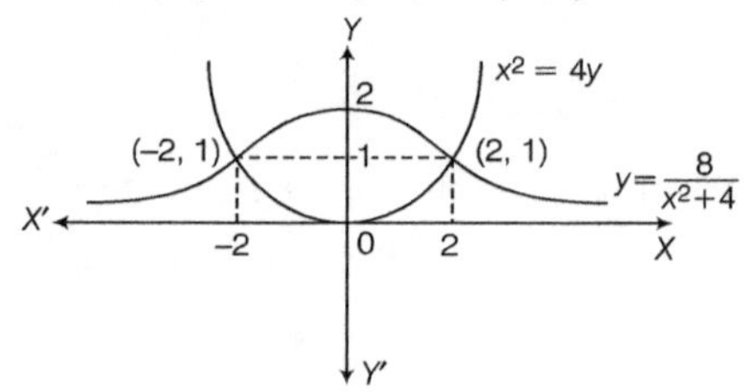


$\therefore$ Required area $= \int_{-2}^2 \left(\frac{8}{x^2 + 4} - \frac{x^2}{4}\right) dx$

$$= 8\left[\tan^{-1}\frac{x}{2}\right]_{-2}^2 - \frac{1}{12}[x^3]_{-2}^2$$

$$= 8[\tan^{-1} 1 - \tan^{-1}(-1)] - \frac{1}{12}(8 + 8)$$

$$= 8\left(\frac{\pi}{4} + \frac{\pi}{4}\right) - \frac{16}{12} = \frac{8\pi}{2} - \frac{16}{12}$$

$$= 4\pi - \frac{16}{12} = 4\pi - \frac{4}{3}$$

Hence, $k = 4$

**1.** (d) Given region is $\{(x, y) : y^2 \le 2x \text{ and } y \ge 4x - 1\}$

$y^2 \le 2x$ represents a region inside the parabola

$$y^2 = 2x \quad \ldots(i)$$

and $y \ge 4x - 1$ represents a region to the left of the line

$$y = 4x - 1 \quad \ldots(ii)$$

The point of intersection of the curves (i) and (ii) is

$$(4x - 1)^2 = 2x$$

$\Rightarrow \quad 16x^2 + 1 - 8x = 2x$

$\Rightarrow \quad 16x^2 - 10x + 1 = 0$

$\Rightarrow \quad x = \frac{1}{2}, \frac{1}{8}$

$\therefore$ The points where these curves intersect, are $\left(\frac{1}{2}, 1\right)$ and $\left(\frac{1}{8}, -\frac{1}{2}\right)$.

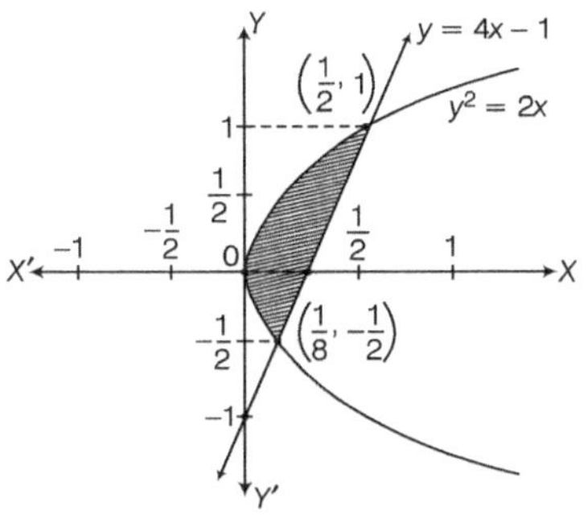


Hence, required area

$$= \int_{-1/2}^{1} \left(\frac{y+1}{4} - \frac{y^2}{2}\right) dy$$

$$= \frac{1}{4}\left[\frac{y^2}{2} + y\right]_{-1/2}^{1} - \frac{1}{6}[y^3]_{-1/2}^{1}$$

$$= \frac{1}{4}\left\{\left(\frac{1}{2} + 1\right) - \left(\frac{1}{8} - \frac{1}{2}\right)\right\} - \frac{1}{6}\left\{1 + \frac{1}{8}\right\}$$

$$= \frac{1}{4}\left\{\frac{3}{2} + \frac{3}{8}\right\} - \frac{1}{6}\left\{\frac{9}{8}\right\}$$

$$= \frac{1}{4} \times \frac{15}{8} - \frac{3}{16} = \frac{9}{32} \text{ sq units}$$

**2.** (a) $A = \{(x, y) : x^2 + y^2 \le 1 \text{ and } y^2 \le 1 - x\}$

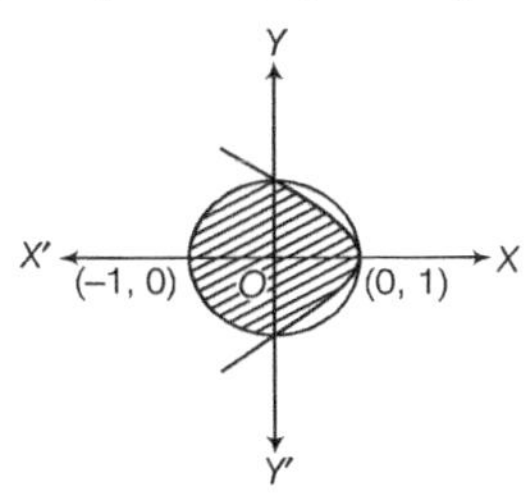


$\therefore$ Required area $= \frac{1}{2}\pi r^2 + 2\int_0^1 (1 - y^2) dy$

$$= \frac{1}{2}\pi(1)^2 + 2\left[y - \frac{y^3}{3}\right]_0^1$$

$$= \left(\frac{\pi}{2} + \frac{4}{3}\right) \text{ sq units}$$

**3.** (a) Given curves are

$$y = \sqrt{x} \quad \ldots(i)$$

and $\quad 2y - x + 3 = 0 \quad \ldots(ii)$

On solving Eqs. (i) and (ii), we get

$$2\sqrt{x} - (\sqrt{x})^2 + 3 = 0$$

$\Rightarrow \quad (\sqrt{x})^2 - 2\sqrt{x} - 3 = 0$

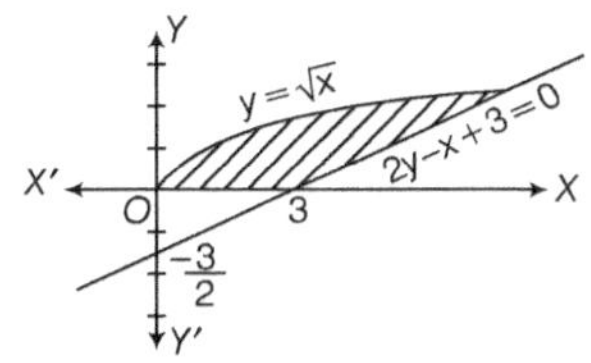


$\Rightarrow \quad (\sqrt{x} - 3)(\sqrt{x} + 1) = 0$

$\Rightarrow \quad \sqrt{x} = 3 \quad [\because \sqrt{x} = -1 \text{ is not possible}]$

$\therefore \quad y = 3$

$\therefore$ Required area $= \int_0^3 (x_2 - x_1) dy = \int_0^3 \{(2y + 3) - y^2\} dy$

$$= \left[y^2 + 3y - \frac{y^3}{3}\right]_0^3$$

$$= 9 + 9 - 9 = 9 \text{ sq units}$$

**4.** (c) **Given** Two parabolas $x^2 = \frac{y}{4}$ and $x^2 = 9y$

**To find** The area bounded between the parabolas and the straight line $y = 2$.

The required area is equal to the shaded region in the drawn figure.

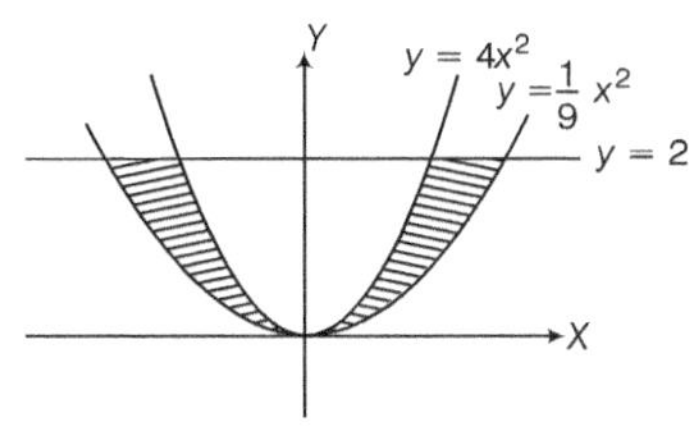


The area of the shaded region (which can be very easily found by using integration) is twice the area shaded in first quadrant.

$\therefore$ Required area $= 2\int_0^2 \left(3\sqrt{y} - \frac{\sqrt{y}}{2}\right) dy$

$$= 2\int_0^2 \left(\frac{5}{2}\sqrt{y}\right) dy = 5\left[\frac{y^{3/2}}{3/2}\right]_0^2$$

$$= \frac{10}{3}[2^{3/2} - 0] = \frac{20\sqrt{2}}{3} \text{ sq units}$$

**5.** (b) Given, $y = x$, $x = e$ and $y = \frac{1}{x}$, $x \ge 0$

Since, $y = x$ and $x \ge 0 \Rightarrow y \ge 0$

$\therefore$ Area to be calculated in I quadrant shown as

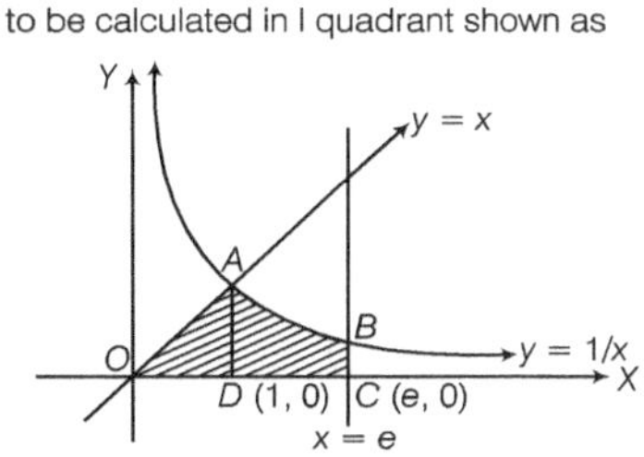

$\therefore$ Required area = Area of $\Delta ODA$ + Area of $ABCD$

$$= \frac{1}{2}(1 \times 1) + \int_1^e \frac{1}{x}\,dx = \frac{1}{2} + [\log|x|]_1^e$$

$$= \frac{1}{2} + \{\log|e| - \log 1\}$$

$$= \frac{1}{2} + 1 = \frac{3}{2} \text{ sq units}$$

**6.** (c) For the point of intersection, solve $y^2 = 4x$ and $x^2 = 4y$.

$$\Rightarrow \quad \left(\frac{x^2}{4}\right)^2 = 4x$$

$$\Rightarrow \quad x^4 = 4^3 x \Rightarrow x = 0, 4$$

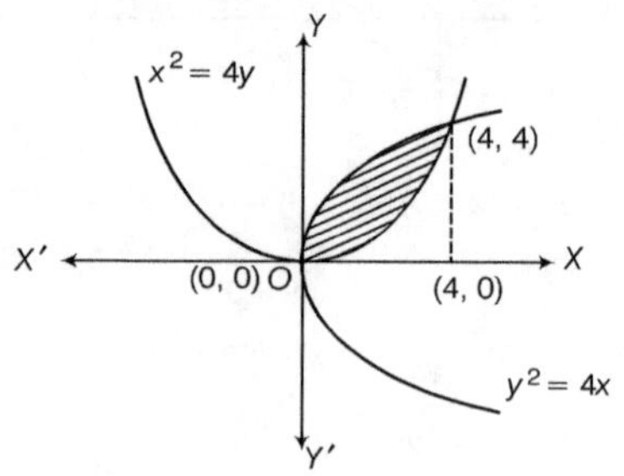


$\therefore$ Area bounded between curves

$$= \int_0^4 \left(\sqrt{4x} - \frac{x^2}{4}\right) dx = \left[2 \cdot \frac{x^{3/2}}{\frac{3}{2}} - \frac{x^3}{12}\right]_0^4$$

$$= \frac{4}{3} \cdot (4)^{3/2} - \frac{(4)^3}{12} = \frac{32}{3} - \frac{16}{3}$$

$$= \frac{16}{3} \text{ sq units}$$

**7.** (a)

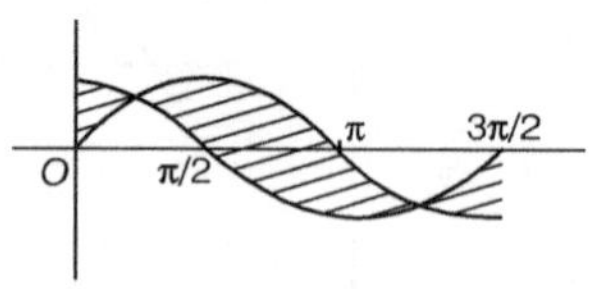


$$\therefore \text{Required area} = \int_0^{\pi/4} (\cos x - \sin x)\,dx + \int_{\pi/4}^{5\pi/4} (\sin x - \cos x)\,dx + \int_{5\pi/4}^{3\pi/2} (\cos x - \sin x)\,dx$$

$$= [\sin x + \cos x]_0^{\pi/4} + [-\cos x - \sin x]_{\pi/4}^{5\pi/4} + [\sin x + \cos x]_{5\pi/4}^{3\pi/2}$$

$$= \left(\frac{1}{\sqrt{2}} + \frac{1}{\sqrt{2}} - 1\right) + \left[\left(\frac{1}{\sqrt{2}} + \frac{1}{\sqrt{2}}\right) - \left(\frac{-1}{\sqrt{2}} - \frac{1}{\sqrt{2}}\right)\right] + \left[(-1 + 0) - \left(\frac{-1}{\sqrt{2}} - \frac{1}{\sqrt{2}}\right)\right]$$

$$= \frac{8}{\sqrt{2}} - 2 = (4\sqrt{2} - 2) \text{ sq units}$$

**8.** (b) The equation of tangent at (2, 3) to the given parabola is

$$x = 2y - 4$$

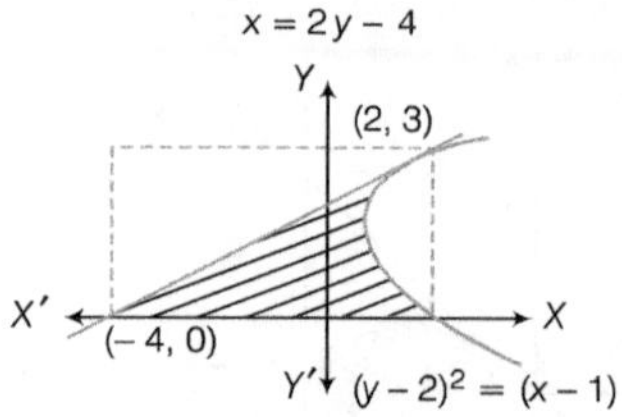


$$\therefore \text{Required area} = \int_0^3 \{(y-2)^2 + 1 - 2y + 4\}\,dy$$

$$= \left[\frac{(y-2)^3}{3} - y^2 + 5y\right]_0^3 = \frac{1}{3} - 9 + 15 + \frac{8}{3}$$

$$= 9 \text{ sq units}$$

**9.** (a) Given curves are

$$x + 3y^2 = 1 \quad \ldots \text{(i)}$$

and

$$x + 2y^2 = 0 \quad \ldots \text{(ii)}$$

On solving Eqs. (i) and (ii), we get

$$y = \pm 1 \text{ and } x = -2$$

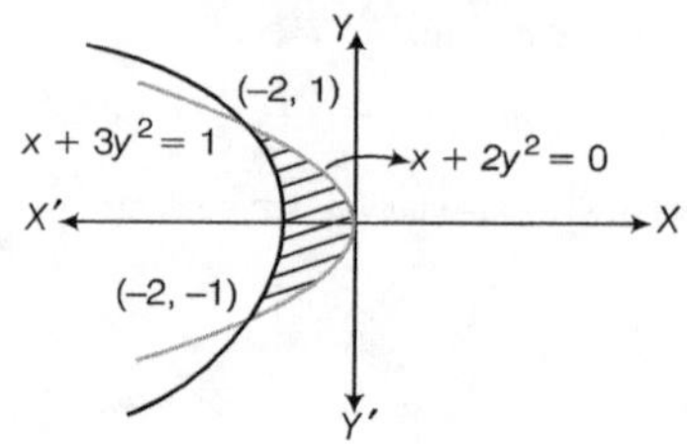


$$\therefore \text{Required area} = \left|\int_{-1}^1 (x_1 - x_2)\,dy\right|$$

$$= \left|\int_{-1}^1 (1 - 3y^2 + 2y^2)\,dy\right|$$

$$= \left|\int_{-1}^1 (1 - y^2)\,dy\right|$$

$$= \left|2\int_0^1 (1 - y^2)\,dy\right|$$

[since, integral is an even function.]

$$= \left|2\left[y - \frac{y^3}{3}\right]_0^1\right| = \left|2\left(1 - \frac{1}{3}\right)\right|$$

$$= \frac{4}{3} \text{ sq units}$$

**10.** (c) Required area,

$$A = \int_0^1 (\sqrt{x} - x)\,dx$$

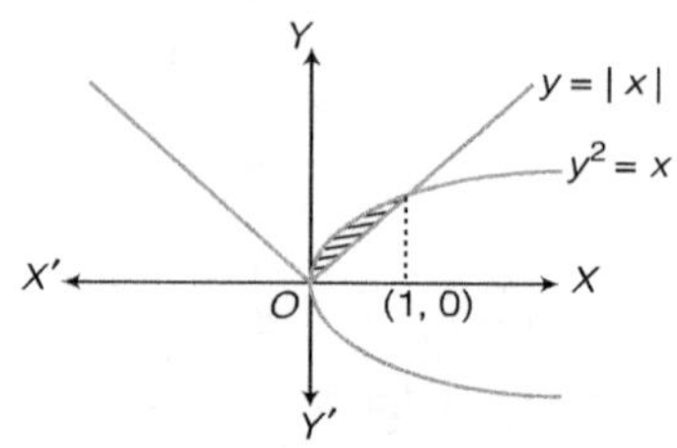


$$= \left[\frac{2}{3}x^{3/2} - \frac{x^2}{2}\right]_0^1$$

$$= \frac{2}{3} - \frac{1}{2}$$

$$= \frac{1}{6} \text{ sq unit}$$

**11.** (a) According to the given condition,

$$\int_{\pi/4}^{\beta} f(x)\,dx = \beta \sin\beta + \frac{\pi}{4}\cos\beta + \sqrt{2}\,\beta$$

On differentiating both sides w.r.t. $\beta$, we get

$$f(\beta) = \sin\beta + \beta\cos\beta - \frac{\pi}{4}\sin\beta + \sqrt{2}$$

$$\therefore \quad f\left(\frac{\pi}{2}\right) = 1 + 0 - \frac{\pi}{4} + \sqrt{2} = 1 - \frac{\pi}{4} + \sqrt{2}$$

**12.** (d) Required area,

$$A = \int_{1-e}^{0} \log_e (x + e)\, dx$$

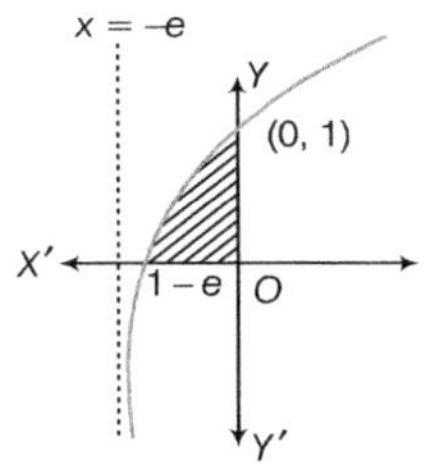


Put $x + e = t$

$\Rightarrow \quad dx = dt$

$\therefore \; A = \int_1^e \log_e t \, dt = [t \log_e t - t]_1^e$

$= (e - e - 0 + 1) = 1$ sq unit

**13.** (a) It is clear from the figure that

$$S_1 = S_3 = \int_0^4 y\, dx = \int_0^4 \frac{x^2}{4} dx = \frac{1}{4}\left[\frac{x^3}{3}\right]_0^4$$

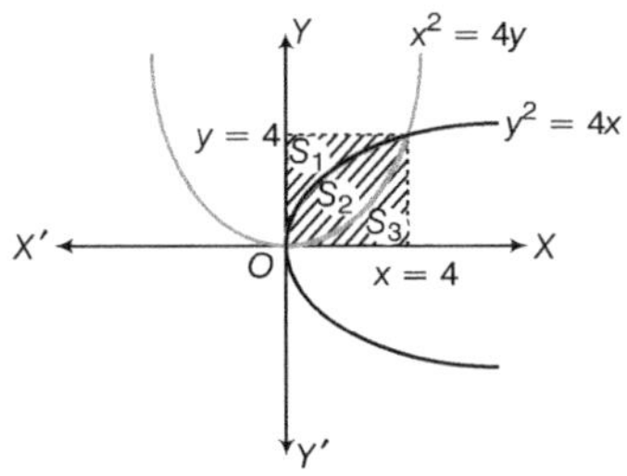


$\Rightarrow \quad S_1 = S_3 = \frac{1}{12} \times 64 = \frac{16}{3}$ sq units ...(i)

and $\quad S_2 + S_3 = \int_0^4 \sqrt{4x}\, dx$

$= 2\left[\frac{x^{3/2}}{3/2}\right]_0^4 = \frac{4}{3} \times 8$

$\Rightarrow \quad S_2 = \frac{32}{3} - \frac{16}{3}$ [from Eq. (i)]

$\Rightarrow \quad S_2 = \frac{16}{3}$ sq units

$\therefore \quad S_1 : S_2 : S_3 = \frac{16}{3} : \frac{16}{3} : \frac{16}{3}$

$= 1 : 1 : 1$

**14.** (a) Required area $= \int_1^3 y dx = \int_1^3 |x - 2|\, dx$

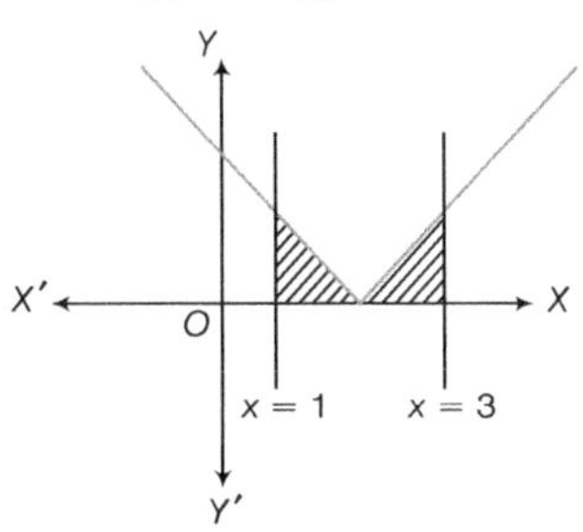


$= \int_1^2 -(x-2)\, dx + \int_2^3 (x-2)\, dx$

$= \int_1^2 (2 - x)\, dx + \int_2^3 (x-2)\, dx$

$= \left[2x - \frac{x^2}{2}\right]_1^2 + \left[\frac{x^2}{2} - 2x\right]_2^3$

$= \left[4 - 2 - \left(2 - \frac{1}{2}\right)\right] + \left[\frac{9}{2} - 6 - (2 - 4)\right]$

$= 2 - \frac{3}{2} - \frac{3}{2} + 2$

$= 1$ sq unit

**15.** (c) Since, $\quad y = |x - 1| = \begin{cases} x - 1, & x > 1 \\ -x + 1, & x \le 1 \end{cases}$

and $\quad y = 3 - |x| = \begin{cases} 3 + x, & x \le 0 \\ 3 - x, & x > 0 \end{cases}$

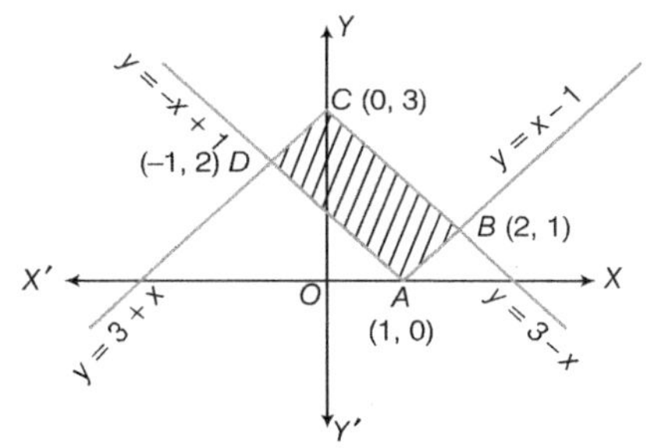


On solving, we get

$y = x - 1$

and $\quad y = 3 - x$

$\Rightarrow \quad x - 1 = 3 - x$

$\Rightarrow \quad x = 2$

and $\quad y = 3 - 2$

$\Rightarrow \quad y = 1$

Now, $\quad AB^2 = (2 - 1)^2 + (1 - 0)^2$

$= 1 + 1 = 2$

$\Rightarrow \quad AB = \sqrt{2}$

and $\quad BC^2 = (0 - 2)^2 + (3 - 1)^2$

$= 4 + 4 = 8$

$\Rightarrow \quad BC = 2\sqrt{2}$

$\therefore$ Area of rectangle $ABCD = AB \times BC$

$= \sqrt{2} \times 2\sqrt{2}$

$= 4$ sq units

**16.** (a) The equations of given curve and a line are

$y = 2x - x^2$ ...(i)

and $\quad y = -x$ ...(ii)

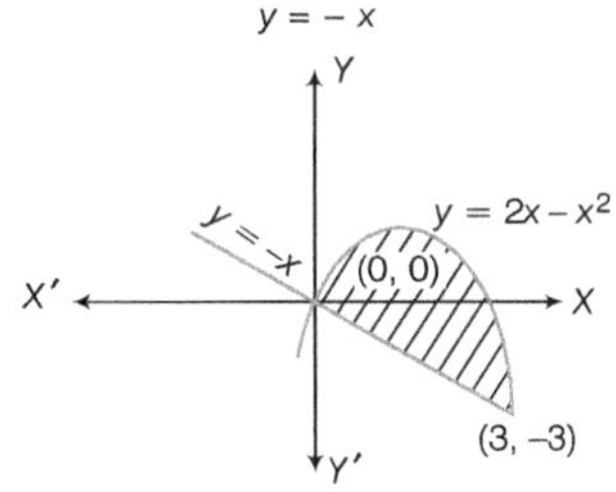


On solving Eqs. (i) and (ii), we get the points of intersection of curves which are (0, 0) and (3, –3).

$\therefore$ Required area $= \int_0^3 [(2x - x^2) - (-x)]\, dx$

$= \int_0^3 (3x - x^2)\, dx = \left[\frac{3x^2}{2} - \frac{x^3}{3}\right]_0^3$

$= \frac{27}{2} - \frac{27}{3} = \frac{9}{2}$ sq units

# JEE Advanced & IIT JEE Archive

**1.** (b) **Concept Involved**

To find the bounded area between $y = f(x)$ and $y = g(x)$ of limits $x = a$ to $x = b$.

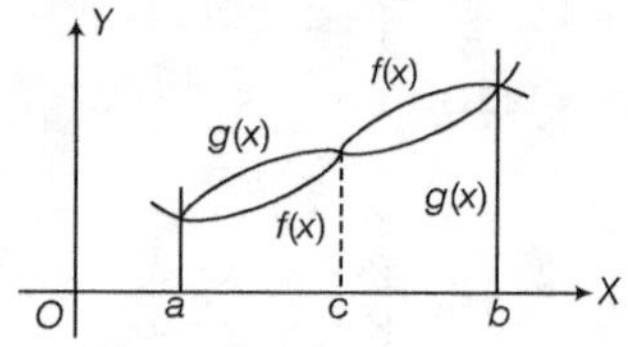


$\therefore$ Area bounded $= \int_a^c [g(x) - f(x)]dx + \int_c^b [f(x) - g(x)]dx$

$= \int_a^b |f(x) - g(x)|dx$

Here, $f(x) = y = \sin x + \cos x$, when $0 \le x \le \frac{\pi}{2}$

and $g(x) = y = |\cos x - \sin x|$

$$= \begin{cases} \cos x - \sin x, & 0 \le x \le \frac{\pi}{4} \\ \sin x - \cos x, & \frac{\pi}{4} \le x \le \frac{\pi}{2} \end{cases}$$

could be shown as

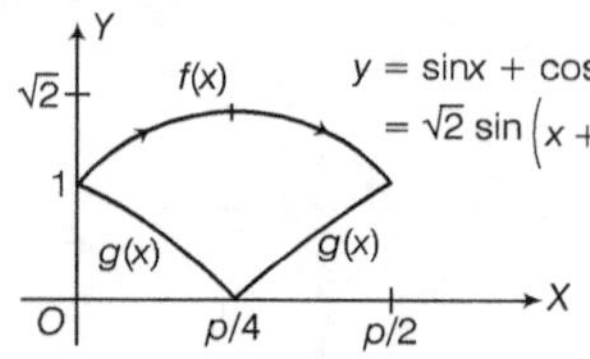


$\therefore$ Area bounded

$= \int_0^{\pi/4} \{(\sin x + \cos x) - (\cos x - \sin x)\}dx$

$+ \int_{\pi/4}^{\pi/2} \{(\sin x + \cos x) - (\sin x - \cos x)\}dx$

$= \int_0^{\pi/4} 2\sin x dx + \int_{\pi/4}^{\pi/2} 2\cos x dx$

$= -2[\cos x]_0^{\pi/4} + 2[(\sin x)]_{\pi/4}^{\pi/2}$

$= 4 - 2\sqrt{2}$

$= 2\sqrt{2}(\sqrt{2} - 1)$

**2.** (d) **Plan**

(i) $y = mx + a/m$ is equation of tangent to the parabola $y^2 = 4ax$.

(ii) A line is a tangent to circle, if distance of line from centre is equal to the radius of circle

(iii) Equation of chord drawn from exterior point $(x_1, y_1)$ to a circle/parabola is given by $T = 0$

(iv) Area of Trapezium $= \frac{1}{2}$ (Sum of parallel sides) $\times$ (Distance between them)

Let equation of tangent to parabola be $y = mx + \frac{2}{m}$

It also touches the circle $x^2 + y^2 = 2$.

$\therefore \quad \left| \frac{2}{m\sqrt{1 + m^2}} \right| = \sqrt{2}$

$\Rightarrow \quad m^4 + m^2 = 2$

$\Rightarrow \quad m^4 + m^2 - 2 = 0$

$\Rightarrow \quad (m^2 - 1)(m^2 + 2) = 0$

$\Rightarrow \quad m = \pm 1, m^2 = -2 \quad [m^2 = -2$ rejected]

So, tangents are $y = x + 2, y = -x - 2$

They intersect at $(-2, 0)$.

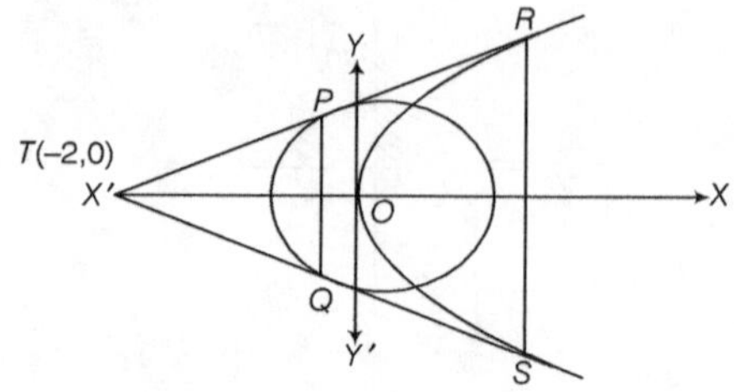


Equation of chord $PQ$ is $-2x = 2 \Rightarrow x = -1$

Equation of chord $RS$ is $0 = 4(x - 2)$

$\Rightarrow \quad x = 2$

$\therefore$ Coordinates of $P$, $Q$, $R$ and $S$ are

$P(-1, 1), Q(-1, -1), R(2, 4), S(2, -4)$

Hence, area of $PQRS = \frac{(2 + 8) \times 3}{2} = 15$ sq units

**3.** (a,b,d) **Concept Involved**

(i) Area of region $f(x)$ bounded between $x = a$ to $x = b$ is

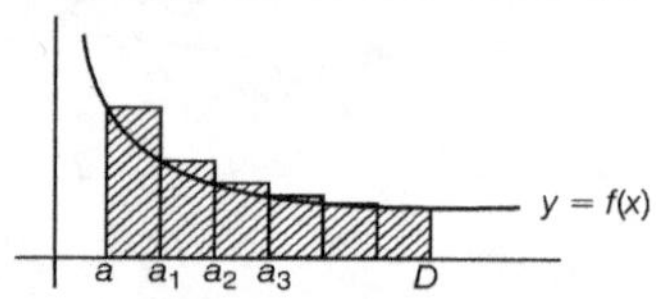


$\int_a^b f(x)dx =$ Sum of areas of rectangle shown in shaded part.

(ii) If $f(x) \ge g(x)$, when defined in $[a, b]$

$\Rightarrow \quad \int_a^b f(x)dx \ge \int_a^b g(x)\,dx$

**Description of Situation** As the given curve $y = e^{-x^2}$ cannot be integrated thus we have to bound this function by using above mentioned concept.

Graph for $y = e^{-x^2}$,

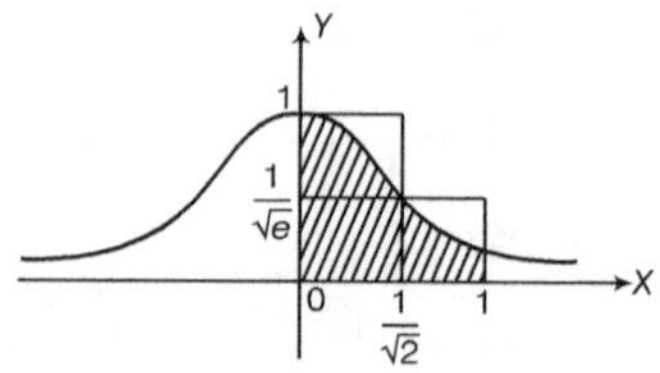


Since, $x^2 \le x$, when $x \in [0, 1]$

$\Rightarrow \quad -x^2 \ge -x$ or $e^{-x^2} \ge e^{-x}$

$\therefore \quad \int_0^1 e^{-x^2}dx \ge \int_0^1 e^{-x}dx$

$\Rightarrow \quad S \ge -(e^{-x})_0^1 = 1 - \frac{1}{e} \quad \ldots(i)$

Also, $\int_0^1 e^{-x^2}dx \le$ Area of two rectangles

$\le \left(1 \times \frac{1}{\sqrt{2}}\right) + \left(1 - \frac{1}{\sqrt{2}}\right) \times \frac{1}{\sqrt{e}}$

$\le \frac{1}{\sqrt{2}} + \frac{1}{\sqrt{e}}\left(1 - \frac{1}{\sqrt{2}}\right) \quad \ldots(ii)$

$\therefore \quad \frac{1}{\sqrt{2}} + \frac{1}{\sqrt{e}}\left(1 - \frac{1}{\sqrt{2}}\right) \ge S \ge 1 - \frac{1}{e}$ [from Eqs. (i) and (ii)]

**4.** (b) Here, area between 0 to $b$ is $R_1$ and $b$ to 1 is $R_2$.

$$\therefore \quad \int_0^b (1-x)^2\,dx - \int_b^1 (1-x)^2\,dx = \frac{1}{4}$$

$$\Rightarrow \quad \left[\frac{(1-x)^3}{-3}\right]_0^b - \left[\frac{(1-x)^3}{-3}\right]_b^1 = \frac{1}{4}$$

$$\Rightarrow \quad -\frac{1}{3}\{(1-b)^3 - 1\} + \frac{1}{3}\{0-(1-b)^3\} = \frac{1}{4}$$

$$\Rightarrow \quad -\frac{2}{3}(1-b)^3 = -\frac{1}{3} + \frac{1}{4} = -\frac{1}{12}$$

$$\Rightarrow \quad (1-b)^3 = \frac{1}{8}$$

$$\Rightarrow \quad (1-b) = \frac{1}{2}$$

$$\Rightarrow \quad b = \frac{1}{2}$$

**5.** (c) $R_1 = \int_{-1}^{2} x\, f(x)\,dx$ ...(i)

Using $\int_a^b f(x)\,dx = \int_a^b f(a+b-x)\,dx$

$$R_1 = \int_{-1}^{2} (1-x)\, f(1-x)\,dx \quad [\text{given, } f(x) = f(1-x)]$$

$$\therefore \quad R_1 = \int_{-1}^{2} (1-x)\, f(x)\,dx \quad \text{...(ii)}$$

Given, $R_2$ is area bounded by $f(x)$, $x = -1$ and $x = 2$

$$\therefore \quad R_2 = \int_{-1}^{2} f(x)\,dx \quad \text{...(iii)}$$

On adding Eqs. (i) and (ii), we get

$$2R_1 = \int_{-1}^{2} f(x)\,dx \quad \text{...(iv)}$$

From Eqs. (iii) and (iv), we get

$$2R_1 = R_2$$

**6.** (b,c,d) Shaded area $= e - \left(\int_0^1 e^x\,dx\right) = 1$

Also, $\int_1^e \ln(e+1-y)\,dy$

[put $e + 1 - y = t \Rightarrow -dy = dt$]

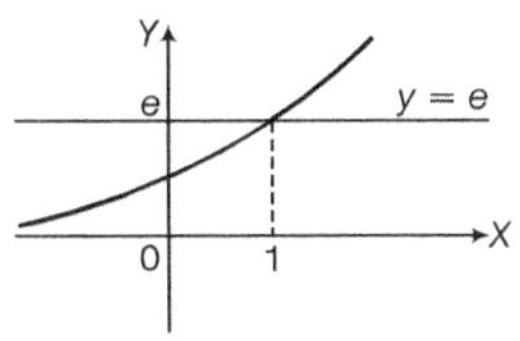


$$= \int_e^1 \ln t\,(-dt) = \int_1^e \ln t\,dt$$

$$= \int_1^e \ln y\,dy = 1$$

**7.** (b) Required area $= \int_0^{\pi/4}\left[\sqrt{\frac{1+\sin x}{\cos x}} - \sqrt{\frac{1-\sin x}{\cos x}}\right]dx$

$$\left[\because \frac{1+\sin x}{\cos x} > \frac{1-\sin x}{\cos x} > 0\right]$$

$$= \int_0^{\pi/4}\left(\sqrt{\frac{1+\dfrac{2\tan\frac{x}{2}}{1+\tan^2\frac{x}{2}}}{\dfrac{1-\tan^2\frac{x}{2}}{1+\tan^2\frac{x}{2}}}} - \sqrt{\frac{1-\dfrac{2\tan\frac{x}{2}}{1+\tan^2\frac{x}{2}}}{\dfrac{1-\tan^2\frac{x}{2}}{1+\tan^2\frac{x}{2}}}}\right)dx$$

$$= \int_0^{\pi/4}\left(\sqrt{\frac{1+\tan\frac{x}{2}}{1-\tan\frac{x}{2}}} - \sqrt{\frac{1-\tan\frac{x}{2}}{1+\tan\frac{x}{2}}}\right)dx$$

$$= \int_0^{\pi/4}\frac{1+\tan\frac{x}{2} - 1 + \tan\frac{x}{2}}{\sqrt{1-\tan^2\frac{x}{2}}}\,dx$$

$$= \int_0^{\pi/4}\frac{2\tan\frac{x}{2}}{\sqrt{1-\tan^2\frac{x}{2}}}\,dx$$

$$= \int_0^{\tan\frac{\pi}{8}}\frac{4t\,dt}{(1+t^2)\sqrt{1-t^2}}$$

$$\left[\text{put } \tan\frac{x}{2} = t \Rightarrow \frac{1}{2}\sec^2\frac{x}{2}\,dx = dt\right]$$

as, $\tan\dfrac{\pi}{8} = \sqrt{2}-1$

$$= \int_0^{\sqrt{2}-1}\frac{4t}{(1+t^2)\sqrt{1-t^2}}\,dt$$

**8.** (a) Since, tangents drawn from external point to the circle subtends equal angle at the centre.

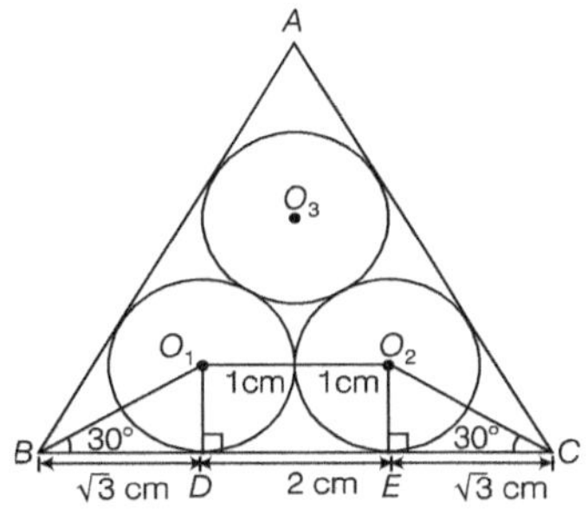


$$\therefore \quad \angle O_1BD = 30^\circ$$

In $\Delta O_1BD$, $\tan 30^\circ = \dfrac{O_1D}{BD}$

$$\Rightarrow \quad BD = \sqrt{3}\text{ cm}$$

Also, $DE = O_1O_2 = 2$ cm

and $EC = \sqrt{3}$ cm

Now, $BC = BD + DE + EC$

$$= 2 + 2\sqrt{3}$$

$\therefore$ Area of $\Delta ABC = \dfrac{\sqrt{3}}{4}(BC)^2 = \dfrac{\sqrt{3}}{4}\cdot 4(1+\sqrt{3})^2$

$$= (6 + 4\sqrt{3})\text{ sq cm}$$

**9.** (a) The curves $y = (x-1)^2$, $y = (x+1)^2$ and $y = 1/4$ are shown as

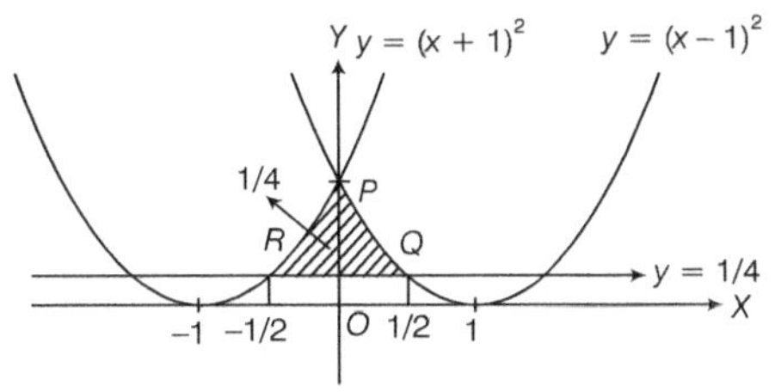

where point of intersection are

$$(x-1)^2 = \frac{1}{4}$$

$$\Rightarrow \quad x = \frac{1}{2}$$

and $$(x+1)^2 = \frac{1}{4}$$

$$\Rightarrow \quad = -\frac{1}{2}$$

So, $Q\left(\frac{1}{2}, \frac{1}{4}\right)$ and $R\left(-\frac{1}{2}, \frac{1}{4}\right)$

$\therefore$ Required area $= 2\int_0^{1/2} \left[(x-1)^2 - \frac{1}{4}\right] dx$

$$= 2\left[\frac{(x-1)^3}{3} - \frac{1}{4}x\right]_0^{1/2}$$

$$= 2\left[-\frac{1}{8\cdot 3} - \frac{1}{8} - \left(-\frac{1}{3} - 0\right)\right]$$

$$= \frac{8}{24} = \frac{1}{3} \text{ sq unit}$$

**10.** The region bounded by the curves $y = x^2$, $y = -x^2$ and $y^2 = 4x - 3$ is symmetrical about $X$-axis, where $y = 4x - 3$ meets at (1, 1).

$\therefore$ Area of curve ($OABCO$)

$$= 2\left[\int_0^1 x^2\,dx - \int_{3/4}^1 (\sqrt{4x-3})\,dx\right]$$

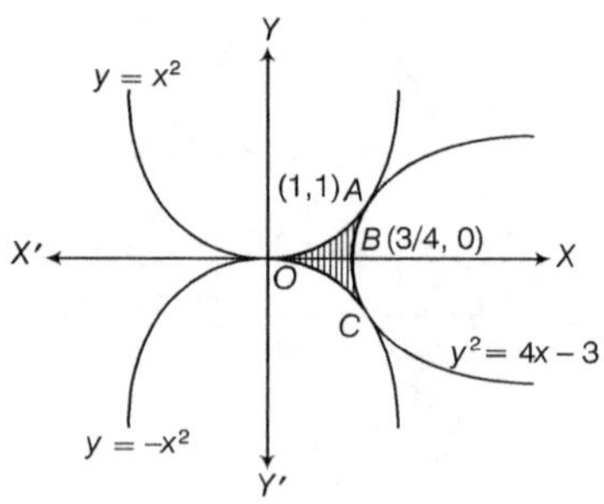


$$= 2\left[\left(\frac{x^3}{3}\right)_0^1 - \left(\frac{(4x-3)^{3/2}}{3\cdot 4/2}\right)_{3/4}^1\right]$$

$$= 2\left(\frac{1}{3} - \frac{1}{6}\right)$$

$$= 2\cdot\frac{1}{6} = \frac{1}{3} \text{ sq unit}$$

**11.** Given, $\begin{bmatrix} 4a^2 & 4a & 1 \\ 4b^2 & 4b & 1 \\ 4c^2 & 4c & 1 \end{bmatrix}\begin{bmatrix} f(-1) \\ f(1) \\ f(2) \end{bmatrix} = \begin{bmatrix} 3a^2 + 3a \\ 3b^2 + 3b \\ 3c^2 + 3c \end{bmatrix}$

$$\Rightarrow \quad 4a^2 f(-1) + 4a\,f(1) + f(2) = 3a^2 + 3a \quad \text{...(i)}$$

$$4b^2 f(-1) + 4b\,f(1) + f(2) = 3b^2 + 3b \quad \text{...(ii)}$$

and $$4c^2 f(-1) + 4cf(1) + f(2) = 3c^2 + 3c \quad \text{...(iii)}$$

where, $f(x)$ is quadratic expression given by,

$f(x) = ax^2 + bx + c$ and (i), (ii) and (iii)

$$\Rightarrow \quad 4x^2 f(-1) + 4x\,f(1) + f(2) = 3x^2 + 3x$$

or $$\{4f(-1) - 3\}x^2 + \{4f(1) - 3\}x + f(2) = 0 \quad \text{...(iv)}$$

As above equation have 3 roots $a$, $b$ and $c$.

So, above equation is identity in $x$.

i.e. coefficients must be zero.

$$\Rightarrow \quad f(-1) = 3/4,\ f(1) = 3/4,\ f(2) = 0 \quad \text{...(v)}$$

$$\because \quad f(x) = ax^2 + bx + c$$

$\therefore$ $a = -1/4$, $b = 0$ and $c = 1$, using Eq. (v)

Thus, $f(x) = \frac{4 - x^2}{4}$ shown as,

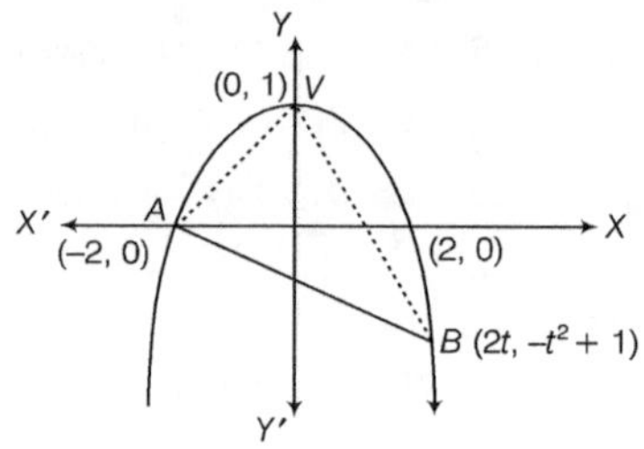


Let $A(-2, 0)$, $B = (2t, -t^2 + 1)$

Since, $AB$ subtends right angle at vertex $V$ (0, 1).

$$\Rightarrow \quad \frac{1}{2}\cdot\frac{-t^2}{2t} = -1 \Rightarrow t = 4$$

$$\therefore \quad B(8, -15)$$

Equation of chord $AB$ is $y = \frac{-(3x+6)}{2}$

$\therefore$ Required area $= \left|\int_{-2}^{8}\left(\frac{4 - x^2}{4} + \frac{3x+6}{2}\right)dx\right|$

$$= \left|\left(x - \frac{x^3}{12} + \frac{3x^2}{4} + 3x\right)_{-2}^{8}\right|$$

$$= \left|\left[8 - \frac{128}{3} + 48 + 24 - \left(-2 + \frac{2}{3} + 3 - 6\right)\right]\right|$$

$$= \frac{125}{3} \text{ sq units}$$

# Last 3 Years' Questions
# JEE Main & Advanced
# (2016-2018)

## JEE Main

### 2016

1. The integral $\int \frac{2x^{12}+5x^9}{(x^5+x^3+1)^3}\,dx$ is equal to **[Indefinite Integral]**

(a) $\frac{-x^5}{(x^5+x^3+1)^2}+C$
(b) $\frac{x^{10}}{2(x^5+x^3+1)^2}+C$
(c) $\frac{x^5}{2(x^5+x^3+1)^2}+C$
(d) $\frac{-x^{10}}{2(x^5+x^3+1)^2}+C$

where, $C$ is an arbitrary constant.

2. The area (in sq units) of the region $\{(x, y): y^2 \geq 2x \text{ and } x^2+y^2 \leq 4x, x \geq 0, y \geq 0\}$ is **[Area]**

(a) $\pi - \frac{4}{3}$
(b) $\pi - \frac{8}{3}$
(c) $\pi - \frac{4\sqrt{2}}{3}$
(d) $\frac{\pi}{2} - \frac{2\sqrt{2}}{3}$

3. If a curve $y = f(x)$ passes through the point $(1, -1)$ and satisfies the differential equation, $y(1+xy)dx = x\,dy$, then $f\left(-\frac{1}{2}\right)$ is equal to **[Differential Equations]**

(a) $-\frac{2}{5}$ (b) $-\frac{4}{5}$ (c) $\frac{2}{5}$ (d) $\frac{4}{5}$

### 2017

4. The area (in sq units) of the region $\{(x, y): x \geq 0, x+y \leq 3, x^2 \leq 4y \text{ and } y \leq 1+\sqrt{x}\}$ is **[Area]**

(a) $\frac{59}{12}$ (b) $\frac{3}{2}$ (c) $\frac{7}{3}$ (d) $\frac{5}{2}$

5. If $(2+\sin x)\frac{dy}{dx}+(y+1)\cos x = 0$ and $y(0)=1$, then $y\left(\frac{\pi}{2}\right)$ is equal to **[Differential Equations]**

(a) $\frac{1}{3}$ (b) $-\frac{2}{3}$ (c) $-\frac{1}{3}$ (d) $\frac{4}{3}$

6. Let $I_n = \int \tan^n x\,dx \; (n > 1)$. If $I_4 + I_6 = a\tan^5 x + bx^5 + C$, where $C$ is a constant of integration, then the ordered pair $(a, b)$ is equal to **[Indefinite Integral]**

(a) $\left(-\frac{1}{5}, 1\right)$
(b) $\left(\frac{1}{5}, 0\right)$
(c) $\left(\frac{1}{5}, -1\right)$
(d) $\left(-\frac{1}{5}, 0\right)$

7. $\int_{\pi/4}^{3\pi/4} \frac{dx}{1+\cos x}$ is equal to **[Definite Integral]**

(a) $-2$ (b) 2 (c) 4 (d) $-1$

## 2018

**8.** The integral

$$\int \frac{\sin^2 x \cos^2 x}{(\sin^5 x + \cos^3 x \sin^2 x + \sin^3 x \cos^2 x + \cos^5 x)^2} dx \text{ is equal to}$$

**[Indefinite Integral]**

(a) $\frac{1}{3(1+\tan^3 x)} + C$

(b) $\frac{-1}{3(1+\tan^3 x)} + C$

(c) $\frac{1}{1+\cot^3 x} + C$

(d) $\frac{-1}{1+\cot^3 x} + C$

(where $C$ is a constant of integration)

**9.** The value of $\int_{-\pi/2}^{\pi/2} \frac{\sin^2 x}{1+2^x} dx$ is

**[Definite Integral]**

(a) $\frac{\pi}{8}$

(b) $\frac{\pi}{2}$

(c) $4\pi$

(d) $\frac{\pi}{4}$

**10.** Let $g(x) = \cos x^2$, $f(x) = \sqrt{x}$ and $\alpha, \beta\ (\alpha < \beta)$ be the roots of the quadratic equation $18x^2 - 9\pi x + \pi^2 = 0$. Then, the area (in sq units) bounded by the curve $y = (gof)(x)$ and the lines $x = \alpha$, $x = \beta$ and $y = 0$, is **[Area]**

(a) $\frac{1}{2}(\sqrt{3} - 1)$

(b) $\frac{1}{2}(\sqrt{3} + 1)$

(c) $\frac{1}{2}(\sqrt{3} - \sqrt{2})$

(d) $\frac{1}{2}(\sqrt{2} - 1)$

**11.** Let $y = y(x)$ be the solution of the differential equation $\sin x \frac{dy}{dx} + y \cos x = 4x$, $x \in (0, \pi)$.

If $y\left(\frac{\pi}{2}\right) = 0$, then $y\left(\frac{\pi}{6}\right)$ is equal to

**[Differential Equations]**

(a) $\frac{4}{9\sqrt{3}}\pi^2$

(b) $\frac{-8}{9\sqrt{3}}\pi^2$

(c) $-\frac{8}{9}\pi^2$

(d) $-\frac{4}{9}\pi^2$

# JEE Advanced

## 2016

**1.** The total number of distincts $x \in [0, 1]$ for which

$\int_0^x \frac{t^2}{1+t^4} dt = 2x - 1$ is

**[Definite Integral, Type 3]**

**2.** Area of the region $\{(x, y)\} \in R^2 : y \geq \sqrt{|x+3|}, 5y \leq (x+9) \leq 15\}$ is equal to **[Area, Type 1]**

(a) $\frac{1}{6}$

(b) $\frac{4}{3}$

(c) $\frac{3}{2}$

(d) $\frac{5}{3}$

**3.** The value of $\int_{-\pi/2}^{\pi/2} \frac{x^2 \cos x}{1+e^x} dx$ is equal to

**[Definite Integral, Type 1]**

(a) $\frac{\pi^2}{4} - 2$

(b) $\frac{\pi^2}{4} + 2$

(c) $\pi^2 - e^{-\pi/2}$

(d) $\pi^2 + e^{\pi/2}$

**Note** ***Type-1*** *Only one correct option,* ***Type-2*** *More then one correct option,* ***Type-3*** *Single digit Answer,* ***Type-4*** *Numerical value based*

## 2017

**4.** Let $f: R \to (0, 1)$ be a continuous function. Then, which of the following function(s) has (have) the value zero at some point in the interval $(0, 1)$? **[Definite Integral, Type 2]**

(a) $e^x - \int_0^x f(t) \sin t \, dt$

(b) $f(x) + \int_0^{\frac{\pi}{2}} f(t) \sin t \, dt$

(c) $x - \int_0^{\frac{\pi}{2} - x} f(t) \cos t \, dt$

(d) $x^9 - f(x)$

**5.** Let $f: R \to R$ be a differentiable function such that $f(0) = 0$, $f\left(\frac{\pi}{2}\right) = 3$ and $f'(0) = 1$.

If $g(x) = \int_0^{\frac{\pi}{2}} f'(t) \operatorname{cosec} t - \cot t \operatorname{cosec} t \, f(t)] \, dt$

for $x \in \left(0, \frac{\pi}{2}\right]$, then $\lim_{x \to 0} g(x) =$ **[Definite Integral, Type 3]**

**6.** If $I = \sum_{k=1}^{98} \int_k^{k+1} \frac{k+1}{x(x+1)} dx$, then **[Definite Integral, Type 2]**

(a) $I > \log_e 99$

(b) $I < \log_e 99$

(c) $I < \frac{49}{50}$

(d) $I > \frac{49}{50}$

**7.** If the line $x = \alpha$ divides the area of region $R = \{(x, y) \in R^2 : x^3 \le y \le x, 0 \le x \le 1\}$ into two equal parts, then **[Area, Type 2]**

(a) $2\alpha^4 - 4\alpha^2 + 1 = 0$

(b) $\alpha^4 + 4\alpha^2 - 1 = 0$

(c) $\frac{1}{2} < \alpha < 1$

(d) $0 < \alpha \le \frac{1}{2}$

**8.** If $g(x) = \int_{\sin x}^{\sin(2x)} \sin^{-1}(t) \, dt$, then **[Definite Integral, Type 2]**

(a) $g'\left(-\frac{\pi}{2}\right) = 2\pi$

(b) $g'\left(-\frac{\pi}{2}\right) = -2\pi$

(c) $g'\left(\frac{\pi}{2}\right) = 2\pi$

(d) $g'\left(\frac{\pi}{2}\right) = -2\pi$

## 2018

**9.** Let $f: [0, \infty) \to R$ be a continuous function such that $f(x) = 1 - 2x + \int_0^x e^{x-t} f(t) \, dt$ for all $x \in [0, \infty)$. Then, which of the following statement(s) is (are) TRUE? **[Definite Integral, Type 2]**

(a) The curve $y = f(x)$ passes through the point $(1, 2)$

(b) The curve $y = f(x)$ passes through the point $(2, -1)$

(c) The area of the region $\{(x, y) \in [0, 1] \times R : f(x) \le y \le \sqrt{1 - x^2}\}$ is $\frac{\pi - 2}{4}$

(d) The area of the region $\{(x, y) \in [0, 1] \times R : f(x) \le y \le \sqrt{1 - x^2}\}$ is $\frac{\pi - 1}{4}$

**10.** A farmer $F_1$ has a land in the shape of a triangle with vertices at $P(0, 0)$, $Q(1, 1)$ and $R(2, 0)$. From this land, a neighbouring farmer $F_2$ takes away the region which lies between the sides $PQ$ and a curve of the form $y = x^n$ $(n > 1)$. If the area of the region taken away by the farmer $F_2$ is exactly 30% of the area of $\Delta PQR$, then the value of $n$ is .................... . **[Area, Type 4]**

**11.** The value of the integral $\int_0^{1/2} \frac{1 + \sqrt{3}}{((x+1)^2 (1-x)^6)^{1/4}} dx$ is ....... **[Definite Integral, Type 4]**

# Answer *with* Explanations

## JEE Main

**1.** *(b)* Let $I = \int \frac{2x^{12} + 5x^9}{(x^5 + x^3 + 1)^3} dx$

$$= \int \frac{2x^{12} + 5x^9}{x^{15}(1 + x^{-2} + x^{-5})^3} dx$$

$$= \int \frac{2x^{-3} + 5x^{-6}}{(1 + x^{-2} + x^{-5})^3} dx$$

Now, put $1 + x^{-2} + x^{-5} = t$

$\Rightarrow \quad (-2x^{-3} - 5x^{-6})\, dx = dt$

$\Rightarrow \quad (2x^{-3} + 5x^{-6})\, dx = -dt$

$\therefore \quad I = -\int \frac{dt}{t^3} = -\int t^{-3}\, dt$

$$= -\frac{t^{-3+1}}{-3+1} + C = \frac{1}{2t^2} + C$$

$$= \frac{x^{10}}{2(x^5 + x^3 + 1)^2} + C$$

**2.** *(b)* Given equations of curves are

$y^2 = 2x$ ...(i)

which is a parabola with vertex (0, 0) and axis parallel to $X$-axis.

and $x^2 + y^2 = 4x$

which is a circle with centre (2, 0) and radius = 2 ...(ii)

On substituting $y^2 = 2x$ in Eq. (ii), we get

$x^2 + 2x = 4x$

$\Rightarrow \quad x^2 = 2x$

$\Rightarrow \quad x = 0$

or $\quad x = 2$

$\Rightarrow \quad y = 0$ or $y = \pm 2$ [using Eq. (i)]

Now, the required area is the area of shaded region, i.e.

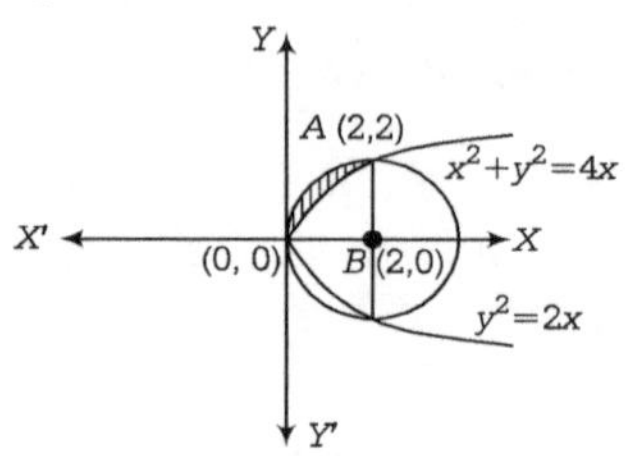


Required area $= \frac{\text{Area of a circle}}{4} - \int_0^2 \sqrt{2x}\, dx$

$$= \frac{\pi(2)^2}{4} - \sqrt{2}\int_0^2 x^{1/2} dx$$

$$= \pi - \sqrt{2}\left[\frac{x^{3/2}}{3/2}\right]_0^2$$

$$= \pi - \frac{2\sqrt{2}}{3}[2\sqrt{2} - 0]$$

$$= \left(\pi - \frac{8}{3}\right) \text{ sq unit}$$

**3.** *(d)* Given differential equation is

$$y(1 + xy)\, dx = x\, dy$$

$\Rightarrow \quad y\, dx + xy^2\, dx = x\, dy$

$\Rightarrow \quad \frac{x\, dy - y\, dx}{y^2} = x\, dx$

$\Rightarrow \quad -\frac{(y\, dx - x\, dy)}{y^2} = x\, dx$

$\Rightarrow \quad -d\left(\frac{x}{y}\right) = x\, dx$

On integrating both sides, we get

$$-\frac{x}{y} = \frac{x^2}{2} + C \quad \text{...(i)}$$

$\because$ It passes through (1, −1).

$\therefore \quad 1 = \frac{1}{2} + C \Rightarrow C = \frac{1}{2}$

Now, from Eq. (i)

$$-\frac{x}{y} = \frac{x^2}{2} + \frac{1}{2}$$

$\Rightarrow \quad x^2 + 1 = -\frac{2x}{y}$

$\Rightarrow \quad y = -\frac{2x}{x^2 + 1}$

$\therefore \quad f\left(-\frac{1}{2}\right) = \frac{4}{5}$

**4.** *(d)* Required area $= \int_0^1 (1 + \sqrt{x})dx + \int_1^2 (3 - x)dx - \int_0^2 \frac{x^2}{4}dx$

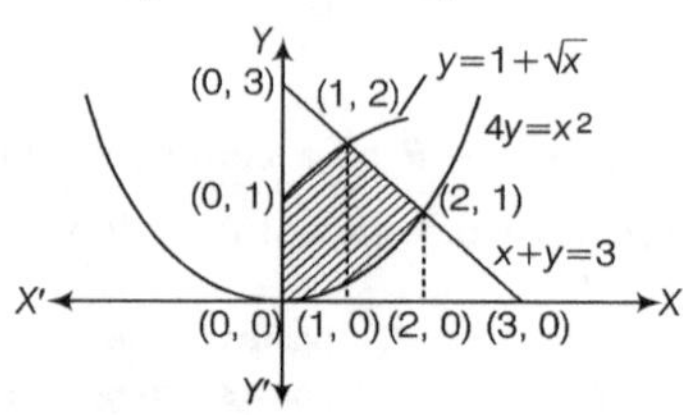


$$= \left[x + \frac{x^{3/2}}{3/2}\right]_0^1 + \left[3x - \frac{x^2}{2}\right]_1^2 - \left[\frac{x^3}{12}\right]_0^2$$

$$= \left(1 + \frac{2}{3}\right) + \left(6 - 2 - 3 + \frac{1}{2}\right) - \left(\frac{8}{12}\right)$$

$$= \frac{5}{3} + \frac{3}{2} - \frac{2}{3} = 1 + \frac{3}{2} = \frac{5}{2} \text{ sq units}$$

**5.** *(a)* We have,

$$(2 + \sin x)\frac{dy}{dx} + (y + 1)\cos x = 0$$

$\Rightarrow \quad \frac{dy}{dx} + \frac{\cos x}{2 + \sin x}y = \frac{-\cos x}{2 + \sin x}$

which is a linear differential equation.

$\therefore \quad \text{IF} = e^{\int \frac{\cos x}{2 + \sin x}dx} = e^{\log(2 + \sin x)}$

$$= 2 + \sin x$$

$\therefore$ Required solution is given by

$$y\cdot(2+\sin x)=\int\frac{-\cos x}{2+\sin x}\cdot(2+\sin x)dx+C$$

$\Rightarrow \quad y(2+\sin x)=-\sin x+C$

Also, $\quad y(0)=1$

$\therefore \quad 1(2+\sin 0)=-\sin 0+C$

$\Rightarrow \quad C=2$

$\therefore \quad y=\dfrac{2-\sin x}{2+\sin x}$

$\Rightarrow \quad y\left(\dfrac{\pi}{2}\right)=\dfrac{2-\sin\frac{\pi}{2}}{2+\sin\frac{\pi}{2}}=\dfrac{1}{3}$

6. *(b)* We have, $I_n=\int\tan^n x\,dx$

$\therefore \quad I_n+I_{n+2}=\int\tan^n x\,dx+\int\tan^{n+2}x\,dx$

$=\int\tan^n x(1+\tan^2 x)\,dx$

$=\int\tan^n x\sec^2 x\,dx$

$=\dfrac{\tan^{n+1}x}{n+1}+C$

Put $n=4$, we get $I_4+I_6=\dfrac{\tan^5 x}{5}+C$

$\therefore \quad a=\dfrac{1}{5}$ and $b=0$

7. *(b)* Let $I=\displaystyle\int_{\pi/4}^{3\pi/4}\frac{dx}{1+\cos x}$

$=\displaystyle\int_{\pi/4}^{3\pi/4}\frac{1-\cos x}{1-\cos^2 x}dx$

$=\displaystyle\int_{\pi/4}^{3\pi/4}\frac{1-\cos x}{\sin^2 x}dx$

$=\displaystyle\int_{\pi/4}^{3\pi/4}(\text{cosec}^2 x-\text{cosec}\,x\cot x)dx$

$=[-\cot x+\text{cosec}\,x]_{\pi/4}^{3\pi/4}$

$=[(1+\sqrt{2})-(-1+\sqrt{2})]=2$

**8.** (b) We have,

$$I=\int\frac{\sin^2 x\cdot\cos^2 x}{(\sin^5 x+\cos^3 x\cdot\sin^2 x+\sin^3 x\cdot\cos^2 x+\cos^5 x)^2}dx$$

$$=\int\frac{\sin^2 x\cos^2 x}{\{\sin^3 x(\sin^2 x+\cos^2 x)+\cos^3 x(\sin^2 x+\cos^2 x)\}^2}dx$$

$$=\int\frac{\sin^2 x\cos^2 x}{(\sin^3 x+\cos^3 x)^2}dx$$

$$=\int\frac{\sin^2 x\cos^2 x}{\cos^6 x(1+\tan^3 x)^2}dx$$

$$=\int\frac{\tan^2 x\sec^2 x}{(1+\tan^3 x)^2}dx$$

Put $\tan^3 x=t \Rightarrow 3\tan^2 x\sec^2 x dx=dt$

$\therefore \quad I=\dfrac{1}{3}\displaystyle\int\frac{dt}{(1+t)^2}$

$\Rightarrow \quad I=\dfrac{-1}{3(1+t)}+C$

$\Rightarrow \quad I=\dfrac{-1}{3(1+\tan^3 x)}+C$

**9.** (d) Use property $=\displaystyle\int_a^b f(x)dx=\int_a^b f(a+b-x)dx$

Let $\quad I=\displaystyle\int_{-\pi/2}^{\pi/2}\frac{\sin^2 x}{1+2^x}dx$

$\Rightarrow \quad I=\displaystyle\int_{-\pi/2}^{\pi/2}\frac{\sin^2\left(-\frac{\pi}{2}+\frac{\pi}{2}-x\right)}{1+2^{-\frac{\pi}{2}+\frac{\pi}{2}-x}}dx$

$\left[\because \displaystyle\int_a^b f(x)dx=\int_a^b f(a+b-x)dx\right]$

$\Rightarrow \quad I=\displaystyle\int_{-\pi/2}^{\pi/2}\frac{\sin^2 x}{1+2^{-x}}dx$

$\Rightarrow \quad I=\displaystyle\int_{-\pi/2}^{\pi/2}\frac{2^x\sin^2 x}{2^x+1}dx$

$\Rightarrow \quad 2I=\displaystyle\int_{-\pi/2}^{\pi/2}\sin^2 x\left(\frac{2^x+1}{2^x+1}\right)dx$

$\Rightarrow \quad 2I=\displaystyle\int_{-\pi/2}^{\pi/2}\sin^2 x\,dx$

$\Rightarrow \quad 2I=2\displaystyle\int_0^{\pi/2}\sin^2 x\,dx \qquad$ [$\because \sin^2 x$ is an even function]

$\Rightarrow \quad I=\displaystyle\int_0^{\pi/2}\sin^2 x dx$

$\Rightarrow \quad I=\displaystyle\int_0^{\pi/2}\cos^2 x dx \qquad \left[\because \int_0^a f(x)dx=\int_0^a f(a-x)dx\right]$

$\Rightarrow \quad 2I=\displaystyle\int_0^{\pi/2}dx$

$\Rightarrow \quad 2I=[x]_0^{\pi/2} \Rightarrow I=\dfrac{\pi}{4}$

**10.** (a) We have,

$\Rightarrow \quad 18x^2-9\pi x+\pi^2=0$

$\Rightarrow \quad 18x^2-6\pi x-3\pi x+\pi^2=0$

$(6x-\pi)(3x-\pi)=0$

$\Rightarrow \quad x=\dfrac{\pi}{6},\dfrac{\pi}{3}$

Now, $\alpha<\beta \quad \alpha=\dfrac{\pi}{6},\beta=\dfrac{\pi}{3}$

Given, $g(x)=\cos x^2$ and $f(x)=\sqrt{x}$

$y=gOf(x)$

$\therefore \quad y=g(f(x))=\cos x$

Area of region bounded by $x=\alpha, x=\beta$, $y=0$ and curve $y=g(f(x))$ is

$A=\displaystyle\int_{\pi/6}^{\pi/3}\cos x\,dx$

$A=[\sin x]_{\pi/6}^{\pi/3}$

$A=\sin\dfrac{\pi}{3}-\sin\dfrac{\pi}{6}=\dfrac{\sqrt{3}}{2}-\dfrac{1}{2}$

$A=\left(\dfrac{\sqrt{3}-1}{2}\right)$

**11.** (c) We have,

$$\sin x\frac{dy}{dx} + y\cos x = 4x$$

$$\Rightarrow \quad \frac{dy}{dx} + y\cot x = 4x \operatorname{cosec} x$$

This is a linear differential equation of form

$$\frac{dy}{dx} + Py = Q$$

where $P = \cot x$, $Q = 4x \operatorname{cosec} x$

Now, $IF = e^{\int P dx} = e^{\int \cot x dx} = e^{\log \sin x} = \sin x$

Solution of the differential equation is

$$y \cdot \sin x = \int 4x \operatorname{cosec} x \sin x dx + C$$

$$\Rightarrow \quad y\sin x = \int 4x dx + C = 2x^2 + C$$

Put $x = \frac{\pi}{2}$, $y = 0$, we get

$$C = -\frac{\pi^2}{2} \quad \Rightarrow \quad y\sin x = 2x^2 - \frac{\pi^2}{2}$$

Put $\quad x = \frac{\pi}{6}$

$$\therefore \quad y\left(\frac{1}{2}\right) = 2\left(\frac{\pi^2}{36}\right) - \frac{\pi^2}{2}$$

$$\Rightarrow \quad y = \frac{\pi^2}{9} - \pi^2 \quad \Rightarrow \quad y = -\frac{8\pi^2}{9}$$

**Alternate Method**

We have, $\sin x\frac{dy}{dx} + y\cos x = 4x$, which can be written as

$$\frac{d}{dx}(\sin x \cdot y) = 4x$$

On integrating both sides, we get

$$\int \frac{d}{dx}(\sin x \cdot y) \cdot dx = \int 4x \cdot dx$$

$$\Rightarrow \quad y \cdot \sin x = \frac{4x^2}{2} + C$$

$$\Rightarrow \quad y \cdot \sin x = 2x^2 + C$$

Now, as $y = 0$ when $x = \frac{\pi}{2}$

$$\therefore \quad C = -\frac{\pi^2}{2} \quad \Rightarrow \quad y \cdot \sin x = 2x^2 - \frac{\pi^2}{2}$$

Now, putting $x = \frac{\pi}{6}$, we get

$$y\left(\frac{1}{2}\right) = 2\left(\frac{\pi^2}{36}\right) - \frac{\pi^2}{2}$$

$$\Rightarrow \quad y = \frac{\pi^2}{9} - \pi^2 = -\frac{8\pi^2}{9}$$

# JEE Advanced

1. (1) Let $f(x) = \int_0^x \frac{t^2}{1 + t^4} dt$

$$\Rightarrow f'(x) = \frac{x^2}{1 + x^4} > 0, \text{ for all } x \in [0, 1]$$

$\therefore f(x)$ is increasing.

At $x = 0$, $f(0) = 0$ and at $x = 1$, $f(1) = \int_0^1 \frac{t^2}{1 + t^4} dt$

Because, $0 < \frac{t^2}{1 + t^4} < \frac{1}{2}$

$$\Rightarrow \quad \int_0^1 0 \cdot dt < \int_0^1 \frac{t^2}{1 + t^4} dt < \int_0^1 \frac{1}{2} \cdot dt \quad \Rightarrow \quad 0 < f(1) < \frac{1}{2}$$

Thus, $f(x)$ can be plotted as

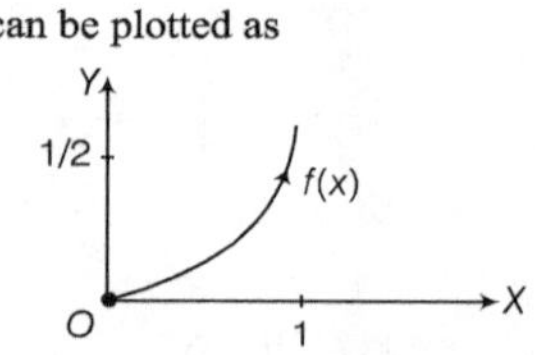


$\therefore y = f(x)$ and $y = 2x - 1$ can be shown as

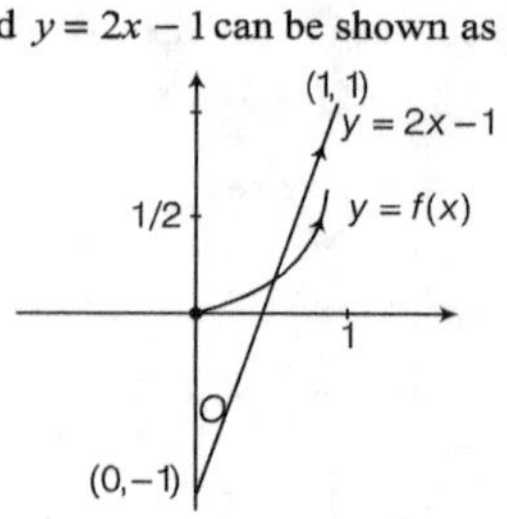


From the graph, the total number of distinct solutions for $x \in (0, 1] = 1$.

[as they intersect only at one point]

2. (c) Here, $\{(x, y) \in R^2 : y \geq \sqrt{|x + 3|}, 5y \leq (x + 9) \leq 15\}$

$$\therefore \quad y \geq \sqrt{x + 3}$$

$$\Rightarrow \quad y \geq \begin{cases} \sqrt{x + 3}, \text{ when } x \geq -3 \\ \sqrt{-x - 3}, \text{ when } x \leq -3 \end{cases}$$

or $\quad y^2 \geq \begin{cases} x + 3, \text{ when } x \geq -3 \\ -3 - x, \text{ when } x \leq -3 \end{cases}$

Shown as

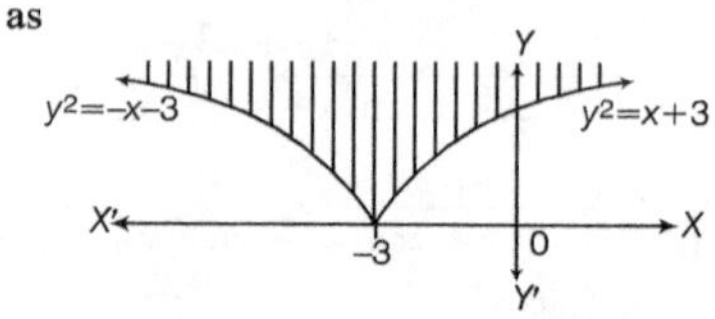


Also, $\quad 5y \leq (x + 9) \leq 15$

$$\Rightarrow \quad (x + 9) \geq 5y \text{ and } x \leq 6$$

Shown as

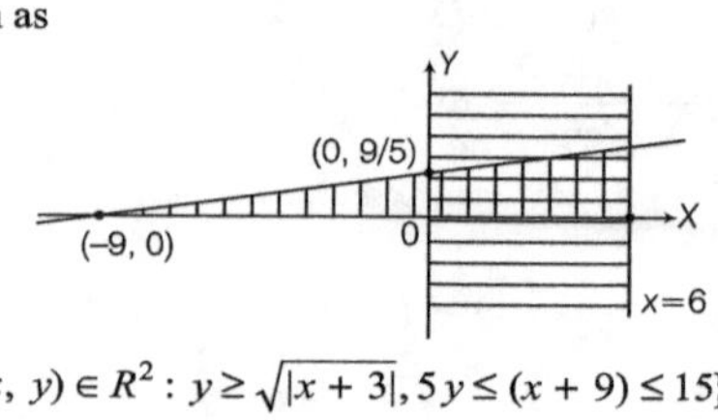


$$\therefore \quad \{(x, y) \in R^2 : y \geq \sqrt{|x + 3|}, 5y \leq (x + 9) \leq 15\}$$

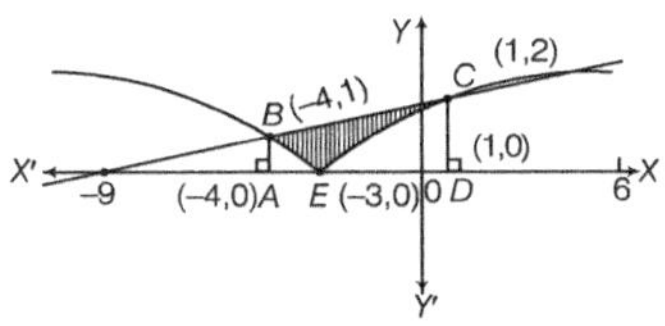


$\therefore$ Required area = Area of trapezium $ABCD$

– Area of $ABE$ under parabola

– Area of $CDE$ under parabola

$$= \frac{1}{2}(1+2)(5) - \int_{-4}^{-3}\sqrt{-(x+3)}\,dx$$

$$-\int_{-3}^{1}\sqrt{(x+3)}\,dx = \frac{15}{2} - \left[\frac{(-3-x)^{3/2}}{-\frac{3}{2}}\right]_{-4}^{-3} - \left[\frac{(x+3)^{3/2}}{\frac{3}{2}}\right]_{-3}^{1}$$

$$= \frac{15}{2} + \frac{2}{3}[0-1] - \frac{2}{3}[8-0]$$

$$= \frac{15}{2} - \frac{2}{3} - \frac{16}{3} = \frac{15}{2} - \frac{18}{3} = \frac{3}{2}$$

**3.** *(a)* Let $I = \int_{-\pi/2}^{\pi/2} \frac{x^2 \cos x}{1+e^x}\,dx$ ...(i)

$$\left[\because \int_a^b f(x)dx = \int_a^b f(a+b-x)\,dx\right]$$

$\Rightarrow$ $I = \int_{-\pi/2}^{\pi/2} \frac{x^2 \cos(-x)}{1+e^{-x}}\,dx$ ...(ii)

On adding Eqs. (i) and (ii), we get

$$2I = \int_{-\pi/2}^{\pi/2} x^2 \cos x \left[\frac{1}{1+e^x} + \frac{1}{1+e^{-x}}\right] dx$$

$$= \int_{-\pi/2}^{\pi/2} x^2 \cos x \cdot (1)\,dx$$

$$\left[\because \int_{-a}^{a} f(x)dx = 2\int_0^a f(x)dx, \text{when } f(-x) = f(x)\right]$$

$\Rightarrow$ $2I = 2\int_0^{\pi/2} x^2 \cos x\,dx$

Using integration by parts, we get

$$2I = 2\,[x^2(\sin x) - (2x)(-\cos x) + (2)(-\sin x)]_0^{\pi/2}$$

$\Rightarrow$ $2I = 2\left[\frac{\pi^2}{4} - 2\right]$

$\therefore$ $I = \frac{\pi^2}{4} - 2$

**4.** *(c,d)*

(a) $\because$ $e^x \in (1, e)$ in $(0, 1)$ and $\int_0^x f(t)\sin t\,dt \in (0, 1)$ in $(0, 1)$

$\therefore$ $e^x - \int_0^x f(t)\sin t\,dt$ cannot be zero.

So, option (a) is incorrect.

(b) $f(x) + \int_0^{\frac{\pi}{2}} f(t)\sin t\,dt$ always positive

$\therefore$ Option (b) is incorrect.

(c) Let $h(x) = x - \int_0^{\frac{\pi}{2}-x} f(t)\cos t\,dt$,

$$h(0) = -\int_0^{\frac{\pi}{2}} f(t)\cos t\,dt < 0$$

$$h(1) = 1 - \int_0^{\frac{\pi}{2}-1} f(t)\cos t\,dt > 0$$

$\therefore$ Option (c) is correct.

(d) Let $g(x) = x^9 - f(x)$

$$g(0) = -f(0) < 0,\ g(1) = 1 - f(1) > 0$$

$\therefore$ Option (d) is correct.

**5.** *(2)* $g(x) = \int_x^{\frac{\pi}{2}} (f'(t)\operatorname{cosec} t - \cot t \operatorname{cosec} t\, f(t))dt$

$\therefore$ $g(x) = f\left(\frac{\pi}{2}\right)\operatorname{cosec}\frac{\pi}{2} - f(x)\operatorname{cosec} x$

$\Rightarrow$ $g(x) = 3 - \frac{f(x)}{\sin x}$

$$\lim_{x\to 0} g(x) = \lim_{x\to 0}\left(\frac{3\sin x - f(x)}{\sin x}\right)$$

$$= \lim_{x\to 0}\frac{3\cos x - f'(x)}{\cos x}$$

$$= \frac{3-1}{1} = 2$$

**6.** *(b,d)* $I = \sum_{k=1}^{98}\int_k^{k+1}\frac{(k+1)}{x(x+1)}dx$

Clearly, $I = \sum_{k=1}^{98}\int_k^{k+1}\frac{(k+1)}{x(x+1)^2}dx$

$\Rightarrow$ $I > \sum_{k=1}^{98}(k+1)\int_k^{k+1}\frac{1}{(x+1)^2}dx$

$\Rightarrow$ $I > \sum_{k=1}^{98}(-(k+1))\left[\frac{1}{k+2} - \frac{1}{k+1}\right]$

$\Rightarrow$ $I > \sum_{k=1}^{98}\frac{1}{k+2}$

$\Rightarrow$ $I > \frac{1}{3} + \ldots + \frac{1}{100} > \frac{98}{100}$

$\Rightarrow$ $I > \frac{49}{50}$

Also, $I < \sum_{k=1}^{98}\int_k^{k+1}\frac{k+1}{x(k+1)}dx$

$$= \sum_{k=1}^{98}[\log_e(k+1) - \log_e k]$$

$$I < \log_e 99$$

**7.** *(a,c)* $\int_0^1 (x - x^3)dx = 2\int_0^{\alpha}(x - x^3)dx$

$$\frac{1}{4} = 2\left(\frac{\alpha^2}{2} - \frac{\alpha^4}{4}\right)$$

$$2\alpha^4 - 4\alpha^2 + 1 = 0$$

$$\alpha^2 = \frac{4 - \sqrt{16-8}}{4} \quad (\because \alpha \in (0,1))$$

$\because$ $\alpha^2 = 1 - \frac{1}{\sqrt{2}}$

**8.** *(*)* $g(x) = \int_{\sin x}^{\sin 2x} \sin^{-1}(t)dt$

$$g'(x) = 2\cos 2x \sin^{-1}(\sin 2x) - \cos x \sin^{-1}(\sin x)$$

$$g'\left(\frac{\pi}{2}\right) = -2\sin^{-1}(0) = 0$$

$$g'\left(-\frac{\pi}{2}\right) = -2\sin^{-1}(0) = 0$$

No option is matching.

**9.** (b, c) We have,

$$f(x) = 1 - 2x + \int_0^x e^{x-t} f(t)\, dt$$

On multiplying $e^{-x}$ both sides, we get

$$e^{-x} f(x) = e^{-x} - 2xe^{-x} + \int_0^x e^{-t} f(t)\, dt$$

On differentiating both side w.r.t. $x$, we get

$$e^{-x} f'(x) - e^{-x} f(x) = -e^{-x} - 2e^{-x} + 2xe^{-x} + e^{-x} f(x)$$

$\Rightarrow \quad f'(x) - 2f(x) = 2x - 3 \quad$ [dividing both sides by $e^{-x}$]

Let $\quad f(x) = y$

$\Rightarrow \quad f'(x) = \dfrac{dy}{dx}$

$\therefore \quad \dfrac{dy}{dx} - 2y = 2x - 3$

which is linear differential equation of the form $\dfrac{dy}{dx} + Py = Q$.

Here, $P = -2$ and $Q = 2x - 3$.

Now, IF $= e^{\int P\, dx} = e^{\int -2\, dx} = e^{-2x}$

$\therefore$ Solution of the given differential equation is

$$y \cdot e^{-2x} = \int \underset{\text{I}}{(2x-3)}\, \underset{\text{II}}{e^{-2x}}\, dx + C$$

$$y \cdot e^{-2x} = \frac{-(2x-3) \cdot e^{-2x}}{2} + 2\int \frac{e^{-2x}}{2}\, dx + C$$

[by using integration by parts]

$\Rightarrow \quad y \cdot e^{-2x} = \dfrac{-(2x-3)\, e^{-2x}}{2} - \dfrac{e^{-2x}}{2} + C$

$\Rightarrow \quad y = (1 - x) + Ce^{2x}$

On putting $x = 0$ and $y = 1$, we get

$1 = 1 + C \Rightarrow C = 0$

$\therefore \quad y = 1 - x$

$y = 1 - x$ passes through $(2, -1)$

Now, area of region bounded by curve $y = \sqrt{1 - x^2}$ and $y = 1 - x$ is shows as

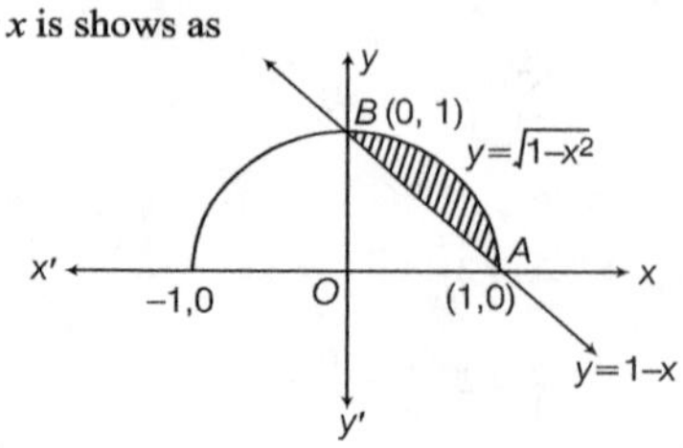


$\therefore$ Area of shaded region = Area of 1st quadrant of a circle − Area of $\Delta OAB$

$$= \frac{\pi}{4}(1)^2 - \frac{1}{2} \times 1 \times 1 = \frac{\pi}{4} - \frac{1}{2} = \frac{\pi - 2}{4}$$

Hence, options $b$ and $c$ are correct.

**10.** (4) We have,

$y = x^n, n > 1$

$\because$ $P(0, 0)$ $Q(1, 1)$ and $R(2, 0)$ are vertices of $\Delta PQR$.

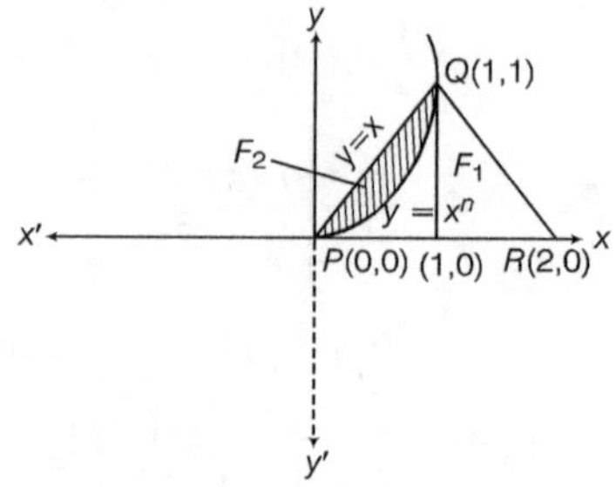


$\therefore$ Area of shaded region = 30% of area of $\Delta PQR$

$\Rightarrow \quad \displaystyle\int_0^1 (x - x^n)\, dx = \frac{30}{100} \times \frac{1}{2} \times 2 \times 1$

$\Rightarrow \quad \left[\dfrac{x^2}{2} - \dfrac{x^{n+1}}{n+1}\right]_0^1 = \dfrac{3}{10}$

$\Rightarrow \quad \left(\dfrac{1}{2} - \dfrac{1}{n+1}\right) = \dfrac{3}{10}$

$\Rightarrow \quad \dfrac{1}{n+1} = \dfrac{1}{2} - \dfrac{3}{10} = \dfrac{2}{10} = \dfrac{1}{5}$

$\Rightarrow \quad n + 1 = 5 \Rightarrow n = 4$

**11.** (2) Let $I = \displaystyle\int_0^{1/2} \frac{1 + \sqrt{3}}{[(x+1)^2 (1-x)^6]^{1/4}}\, dx$

$\Rightarrow \quad I = \displaystyle\int_0^{1/2} \frac{1 + \sqrt{3}}{(1-x)^2 \left[\left(\dfrac{1-x}{1+x}\right)^6\right]^{1/4}}\, dx$

Put $\dfrac{1-x}{1+x} = t \Rightarrow \dfrac{-2\, dx}{(1+x)^2} = dt$

when $x = 0$, $t = 1$, $x = \dfrac{1}{2}$, $t = \dfrac{1}{3}$

$\therefore \quad I = \displaystyle\int_1^{1/3} \frac{(1+\sqrt{3})\, dt}{-2(t)^{6/4}}$

$\Rightarrow \quad I = \dfrac{-(1+\sqrt{3})}{2}\left[\dfrac{-2}{\sqrt{t}}\right]_1^{1/3}$

$\Rightarrow \quad I = (1+\sqrt{3})(\sqrt{3} - 1) \Rightarrow I = 3 - 1 = 2$

www.ingramcontent.com/pod-product-compliance
Lightning Source LLC
Chambersburg PA
CBHW080210170726
48282CB00030B/1058